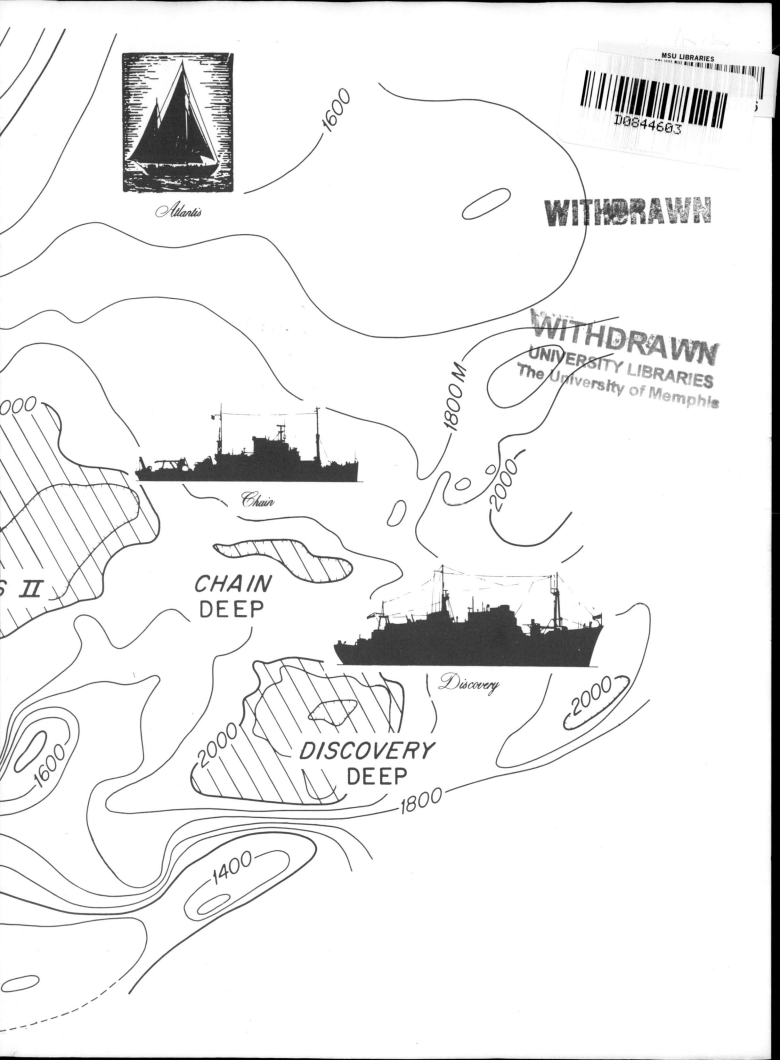

Atlantis

1600

1800 M

2000

Chain

S II

CHAIN
DEEP

2000

Discovery

2000

DISCOVERY
DEEP

2000

1600

1800

1400

Hot Brines and Recent Heavy Metal Deposits in the Red Sea

Hot Brines and Recent Heavy Metal Deposits in the Red Sea

A Geochemical and Geophysical Account

Edited by
EGON T. DEGENS AND DAVID A. ROSS
Woods Hole Oceanographic Institution
Woods Hole, Massachusetts

Springer-Verlag New York Inc.
1969

Contributors

MONEM ABDEL-GAWAD
  Science Center, North American Rockwell Corporation, Thousand Oaks, California

FRANK C. ALLSTROM
  Alpine Associates Inc., Norwood, New Jersey

WILLIAM A. BERGGREN
  Department of Geology and Geophysics, Woods Hole Oceanographic Institution, Woods Hole, Massachusetts

J. L. BISCHOFF
  Department of Geological Sciences, University of Southern California, Los Angeles, California

ANNE BOERSMA
  Department of Experimental Astronomy, Massachusetts Institute of Technology, Cambridge, Massachusetts

CARL O. BOWIN
  Department of Geology and Geophysics, Woods Hole Oceanographic Institution, Woods Hole, Massachusetts

PETER G. BREWER
  Department of Chemistry, Woods Hole Oceanographic Institution, Woods Hole, Massachusetts

R. R. BROOKS
  Department of Chemistry, Massey University, Palmerston North, New Zealand

PETER J. BUREK
  Southwest Center for Advanced Studies, Dallas, Texas

RICHARD L. CHASE
  Department of Geology, University of British Columbia, Vancouver, B.C., Canada

CHIN CHEN
  Lamont Geological Observatory of Columbia University, Palisades, New York

J. A. COOPER
  Department of Geophysics and Geochemistry, Australian National University, Canberra, Australia

HARMON CRAIG
  Scripps Institution of Oceanography, La Jolla, California

EGON T. DEGENS
  Department of Chemistry, Woods Hole Oceanographic Institution, Woods Hole, Massachusetts

C. D. DENSMORE
  Department of Physical Oceanography, Woods Hole Oceanographic Institution, Woods Hole, Massachusetts

WERNER G. DEUSER
  Department of Chemistry, Woods Hole Oceanographic Institution, Woods Hole, Massachusetts

GÜNTER DIETRICH
  Institut für Meereskunde, Kiel, Germany

K. O. EMERY
  Department of Geology and Geophysics, Woods Hole Oceanographic Institution, Woods Hole, Massachusetts

ALBERT J. ERICKSON
  Department of Geophysics, Massachusetts Institute of Technology, Cambridge, Massachusetts

MAURICE EWING
  Lamont Geological Observatory of Columbia University, Palisades, New York

GUNTER FAURE
  Department of Geology, Ohio State University, Columbus, Ohio

R. W. GIRDLER
  Department of Physics, University of Newcastle, Newcastle-Upon-Tyne, England

R. GOLL
  Lamont Geological Observatory of Columbia University, Palisades, New York

WILLIAM L. GRIFFIN
  Suite 525, 1100 Connecticut Ave. N.W., Washington, D.C.

MARTIN HARTMANN
  Geologisch—Paläontologisches, Institut der Universität Kiel, Germany

EARL E. HAYS
  Department of Geology and Geophysics, Woods Hole Oceanographic Institution, Woods Hole, Massachusetts

RUTH L. HENDRICKS
  Kennecott Copper Corporation, Exploration Services, Salt Lake City, Utah

v

YVONNE HERMAN
Department of Geology, Washington State University, Pullman, Washington

JOHN M. HUNT
Department of Chemistry, Woods Hole Oceanographic Institution, Woods Hole, Massachusetts

H. L. JAMES
U.S. Geological Survey, Washington, D.C.

LOIS M. JONES
Department of Geology, Ohio State University, Columbus, Ohio

I. R. KAPLAN
Department of Geology and Institute of Geophysics and Planetary Physics, University of California, Los Angeles, California

G. KRAUSE
Institut für Meereskunde, Kiel, Germany

TEH-LUNG KU
Department of Chemistry, Woods Hole Oceanographic Institution, Woods Hole, Massachusetts

V. J. LINNENBOM
Ocean Science Division, U.S. Naval Research Laboratory, Washington, D.C.

EDWIN J. MAHAFFEY
Kennecott Copper Corporation, Exploration Services, Salt Lake City, Utah

FRANK T. MANHEIM
U.S. Geological Survey, Woods Hole, Massachusetts

GUY G. MATHIEU
Lamont Geological Observatory of Columbia University, Palisades, New York

ANDREW MCINTYRE
Lamont Geological Observatory of Columbia University, Palisades, New York

B. E. MCMAHON
Massachusetts Institute of Technology, Cambridge, Massachusetts

A. R. MILLER
Department of Physical Oceanography, Woods Hole Oceanographic Institution, Woods Hole, Massachusetts

R. G. MUNNS
Department of Physical Oceanography, Woods Hole Oceanographic Institution, Woods Hole, Massachusetts

ARIE NISSENBAUM
Department of Geology and Institute of Geophysics and Planetary Physics, University of California, Los Angeles, California

FEODOR OSTAPOFF
Atlantic Oceanographic Laboratories, Sea-Air Interaction Laboratory, Miami, Florida

MELVIN N. A. PETERSON
Scripps Institution of Oceanography, La Jolla, California

JOSEPH D. PHILLIPS
Department of Geology and Geophysics, Woods Hole Oceanographic Institution, Woods Hole, Massachusetts

DAVID T. PUGH
Admiralty Research Laboratories, Teddington, Middlesex, England

FREDRICK B. REISBICK
Kennecott Copper Corporation, Exploration Services, Salt Lake City, Utah

JOHN R. RICHARDS
Department of Geophysics and Geochemistry, Australian National University, Canberra, Australia

D. BLAIR ROBERTS
Kennecott Copper Corporation, Exploration Services, Salt Lake City, Utah

P. E. ROSENBERG
Department of Geology, Washington State University, Pullman, Washington

DAVID A. ROSS
Department of Geology and Geophysics, Woods Hole Oceanographic Institution, Woods Hole, Massachusetts

WILLIAM B. F. RYAN
Lamont Geological Observatory of Columbia University, Palisades, New York

RUSHDI SAID
22 Road 6, Maadi, Cairo, Egypt

CLIFFORD E. SCHATZ
Bear Creek Mining Company, San Diego, California

GEROLD SIEDLER
Institut für Meereskunde, Kiel, Germany

GENE SIMMONS
Department of Geophysics, Massachusetts Institute of Technology, Cambridge, Massachusetts

DEREK W. SPENCER
Department of Chemistry, Woods Hole Oceanographic Institution, Woods Hole, Massachusetts

R. J. STANLEY
Department of Physical Oceanography, Woods Hole Oceanographic Institution, Woods Hole, Massachusetts

J. D. STEPHENS
MMD Research, Kennecott Copper Corporation, Salt Lake City, Utah

DAVID W. STRANGWAY
Department of Geophysics, Massachusetts Institute of Technology, Cambridge, Massachusetts

JOHN C. SWALLOW
National Institute of Oceanography, Wormley, Godalming-Surrey, England

R. E. SWEENEY
Department of Geology and Institution of Geophysics and Planetary Physics, University of California, Los Angeles, California

J. W. SWINNERTON
Ocean Sciences Division, U.S. Naval Research Laboratory, Washington, D.C.

EDWARD M. THORNDIKE
Lamont Geological Observatory of Columbia University, Palisades, New York

DAVID L. THURBER
Lamont Geological Observatory of Columbia University, Palisades, New York

HANS G. TRÜPER
Institut für Mikrobiologie, Universität Gottingen, Gottingen, Germany

J. S. TURNER
Department of Applied Mathematics and Theoretical Physics, University of Cambridge, Cambridge, England

DAVID A. WALL
Department of Geology and Geophysics, Woods Hole Oceanographic Institution, Woods Hole, Massachusetts

THOMAS N. WALTHIER
International Exploration, Occidental Minerals Corporation, Denver, Colorado

JOHN S. WARREN
University of Cincinnati, Cincinnati, Ohio

JOHN B. WATERBURY
Department of Biology, Woods Hole Oceanographic Institution, Woods Hole, Massachusetts

STANLEY W. WATSON
Department of Biology, Woods Hole Oceanographic Institution, Woods Hole, Massachusetts

RAY WEISS
Scripps Institution of Oceanography, La Jolla, California

R. W. WITTKOPP
MMD Research, Kennecott Copper Corporation, Salt Lake City, Utah

JOHN WOODSIDE
Bedford Institute of Oceanography, Dartmouth, Nova Scotia, Canada

Preface

Cooperative research ventures between oceanographic institutions and nations today frequently start with a series of official meetings, councils, and so forth, followed by several years of research, and finally a group of papers emerging in various technical journals. The study of the Red Sea is an exception to this procedure. It is a good example of the kind of spontaneous cooperation that can occur when individual scientists get excited about a unique problem and work together exchanging samples and data and publishing their final results in a single volume. The problem of the hot holes of the Red Sea required real teamwork from scientists of many different disciplines as well as different nationalities.

The first indication of anomalously high temperatures near the bottom of the Red Sea was recorded by the Swedish vessel, **ALBATROSS,** in 1948. It was not until the years 1963–1965 that the combination of stratified high temperature and high saline waters was recognized in cruises by Charnock and Swallow on the **R. V. DISCOVERY,** Dietrich on the **R. V. METEOR** and Miller on the **R. V. ATLANTIS II.** The latter cruise was the first to core the sediments and discover their high metal content. In the summer of 1966, J. M. Hunt of the Woods Hole Oceanographic Institution received a grant from the National Science Foundation to conduct a detailed six-week geophysical, geological, and geochemical study of the Red Sea. That October, **R. V. CHAIN** left Beirut with a scientific crew of 25 representing 7 different universities and organizations, and including geophysicists, geologists, chemists, biologists, paleontologists, and physical oceanographers. More than 100 traverses were made of the rift valley and over 200 water and sediment samples were taken in and around the hot hole area. Practically every oceanographic instrument was utilized plus some new ones, such as a telemetering pinger for getting a continuous temperature profile and a thermoprobe accurate to 0.005°C.

When the **R. V. CHAIN** returned home, there were requests from many laboratories around the world for sediment samples to analyze. These requests were filled insofar as was possible without exhausting all available sample.

Although the hot brine and heavy metal deposits cover an area of less than 100 square miles, they are clearly part of a larger geological scheme. The world-wide interest in rift valleys and sea floor spreading with their attendant hydrothermal and volcanic activity can be studied in a mini-scale in the Red Sea. The data reported in this book cover both the regional and local geology of the heavy metal deposits. They suggest that such deposits may be more commonly found in the decades ahead as scientists study the rift valleys of the world. The similarity of the heavy metal deposits with some ancient ore formations indicates that the conditions observed in the Red Sea may be a counterpart of some ancient ore forming processes.

The papers represent a broad international authorship and cover quite a variety of subject matter. Mathematicians, physicists, chemists, biologists, geologists, oceanographers, lawyers, and economists from numerous countries have participated in this endeavor, and the editors are grateful to them for their enthusiastic support. We also wish to express our appreciation to Dr. K. Springer of the Springer-Verlag for his help in arranging this publication and making it available in such a short period of time. To many scientific colleagues we are indebted beyond measure for their scientific papers and personal communications, without which this work would only be fragmentary. The help and enthusiasm of the officers and crew of the R. V. CHAIN was one of the most satisfying aspects of the Red Sea work. In particular we would like to thank Captain C. Davis, Boatswain Jerry Cotter, Earl M. Young Jr., and Frank B. Wooding. Our thanks go to Miss Louise Langley and Mrs. Belinda Collins for their secretarial help in the final editing of the manuscript. Also, we would like to thank the National Aeronautics and Space Administration for the use of the jacket photograph. In conclusion, we give our special appreciation to the Oceanography Section of the National Science Foundation whose generous support made this work possible.

The editors will contribute all the royalties from the sale of this book to the libraries of the ships that have explored the hot brine region of the Red Sea.

EGON T. DEGENS
DAVID A. ROSS

Contents

Introduction

History of the Exploration of the Hot Brine Area of the Red Sea: **DISCOVERY** Account

J. C. SWALLOW

National Institute of Oceanography
Wormley, Godalming,
Surrey, England

Abstract

The chief contribution made by the **RRS DIS-COVERY,** in exploring the hot brine pools of the Red Sea, was the delineation in 1964 of a small area (Discovery Deep) within which a brine with a temperature over 44°C was found. In addition, acoustic reflecting layers were observed in mid-water in the brine areas during three passages of the **DISCOVERY** through the Red Sea. These observations include the first known indication of the layering in Atlantis II Deep, in November 1963, and a layer observed in April 1967 in a small depression to the south of the known brine area not previously described. In this chapter, the circumstances surrounding the **DISCOVERY's** observations are narrated, and the observations themselves are outlined.

Introduction

The first conscious attempt at observing the hot brines of the Red Sea from the **DIS-COVERY** was made in March 1964 when water samples were taken in a deep basin to the south of the brine area. They showed traces of the hot brine mixed with normal, deep Red Sea water (Charnock, 1964). In an earlier cruise, however, in November 1963, the **DISCOVERY** had passed over what is now known as Atlantis II Deep, and reflecting layers in mid-water had been noted on the echo-sounding record though their significance was not appreciated at the time.

On returning through the Red Sea in September 1964, a more deliberate attempt was made to locate the hot salty water. A small bathymetric survey revealed two depressions with depths exceeding 2,000m, and water samples from below the sill of one of these depressions had the remarkable temperature of over 44°C and nearly seven times the normal salt concentration.

In April 1967, when the **DISCOVERY** was again passing through the Red Sea returning from the Gulf of Aden, the mid-water reflecting layers in Atlantis II Deep were again observed, and a further reflecting layer was noted in yet another small basin to the south of Discovery Deep.

In describing these investigations more fully below, it will be convenient to deal first with the water sampling and bathymetry, then with the reflecting layers.

Water Sampling

March 1, 1964

There had been three reports of the hot salty water up to that time, the most recent (and hottest) being Atlantis II station 42 in July 1963 (Miller, 1964). On passage out to the Arabian Sea, one of our aims was to reoccupy that Atlantis II station. When the **DISCOVERY** passed near the intended position, however, the observed water depths were less than expected, only a little over 1,900m, and too shallow for the anomalous water. The track was later found to have passed slightly east of the intended position, but comparison of the echo-sounding profile with the **CHAIN** contour chart (Ross *et al.,* 1969) suggests that it must have been well within a mile of Atlantis II Deep. Time was too short to allow any diversion, to search for deeper water, but while continuing southwards depths exceeding 2,000m were encountered not far from the position of Albatross station 254, where abnormal water had first been found in 1948 (Figs. 1, 3). One station (Discovery No. 5247) was worked near there (Figs. 1, 2) in a water depth of 2,394m. As shown by Charnock (1964) the θ-S curve

for samples below 1,500m at this station diverged from that for normal deep Red Sea water, indicating a slight admixture of the abnormal water.

September 11, 1964

On approaching the central Red Sea area during the return passage from the Indian Ocean, Discovery station No. 5579 was worked in 2,339m water depth, not far from the previous station 5247 and possibly in the same basin (Fig. 1). Again, traces of the abnormal water were detected at depths below 1,500m (Fig. 2). This time twelve hours had been kept in hand in which to search for more concentrated abnormal water. Continuing northward from

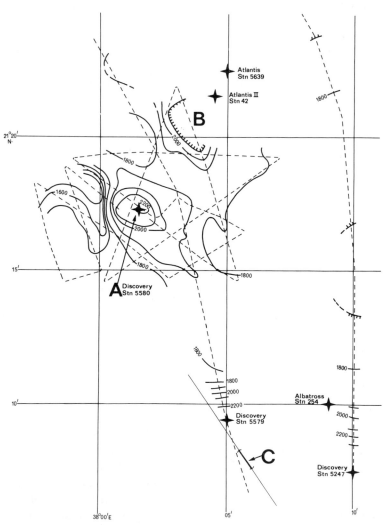

Fig. 1. Station positions, and bathymetry as revealed by **DISCOVERY** soundings in 1964. Note A—Discovery Deep. B—part of Atlantis II Deep. C—position of new reflecting layer (see Fig. 5). Depth contours in meters corrected according to Matthews' table for Area 51.

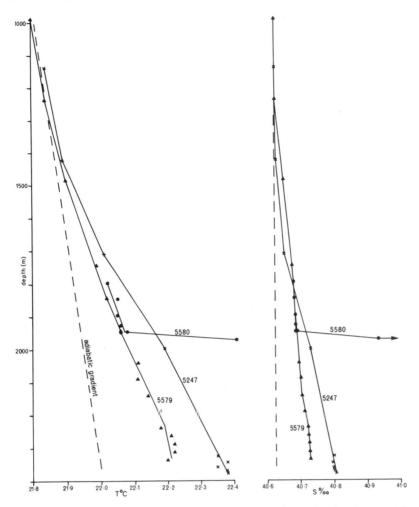

Fig. 2. Temperature-depth and salinity-depth curves for three stations, showing traces of the hot brine of Discovery Deep (station 5580) in the deeper water to the south (stations 5247, 5579).

station 5579, a small depression was encountered near 21° 17'N 38° 03'E, and a buoy with a radar reflector was anchored to provide a reference point for a small bathymetric survey. Any echo-sounding profile through that part of the Red Sea shows relief of several hundred meters amplitude, with a horizontal dimension of the order of 1 or 2 miles. Clearly there was a need for relative navigation to better than a mile to determine whether any depressions were closed basins or not and to enable water sampling to be done in realistic relationship to the bathymetry. Fortunately the buoy was anchored before morning stars so that a reliable absolute position could be assigned to the relative survey. This took the form of a 5-pointed star, out to 4 miles from the buoy. It revealed two depressions in

which the water depth exceeded 2,000m, one of which appeared to be roughly circular and about 1½ miles in diameter. A cast of water bottles (Discovery station 5580) was lowered into the deepest part of this depression, revealing normal, deep Red Sea water (about 22°C) down to 1,960m wire out, over 26°C at 2,010m, and off scale (i.e. over 35°C) on the protected thermometers at 2,060m out and below. From the readings of two unprotected thermometers below that depth, it was evident that the water near the bottom had a temperature of over 44°C. A second cast was lowered, using more 60°C-range unprotected thermometers, and two 7-liter water bottles to collect a larger sample of the 44°C water. The unusually high temperatures were confirmed; even after being hauled up

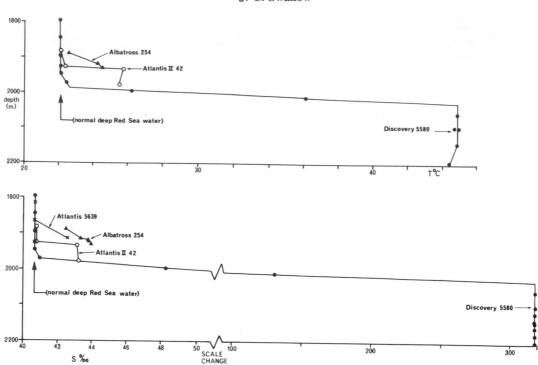

Fig. 3. Temperature-depth and salinity-depth curves for Discovery station 5580, compared to previously reported examples of abnormal water. Albatross 254 – Bruneau *et al.* (1948); Atlantis 5639 – Neumann and Densmore (1959); Atlantis II 42 – Miller (1964). "Salinities" are shown as derived from conductivity after dilution in known volume ratio, *cf.* Munns *et al.* (1967).

through nearly 2,000m of 22°C water, the large water samples when drawn off on deck still had a temperature exceeding 37°C. Fig. 3 shows the results of this station compared to the earlier reports of anomalous water.

In the short time remaining for observations the bathymetric survey was filled in so as to define more clearly the shape of the depression containing the 44°C water, subsequently named Discovery Deep. We could have chosen instead to take samples in the other depression revealed by the survey and certainly would have done so if the mid-water reflecting layers, which were present on the echo-sounding record, had been noticed. We therefore missed sampling the even hotter water (56°C) in what has become known as Atlantis II Deep. The salinity of the 44°C water was measured on board by diluting it with pure water in a known volumetric proportion so that its conductivity came within the range of an Autolab salinometer. The apparent salinity obtained in this way was approximately 318 per mill, more than seven times as high as

any previously reported oceanic salinity. Though there was certainly a large quantity of dissolved salts in the water, the precise number, 318 per mill, was of little significance. Not only should the dilution have been by weight but the ionic composition of the brine differed from that of sea water so that the available conductivity tables were not strictly applicable.

The hydrographic data are described more fully in another chapter, but it seems appropriate here to explain how the temperatures and depths for Discovery station 5580 were calculated and to point out that the provisional temperatures first quoted for the 44°C water (Swallow and Crease, 1965) are about 0.5°C too low. Since only unprotected thermometers had been available for measuring the high temperatures, their readings had to be corrected for pressure. The appropriate pressures were estimated in the following manner: in the first cast there were three bottles with unprotected thermometers in the normal Red Sea water, at 1,810, 1,910 and 2,010m wire out. Depths calculated for these bottles in the normal way showed a mean loss of depth (wire out minus depth) of 14m. Since the wire angle was negligible, it was assumed that the same loss of depth could be applied to the deeper bottles, down to 2,210m. The depths so derived were converted to pressures assuming a density of $1.2gm/cm^3$ in the hot brine in the bottom 150m of the cast, and normal Red Sea water above.

For the second cast there was a 20° wire angle and only two unprotected thermometers above the hot brine, and correction of the deeper temperatures was not so straightforward. From the first cast, however, it was possible to make a crude estimate of the velocity of sound in the brine, which could be used in calculating the depth of the second cast. There was a discrepancy of 25m on the first cast, between the depth of a pinger as estimated from its position on the wire (5m below the lowest bottle) and as obtained from the echo-sounding record, corrected according to Matthews' tables (1939). This could be accounted for by assuming a velocity of sound of 1,800m/sec in the hot brine, and this velocity was used in correcting the pinger depth on the second cast. The depths, and thence ambient pressures, for the water bottles in the hot brine were then obtained by interpolation between the pinger depth and the depths from the unprotected thermometers in the normal Red Sea water above.

Mid-water Reflecting Layers

November, 1963

The **DISCOVERY** passed over what is now known as Atlantis II Deep on November 19, 1963. Reflecting layers are clearly visible in mid-water on the echo-sounding record (Fig. 4) and were noticed at the time. There is a note in the echo-sounding log book alongside the sounding for 1,620/19.XI1.63 saying "horizontal layering in water at c.1100 fms," unsigned but recognizably in the handwriting of the late Dr. M. N. Hill, chief scientist on that cruise. The ship was not stopped for any related sampling, and the observation went unnoticed until mid-1967 when it came to light during a search for other mid-water reflecting layers.

September 11, 1964

As mentioned above, mid-water reflections can be seen in the part of the bathymetric survey that extends over Atlantis II

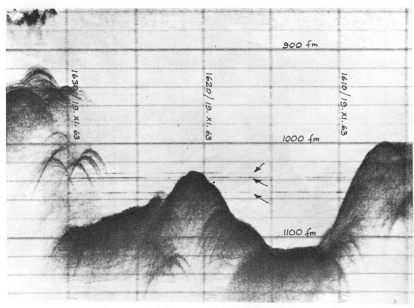

Fig. 4. Reflecting layers (arrows) observed by **DISCOVERY** on November 19, 1963, in what is now known as Atlantis II Deep.

Deep. They were not noticed at the time but were found later after the **METEOR** (G. Dietrich, personal communication, 1965; Krause and Ziegenbein, 1966) reported finding reflecting layers in the hot brine area. They do not differ significantly in depth from those recorded in November, 1963.

April 21, 1967

This third passage over Atlantis II Deep again revealed the reflecting layers, at the same depths. The depths are 1,032, 1,038 and 1,054fms (uncorrected), which convert to 1,994, 2,006 and 2,037m on correction using Matthews' table 51 (1939). These depths correspond fairly well to the discontinuities at approximately 1,990, 2,010 and 2,040m in the published temperature-depth and salinity-depth curves (Munns *et al.*, 1967; Brewer *et al.*, 1969; Ross, 1969).

The constancy in depth of these reflecting layers, to within 1m over the period November, 1963–April, 1967, seems remarkable considering the variability in the transition layer noted by Munns *et al.* (1967). Perhaps the shallowest reflector may be related to a sill, but this seems unlikely in the case of the deepest one, which appears to belong to the boundary between the 44°C and 56°C water of Atlantis II Deep. Munns *et al.* (1967) report an average increase in temperature of 0.56°C in the "56°C" Atlantis II water between February, 1965 and October, 1966. Assuming this to be due to heat flow, and assuming an average thickness of 50m for the 56°C brine, an average heat flow of about 50×10^{-6} cal/cm²/sec would be needed, which seems exceedingly high. If all the heat were brought in by a flow of hot water through the sea floor, the incoming water must have had a temperature of at least 90°C to cause the observed change of mean temperature without detectable change of volume.

A reflecting layer was observed in the water of another depression (Fig. 5) some

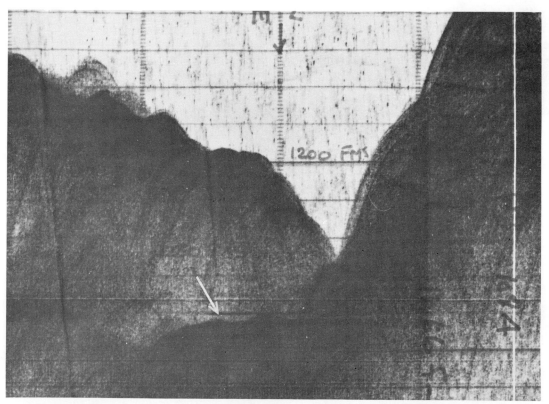

Fig. 5. Reflecting layer (arrow) observed by **DISCOVERY** on April 21, 1967, near 21°08′N, 38°06′E, not previously described.

9 miles SSW of Discovery Deep while the **DISCOVERY** was approaching the known hot brine area on April 21, 1967 (A. S. Laughton, personal communication, 1967). This reflector, at 1,280fms (uncorrected) or 2,486m (corrected according to Matthews, 1939), is considerably deeper than those of Atlantis II Deep. No water sampling or coring was attempted in relation to this new reflecting layer, and its extent cannot be determined from the available sounding tracks. It was proposed, by those on board **DISCOVERY** at the time that if this should turn out to be a new hot brine region, it should be named Albatross Deep after the ship that first sampled anomalous water near that area.

References

Brewer, P. G., C. D. Densmore, R. G. Munns, and R. J. Stanley: Hydrography of the Red Sea brines. *In; Hot brines and recent heavy metal deposits in the Red Sea,* E. T. Degens and D. A. Ross (eds.). Springer-Verlag New York Inc., 138–147 (1969).

Bruneau, L., N. G. Jerlov, and F. Koczy: Physical and chemical methods. Rep. Swedish Deep Sea Expedition, **3,** 99 (1953).

Charnock, H.: Anomalous bottom water in the Red Sea. Nature, **203,** 591 (1964).

Krause, G., and J. Ziegenbein: Die Struktur des heissen salzreichen Tiefenwassers im zentralen Roten Meer. METEOR Forschungsergebnisse, Reihe A, **1,** 53 (1966).

Matthews, D. J.: Tables of the velocity of sound in pure water and sea water for use in echo-sounding and sound-ranging, 2nd ed. London Hydrographic Dept., Admiralty, 52 p. (1939).

Miller, A. R.: Highest salinity in the world ocean? Nature, **203,** 590 (1964).

Munns, R. G., R. J. Stanley, and C. D. Densmore: Hydrographic observations of the Red Sea brines. Nature, **214,** 1215 (1967).

Neumann, A. C., and C. D. Densmore: Oceanographic data from the Mediterranean Sea, Red Sea, Gulf of Aden and Indian Ocean. ATLANTIS cruise 242 for the International Geophysical Year 1957–8. Unpublished manuscript, Ref. 60–2, 44 p. Woods Hole Oceanographic Institution (1959).

Ross, D. A.: Temperature structure of the Red Sea brines. *In: Hot brines and recent heavy metal deposits in the Red Sea,* E. T. Degens and D. A. Ross (eds.). Springer-Verlag New York Inc., 148–152 (1969).

——, E. E. Hays, and F. C. Allstrom: Bathymetry and continuous seismic profiles of the hot brine region of the Red Sea. *In: Hot brines and recent heavy metal deposits in the Red Sea,* E. T. Degens and D. A. Ross (eds.). Springer-Verlag New York Inc., 82–97 (1969).

Swallow, J. C., and J. Crease: Hot salty water at the bottom of the Red Sea. Nature, **205,** 165 (1965).

The Observations of the Vertical Structure of Hot Salty Water by **R.V. METEOR**

G. DIETRICH AND G. KRAUSE

Institut für Meereskunde
Kiel, Germany

Abstract

On her cruise to the Indian Ocean, **R.V. METEOR** worked a section through the Red Sea with stations in several deep holes. Hot salty water occurred only in the Discovery Deep (st. 28) and in the area of the Atlantic Deep (st. 384). The upper limits of the abnormal water were indicated by sound scattering layers. In both deeps continuous temperature and transparency measurements were carried out. At station 28 the temperature increased with depth within a thermocline of 45-m thickness from 21.6°C to 44.8°C. At station 384 three layers were observed; the temperature first increased from 21.6°C to 41.2°C, then to 57.7°C and finally to 58.4°C near the bottom. On both stations the water was found to be very turbid. Some chemical properties of the water were determined.

On October 29, 1964, **R.V. METEOR** left Hamburg on her maiden cruise. This cruise led to the Arabian Sea as part of the International Indian Ocean Expedition (IIOE).

Before this cruise our attention was drawn to articles by Charnock (1964) and by Miller (1964) who both reported anomalous salinities at a certain position in the Red Sea. No explanation was given at the time. Just before **R.V. METEOR** left for her cruise, we received a letter dated September 12, 1964, from J. C. Swallow aboard the **R.R.S. DISCOVERY** in Port Said. He informed us that near the hole reported by **R.V. ATLANTIS II** a somewhat deeper hole was found. Furthermore, he pointed out that we ought to prepare for temperatures of well over 40°C and of salinities of over 300 per mill if we intended to investigate these phenomena. These facts were later published by Swallow and Crease (1965).

Thanks to this information we could prepare for a special short investigation. We intended to scan the area for other holes and to check whether they too contained hot brine. The likelihood of the occurrence of other holes with hot salty water is confirmed by the report about such brines in the Sea of Galilee, even though no brine has been found in the Dead Sea (Neev and Emery, 1967). From a detailed depth chart prepared by Pfannenstiel, a member of our expedition, we obtained detailed information about the positions of other deep holes.

To detect the hot salty water in the holes some special devices were used in addition to the normal equipment. Sound reflections could be expected at the upper boundary of the hot dense water in the holes. The instrument used was the ELAC precision depth recorder which has an extremely narrow beam and hence high accuracy and sensitivity. Furthermore, a transparency meter (Krause, 1963) provided an additional means of measuring a parameter which would help to characterize the nature of this hot salty water. This apparatus carried a bottom contact device (Holzkamm, 1964). In addition, a thermistor was prepared that could be used for the recording of high temperatures with an accuracy of ±0.2°C and

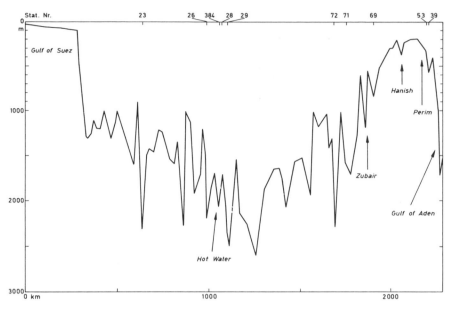

Fig. 1. Bottom topography along the axis of the Red Sea Trench according to the depth chart of M. Pfannen-stiel. Meteor stations and the small holes with hot salty water (station 28 and station 384) are indicated.

which could be used to record the fine structure of the vertical temperature distribution. It was also intended to take some water samples. Finally, we planned to take cores in the holes on our return from the IIOE. At that time the geologist Seibold would be on board the **R.V. METEOR.**

With the transparency-temperature recorder, measurements were made at the stations indicated in Fig. 1. Several other deep holes were investigated but no anomalous water was observed (Dietrich *et al.*, 1966), except at the already known positions at stations 28 and 384.

Station 28 at 21°17.2′N, 38°2.5′E, was investigated on the way to the Indian Ocean. The transparency recorder went off scale while passing into the hot water so that only a temperature record was obtained (Fig. 2b). The thickness of the main thermocline in the deep water was 45m. The highest temperature found was 44.8°C. The echo-sounder record (Fig. 2a) shows that a sound scattering layer coincides with the recorded temperature discontinuity layer. It should be mentioned here that in the fall of 1966 **R.V. CHAIN** found another hole near Discovery Deep with a temperature of 34°C (Ross and Hunt, 1967). However, the deepest part of this hole was not investigated at that time.

On March 23, 1965, **METEOR** met **R.V. ATLANTIS II** in the Arabian Sea. **ATLANTIS II** had just found another hot hole in the Red Sea (Miller *et al.*, 1966). After a comparison of the results obtained on the two ships, it was decided that **R.V. METEOR** would briefly repeat measurements at the Atlantis II Deep rather than at the Discovery Deep. These measurements were carried out on Meteor station 384 at 21°33′N, 38°3.8′E on April 27, 1965.

Three well separated layers were observed. The echogram, temperature and the transparency records are shown in Fig. 3. The temperature increased from 21.6°C to 41.2°C, then to 57.7°C and finally to 58.4°C near the bottom. The turbidity was so high in the first and near the third layer that it caused the instrument to run off scale.

In the intermediate layer the turbidity was smaller and could be measured. Ryan *et al.* (1969) have studied these layers using a nepholometer and noted similar features. It should be noted that the depth of the upper scattering layer corresponds to the first minimum in transparency and not to the depth of the thermocline. From Fig. 3b it is apparent that the gradient of the first thermocline is not as steep as those of the subsequent ones, and this may be the rea-

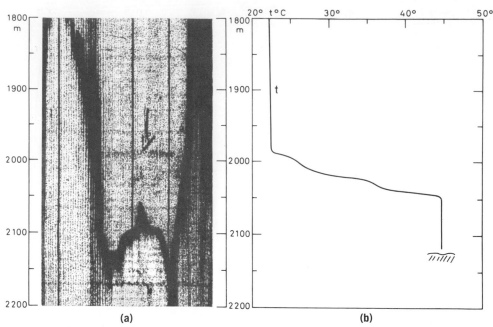

Fig. 2. Meteor station 28. (a) Echo-sounder record with the deep scattering layer at the depth of the deep thermocine. (b) Temperature record for a distance of 200m from the bottom.

son why in this case the scattering layer can be traced back to the layer with maximum turbidity.

Chemical analysis has been carried out on water samples from stations 28 and 384. These samples were taken by 5-liter water bottles which were attached to the trans-parency meter and which were closed about 10m above the bottom. The results of the analysis are presented in Table 1.

Chemical studies on the water of the different layers are of fundamental importance for a theory of the origin of the abnormal water (Riley, 1967). However, only a small

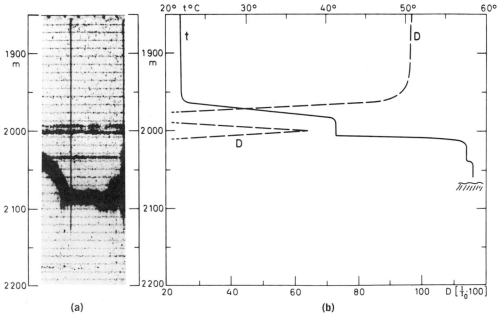

Fig. 3. Meteor-station 384. (a) Echo-sounder record with three deep scattering layers. (b) Temperature (*t*) and transparency (*D*) record. Scale for transparency is given in per cents of the transparency of clear water.

Table 1 Chemical Analysis of Red Sea Brines *

station 28			station 384		
NaCl	280.4	g/l	Cl	18.1	%
CaCl$_2$	16.0	"	NO$_2$-N	0.36	μgat/l
CaSo$_4$	1.2	"	NO$_3$-N	0.90	"
KCl	4.9	"	SiO$_2$	271	"
MgCl$_2$	2.7	"	PO$_4$-P	+	
Br$^-$	0.14	"	Mn	++++	
Cl$^-$	184.5	"			
Al$^+$					
SiO$_2$	128	μgat/l			
PO$_4$-P	0.19	"			

* The authors wish to thank Mr. M. R. Bloch and Dr. K. Grasshoff for providing the chemical data.

number of water samples could be obtained on this cruise because of the restricted time for these investigations. Nevertheless they can be used to give a survey over some main chemical properties.

The color of the water sample from station 384 when it arrived on deck was light brown; however, within a few minutes after the sample was drawn a brown precipitate formed. After six weeks, when the sample was analysed in our laboratory in Kiel, a thick brownish-white layer had formed on the bottom of the sample bottle, which was analysed separately. The "salinity" of the supernatant water was about 32.65 per cent on the basis of a conductivity measurement on a sample that was diluted with distilled water to 10 per cent of its concentration.

The residue is not due to precipitation from a saturated solution. In the bottom water reductive conditions exist, which suggests that after the sample was taken the oxygen which entered the sample caused a change in the oxidation state of the Fe^{++}, thus causing a precipitate of Fe^{+++} hydroxide to form. The precipitate did not dissolve in 0.01 N hydrochloric acid but did in a 0.1 N solution. A qualitative analysis indicated that next to the main components Fe^{+++}, zinc and manganese there were also traces of aluminum and magnesium in the form of oxides or carbonates. The precipitation of Fe^{+++} hydroxide at the surface of the brine layers may well explain the enormous rise in turbidity and the subsequent scattering layer.

The brine at station 28 not only shows a different pattern than at station 384, but also the water itself seems to have a different character. No formation of a precipitate was observed. Analysis by Brewer *et al.* (1965) on the water from the Discovery Deep also showed a marked difference compared to the water at station 384, particularly since these authors did not observe any large quantity of Fe^{++}. From the chemical analysis it seems clear that the lower parts of the holes are not connected to each other.

Hartmann (1969) made geological-geochemical analyses on the sediments from these holes. The measurements were sufficient to describe several features of the phenomena, but they were not detailed enough to lead to a theory of the origin of this abnormal water. We cannot decide from these results whether the formation of several layers is caused by the Turner-Stommel mechanism (Turner and Stommel, 1964) or by a mechanism where water from different sources flow over each other horizontally (Krause and Ziegenbein, 1965; Hunt *et al.*, 1967).

References

Brewer, P. G., J. P. Riley, and F. Culkin: The chemical composition of the hot salty water from the bottom of the Red Sea. Deep-Sea Res., **12**, 497 (1965).

Charnock, H.: Anomalous bottom water in the Red Sea. Nature, **203**, 590 (1964).

Dietrich, G., G. Krause, E. Seibold, and K. Vollbrecht: Reisebericht der Indischen Ozean Expedition mit dem Forschungsschiff METEOR 1964–1965. METEOR-Forschungsergebnisse, Reihe A, **1**, Berlin (1966).

Dietrich, G., W. Düing, K. Grasshoff, and P. H. Koske: Physikalische und chemische Daten nach Beobachtungen des Forschungsschiffes METEOR im Indischen Ozean 1964–1965. METEOR-Forschungsergebnisse, Reihe A, **2**, Berlin (1966).

Hartmann, M.: Investigation of the Atlantis II Deep samples, taken by the F.S. METEOR. *In: Hot brines and recent heavy metal deposits in the Red Sea*, E. T. Degens and D. A. Ross (eds.). Springer-Verlag New York Inc., 204–207 (1969).

Holzkamm, F.: Bodenberührungsschalter für Geräte mit Einleiterkabeln. Kieler Meeresforsch., **20**, 136 (1964).

Hunt, J. M., E. E. Hays, E. T. Degens and D. A. Ross: Red Sea: detailed survey of hot-brine areas. Science, **156**, 514 (1967).

Krause, G.: Eine Methode zur Messung optischer Eigenschaften des Meerwassers in grossen Tiefen. Kieler Meeresforsch., 19, 175 (1963).

—— and J. Ziegenbein: Die Struktur des heissen salzreichen Tiefenwassers im zentralen Roten Meer. METEOR-Forschungsergebnisse, Reihe A, 1, Berlin (1966).

Miller, A. R.: Highest salinity in the world ocean? Nature, 203, 590 (1964).

——, C. D. Densmore, E. T. Degens, J. C. Hathaway, F. T. Manhein, P. F. McFarlin, R. Pocklington, and A. Jokela: Hot brines and recent iron deposits in deeps of the Red Sea. Geochim. et Cosmochim. Acta, 30, 341 (1966).

Neev, D. and K. O. Emery: The Dead Sea. Israel Geological Survey, Bull. 41, 1 Jerusalem (1967).

Riley, J. P.: The hot saline waters of the Red Sea bottom and their related sediments. Oceanogr. Mar. Biol. Ann. Rev., George Allen and Unwin Ltd., London, 5, 141 (1967).

Ross, D. A. and J. M. Hunt: Third brine pool in the Red Sea. Nature, 213, 687 (1967).

Ryan, W. B. F., E. M. Thorndike, M. Ewing, and D. A. Ross: Suspended matter in the Red Sea brines and its detection by light scattering. *In: Hot brines and recent heavy metal deposits in the Red Sea*, E. T. Degens and D. A. Ross (eds.). Springer-Verlag New York Inc., 153–157 (1969).

Swallow, J. C. and J. Crease: Hot salty water at the bottom of the Red Sea. Nature, 205, 165 (1965).

Turner, J. S. and H. Stommel: A new case of convection in the presence of combined vertical salinity and temperature gradients. Proc. Nat. Acad. Sci., 52, 49 (1964).

ATLANTIS II Account *

ARTHUR R. MILLER

Woods Hole Oceanographic Institution
Woods Hole, Massachusetts

Abstract

A brief account is given of the events leading to and occurring during the 1965 **ATLANTIS II** trip to the hot brine region of the Red Sea.

At the Intergovernmental Oceanographic Commission in July, 1965 in Paris the findings of **ATLANTIS II** in the Red Sea were announced as a mutual discovery resulting from direct cooperation, interest and exchange between participating scientists and ships in the International Indian Ocean Expedition (1963–1965).

During the Expedition none of the ships was free to explore fully the anomalous condition in the Red Sea because of the commitment to investigations in the Indian Ocean according to plan. Yet logistically there were normal requirements to do productive research while in transit to the Indian Ocean. Accordingly, in 1963, **ATLANTIS II** carried out a series of hydrographic stations along the length of the Red Sea. It was only by circumstance that station 42 along the profile was coincident with the previously reported stations of **ATLANTIS,** station 5639 (Neumann and Densmore, 1959), and of **ALBATROSS,** station 254 (Bruneau et al., 1953). At the suggestion of Mr. Densmore of the scientific team, the previous station position of **ATLANTIS** was occupied, not only to fill in the profile but also to check out the solitary anomaly given in the data of station 5639 obtained five years earlier.

The observations of unusually warm saline deep water from station 42 of **ATLANTIS II** began the chain of subsequent findings with the recognition that the anomalous results were factual (Miller, 1964). These results were transmitted to the **DISCOVERY** scientists and on **DISCOVERY's** first passage through the Red Sea anomalies were found and reported (Charnock, 1964). On her return voyage, re-occupying the station, she made the startling discovery of the hot, 44°C, brines (Swallow and Crease, 1965). The **METEOR** was the next ship to go to the Indian Ocean by way of the Red Sea route and it confirmed the results of the **DISCOVERY** (Dietrich et al., 1966).

Concern was expressed in Woods Hole for the limited range of ordinary deep-sea reversing thermometers and the amount of extrapolation required to arrive at the 44° determination. The maximum range of thermometers protected from pressure is only about 0°–30°C with even smaller ranges, 0°–6°C, normally used for very deep water. Depths are obtained by pairing a protected thermometer with one that is unprotected from pressure. Since they are both measuring the same temperature, the difference in the two readings gives the effect of pressure which is ultimately computed to give the depth of observation. The range of unprotected thermometers may correspond to about 0°–64°C to allow for the additive effect of pressure on the mercury column. A solution for measuring the temperature of the hot, deep brines was suggested by G. G. Whitney, Jr., i.e., unprotected thermometers converted to protected thermometers with a range of 0°–64°C and calibrated, thereby affording a

* Woods Hole Oceanographic Institution Contribution No. 2180.

15

standard of measurement commensurable with other deep ocean observations.

Planning for the 1965 cruise of **ATLANTIS II** to the Indian Ocean involved multi-disciplinary investigations but, unlike many other cruises, geological and physical oceanographic observations were to proceed concurrently. Consequently, the geologists and oceanographers were working together in the Red Sea aboard **ATLANTIS II.** In fact, sedimentary cores were obtained along with the hydrographic cast, a symbiotic arrangement that provided the oceanographer with bottom water. Piston cores and a number of free-fall cores were aboard. There were, however, no chemicals aboard for analysis other than those deemed necessary for various types of planned observations, having little to do with geochemical objectives associated with the brines. So when **ATLANTIS II** entered the Red Sea in 1965 she was prepared to measure the excessive temperatures, to test the brine for conductivity by volume dilution, to obtain geological cores in various ways, to make standard chemical analyses for oxygen and phosphorus, but not equipped to perform other analytical chemistry.

It was our intention to examine other depressions along the length of the Red Sea for the possibility of brine sources, for there were several major depressions north and south of the brine area. Nothing was observed to warrant further investigation beyond the hot brine area. Our navigation was enhanced by the addition of a Satellite Navigation System and a computer to convert the satellite message to fixed coordinates. On approaching the hot brine site we intended to pass over it briefly, towing a magnetometer, record the depths, and then to double back and stop on station. This proceeded according to plan.

At 1245 on February 17, 1965, we were on station but the ship quickly drifted away from the deep soundings. It took two hours to find our way back to the hole. Although there was very little wind, ship's drift began to plague our operations. On this station we had the excitement of getting our first samples of the hot, 44° water and, along with it, a core of black material. The rumor began to circulate about the ship that we had found oil, for its appearance was more like sludge from an automotive repair shop rather than the deep sea cores the crew were accustomed to. At the end of the station we drifted away again and spent another two hours getting back. Another station was made with a deep maximum temperature of only 27.8°C. Rather than repeating it immediately, the piston corer was put over the side. On retrieval of the piston corer, we found we had drifted away once more, and again found it necessary to search for deep soundings.

At 0040, February 18, a free-fall corer was put overboard. It was not recovered. The hydrographic station was made. To our surprise, the deep, hot brine was warmer than the first by 12°. At 0345 another deep cast was made confirming the first. It was evident that the **ATLANTIS II,** in the search to recover position in the 44° water, had come upon a different brine area where the temperature was 56°C. The satellite fix at 0642 and astronomical fixes showed that we had moved 4 to 5 miles north of the position of the 44° water.

The searching procedure that was used was simple enough, though a navigator's nightmare. The echosounder demonstrated that the sides of the hole were very precipitous. With an unknown current, the direction of drift was indeterminate. Navigational fixes were too infrequent to estimate ship's movement. In order to return to the hole, the procedure was to move slowly in any direction and to perform a series of circling maneuvers governed by the trend of the soundings. Thus, the ship was steered with 10 degrees right rudder and, before coming full circle, shifted to 10 degrees left rudder. As soundings perceptibly deepened it was possible to follow the trend with this circling procedure by timing the rudder shift to the subsequent exploratory circle. It was this type of "blindman's buff" that put us in what is now known as Atlantis II Deep.

After recovering position again to counteract the set of the current, another station was made in Atlantis II Deep with a coring device in place of the hydrographic lead weight. A new shiny brass swivel was attached above the corer. When it was retrieved the brass swivel had turned a dull black with a successful core achieved. The time was now 1045 and we prepared to lower the piston corer after an interval of one hour to recover position. Our position, however, was not secure and it was not until 1400 that the piston corer could be lowered. Only a 2m core was obtained.

There was no scheduled time available to continue with this investigation. With our commitment in the Indian Ocean and the difficulties in maintaining position and sampling, it was decided that enough new information had been obtained to warrant significant attention in the future with better preparation. There still remained another hole to the south to investigate. This latter deep, however, showed no results relative to the hot brine problem.

During the passage between the area of Atlantis II Deep and Aden, ways were devised to analyze the content of the brines, not only their salt content but also the individual chemical properties. Samples of core material were also analyzed. The ingenuity of Mr. Pocklington and Dr. Degens was put to task to provide analytical chemicals out of such material that one could find in galley provisions and other ships' stores (Densmore, 1965). It is remarkable that the significance of the samples was recognized through these improvised analyses before we reached Aden. Consequently, on arrival at the port of Aden, samples were carried by hand to Woods Hole by Mr. Robert Alexander in order to confirm the shipboard results.

It is difficult to describe the feelings aboard ship with the recognition of a finding which was significant scientifically and perhaps economically. At first we thought the investigation was purely a problem in physical oceanography but it turned out to be a geological phenomenon of greater importance. Speculation was rife as to the economic value of the metals, the means of recovery and the existence of similar phenomena elsewhere. As for the latter, assuming uniqueness, there was no question that broader scientific inquiry would soon follow. Detailed results of this cruise were published in 1966 (Miller *et al.*, 1966).

References

Bruneau, L., N. G. Jerlov, and F. Koczy: Rep. Swedish deep-sea expedition, **3**, 4. Appendix, Table 2, XIV–XXX (1953).

Charnock. H.: Anomalous bottom water in the Red Sea. Nature, **203**, 591 (1964).

Densmore, C. D.: Missing ingredients found. Oceanus, **11**(3), 6 (1965).

Dietrich, G., E. Krause, E. Seibold, and K. Vollbrecht: METEOR Forschungsergebnisse, Reihe A, **1**, Deutsche Forschungsergebnisse (1966).

Miller, A. R.: Highest salinity in the world ocean? Nature, **203**, 590 (1964).

——, C. D. Densmore, E. T. Degens, J. C. Hathaway, F. T. Manheim, P. F. McFarlin, R. Pocklington, and A. Jokela: Hot brines and recent iron deposits in deeps of the Red Sea. Geochimica et Cosmochimica Acta, **30**, 341 (1966).

Neumann, A. C. and C. D. Densmore: Oceanographic data from the Mediterranean Sea, Red Sea, Gulf of Aden and Indian Ocean. Woods Hole Oceanographic Institution, Ref. No. 60-2 (Unpublished manuscript) (1959).

Swallow, J. C. and J. Crease: Hot salty water at the bottom of the Red Sea. Nature, **205**, 165 (1965).

A Fourth Brine Hole in the Red Sea?

FEODOR OSTAPOFF

Atlantic Oceanographic Laboratories
Sea Air Interaction Laboratory
Miami, Florida

Abstract

This paper discusses some of the results of the **OCEANOGRAPHER** (USC & GSS) crossing of the Red Sea. A possible new hot hole was observed 617km north-northwest of the original hot brine area. Three reflecting layers were observed in the Atlantis II Deep that are apparently related to the different brine layers.

Recently a number of publications (Swallow and Crease, 1965; Krause and Ziegenbein, 1966; Miller *et al.,* 1966; Hunt *et al.,* 1967; Ross and Hunt, 1967; and Munns *et al.,* 1967) have dealt with the interesting problem of the hot brine holes in the Red Sea. Three such hot brine holes were discovered, documented and subsequently named Atlantis II Deep, Discovery Deep and Chain Deep. All three are located in a small area of less than 10 by 10 nautical miles.

In May, 1967 the USC & GSS **OCEANOGRAPHER** traversed the Red Sea on her scientific Global Expedition. A station in the Atlantis II Deep was occupied at the position 21°25′02″N and 38°03′03″E to obtain large volume water samples and cores for specialized analysis. The depth recorder was operating all the time. Positions were determined by satellite fixes taken every few hours.

The **OCEANOGRAPHER** is equipped with a General Electric narrow beam (2.66° at 3db) mechanically stabilized sound transducer system and operates on 19KHz. The recording is made on a precision fathometer recorder (Raytheon PFR-193A).

The calibration is set at 1,463msec⁻¹. Fig. 1 is a sample of the PFR record, recorded while on station in the Atlantis II Deep. Three characteristic sound reflections indicate the presence of the hot brine. The weakest reflection at 1,034 fathoms (1,891m), and two stronger reflections are found at 1,041 and 1,057 fathoms (1,904 and 1,933m), respectively. All depths quoted in this paper are uncorrected.

The **OCEANOGRAPHER's** fathogram may be compared with a number of similar records which were obtained by **METEOR**

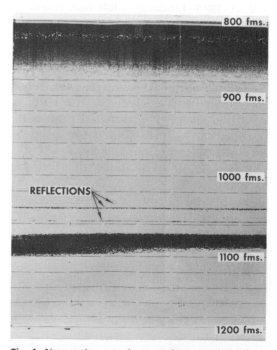

Fig. 1. Narrow-beam echo-sounder record obtained on **OCEANOGRAPHER** while on station in Atlantis II Deep at 21°25′2″N and 38°13′13″E. Depth scale uncorrected.

and published by Krause and Ziegenbein (1966). The **METEOR** carried the ELAC narrow-beam sounder 1 CO operating at the frequency of 30KHz with a beam width of 2.8° at 3db. The transducer is mechanically stabilized and calibrated for a sound speed of 1,500msec⁻¹.

Considering the instrumental limitations and differences, the agreement between the **OCEANOGRAPHER's** PFR record and the **METEOR's** PDR record is rather good. The travel time of the sound waves between the surface and the first reflection for the **METEOR** instrument and the **OCEANOGRAPHER** instrument are about the same (estimated as less than 4m). The **METEOR** station was located less than 4km to the south-southwest of the **OCEANOGRAPHER's** station. Thus, the **OCEANOGRAPHER** found, about two years later, conditions similar to those observed by the **METEOR** as far as the layering structure is concerned.

Table 1 presents scaled distances in meters between the identifiable reflectors. Individual differences appear to be relatively large, although the distance between the first and the third reflecting surface is very little. The differences may arise from the fact that one echo-sounder operates on 19KHz, the other on 30KHz; thus, different scatterers may have been involved.

The three different reflecting layers correlate well with the temperature structure observed by Ross (1969). These layers occur at depths of large temperature and salinity changes (Table 2) and are probably due to the increase in density associated with these changes.

The narrow-beam deep echo-sounding systems used on the **METEOR** and **OCEANOGRAPHER** thus proved to be useful tools in detecting, among other things, the hot brine holes. Scanning the PFR

Table 2 Comparison of Depths of Reflecting Layers and Temperature Structure as Observed by Ross, 1969

(All depths uncorrected fathoms)

Reflecting Layers	Depth	Temperature Structure	Depth
1	1,034	Start of hot brine	1,025
2	1,041	Start of 44°C water	1,037
3	1,057	Start of 56°C water	1,054

records for specific characteristic features revealed an interesting reflection in a very narrow hole located some 617km north-northwest of the original hot brine area. It was named Oceanographer Deep. A reproduction of the original record is presented in Fig. 2. The bottom of the hot hole was observed at 785 fathoms (1,446m). The Oceanographer Deep measures on top 5.5km and is at the bottom about 1.0km wide along the trackline. East-west dimensions are unknown. Possibly the course of the **OCEANOGRAPHER** may have crossed the hole as its outer edge. The reflection can

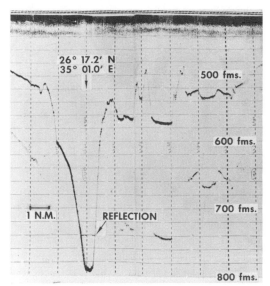

Fig. 2. Narrow-beam echo-sounder record obtained on **OCEANOGRAPHER** showing one reflecting layer. Depth scale uncorrected.

Table 1 Distances in Meters Between Reflecting Layers for the METEOR and OCEANOGRAPHER Records

	1st–2nd layer	2nd–3rd layer	1st–3rd layer
METEOR	8m	31m	39m
OCEANOGRAPHER	13m	29m	42m

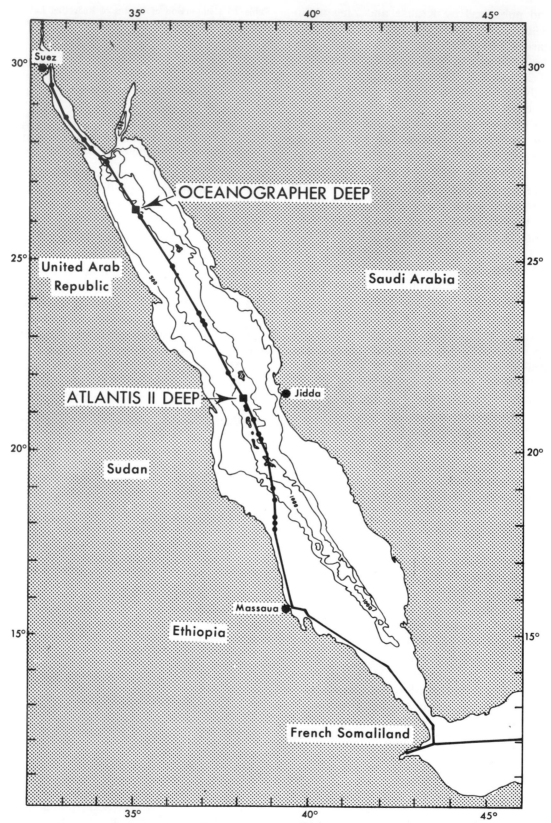

Fig. 3. Trackline of the **OCEANOGRAPHER** through Red Sea showing geographic locations of Atlantis II Deep and Oceanographer Deep; 500m and 1,000m contour lines after Drake and Girdler (1964).

be clearly identified and seems rather strong, raising the suspicion of a possible hot hole in an unsuspected area and at a shallower depth than the other three known holes.

Most of the available records from the other holes show multiple reflections unlike the one from the Oceanographer Deep, which reveals only one reflecting layer at an uncorrected depth of 741 fathoms or 1,355m (Fig. 2). Krause and Ziegenbein (1966) published a record obtained at Meteor station 28 (21°17.2′N, 38°02.5′E) from the eastern edge of the Discovery Deep with only one reflecting layer.

Fig. 3 shows the trackline of the **OCEANOGRAPHER,** the location from which the PFR record (Fig. 2) was obtained in relation to the Atlantis II Deep where the record (Fig. 1) was obtained. Indeed, if the Oceanographer Deep does contain hot brine, for the proof of which we must await further research, then a new area of hot holes has been found, probably belonging to a separate system.

Acknowledgment

The author would like to thank participating scientists and the crew of the USC & GSS **OCEANOGRAPHER** for their efforts; in particular, Mr. W. Moore of New York State University at Stony Brook, Long Island and Dr. J. Swinnerton of the Naval Research Laboratory, Washington, D.C., for their assistance searching the voluminous records.

References

Drake, C. L. and R. W. Girdler: A geophysical study of the Red Sea. Geophys. J. Royal Astro. Soc., **8,** 473 (1964).

Hunt, J. M., E. E. Hays, E. T. Degens, and D. A. Ross: Red Sea: detailed survey of hot brine areas, Science, **156,** 514 (1967).

Krause, G. and J. Ziegenbein: Die structur des heissen salzreichen Tiefenwassers im Zentralen Rotem Mccr. METEOR-Forschungsergebnisse Reihe A. **1,** Gebr. Borntraeger, 53 (1966).

Miller, A. R., C. D. Densmore, E. T. Degens, J. C. Hathaway, F. T. Manheim, P. F. McFarlin, R. Pocklington, and A. Jokela: Hot brines and recent iron deposits in deeps of the Red Sea. Geochimica et Cosmochimica Acta, **30,** 341 (1966).

Munns, R. G., R. J. Stanley, and C. D. Densmore: Hydrographic observations of the Red Sea brines. Nature, **214,** 1215 (1967)

Ross, D. A.: Temperature structure of the Red Sea brines. *In: Hot brines and recent heavy metal deposits in the Red Sea,* E. T. Degens and D. A. Ross (eds.). Springer-Verlag New York Inc., 148–152 (1969).

Ross, D. A. and J. M. Hunt: Third brine pool in the Red Sea. Nature, **213,** 687 (1967).

Swallow, J. C. and J. Crease: Hot salty water at the bottom of the Red Sea. Nature, **205,** 165 (1965).

Geological and Geophysical Setting

Geological Structures of the Red Sea Area Inferred from Satellite Pictures

MONEM ABDEL-GAWAD

Science Center, North American Rockwell Corporation
Thousand Oaks, California

Abstract

Photographs of the Red Sea area taken during the Gemini earth-orbiting missions show the gross structural relationship between the highlands of the Arabian-Nubian shield, the sediment-filled Red Sea "graben" system and the marginal sedimentary cover with large areas of volcanic activity.

Among five major systems of faults affecting the Arabian-Nubian massif, three systems trending N, NW, and WNW appear to have a profound influence on the structure of the Red Sea. The other systems oriented NE and E are more evident in the Gulf of Aden area. The fissure systems control the coastal outlines, and major physiographic features such as depressions, drainage lines, and much of the Nile Valley. The relative positions of two pairs of shear zones intersecting the northern part of the Red Sea provide additional geological evidence consistent with the relative northward movement of Arabia some 150km.

The regional fissure system disposition may be interpreted in terms of a combination of shear and tension producing conjugate fissures and gravity faults.

Introduction

The Red Sea extends in a northwesterly direction from the narrow straits of Bab El Mandeb at its southern end to the southern tip of Sinai Peninsula, a distance of about 1,930km. It cuts across a huge dome of Precambrian basement rocks (Arabian-Nubian massif) flanked by epicontinental and marine sediments (Picard, 1939). At its southern end, the Red Sea is connected to the Gulf of Aden which trends ENE-WSW, opening into the Indian Ocean and separating the southern part of the Arabian Peninsula from the Somali Plateau. At the northern end, the Red Sea bifurcates into the Gulfs of Suez and Aqaba creating the triangular Sinai Peninsula.

During the earth-orbiting Gemini missions (1965–1966), the United States' National Aeronautics and Space Administration (NASA) conducted a series of synoptic terrain photography experiments, which provided many high-quality color photographs of the Red Sea basin and surrounding terrain in Africa and Arabia. Most pictures were taken by a hand-held modified Hasselblad camera (Model 500-A) on Eastman Kodak Ektachrome film. They varied considerably in ground coverage, resolution and color qualities, depending on many factors, including the orbital altitude (100–850 statute miles), focal length (80mm), camera tilt and weather conditions.

On standard prints (8 × 8 inches) the photographic scale ranged from about 1:500,000 in nearly vertical photographs to less than 1:5,000,000.

Of more than thirty photographs used in this study, five were selected as illustrations. Figs. 1–5 are black and white reproductions of color transparencies which cover the entire Red Sea basin from the Mediterranean to the Gulf of Aden. Although most of the geological features described could be identified in at least one of these photographs, some features are better observed in lower altitude and nearly vertical pictures, and in early generation color reproductions. Those familiar with satellite pictures realize that observations made purely by visual inspection may be quite misleading and that careful correlation to topographic and geological maps and available geological literature is often necessary.

Fissures or faults are often recognized by lineaments in drainage, geological contacts, geomorphic alignments or subtle differences in color and tones along straight or gently curved lines. Since most satellite photographs show gross distortion, triangulation control of both photograph and a suitable base map is necessary for plotting the right feature in the right place.

This article describes the geological structures of the Red Sea area as they appear in each photograph, sequentially from north to south. For each picture a brief description of the regional geological setting as it is related to the physiography is followed by specific observations on prominent fissure systems and their relation to the Red Sea structures. In the final section a summary of the regional disposition of fissure systems is given along with the author's view on the nature of the Red Sea rift.

Stated briefly, geological evidence that a northward movement of Arabia relative to Africa was found along the sides of the northern segment of the Red Sea. It is suggested that as a result of a principal left-lateral shearing force of Erythrean trend (NW-SE) combined with NE-SW tension, conjugate fissures of shear trending N-S and WNW-ESE were produced along with principal Erythrean fissures. These fissures appear to have a profound effect on the development of the Red Sea rift.

Gulfs of Suez and Aqaba

Two Gemini photographs (Figs. 1 and 2) show the regional structure of that area in different perspectives. Fig. 1 shows the northern part of the Arabian-Nubian massif (Precambrian shield, PC) which is extensively developed in the Egyptian Eastern Desert, the Sinai Peninsula and the western Arabian highlands. In Sinai the Precambrian shield (PC) occupies the southern highlands. It is essentially a triangular horst block tilted to the north, towards the Mediterranean, and overlapped by a sedimentary plateau of Carboniferous (C), Cretaceous (K), and Tertiary strata (E).

West of the Gulf of Suez, the shield occupies an elongated fault block (Red Sea Range) which widens considerably inland south of Ras Benas. The shield is overlapped on the western margin by sediments of the Egyptian basin, notably epicontinental Nubian sandstone (N), Cretaceous (K) shales, limestone, chalk, Paleocene-Eocene (E) shale and limestones. The latter forms the extensive plateau between Wadi Qena and the Nile Valley. The thickness of these sediments increases sharply towards the axis of the Egyptian basin west of the Nile (Said, 1962).

The Gulf of Suez is bordered by coastal strips covered by Miocene and younger sediments (M), which represent the Miocene "graben" that connected the Red Sea to the Mediterranean. The Gulf of Aqaba, which was not fully developed until the Pliocene, lacks the Miocene coastal strips and is directly bordered by shield rocks. On the eastern side of the gulf, the Precambrian shield widens considerably inland further south, forming the Arabian highlands. It is overlapped by Paleozoic to Mesozoic sediments with a gentle northerly dip.

Alpine Foreland Folding

The slope of the Arabian-Nubian massif is indicated by the regional dip of stratified rocks overlapping its margins. The dip changes from westerly on the western side of the Gulf of Suez to northerly in Sinai and the northeastly in the northern part of Arabia. This area, set in the foreland of the Alpine orogenic belt, is affected by gentle northeast trending folds (Syrian Arc). A prominent example of this fold system is the huge Wadi Araba (Fig. 2) anticline on the western side of the Gulf of Suez, but similar folds are known to occur in northern Sinai (Said, 1962).

Fig. 1. Gemini XII photograph (looking east) showing northern part of the Red Sea and the Gulfs of Suez and Aqaba and southern part of Sinai Peninsula. Fault-controlled Nile Valley separates the Eastern and Western Deserts in Egypt. The Arabian Highlands are seen beyond the Red Sea in background. Photo number S-66-63479, courtesy of NASA.

Symbols

C = Carboniferous strata	N = Nubian Sandstone formation, age variable
E = Tertiary (mostly Eocene) strata	PC = Precambrian shield
F = Fault, arrow indicates fault trace	V = Volcanics, mostly basic extrusives
K = Cretaceous strata	⊤ = Approximate strike and dip of strata
M = Miocene-Quaternary sediments	⇌ = Relative strike-slip movement

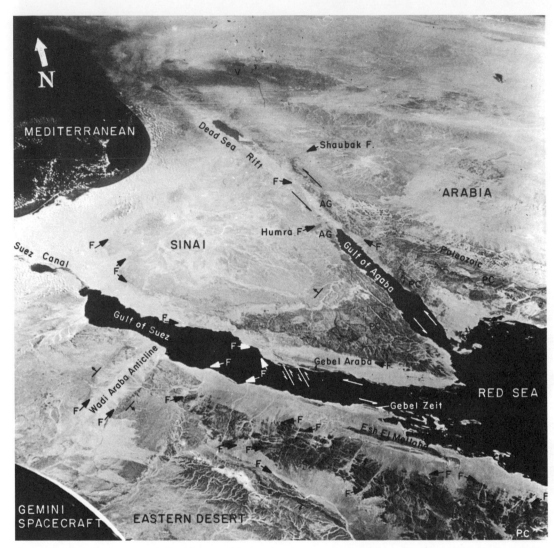

Fig. 2. Gemini XI photograph (looking north) showing transcurrent faults in the Dead Sea-Gulf of Aqaba and Gulf of Suez regions. Note the influence of faults on coastal lines, drainage and the Precambrian shield. Photo number S-66-54893, courtesy of NASA.

Symbols

AG = Aqaba Granite   V = Volcanics, mostly basic extrusives
 F = Fault, arrow indicates fault trace   ⊤ = Approximate strike and dip of strata
PC = Precambrian shield   ⇌ = Relative strike-slip movement

Transcurrent Movements

Gemini photographs show that the Gulfs of Suez and Aqaba are more than simple graben structures. Along the Gulf of Aqaba we note evidence supporting Quennell's argument that the Arabian side of the Gulf of Aqaba-Dead Sea rift has moved northward relative to Sinai cumulatively more than 100km (Quennell, 1958). One piece of his evidence, which concerns the general displacement of the shield rocks bordering the Gulf, is supported by our observations on marked differences in photographic color and tone of Precambrian rocks. We found by examination of Gemini photographs and geological maps * that distinctive rock types could be correlated on both sides and that reconstruction will bring these rocks into a more reasonable structural position. We also note in Fig. 2 the displacement of Quennell's Shaubak and Humra faults which mark the northern limits of the Aqaba Granite (AG) blocks (Fig. 2).

The Gulf of Suez area has generally been regarded as a graben structure controlled by NW-SE normal faults. There is considerable evidence, however, that it is profoundly influenced by intersecting systems of faults trending E-W, NE-SW, N-S, NW-SE, and WNW-ESE. The N-S and WNW-ESE faults are particularly responsible for the zigzag shape of the coastlines, the outline of the Precambrian shield, and major drainage lines west of Suez Gulf. The influence of these faults on the shield rocks of Sinai is equally evident. Along many of these faults lateral displacements are observed: a notable example is the left-lateral off-set of Gebel Araba block in western Sinai along N-S faults. Small displacements along WNW faults are inferred across Esh El Mellaha range west of the Gulf of Suez.

The relationship between the Precambrian blocks in Gebel Araba and Gebel Zeit along the coasts of the Gulf of Suez is rather intriguing because of their similar-

* Rocks on the Arabian side mapped by Bramkamp et al. (1963) as Giddah Greenstone (gd), Granite-Gneiss Complex (gg), and Hornblende Granite (gr) correspond in Sinai to Schist-Andesite, Red Granite-Gneiss and Hornblende Granite (Hume, 1934, Plate LII).

ity in composition and structure. The similarity may be coincidental, but on the other hand a left-lateral movement of some 50km along the Gulf of Suez could not be ruled out, particularly when we note that the Precambrian terrain of the northern Red Sea Range west of the Gulf shows a striking similarity to the northwestern corner of the basement rocks in Sinai.

Northern Red Sea

From the southern tip of Sinai to Ras Benas and Ras Abu Madd, the Red Sea is characterized by relatively straight and parallel coastal lines. On the African side (lower left corner, Fig. 3) the Precambrian shield occupies about two-thirds the width of the Egyptian Eastern Desert, east of the Nile Valley. The western and eastern edges of the shield are controlled by Erythrean faults parallel to the main direction of the Red Sea, but the influence of N-S, WNW-ESE, NE-SW and E-W faults is also quite evident. The thin coastal strip between the shield and the Red Sea is occupied by Miocene and younger sediments.

On the Arabian side (upper part, Fig. 3) the inland limit of the shield is less definite and is affected by enormous shear zones undulating in a WNW-ESE to NW-SE sinuous pattern. The shield is overlapped on the east by cuestas of terrestrial to epicontinental Paleozoic and marine Mesozoic-Paleogene sediments. These are partly covered by extensive Tertiary and Quaternary basaltic lava flows (V). Towards the Red Sea an irregular fault-controlled strip is occupied by Miocene sediments and Quaternary alluvial and gravel cover (M).

New Evidence of Displacement

The geological evidence for the relative movement of Arabia some 100km along the Gulf of Aqaba-Dead Sea rift (Quennell, 1958) suggests that "the horizontal displacement must also have affected the most northern part of the Red Sea" (Girdler, 1965). Although this extrapolation has been largely deductive, we found geological evi-

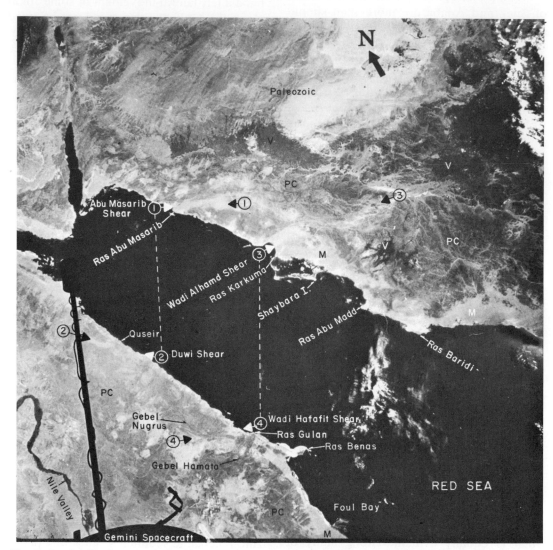

Fig. 3. Gemini XI photograph (looking north) showing geological evidence for a relative northward movement of Arabia (background) from Africa (foreground). Note the displacement of Abu Masarib shear (1-1) from Duwi shear (2-2), and Wadi Alhamd shear (3-3) from Wadi Hafafit shear (4-4). Photo number S-66-54664, courtesy of NASA.

Symbols

M = Miocene-Quaternary sediments
PC = Precambrian shield
V = Volcanics, mostly basic extrusives

dence that it may, indeed, be the case. This evidence is derived from the position of two pairs of shear zones in Arabia and Africa which indicate a lateral displacement consistent with that observed along the Gulf of Aqaba-Dead Sea rift (Abdel-Gawad, 1969).

Abu Masarib-Duwi shear zones (Fig. 3) constitute a distinctive pair intersected by eastern and western coasts of the Red Sea. On the Arabian side the shear zone ending near Ras Abu Masarib is a prominent structure which appears in the U.S. Geological Survey Maps of the Arabian Peninsula (Brown *et al.*, 1963). The zone is characterized by enormous wrench faults trending NW-SE and intersecting the eastern Red Sea coast at about 30°. Abu Masarib shear is remarkably well displayed in Gemini pictures (1-1, Fig. 3) because the conforming drainage and associated light colored alluvium stand out in contrast to the surrounding dark schistose terrain. It is perhaps very significant that Miocene evaporite sediments and raised coral reefs occur along this structure for a considerable distance inland, which indicates that the zone was under the water of an elongated gulf during the Miocene and was subsequently raised above the sea level with the general taphrogenic uplift of the Arabian highlands. There are indications from bathymetric contours in the geological map of the area (Brown *et al.*, 1963) that the structure is continued for some 70km off shore.

In Egypt we find a similar and most probably related structure in Gebel Duwi area near Quseir. The Duwi shear zone intersects the Red Sea south of Quseir at about 30° and is interrupted by NE-SW faults (2-2, Fig. 3). The structure of that area as described by Said (1962, Fig. 16) is "characterized by strike faults where the fundamental complex of ancient crystalline rocks with its overlying mantle of Cretaceous and Eocene sediments were subjected then to intense deformation which resulted in a series of highly tilted fault blocks running along axes trending north-northwest or northwest." On top there are relatively undisturbed Miocene evaporites and raised beach sediments.

The similarity of the two structures, now separated by the Red Sea rift, is striking and it seems very probable that the two structures were once a continuous shear zone. The age of this deformation is not known and may be quite old. Strike-slip deformation of Eocene rocks in Gebel Duwi area indicates that shearing was active during the Eocene. The Abu Masarib-Duwi shear began to be displaced along north-south left-lateral faults related to the Gulf of Aqaba-Dead Sea movement, which according to Quennell (1958) started in the Cenozoic time. Miocene sediments along these structures indicate that they were flooded by the Miocene Sea and had perhaps continued to separate until their present position.

Wadi Alhamd-Wadi Hafafit Shear Zones are a second pair of apparently related structures. On the Arabian side Wadi Alhamd shear (3-3, Fig. 3) is a zone of great wrench faults (WNW-ESE) which were mapped by the U.S. Geological Survey (Brown *et al.*, 1963). The structure is generally marked by the drainage of Wadi Alhamd and can be inferred to extend below the alluvial cover intersecting the Red Sea at about 45° near Ras Karkuma.

It is perhaps no coincidence that on the African side there is an equally impressive shear zone of a similar trend which intersects the coastal line near Ras Gulan (4-4, Fig. 3). This Wadi Hafafit shear zone coincides with and is most probably responsible for one of the greatest structural and topographic breaks of the Red Sea Precambrian Range in Egypt. The magnitude of this break can be judged from its elevation (about 300m) in contrast to the high peaks of Gebel Nugrus block to the north (1,505m) and Gebel Hamata block to the south (1,977m). Impressive evidence of the tectonic complexity of this wrench zone is observed in nearly vertical Gemini photographs,* showing sinuous folds and wrench faults in the Hafafit-Nugrus area north of Ras Benas Peninsula.

The relative positions of Masarib-Duwi

* Not shown in this article, NASA Gemini IV photo number S-65-34784.

and Alhamd-Hafafit shear zones provide geological evidence consistent with a northward displacement of Arabia some 150km. Reconstruction will not only bring the shear zones in compatible positions but will also bring the embayment of Foul Bay and Ras Abu Maad-Ras Baridi "bulge" in structural harmony (Fig. 3).

Coastal Lines and Fissures

Most of the fissure systems observed inland are generally reflected in the coastal lines. Along the northern part of the Red Sea where coastal lines are generally straight, they conform to widespread Erythrean (NW-SE) fissures in Egypt and Arabia. Details of coastal lines and basement outlines, however, show small but distinct deflections, particularly along N-S and WNW-ESE lineaments. The effect of these systems on the Arabian shield is quite evident in the down-faulted Miocene blocks (M) along the coast: e.g. the polygonal inland block opposite Shaybara Island (Fig. 3).

Further south the coastal lines show major swings whereby Foul Bay appears to fit the bulge south of Ras Abu Madd. It is in these areas adjacent to major coastal swings that we observe a most profound tectonic complexity. Detailed tectonic maps prepared by the author from vertical Gemini photographs (Abdel-Gawad, 1967) show that the Precambrian shield is broken by at least five sets of faults trending N-S, NW-SE, WNW-ESE, E-W, and NE-SW. Some of these faults extend hundreds of kilometers inland and control major drainage lines, the Nile Valley, and large tectonic depressions in the Eastern and Western Deserts of Egypt. The same complexity is observed in corresponding areas in the Arabian Shield.

Southern Red Sea

Sinuosity and Faulting

The Red Sea between Ras Benas-Ras Abu Madd and the Dahlac-Farasan Islands is characterized by sinuous coastal lines changing from NNW to WNW (Fig. 4). Here, the rift cuts across the widest

and central part of the Arabian-Nubian massif and is noted for a well-developed central trench, associated positive gravity and magnetic anomalies (Girdler, 1969). The hot brine deeps are located approximately midway between Ras Abu Shaghara Peninsula and the Arabian coast.

On the African side the major physiographic features are essentially a continuation of structures observed in the north, where the high Red Sea chains of crystalline rocks are bound to the east by steep fault scarps separating the high mountains from the low coastal strip. The western slope of the shield is gentle, extending further inland beyond the Nile, and is overlapped by an extensive sheet of Nubian sandstone (e.g. the area between the River Atbara and the Spacecraft in Fig. 4). Further south the great horst of the Ethiopian Plateau, which rises to elevations greater than 1,500m, is largely obscured by cloud cover.

In the Arabian Peninsula the dark shield rocks extend deep inland but narrow considerably towards the south in Yemen. The southeastern margin of the massif is overlapped by a succession of Paleozoic-Mesozoic and Cenozoic strata.

The sinuosity and parallelism of the Red Sea coastal lines are among the salient features which have attracted the attention of all students of the Red Sea problem and have been for a long time a central argument for the split of Africa from Arabia along a central sinuous fissure. The regular N-S and WNW swings of the coastal lines and the central trench are determined by major breaks of similar trends in the Precambrian shield.

For example, south of Suakin the coast changes from N-S to WNW-ESE but the N-S faulting is continued deep inland and is clearly marked by the drainage trunk of Wadi Barca (F_1) and related geomorphic lineaments (F_2). The WNW-ESE fault lineaments in the basement are also evident (F_3-F_3, Fig. 4).

East African Rift

Figs. 4 and 5 show the critical junction of the Red Sea, the Gulf of Aden and the East African rifts. South of Dahlac-

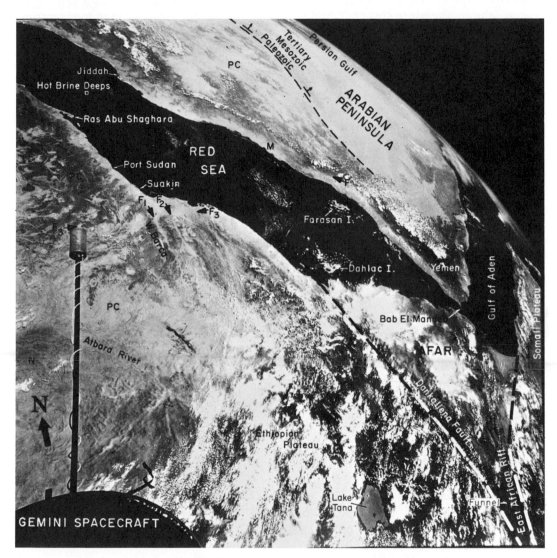

Fig. 4. Gemini XI photograph (looking northeast) showing the sinuous coast line of the Red Sea in relation to faults affecting the Precambrian shield. The lower right quadrant shows the junction of the Red Sea, East African and Gulf of Aden rifts (Funnel). Note the location of the Hot Brine Deeps midway between Ras Abu Shaghara and Jiddah. Photo number S-66-54533, courtesy of NASA.

Symbols

F = Fault, arrow indicates fault trace   PC = Precambrian shield
M = Miocene-Quaternary sediments   ⊤ = Approximate strike and dip of strata
N = Nubian Sandstone formation, age variable

Farasan Islands the converging coasts of the Red Sea join the Gulf of Aden through the narrow strait of Bab El Mandeb which separates the southeastern corner of Arabia from Africa. The eastern branch of the Great Rift Valley is discernible through a break in the cloud cover. In this area, called the Funnel, the East African rift widens into a triangular tectonic depression enclosing the Afar plains covered mostly by Cretaceous to Quaternary basaltic lava flows (V). The western edge of the triangle (Dankaliena fault) extends from the Red Sea coast near Dahlac Island in a N-S direction, forming a funnel neck with the southern edge of the rift. The Dankaliena fault separates the Afar volcanic province from the Cretaceous volcanic highlands of the Ethiopian Plateau (Fig. 4).

The southern edge of the rift separates the Somali Plateau (Jurassic-Tertiary marine strata) from the Afar volcanic province and, across the Gulf of Aden, from Southern Arabia. Beyond the narrow coastal plains which border the Red Sea and the Gulf of Aden, there is an essential continuity of composition and structure.

In Arabia we can distinguish four major physiographic provinces. Along the Red Sea coast there is a maritime plain covered by deltaic stream deposits and recent eolian sands (M, Fig. 5). This is followed inland by the mountainous Yemen horst, complicated by many secondary faults. The terrain consists (Geukens, 1966) mostly of tuffs and horizontal lava flows of Cretaceous to Quaternary age (V), and in part by Precambrian shield rocks. The inland region consists of sedimentary plateaus of Jurassic to Tertiary age. The most conspicuous is the great arch of the Hadramawt Plateau, which grades inland into the vast sand dune desert of the Empty Quarters (Ar Rub Al Khali).

Gulf of Aden

The continuity of geological provinces in Africa and Arabia and the configuration of coastal lines of the Gulf of Aden led to the belief that the land masses must have once been connected. We note from satellite pictures that abrupt swings in coastal lines are often associated with major fault lines which could be traced for considerable distances inland. The most prominent and perhaps the least noted in the literature are great lineaments marking fissure lines trending WNW which cut diagonally across Arabia from the Gulf of Aden to the Red Sea and across the "horn of Africa" from the Gulf of Aden to the Indian Ocean. Important examples of these major trends were described in Southern Arabia (Beydoun, 1966) and in the Somali Plateau (Somaliland Oil Exploration Co., 1954). Along with these old fissures there are NE-SW and E-W faults, all shaping the zigzag coastal outlines of the Gulf of Aden. As we approach the junction with the Red Sea, however, two additional fault systems become prominent: N-S, parallel to the eastern coast of the Red Sea; NW-SE, such as those cutting the Afar Plains near the western end of the Gulf of Tajura.

Large areas of the southern Red Sea region are concealed under cloud cover and many critical structures are obscured by the lava fields. We could, nevertheless, observe many geomorphic lineaments consistent with trends observed all over the Red Sea and Gulf of Aden rifts.

Fissure Systems in the Red Sea Area

Gemini photographs show that throughout the entire region surrounding the Red Sea five major fissure systems are developed.

North-South (East African). This system is parallel to the N-S trend of the East African rift and probably includes NNE-SSW aualitic trends (parallel to the Gulf of Aqaba). Prominent examples occur in the Aqaba-Dead Sea region along N-S trends of the Red Sea coast and the Nile River system, the Dankaliena fault zone marking the western side of Afar triangle and the western side of the Yemen horst.

Northwest-Southeast (Erythrean). This system is parallel to the prominent NW lineaments and fissures in Erythrea and marks the general trend of the Red Sea. Prominent examples control the coastal lines of the Red Sea north of latitude 24°

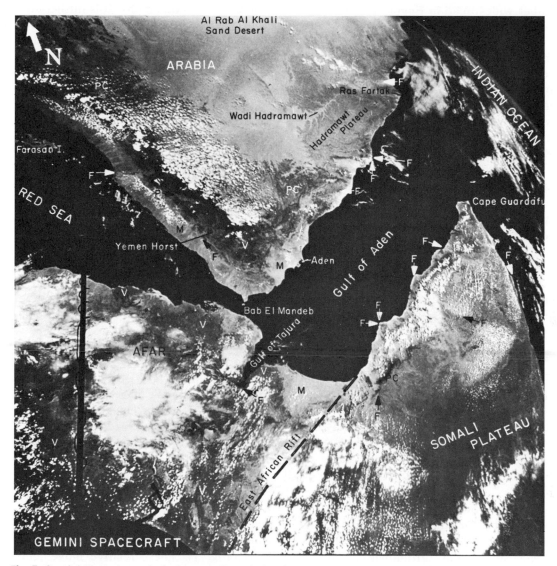

Fig. 5. Gemini XI photograph (looking northeast) showing the converging coasts of the Red Sea joining the Gulf of Aden through the strait of Bab El Mandeb. Note the continuity of geological features across the seaways. Photo number S-66-54536, courtesy of NASA.

Symbols

F = Fault, arrow indicates fault trace   PC = Precambrian shield
M = Miocene-Quaternary sediments   V = Volcanics, mostly basic extrusives

and the general direction of the Gulf of Suez. This system also coincides with NW trends in the Nile River system and is rather widespread in Egypt, Sudan and the Arabian Peninsula.

WNW-ESE. This important system has so far received little attention. It appears to be responsible for most of the westerly swings of the Red Sea and the Gulf of Suez and marks major breaks affecting large basement blocks on both sides of the Red Sea and in southern Arabia and the Somali Plateau.

In Southern Arabia, Beydoun (1966) states "that the most numerous and prominent faults are not strictly on the Red Sea trend (northwest) in The East Aden Protectorate but follow a west-northwest trend. . . ." He continues to describe their profound effect on the basement of Southern Arabia and relates them to similar structures in the Somali Plateau (Somali Oil Exploration Co., Ltd., 1954). Beydoun (1966) suggests that these long and straight lines are involved in the transcurrent movement along the Gulf of Aden and describes structures indicative of "block jostling" along these lines.

East-West (Tethyan). This system marks Tethyan trends and may include ENE trends in the rift system: Gulf of Aden, southern end of Afar triangle, east-west breaks in northern Egypt and Sinai Peninsula.

Northeast-Southwest. This system appears to parallel the Syrian Arc structures and is well developed in Sinai and the Dead Sea region. Prominent lineaments along this direction appear to influence NE swings of the "great bend" of the Nile in Sudan. Similar trends occur within the East African rift and along NE coastal swings in the Gulf of Aden.

Summary

Among five major systems of faults affecting the Arabian-Nubian massif, three systems trending N, NW, and WNW appear to have a profound influence on the structure of the Red Sea. The other systems oriented NE and E (also ENE) are more evident in the Gulf of Aden area.

Evidence of a left-lateral transcurrent movement along the Gulf of Aqaba-Dead Sea rift (Quennell, 1958) could be inferred from Gemini photographs. It is further corroborated by our correlation of distinctive rock types in the basement complex bordering the Aqaba Gulf. We also found geological evidence that the movement has affected the northern part of the Red Sea. If we consider that the shear zone pairs, Abu Masarib (Arabia)-Duwi (Africa) and Alhamd (Arabia)-Hafafit (Africa), were once two continuous structures and were displaced along N-S faults, a relative northward movement of Arabia from Africa amounting to about 150km is indicated.

There is considerable evidence that the major breaks affecting the Arabian-Nubian massif and coastal lines occur along faults trending N, NW, and WNW. Since the N-S and WNW-ESE fissures are diagonal to the NW-SE suture of the Red Sea, they may be considered as conjugate fissures of shear associated with it. Opening of chasms consistent with the Red Sea structure is possible if the N-S fissures are associated with left-lateral and WNW-ESE with right-lateral movements. We have clear examples of left-lateral displacements along N-S faults (Gulf of Aqaba-Dead Sea and Gebel Araba in Sinai). The right-lateral displacement could so far only be inferred from the relative positions of opposing blocks in Arabia and Africa where the Red Sea takes westerly swings. It is perhaps significant that E-W faults in northern Sinai invariably show right lateral displacement (Vroman, 1961, Fig. 7). These relative movements are consistent with experiments on a shearing-tensional model (Belousov and Gzovsky, 1965, Fig. 24, d). If we apply this model to the Red Sea, a tensional force of NE-SW trend with a left-lateral shearing couple of Erythrean trend will produce a sinuous chasm and lateral displacements consistent with the observed structures in the Red Sea area.

Acknowledgment

Gemini photographs used in this study were made available through the courtesy of the U.S. National Aeronautics and Space Administration. The author is particularly indebted to Dr. Paul D. Lowman, Jr., Principal Investigator of Synoptic Terrain Photography, NASA Goddard Space Flight Center, Greenbelt, Maryland, for his cooperation and interest.

References

Abdel-Gawad, M.: Geologic exploration and mapping from space. Presented in AAS Meeting, Boston, May 1967.

——: New evidence of transcurrent movements in Red Sea area and petroleum implications. Amer. Assoc. Petroleum Geologists Bull. (in press, 1969).

Belousov, V. V., and M. V. Gzovsky: Experimental tectonics. *In: Physics and Chemistry of the Earth,* Ahrens, Press, Runcorn, and Urey (eds.), Pergamon Press, **6,** 410 (1965).

Beydoun, Z. R.: Geology of the Arabian Peninsula, Eastern Aden Protectorate and part of Dhufar. U.S. Geological Survey Professional Paper 560-H, 49 (1966).

Bramkamp, R. A., G. F. Brown, D. A. Holm, and M. Layne Newton, Jr.: Geology of the Wadi As Sirhan Quadrangle, Kingdom of Saudi Arabia. U. S. G. S. Miscellaneous Geological Investigations Map 1-200A, Scale 1:500,000 (1963).

Brown, G. F., R. O. Jackson, R. G. Bogue, and E. L. Elberg, Jr.: Geology of the Northwestern Hijaz Quadrangle, Kingdom of Saudi Arabia. U. S. G. S. Miscellaneous Geological Investigations Map I-204A, Scale 1:500,000 (1963).

Geukens, F.: Geology of the Arabian Peninsula, Yemen. U. S. G. S. Professional Paper 560-B, 23 (1966).

Girdler, R. W.: The role of translational and rotational movements in the formation of the Red Sea and Gulf of Aden. *In: The World Rift System,* Geological Survey of Canada Paper 66-14, 65 (1965).

——: The Red-Sea — A geophysical background. *In: Hot brines and recent heavy metal deposits in the Red Sea,* E. T. Degens and D. A. Ross (eds.). Springer-Verlag New York Inc., 38–58 (1969).

Hume, W. F.: Geology of Egypt. Egyptian Survey Department, Cairo, **2,** Part 1, 300 (1934).

Picard, L.: Outline of the tectonics of the earth with special emphasis upon Africa. Bull. Geol. Dept., Hebrew University, II, **3–4** (1939).

Quennell, A. M.: The structural and geomorphic evolution of the Dead Sea rift. Quart. J. Geol. Soc., London, **114,** 1 (1958).

Said, R.: *In: The Geology of Egypt.* Elsevier Publishing Co., Amsterdam and New York, 377 (1962).

Somaliland Oil Explorations Co., Ltd.: A geological Reconnaissance of the Sedimentary Deposits of the Protectorate of British Somaliland. Pub. by Crown Agents for the Colonies for the Government of the Somaliland Protectorate, Millbank, London, 41 (1954).

Vroman, A. J.: On the Red Sea Rift problem. Bull. Res. Counc. of Israel, **10G,** 321 (1961).

United States Geological Survey: Geologic map of the Arabian Peninsula. U. S. G. S. Miscellaneous Geological Investigations Map I-270A, Scale 1:2,000,000.

United States National Aeronautics and Space Administration. Gemini Photographs S-66-54533, S-66-54536, S-66-54664, S-66-54893, S-66-63479, Gemini IV-XII (1965–1966).

The Red Sea—A Geophysical Background

R. W. GIRDLER

School of Physics
University of Newcastle upon Tyne
United Kingdom

Abstract

The history and evolution of the Red Sea are considered in the light of the continental movement of Africa and Arabia. The relevant palaeomagnetic data pertaining to the latter are summarized. The Red Sea developed in phases, and it is likely that these correspond in time to phases of continental drift. The geophysical data for the Red Sea are summarized in a general résumé and in more detail by considering the Red Sea in three parts, viz. the northern Red Sea (north of 25°N), the central Red Sea (17° to 25°N) and southern Red Sea (south of 17°N). Structural differences from north to south are noted. Finally, a summary is given for the evolution of the Red Sea taking into account geophysical and geological data for the whole Afro-Arabian region.

Introduction

The Red Sea occupies part of a big crack in the continental crust of Africa and Arabia. Its length from the tip of the Sinai Peninsula in the north to the Strait of Bab el Mandeb in the south is 1,680km. Its shore to shore width increases from north to south being 180km in the north and 370km at 16°N where it is widest. The land on either side reaches heights of more than 2km, the highest land being in the south. The deepest part of the sea, in an axial trough, has depths exceeding 2km. Other parts of the big crack include the Gulf of Aden which extends eastward for about 1,000km, the East African rift which extends south through Ethiopia, Kenya and Tanzania for 3,500km and the Wadi Aqaba-Dead Sea— Jordan rift which extends northward for more than 450km from the Gulf of Aqaba. Beyond the Gulf of Aden the crack joins the Carlsberg rift and the world oceanic rift system after being offset at the Alula-Fartak trench and Owen fracture zone.

The Red Sea and Gulf of Aden is by far the widest part of the crack. The parallelism of the Red Sea shores and Gulf of Aden shores have fascinated many, including Wegener. Wegener considered the Red Sea to be a place where the continental crust has completely separated and where the early stages of continental drift are seen. There is now geophysical evidence for some separation of the crust with the evolution of oceanic crust in between. Further, there is palaeomagnetic evidence for continental drift and geophysical evidence for sea floor spreading away from the oceanic rifts.

Such a large crack extending for nearly 7,000km through the continental crust of Africa is most likely related to the movement of the African-Arabian landmass and to sub-crustal movements. The development of the Red Sea is therefore first considered in the light of palaeomagnetic evidence for the continental drift of Africa.

Continental Movement of Africa and Arabia

African Virtual Magnetic Poles for the Last 600 Million Years

The virtual palaeomagnetic poles for Africa for the last 600 million years are listed in Table 1. Many of the ages are from potassium-argon determinations. Where

Table 1

Age (myr)	Rock Type	Polar-ity	North Pole		Az.	Lat.	Ref.
<2.0	Kenya Trachytes	R	77.1°N	127.1°E	+13°	2°S	Nairn, 1964
1.8	Olduvai basalts	N	84.7°N	101.0°E	+5	0	Grommé & Hay, 1963
1.9	Olduvai trachyandesites	N	89.0°N	107.4°E	0	2°S	Grommé & Hay, 1967
4±3	Nairobi basalts	N,R	87.6°N	115.7°W	−1	4°S	Mussett et al., 1964; Raja et al., 1966
4 ± 3	Rungwe lavas	N,R	73.1°N	146.9°E	+16	7°S	Nairn, 1964
4.5 ± 2.5	Kerichwa tuffs	N	78.0°N	140.8°E	+12	5°S	Nairn, 1964
16 ± 1.0	Basalts & phonaolites	N,R	82.7°N	105.3°E	+7	0	Nairn, 1964
22.5 ± 10.5	Turkana lavas	N,R	85.4°N	170.5°W	+2	6°S	Raja et al., 1966
45 ± 18	Ethiopian lavas	N	85.8°N	173.3°E	+3	5°S	Grasty, 1965
>109	Lupata volcanics	N	61.8°N	101.0°W	−20	22°S	Gough & Opdyke, 1963
122 ± 12	Mlanje syenite	N	59.5°N	96.9°W	−23	22°S	Briden, 1967
168 ± 4	Mateke Hills gabbro	R	58.6°N	100.3°W	−23	24°S	Gough et al., 1964; Brock, 1968
172 ± 18	Karroo dolerites	N	65.4°N	104.9°W	−16	21°S	McElhinny & Jones, 1965; Graham & Hales, 1957; Nairn, 1964
172 ± 18	Stormberg lavas	N,R	70.5°N	91.4°W	−17	14°S	Van Zijl et al., 1962
189 ± 7	Marangudzi complex	N,R	70.2°N	74.4°W	−18	10°S	Gough et al., 1964; Brock, 1968
205 ± 10	Red s.s. Zambia	N	68.1°N	130.4°W	−7	23°S	Opdyke, 1964
209 ± 16	Shawa ijolite	N	64.2°N	94.4°W	−20	19°S	Gough & Brock, 1964
260 ± 20	Ecca Red ss.	R	58°N	106°W	−21	26°S	Nairn, 1964
260 ± 20	Ketewaka & Songwe Red Beds	R	27.1°N	90.8°W	−58	35°S⎤	Opdyke, 1964; McElhinny
303 ± 23	Galula Red Beds	R	45.5°N	139.8°W	−4	47°S⎦	& Opdyke, 1968
?335 ± 10	Dwyka Varves	N,R	26.7°N	153.8°W	+20	64°S	McElhinny & Opdyke, 1968
?335 ± 10	* Dwyka Varves	N,R	14.5°N	158.5°W	+48	70°S	Nairn, 1960
415 ± 20	* Lower Table Mt. Series	R	51.1°S	168.4°E	+148	25°S	Graham & Hales, 1961
500 ± 17	Hook Intrusives	R	14.1°S	156.8°E	+107	28°S	Brock, 1967
600 ± ?	Ntonya Ring Complex	R	27.7°S	164.9°E	+124	32°S	Briden, 1968

Az. = Palaeodeclination of Nairobi. + = Measured clockwise from present north.
− = Measured anticlockwise from present north.

Lat. = Palaeolatitude of Nairobi
* May be unreliable due to lack of cleaning.

radiometric ages are not available, the absolute ages have been estimated using the Geological Society of London (1964) time scale. Most of the rocks are igneous and have been subjected to alternating field or thermal demagnetization to remove isothermal components of magnetization. The results should therefore be reliable. The pole positions have been computer checked. The method of computing poles from site mean poles has been preferred where the data are available.

The north pole positions and provisional polar wander curve for Africa are shown in Fig. 1. Four deductions may be made. First, the poles for the interval 0 to 50 million years (myrs) are somewhat scattered and mostly lie to the east of the 0–180° meridian. Secondly, the poles for the longer time

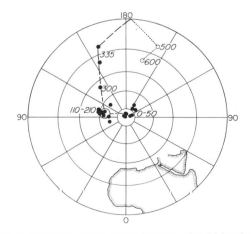

Fig. 1. Provisional polar wander curve for Africa for the last 600 myrs. The mean pole for the period 110 to 210 myrs is at 65.4°N, 99.5°W with 95% circle of confidence $\alpha_{95} = 5.2°$.

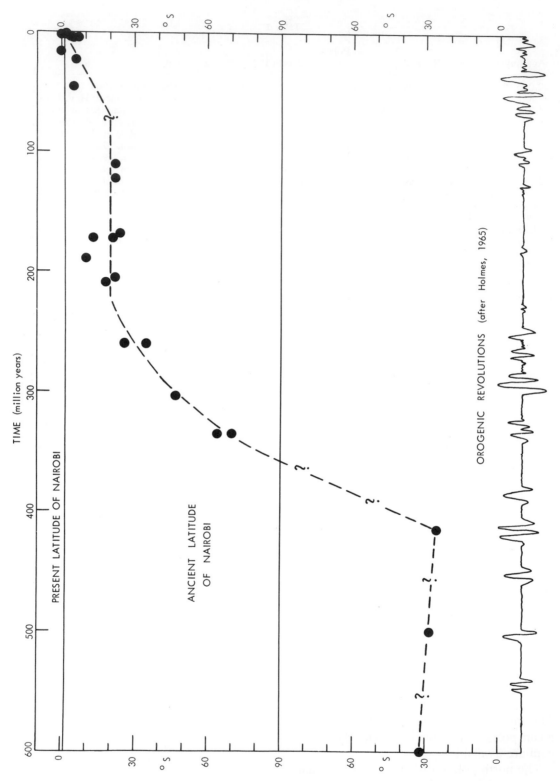

Fig. 2. Paleolatitudes of Nairobi (present position 1.27°S, 36.80°E). The site of Nairobi crossed over the polar regions about 360 myrs ago. From about 250 myrs ago to possibly 70 myrs ago the African continent was stationary. The site of Nairobi then moved north from 20°S to its present position. The paleomagnetic data to date confirm there were at least two phases of continental drift; these correspond in time to the Hercynian and Alpine orogenies.

interval 110 to 210 (?240) myrs are relatively closely grouped, six of the eight falling extremely close together. It seems there was little or no polar wander (and hence no continental drift of Africa) during this interval, the mean pole position being 65.4°N, 99.5°W with 95 per cent circle of confidence (Fisher, 1953) $\alpha_{95} = 5.2°$. Thirdly, there was rapid polar movement during the interval 240 to 340 myrs probably corresponding to rapid phases of continental drift. Fourthly, there was much less polar movement in the interval ?400 to 600 myrs.

Palaeolatitudes of Africa

For some purposes such as relating continental movement to rifting and other aspects of geological history, it is useful to express the palaeomagnetic results in terms of palaeolatitudes and palaeoazimuths. For this, Nairobi with present lat. 1.3°S, long. 36.8°E is chosen as reference. Fig. 2 shows the palaeolatitudes of Nairobi computed from the poles of Table 1.

For the interval before 340 myrs ago there is a shortage of data. The site of Nairobi was probably at latitude 30°S about 600 to 400 myrs ago and then crossed over the polar region about 360 myrs ago. From 210 to 110 myrs ago the site was stationary at 20°S and this was followed by a northward movement to its present latitude of 1.3°S.

Fig. 2 also shows the orogenic revolutions of Holmes (1965). It is seen that the time interval 110 to 210 myrs lies within the quiet period from 70 to 250 myrs ago. It is tempting to suggest that Africa remained stationary during this period. Two phases of rapid movement of the African continent occur at the same times as the Hercynian and Alpine orogenies, i.e., before 250 myrs ago and during the last 70 myrs. The Red Sea and African rift system have largely developed during this last phase of northward movement of the African continent.

Palaeoazimuths: Rotation of Africa

Evidence for a possible rotation of Africa is especially important for theories of origin of the Red Sea. The possible rotation of Nairobi is represented in diagrammatic form in Fig. 3. To illustrate the direction of rotation, the palaeomagnetic azimuths are plotted with opposite sign.

Owing to the fact that Nairobi crossed the polar region between 340 and 400 myrs ago, the azimuths for rocks older than this are reversed in sign and the rotation for 600 to 500 myrs is shown separately. For the period 110 to 210 myrs, the azimuths are shown by the shaded zone, there being little significant differences for this period. By analogy with Fig. 2 it seems likely that the azimuth of Nairobi (N 19 W) remained constant from 250 to 70 myrs ago when Africa began to rotate anticlockwise and move northwards. It seems the anticlockwise rotation continued to a point beyond (D = 0) as eight of the nine azimuths for the past 50 myrs range from 0 to +16° (i.e., east of north). There is therefore some suggestion of a change in direction of rotation from anticlockwise to clockwise for Africa to reach its present orientation. It is difficult to ascertain when this change in direction of rotation took place, but other evidence such as the interpretation of magnetic anomalies over the Red Sea suggests, it may have been between 5 and 15 myrs ago.

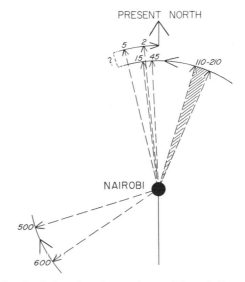

Fig. 3. Paleoazimuths and possible rotation of Nairobi. The paleoazimuths of Table 1 are plotted with opposite sign so as to give an impression of the way in which Africa has rotated. During the past 70 myrs Africa rotated anticlockwise and over the past 10 myrs or so may have rotated clockwise.

Rotation of Arabia

Irving and Tarling (1961) and Tarling *et al.* (1967) have obtained palaeomagnetic results for volcanic rocks from the southern part of the Arabian peninsula. The declinations are found to be west of north in contrast to the easterly declinations for rocks from Africa. The rocks range in age from 5 to 10 myrs, suggesting an anticlockwise rotation of Arabia during and since this time.

Continental Movement of Africa and Arabia and the Development of the Red Sea

The difficulty in interpreting palaeomagnetic results is to separate continental drift from polar wander. The African results indicate there was no polar wander and no continental drift of Africa from 210 to 110 myrs and possibly longer. Before and after this period, there was continental movement of Africa correlating with other geological evidence. There is therefore some doubt about the reality of polar wander, although it is impossible to exclude the possibility of polar wander during periods of continental drift. Briden (1967a) has compared palaeomagnetic results for Africa with the other Gondwana continents and suggests there has been possibly four episodes of rapid drift during the last 600 myrs with little or no drift and polar wander in between. If it is assumed, that continental drift is more important than polar wander in explaining the virtual magnetic poles, the following historical record may be deduced for the movement of Africa and Arabia.

(i) *600 to 400 myrs.* As seen from Fig. 2 there is a shortage of data from 600 to 350 myrs. From 600 to 400 myrs, Africa was relatively stationary with Nairobi having latitude 30°S and azimuth 110 to 125°.

(ii) *400 to 230 myrs.* During this period there was large scale continental movement when Africa crossed the polar regions.

(iii) *220 to 100 myrs.* A period of no continental movement and no polar wander with Nairobi stationary at latitude 20°S, azimuth −20°.

(iv) *?70 to ?10 myrs.* During this period Africa moved northwards and rotated anticlockwise, Nairobi reaching its present latitude of 1.3°S. The main rifting of the Red Sea took place in this interval.

(v) *?10 myrs to present.* For this interval continental and rift movements continued, and there is a possibility that Africa started to rotate clockwise whereas Arabia continued rotating anticlockwise. The situation is illustrated diagrammatically in Fig. 4, the axial trough of the Red Sea developing in this interval.

With this background of continental movement of Africa and Arabia, the historical development of the Red Sea may be considered. A recent geological review has been given by Heybroek (1965), and a summary from various sources is shown in Fig. 5. It seems that some kind of depression may have existed 345 to 325 myrs ago (Lower Carboniferous). This was during the earlier phase of continental movement when Africa was nearer the polar regions. After this there is little record of rifting until 50 myrs ago; i.e., it seems when there was no drift of Africa, there was no rifting. The most spectacular development of rifting has taken place during the last phase of northward movement and anticlockwise rotation of Africa. Thus the main Red Sea graben was formed about 26 myrs ago following a period of basaltic volcanicity. This was followed by a relatively quiet period for about 14 myrs during which time, large quantities of clastics and evaporites were deposited. A further phase of activity began about 8 myrs ago when movements connected the Red Sea to the Gulf of Aden and Indian Ocean. After a short quiet period, there have been further rift movements continuing to the present day.

The last phase of rifting over the last 3 myrs or so is naturally the best documented. First, the accurately determined epicenters of earthquakes recorded since the introduction of the Worldwide Standard Seismograph Net in 1963 show that the center of the Red Sea, especially the southern part, is seismically active. Secondly, Vine (1966) has reinterpreted the magnetic anomalies over the Red Sea in terms of the spreading

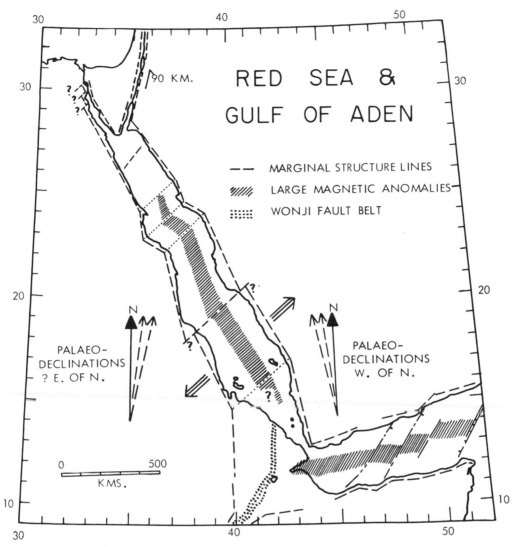

Fig. 4. Map of the Red Sea and Gulf of Aden showing the possible movement of Arabia and Africa as inferred from transform faults and the interpretation of axial magnetic anomalies. The presentation of the paleodeclinations is schematic.

sea floor hypotheses and finds the Red Sea has been opening up along its center at a rate of about 1cm/year over the last 3 myrs. This is consistent with the possible rotations of Africa and Arabia (Fig. 4).

It therefore seems that the Red Sea has developed during periods of continental drift of Africa and Arabia. The development took place in phases, probably correlated to phases of drifting. The earliest phase may have been during Carboniferous drift; the most spectacular phase was in the Lower Miocene when the main Red Sea graben was formed; there was a further phase during the late Miocene/early Plio-

cene and lastly, rift movements have been continuing from the Upper Pliocene, through the Pleistocene to the present day.

Summary of Geophysical Data for the Red Sea

General Résumé

The Red Sea was the subject of geophysical interest in the last decade of the 19th century and the early part of this century. It then received little attention until well after the second world war.

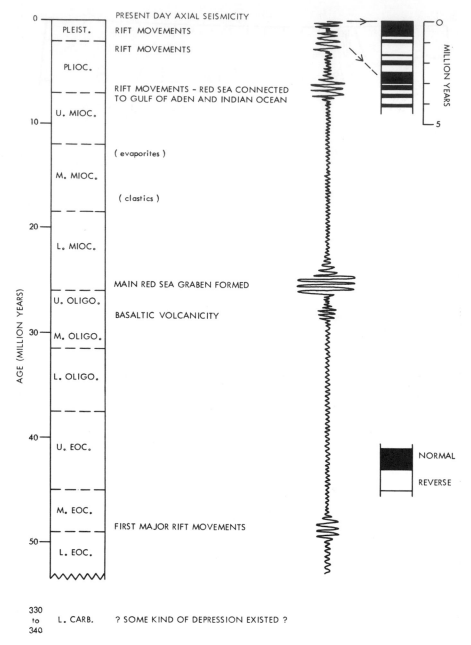

Fig. 5. Geological history of the Red Sea. The reversal timescale on the right is magnified by a factor of two.

One of the most remarkable pieces of work was that of Triulzi (1898, 1901). Triulzi accompanied the German ship **S.S. POLA** and using pendulums measured gravity on the islands and at various places along the Red Sea shores. By interpolation, he produced a Bouguer anomaly map showing the large, positive anomaly centered over the Red Sea. The presence of the anomaly was confirmed by the sea measurements of Hecker (1910) and the submarine pendulum measurements of Vening Meinesz (1934).

It is of interest, that these early gravity measurements indicate that the Red Sea must be in approximate isostatic equilibrium, a fact noticed by Wegener (1922, 1929). Thus, it has been possible to infer for a long time, that at least the deep water of the Red Sea is underlain by oceanic crust, and there must have been some separation of the African-Arabian continental landmass.

In 1955, Harrison made four gravity measurements across the Red Sea at 16° to 16.5°N, using a Vening Meinesz submarine pendulum apparatus (Girdler and Harrison, 1957). These proved so interesting that Girdler (1958) reviewed all Red Sea gravity data available to that date. These enabled a reasonable gravity profile to be constructed across the southern part of the Red Sea. The gravity (Bouguer) maximum of this profile was interpreted as being due to a large basic mass about 60km wide intruded through the downfaulted basement rocks and nearly reaching the surface. A possible mechanism of formation of the Red Sea structure involving crustal separation was given.

In some respects this section is atypical. As the Red Sea remained closed to the Indian Ocean until the late Miocene-early Pliocene, the accumulations of evaporites in the enclosed southern part were particularly voluminous. These deposits obscure the geological structures; the coastlines converge whereas the main fault trough (Fig. 4) continues to widen southwards. In the neighbourhood of the Dahlak Islands record thicknesses of evaporites have been found in exploration drilling for oil. As these have a low specific gravity of about 2.2, there will be an appreciable effect on the Bouguer anomaly profile, the minima on the sides of the Red Sea making the central maximum appear narrower. This means that the width calculated earlier for the axial intrusive zone in the south is a minimum estimate.

In 1958, the Research Vessels **VEMA** and **ATLANTIS** (Lamont Geological Observatory and Woods Hole Oceanographic Institution) rendezvoused in the Red Sea. The total intensity of the Earth's magnetic field was recorded over 4,500km of track and 15 seismic refraction profiles were shot. Large magnetic anomalies were found with amplitudes of up to 1,200 gamma. The extent of the region of large anomalies is shown in Fig. 4. The largest anomalies are found south of 24°N and are over the deep, axial trough. The seismic refraction profiles

are summarized in Figs. 6, 7 and 8. These confirmed the earlier interpretation of the gravity anomalies; high velocities (6.7 to 7.4km/s) being found at shallow depths beneath the deep water. Velocities of 5.5 to 6.4km/s were found nearer the margins, these velocities being typical of basement, crystalline shield rocks. A full account of this work is given in Drake and Girdler (1964).

In 1959, H.M.S. **DALRYMPLE** recorded the total intensity of the Earth's magnetic field over a further 3,000km of track, the location of the track being planned to supplement the data obtained by **R.V. VEMA** the previous year. This work is described by Allan (1964).

In 1961, the **R.V. ARAGONESE** (of the NATO SACLANT ASW Research Center) made a more detailed bathymetric, gravity and magnetic survey of the Red Sea. The work resulted in the publication of a new bathymetric chart by the British Hydrographic Office in 1965. Various aspects of the geophysical work have been described by Plaumann (1963), Allan et al. (1964) and Allan and Pisani (1966). Two continuous recording Askania sea gravimeters were used and a nuclear precession magnetometer. The survey included a total of 53 transverse crossings. As well as some profiles, Plaumann's paper includes a Bouguer anomaly map of the Red Sea. Between 17°N and 26°N there are Bouguer anomalies of 100mgals over the center of the Red Sea with maxima reaching 125 to 150mgals. The positive anomalies extend over the most northern part to just south of the Sinai peninsula, though their magnitudes are less (60 to 70mgal). The **R.V. ARAGONESE** also went into the Gulf of Aqaba where a large negative Bouguer anomaly of −100mgal was confirmed contrasting with the positive anomaly over the Red Sea.

Ten more gravity and magnetic profiles were made over the axial part of the Red Sea north of 17°N by **R.V. CHAIN** in 1964 (Knott et al., 1966). Along seven of these crossings, continuous seismic reflection profiles were obtained using an underwater sparker. The sub-bottom reflectors were found to be most intensely deformed in the deep axial zone. On either side of the axial zone but still in the main trough the deformation was less, with the sediments reaching thicknesses up to 1.8km.

In the early part of 1967, **R.R.S DISCOVERY** returned to the Red Sea, Gulf of Aden and N.W. Indian Ocean. Three new seismic refraction profiles were obtained in the extreme north east of the Red Sea, just south of the Gulf of Aqaba, all giving continental structure. Shore to shore gravity and magnetic traverses were made at latitudes 24°N and 27°N. Much work was done in the Gulf of Aden and some seismic refraction work showed that the outermost part of the Gulf has oceanic structure.

On several cruises, attempts have been made to measure heat flow in the Red Sea. This is difficult, as the bottom sediments are often hard and difficult to penetrate. Successful measurements have been reported by Sclater (1966), Birch and Halunen (1966), and Langseth and Taylor (1967), and are shown in Figs. 6, 7 and 8. Of the seven available measurements, six are higher than the world mean, the highest being 6.37 microcal. cm^{-2} s^{-1}. All are in the central trough and the mean is 3.37 $\pm$ 1.57 microcal. cm^{-2} s^{-1}.

In this general résumé, discussion has been confined to geophysical measurements actually made in the Red Sea. Information of value may also be obtained from Red Sea earthquakes which are recorded at seismograph stations around the world. The seismicity of the Red Sea has been discussed by Gutenberg and Richter (1954), Rothé (1954), Drake and Girdler (1964), and Sykes and Landisman (1964). From the earlier work, it seemed that seismic activity was mainly confined to the southern part of the Red Sea (south of 18°N), and the epicenters plot on the margins as well as in the center. Errors in these determinations may be as great as 100km. Since the introduction of the Worldwide Standard Seismograph Net in 1963 and modern computing techniques, the epicentral determinations have been improved by a factor of ten. Epicenters for earthquakes from January 1955 to the end of 1966 are shown in Figs. 6, 7 and 8 and may be considered to have a 10 to 20km accuracy. It is of interest that

the more recent epicentral determinations (Sykes and Landisman (1964) and Fairhead, personal communication) plot along the center of the Red Sea. This, together with the evidence from bathymetry, seismic reflection profiles and Vine's interpretation of the magnetic anomalies give quite strong support to the idea that the newest part of the Red Sea is down the center where new crust might still be evolving today.

The geophysical data will now be reviewed in more detail. For this, it is convenient to divide the Red Sea into three parts: northern (north of 25°N), central (17 to 25°N), and southern (south of 17°N).

Northern Part including Gulfs of Suez and Aqaba

In the north, the main Red Sea depression is about 200km wide and at latitude 27.8°N it bifurcates into the Gulfs of Suez and Aqaba with the Sinai Peninsula in between. The Gulf of Suez is parallel to the western margin of the Red Sea and the Gulf of Aqaba makes an angle of about 45° to the Red Sea axis (Fig. 6).

Gulf of Suez. The Gulf of Suez fault trough is about 80km wide becoming narrower towards the north. About one third to one half of the fault trough is covered by shallow water averaging 65m depth. The Gulf has a negative gravity anomaly of approximately −50mgal, indicating an infill of light sediments. The magnetic field is relatively smooth reflecting the lack of igneous activity and suggesting the sediments are underlain by downfaulted Pre-Cambrian shield rocks. Both land and marine seismic reflection surveys have been carried out (Masson and Agnich, 1958), connected with oil exploration. Oil has been found in the Miocene sediments along anticlinal flexures and fault zones.

The geology of the Gulf has recently been reviewed by Heybroek (1965). Structural sections show about 2km of sediment (mainly Tertiary) overlying the Pre-Cambrian basement, the sediment thickness occasionally reaching nearly 5km. Numerous faults are shown giving rise to several horsts and grabens within the downfaulted basement. The seismic work shows the top of the Miocene to be a strong reflector and the near surface Pleistocene sediments are distorted. The latter points to fault movements over the last few million years which may still be in progress.

There is no major transcurrent movement along the Gulf (unlike the Gulf of Aqaba and Dead Sea). Arkell (personal communication, 1956) gives three reasons for this. First, the narrow strip of Carboniferous outcropping in Sinai is continued accurately by the Carboniferous outcropping in the Egyptian Wadi Araba; secondly, the boundary of the Archaean shield crosses the Gulf without dislocation and thirdly, there is no appreciable dislocation of the belt of Tertiary folds which cross the Gulf of Suez.

There is no geophysical evidence for an extension of the Gulf of Suez northwards under the Mediterranean. The question arises as to how the Suez rift dies out. One possibility is shown in Fig. 6 where three possible transform faults are shown, the amount of extension becoming progressively less to the north of each one. Northeast-southwest faults beneath the Gulf are referred to by Masson and Agnich (1958), and this seems a major direction of separation throughout the Red Sea area. The Suez faults of Fig. 6 are suggested from the topographic and geological maps and are only hypothetical.

Gulf of Aqaba — Dead Sea Rift. Unlike the Gulf of Suez, the Gulf of Aqaba continues northwards for more than 600km being in continuity with the Dead Sea-Jordan rift (Fig. 6). The understanding of the mechanics of this fault zone is all important to understanding the evolution of the Red Sea.

For a hundred years the Dead Sea-Jordan rift has been considered by some geologists to be a shear zone. Early advocates included Lartet (1869), Dubertret (1932), and Wellings (1938). Estimates for the amount of displacement ranged up to 160km, such that the eastern block (Arabia) moved northwards with respect to the western block (Sinai, Israel and the eastern Mediterranean).

More recent advocates of this transcurrent movement include Quennell (1951, 1958), De Sitter (1962), Freund (1965),

Zak and Freund (1966), and Dubertret (1967). Quennell considers the total horizontal movement to be 107km. He considers the movement to have taken place in two phases, the first intermittently throughout the Miocene and into the Pliocene (62km), and the second throughout the Pleistocene and still continuing today (45km). These movements give average velocities of 0.3cm/yr and 2.2cm/yr, respectively.

De Sitter (1962) has analysed the regional stress pattern over the last 140 myrs. He recognises five phases of movement at about 135 myrs, 70 to 85 myrs, 50 to 70 myrs, 26 to 38 myrs and 0 to 2 myrs. In the first four phases the wrench faulting is considered to be a consequence of compressive stress (largely N.W.-S.E.), and the last phase was accompanied by west-east extension.

The most recent advocate of the horizontal displacement is Freund (1965), who cites further geological evidence including associated drag folding. His best estimate (personal communication) for the total displacement is 90km. Zak and Freund (1966) from a study of air photographs find evidence for a recent horizontal movement of 600m, 150m of which took place within the last 20,000 yrs. This represents an average velocity of more than 0.75cm/yr.

It should be mentioned that not all geologists support horizontal displacement along the Dead Sea rift. For example, Wetzel and Morton (1959) in a full account of the geology of the area find it unnecessary to involve shear, as also does Picard (1966), who has worked in the area for a long time. To the author, the evidence for displacements seems very strong. For example, Quennell (1958) is able to list fifteen different geological features ranging from marine Jurassic strata to ore deposits which come into juxtaposition on restoration of the horizontal displacement of 107km. Further, this movement is a consequence of the northward anticlockwise movement of Arabia which was simultaneously responsible for the opening of the Red Sea and the folding of the Zagros-Taurus mountain range to the north. The average velocities of movement estimated are reasonable and

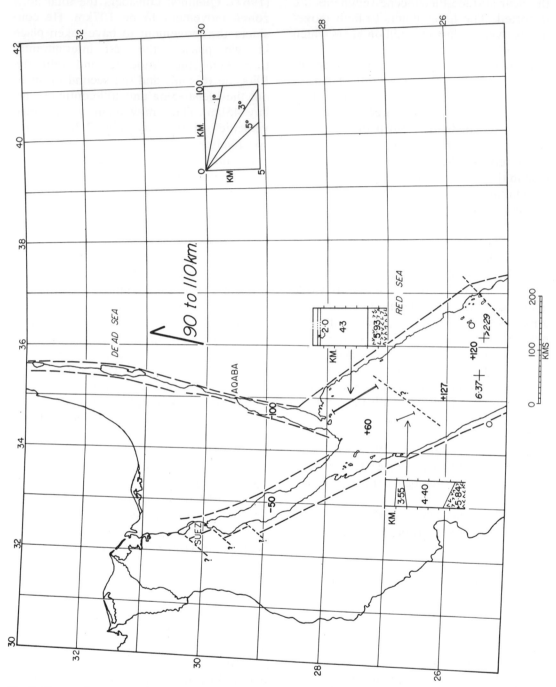

Fig. 6. The northern part of the Red Sea and the Gulfs of Suez and Aqaba. In this and Figs. 7 and 8 long dashed lines indicate marginal structures; short dashed lines connect various structural features across the Red Sea and represent the direction of movement; small circles represent earthquake epicenters; heavy numerals indicate the magnitude of gravity anomalies; italicised numerals are heat flow measurements in microcal. cm^{-2} s^{-1}; 1958 **VEMA-ATLANTIS** seismic refraction profiles ├────┤; 1967 **DISCOVERY** profiles ╞════╡ ; the scales for the reversed profiles are shown separately; for the unreversed profiles, for which horizontal layering is assumed, the horizontal scales are condensed.

similar to estimates from other parts of the world rift system and sea floor spreading.

Few geophysical data exist for the Dead Sea area. Nettleton (private report, 1947, referred to by Picard, 1966) has made a gravity survey of the Dead Sea rift in connection with oil exploration. It is understood that the rift is associated with a large negative Bouguer anomaly. Neev and Emery (1967) cite borehole data indicating more than 4km of Tertiary to Recent sediment infilling the Dead Sea graben. With the exception of Grain Sabt, an olivine basalt intrusive plug 30km north of the Dead Sea, there is no record of intrusive igneous rocks in the rift. A large negative anomaly is therefore to be expected. Knopoff and Belshé (1967) have made two regional traverses which cross the rift to the north of the Dead Sea. They find a positive Bouguer anomaly to the west of the rift and a negative Bouguer anomaly to the east and suggest that the thinner crust associated with the Mediterranean extends as far as the rift zone. The thinner crust to the west is considered to be a bearing surface for the northward movement of Arabia.

To the south, a large negative Bouguer anomaly of about 100mgal exists over the Gulf of Aqaba (Triulzi, 1898; Girdler, 1958; and Allan et al., 1964). The Gulf of Aqaba is much narrower than the Gulf of Suez with a width of 30km compared with 70 to 80km. It is also much deeper, the water depth exceeding 1.8km and the land on either side reaching heights of 0.85km. The total amplitude of the relief is therefore 2.65km making it the most spectacular part of the rift system. The geology of the western margin (Sinai) has been discussed by Beadnell (1926) and the eastern margin (Saudi Arabia) by Mitchell (1957). Both remark on the freshness of the scarps and the latter considers them to be formed by gravity normal and strike slip movement. The Pre-Cambrian on the eastern side of Aqaba is displaced by about 100km northward from the Pre-Cambrian of the southwest tip of Sinai (Fig. 6), the figure being in remarkable agreement with the displacement measured further north along the Jordan valley.

It is concluded that the dominant movement along the Aqaba-Jordan-Dead Sea rift is horizontal, the total displacement being about 100km. The dominant shear might have been associated with minor tension or compression at various times. Recently (last 2 myrs) there has been extension in Aqaba and the Dead Sea. Previously, there may have been compression (De Sitter, 1962). This helps to explain the large, negative gravity anomaly over Aqaba, which is twice as large as the anomaly over Suez which has the greater infill of sediments. It is possible that Aqaba has not recovered from the minor compression phase previous to the Plio-Pleistocene. Finally, if Arabia has moved north and rotated anti-clockwise, the amount of extension should become less and less to the north giving way to compression.

The Northern Red Sea. A 100km displacement along the Gulf of Aqaba must clearly affect the Red Sea. Girdler (1966) has computed the amount of crustal separation in the Red Sea and Gulf of Aden to be expected assuming Arabia has moved north as one unit by 100km and rotated anti-clockwise with respect to Africa by 7° (the divergence of the Red Sea margins). The amount of separation in a direction perpendicular to the Red Sea axis increases from about 70km at 27°N to about 240km at 14.5°N.

A 70km separation in the north is consistent with the bathymetry, deep water (1km) extending as far as Sinai. The contours (Chart C6359) suggest structures parallel to the Gulf of Aqaba. The distance between the 500 fathom (914m) contours is about 60 to 70km, i.e., about the computed amount.

A separation of the continental crust should be accompanied by a positive Bouguer anomaly and such a positive anomaly reaching 60 to 70mgal is observed in the extreme north as far as the Sinai Peninsula and mouth of the Gulf of Aqaba. This is in contrast to the negative anomaly of 100mgal over the Gulf of Aqaba which has a similar water depth. There must therefore be crustal separation or drastic crustal thinning in the extreme northern Red Sea coinciding with the deep water.

In this part of the Red Sea there are no large magnetic anomalies comparable to those found south of 24°N. The anomalies found are smaller and very localised. It therefore seems that the presence of large magnetic anomalies (Drake and Girdler, 1964) cannot be used as a criterion for crustal separation. Large magnetic anomalies are probably only observed where lava flows overlie the region of high density and high seismic velocity.

Unfortunately, there are only two seismic refraction lines in this region of the Red Sea (Fig. 6). The profile in the west (**VEMA-ATLANTIS,** 1958) showed a velocity typical of shield rocks at 6km depth and provisional calculations for the recent **DISCOVERY** profile in the east showed similar velocities. A long seismic refraction line between these two is needed to find the depth of the high density material. Plans for this on **DISCOVERY** cruise 16 were thwarted because of engine trouble.

Further south (Fig. 6) the Bouguer gravity anomaly increases becoming almost double at 26°N. There are two heat flow measurements both of which are high (6.37 and 2.29 microcal. cm^{-2} s^{-1}). These data again lend support to crustal separation in this region which probably expands to about 100km at 24.5°N.

Central Part (17 to 25°N). In this part, the shorelines are no longer straight (*cf.* Figs. 6 and 7). It has long been noticed (*e.g.,* Wegener, 1922) that the shores are remarkably parallel. By comparing the marginal topography of the two sides, the direction in which the center of the Red Sea opened may be ascertained. This is shown by dotted lines in Fig. 7 and is approximately N 45°. At 18 to 20°N the marginal topography is offset on both sides of the Red Sea suggesting the possibility of a transform fault.

The bathymetry of the Red Sea also changes in this region. At about 24 to 25°N a deep, axial trough begins to develop and widens towards the south. The trough reaches depths of more than 1.5km. The Atlantis II, Chain and Discovery Deeps are all in this region (latitude 21.3°N) and have depths greater than 2km. In the south of the region the marginal shelves shoal and widen. This is due to the build up of evaporites and coral reefs.

This region has a positive Bouguer gravity anomaly, the maxima (120 to 140mgal) occurring over the deepest water. The axial trough is also associated with large magnetic anomalies, their complexity increasing towards the south. The rugged topography, and the interpretation of gravity and magnetic anomalies suggest that the axial trough is the youngest part of the Red Sea. In addition, the axis of the trough has been seismically active over the past few years.

The amount of crustal separation perpendicular to the axis of the Red Sea to be expected in this region (Girdler, 1966) is approximately 136km at 22°N and 171km at 19°N. The seismic profiles (Fig. 7) do not conflict with these estimates, four of the profiles showing high velocity material (6.8, 6.9, 7.0 and 7.3km/s) at depths ranging from 4 to 7km. Because of the large magnetic anomalies, the overlying material with velocities 3.5 to 4.5km/s presumably contains large amounts of strongly magnetic volcanics. The two profiles near the western margins reveal the existence of more than 5km of sediment. In the northern profile (23.5°N) the sediment is underlain by material of velocity 6.1km/s which is presumably downfaulted Pre-Cambrian basement. In the southern profile (17.8°N) the underlying velocity is 6.4 km/s which could be either shield rocks or intrusives; it is shown in Fig. 7 as shield rocks as it is well to the west of the region of large gravity and magnetic anomalies. Between these two profiles, there is a borehole (referred to by Heybroek, 1965) on Marghersum Island (latitude 20.8°N, longitude 37.2°E) which gave 0.97km of gypsum, anhydrite and interbedded sandstones overlying 0.46km of salt. The former probably correspond to velocities 3.7 to 4.1km/s and the latter to velocities of 4.4 to 4.8km/s.

In December 1967 and January 1968 a very intensive seismic refraction experiment was carried out by Cambridge University at latitudes 22° to 23°N, longitude 37° to 39°E. Twenty refraction profiles were shot, eighteen of them parallel to the Red Sea and two of them at right angles. An average velocity of 6.6km/s was found at an average depth of 4.6km beneath the main trough (Tramontini and Davies, per-

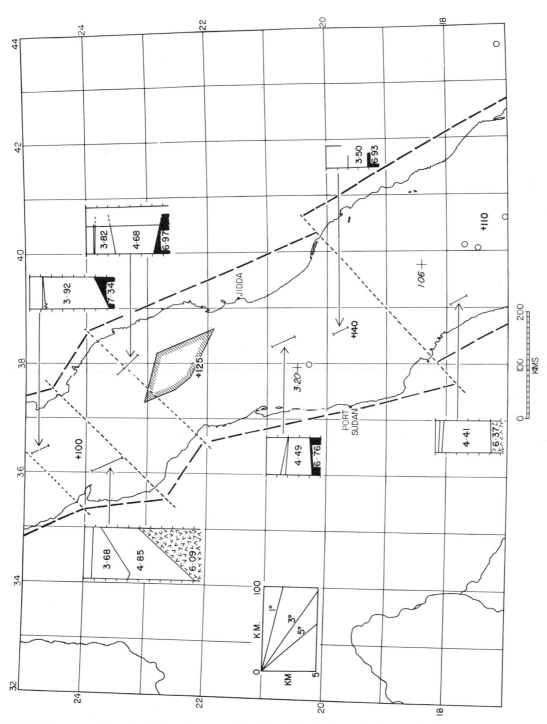

Fig. 7. Various geophysical data for the central Red Sea. (The hachured region marks the boundaries of the 1967/68 Cambridge detailed seismic refraction survey; for explanation of other symbols see Fig. 6.)

sonal communication). A most interesting feature of this work is that the velocities at right angles to the Red Sea trend are not significantly different from the velocities parallel. This suggests that the axial intrusive zone is filled with intrusive rocks rather than a mixture of downfaulted basement and intrusives (*cf.* Drake and Girdler, 1964). The minimum amount of separation perpendicular to the axis of the Red Sea from this work is 130km.

There are two heat flow measurements for this part of the Red Sea, at latitudes 18.4°N and 20.4°N giving 1.06 and 3.20 microcal cm^{-2} s^1. The first is surprisingly low, especially as it is near the center of the Red Sea.

Southern Part (South of 17°N). In discussing this part of the Red Sea, it must be remembered that there was a land barrier until about 6 or 7 myrs ago cutting off the Red Sea from the Gulf of Aden and Indian Ocean. Before this, the Red Sea was connected to the Mediterranean (Montanaro, 1941). The barrier was situated at about the latitude of Perim Island (12.64°N), and is discussed by MacFadyen (1932). The Red Sea was therefore a large Gulf enclosed in the south and conditions were favourable for the deposition of large volumes of evaporites throughout the Miocene. Wide shelves built up eastwards from the Eritrean shore (Dahlak Island), westwards from the Yemen shore (Farsan Island), and northwards from the land barrier in the south. Several kilometers of Miocene evaporites, mostly rock salt, have been found beneath recent coral reefs in exploration wells at latitudes 15° to 17°N on both sides of the Red Sea.

Beneath this vast sedimentary cover, Girdler (1966) estimates the crustal separation perpendicular to the Red Sea axis to be 206km at 16.5°N and 241km at 14.5°N. This implies that much of the Danakil-Afar triangle of Ethiopia is underlain by oceanic rather than continental crust. This is reasonable topographically. Girdler (1958, Fig. 4) suggested that the real structure was best revealed by the 1,500m contour which demarcates the extension of the main Red Sea fault depression in this area. The 1,500m contours for Ethiopia and Arabia

are remarkably parallel. The approximate positions of these are shown in Fig. 8. Some geologists (*e.g.,* Mohr, 1967) disagree with this view and consider the region to be largely underlain by continental material. Some continental material is undoubtedly present as evidenced by the Danakil horst, the foundation of which is Pre-Cambrian. It seems likely, that the fracturing of Arabia and Africa in this region (Fig. 8) was somewhat complex and relatively small fragments of continental crust were left behind when the continental crust separated. This is compatible with the amount of separation estimated.

Geophysical data (Fig. 8) indicate crustal separation at least between the Dahlak and Farsan Islands. The Bouguer anomaly is strongly positive (95 mgal) and there are large magnetic anomalies. At latitude 16°N seismic profiles reveal high velocity material (7.2 to 7.3km/s) at 4km below sea level. However, a seismic profile to the northeast of the Dahlak Islands shows a velocity of 5.9km/s at 2 to 3km below sea level. Continental crust therefore exists near the middle of the sea. To account for a possible 200km separation further intrusive zones must exist. One such possibility is to the south of Dahlak Island where there are active volcanoes along the western margin of the Danakil depression, the depression being below sea level. Gravity and magnetic surveys near here (Holwerda and Hutchinson, 1968) indicate the presence of intrusives and the region is seismically active.

Other seismic refraction profiles also show the presence of continental rocks in the southern part of the Red Sea. Thus, velocities of 5.5 to 5.8km/s at a depth of 3km are found near the Arabian shore at latitude 15°N. Near the entrance of the Red Sea, north of the straits of Babel Mandeb, a velocity of 5.9km/s was found at 2.5 to 3.5km below sea level.

Unlike further north, this region has many volcanoes. The active volcanoes from the catalogue of active volcanoes of the world (Richard *et al.,* 1957) and U.S. Air Force Aeronautical Charts are shown by black triangles in Fig. 8. There are numerous other volcanoes, dormant or extinct over

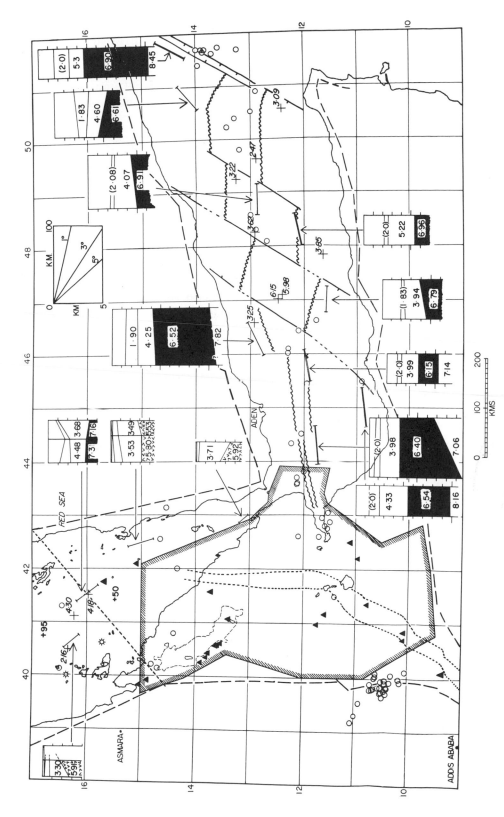

Fig. 8. Various geophysical data for the southern part of the Red Sea and Gulf of Aden (for explanation of symbols see Fig. 6). Solid triangles represent volcanoes. The horizontal and vertical scales for the reversed seismic profiles are shown at the top of the diagram. For the unreversed profiles, for which horizontal layering is assumed, the horizontal scales are condensed. The hachured region marks the limits of the 1968 aeromagnetic survey.

the area. The Red Sea volcanoes of Perim, Zebayir, Jebel Attair and Zukar Hanish have been described by Raisin (1902), Lamare (1930), MacFadyen (1932), and Gass *et al.* (1965). The rocks are all basaltic with basaltic agglomerates, black vesicular basalts and olivine basalts. North of the two active volcanoes shown in Fig. 8, there are three heat flow measurements, all of which give high values, *i.e.* 2.16, 4.18 and 4.30 microcal. cm^{-2} s^{-1}.

The volcanism of the low lying triangle of Ethiopia bordering the Red Sea has been discussed by Lamare (1930a, b), Taylor (1952) and Mohr (1962, 1967). Lamare (1930c) also compares the two sides of the Red Sea and Gulf of Aden and discusses the problems associated with the Afar lavas and fragments of old continental crust left behind. The extensive parallelism of the margins of both the Red Sea and Aden troughs are noted. Various features which can be re-attached are documented. These include fossil beds for the Hedjaz, Sinai and Arabia; various structural units of Central Arabia and Nubia and the Hadramaout of Arabia and Dhofar of Somalia. The presence of acidic lavas in Afar are problematic and Lamare concludes that this region has been formed by both horizontal and vertical movements.

Taylor (1952) discusses the fumaroles, hot springs and recent volcanicity and notes that the near vertical rift scarps denote tension. The volcanicity and rifting is related to the Gulf of Aden and Carlsberg ridge. Mohr (1967) also notes the dominance of the Gulf of Aden trend and describes an intense belt of recent faulting, the "Wonji fault belt" (shown by dotted lines in Figs. 5 and 8). This fault belt reaches 45km wide and has volcanic centers along its western margin. It is possible that this could correspond to the latest phase of activity down the center of the Red Sea.

The most recent work (Holwerda and Hutchinson, 1968) describes the potash deposits in the Danakil Depression. These are Quaternary and deposited from marine brines. The iron content is thought to come from underlying basic intrusives and many igneous plugs are found in this region. The

authors conclude that the Afar Depression is underlain by newly formed oceanic rather than continental crust, the area being largely covered by vast amounts of recent volcanics.

This region is very inaccessible and a full understanding of it is all important to understanding the origin and evolution of the Red Sea. Consequently, an aeromagnetic survey was flown from February to April 1968 with more than 24,000km of track. The object of this work is to find the areal extent of new oceanic crust. The data are being processed at the time of writing.

Discussion of this region would be incomplete without mention of the Gulf of Aden (Fig. 8). Full discussions have been given by Laughton (1966a, b). Similarities with the Red Sea include seismicity, high heat flow (Von Herzen, 1963; and Sclater, 1966), linear magnetic anomalies (Girdler and Peter, 1960; and Laughton 1966a, b) and oceanic structure from seismic refraction measurements (Nafe *et al.*, 1959; and **DISCOVERY** Cruise Rep. 1968). There are some notable differences. First, the central valley becomes associated with a widening ridge towards the east and is more like the Carlsberg and mid-Atlantic ridges than the Red Sea. Secondly, so far, no seismic profiles have shown velocities appropriate to continental rocks unlike the $5.9 \pm 0.2km/s$ velocities for the Red Sea margins; this includes a recent **DISCOVERY** profile fairly close to the Somali shore (Fig. 8). Thirdly, northeast-southwest structures are clearly discernable in the Gulf of Aden unlike the Red Sea. Lastly, the amount of crustal separation (Laughton, 1966b) seems to be more than for the Red Sea. This suggests the possibility of transcurrent movement along the Ethiopian rift to the south of Addis Ababa, the Somali block moving differentially with respect to the Nubian block.

The Evolution of the Red Sea

There are several geological reviews of the history and development of the Red Sea (*e.g.* Owen, 1938; Tromp, 1949, 1950; Shalem, 1954; and Swartz and Arden, 1960), and Vening Meinesz (1950) and

Heiskanen and Vening Meinesz (1958) give a full discussion of the mechanics of rifting. The geological and geophysical data now available enable a comprehensive summary of the evolution of the Red Sea to be made.

(i) The palaeomagnetic data suggest there have been phases of drift of Africa. After a stable period from 220 to 100 myrs ago, Africa has been moving north and rotating anticlockwise during the last 70 myrs. The major features of the Red Sea have all formed during this last period of continental drift.

(ii) The movement of Africa over the last 70 myrs was probably discontinuous and there were corresponding phases of Red Sea rifting. The initial faulting was about 50 myrs ago, the main Red Sea depression formed about 25 myrs ago, a further phase of rifting occurred about 10 myrs ago and the final phase over the last few million years is still continuing.

(iii) During the last phases of rifting the axial trough has been forming. The symmetry of the Red Sea about the axial trough suggests that during this last phase of rifting Africa may have been rotating clockwise and Arabia anticlockwise. The palaeomagnetic data suggests there may have been a change in the direction of rotation of Africa from anticlockwise to clockwise about 10 ± 5 myrs ago.

(iv) The symmetry is complicated by the fact that crustal separation has taken place in a N.E.-S.W. direction. The whole of the Dead Sea-Aqaba-Red Sea-Aden rifting is dominated by N.E.-S.W. movement.

(v) With the possible exception of the Jordan rift, the whole of this part of the rift system is in isostatic equilibrium. The Red Sea has positive Bouguer anomalies indicating the presence of intrusive rocks beneath the deep water. Over the center, there are large magnetic anomalies; these are probably associated with material with seismic velocities of 4.1 ± 0.4km/s (?extrusives) overlying material with velocity 7.1km/s (*cf.* Heirtzler and Le Pichon, 1965).

(vi) Near the margins, seismic velocities of about 6km/s are found suggesting the presence of downfaulted basement. So far,

such velocities have not been found in the Gulf of Aden. The exact amount of crustal separation in the Red Sea is uncertain but seems to be less than in the Gulf of Aden.

(vii) Bathymetry, seismicity and the interpretation of magnetic anomalies all suggest that the newest crust is under the center of the Red Sea. The magnetic anomalies suggest that over the last few million years the center of the Red Sea has been opening symmetrically at about 1cm/year.

(viii) Finally, the opening of the Red Sea and Gulf of Aden, the transcurrent movement of 100km along the Aqaba-Dead Sea-Jordan rift are all consistent with the northward movement and anticlockwise rotation of Arabia with the formation of the compressional features to the north and east, *i.e.* the Taurus-Zagros Range, the Persian and Oman Gulfs. It seems likely that Arabia was moving away from Africa and there was a sequence (in phases) of crustal thinning, rifting, crustal separation accompanied by basic intrusions in the formation of the Red Sea.

Acknowledgments

The author is grateful to Dr. T. D. Allan for showing him his gravity and magnetic maps of the Red Sea and to many others at the Ottawa Meeting on the World Rift System for helpful discussions. For more recent work, sincere thanks are extended to Dr. D. Davies, Mr. C. Tramontini and Dr. A. S. Laughton for seismic refraction data and helpful discussions on the Red Sea and Gulf of Aden respectively.

References

Allan, T. D.: A preliminary magnetic survey in the Red Sea and Gulf of Aden. Boll. di Geof. Teorica ed Applicata, **6**, 199 (1964).

Allan, T. D., H. Charnock, and C. Morelli: Magnetic, gravity and depth surveys in the Mediterranean and Red Sea. Nature, **204**, 1245 (1964).

Allan, T. D. and M. Pisani: Gravity and magnetic measurements in the Red Sea (summary only). The World Rift System. Geol. Surv. Canada Paper 66-14, 62 (1966).

Beadnell, H. J. L.: Central Sinai. Geogr. J. **67**, 387 (1926).

Birch, F. S. and A. J. Halunen: Heat-flow measurements in the Atlantic Ocean, Indian Ocean, Medi-

56   R. W. Girdler

terranean Sea and Red Sea. J. Geophys. Res. **71,** 583 (1966).

Briden, J. C.: Recurrent continental drift of Gondwanaland. Nature, **215,** 1334 (1967a).

――――: A new palaeomagnetic result from the Lower Cretaceous of East-Central Africa. Geophys. J. Roy. Astron. Soc. **12,** 375 (1967b).

――――: Palaeomagnetism of the Ntonya Ring Structure, Malawi. J. Geophys. Res. **73,** 725 (1968).

Brock, A.: Palaeomagnetic results from the Hook Intrusives of Zambia. Nature, **16,** 359 (1967).

――――: Palaeomagnetism of the Nuanetsi Igneous Province and its bearing upon the sequence of Karroo Igneous activity in Southern Africa. J. Geophys. Res. **73,** 1389 (1968).

De Sitter, L. U.: Structural development of the Arabian Shield in Palestine. Geol. en Mijnbouw, **41,** 116 (1962).

Discovery Cruise 16 Report Jan.–May 1967. National Inst. Oceanography Cruise Report Series: **CR16** (1968).

Drake, C. L. and R. W. Girdler: A geophysical study of the Red Sea. Geophys. J. Roy. Astron. Soc., **8,** 473 (1964).

Dubertret, L.: Les formes structurales de la Syrie et de la Palestine; leur origine. C.R. Acad. Sci. Colon., **195,** 66 (1932).

――――: Remarques sur le fosse de la Mer Morte et ses prolongements an Nord jusqu'an Taurus. Rev. de Géogr. Phys. et de Géol. Dyn., **9,** 3 (1967).

Fisher, Sir Ronald: Dispersion on a sphere. Proc. Roy. Soc. Lond., **A 217,** 295 (1953).

Freund, R.: A model of the structural development of Israel and adjacent areas since Upper Cretaceous times. Geol. Mag. **102,** 189 (1965).

Gass, I. G., D. I. J. Mallick, and K. G. Cox: Royal Society Volcanological Expedition to the South Arabian Federation and the Red Sea. Nature, **205,** 952 (1965).

Geological Society of London: The Phanerozoic Time-scale-A Symposium dedicated to Professor Arthur Holmes (1964).

Girdler, R. W.: The relationship of the Red Sea to the East African Rift System. Quart. J. Geol. Soc. Lond. **114,** 79 (1958).

――――: The role of translational and rotational movements in the formation of the Red Sea and Gulf of Aden. The World Rift System. Geol. Surv. Canada Paper 66-14, 65 (1966).

―――― and J. C. Harrison: Submarine gravity measurements in the Atlantic Ocean, Indian Ocean, Red Sea and Mediterranean Sea. Proc. Roy. Soc. **A 239,** 202 (1957).

Girdler, R. W. and G. Peter: An example of the importance of natural remanent magnetization in the interpretation of magnetic anomalies. Geophys. Prosp. **8,** 474 (1960).

Gough, D. I. and A. Brock: The Palaeomagnetism of the Shawa Ijolite. J. Geophys. Res. **69,** 2489 (1964).

Gough, D. I. and N. D. Opdyke: The palaeomagnetism of the Lupata Alkaline Volcanics. Geophys. J. Roy. Astron. Soc. **7,** 457 (1963).

Gough, D. I., A. Brock, D. L. Jones, and N. D. Opdyke: The palaeomagnetism of the Ring Complexes at Marangudzi and the Mateke Hills. J. Geophys. Res. **69,** 2499 (1964).

Graham, K. W. T. and A. L. Hales: Palaeomagnetic measurements on Karroo dolerites. Adv. Phys. **6,** 149 (1957).

――――, ――――: Preliminary palaeomagnetic measurements on Silurian Sediments from S. Africa. Geophys. J. Roy. Astron. Soc. **5,** 318 (1961).

Grasty, R. L. Ph.D. thesis, Univ. of London (1965) quoted by Raja et al. (1966).

Grommé, C. S. and R. L. Hay: Magnetization of basalt of Bed 1, Olduvai Gorge, Tanganyika. Nature, **200,** 560 (1963).

――――, ――――: Geomagnetic polarity epochs: new data from Olduvai Gorge, Tanganyika. Earth & Plan. Sci. Letters, **2,** 111 (1967).

Gutenberg, B. and C. F. Richter: *Seismicity of the Earth.* Princeton University Press (1954).

Hecker, O.: Bestimmung der Schwerkraft auf dem Schwarzen Meere und an dessen Küste sowie neue Ausgleichung der Schwerkraftmessungen auf dem Atlantischen, Indischen und Grossen Ozean. Veröffentl. des Zentralbureaus der Internat. Erdmessung, Neue Folge, 17 (No. 20) Berlin (1910).

Heirtzler, J. R. and X. Le Pichon: Magnetic anomalies over the Mid-Atlantic Ridge. J. Geophys. Res., **70,** 4013 (1965).

Heiskanen, W. A. and F. A. Vening Meinesz: *The earth and its gravity field,* Part 10D, McGraw-Hill, 389 (1958).

Herzen, R. P. von: Geothermal Heatflow in the Gulfs of California and Aden. Science **140,** 1207 (1963).

Heybroek, F.: The Red Sea Miocene Evaporite Basin. Salt Basins around Africa. Inst. of Petr. Lond. (1965).

Holmes, Arthur: *Principles of Physical Geology.* Nelson, London (1965).

Holwerda, J. G. and R. W. Hutchinson: Potash-Bearing Evaporites in the Danakil Area, Ethiopia. Econ. Geol. **63,** 124 (1968).

Hydrographic Office Chart C.6359: Red Sea Bathymetric Chart Compiled by N.A.T.O. Research Centre, La Spezia, Scale 1:2,300,000 (at Lat. 21°). Mercator Projection (1965).

Irving, E. and D. H. Tarling: The Palaeomagnetism of the Aden Volcanics. J. Geophys. Res. **66,** 549 (1961).

Knopoff, L. and J. C. Belshé: Gravity Observations of the Dead Sea Rift. The World Rift System. Geol. Surv. Canada Paper 66-14, 5 (1966).

Knott, S. T., Elizabeth T. Bunce, and R. L. Chase: Red Sea Seismic Reflection. The World Rift System. Geol. Surv. Canada Paper 66-14, 33 (1966).

Lamare, P.: Les Manifestations Volcaniques post-Crétacées de la Mer Rouge et des Pays Limitrophes. Mém. Géol. Soc. France New Series **14,** 21 (1930a).

――――: Nature et Extension des Dépôts Secondaires dans l'Arabie, l'Ethiopie et les Pays Somalis. Mém. Géol. Soc. France New Series, **14,** 49 (1930b).

――――: Observations relatives à la structure des pays

limitrophes de la Mer Rouge et du Golfe d'Aden. Compt. Rend. Somm. Séances Soc. Géol. de France, 72 (1930c).

Langseth, M. G. and P. T. Taylor: Recent heat-flow measurements in the Indian Ocean. J. Geophys. Res. 72, 6249 (1967).

Lartet, L.: La Géologie de la Palestine. Ann. Sci. Géol., 1, 1, 1 (1869).

Laughton, A. S.: The Gulf of Aden. Phil. Trans. Roy. Soc. A 259, 150 (1966a).

———: The Gulf of Aden, in relation to the Red Sea and the Afar Depression of Ethiopia. The World Rift System. Geol. Surv. Canada Paper 66-14, 78 (1966b).

McElhinny, M. W. and D. L. Jones: Palaeomagnetic measurements on some Karroo Dolerites from Rhodesia. Nature, 206, 921 (1965).

McElhinny, M. W. and N. D. Opdyke: Palaeomagnetism of some Carboniferous glacial Varves from Central Africa. J. Geophys. Res. 73, 689 (1968).

MacFadyen, W. A.: The Geology of the Farsan Islands, Gizan and Kamaran Island, Red Sea. Geol. Mag. 67, 310 (1931).

———: On the Volcanic Zebayir Islands, Red Sea. Geol. Mag. 69, 63 (1932).

———: *Geology of British Somaliland.* Gov. of Somali Protectorate, 87 p (1933).

Masson, P. A. A. H. and F. J. Agnich: Seismic survey of Sinai and the Gulf of Suez. Geophysics, 23, 329 (1958).

Mitchell, R. C.: Fault patterns of Northwestern Hegaz. Saudi Arabia. Eclogae Geol. Helvétiae, 50, 257 (1957).

Mohr, P. A.: *The Geology of Ethiopia.* Univ. Coll. Addis Ababa. Press, 268 p (1962).

———: Major volcano-tectonic lineament in the Ethiopian Rift System. Nature, 213, 664 (1967).

Montanaro-Gallitelli, E.: Foraminiferi, posizione stratigrafica e facies di un calcare a "Orperculina" dei Colli di Ebud (Sahel Eritreo). Palaeontogr. Ital., 40, 67 (1941).

Mussett, A. E., T. A. Reilly, and P. K. S. Raja: Palaeomagnetism in East Africa. Proc. E. Afr. Acad., 2, 27 (1964).

Nafe, J. E., J. F. Hennion, and G. Peter: Geophysical measurements in the Gulf of Aden. Preprints Int. Ocean. Congr. New York, 42 (1959).

Nairn, A. E. M.: A palaeomagnetic survey of the Karroo System. Overseas Geol. & Min. Res., 7, 398 (1960).

———: Palaeomagnetic measurements on Karroo and Post-Karroo Rocks; Second Progress Report. Overseas Geol. and Min. Res., 9, 302 (1964).

Neev, D. and K. O. Emery: The Dead Sea: Depositional Processes and Environments of Evaporites. Israel Geol. Surv. Bull 41, 1 (1967).

Opdyke, N. D.: The Palaeomagnetism of Some Triassic Red Beds from N. Rhodesia. J. Geophys. Res. 69, 2495 (1964).

Owen, L.: Origin of the Red Sea Depression. Bull. Amer. Assoc. Petr. Geol. 22, 1217 (1938).

Picard, L.: Thoughts on the graben system in the Levant. The World Rift System. Geol. Surv. Canada Paper 66-14, 22 (1966).

Plaumann, S.: Kontinuierliche Schweremessungen im Roten Meer mit einen Askania-Seagravimeter vom Typ Gas 2 nach GRAF. Zeit für Geophysik, 29, 233 (1963).

Quennell, A. M.: The geology and mineral resources of (Former) Trans-Jordan. Col. Geol. & Min. Res., 2, 85 (1951).

———: The structural and geomorphic evolution of the Dead Sea Rift. Quart. J. Geol. Soc. Lond., 114, 1 (1958).

Raja, P. K. S., T. A. Reilly, and A. E. Mussett: The palaeomagnetism of the Turkana Lavas. J. Geophys. Res., 71, 1217 (1966).

Raisin, C. A.: Perim Island and its relations to the area of the Red Sea. Geol. Mag. 9, 132 and 206 (1902).

Richard, J. J. and M. Neumann van Padang: Catalogue of the active volcanoes of the world including Solfatara Fields; Part IV Africa and the Red Sea. I.A.V. (1957).

Rothé, J. P.: La Zone séismique Médiane Indo-Atlantique. Proc. Roy. Soc. A 222, 387 (1954).

Sclater, J. G.: Heat-flow in the North West Indian Ocean and Red Sea. Phil. Trans. Roy. Soc. Lond. A 259, 271 (1966).

Shalem, N.: The Red Sea and the Erythrean Disturbances. Congr. Géol. Int. Algér 1952, Section XV. La Palaeovolcanologie et ses Rapports avec la Tectonique, 17, 223 (1954).

Swartz, D. H. and D. D. Arden: Geologic history of the Red Sea Area. Bull. Amer. Assoc. Petr. Geol., 44, 1621 (1960).

Sykes, L. R. and M. Landisman: The seismicity of East Africa, the Gulf of Aden and the Arabian and Red Seas. Bull. Seism. Soc. Amer., 54, 1927 (1964).

Tarling, D. H., M. Sanver, and A. M. J. Hutchings: Further palaeomagnetic results from the Federation of South Arabia. Earth & Plan. Sci. Letters, 2, 148 (1967).

Taylor, M. H.: Geologic structures in the Ethiopian Region of East Africa. Trans. New York Acad. Sci. Ser. 11, 14, 220 (1952).

Triulzi, A. E. von: Expedition S.M. Schiff POLA in das Rote Meer, nördliche Hälfte. Wissenschaftliche Ergebnisse 11. Relative Schwerebestimmungen. Denkschr. K. Akad. Wiss. Math-Nat. Classe, 65, 131 (1898).

———: Expedition S.M. Schiff POLA in das Rote Meer, südliche Hälfte. Wissenschaftliche Ergebnisse XII. Relative Schwerebestimmung. Denkschr. K. Akad. Wiss. Math-Nat. Classe, 69, 143 (1901).

Tromp, S. W.: Blockfaulting phenomena in the Middle East. Geol. en Mijn. New Ser. 2, 273 (1949).

———: The age and origin of the Red Sea Graben. Geol. Mag. 87, 385 (1950).

Van Zijl, J. S. V., K. W. T. Graham, and A. L. Hales: The palaeomagnetism of the Stormberg Lavas 11. Geophys. J. Roy. Astron. Soc., 7, 169 (1962).

Vening Meinesz, F. A.: Gravity expeditions at sea II, 1923–1932. Delft (1934).

———: Les Graben Africains, résultat de compression ou de tension dans la croûte terrestre. Bull. des Séances Inst. Roy. Colon. Bélge., 21, 539 (1950).

Vine, F.: Spreading of the ocean floor: new evidence. Science, **154,** 1405 (1966).

Wegener, A.: "Die Entstehung der Kontinente und Ozeane." English Translations of 3rd (1922) and 4th (1929) Editions, Methuen, Lond. (1924 and 1967).

Wellings, F. E.: Observations of Dead Sea Structure. Bull. Geol. Soc. Amer. **49,** 659 (1938).

Wetzel, R. and D. M. Morton: Contribution a la Géologie de la Transjordanie. Notes et Mém. sur le Moyen-Orient, **7,** 95 (1959).

Zak, I. and R. Freund: Recent strike slip Movements along the Dead Sea Rift. Israel J. Earth-Sciences, **15,** 33 (1966).

Structural Effects of Sea-floor Spreading in the Gulf of Aden and the Red Sea on the Arabian Shield *

PETER J. BUREK

Southwest Center for Advanced Studies, Dallas, Texas

Abstract

The Carlsberg Ridge of the Indian Ocean can be traced and connected with the ridge in the Gulf of Aden (Laughton, 1965). In spite of the structural complexities in the southern Red Sea and Afar Depression (Mohr, 1967), it seems reasonable to infer that the Aden extension of the Indian mid-oceanic ridge splits in this area into two branches: the Red Sea branch to the NNW (Girdler, 1965) and the East African Rift System to the SSW (Mohr, 1968).

Linear magnetic anomalies over the rift troughs can be used to infer that sea-floor spreading has occurred (Vine, 1966). The emplacement of a quasi-oceanic crust in the Gulf of Aden and Red Sea rifts resulted in the separation of the former Arabian-Nubian Shield into two continental blocks. The aim of this paper is to consider the tectonic effects of the rifting on the Arabian Shield and the surrounding mobile belt. An attempt of structural and paleogeographic reconstructions is shown in Figs. 1, 2 and 3. These figures also indicate the important geographical features.

Rift Structures Surrounding the Southern Arabian Shield

Traditionally, the Gulf of Aden and the Red Sea were considered as extensive graben features. However, data obtained during the International Indian Ocean Expedition now make it possible to delineate the main geophysical characteristics of the Gulf of Aden and the Red Sea which suggest a different interpretation. The Gulf of Aden and the Red Sea have the following features in common.

Physiography

The continental shelves terminate with steep continental slopes (Fig. 3) which have similar outlines on opposite sides of the basins and do not overlap significantly when fitted together (reconstruction attempts in Figs. 1 and 2). Characteristically the main troughs are mildly deformed with occasional deeper basins along strike, but there is an intensively fractured central zone composed of a series of ridges and valleys of considerable relief often displaced by transcurrent faults.

Seismology

Nafe *et al.* (1959) and Drake and Girdler (1964) reported seismic refraction data for the troughs that indicate layers with P-wave velocities 6.4–7.8km/sec for the Gulf of Aden and 6.7–7.4km/sec for the Red Sea in the basement of the central trough, reflecting the presence of basic material. They interpret P-wave velocities of 5.6 to 6.1km/sec as indicative of granitic material, which although observed on the shelves, were not found in the interior rift troughs. On the shelves as well as in the troughs the basement layers are overlain by materials with mean velocities varying from 3.5–4.7km/sec, which has been interpreted as sedimentary sequences with thicknesses up to 3km. The results of Sykes and Landisman (1964) position recent earthquake epicenters (accuracy ±10km)

* Contribution No. 71, Southwest Center for Advanced Studies.

REGRESSIONS AND VOLCANISM DURING EOCENE

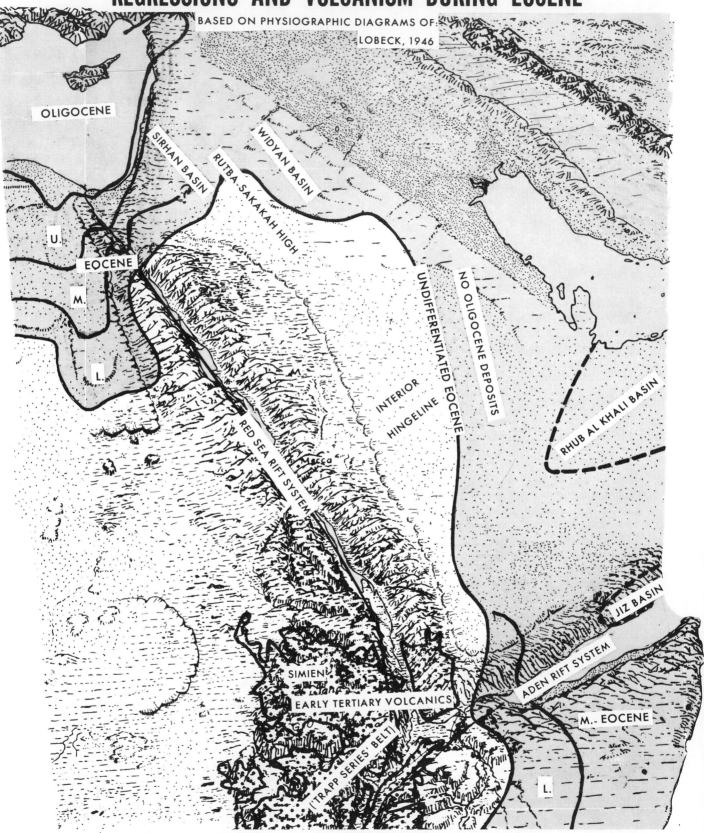

BASED ON PHYSIOGRAPHIC DIAGRAMS OF:
LOBECK, 1946

OLIGOCENE

SIRHAN BASIN

RUTBA-SAKAKAH HIGH

WIDYAN BASIN

U.

EOCENE

M.

L.

RED SEA RIFT SYSTEM

Mecca

INTERIOR

HINGELINE

UNDIFFERENTIATED EOCENE

NO OLIGOCENE DEPOSITS

RHUB AL KHALI BASIN

JIZ BASIN

SIMIEN

EARLY TERTIARY VOLCANICS

(TRAPP SERIES BELT)

ADEN RIFT SYSTEM

M.. EOCENE

L.

Fig. 1

within the central ridge and trough zone, along transcurrent faults or major faults on the shields as shown on Fig. 3.

Gravity Anomalies

Allan and Pisani (1965), Knott *et al.* (1965) and Talwani *et al.* (1965) have confirmed that positive Bouguer gravity anomalies (150mgal) are associated with both rift troughs. The lateral extent of the anomalies coincides with the width of the interior troughs and covers their entire length. However, on entering the Gulf of Aqaba-Dead Sea Fault System a pronounced change in the character of the Red Sea Gravity Anomaly is observed (Knopoff and Belshé, 1965).

Magnetic Anomalies

The axes of the central ridge zones are marked by large magnetic anomalies. In contrast, the magnetic anomalies in the basins on both sides of the central ridge zone are less pronounced and vanish towards the shelves. Individual magnetic anomalies can be correlated and generally follow the strike of the ridges. This pattern is similar to that in the ocean basins and can be used to support the concept of sea-floor spreading (Vine and Matthews, 1963).

In summary, the geophysical data clearly demonstrate that the central portions of both the Gulf of Aden and the Red Sea can no longer be considered as simple graben structures, but instead are rift troughs with basic basements and thin crusts characteristic of oceanic basins.

Although both the Gulf of Aden and the Red Sea have many features in common, some pronounced differences exist which are indicative of their individual stage of development. The Gulf of Aden is wider; the basins are deeper; the secondary transcurrent faults are better developed. The Red Sea, by contrast, has only a narrow main trough and wider shelves. If one considers the geophysical data of the Gulf of Aden, it generally does not involve major interpretation problems. However, the Red Sea data are confused south of 17°N, where they are complex and difficult to understand

because of the fragmentation of continental blocks by extensive basic intrusions (Mohr, 1967). An alternate explanation may be the ±35° rotation of the Danakilia Horst (Laughton, 1965). North of 23°N the magnetic anomalies die out, possibly due to deep burial of igneous material resulting in temperatures exceeding the Curie Point or to the lack of extensive volcanic activity, i.e., sea-floor spreading. Further to the north, the Gulf of Suez area, though clearly characterized by grabens and strike slip faults (Youssef, 1968), is believed to be in a more youthful stage as compared to the Red Sea.

A comparison between the Gulf of Aden and Red Sea troughs suggests that in the Gulf of Aden area the amount of spreading is greater. Thus, spreading is in a more advanced stage in the Gulf of Aden than in the Red Sea.

Volcanic Phases

Extensive areas of West-Arabia, Yemen, Ethiopia and Somalia are intruded or covered by igneous rocks. These volcanic activities are associated with tectonic events in the adjacent rift troughs, as is clearly demonstrated by the close connection of eruption centers to regional fault trends. Though at present radiometric dates are lacking, geological evidence indicates two major eruption phases here grouped as: (a) the Upper Cretaceous-Oligocene Trap Series and (b) the Miocene-Recent Aden Volcanics phase.

(a) Trap Series (Fig. 2): in Yemen, basalts of the Trap Series are described as being interbedded with Cretaceous strata. From the Danakil region Cretaceous andesites have been reported (Filjak *et al.*, 1959). The majority of eruptions started during the late Paleocene with maximum activity during the late Eocene and Oligocene. The assignment of an upper age limit is difficult, for interbedded fossiliferous sediments in the Yemen (Geukens, 1966) indicate some eruptions as late as Miocene. Mohr (1967) reports an apparently conformable passage of the Ethiopian Trap Series (Paleocene-Oligocene) into the still

SCHEMATIC PHYSIOGRAPHIC SKETCH UPPER EOCENE TO OLIGOCENE
I. PHASE

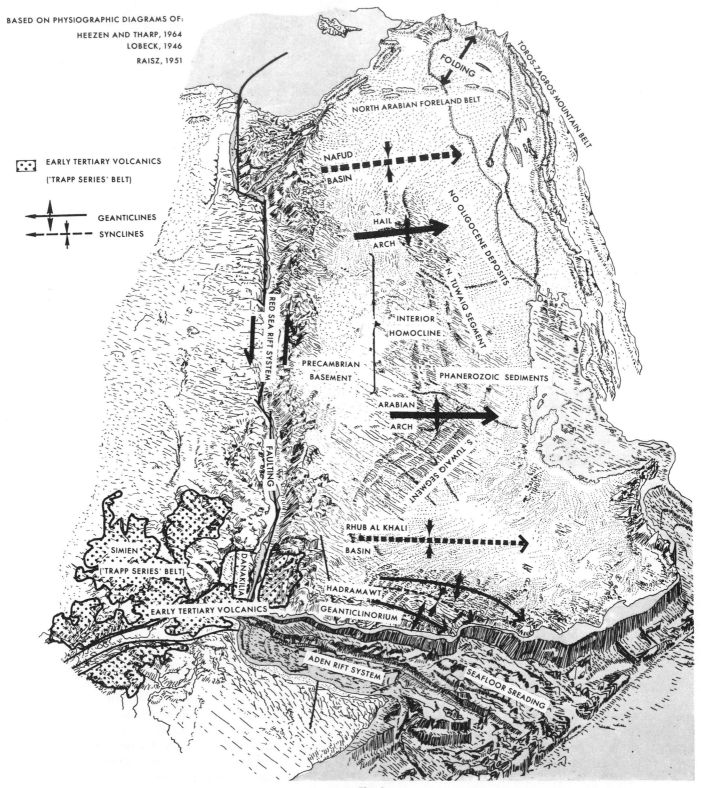

BASED ON PHYSIOGRAPHIC DIAGRAMS OF:
HEEZEN AND THARP, 1964
LOBECK, 1946
RAISZ, 1951

EARLY TERTIARY VOLCANICS
('TRAPP SERIES' BELT)

GEANTICLINES
SYNCLINES

FOLDING

NORTH ARABIAN FORELAND BELT

TOROS-ZAGROS MOUNTAIN BELT

NAFUD BASIN

NO OLIGOCENE DEPOSITS

HAIL ARCH

N. TUWAIQ SEGMENT

RED SEA RIFT SYSTEM

INTERIOR HOMOCLINE

PRECAMBRIAN BASEMENT

PHANEROZOIC SEDIMENTS

FAULTING

ARABIAN ARCH

S. TUWAIQ SEGMENT

SIMIEN ('TRAPP SERIES' BELT)

DANAKLIA

RHUB AL KHALI BASIN

HADRAMAWT GEANTICLINORIUM

EARLY TERTIARY VOLCANICS

ADEN RIFT SYSTEM

SEAFLOOR SREADING

Fig. 2

younger Shield Group of the Simien Center.

Mohr (1967) and Greenwood and Bleackley (1967) emphasize the generally strong alkaline character of the Trap Series although some tholeitic basalts were found. From Yemen, Karrenberg (1956) reported alkaline granites intruding into the Trap Series. In general, the early Tertiary volcanism started with dominantly mafic basaltic effusives, followed by more silicic lavas. Granitic intrusions were emplaced during this later phase. It is interesting to note that the older Tertiary "Trap Series" occurs only in areas surrounding the eastern Gulf of Aden (Figs. 1 and 2), thus indicating a late Eocene-Oligocene period of tectonic unrest in that area.

(b) Aden Volcanics (Fig. 3): in S.W. Arabia (Yemen and Aden Protectorate) the term Aden Volcanics is used to describe volcanic material of post-Trap age (Miocene-Recent). In Ethiopia this definition is certainly somewhat arbitrary since the volcanic activity was much more continuous and extensive than on the Arabian side.

The earliest silicic lavas in Ethiopia are of Upper Oligocene to Middle Miocene age and were described from the Shield Group of the Simien Mountains in northern Ethiopia (Mohr, 1967). This could indicate that they represent a local connection between the late silicic Trap Series and early silicic Aden Volcanics. However, the most intensive volcanism was during Pliocene to Pleistocene times.

The post-Shield Group volcanics of Ethiopia, with the exception of the Pliocene ignimbrites, are concentrated around the Afar Depression. In contrast to the occurrence of the older Tertiary Trap Series around the Gulf of Aden, the younger Cenozoic belt of Aden Volcanics (Fig. 3) extends from the northern part of the East African Rift System into areas mainly east of the Red Sea (Aden Volcanoes, Saudi-Arabian Harrats) and occurs even in the northernmost parts of the Arabian Shield (Jebel Hauran/Syria and Karaca Dag/Turkey). Except for a few relatively small outcrops, no major Tertiary volcanics are mapped on the Egyptian side of the Red Sea.

In summary, we can separate two main belts of volcanism that differ in time and position: the Trap Series which are to be associated with events in the Gulf of Aden and the younger Aden Volcanics related to the Red Sea-East African Rift System.

Structures on the Arabian Shield

The Arabian Shield, a deeply eroded, virtually peneplained complex of igneous and metamorphic rocks (Brown and Jackson, 1960), is the relatively stable base for an almost flat-lying Phanerozoic sedimentary cover which in places measures 5,500m thick. Epirogenic movements that can hardly be detected in the Precambrian basement are clearly reflected in its sedimentary mantle. Geologic and topographic maps (for example, U.S.G.S. Maps, 1:2,000,000, 1963) show two sets of gentle, broad geanticlines and synclines which differ in structural trend, location and age.

Fig. 2 shows structures with fold axes that parallel the Gulf of Aden: these are the Hadramawt Geanticlinorium [Southern Hadramawt Arch, Wadi Hadramawt-Jiz Syncline, Northern Hadramawt Arch (Fig. 3)], Rub Al Khali, Interior Homocline (Central Arabian Arch, Hail Arch) and Nafud Basin (Fig. 2). Eocene sedimentary rocks are affected by the warping in all these structures, indicating that the last major epirogenic phases related to the Gulf of Aden trend occurred during the late Eocene and probably during the Oligocene. It is interesting to note that the width of this wave like pattern increases from S to N, i.e., with growing distance from the Gulf of Aden and appears to die out north of the Nafud desert.

The epirogenic warps are interpreted as compressional features, as inferred from the following observations: faulting occurs only close to the oceanic basins; the distance of separation between individual arches and basins increases significantly with distance from the rifting zones (Figs. 2 and 3); finally, in the Interior Homocline and in the Rutba-Sakakah Arch "Scheitelgraben" (such as Wadi Nisah-Wadi as Sahba Graben and Khwar Umm Wu'al

SCHEMATIC PHYSIOGRAPHIC SKETCH MIOCENE TO RECENT
II. PHASE

BASED ON PHYSIOGRAPHIC DIAGRAMS OF:
HEEZEN AND THARP, 1964
LOBECK, 1946
RAISZ, 1951

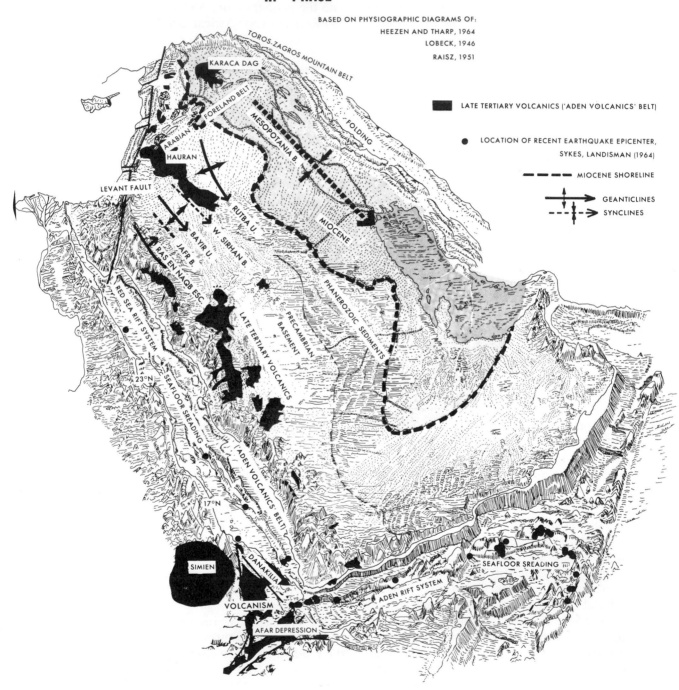

Fig. 3

Graben) in the sense of Cloos (1936) were developed (Powers *et al.,* 1966).

Furthermore, close to the edges of the Arabian Shield, near the more unstable, former Tethys Geosyncline, several zones of more tightly folded belts were formed, apparently contemporaneously with the broad arching of the shield itself (Fig. 2). These are: (a) the folded foreland belt which included the Sinai-Palestine folds, Lebanon, M. Hermon, Antilebanon, Palmyra Arch, Jebel Sinjar, etc., which were involved in a Late Eocene to Oligocene phase of folding (Knetsch, 1957); (b) the Alpine Toros-Zagros belt which underwent a pronounced orogenic phase during the same time interval (Ilhan, 1967).

Fig. 3 shows a younger set of epirogenic warps with fold axes parallel to the Red Sea: Ras En Nagb Escarpment, Jafr Depression, Bajir Uplift, Wadi Sirhan Basin, Rutba-Sakakah Arch and the Widyan-Mesopotamian Basin. These structures occur only on the northern part of the Arabian Shield and in Egypt, i.e., in areas where epirogenic warps trending parallel to the Gulf of Aden die out. In the Widyan and the Wadi Sirhan Basin Pliocene sediments uncomformably overlie Late Eocene or older sedimentary units which suggests that these structures reached their present form during Mio-Pliocene times. Some faults suggest Pleistocene movements.

Movements along the Levant (Aqaba-Dead Sea) Fault System are associated with these Miocene to Pleistocene phases (Quennell, 1958) of tectonic activity, as well as time correlative folding occurs in the Zagros Mountains (Naqib, 1967). Furthermore, seismic reflection studies of Red Sea sediments (Knott *et al.,* 1965) show deformation resulting from two periods of rifting, one before the beginning of the Pliocene, and one in late Plio-Pleistocene time.

Structural Evolution on the Arabian Shield

It has been shown above that the volcanism and epirogenic movements associated with the Gulf of Aden trend reached a period of maximum intensity during the late Eocene and Oligocene, while the volcanism and the warping of structures parallel to the Red Sea are more recent, with activity peaks that lasted from Miocene well into Pleistocene times. These dates correspond only to the last pronounced phases associated with the two trends, yet it is important to study their evolution in order to learn when they first appeared. Epirogenic movements that control the paleogeographic evolution are best displayed by changes in sedimentation rates, unconformities, shifts of shorelines, etc. (Summarizing Literature: U.S.G.S. Prof. Paper Ser. 560, 1967, Lex. Strat. Int., Vol. III, Fasc. 10, 1963, Wolfart, 1967, etc., Burdon, 1959).

Paleogeology of Central and South Arabia

The Phanerozoic sedimentary cover of the Assyntian Basement records four major episodes related to its paleogeographic development: (a) the Paleozoic formation of the "Arabian High" in Central Arabia, (b) its Upper Permian to Jurassic complete inversion to the "Central Arabian Embayment" (Fig. 4), (c) the Lower Cretaceous to Oligocene evolvement of the Interior Homocline and the Hadramawt Geanticlinorium (Fig. 2), (d) the Upper Cretaceous to Pliocene development of the "Rutba-Sakakah Arch" with its adjoining basins (Figs. 1 and 3).

The stratigraphic N-S cross section from the Rub Al Khali to the Nafud through the Phanerozoic sedimentary mantle of the Saudi part of the Arabian Shield (Fig. 4) is based on ARAMCO field work (U.S.G.S. Prof. Paper 560-D, pt. 1). This section clearly shows the location of the "Central Arabian High" and reflects the later development of the "Central Arabian Embayment." These first two major paleogeographic episodes demonstrate that the present morphological pattern must be younger than late Jurassic and definitely is not an old, inherited structure.

During the late Jurassic (Tithonian) and Early Cretaceous (Berriasian, Hauterivian) epirogenic movements resulted in the for-

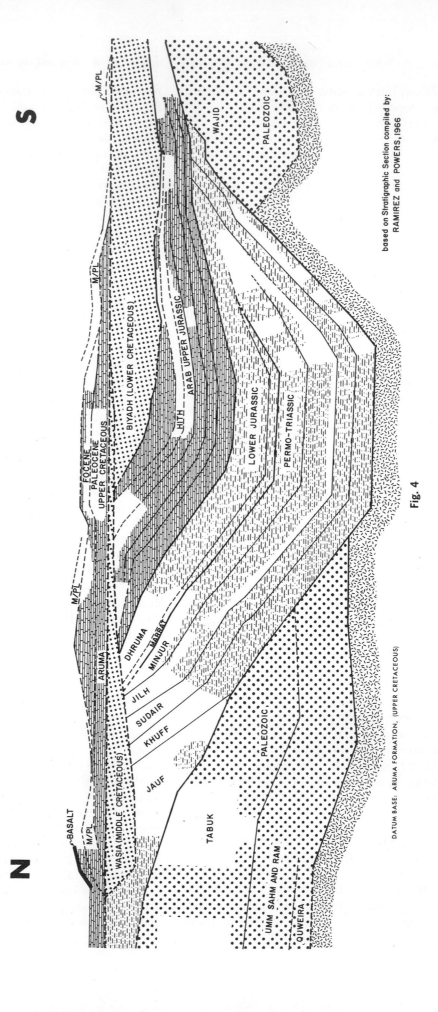

Regional Stratigraphic N-S Section
NAFUD - RUB AL KHALI

N

S

BASALT

M/PL

WASIA (MIDDLE CRETACEOUS)

ARUMA

DHRUMA

MARRAT

MINJUR

JILH

SUDAIR

KHUFF

JAUF

TABUK

UMM SAHM AND RAM

QUWEIRA

PALEOZOIC

EOCENE

PALEOCENE

UPPER CRETACEOUS

BIYADH (LOWER CRETACEOUS)

HITH

ARAB UPPER JURASSIC

LOWER JURASSIC

PERMO-TRIASSIC

PALEOZOIC

WAJID

M/PL

DATUM BASE: ARUMA FORMATION, (UPPER CRETACEOUS)

based on Stratigraphic Section compiled by:
RAMIREZ and POWERS, 1966

Fig. 4

mation of a large hinge line (Fig. 1) roughly coinciding with the axis of the "Central Arabian Arch" as interpreted from unconformities and shifts in areas of sedimentation. During the Lower Cretaceous (Aptian, Albian) the Northern Tuwaiq Segment (escarpment between Central Arabian and Hail Arch, Fig. 2) was first uplifted, while the south continued to receive sediments (Biyad Sandstone, Fig. 4). During the Upper Cretaceous (Cenomanian-Maastrichtian) the Southern Tuwaiq Segment (area between Central Arabian Arch and Rub Al Khali, Fig. 2) was slightly raised, while the north subsided (Wasia-Aruma Formations, Fig. 4). Thus, during the Cretaceous we find the first indication of the evolving "Interior Homocline" (Powers *et al.,* 1966). During Paleocene and Eocene the general pattern inherited from the Cretaceous was further emphasized (Fig. 1). The area north of the "Interior Homocline" continued to sag and the Rub Al Khali basin became an area of active subsidence as well (Fig. 1), thus roughly outlining the present structural pattern of the Nafud Basin, Interior Homocline, Rub Al Khali (Fig. 2).

This epirogenic evolvement in Central Arabia corresponds with the evolution of the South Arabian "Hadramawt Geanticlinorium." During the Upper Jurassic South Arabia and Somalia subsided for the first time since the late Precambrian. Jurassic sedimentary basins were separated by the northerly trending "Mukalla" and "Dufar Highs" (Beydoun, 1966) which are also recognized on the African horn. As in Central Arabia Late Jurassic and Early Cretaceous unconformities are again observed and can be used to infer tectonic disturbances. It may be more than mere coincidence that at the same time volcanism and faulting was taking place in the Levant (Swartz and Arden, 1960).

During the Lower Cretaceous the paleogeography of this region changed profoundly. The Jurassic highs became less pronounced. An east-west striking basin, parallel to the present Gulf of Aden and situated almost on the axis of the present Wadi Hadramawt-Jiz syncline, appeared for the first time. Since the Senonian, the basin was leveled and facies differences became far less pronounced (Mukalla Sandstone), indicating a period of tectonic quiescence. However, during the Paleocene and Lower Eocene the subsidence of the same basin reappeared, involving a greater area than in pre-Senonian time. This rejuvenated trough (Fig. 1) covered most of the area of the Hadramawt-Jiz syncline. Tilting and faulting of Eocene sediments show that the whole Hadramawt System had developed by Late Eocene times and was subsequently uplifted during Oligocene times.

Paleogeology of North Arabia

Since we find on the northern part of the Arabian Shield a great variety of structures (Fig. 3): the S-shaped foreland belt, extending from Sinai to North Iraq; the Toros-Zagros mountain chain which surrounds the shield; epirogenic warps parallel to the Red Sea and fault zones such as the Levant fault, the geological evolution of North Arabia is much more difficult to unravel than that of the main block of Arabia.

Schürmann (verbal communication) suggests that one of the many parallel faults which existed by Precambrian time was the tectonic element along which the Levant (Aqaba-Dead Sea) fault system, as well as the Red Sea rift, later formed (Youssef, 1968). The Levant trend (NNE-SSW) played a major role in the paleogeographic development of the region until Early Cretaceous times (Bender, 1965; Burdon, 1959; Picard, 1943).

During the Late Cretaceous a new paleogeographic element with Red Sea trend appeared on the northern shield. The Sakakah-Rutba Arch (Fig. 1) separated marine sedimentation basins which trend parallel to the Red Sea in Jordan (W. Sirhan Basin) and Mesopotamia (Widyan Basin). The situation prevailed and was accentuated throughout the Paleocene and Eocene (Fig. 1). During the Late Eocene, North Arabia was uplifted. Pliocene deposits which unconformably overlie older strata in the Wadi Sirhan and the Widyan Basin and Quaternary mud flats in the Jafr Depression imply that the warps of the Ras

En Nag Escarpment, Jafr-Tayma Depression, Bayir-Al Hubrah Uplift, Wadi Sirhan Basin, Rutba-Sakakah Arch, Widyan-Mesopotamian Basin were developed during Mio-Pliocene times (Fig. 2). Quennell (1958) and Freund (1965) showed that faulting along the Levant fault zone occurred mainly during the Pliocene to Pleistocene, but movements commenced in the Miocene and Upper Cretaceous.

Conclusion

A review of recent oceanographic research shows that the Gulf of Aden and Red Sea can no longer be considered as simple graben structures but are rift troughs in an early stage of development. A comparison between both troughs demonstrates that the Aden rift trough is the more mature one. Intrusion and extrusion of basic igneous material which is interpreted as the emplacement of a quasi-oceanic crust resulted in an expansion of the rift troughs, i.e., ocean-floor spreading, thus separating the Arabian block from Africa.

Volcanism, epirogenic and orogenic events as well as movement along pronounced fault systems on the shield are closely associated with the events in the rift troughs. Four major regional features are related in age and trend to the Gulf of Aden and Red Sea:

1. Two volcanic episodes which differ in age and geographic locality can be separated into: (a) the Trap Series that occurs only in the Gulf of Aden area and ranges in age from Late Cretaceous into the Early Miocene. The most intense volcanism, however, occurred during Eocene to Oligocene time, (b) the Aden Volcanics Belt, which extends from the Afar Depression over the West Arabian Shield to South-East Turkey. The age of these volcanics span Miocene to Recent time with eruption peaks during the Plio-Pleistocene.
2. Two different sets of epirogenic warps which are essentially parallel and adjacent to the rift troughs can be recognized: (a) Structures related to the Gulf of Aden (Fig. 2) appeared during the Lower Cretaceous. They reached their present configuration during the Late Eocene and Oligocene, (b) Epirogenic structures (Fig. 3) that parallel the Red Sea are younger and occur only on the North Arabian Shield where the structures associated with the Gulf of Aden die out. During Late Cretaceous time we find the first indications of their appearance, but their present morphology was developed during the Plio-Pleistocene times.
3. Orogenic structures surround the north and east Arabian Shield: the Toros-Zagros mountains and its S-shaped foreland belt. Orogenic phases are apparently correlative with epirogenic movements observed on the shield.
4. Pronounced fault systems such as: (a) the initial Red Sea fault system, (b) the Levant fault zone, are associated with the different movements of the shield.

The time sequence of the structural evolution of Arabia can be separated in two phases:

Phase I – The Gulf of Aden extension of the Carlsberg ridge first became active during the Early Cretaceous and continued to develop in several phases that culminated in Late Eocene time. The spreading in the Gulf of Aden resulted in a NNW-movement of the Arabian Shield (Fig. 2). Left-lateral movement along the initial Red Sea-Suez fault system allowed the separation of Arabia from the Nubian Shield. The shield area itself suffered repeated compressional stresses resulting in epirogenic warping (Fig. 2). Several phases of orogenic folding and thrusting in the unstable shield margin are the effects of the northward drift of Arabia. Volcanism around the Western Aden rift trough; margins are associated with faulting.

Phase II – The evolution of the Red Sea Basin is younger (Fig. 3) than the Gulf of Aden. Although paleogeographic evidence indicates Late Cretaceous movements, the major tectonic activity peaks date from Miocene to Recent. As it is believed that the estimated tension in the Suez Graben cannot account for the whole amount of separation in the Red Sea, it seems reasonable to suggest that the NE-movement of Arabia which is associated with the spreading in the Red Sea occurred along the Levant (Aqaba-Orontes) fault zone. The associated volcanic, epirogenic and orogenic development is very similar to that associated with the spreading in the Gulf of Aden.

The separation of the Arabian Shield from the African continents is our most recent example of continental drift and thus most suitable for the study of the tectonic effects of ocean-floor spreading. The structural evolution on the Arabian Shield implies that the tectonic forces that caused epirogenic, orogenic movements and volcanism originated in the rift troughs, i.e., are related to sea-floor spreading.

Acknowledgments

I am greatly indebted to Deutsche Forschungs Gemeinschaft, Bad Godesberg, Germany, for awarding grants for 2 years of field work in the Middle East. In Turkey I received valuable help from Dr. S. Alpan, M. T. A. Ankara and Prof. Dr. R. Brinkmann, EGE UNIVERSITESI, Jzmir, and sincere gratitude is expressed to Dr. G. F. Brown, U.S.G.S., who was most helpful and was responsible for obtaining support for my program from His Excellency Dr. F. Kahbani, Ministry of Mineral Resources of the Kingdom of Saudi Arabia and Dr. Ranoux, BRGM, Jeddah. Dr. M. Barber, UNESCO and Dr. F. Bender, Deutsche Geol. Mission, Amman, substantially aided my paleomagnetic field work in Jordan. This paper was prepared at the Southwest Center for Advanced Studies, Dallas, and was supported under NASA Grant NSA 269-62 and NSF Grant GA 1601 and GA 590. Drs. A. L. Hales, M. Halpern, C. E. Helsley and I. MacGregor, SCAS, Dallas, reviewed the paper. Last, but not least, I would like to thank my wife who turned out to be an excellent field assistant.

References

Allan, T. D. and M. Pisani: Gravity and magnetic measurements in the Red Sea. The World Rift System, Geol. Survey Canada Paper 66-14, 62 (1965).

Bender, F.: Zur Geologie der Kupfererzvorkommen am Ostrad des Wadi Araba, Jordanien. Geol. Jb., **83,** 181 (1965).

Beydoun, Z. R.: Geology of the Arabian Peninsula, Eastern Aden Protectorate and Part of Dhufar. U.S. Geol. Survey Prof. Paper 560-H, 1 (1966).

Brown. G. F. and R. O. Jackson: The Arabian Shield. Internat. Geol. Cong. Rept. 21, Copenhagen, **9,** 69 (1960).

Burdon, D. J.: *Handbook of the Geology of Jordan,* to accompany and explain the three sheets of the 1:250,000 geological map of the Rift, by A. M. Quennell, Colchester (Berhan & Co.), 1 (1959).

Cloos, H.: *Einführung in die Geologie,* Berlin (1936).

Drake, C. L. and R. W. Girdler: A geophysical study of the Red Sea. Geophys. J., **8,** 473 (1964).

Filjak, R., N. Glumicic, and Z. Zagorac: *Oil possibilities of the Red Sea region in Ethiopia,* Naftaplin, Zagreb, 1 (1959).

Freund, R.: A model of the structural development of Israel and adjacent areas since Upper Cretaceous times. Geol. Mag., **102,** 189 (1965).

Geukens, F.: Geology of the Arabian Peninsula, Yemen. U.S. Geol. Survey Prof. Paper 560-B, 1 (1966).

Girdler, R. W.: The role of translational and rotational movements in the formation of the Red Sea and Gulf of Aden. The World Rift System, Geol. Survey Canada Paper 66-14, 65 (1965).

Greenwood, G. C. G. W. and D. Bleakley: Geology of the Arabian Peninsula, Aden Protectorate. U.S. Geol. Survey Prof. Paper 560-C, 1 (1967).

Ilhan, E.: Toros-Zagros folding and its relation to Middle East oil fields. Amer. Assoc. Petrol. Geol., **51,** 651 (1967).

Karrenberg, H.: Junger Magmatismus and Vulkanismus in Südwestarabien (Jemen). Internat. 20th Geol. Cong., Mexico City, Résumé, **1,** 171 (1956).

Knetsch, G.: Eine Struktur-Skizze Ägyptens und einiger seiner Nachbargebiete. Geol. Jb., **74,** 75 (1957).

Knopoff, L. and J. C. Belshé: Gravity observations of the Dead Sea Rift. The World Rift System, Geol. Survey Canada Paper 66-14, 5 (1965).

Knott, S. T., E. T. Bunce, and R. T. Chase: Red Sea seismic reflection studies. The World Rift System, Geol. Survey Canada Paper 66-14, 33 (1965).

Laughton, A. S.: The Gulf of Aden, in relation to the Red Sea and Afar Depression of Ethiopia. The World Rift System, Geol. Survey Canada Paper 66-14, 78 (1965).

Mohr, P. A.: The Ethiopian Rift System. Bull. Geophys. Obs., **11,** 1 (1967).

———: Transcurrent faulting in the Ethiopian Rift System. Nature, **218,** 938 (1968).

Nafe, J. E., J. F. Hennion, and G. Peter: Geophysical measurements in the Gulf of Aden. Preprints I.O.C., 42 (1959).

Naqib, K. M. Al.: Geology of the Arabian Peninsula, Southwestern Iraq. U.S.G.S. Prof. Paper 560-G, 1 (1967).

Picard, L.: Structure and evolution of Palestine with comparative notes on neighboring countries. Bull. Geol. Dept., Hebr. Univ., Jerusalem, **4,** 2–4, 1 (1943).

Powers, R. W., L. T. Ramirez, C. D. Redmond, and E. L. Elberg, Jr.: Geology of the Arabian Peninsula, sedimentary geology of Saudi Arabia. U.S. Geol. Survey Prof. Paper 560-D, 1 (1966).

Quennell, A. M.: The structural and geomorphic evolution of the Dead Sea Rift. Quart. Journ. Geol. Soc., 114, 1 (1958).

Sykes, L. R. and M. Landisman: The seismicity of East Africa, the Gulf of Aden, and the Arabian and Red Seas. Bull. Seism. Soc. Amer., 54, 1927 (1964).

Swartz, D. H. and S. S. Arden: Geologic history of the Red Sea. Amer. Assoc. Petrol. Geol., 44, 1621 (1960).

Talwani, M., X. Le Pichon, and M. Ewing: Crustal structure of the Mid-Ocean Ridges, Pt. 2, computed model from gravity and seismic refraction data. J. Geophys. Res., 70, 341 (1965).

Vine, F. J.: Spreading of the ocean floor—new evidence. Science, 154, 1405 (1966).

Vine, F. J. and D. H. Matthews: Magnetic anomalies over oceanic ridges. Nature, 199, 947 (1963).

Wolfart, R.: *Geologie von Syrien und dem Libanon,* Gebr. Bontraeger, Berlin, 1 (1967).

Youssef, M. I.: Structural pattern of Egypt and its interpretation. Amer. Assoc. Petrol. Geol., 52, 601 (1968).

General Stratigraphy of the Adjacent Land Areas of the Red Sea

RUSHDI SAID

Cairo, Egypt

Abstract

The general stratigraphy of the Red Sea and adjacent land areas is presented. Stratigraphical correlations are made between the geological formations in Arabia, Egypt, Ethiopia, and the Sudan.

Introduction

The Red Sea is about 1,940km in length and between 280 and 320km in width stretching out in a northwest-southeast direction. At its southeastern extremity the Red Sea joins the Indian Ocean through the Gulf of Aden, and at its northwestern end the Red Sea bifurcates into the Gulf of Suez, extending in the same northwest direction. The Gulf of Aqaba strikes off in a northeast direction.

The Gulf of Suez, the Gulf of Aqaba and the Red Sea proper represent three different structural elements that are genetically related to one of the principal structural features of the African continent, the Great Rift Valley system of East Africa. This system consists of a large number of valleys running generally in a north-south direction and extending over 5,000km from the Zambesi River in the south to Jordan, Syria and Lebanon in the north.

The Red Sea may be viewed as a great elongated depression, separating the massifs of the Arabian subcontinent from those forming the backbone of the Eastern Desert of Egypt, the Sudan and Erithrea (Red Sea Hills). While the center of this depression is filled with the waters of the sea, the comparatively low country on either side is chiefly made up of rocks of Miocene and younger age. It is only along both sides of the Gulf of Suez, the northern end of the Red Sea and a small area along the coast of Jiddah that the relief is broken by elongated ridges composed of Cretaceous or Eocene strata with occasional outcrops of basement rocks or older sediments.

This paper deals only with the stratigraphy of this coastal plain which fringes the entire sea on both sides and forms part of the marginal zone of the main trough of the Sea itself (Drake and Girdler, 1964; Heezen and Tharp, 1964), a zone which slopes from the shore to a depth of 300 to 600m and which is mildly deformed. The following paragraphs give a description of the stratigraphy of the coastal plains of the three elements that constitute the Red Sea.

The Gulf of Suez Coastal Plain

Situated within the stable shelf of Egypt, the Gulf of Suez constitutes a zone of subsidence throughout its geological history and has been the site of an immense accumulation of sediments (Fig. 1). The cumulative thickness of the sedimentary section is 10,000m, and the Gulf is essentially a taphrogeosyncline. The region possesses a unique sedimentary history when compared with the bordering regions or with the Red Sea proper of which it is today a part. The Gulf of Suez basin started at least as early as the Upper Paleozoic in the form of a graben that probably encompassed present-day southern Sinai. The basin devel-

71

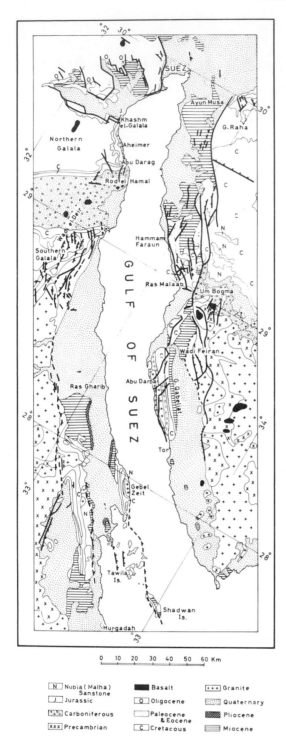

0   10   20   30   40   50   60 Km

N — Nubia (Malha) Sanstone
J — Jurassic
Carboniferous
xxx — Precambrian
Basalt
O — Oligocene
Paleocene & Eocene
C — Cretaceous
+++ Granite
Quaternary
Pliocene
Miocene

Fig. 1. Schematic geologic map of the Gulf of Suez region.

oped as a result of a master fault pattern along the oversteepened sides of two interbasin arches that are assumed to have penetrated into the major sedimentary basin of the unstable shelf of Egypt and the Le-

vant. This situation continued to the end of Lower Eocene time. During this interval the intermittent activation of the blocks that constitute the sides kept rejuvenating the graben. The eastern interbasin arch probably occupied the northern part of the Eastern Desert of Egypt (Cairo-Suez district). It apparently existed as a topographic high for most of the time from its elevation to the Lower Eocene. This high is still expressed in the Cairo-Suez district, whose sediments appear to be thin, apparently responding to subcrustal deformation by faulting movements on the present surface of this district (Said, 1963). The western interbasin arch seems to have occupied the "tilted shelf" of Jordan and Syria (Bender, 1963), extending southward into Saudi Arabia and including the Gulf of Aqaba. All known occurrences of Lower Paleozoic rocks in this region are found along this shelf. The areal distribution of these sediments indicates that this "tilted shelf" constituted an interbasin arch in Upper Paleozoic time: Upper Paleozoic sediments only occur along both flanks of this block.

The birth of the Gulf of Suez graben, in its present form, dates back to the Oligocene when the Gulf of Suez represented an uplifted region on top of which the modern graben was formed in late Oligocene time. By contrast, the Red Sea proper formed part of the stable shelf throughout its history and is a feature which developed in Lower Miocene time.

The movements that affected the Gulf were tensional. The Gulf is bordered by and is made up of a large number of blocks that were continuously rising and sinking with varying magnitudes and intensities. Some of these blocks are so small and old that at one locality (e.g. 82km along the Suez-Ras Gharib road) two adjacent blocks have different stratigraphic successions (Said, 1964). One block does not show the Malha (Lower Cretaceous) and Galala (Cenomanina) Formations while the other shows a complete stratigraphic succession. The structure of the Gulf is exceedingly complex, for these faulting movements not only occurred with different intensities but they also affected a lithologically heterogenous section, so that one commonly finds almost vertical strata on the upthrown side of some of the intense and sudden faults (e.g. Wadi el-Deir, next to St. Anthony's Monastery).

The stratigraphic section in the Gulf of Suez is thick, and deposits of every age from the Upper Paleozoic on are represented. Sediments belonging to the Lower Paleozoic lie beyond the realm of the Gulf of Suez taphrogeosyncline, either in the northern reaches of Sinai which constituted part of the unstable shelf or on top of the western interbasin arch that bordered the taphrogeosyncline. The recently discovered exposure of Lower Paleozoic on the Island of Tiran at the entrance of the Gulf of Aqaba, which can be correlated with the Tawil Formation of Saudi Arabia, belongs to the borderlands of this old Gulf of Suez. The oldest formations so far described from the Gulf proper date back to the Carboniferous (Rod el-Hamal, Aub Darag and Um Bogma Formations). Marine Permo-Carboniferous (Aheimer Formation) and Permo-Triassic (Qusaib Redbeds) have been recently described from the western coastal plain of the Gulf of Suez (Abdallah and Abindany, 1965). Recent mapping by Hassan (1967) of the Paleozoic clastic section of the eastern coastal plain of the Gulf of Suez, which overlies the basement rocks, has shown that the section can be divided into a lower unit made up of non-fossiliferous coarse variegated sands with occasional basal conglomerates, a middle unit of white friable sands and an upper unit made up of multicolored to black compact shales. It is in this last unit that Hassan discovered fossil-bearing beds that belong to the Carboniferous in Abu Durba and Wadi Feiran and which can be correlated with the Sinai clastic carboniferous section known from this coastal strip in the Abu Weneima area.

The Jurassic is represented in the Gulf by the Bathonian clastic section of Khashm el-Galala (Sadek, 1926) and by Callovian and Kimmeridgian exposures that have been recently discovered by Abdallah on the west coast of the Gulf in the form of small elliptical down-faulted blocks ap-

pearing from within the Carboniferous or Permo-Carboniferous of the western coastal strip of the Gulf. Oxfordian rocks have also been recognized from the subsurface in the Ayun Musa wells on the Sinai coastal plain of the Gulf to the south of Suez.

Lower Cretaceous rocks are represented by fine grained variegated sands and fissile shales which are plant-bearing and in places carbonaceous. An examination of the fossils found in these proved their age to be Aptian or Lower Albian. These continental to paralic deposits were previously described as forming part of the Nubia Sandstone (Said, 1962), but they are now recognized as a new formation (Malha) because of their lithological identity and stratigraphic relations. The term Nubia Sandstone is being restricted to those rocks having a facies similar to that of Nubia. The Cenomanian is represented in the coastal plain of the Gulf of Suez by a uniform and richly fossiliferous marl-shale section with few limestone interbeds (Galala and Raha Formations). This uniform and extensive formation overlaps older formations. In the south, along the Sinai coastal plain, the Cenomanian overlies the upper variegated to black shale unit of Hassan (*vide supra*) while to the north it overlies the Qusaib Redbeds, and still farther north it overlies the Lower Cretaceous Malha Formation. The Turonian is known in many blocks of the Gulf of Suez and assumes a dolomite-limestone facies (Wata Formation). The Coniacian is represented by a red clastic section while the Santonian shows variations in lithological composition, being of a chalk facies with medium-sized *Pycnodonta vesicularis* (Wadi el-Deir) or of a marl facies (Matulla Formation). The Campanian is known in the Gulf of Suez in several blocks (best represented in southern Galala). It has a yellow marl facies. The Mastrichtian is represented by a white chalk section with *Orbitoides* (southern Galala) or with planktonics (Gebel Qabeliat). Paleocene and Eocene sections in the Gulf of Suez are thick and form many of the plateaus that border the coastal plain of the Gulf or make up the occasional ridges that cut across the plain.

The Upper Eocene section in east central Sinai has been divided into a number of rock units, which from top to bottom are: Tayiba Redbeds, Tanka beds, Gypseous Marls, Green beds and *Cardita* beds. Oligocene "fluviatile" deposits, such as those known in many other parts of Egypt, are recorded only in the topographic lows to the west of the Gulf. The Miocene of the Gulf of Suez is a basin deposit which transgressed older rocks on all sides of the Gulf. It is represented by marine to paralic deposits: the Gharandal and the Ras Malaab (Evaporite) groups. Exceptionally thick post-Miocene sediments are also known along the coastal plain of the Gulf of Suez. These have been accumulating at a rate that would indicate a subsidence at a rate of 1m per thousand years.

Despite the fact that the Miocene section of the Gulf of Suez shows lateral variations, it is possible to divide this section into several rock units that can be traced for long distances on both the eastern and western coastal plains of the Gulf of Suez. The Miocene section of the Gulf is made up of two distinct units of group status (Table 1). The upper Evaporite (Ras Malaab) Group is made up of a succession of evaporites alternating with sands, shales and

Table 1   The Classification of the Miocene Section of the Gulf of Suez, According to Different Workers

Age (Said and El-Heiny, 1967)	Moon and Sadek, 1923	Lexique, Said and El-Heiny, 1967		Stratigraphic Commission General Petroleum Organization, 1964		
Helvetian?	Gypsum 5	EVAPORITE GROUP	Upper Evaporite	RAS MALAAB GROUP		Zeit
			Evaporite V			South Gharib
Burdigalian	Nullipore rock		Δ Interevaporite marls		Belayim	Hamman Faraun
	Gypsum 4		Evaporite IV			Feiran
	Upper intergypsiferous marls		γ Interevaporite marls			Sidri
	Gypsum 3		Evaporite III			Baba
	Lower intergypsiferous marls and gypsum		β Interevaporite marls		Karim	Shagar
			Evaporite II			Markha
			α Interevaporite marls			
	Gypsum 1		Evaporite I			
Aquitanian	Miocene clays	GHARANDAL GROUP	*Globigerina* marls	GHARANDAL GROUP	Rudeis	Mreir
	Miocene grits					Asl
						Hawara
	Miocene clays					Mheirerat
	Flint and basal conglomerate		Nukhul			Nukhul

minor carbonate beds which reach a thickness of about 700m in Wadi Gharandal, Sinai. This unit overlies the Gharandal Group. In outcrop the evaporite members of the Ras Malaab Group are almost always made up of gypsum. Below the surface, however, these members occur in the form of anhydrite, and there are also some halite beds. The Evaporite (Ras Malaab) Group falls into a number of distinct units which are from top to bottom:

Zeit (Upper Evaporite) Formation: This unit is made up of alternating beds of evaporites and clastics.

South Gharib (Evaporite V) Formation: This unit is made up of a number of anhydrite and halite beds alternating with comparatively thin shale fossil-bearing interbeds with *Orbulina universa, Biorbulina bilobata,* etc. This is the unit that was referred to as the "Evaporite Series" by Said (1962).

Belayim Formation: This unit is made up of the following members: Hammam Faraun (= Δ interevaporite marls); Feiran (= Evaporite IV); Sidri (= γ interevaporite marls); and Baba (= Evaporite III). The Hammam Faraun member is calcareous and reefal in the Gharandal surface section and is sometimes referred to as the "Nullipore rock." Evaporite III is a solid anhydrite-halite unit.

Karim Formation: This unit is made up of the Shagar and Markha members. The lower Gharandal Group is divided into the Rudeis and Nukhul Formations. The Rudeis Formation is variable in thickness while the Nukhul Formation is made up of a basal conglomerate, sandstone and a few marl beds. The environment of deposition of the Rudeis Formation is open marine; the Nukhul is shallow to brackish.

The survey of the stratigraphy of the coastal plains of the Gulf of Suez shows that this region formed a zone of subsidence since the Upper Paleozoic. This zone was in the form of a wide graben encompassing the larger part of southern Sinai from the time of its inception to post Lower Eocene time when the elevation of the southern Sinai horst block delimited the Gulf in more or less its modern shape during late Oligocene time. Miocene marine to paralic sediments filled the modern graben, which was elevated into land during the Pliocene and was flooded during several stages of the Pleistocene by the waters of the then formed Red Sea proper.

The Gulf of Aqaba Coastal Plain

The Gulf of Aqaba has a narrow coastal plain as the bordering cliffs rise abruptly from its waters. The early geological history of the Gulf is unknown, but judging from the distribution of the Paleozoic sediments in Sinai and the Levant, it is possible that the site of the Gulf of Aqaba formed part of the tilted shelf that became high during and since Upper Paleozoic time, forming the eastern interbasin arch that delimited the early Gulf of Suez basin. Since its elevation, there is evidence to believe that this shelf remained above water until late Pliocene or early Pleistocene time when the modern graben of the Gulf of Aqaba formed. No Miocene or Pliocene sediments along the coastal plain of the Gulf of Aqaba are known, indicating that the Gulf was an upland plateau during these periods. The Miocene occurrences recorded along the southern tip of the Gulf and in the Island of Tiran lie outside the realm of the Gulf.

The sediments that occur along the coastal plain of the Gulf of Aqaba indicate that the Gulf was only flooded during the late Pliocene or early Pleistocene time. Pleistocene sediments are found at the mouths of several of the wadis that drain into the Gulf. The presence of igneous and metamorphic pebbles as well as sedimentary pebbles of Cretaceous and Eocene rocks show that the southern land of Sinai was uncovered before the beginning of the Pleistocene. Gravel terraces, at least at two levels (23 and 30m), can also be traced along the sides of many of the major wadis. Other Pleistocene deposits that border the coastal strip of the Gulf are in the form of coral reefs recognisable in at least four levels (2, 10, 15 and 30m). These have been studied by Walther (1888) and Hume (1906).

The Red Sea Proper Coastal Plain

The narrow coastal plain of the Red Sea proper lies between the high, fringing mountains consisting mostly of crystalline rocks and the waters of the sea (Figs. 2 and 3). Along the shores there is an almost continuous band of emergent reef terraces between .5 to 10km wide. Between these and the foot of the crystalline hills extends a sand gravel surface which is inclined toward the sea with gradients that range from 1:80 to 1:200 (Sestini, 1965; Akkad and Dardir, 1966). The width of this plain ranges from less than one km to over 20km.

This plain is covered mainly with Middle Miocene and later sediments. Pre-Middle Miocene sediments are of restricted occurrence and peculiar facies. In Egypt, at the extreme northern end of the Red Sea proper in the Safaga-Quseir reach, strike faults give rise to remarkable topographical complexity in which the pre-Miocene (Cretaceous and Eocene) strata are preserved, forming limestone plateaus.

The detailed stratigraphy of the pre-Miocene strata of the Quseir-Safaga reach is given elsewhere (Said, 1962). It is sufficient here to mention that the Cretaceous and Eocene strata of this district are very similar to those of the area to the west of the Red Sea range, indicating that they were most probably deposited in the advancing sea that transgressed over the surface of Egypt from the north rather than in a differentiated basin that occupied more or less the modern Red Sea. The Nubia Sandstone in this area is identical to that of other stable shelf areas of Egypt and is different in many aspects from that of the neighboring Gulf of Suez.

Formations older than the Miocene along the coastal plain of the Sudan are of limited extent and are known only from the subsurface. Carella and Scarpa (1962) describe a clastic section with minor carbonate beds (Mukawar Formation) from a well drilled in the northern part of the Maghersum Island. The section, made up mainly of dark grey indurated silty shales, is poor in fossils, but some ostracodes found in a bed of marl indicate an uppermost Cretaceous or possibly Paleocene age and a brackish shal-low marine environment. The Mukawar Formation is 190m thick. It overlies, with probable unconformity, a basaltic tuffaceous breccia of unknown age and underlies conformably red quartzitic to arkosic sandstone referred to as the Hamamit Formation. On the eastern coast of the Red Sea proper, along the coast of Jiddah, a marl-limestone section with marine fossils of Maastrichtian age has been described by Karpoff (1957).

The Hamamit section described by Carella and Scarpa from the coastal plain of the Sudan is made up of clastic non-fossiliferous beds that could be of Lower Miocene to Eocene age according to stratigraphic relations. Karpoff (1957) describes a comparable section in Saudi Arabia several hundred meters thick made up of sandstone, grits, shales and conglomerates (the Shumazi Group) overlying unconformably the Maastrichtian limestones. This group could well range from the Eocene into the Lower Miocene.

Apart from these scattered pre-Miocene occurrences which were probably deposited in older basins, the entire plain of the Red Sea is made up of Middle Miocene and younger sediments. These are in the form of disconnected patches that abut with depositional dips on the slopes of the high ranges that delimit the Red Sea, indicating that these were deposited after the graben took shape.

The Miocene and younger rocks of the Red Sea coastal plain of Egypt have been the subject of numerous studies (Beadnell, 1924; Cox, 1929; Said, 1962; Souaya, 1963; and Akkad and Dardir, 1966). The following table gives the classification of the Miocene and younger sediments of the Egyptian part of the Red Sea coastal plain.

The Gebel Rusas Formation is made up of sands, gravels, coral limestones and limegrits. These beds lie with an angular unconformity over the basement rocks. Apart from several corals, *Pecten* and *Lithodomus sp.* which are non-diagnostic, Souaya (1963) recorded from these beds *Neoalveoline melo,* which is a diagnostic Middle Miocene fossil in Northern Africa. This indicates that these basal beds are younger than the basal Miocene beds of the Gulf

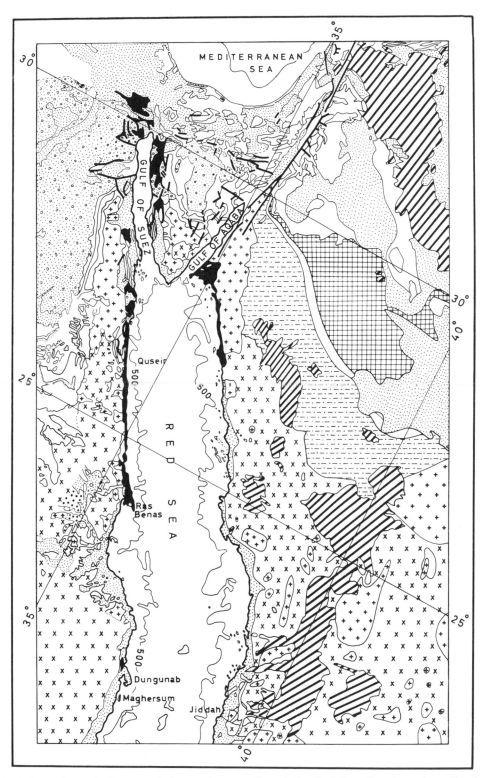

Fig. 2. Schematic geologic map of the northern portion of the Red Sea. See legend on Fig. 3 (scale 1:5,000,000).

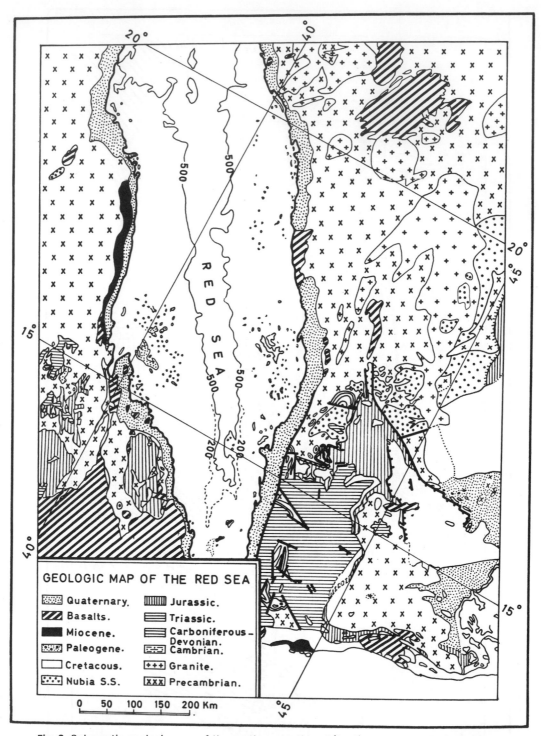

Fig. 3. Schematic geologic map of the northern portion of the Red Sea (scale 1:5,000,000).

Table 2   The Classification of the Miocene Section of the Egyptian Red Sea Coast, According to Different Authors

Age (Cox, 1929)	Beadnell, 1924			Said, 1962	Akkad and Dardir, 1966
Pleisto-cene	*Laganum-Clypeaster* series		Raised beaches of coral limegrits and gravels		Pleistocene Organic Reefs
Pliocene			Coral & nullipore limegrits, sands, sandstones & gravels	*Clypeaster-Laganum* series	Ras Shagara
	Ostrea-Pecten series		Arenaceous series with marls	Oyster and cast beds	Gabir
Middle Miocene			Brackish water marls	Brackish water marls	Samh
	Gypseous series		Massive gypsum	Evaporite Formation	Gypsum Formation
			Basal group	Basal limegrits	Gebel Rusas

of Suez region. The Evaporite (Gypsum) Formation which overlies the Gebel Rusas Formation is made up of massive non-fossiliferous gypsum deposits that have a patchy distribution and variable thickness. The age is accepted as Middle Miocene on the basis of their similarity to the well dated Evaporites of the Gulf of Suez although exact correlation with these is difficult to ascertain. Akkad and Dardir (1966) show that many of the Red Sea evaporites are in the form of domes which seem to have resulted from the plastic flow of these deposits rather than from forces of deep origin. These authors mention several evidences for their conclusion such as the disposition of the Evaporite formation, its isopachous variations, relations with the bordering basement rocks and the non-deformation of the underlying Gebel Rusas formation.

The Samh Formation overlies conformably the evaporites and is made up of marls, arkosic sandstones, shales, clays and intraformational conglomerates. The rare fossils described from this formation by Akkad and Dardir (1966) led these authors to advocate a marine origin for this formation rather than a brackish environment as was previously held. The Gabir and Shagara Formations are richly fossiliferous including a fauna which, according to Cox (1929), is of Pliocene age and of Indo-Pacific origin.

In the south, along the Sudan coastal plain, Carella and Scarpa (1962) and Sestini (1965) describe a Miocene-Pliocene section. The Miocene succession is made up of a lower sandstone-evaporite unit (The Maghersum Formation) which crops out in the Island of Maghersum. This formation consists essentially of an irregular alternation of coarse, gravel-like sands, fine sands and clays and shows great thickness variations, reaching 1,435m in thickness in the type locality. The upper members of this formation are dominantly clastic, rich in marine fossils of Middle Miocene age. The middle members are made up of evaporites interbedded with sandstones and sandy shales while the lower member is dominantly halite. Overlying the Maghersum is a unit (Khor Eit Formation) which was suggested by Sestini (1965) for the succession of mixed carbonates and clastics, which is transitional in character and which intervenes between the lower clastic and the upper carbonate parts of the Miocene section. The fossils described from the formation indicate a Middle Miocene age. The Abu Imama Formation follows on top of the Khor Eit Formation. The thickness of the Abu Imama at its type locality at Gebel Abu Imama is 137m. It is made up of a lower member of conglomerates, a middle member of calcarenites and an upper member of reefal limestone. The age is Middle Miocene. The Dungunab Formation follows on top of the Abu Imama. It is made up of massive halites and gypsum with interbedded clays and sandstones. The formation assumes a thickness of 722m in its type locality. No fossils are found in this formation, but the age is considered, on stratigraphic grounds, to be post-Middle Miocene and pre-Pleistocene.

The Pleistocene deposits of the Red Sea coastal plain are varied. They include the emergent thick sections made up of alternating beds of massive coral reefs and gravels. Akkad and Dardir (1966) describe the presence of four coral reefs alternating with gravel beds along the Egyptian coastal plain. A similar succession is also recorded from the Sudan coastal plain by Carella and Scarpa (1962) and is given the name Abu Shagara Group. In addition to these old interbedded coral reefs that seem to have been deposited in a rising sea, there are a series of late Pleistocene reefs that abut against the old cliffs and stand as wave-cut terraces of different heights above the modern sea. The most consistent of these heights is the 1 to 3m terrace, the 7m terrace, and the 12m terrace. These terrace elevations have also been noted in the Sudan and Erithrea (Sestini, 1965; and Mohr, 1962). Nestroff (1955, 1959) gives a description of these older coral reefs and gives the data of a 30m terrace from the Abulat Island off the Arabian Coast near Lith as 35,000 years, which is above the reliability limit of the C14 method.

Continental deposits in the form of thick wadi and alluvial deposits and gravel terraces are recorded from all over the coastal plain. Lacustrine and glacial deposits are also known from the coastal plain of Erithrea (Mohr, 1962).

The volcanic activity along the Red Sea basin is treated by Chase (1969), and the tectonic events by Girdler (1969). It is sufficient to mention here the continuous activity that has beset the graben since its inception in Lower Tertiary time. The numerous basalts and other effusive rocks are classified as the Trap and Aden Volcanic series.

References

Abdallah, A. M. and A. Abindany: Stratigraphy of Upper Paleozoic rocks western side of the Gulf of Suez. Egypt Geol. Surv. Paper 25, 18 p. (1965).

Akkad, S. and A. A. Dardir: Geology of Red Sea coast between Ras Ghagara and Mersa Alam. Egypt Geol. Surv. Paper 35, 67 p. (1966).

Beadnell, H. J. L.: Report on the geology of the Red Sea coast between Quseir and Wadi Ranga. Petrol. Res. Bull., Government Press, Cairo, **13,** 37 p. (1924).

Bender, F.: Extrême sud de la Jordanie. Lexique stratigraphique Internat., Asie, Fasc. 10 cl. (1963).

Carella, R. and M. Scarpa: Geological results of exploration in Sudan by AGIP Mineraria. 4th Arab Petrol. Congress, Beirut, 23 p. (1962).

Chase, R. L.: Basalt from the axial trough of the Red Sea. *In: Hot brines and recent heavy metal deposits in the Red Sea*, E. T. Degens and D. A. Ross (eds.). Springer-Verlag New York Inc., 122–128 (1969).

Cox, L. R.: Notes on the post-Miocene Osteridae and Petenidae of the Red Sea region with remarks on the geological significance of their distribution. Proc. Malacol. Soc., **18,** 165 (1929).

Drake, C. L. and R. W. Girdler: A geophysical study of the Red Sea. Geophys. Jour. Royal Astron. Soc., **8,** 473 (1964).

Girdler, R. W.: The Red Sea—A geophysical background. *In: Hot brines and recent heavy metal deposits of the Red Sea*, E. T. Degens and D. A. Ross (eds.). Springer-Verlag New York Inc., 38–58 (1969).

Hassan, A. A.: A new carboniferous occurrence in Abu Durba-Sinai-Egypt. 6th Arab Petrol. Congress, Baghdad, 16 p. (1967).

Heezen, B. C. and M. Tharp: Physiographic diagram of the Indian Ocean. Geol. Soc. America, Special Paper 65 (1964).

Heybrock, F.: The Red Sea Miocene Evaporite basin. *In: Salt Basins around Africa.* Inst. of Petrol., London, 17 (1965).

Hume, W. F.: The topography and geology of the Peninsula of Sinai (South-Eastern Portion). Egypt. Survey Dept., Cairo, 280 p. (1906).

Karpoff, R.: Sur l'existence du Maestrichtien au nord de Djeddah. Acad. Sciences, C.R., **245**, 1322 (1957).

Mohr, P. A.: *The Geology of Ethiopia.* University College of Addis Ababa Press, 268 p. (1962).

Moon, F. W. and H. Sadek: Preliminary geological report on Wadi Gharandel area. Petrol. Res. Bull., Government Press, Cairo, **10**, 42 p. (1923).

Nesteroff, W. D.: Les recifs corallines du Banc Farsan Nord (Mer Rouge). Inst. Ocean. Ann., **30**, 7 (1955).

———: Age des derniers mouvements du graben de la Mer Rouge determinés par la méthode du C^{14} appliqueé aux recifs fossils. Bull. Soc. Géol. France, **7**, 415 (1959).

Sadek, H.: The geography and geology of the district between Gebel Ataqa and El-Galala El-Bahariya (Gulf of Suez). Geol. Survey, Egypt, Cairo, 120 p. (1926).

Said, R.: *The Geology of Egypt.* Elsevier Publishing Co., Amsterdam, New York, 377 p. (1962).

———: Structural setting of Gulf of Suez, Egypt. 3rd World Petrol. Congress, Frankfurt, 201 (1963).

———: Trip to Gulf of Suez. Petrol. Exploration Soc. Libya, 6th Annual Field Conference, 141 (1964).

———: Egypt. Lexique stratigraphique Internat., Afrique (1965).

——— and I. El-Heiny: Planktonic foraminifera from the Miocene rocks of the Gulf of Suez region, Egypt. Contr. Cushman Found. Foram. Res., **18**, 14 (1967).

——— and Z. S. Soliman: The distribution of the Miocene rock units of the East Belayim oilfield, East Coast, Gulf of Suez, Egypt. 6th Arab Petrol. Congress, Baghdad, 12 p. (1967).

Sestini, J.: Cenozoic stratigraphy and depositional history, Red Sea coast, Sudan. Bull. Amer. Assoc. Petrol. Geol., **49**, 1453 (1965).

Souaya, F.: Micropaleontology, of four sections south of Quseir, Egypt. Micropaleontology, **9**, 233 (1963).

Walther, J. K.: Die Korallenriffe der Sinaihalbinsel; Geologische und biologische Beobachtungen. Abh. sächs. Akad. Wiss. Leipzig, Math.-Naturw. Kl., **14**, 439 (1888).

Bathymetry and Continuous Seismic Profiles
of the Hot Brine Region of the Red Sea *

DAVID A. ROSS AND EARL E. HAYS

Woods Hole Oceanographic Institution
Woods Hole, Massachusetts

FRANK C. ALLSTROM

Alpine Associates Inc.
Norwood, New Jersey

Abstract

A detailed bathymetric and geophysical study has been made of the three known deeps in the hot brine region of the central rift valley of the Red Sea. The largest, the Atlantis II Deep, is an elongate basin with an irregular and sometimes tilted bottom. The Discovery Deep is circular and flat-floored, and the Chain Deep is relatively narrow and V-shaped.

Continuous seismic profiling produced a distinct reflection from the hot brine-normal sea water contact in the Atlantis II and Discovery Deeps. In contrast, reflections from the sediment-brine interface are commonly indistinct.

Continuous seismic profiling, acoustic returns from the lowering of a pinger to the sea bottom, and coring indicate that the deeps are covered with a thin veneer of heavy-metal rich sediments. Evidence from the pinger lowerings suggests that a major portion of the Atlantis II Deep contains more than 20m of heavy metal deposits. The sediment distribution in the brine area is compatible with a brine source in the Atlantis II Deep and periodic overflows into Chain and Discovery Deeps at some time in the past.

A subsurface seismic reflector postulated to be an unconformity of late Miocene or early Pliocene age is common to records from the flanks of the deeps. If the age of this unconformity is correct, it suggests that the central rift valley of the Red Sea, in the studied area, has formed prior to this time.

Introduction

One objective of the **R/V CHAIN** cruise 61 to the Red Sea was a detailed bathymetric and geophysical study of the hot brine region. This paper describes previous surveys of the area and the results of the **CHAIN** expedition.

Previous Work

A bathymetric contour chart of the Red Sea at a scale of 1:2,300,000 was prepared by the Hydrographer of the British Navy (Admiralty Chart No. C6359) from all soundings available to the British Hydrographic Department of the Navy. Navigation control was by radar, supplemented by celestial sights and dead-reckoning when out of radar range. Accuracy of the radar positions was estimated by Allan (1966) at ±2–3 miles (about 4 to 6km). The small scale of the chart does not clearly indicate the hot brine region.

Indications of abnormal water in the central portion of the Red Sea were first noted during the Swedish **ALBATROSS** expedi-

* Woods Hole Oceanographic Institution Contribution No. 2181.

tion of 1947–48 (Bruneau *et al.,* 1953) and from an **ATLANTIS** cruise (Neumann and Densmore, 1959). In 1963 the **DISCOVERY** (Charnock, 1964) and **ATLANTIS II** (Miller, 1964) reinvestigated and definitely established the presence of warm, saline water. None of the above expeditions mapped the area. The **DISCOVERY** returned to the brine area in 1964 and made a local bathymetric survey (Swallow and Crease, 1965), mainly in the area now called the Discovery Deep. Subsequent visits to the brine area by **METEOR** (Dietrich *et al.,* 1966) and **ATLANTIS II** (Miller *et al.,* 1966) confirmed the hydrographic observations but did not detail the bathymetry.

The presence of the Red Sea in the World Rift System has encouraged considerable geophysical investigation of the area. Previous gravity and magnetic work is discussed in this book by Phillips *et al.* (1969). Seismic refraction profiles have been interpreted by Drake and Girdler (1964), indicating that the Red Sea has been formed as a tension feature, the deep axial trough representing the lateral separation of Africa and Arabia. The basic intrusives beneath the axial trough cause the magnetic anomalies associated with the rift and permit it to remain in isostatic equilibrium.

Seven continuous seismic reflection profiles were made by Knott *et al.* (1965) north of 17°N latitude. They divided the main trough of the rift into a mildly deformed marginal zone and a deeper intensely fractured axial zone. No reflection data was obtained from the brine area.

The CHAIN Expedition

During **CHAIN** 61, about two days were spent surveying the brine area. These results were supplemented by crossings made while obtaining other geophysical data and going from one station to another. Over 100 traverses were made of the rift valley between 21° 10′ and 21° 30′N. Positions were determined by radar fixes on anchored buoys. In most instances the radar range was less than 5 miles; the radar accuracy

was estimated at about ±0.2 miles. The good control and generally excellent fit of the sounding crossings suggest that the accuracy of the bathymetric survey is about ±0.5 miles. Location of the hot brine area was determined by celestial sights and dead-reckoning and probably is correct to within ±1 mile.

During the **CHAIN** cruise, depths were recorded in fathoms on a Precision Graphic Recorder (PGR) using a short (about 0.5 millisecond) with an EDOUQNIB transducer (±30° beam). Positions were obtained every 5 or 10 minutes and depths were plotted every minute or less. Course changes were generally made outside of the surveying area. A preliminary map prepared on ship was used as a navigational aid for additional surveying and sampling. At the conclusion of the **CHAIN** cruise, one of the authors (F. C. Allstrom) collected and correlated all the surveying lines, converted the soundings to meters, corrected for sound velocity according to Matthews (1939) and prepared the bathymetric map (Fig. 1). No correction has been applied for the increase in sound velocity in the hot brine.

Sixteen continuous seismic reflection profiles were made over the hot brine region during the cruise. Because of radar difficulties, good navigational control was available for only 11 of the crossings. In addition two tracks approximating equilateral triangles were made, one east and one west of the brine region.

The sound source of the continuous seismic profiler was an underwater spark discharged at about 60,000 joules of stored energy. The sparker was towed behind the ship at a depth of about 8m and discharged every 10 seconds. The receiver was a 20m linear hydrophone array of five elements towed 91m astern. Received signals were recorded both on a Precision Graphic Recorder and magnetic tape. Recording bandwidth was controlled by variable filters, usually a bandwidth of 37.5 to 150 (or 180) cps was used. Ship speed during the seismic surveys was about 6 knots (11km/hour).

Sediment thickness as determined from the seismic profiles is expressed in seconds

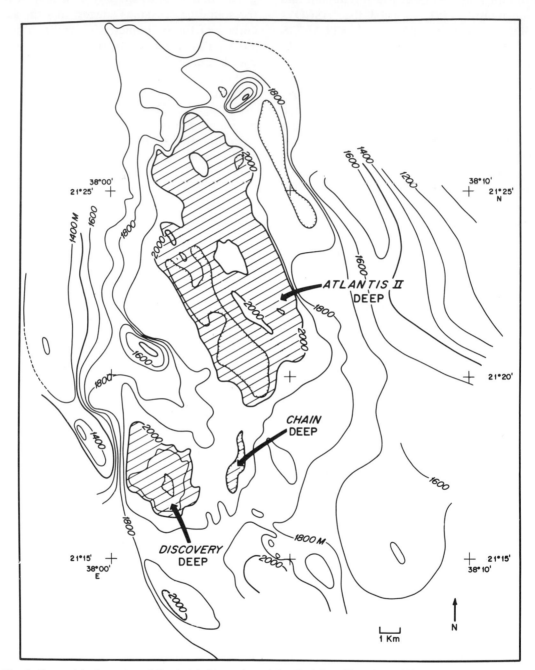

Fig. 1. General bathymetry of the hot brine region of the Red Sea. The slanted lines indicate areas of hot brine (modified from Hunt *et al.*, 1967). Contours based on soundings made during **CHAIN** cruise 61. Contour interval in meters corrected for changes in sound velocity according to Matthews (1939).

of penetration. 2 to 3km per second is the typical range for sediment velocities in the region.

Results of the CHAIN Expedition

Bathymetry

Three distinct and apparently separated deeps are evident from the bathymetric map of the brine area (Fig. 1). The largest deep, the Atlantis II Deep, in the area sampled by **ATLANTIS** in 1959 is an elongate basin trending slightly NW-SE. As defined by the 2,000m contour interval, it is about 14km long and 5km wide. A maximum depth of 2,170m occurs in the relatively flat-floored southwestern portion of the deep. The flanks of the Atlantis II Deep have slopes that range from about 35° on a portion of the eastern flank to about 5° to 15° for the other areas.

Three small topographic highs trending NW-SE are situated along the central part of the deep. The middle one of these highs is the shallowest, having a depth of 1,902m, about 150m above the surrounding sea floor. The position of these topographic highs within the basin is suggestive of an ancient caldera (Hunt *et al.*, 1967). A dredging attempt on the southern high, however, obtained only some semi-consolidated clay. A dredge on the flank of the southwestern part of the Atlantis II Deep obtained a small piece of anhydrite.* A small basin northeast of the Atlantis II Deep with closure at 1,900m did not contain any hot brine.

The Chain Deep, discovered during this expedition (Ross and Hunt, 1967), is very narrow and elongated with a maximum depth of 2,066m. This deep is located on the eastern side of a saddle between the Atlantis II and Discovery Deeps. A detailed survey of the area shows the Chain Deep to be a narrow rift with a V-shaped floor. Its flanks are relatively steep, with slopes of about 20°. This deep, as defined

by the 2,000m contour interval, is about 3km long and 0.7km wide.

The Chain Deep is separated from the Discovery Deep by a sill at 1,980m (Ross and Hunt, 1967). The presence of a sill between the Atlantis II and Chain Deep was not established by the bathymetric survey, but a sill shallower than 2,009m is indicated by the water temperature structure in the two deeps (Ross and Hunt, 1967, Fig. 4).

The Discovery Deep is circular in shape and generally flat-floored. As defined by the 2,000m contour interval, it is about 4km long and 2.5km wide. Its depth is greater than the other three, known deeps, having a maximum depth of 2,220m. The flanks of the Discovery Deep slope about 15° or less, except for the western flank which is about 40°.

A small depression southeast of Discovery Deep does not contain hot brine at present, although Bischoff (1969) reported a layer of goethite at a depth of 220 to 310cm in a core taken from this area. The goethite is probably related to the hydrothermal events in the hot holes, and it suggests that the distribution of hot saline water may have been more extensive in the past.

Continuous Seismic Profiles

The location of the continuous seismic profiles across the brine areas is shown in Fig. 2. Profiles were made parallel (Fig. 3) and transverse to the rift valley (Fig. 4). Two equilateral triangle traverses outside of the brine area were made on each side of the rift valley.

Continuous Seismic Profiles Parallel to the Rift Valley. Profiles B-A, B-C and D-C were made across the Atlantis II Deep (Fig. 3). In these profiles (as in all profiles across the Atlantis II and Discovery Deeps) a distinct reflection was obtained from the hot brine. This reflection at a depth of about 1,960m (assuming a water velocity of 1,500m/sec) is probably due to the density difference between the hot brine and the overlying Red Sea bottom water. The depth of this reflector, cor-

---

* Some basaltic and volcanic glass fragments were in the core catcher of some piston cores (Chase, 1969).

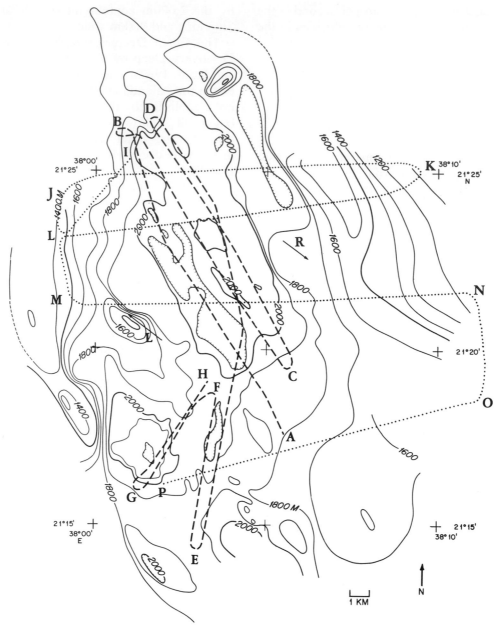

Fig. 2. Location of continuous seismic profiles made across the hot brine region of the Red Sea during **CHAIN** 61. The dashed lines are profiles parallel to the rift valley; the dotted lines are transverse to the rift valley. Arrows indicate approximate starting position of triangular traverses made adjacent to the brine area. R refers to the right or eastern triangle, L to the left or western triangle.

rected for sound velocity according to Matthews (1939), is about 2,037m, which is near the transition from 44°C water to 56°C water for the Atlantis II Deep and near the start of the 44°C water in the Discovery Deep.

Profile B-A crosses the deepest and generally flattest part of the Atlantis II

Deep. Subsurface reflections are not clear but the traces suggest that more than a superficial layer of sediments may be present (*i.e.,* more than 0.03 seconds or 30m) along the flanks of the depression. Side slopes continue below the flat bottom, suggesting some thickness to the deposit in the center of the depression. A distinct,

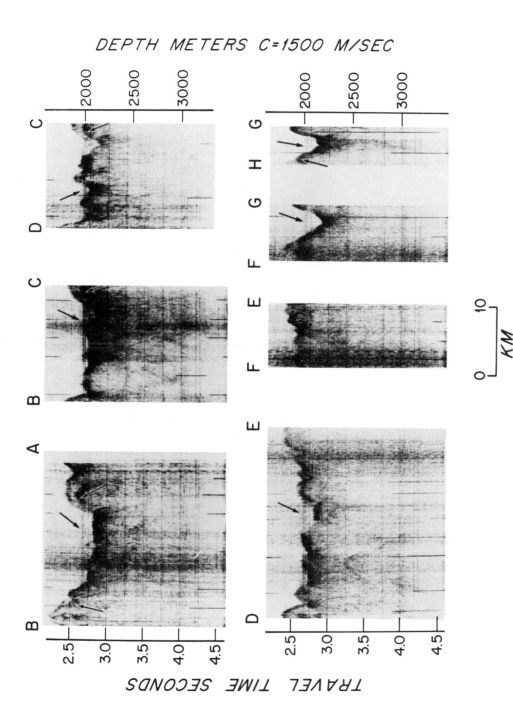

Fig. 3. Continuous seismic profiles made parallel to rift valley. Profiles run from north to south (see Fig. 2 for location). Travel time is two way time. Distance scale is approximate. Arrows pointing down toward bottom of page indicate the brine reflection; arrows pointing up toward top of page indicate the strong subsurface reflector.

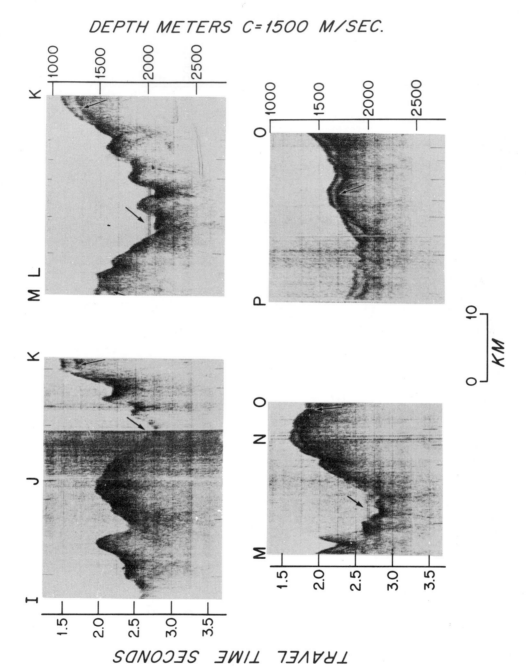

Fig. 4. Continuous seismic profiles made transverse to rift valley. Profiles run west to east (see Fig. 2 for location). Travel time is two way time. Distance scale is approximate. Arrows pointing down toward bottom of page indicate the brine reflection; arrows pointing up toward top of page indicate the strong subsurface reflector.

strong, subsurface reflector, unconformable to the overlying material, is observed from the flanks of the deep. This strong, subsurface reflector is common to many records from both the brine area and areas to the north and south.

Profile B-C crosses some of the topographic highs that extend above the brine interface. Reflections from the brine appear on all sides of the highs (see also profiles D-C, D-E, L-K and M-N). Little delineation of sediment is observed on this profile. The strong, subsurface reflector extends below the brine interface on the southeastern part of this crossing (see also D-C).

Profile D-C, slightly northeast of B-C, also crosses some of the topographically high areas, one of which is partially buried. As in the other profiles, there is generally little indication of a large sediment accumulation in the bottom of the basin. Some internal structure is observed at about 3.2 seconds. In the southeastern part of profiles B-C, D-C, the sediment surface reflection is unclear and may be within the brine reflection. If so, a sediment thickness of about 0.1 seconds is indicated.

Profile D-E crosses the Atlantis II Deep and continues south, passing slightly east of Chain Deep. This profile shows little indication of subsurface structure or sediment accumulation. The Chain Deep was not known at the time of the survey and thus was not examined.

Profile F-E, east of Discovery Deep, shows little indication of structure or sediment accumulation. Profiles F-G and H-G, across the southeastern part of Discovery Deep, show a U-shaped bottom. Further to the northwest, however, in the deeper part of the basin, the bottom is relatively flat. Very little sediment accumulation is indicated from either of the profiles. The southwestern flank on profile H-G is steep and suggestive of a fault. Water reflections from the brine are observed in both profiles and are at a depth similar to the crossings in the Atlantis II Deep.

Continuous Seismic Profiles Transverse to the Rift Valley. Profile I-J-K (Fig. 4) crosses the northern part of the Atlantis II Deep.

Section I-J, from the western flank of the brine area, shows a hilly topography with little sediment cover. Section J-K crosses a relatively steep, V-shaped part of the Atlantis II Deep. The reflection from the brine is faintly visible at the apex of the V and may be slightly deeper than in other records. Little sediment is evident in the bottom of the basin. A small basin, on the eastern flank, has less than 0.1 seconds of sediment and is tilted down to the east. The strong subsurface reflector is present on the eastern flank of the depression.

Profile M-L-K, about 2km south of the previous profile, crosses one of the topographic highs in the central part of the Atlantis II Deep. The western flank is hilly with little sediment cover and has a slight suggestion of the strong, subsurface reflection. This strong subsurface reflector is more evident on the eastern flank. The eastern flank descends in a series of steps, the deepest of which in the brine pool appears to be a fault. The bottom of the deep is relatively flat although some tilting down to the east is indicated. Sediment accumulation at the bottom appears larger than usual and may, although the records are not clear, approach 0.1 seconds.

Profile M-N-O, 4km south of the previous profile, shows a small hill on the western flank that leads to a flat platform area above the brine area. The western part of the deep is narrow, flat-floored and slopes down to the east; the eastern part is hilly. There is little indication of sediment accumulation in the deep along this profile. The strong, subsurface reflector begins on the eastern flank at about 2.3 seconds and continues to the end of the profile.

Profile P-O crosses the rift valley south of the Chain and Discovery Deeps. The strong subsurface reflector is present on the entire record covered by sediments 0.1 to 0.2 seconds thick. The character of the bottom is generally smooth, except immediately south of the Chain Deep where the bottom topography is hilly.

Continuous Seismic Profiles Adjacent to the Brine Area. Two traverses, each in the approximate shape of an equilateral triangle with 25-km sides, were run outside

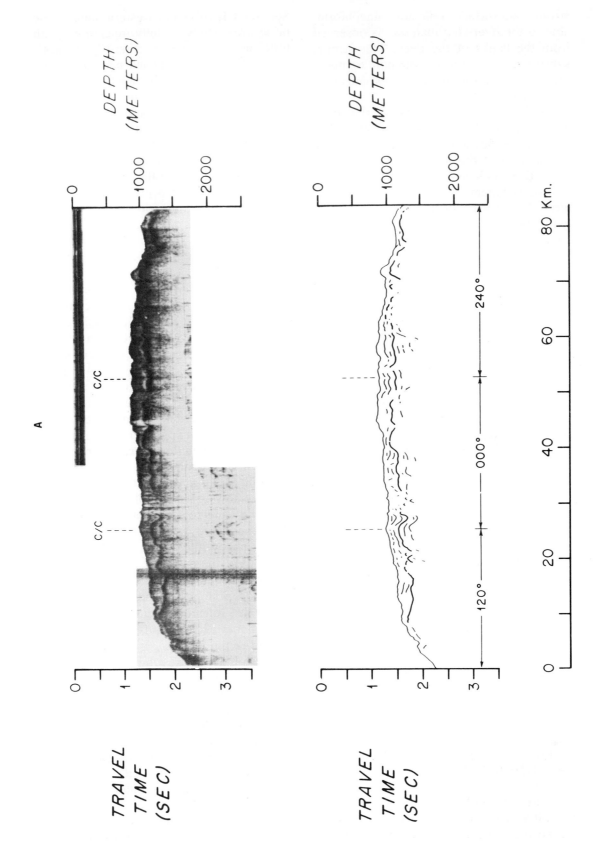

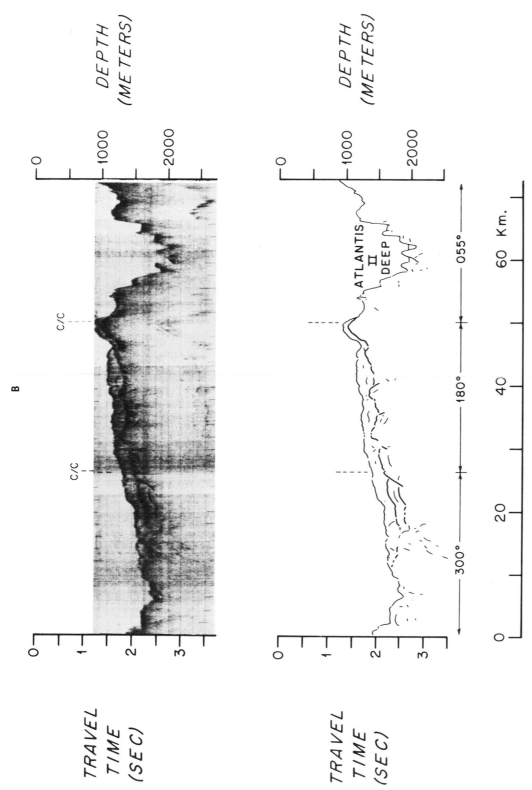

Fig. 5. Continuous seismic profiles made adjacent to the brine area. See Fig. 2 for the approximate starting position of the traverses. Travel time is two way time. A is the right or eastern triangle; B is the left or western triangle.

the brine area in the mildly deformed marginal zone of the rift valley (Fig. 5). Navigation for the triangles was dead-reckoning and thus is not as accurate as the previously discussed surveys.

The right-hand or eastern triangle starts near the Atlantis II Deep (Fig. 2), with a 120° leg followed by a 000° leg and a 240° leg that ends a few km east of the starting point. The strongly reflecting subsurface layer is observed on most of the traverse although it is not always continuous and has numerous small displacements, especially on the 000° and 240° legs. Overlying the strongly reflecting subsurface layer is a variable, sometimes relatively acoustically transparent (at the frequencies we examined), irregular sediment accumulation. The sediments are generally thicker in the landward (eastern) direction, having a maximum thickness of 0.3 seconds on this end of the traverse. Sediments are strongly folded and possibly faulted at the beginning of the 000° leg. Folding is relatively gentle on the remainder of the traverse.

The left-hand or western triangle starts west of the Atlantis II Deep with a 300° leg followed by 180° leg and a 055° leg. Because of currents, this triangle did not end at its starting point but extended across the Atlantis II Deep.

The strong subsurface reflector is not evident on this traverse until the middle of the 300° leg. When the reflector is present, it is irregular, discontinuous, folded and probably faulted. Near the end of the 300° leg two of these reflectors appear, both of which are topographically similar to the surface structure. One of the two reflectors is probably a side echo. There is some indication of structure below this strong reflecting layer.

The sediment thickness overlying this layer is generally less than that found on the eastern or right traverse, but it is also greater near land toward the west. Sediments are relatively transparent but not as folded and faulted as on the eastern traverse. The delineated sediments end abruptly at the beginning of the 055° leg at the start of the more intensely fractured axial zone. The crossing of the Atlantis II

Deep is similar to previous profiles (see Figs. 3 and 4), except for the absence of the strong subsurface reflector on the eastern flank of the deep.

Discussion

General Structure

The three known hot brine deeps of the Red Sea, although very close to each other, differ somewhat in their topography. The Chain Deep is narrow and V-shaped; the Discovery Deep is circular and flat-floored; the Atlantis II Deep is elongated with an irregular bottom.

The continuous seismic profiler records of the hot hole region clearly indicate the dense, saline water. No similar reflections, suggestive of brine, were observed in the other surveyed areas of the Red Sea.

Reflections from the sediment-brine interface are sometimes poorly defined, especially in the areas of rugged topography. Subsurface reflections in the deeps are rare and the records generally show little indication of significant sediment accumulation (greater than 0.03 seconds or 30m) although in some isolated places as much as 0.1 seconds may be present.

A strong subsurface reflector occurs on many records, especially those outside of the brine area. Sediments above this reflector range in thickness from 0.35 to 0 seconds and are relatively transparent. This subsurface reflector is topographically similar to the sea floor in the records transverse and parallel to the brine area, but it is generally unrelated to the surface structure on the triangular traverses made outside the brine area (Fig. 5). A similar reflector has been noted by Knott et al. (1966) from seven reflection profiles north of 17°N latitude and by Ross (1969a) from 13 profiles north of 19°N latitude. Knott and his co-workers postulate that this reflector is an unconformity of late Miocene or early Pliocene age and that most of the deformation in the axial zone took place in two periods, one before the beginning of the Pliocene, the other in late Plio-Quaternary

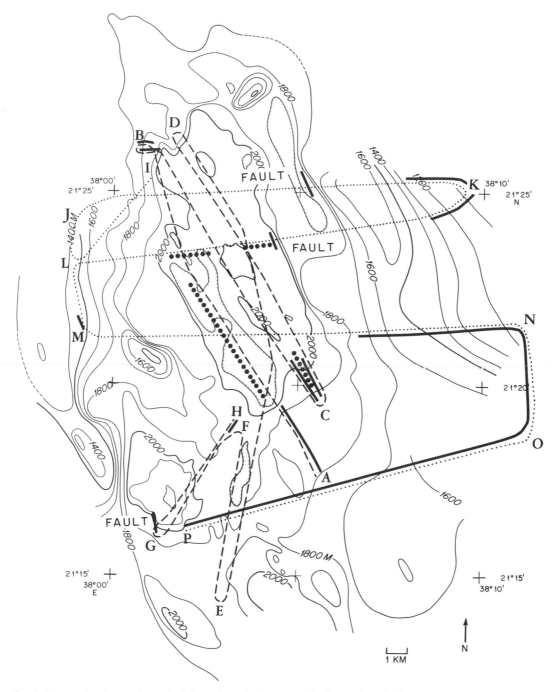

Fig. 6. Some structural characteristics of the brine area. The heavy lines indicate areas where the strong subsurface reflector is present. Three possible faults are shown. Their direction of strike is drawn parallel to the rift valley. The filled hexagons on profiles B-A, B-C, D-C and L-K are areas of sediment accumulations of about 100m as indicated from the seismic profiles.

time. Some deformation prior to this reflector or pre-Pliocene is indicated on the traverses outside of the brine area. Post-Pliocene deformation is indicated by the irregularity of the unconformity to the overlying beds. If the age of the unconformity is correct, this part of the rift valley was formed after late Miocene or early Pliocene age. Records north and south of the hot brine area (Ross, 1969a) likewise show a general absence of this unconformity in the axial zone.

The strong subsurface reflector appears more often and generally at greater depths on the eastern part of the rift valley than on the western part (Fig. 6), suggesting that some tilting down to the east has occurred since the Pliocene. A small basin on the eastern flank of profile I-J-K also appears tilted down to the east. Another indication of tilting is shown by some seismic profiles from within the brine area: the bottom of the deeps in profiles K-L-N and M-N-O is tilted down to the east. Normal or block faulting, as suggested by Drake and Girdler (1964), could cause such tilting.

Identification of faults within the intensely fractured axial zone is difficult; however, three small faults may be present (Fig. 6). Two occur on the east flank of the Atlantis II Deep, the third near the western flank of the Discovery Deep. The strike of the faults cannot be determined from the information available and the direction indicated on Fig. 6 is that of the trend of the rift valley in the area of the presumed fault.

Sediment Thickness

Since there is little indication of rock outcrop within the brine area, it is assumed that the sea floor is covered with a layer of sediment. This assumption is based on the echo-sounding records, the continuous seismic profiles, the successful coring attempts and two dredge attempts that did not obtain any hard rock.

The amount of iron-rich sediment in the brine area is a very important question for the understanding of the geologic history of the area and for evaluating its economic value. Records obtained by continuous seismic profiling do shed some light on this question but are not definitive since the instrument used could not detect sediment less than about 0.03 seconds (about 30m) thick. In addition, in areas where larger sections may be present the records are usually of poor quality. Coring with long piston corers can give only a minimum value of the iron-rich sediment present.

An estimate of sediment thickness was made by examining acoustic returns from pinger lowerings. A 12 kHz temperature telemetering pinger (Ross and Tyndale, 1967) was lowered to within a few fathoms of the bottom to study the water temperature structure of the brines (Ross, 1969b). Records from a pinger lowering can show good resolution of subsurface reflectors when the device is very close to the bottom, where confusion due to side echoes is at a minimum and the cone of sound from the pinger covers a relatively small area. A similar technique has been used to trace coarse-grained sediment layers on the Tyrrhenian Abyssal Plain (Ryan et al., 1965).

Three main types of reflectors can be recognized from the pinger recordings:

1. Two thin layers having a thickness of 0.01 to 0.03 seconds.
2. A thick layer, about 0.02 to 0.03 seconds thick.
3. A strong surface reflector, with many subsurface reflectors, having a maximum thickness of 0.20 seconds.

Graduations between and within some of the above types were observed. Reflection Type 1 sometimes had the lower layer more intense than the upper layer (Type 1A). Types 1 and 2 were sometimes indistinguishable. Type 3 was very distinct and did not grade into either of the other types.

The areal distribution and thickness of the different reflector types are shown in Fig. 7. Reflector Types 1, 1A, and 1-2 are confined mainly to the deepest part of the Atlantis II Deep. Type 2 is common throughout the remainder of the Atlantis II and Discovery Deeps. Type 3 is found outside of the brine region and in the Dis-

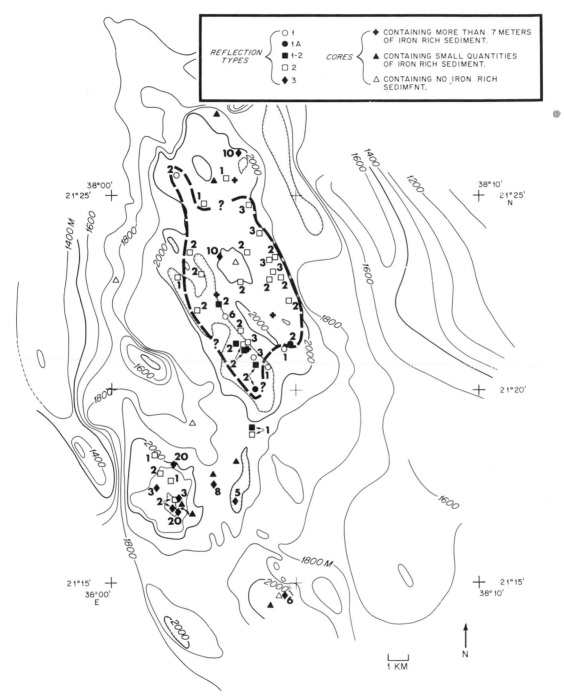

Fig. 7. Distribution of different sediment reflector types as determined by pinger lowerings. The numbers below the reflector types indicate the amount of sediment (assuming a sediment velocity of 2 km/sec) present in tens of meters. Dashed line encloses (for areas deeper than 2,000m) areas having more than 20m of heavy metal sediment (see discussion in the text).

covery Deep where the heavy metal deposits are relatively thin.

It appears that these reflectors are related to the sediment type present. Type 3 may be the calcareous ooze common outside the brine area and underlying the heavy metal deposits in the Discovery Deep. Types 1 and 2 probably are due to slight changes within the heavy metal deposits, possibly porosity differences or slight differences in the mineralogy or salt content of the interstitial waters. The occurrence of Type 1 and its subspecies in the deepest part of the Atlantis II Deep, an area suspected of being a source for the brine solution, is especially interesting.

Data from the temperature pinger lowerings combined with the type and amount of material obtained by cores can be used to indicate minimal sediment thickness in the brine area (Fig. 7). Apparently, less heavy metal sediments are present in the Chain and Discovery Deeps than in the Atlantis II Deep; however, the data from the former two are not as extensive as that of the Atlantis II Deep. The intermediate areas between the known brine areas have a thin veneer of heavy metal sediment. These two observations are consistent with a brine source in the Atlantis II Deep and periodic overflows into Chain and Discovery Deeps at some time in the past. A heavy metal sediment thickness of 20 to 30m is common in a large part of the Atlantis II Deep. The sediment thickness in this deep apparently decreases towards the north and near the topographic highs. Continuous seismic profiles indicate a maximum thickness in the southwestern part of the deep (see Figs. 3 and 4, especially profile B-A). It should be restated that these sediment thicknesses are minimal values based on an admittedly imperfect technique. Additional coring and more sophisticated geophysical techniques are needed to fully evaluate the thickness of the heavy metal deposit.

Summary

A detailed bathymetric and geophysical survey of the hot brine region of the Red Sea has defined three topographic deeps in the area 21° 10'N to 21° 30'N. The deeps, located in an intensely fractured axial zone, apparently are isolated from each other at present, and there is no exchange of hot brines between them. This part of the axial zone differs from the axial zone of most other parts of the Red Sea in that it trends north-south, whereas the other areas trend northwest-southeast.

Individual deeps differ in their topography: Atlantis II Deep is elongated with an irregular and sometimes tilted bottom. Chain Deep is narrow and V-shaped. Discovery Deep is circular and generally flat-floored.

Continuous seismic profiling, acoustic returns from lowerings of a pinger near sea bottom and coring indicate that the deeps are floored with at least a thin layer of sediment. Some tentative conclusions about the thickness of the heavy metal deposits can be made:

1. Both the Chain and Discovery Deeps have relatively small amounts (probably less than 10m) of heavy metal deposits.
2. Some areas outside of the known deeps have a thin veneer of heavy metal sediments that apparently are related to an overflow of the brines in the past.
3. Evaluation of pinger returns suggests that a large portion of the Atlantis II Deep contains more than 20m of heavy metal deposits.

A subsurface reflector, postulated to be an unconformity of late Miocene or early Pliocene age (Knott *et al.*, 1966), is common to many of the records but is not present in the deeps themselves. If this is the age of the unconformity, this part of the rift area was formed in late Miocene or early Pliocene.

Acknowledgments

We appreciate the aid and cooperation of the officers and crew of the **R/V CHAIN.** Elazar Uchupi and S. T. Knott reviewed the manuscript. This work was supported by NSF grant GA-584.

References

Admiralty Chart C 6359. Compiled by the N.A.T.O. Research Centre, La Spezia, Scale 1:2,300,000 (1965).

Allan, T. D.: A bathymetric chart of the Red Sea. Intern. Hydrographic Review, XLIII, 33 (1966).

Bischoff, J. L.: Red Sea geothermal brine deposits: their mineralogy, chemistry, and genesis. *In: Hot brines and recent heavy metal deposits in the Red Sea,* E. T. Degens and D. A. Ross (eds.). Springer-Verlag New York Inc., 368–401 (1969).

Bruneau, L., N. G. Jerlov, and F. F. Koczy: Reports of the Swedish Deep Sea Exped., Physics and Chemistry, Fasc. II, 3, Appendix 29 (1953).

Charnock, H.: Anomalous bottom water in the Red Sea. Nature, **203,** 590 (1964).

Chase, R. L.: Basalt from the axial trough of the Red Sea. *In: Hot brines and recent heavy metal deposits in the Red Sea,* E. T. Degens and D. A. Ross (eds.). Springer-Verlag New York Inc., 122–128 (1969).

Dietrich, G., E. Krause, E. Seibold, and K. Volbrecht: METEOR Forschungsergebnisse, Reihe A, **1,** Deutsche Forschungsergebnisse (1966).

Drake, C. L., and R. W. Girdler: A geophysical study of the Red Sea. Geophys. Jour. Roy. Astronom. Soc., **8,** 473 (1964).

Hunt, J. M., E. E. Hays, E. T. Degens, and D. A. Ross: Red Sea, detailed survey of hot-brine areas. Science, **156,** 514 (1967).

Knott, S. T., E. T. Bunce, and R. L. Chase: Red Sea seismic reflection studies. Geol. Surv. Canada, **66-14,** 33 (1966).

Matthews, D. J.: Tables of the velocity of sound in pure water and sea water for use in echo-sounding and sound-ranging, 2nd Ed. London Hydrographic Dept., Admiralty 52 p. (1939).

Miller, A. R.: Highest salinity in the world ocean? Nature, **203,** 590 (1964).

———, C. D. Densmore, E. T. Degens, J. C. Hathaway, F. T. Manheim, P. F. McFarlin, R. Pocklington, and A. Jokela: Hot brines and recent iron deposits in deeps of the Red Sea. Geochim. et Cosmochim. Acta, **30,** 341 (1966).

Neumann, A. C., and C. D. Densmore: Oceanographic data from the Mediterranean Sea, Gulf of Aden and Indian Ocean. Woods Hole Ocean. Inst., Ref. 60-2, unpub. manuscript (1959).

Phillips, J., J. Woodside, and C. O. Bowin: Magnetic and gravity anomalies in the central Red Sea. *In: Hot brines and recent heavy metal deposits in the Red Sea,* E. T. Degens and D. A. Ross (eds.). Springer-Verlag New York Inc., 98–113 (1969).

Ross, D. A.: Seismic reflection profiles from the Red Sea, in preparation (1969a).

———: Temperature structure of the Red Sea brines. *In: Hot brines and recent heavy metal deposits in the Red Sea,* E. T. Degens and D. A. Ross (eds.). Springer-Verlag New York Inc., 148–152 (1969b).

———, and J. M. Hunt: Third brine pool in the Red Sea. Nature, **213,** 687 (1967).

———, and C. Tyndale: Hot hole-revisited. *Geo-Marine Technology,* **3,** 7 (1967).

Ryan, W. B. F., F. Workum, Jr., and J. B. Hersey: Sediments on the Tyrrhenian Abyssal Plain. Bull. Geol. Soc. Amer., **76,** 1261 (1965).

Swallow, J. C., and J. Crease: Hot salty water at the bottom of the Red Sea. Nature, **205,** 165 (1965).

Magnetic and Gravity Anomalies in the Central Red Sea *

J. D. PHILLIPS
J. WOODSIDE
C. O. BOWIN
Woods Hole Oceanographic Institution
Woods Hole, Massachusetts

Abstract

Marine magnetic and gravity profile results over the hot brine area (21°30′N, 38°E) and axial trough of the Red Sea between 25°N and 18°N latitudes confirm previous hypotheses that dense, strongly magnetic rocks underlie these areas. The gravity and magnetic anomalies in the immediate vicinity of the hot brine area appear to be part of broad, regional anomaly trends subparallel to the axial trough. It is unlikely that the heavy metal deposits of the hot brine area contribute significantly to the observed anomalies. Model studies show the configuration of bodies which could account for the axial trough anomalies to be compatible with a rifting origin associated with seafloor spreading. There appear to be at least two distinct zones of active seafloor spreading in the axial trough. One region is located beneath the hot brine area; its spreading axis trends about N70°E with an apparent rate of 1.5cm/yr. The other zone, extending between 20°N and 16°N, has its spreading axis trending about N35°E near 18°N, 40°E with an apparent rate of 1.0cm/yr. Assuming a nearly N-S direction of separation for Africa and Arabia, the true spreading rates are 1.6 and 1.7cm/yr, respectively. The location of the hot brine area over the northern seafloor spreading center (near 21°30′N) and the known volcanism associated with seafloor spreading zones on other mid-ocean ridges strongly suggest the heavy metal deposits and hot brine waters result from submarine volcanic emanations.

Introduction

Detailed magnetic and gravity measurements were made over the hot brine area of the Red Sea during **CHAIN** cruise 61. Although these observations are primarily useful for delineating the geologic structure beneath the hot brine area, their significance is better understood when considered as part of the regional magnetic and gravity anomaly pattern of the central Red Sea. In this way the deep crustal structure associated with the hot brine area can be compared with other regions of the Red Sea where hypotheses have been proposed for the nature and origin of the crustal rocks. This comparison will be shown to be especially relevant in view of recent hypotheses concerning continental drift and seafloor spreading as the main phenomena responsible for the formation of the Red Sea (Vine, 1966; Girdler, 1966). Magnetic evidence to be presented here strongly supports these ideas and suggests that the origin of the hot brine deposits is intimately associated with the seafloor spreading process.

Primary geophysical data for the hot brine area comes from detailed magnetic survey measurements made during

* Woods Hole Oceanographic Institution Contribution No. 2182.

98

CHAIN cruise 61 (Fig. 1). In addition, magnetic profile data previously collected aboard **CHAIN, VEMA, CONRAD** and **DALRYMPLE** in the regions adjacent to

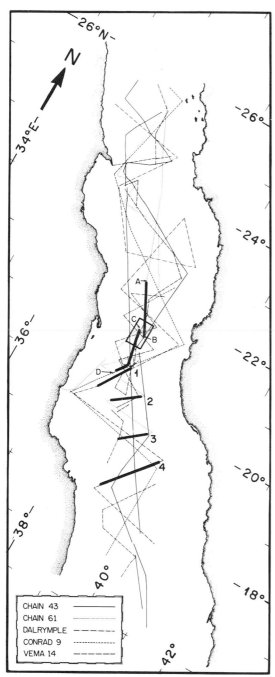

Fig. 1. Index map showing magnetic profile measurements in the central Red Sea. The location of the detailed magnetic survey over the hot brine area is indicated by the rectangle near 21°30′N, 38°E. The locations of certain profiles over the hot brine area and southern Red Sea are indicated by letters (A–D) and numbers (1–4), respectively.

the hot brine area are presented (Fig. 1). Extensive magnetic investigations have also been made aboard **ARAGONESE** (Allan *et al.,* 1964). Unfortunately these data are unavailable at this time.

The gravity data for the survey area are of lesser value because the relatively short survey lines (15–20 minutes duration) and heavy ship traffic required frequent course and speed changes. Since about 5–10 minutes are needed to stabilize the gravity meter after such changes few data were obtained. Although useful gravity data were obtained on several longer tracks crossing the hot brine area, the limited coverage did not warrant their contouring.

Instrumentation and Data Analysis

A Varian proton precession magnetometer system was employed for measurements of the total geomagnetic field intensity during **CHAIN** cruise 61. The sensor was towed 250m astern. Similar proton precession magnetometers were used aboard previous cruises of **CHAIN, CONRAD** and **DALRYMPLE**. The **VEMA** data were obtained using a fluxgate magnetometer.

The magnetic anomalies along each track (Fig. 1) were computed and the anomaly amplitude (gammas) plotted normal to the track (Figs. 3–5). The anomaly profiles along **CHAIN** and **DALRYMPLE** tracks were computed by removing the regional field proposed by Allan (1964) for the Red Sea, and updated for secular variations according to estimates provided by U.S. Hydrographic Office Chart No. 1703. The **CONRAD** and **VEMA** anomaly profiles were computed by using a spherical harmonic expansion estimate of the geomagnetic regional field (Cain *et al.,* 1964). The **VEMA** fluxgate data were calibrated for absolute intensity using 18 crossings of the **DALRYMPLE** profiles (Drake and Girdler, 1964). It should also be noted that all profiles were adjusted an arbitrary amount along the gamma scale of each straight track segment for purposes of clear identification of the anomalies with their respective track. Thus, the anomaly amplitudes shown in Figs. 3, 4 and 5 should only be taken to indicate relative amplitudes. No attempt has been made to correct the data for temporal variation since the amplitudes of anomalies associated with the Red Sea axial trough are much larger than expected diurnal changes.

A LaCoste-Romberg gimbal-mounted sea gravimeter was used for making gravity measurements aboard **CHAIN**. A real-time data acquisition and computer system (Bowin *et al.*, 1967) was employed to process and display the magnetic, gravity and navigational information. The simple Bouguer anomaly was computed using an assumed crustal density of 2.67gm/cm^3.

Navigational control for the **CHAIN** survey of the hot brine area was provided by moored buoys equipped with radar reflectors. Celestial fixes and radar bearings on landmarks were the primary navigational methods for the longer **CHAIN** cruise 61 profiles and for the previous **CHAIN**, **VEMA**, **CONRAD** and **DALRYMPLE** profiles.

Magnetic Investigations

Previous Work

Magnetic anomalies associated with the axial trough of the Red Sea have been described in several papers (Drake and Girdler, 1964; Allan, 1964; Allan *et al.*, 1964; and Knott *et al.*, 1966). Drake and Girdler (1964) first noted that maxima and minima trends of the total magnetic field intensity appear to parallel in a general way the axis of the trough and shorelines, although they admitted some difficulty in correlating detailed features from profile to profile. Allan's (1964) preliminary magnetic anomaly contour chart of the Red Sea also delineated magnetic trends parallel to the axial trough. These trends were especially evident in the southern Red Sea between latitudes 16°N and 20°N. Here the magnetic anomalies and axial trough trend about N35°W. Between 20°N and 24°N contours shown in both Allan's and Drake

and Girdler's maps suggest that the magnetic trends strike more northerly, as does the axial trough. North of 24°N latitude the amplitudes of the magnetic trends decrease to less than about 500 gammas (as compared with amplitude of nearly 2,000 gammas farther south), and the axial trough is absent. However, the magnetic trends still continue to parallel the main trough. Drake and Girdler (1964) and Girdler (1963; 1966) proposed that the magnetic anomalies of the axial trough are due to the emplacement of strongly magnetic basaltic rock along tensional cracks in the continental crust as Africa and Arabia rifted apart. Girdler (1966) further proposed that the true crustal separation (drift) of the Arabian peninsula away from Africa has been primarily northward with a 7° anticlockwise rotation. Vine's (1966) analysis of magnetic profiles provided further support to a continental drift origin for the Red Sea. He interpreted certain magnetic anomaly profiles of the southern Red Sea to represent seafloor spreading-type anomalies.

Results

During October and November, 1966 (**CHAIN** Cruise 61) several long magnetic profiles were obtained across the axial trough of the Red Sea between 20°N and 24°N latitudes as part of an effort to determine the extent of the known hot brine area near 21°N and to find other brine areas that might exist along the trough. Fig. 2 shows the total magnetic intensity contour map developed from a detailed magnetic survey over the hot brine area. Figs. 3, 4 and 5 show the magnetic anomalies plotted normal to the ship's track for the long profiles made during **CHAIN** cruise 61 as well as other magnetic profiles made previously aboard **CHAIN** (cruise 43; Knott *et al.*, 1966; and largely unpublished); **VEMA** 14 (Drake and Girdler, 1964; and largely unpublished); **CONRAD** 9 (unpublished) and **DALRYMPLE** (Allan, 1964).

The large horizontal gradients (over 200 gammas/km) and short spatial wavelengths of the anomalies associated with the axial trough made it difficult to contour accurately the magnetic data away from the hot brine area, as Drake and Girdler (1964) have already noted. In any event, contouring of such limited magnetic data tended to smooth the detailed features of the profiles and did not reveal any more information than is already shown by the earlier maps (Drake and Girdler, 1964; Allan, 1964). It was found more illuminating simply to correlate distinctive anomaly features, shown along the tracks in Figs. 3, 4 and 5, from profile to profile. The proposed correlations will be considered later.

Examination of the total magnetic intensity map (Fig. 2) over the hot brine area suggests there is a positive magnetic anomaly associated with the Chain and Discovery Deeps and a negative anomaly with the Atlantis II Deep (see hot brine area bathymetric maps of Ross *et al.*, (1969). The trend of these magnetic features is about N70°E. This trend is significantly different from the general northerly trend previously suggested for this portion of the Red Sea axial trough (Drake and Girdler, 1964; Allan, 1964). To determine if this trend was of only local significance, related primarily to the distribution of magnetic materials associated with the heavy metal deposits in the hot brine deeps, laboratory magnetic analyses of core materials from the hot brine deep were made. Measurements of magnetic susceptibility and remanence were made on 5cc cylindrical specimens. The specimens were taken 10cm apart along the length of cores obtained during **CHAIN** cruise 61 (station 85 — Chain Deep; station 143 — Atlantis II Deep; and station 154 — outside hot brine area). The natural remanence was measured with a spinner magnetometer (Phillips and Kuckes, 1967). The susceptibility was measured with a commercial inductance bridge. The average susceptibility and remanence for each core, giving each specimen unit weight, is shown in Table 1. It is suggested from these measurements that the maximum total magnetization (remanence + induced) of the heavy metal deposits is on the order of 1×10^{-4} emu/cc (station 143). This is only slightly more than the magnetization of the more normal sediments obtained outside

Table 1

Station No.	Average Remanence (emu/cc)	Average Susceptibility (emu/cc/oe)
85	$9.5(\pm 8.0) \times 10^{-5}$	$8.0(\pm 5.3) \times 10^{-5}$
143	$8.2(\pm 5.1) \times 10^{-4}$	$2.3(\pm 2.9) \times 10^{-5}$
154	$2.0(\pm 3.0) \times 10^{-4}$	$4.5(\pm 6.0) \times 10^{-5}$

Induced magnetization taken to be .35 oersted (oe) × average susceptibility. The + and − values refer to the standard deviation of the mean.

the hot brine area (station 154). Assuming the heavy metal deposits are on the order of 20m thick (Ross *et al.*, 1969) and the water depth of the deposits is about 2km, appropriate computer model studies were made to simulate the observed anomaly. These studies clearly showed that the observed anomaly amplitude (greater than 1,000 gammas) cannot be accounted for by such weakly magnetized materials. Whether these core materials are truly representative of all the hot brine deposit is not known.

The significance of the N70°E magnetic trend of the hot brine area is better understood when it is considered in terms of the overall anomaly trends of the central Red Sea region. In order to illustrate the regional anomaly trends, heavy bar lines have been used to connect the most likely correlative anomalies along each track in Figs. 3, 4 and 5. It must be recognized that this is a subjective technique and is strongly influenced by the orientation of ship tracks in a given area. However, in regions where various track orientations are available, it is a useful technique. Most of the correlations shown in Figs. 3 and 5 have several track orientations and are considered firm. It must be noted that the trends indicated by the heavy bar lines in Fig. 4 and the northern part of Fig. 3 are considered tentative in that most of the tracks here are oriented NW-SE. This orientation unduly favors a NE-SW trend assignment. Other possible trend correlations are indicated by a dashed bar line in such areas.

Inspection of Fig. 3 suggests that the hot brine anomaly trend, indicated by the arrow-tipped bar line, is one of several anomaly features which trend approximately east-northeast between 20°N–23°N

J. D. Phillips, J. Woodside, and C. O. Bowin

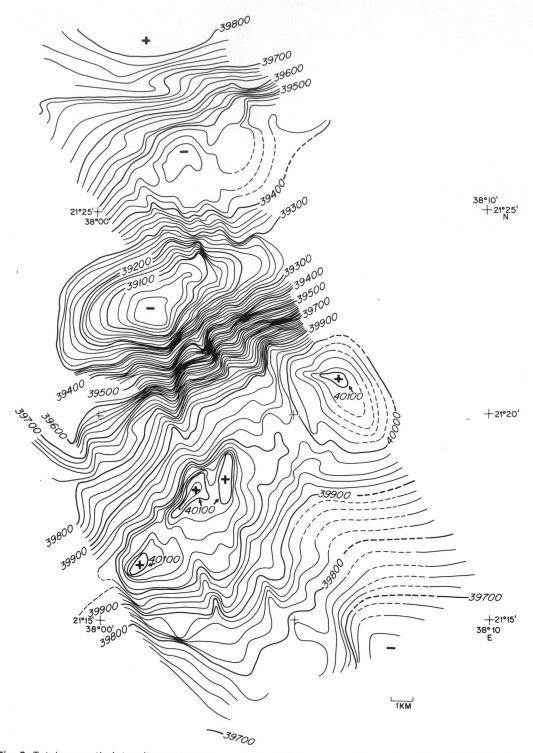

Fig. 2. Total magnetic intensity contour map over the hot brine area. The contour interval is 20 gammas. The location of the survey is shown in Fig. 1.

latitude. North of 23°N latitude the trends appear to be more northerly. However, as mentioned above, an alternative trend (dashed line) striking northwest (N45°W) is also a reasonable correlation (Fig. 4). This is nearly parallel to that in the southern Red Sea, discussed below. NE-SW oriented tracks are needed in the area to confirm this possibility. South of 20°N latitude the anomaly trends are clearly NW-SE, striking at about N35°W, as noted by previous investigators (Fig. 5). This is essentially normal to that of the N70°W trend of the hot brine area. It is this southern Red Sea trend (N35°W) which Vine (1966) interpreted to result from the seafloor spreading process.

Careful examination of certain profiles near 20°N latitude showed that the anomalies (arrow-tipped bar lines in Fig. 5), which correlate with the simulated central anomaly of Vine's seafloor spreading model,

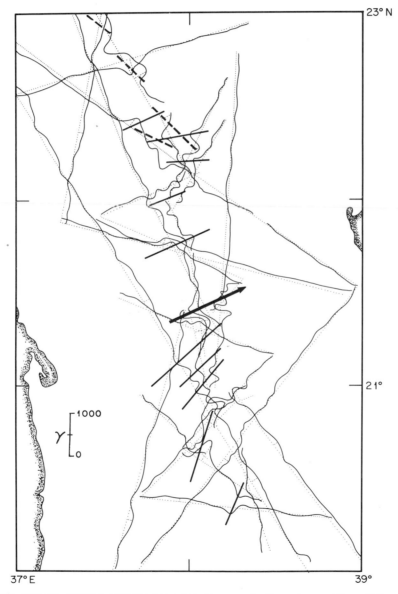

Fig. 3. Magnetic anomaly profiles over the hot brine deeps and adjacent area. The relative anomaly amplitudes have been plotted normal to the respective ship tracks shown in Fig. 1. The heavy bar lines indicate anomaly trends. The arrow-tipped bar is over the hot brine area. The dashed bar line shows an alternative interpretation of certain anomaly trends. A dotted line connects the trace of the positive anomalies to the respective ship tracks.

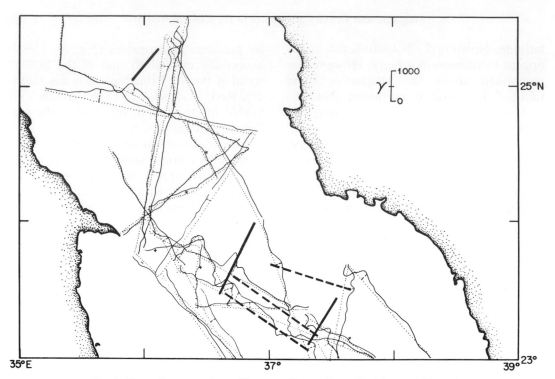

Fig. 4. Magnetic anomaly profiles over the northern Red Sea axial trough.

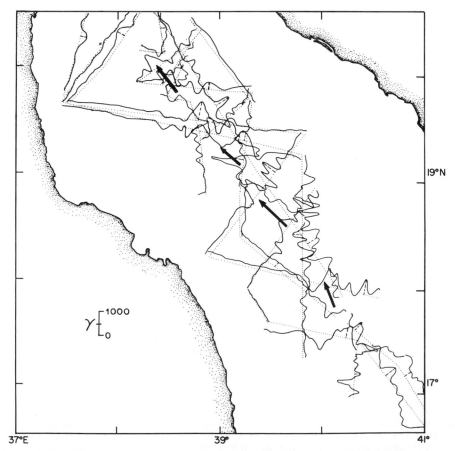

Fig. 5. Magnetic anomaly profiles over the southern Red Sea axial trough.

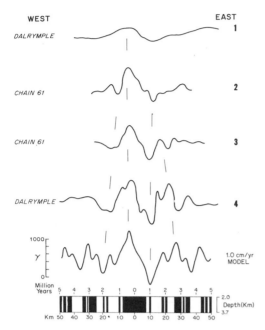

Fig. 6. The top four curves show magnetic anomaly profiles projected normal to the trend (N35°W) of the axial trough of the southern Red Sea. The location of the profiles is shown in Fig. 1. The bottom curve is a simulated profile generated by the sea-floor spreading model beneath it. The black blocks indicate normal magnetization. The open blocks represent reversed magnetization. The magnetization direction is with respect to an axial dipole vector. Effective susceptibility is .01 cgs units; except for the central block which is .02. The assumed model parameters are: intensity and dip of Earth's field 35,000 gammas and 36° respectively. The strike of the profile is N55°E. The reversal time scale was taken from Vine (1968). The simulated magnetic profile here and in Fig. 9 was calculated using a computer program similar to that described by Talwani and Heirtzler (1964).

decrease in amplitude to the north (Fig. 6). It should be pointed out that these central-type anomalies do not appear to be continuous with the NE trending hot brine area anomalies (Fig. 5). The central anomaly pattern seems to terminate abruptly near the Dalrymple profile 1 (Fig. 1). It should be further noted that the axial trough shows an abrupt change in strike in this area to a more northerly direction.

Gravity Investigations

Seven submarine pendulum gravity measurements in the Red Sea were reported by Vening Meinesz (1934) and Girdler and Harrison (1957). These meas-

urements suggested that the center of the Red Sea is associated with positive Bouguer anomalies. Girdler (1958) tabulated 76 gravity measurements from the Red Sea shores and islands. Plaumann (1963) reported on the results of extensive sea gravity observations in the Red Sea made aboard the **ARAGONESE** in 1961. This survey confirmed the occurrence of a Bouguer anomaly high associated with the central trough along the center line of the Red Sea. Girdler (1958), Plaumann (1963) and Drake and Girdler (1964) interpreted this Bouguer high to be due to intrusion of dense subcrustal magma.

Results

The location of gravity observations in the vicinity of the hot brine deeps together with profiles of gravity anomaly, total magnetic field intensity and bathymetry are shown in Fig. 7. Profiles A-A', B-B' and C-C' cross the median valley of the Red Sea south of the hot brine deeps; D-D', a composite of two lines made in opposite directions, crosses the deep; E-E' is a north-northwest traverse along the median valley. The gravity values of profile C-C' are similar to those of profile A-A', about 100km to the south (Fig. 7). Moreover, the maximum Bouguer anomaly across the hot brine deeps (profile D-D') is only slightly lower than the maximum Bouguer anomalies found along profiles A-A' and C-C'.

The free-air anomalies in the central Red Sea typically average to nearly zero along more transverse profiles, suggesting that the Red Sea Rift is approximately in isostatic equilibrium. The Bouguer anomalies attain their highest values (about 110 to 135mgals in this portion of the Red Sea) over the axial trough, and they decrease to near zero at the margins of the Red Sea. The Bouguer high suggests the occurrence of dense material underlying the central part of the Red Sea.

The gravity effect of the inferred small volume (Ross et al., 1969) of heavy metal-bearing sediments in the hot brine deeps should be relatively small and probably masked within the error for the gravity

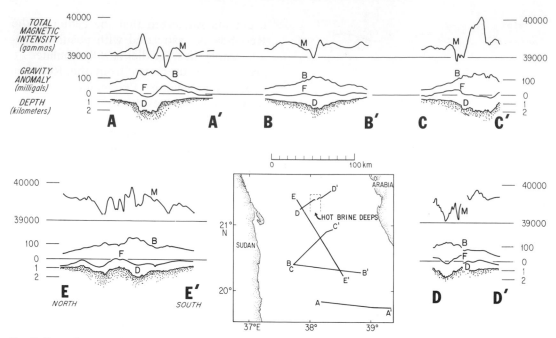

Fig. 7. Free-air anomaly, Bouguer anomaly, and total magnetic intensity profiles in the vicinity of the hot brine deeps. Profile E–E' from **CHAIN** 43, others from **CHAIN** 61.

measurements. The estimated absolute error is about ±10mgals; relative precision along a profile is better, perhaps ±5mgals. The central trough of the Red Sea is sufficiently complex geologically that it is unlikely that such fine structures could be distinguished and identified.

Discussion

The location of the hot brine deposits within the zone of NE trending magnetic anomalies and their association with a positive Bouguer gravity anomaly suggest that the origin of these hot brine deposits is closely related to the origin of the axial trough of the Red Sea which in general shows similar features.

A widely accepted hypothesis proposed for the origin of the Red Sea concerns the rift-like nature of the axial trough and its similarity of the mid-ocean ridge system. Drake and Girdler (1964) and Girdler (1966) have pointed out various petrologic, thermal, seismic, gravity and magnetic characteristics which provide strong evidence for a tensional origin of the trough associated with the drifting apart of Africa and Arabia.

Gravity models based on the data of this investigation generally support this idea. The configuration of dense rocks which appear to underlie the axial trough can be interpreted to result from intrusions which have rifted the lighter continental crustal apart.

Two structure models across the Red Sea Rift were constructed utilizing seismic refraction, bathymetric and gravity measurements (Figs. 8 and 9). These models are both along profile A-A' (Fig. 7), which was chosen because this is one of the longest of the four profiles across the rift, has characteristic gravity and magnetic anomalies and is located near Vema refraction profile 178 (Drake and Girdler, 1964). Structural control for the shelf areas flanking the axial trough was obtained from the generalized sections presented by Drake and Girdler (1964, Fig. 8) and their seismic sections (Fig. 7) for other locations in the Red Sea. Profile 178 is unreversed and indicates the presence in the axial trough of a thick (almost 2km) layer of material with a low compressional wave velocity (about 3.5km/sec) that rests directly upon a basement having velocity of 6.9km/sec. Seismic profile 174 similarly shows a layer of 3.8km/sec material above basement of 7.2km/sec veloc-

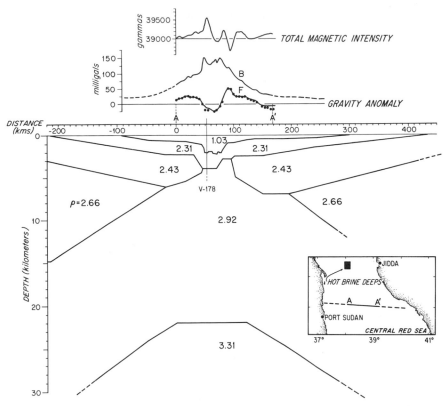

Fig. 8. Structure model near 20°N. Solid lines are observed free-air and Bouguer anomalies, dashed lines are inferred Bouguer values, and dots are calculated free-air values.

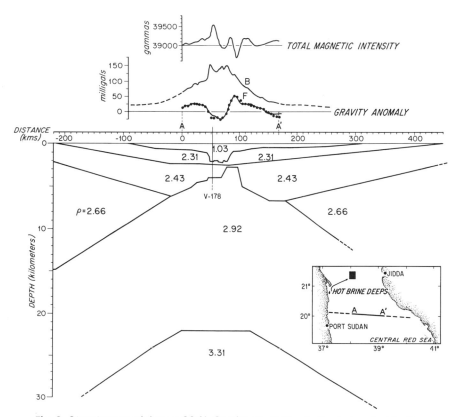

Fig. 9. Structure model near 20°N. Gravity anomalies indicated as in Fig. 8.

ity. In two other refraction profiles in the axial trough, a layer of 4.5km/sec velocity material overlies basement, 6.8 to 7.3km/sec velocity. Both the 3.5 and 4.5km/sec velocity layers are interpreted by Drake and Girdler to be sedimentary rocks, evaporites and/or pyroclastics. Material of 3.3 to 3.8km/sec velocity was found to overlie material of 4.5 to 5.9km/sec velocity in most of the seismic refraction profiles in locations other than the central trough. Because it is not certain whether the material having velocity of near 3.5km/sec retains its 2km thickness, thins or is absent in the central trough, models were constructed for both cases (Figs. 8 and 9).

The computed gravitational attraction closely matches the observed free-air gravity anomaly profiles for both models. These were computed by the polygon technique described by Talwani *et al.* (1959a). The density values used were the differences between the assumed density for each layer, shown in Fig. 8, and a reference density of 2.67gm/cm³. The reference point for the calculated gravitational attraction is indicated on the figures. The empirical curve of Nafe and Drake was used to assign densities to the seismically determined layers. The density of the 3.5km/sec layer is 2.31gm/cm³, that of the 4.5km/sec layer is 2.43gm/cm³, thus a density contrast of 0.12gm/cm³ was used in making the minor adjustment of structure between the two models.

If the axial trough formed by simple downfaulting of the 2.31gm/cm³ layer, then the model shown in Fig. 8 might be the more reasonable of the two. However, the magnetic data indicate that seafloor spreading has occurred during the last 2–4 million years, and it is reasonable to expect that some thinning has occurred. Thus the model of Fig. 9 is considered to be more representative of the probable crustal structure. In this model, both the 2.31 and the 2.43gm/cm³ layers are thinner beneath the axial trough and their absence is quite possible. This model (Fig. 9) also suggests that in different parts of the same crossing of the axial trough either the 3.5km/sec or the 4.5km/sec material may be the dominant lower velocity layer and directly overlie

the basement. Magnetic model studies discussed later indicate that material at the depth of these layers is probably responsible for the magnetic anomalies.

The material of density 2.66gm/cm³ (Fig. 8) attempts to represent the material having seismic velocities of 5.5 to 6.4km/sec encountered in the seismic refraction profiles beneath shelf portions of the Red Sea. Drake and Girdler (1964) interpreted this material to be shield rocks. The approximately 250km separation of this material in the crustal models is inferred to have resulted from both the present and past spreading apart of the Red Sea region.

The Mohorovičić discontinuity present in both models has been added to complete the structural picture. There is no seismic evidence to substantiate the presence of a distinct or a transitional boundary between the crust and upper mantle. The probable depth to the mantle layer shown was simply calculated assuming that the crustal mass per square centimeter at a depth of 40km is common along the entire profile. If the upper layers of the crust are thought to be present in the configuration as shown, the position of the Mohorovičić discontinuity may be set so that the crustal mass 40km beneath any point at sea level is about 11,800kg per square centimeter. This value is the average of 468 crustal mass computations made from seismic refraction profiles around the world (from compilation of McConnell and McTagart-Cowan, 1963), using the Nafe-Drake empirical relationship as before. Thus the upper layers of the model are determined from seismic refraction measurements and the lower section is added on the premise that equilibrium is attained between adjacent columns of crust and upper mantle material at a depth of 40km and that the crustal mass to this depth is the same. The gravity effect of the mantle layer is very broad compared with the effect of the more shallow structure. Even if the calculations were extended only to shallow depths, they would not substantially alter the crustal configuration shown.

The rifting hypothesis for the origin of the Red Sea axial trough has probably received its strongest support to date from Vine's recent analysis of two magnetic profiles across the southern Red Sea (Vine, 1966). He suggested that seafloor spreading-type magnetic models, which have been used to account for the magnetic anomalies over the mid-ocean ridges, can also provide an explanation of certain anomalies in the Red Sea. Vine found that a spreading rate of about 1cm/yr in a N55°E direction generated a simulated profile which best fit the observed data near both 16°N and 20°N latitudes. The results of a similar comparison of four other profiles between 20°N and 18°N latitudes with a simulated 1cm/yr profile are shown in Fig. 6. The similarity of peak positions to the simulated profile is considered good, especially for the Dalrymple profile 4. In this profile even certain shape details of the eastern side of the profile are represented in the simulated profile. However, the lack of essential elements of symmetry about the central anomaly and the decreased amplitude and smoother character of the northern three profiles suggest that a uniform seafloor spreading process cannot be used to account for the entire area. This change in character may indicate that seafloor spreading-type anomalies have only recently begun to form toward the north. This explanation would give rise to a wider, more distinctive symmetrical pattern in the south where spreading first began. A simple single block model, representing only the present normal polarity interval, could easily account for the northernmost profile. Vine (1966) also implied such an age difference by using a less complex block model for his "fit" to the Aragonese 20°N latitude profile. However, alternative explanations are possible. For example, a decreased spreading rate near 20°N as compared to that at 18° and 16°N could account for the subdued nature of the Dalrymple #1 profile.

In light of the reasonably strong evidence which supports seafloor spreading as an origin of the magnetic anomalies of the southern Red Sea, a similar investigation of profiles across the N70°E anomaly trends of the hot brine area was made. Fig. 10 shows a comparison of a composite profile (**CHAIN** Cruise 61, A, B, C, D of Fig. 1), projected normal to the N70°E anomaly trend, with a simulated profile generated by

a spreading rate of 1.5cm/yr. The fact that the observed profile has several symmetry elements in itself justifies use of a seafloor spreading magnetic model. The most obvious symmetry elements are the two large positive anomalies which flank the central positive anomaly at a distance of about 50km on each side. In addition, two smaller positive anomalies are located at a distance of 20km. The detailed survey over the hot brine area (Fig. 2) encompasses the central positive anomaly and the negative anomaly trough to the north. The approximate position of the Atlantis II Deep, indicated by the cross in Fig. 10, appears to be over the center of the central block of the seafloor spreading model. The extrusion of basaltic lava flows are well known on

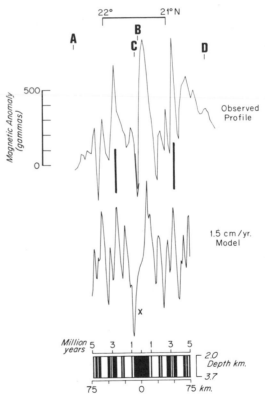

Fig. 10. The top curve is a composite magnetic profile projected normal to anomaly trends (N70°E) over the hot brine area. The location of the profile segments AB and CD are shown in Fig. 1. The bottom curve is a simulated magnetic anomaly profile generated by the seafloor spreading model shown beneath it. The model parameters are the same as those in Fig. 6 except the strike of the profile is N20°W and the dip of the Earth's field is 35°. The approximate position of the Atlantis II hot brine deep with respect to these profiles is indicated by a cross above the block model.

most parts of mid-ocean ridges which are believed to be actively spreading. This association strongly supports an origin related to submarine volcanism for the hot brine deposits.

The "fit" of the 1.5cm/yr simulated profile is considered good in view of the complication which might be presented by fragments of crustal blocks being incorporated in the axial zone of such an embryonic mid-ocean rift. It is also likely that a newly initiated spreading zone may not be organized sufficiently along its length to develop uniform linear bands in a consistent direction. The very fact that the Red Sea anomaly trends appear to be different along the length of the axial trough and different from the Gulf of Aden anomaly trend suggest that similar disorganization may exist on a smaller scale within each spreading zone.

The spreading rates, determined in this paper for the hot brine area and by Vine (1966) for the southern Red Sea, should be considered the minimum amounts for their respective regions in that it is unlikely that crustal separation across the Red Sea has been normal to the axis of the axial trough or anomaly trends. Morgan (1968) has shown that the true direction and rate of spreading can only be deduced from seafloor spreading-type anomalies when the orientation of associated transform faults (Wilson, 1965) is known. Transform faults in the Red Sea have not yet been described. However, it is reasonable that such a fault zone separates the seafloor spreading-type anomalies which trend N35°W, south of 20°N and those which trend N70°W, north of 20°N: the loci of the respective central anomalies are displaced; the anomaly trends are nearly normal; the apparent spreading rates are significantly different. The approximate orientation of such a fault can be inferred from the evidence which indicates the relative movement of Arabia away from Africa. Girdler (1966) has summarized this evidence and concludes that for Arabia a northward translation of 100km with a minor anticlockwise rotation of 7° is the most reasonable estimate. Hence, the orientation of transform faults and the spreading direction should be nearly north-

south. Girdler has also computed the separation in the direction of movement for various latitudes along the Red Sea axis.

If we assume that a north-south movement represents the true direction of spreading, then the true spreading rate for the N70°E hot brine spreading axis is about 1.6cm/yr while the N35°E southern Red Sea spreading axis has a true rate of about 1.7cm/yr. The proposed spreading units of the hot brine area and southern Red Sea with their respective true spreading rates are shown in Fig. 11. The orientation of the northernmost spreading unit near 24°N is speculative in that insufficient data are available to determine an apparent rate or direction.

To bring the true spreading rates at 22°N and 19°N into coincidence, the true direction of spreading would have to be N3°E, near Girdler's direction of movement N7°E (1966, Fig. 2). The adjusted rates would be 1.62cm/yr. However, a slightly faster spreading rate near 19°N than that near 22°N is expected for a pole of rotation situated to the west of the Red Sea (Morgan, 1968). Le Pichon and Heirtzler (1968) and Heirtzler *et al.* (1968) have determined a pole of rotation for Africa and Arabia at about 26°N and 21°E (Southern Libya) from fault trends in the Gulf of Aden. They presented a comparison of calculated and inferred fault strikes (separation directions) and spreading rates resulting from rotation about this pole for locations from the Red Sea to the south Indian Ocean. The comparison results were poorest for the Red Sea, also their pole of rotation requires a true spreading direction of N25°E in the Red Sea at 19°N. This suggests that the pole determined by Le Pichon and Heirtzler is not applicable to the Red Sea.

A qualitative test of the proposed spreading rate and spreading axes variations between the hot brine area and southern Red Sea can be made if it is assumed that Africa and Arabia have indeed rifted apart with both translation and rotation movements as Girdler (1966) has suggested. The separation distance in the vicinity of the hypothesized fault line (Fig. 11) in the direction of movement, divided by twice the inferred spreading rate of about 1.62cm/yr, should indicate the time at which separation began at a particular latitude. Further, if Africa and Arabia are considered to have moved as a rigid block, the time of separation should be the same regardless of the ap-

parent spreading rates, spreading axis orientations or latitude. This computation has been carried out using Girdler's proposed separation distance near the hot brine and southern Red Sea spreading units (230km at 20°30′N; Girdler, 1966; Table 1). The inferred time of separation is about 7.0 million years b.p. This result is approximate since the uncertainties in the apparent spreading rate and anomaly trend determinations are on the order of 0.1cm/yr and 10° respectively. Such uncertainties could introduce errors of as much as 1 million years. Therefore, a time for separation of about 6–8 million years b.p. is most probable. This corresponds roughly to Late Miocene time (Tongiorgi and Tongiorgi, 1964; W. A. Berggren, personal communication) and is in reasonable agreement with other geologic evidence for the time of initiation of deformation in the Red Sea sediments (Knott *et al.*, 1966) and the opening of the Gulf of Aden (Laughton, 1966).

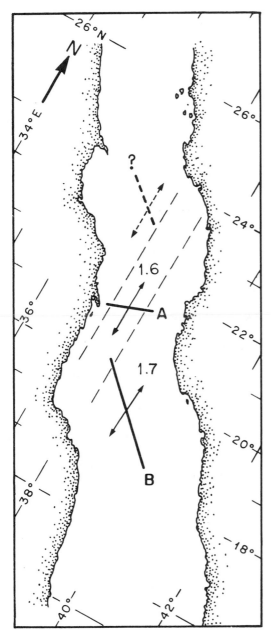

Fig. 11. Zones of seafloor spreading in the Red Sea. Zone A beneath the hot brine area has a spreading axis oriented N70°E. It is indicated by the heavy bar marked A. Zone B near 18°N, 40°E has its spreading axis oriented N35°W. The directions and magnitudes of true spreading are indicated by the arrows, parallel to the inferred north-south direction of movement (dashed lines). The orientation of the northernmost spreading zone, indicated by dotted lines, is considered tentative.

Conclusions

The data presented here strongly suggest that symmetrical magnetic anomaly trends associated with the axial trough and hot brine area of the Red Sea can be explained by the seafloor spreading hypothesis. The positive Bouguer gravity high that is observed over the hot brine area and parallel to the axial trough provides further evidence that dense rocks have been emplaced beneath these regions. Magnetic and gravitational model studies also indicate that the configuration of causative bodies which accounts for the observed anomalies is compatible with a rifting origin for the axial trough. On the assumption that the seafloor spreading hypothesis is applicable, certain conclusions regarding the geologic significance of the anomalies in regard to the origin of the axial trough and the nature of the hot brine deposits can be drawn:

1. There appear to be at least two distinct zones of active seafloor spreading in the axial trough. One zone (A)

is located beneath the hot brine area near 20°30'N, 38°E. Its spreading axis trends about N70°E. The other zone (B) extends between 20°N and 16°N latitude with the axis of spreading trending about N35°E near 18°N, 40°E.

2. The apparent spreading rates are about 1.5cm/yr and 1.0cm/yr, respectively.

3. If the direction of separation (true spreading) is taken to be nearly N-S, the true spreading rates are about 1.6cm/yr and 1.7cm/yr.

4. Comparison of these calculated true spreading rates with inferred separation distances across the Red Sea suggest Arabia and Africa began rifting apart about 6–8 million years ago.

5. The hot brine area appears to be located directly over the center of spreading for the northern seafloor spreading unit (A) near 21°30'N latitude.

6. Since similar seafloor spreading zones on deep ocean ridges (i.e., Mid-Atlantic and Reykjanes) are known areas of submarine volcanism, it is probable that the origin of the heavy metal deposits and hot brine waters are related to volcanic emanations. It is not clear why the hot brine deposits are restricted to the 21°30'N area when it is considered that the entire length of the axial trough is probably spreading.

7. The association of the hot brine area magnetic and gravity anomalies with regional seafloor spreading-type anomalies and the relatively weak magnetization of the heavy metal deposits make it unlikely that the deposits themselves contribute significantly to the observed anomalies.

Acknowledgments

We thank E. T. Bunce for her comments and critical review of the manuscript; W. C. Pitman and W. Ryan for kindly providing certain unpublished VEMA and CONRAD magnetic profile data; C. A. Davis, Master and the officers and crew of CHAIN for their cooperation and assistance during CHAIN cruises 43 and 61.

CHAIN cruise 61 was supported under Contract Nonr 4029(00) NR 260-101 with the Office of Naval Research. Laboratory analyses were also supported by this contract and a grant from the National Science Foundation (GP 5353).

References

Allan, T. D.: A preliminary magnetic survey in the Red Sea and Gulf of Aden. Bollettino di Geofisica Teorica ed Applicata VI, **23**, 199 (1964).

——, H. Charnock, and C. Morelli: Magnetic, gravity and depth surveys in the Mediterranean and Red Sea. Nature, **204**, 1245 (1964).

Bowin, C. O., R. Bernstein, E. D. Ungar, and J. R. Madigan: A shipboard oceanographic data processing and control system. IEEE Transactions on Geoscience Electronics, **GE-5,** 41 (1967).

Cain, J. C., S. Hendricks, W. E. Daniels, and D. C. Jensen: Computation of the main geomagnetic field from spherical harmonic expansions. NASA Publication, X-611-64-316, Greenbelt, Maryland (1964).

Drake, C. L. and R. W. Girdler: A geophysical study of the Red Sea. Geophys. Journ. Roy. Astron. Soc., **8**, 473 (1964).

Girdler, R. W.: The relationship of the Red Sea to the East African Rift System. Quarterly Journal of the Geol. Soc. of London, **CXIV**, 79 (1958).

——: Geophysical evidence on the nature of magmas and intrusions associated with rift valleys. Bull. Volcanologique, **26**, 37 (1963).

——: The role of translational and rotational movements in the formation of the Red Sea and Gulf of Aden. Geol. Surv. Canada, Paper 66-14, 65 (1966).

—— and J. C. Harrison: Submarine gravity measurements in the Atlantic Ocean, Indian Ocean, Red Sea, and Mediterranean Sea. Proc. Roy. Soc. A, **239**, 202 (1957).

Heirtzler, J. R., G. O. Dickson, E. M. Herron, W. C. Pitman, III, and X. Le Pichon: Marine magnetic anomalies, geomagnetic field reversals, and motions of the ocean floor and continents. Journal of Geophysical Research, **73**, 2119 (1968).

Knott, S. T., E. T. Bunce, and R. L. Chase: Red Sea seismic reflection studies. Geol. Surv. Canada, Paper 66-14, 33 (1966).

Laughton, A. S.: The Gulf of Aden. Phil. Trans. Roy. Soc. Ser. *A*, **259**, 150 (1966).

Le Pichon, X. and J. R. Heirtzler: Magnetic anomalies in the Indian Ocean and sea floor spreading. Journal of Geophysical Research, **73**, 2101 (1968).

McConnell, R. K., Jr., and G. H. McTaggart-Cowan: Crustal seismic refraction profiles, a compilation. Scientific Report No. **8**, Institute of Earth Sciences, University of Toronto (1963).

Morgan, W. J.: Rises trenches, great faults, and crustal blocks. Journ. Geophys. Res., **73**, 1959 (1968).

Nafe, J. E. and C. L. Drake: Physical properties of marine sediments. *In: The Sea.* M. N. Hill (ed.). Interscience Publishers New York, **3**, 794 (1963).

Phillips, J. D. and A. F. Kuckes: A spinner magnetometer. Journ. Geophys. Res., **72**, 2209 (1967).

Plaumann, V. S.: Kontinuierliche Schweremessungen im Roten Meer mit einem Askania-Seegravimeter vom Typ Gss 2 nach GRAF. Zeitschrift fur Geophysik, **5**, 233 (1963).

Ross, D. A., E. E. Hays, and F. C. Allstrom: Bathymetry and continuous seismic profiles of the hot brine region of the Red Sea. *In: Hot brines and recent heavy metal deposits in the Red Sea.* E. T. Degens and D. A. Ross (eds.). Springer-Verlag New York Inc., 82–97 (1969).

Talwani, M. and J. R. Heirtzler: *Computers in the Mineral Industries.* G. A. Parks (ed.). Stanford University Press, 464 (1964).

Talwani, M., J. L. Worzel, and M. Landisman: Rapid gravity computations for two-dimensional bodies with application to the Mendocino Submarine Fracture Zone. Jour. Geophys. Res., **64**, 49 (1959a).

Talwani, M., G. H. Sutton, and J. L. Worzel: A crustal section across the Puerto Rico Trench. Jour. Geophys. Res., **64**, 10, 1545–1555 (1959b).

Tongiorgi, E. and M. Tongiorgi: Age of Miocene-Pliocene limit in Italy. Nature, **201**, 365 (1964).

Vening Meinesz, F. A.: Gravity Expeditions at Sea 1923–1932, Delft, **II**, 208 (1934).

Vine, F. J.: Spreading of the ocean floor: new evidence. Science, **154**, 1405 (1966).

———: Magnetic anomalies associated with mid-ocean ridges. *In: History of the Earth's Crust,* NASA Symposium, R. A. Phinny (ed.). Princeton University Press (1968).

Wilson, J. T.: Transform faults, oceanic ridges, and magnetic anomalies southwest of Vancouver Island. Science, **150**, 482 (1965).

Thermal Measurements in the Red Sea Hot Brine Pools

ALBERT J. ERICKSON AND GENE SIMMONS

Department of Geology and Geophysics
Massachusetts Institute of Technology
Cambridge, Massachusetts

Abstract

Fourteen thermal gradient measurements in the sediments beneath the Atlantis II Deep show large temperature variations both with location in the deep and vertically within individual measurements. Maximum sediment temperature observed was 62.3°C. Thermal gradients ranging from 3.75°C/m (hotter at depth) to −0.87°C/m (cooler at depth) were found, with a tendency for gradients in regions of high absolute temperature to be small or slightly negative.

The average measured conductive heat flux into the 56°C water of 15 to 20μcal/cm² sec is inadequate to provide the heat lost to the 44°C water above or to warm the 56°C water. Of possible heat transfer mechanisms considered, the uniform convection of hot brine through the sediment beneath the Atlantis II Deep appears most likely to be active now, although former periods of hydrothermal discharge are probable.

Introduction

The Red Sea Hot Brine Pools, consisting of the Atlantis II Deep, the Discovery Deep and the recently discovered Chain Deep (Ross and Hunt, 1967) are located near 21°21′N latitude and 38°05′E longitude, midway between Port Sudan and Mecca. The pools are associated with an anomalous magnetic gradient of 125γ/km and a Bouguer gravitational anomaly of over 120mgal (Hunt et al., 1967). Various aspects of this general area are summarized elsewhere in this volume. General discus-

sions of heat flow which include measurement techniques, equipment and the geophysical significance of such work are found in Lee (1965).

During 1966, geothermal data were obtained in the area of the Red Sea Hot Brine Pools as one part of an extensive survey aboard the **R/V CHAIN**. The temperature and salinity structures of the brine pools were obtained from hydrographic stations (Munns et al., 1967; Brewer et al., 1969), from temperatures telemetered by pingers (Ross, 1969) and from the thermal data collected during the passage of the marine heat flow equipment through the near-bottom water. In addition absolute temperatures and thermal gradients in the sediment were determined at 15 locations.

Experimental Techniques

Temperatures in the sediments were obtained using the equipment shown schematically in Fig. 1. Five outrigger type thermistor probes, located more or less evenly along the core barrel (which was 2 to 10m in length), sensed the local temperature relative to the absolute temperature determined by the uppermost probe. Absolute temperature measurements in the sediment as well as in the overlying water were obtained.

Because the thermistor probes have a short time constant (less than half a minute) and the initial thermal perturbation due to frictional heating becomes negligible after about one minute, the determination of temperature, and hence thermal gradient, requires only a few minutes in the sediment.

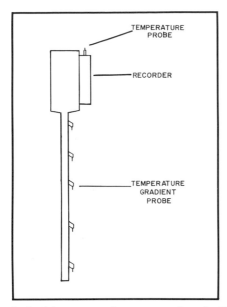

Fig. 1. A piston or gravity corer equipped with a thermal gradient recording instrument, an upper temperature sensor and five outrigger thermistor probes for the measurement of temperatures in the near-bottom water and sediment.

The recorder, identical to the one described in detail by Langseth (1965), consists of a motor, a rotary switch, Wheatstone bridge, amplifier and a galvanometer-optical system. The thermistor resistances are compared with precision resistors of low temperature coefficients. Thermistor calibrations were obtained by us with a platinum thermometer and a Mueller G-2 bridge, and are accurate to at least .03°C. A very rough check on our ability to measure absolute temperature is obtained by comparing the temperature measured by our equipment with that measured by reversing thermometers. In the Atlantis II Deep an average water temperature of 56.38°C was indicated by our equipment; a value of 56.48°C was obtained with reversing thermometers. The thermal gradient measurements are believed to be accurate to about 0.05°C/m, an accuracy considerably poorer than the usual 0.002°C/m. The inaccuracy was caused by the penetration of the entire corer beneath the sediment with a resultant shift in the reference resistance provided by the upper probe.

Both the thermal gradient and thermal conductivity are required to evaluate the amount of heat flowing through the water-sediment interface. $Q = -K \, dT/dz$, where Q is the heat flux per unit area per unit time, K the thermal conductivity, T temperature and z depth; the usual units are HFU (HFU, heat flow unit, is 1×10^{-6} cal/cm^2 sec). The conductivities were measured

with the needle probe described by Von Herzen and Maxwell (1959). The cores were not available on board ship for conductivity measurements. When actually measured about four months later, the cores showed varying degrees of mechanical disturbance. Thus errors of as much as 20 per cent may be possible in some of the thermal conductivity values.

Data

Locations of thermal gradient measurements are shown in Fig. 2; Table 1 summarizes the results of these fifteen measurements. Although the data obtained by Pugh (1967) and Munns *et al.* (1967) implied that the 56°C water mass in the Atlantis II Deep was nearly isothermal, our measurements of the absolute temperatures and thermal gradients within the underlying sediment show extreme vertical and horizontal variation. Fig. 3 shows the individual temperature profiles. Note the inverse correlation for several stations between absolute sediment temperatures and thermal gradients: stations with high temperatures have low gradients and those with low temperatures tend to have high gradients. Indeed, negative thermal gradients (temperature decreasing with depth) can be observed in the sediments at several sta-

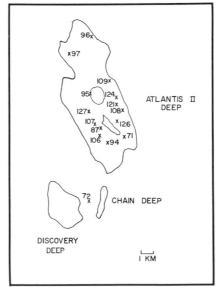

Fig. 2. Positions of heat flow stations in the hot brine pools.

Table 1   Thermal Measurements in the Red Sea Hot Brine Pools

Station Number	Thermal Gradient (°C/m)			Temperature (°C)			Depth (corr. m)
	Upper	Middle	Lower	Water	Max. Sed.	Min. Sed.	
71	(+)	(+)	(+)	56.46	>59.00	—	2,095
72	0.77	—	1.69	44.68	44.76	44.21	1,930
87	0.93	0.41	−0.22	56.45	57.93	57.47	2,073
94	0.24	0.27	0.63	56.40	57.50	56.55	2,170
95	(+)	(+)	(+)	44	—	—	1,910
96	0.01	0.00	0.01	55.29	55.14	55.06	2,047
97	2.33	0.82	−0.22	56.33	59.18	57.34	2,042
106	3.75	1.20	0.95	56.40	61.87	57.84	2,164
107	—	—	—	56.39	—	—	2,151
108	−0.24	−0.10	−0.22	56.33	60.77	60.23	2,071
109	0.72	−0.87	0.20	—	62.26	60.27	2,077
121	0.57	0.07	—	56.32	60.95	60.29	2,064
124	0.27	0.40	0.09	56.34	58.39	55.60	2,056
126	−0.03	−0.09	−0.18	56.37	61.27	60.32	2,066
127	0.95	−0.15	−0.24	56.41	59.97	57.55	2,106

tions. Fig. 4 is a plot based on the same data, but allowance has been made for the tendency of the entire corer to penetrate the sediment layer. Superpenetration results in ambiguity regarding the depths below the sediment-water interface over which the thermal gradient is measured. In an effort to remove this ambiguity the depth of penetration was estimated from the absolute temperature at the top of the core, together with the assumption that the temperature in the upper few meters was $T = 56°C + Az$, where z is depth in meters and A is the gradient in the upper few meters of 1°C/m, an unusually high value for oceanic as well as continental areas. Justification

for this method of determining the depth of penetration is suggested in Fig. 3 which shows that the temperature of the upper probe while in the sediment is as high as 62°C. Yet the temperature of the water immediately above the interface is 56°C.

A difference appears to exist between temperature profiles obtained in the eastern

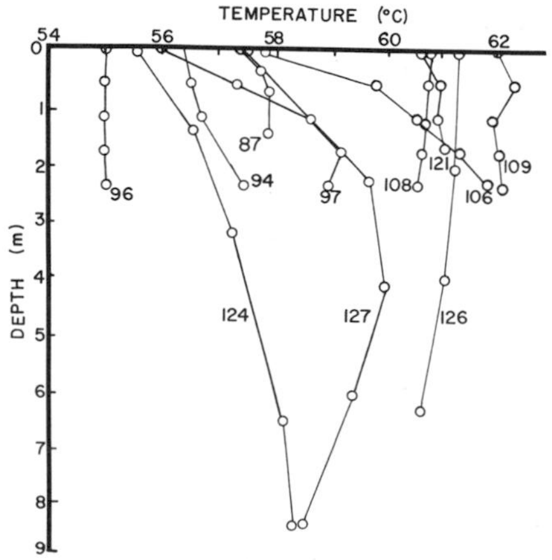

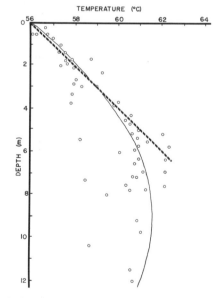

Fig. 4. A composite temperature profile of all thermal data obtained in the sediment beneath the 56.4°C water. Sediment temperature is plotted against the estimated depth below the sediment surface; see text for discussion. Also shown is the thermal gradient (dashed line) chosen to represent the average gradient at the surface and the theoretical thermal gradient (solid line) for a 10m thick layer, initially at 74°C, which has been cooling by conduction for six or seven years.

Fig. 3. Temperature profiles obtained beneath the 56.4°C water in the Atlantis II Deep. Absolute sediment temperature is plotted against the depth beneath the uppermost probe (NOT the depth beneath the sediment surface).

part of the Atlantis II Deep (stations 108, 109, 121, 124 and 126) and profiles obtained elsewhere in the deep. The five eastern profiles are, with one exception, characterized by consistently high absolute temperatures and small or negative thermal gradients. The remaining profiles show considerably greater variations both between each other and within a given profile than are observed at the eastern stations.

The value of the heat flowing into the 56°C water by ordinary molecular and lattice conduction processes (rather than convective transport) is given by the product of thermal gradient and thermal conductivity measured immediately below the sediment-water interface. From Fig. 4 we take dT/dz to be 1.0°C/m. The thermal conductivity, measured on the cores, was approximately 1.5 to 2.0 mcal/cm sec-°C. Thus the average flux into the water is 15 to 20 HFU.

The solid line in Fig. 4, constructed from the composite temperature-depth curves, can be treated as a cooling curve. The model, not illustrated, is that of a layer 10m thick, of initial temperature 74°C, deposited 6–7 years ago. This model, though purely hypothetical, is consistent with the observations. It can, of course, be tested against future data.

Variations with time of the temperature of the bottom water were shown by some of our data. These data were obtained at three stations in the Atlantis II Deep while our equipment was suspended at a fixed depth in the 56°C water. They showed that the absolute temperature varied with a period of about three minutes and a maximum amplitude of 0.2°C. The amplitude appeared to decrease as depth below the 56°C–44°C interface increased. In at least one instance the oscillations are reflected by thermal gradients increasing and decreasing along the entire barrel with the same period. A qualitative analysis of these fluctuations suggests that they may be caused by internal waves propagating along the density discontinuity.

In addition to thermal gradient measurements in the sediments and water of the brine pools, four heat flow measurements were made outside of the brine region in an effort to clarify the relationship of the brine pools to the regional geothermal environment. The four new heat flow values, as well as other values available for the Red Sea, are plotted in Fig. 5 and listed in Table 2. The new values, all situated in the long median trough in which the brine pools are found, range from 1.5 to 8.0 HFU, in good general agreement with other heat flow data available for the Red Sea.

In contrast to the many measurements obtained in the Atlantis II Deep, only one value was obtained in the Discovery Deep. The value seems reliable, as judged from the experimental data. Because of its bearing on the origin of the hot water, we report this single data point. The thermal gradient in the sediment is positive (temperature increases with depth) and therefore indicates that heat is flowing from the sediment into the water, a result that is not consistent with the hypothesis that the 44.7°C water has spilled over from the Atlantis II Deep, suggested by Ross and Hunt (1967), Pugh (1967) and Turner (1969) and is cooling by conductive heat loss into the sediment. We note, however, that the thermal structure in the water is consistent with the hypothesis. Our single measurement indicated a decrease in water temperature near the floor of the deep of about 0.1°C, a value smaller than that observed by Pugh (1967).

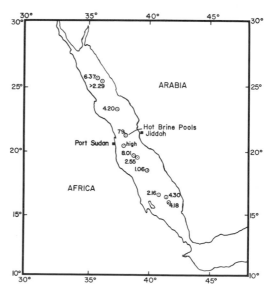

Fig. 5. A map showing the location of heat flow measurements in the Red Sea. The heat flux at each location is given in units of 10^{-6} cal/cm² sec.

Table 2   Heat Flow Measurements in the Red Sea

Station	Position Lat. (N)	Position Long. (E)	Depth (corrected)	Heat Flux (μcal/cm^{2}sec)	Reference
CH61-153	19°43′	38°41′	2,704	8.0	1
CH61-154	19°34′	38°59′	1,276	2.5	1
CH61-155	19°23′	38°54′	2,027	1.5	1
CH61-167	23°20′	37°20′	826	4.2	1
CH43-24	25°24′	36°10′	2,205	>2.29	2
5231	15°58′	41°31′	1,735	4.18	3
5232	18°24′	39°47′	1,480	1.06	3
5234	20°27′	37°55′	870	High	3
C9-111	16°13′	41°13′	941	4.30	4
C9-112	16°20′	40°28′	1,223	2.16	4
C9-114	25°28′	35°29′	1,335	6.37	4
CH-61 *	21°22′	38°05′	≈2,000	>79.0	1

* Heat flux calculated on the basis of hydrographic data in the Atlantis II Deep.
1. This paper.
2. Birch and Halunen (1966).
3. Sclater (1965).
4. Langseth and Taylor (1967).

The sparsity of data and the location of this single station near the edge of the deep preclude further interpretation.

Interpretation

Because most of the geothermal measurements were obtained in the Atlantis II Deep, our discussion is limited to that area. An understanding of the nature of the thermal source and transfer mechanism in and beneath the Atlantis II Deep is important to many of the related mineralogical, geochemical and hydrographic studies now in progress. Turner (1964, 1965, 1969) has shown that the isothermal layers characterized by temperatures of 44°C and 56°C respectively, and separated by a sharp interface, could result from thermal convection in a salinity-stabilized water mass that is heated uniformly from below. Turner also suggests that for a density difference as large as that between the 56°C and 44°C water, the transfer of heat between the layers should be due almost entirely to molecular conduction. Any transfer by eddy conduction induced by convection or internal waves will increase the effective conductivity.

Molecular conduction through the sediments is inadequate to provide the thermal energy necessary either to warm the water 0.56°C in 20 months (Munns *et al.,* 1967) or to replace the heat lost to the 44°C

water. With the area and volume of the 56°C water given by Ross (1969), the vertical flux into the water must average 42 HFU merely to heat the water at the required rate. Although the amount of heat loss into the 44°C water is uncertain, a minimum value of 37 HFU can be calculated by assuming a maximum "boundary thickness" of 5m between the 56°C and 44°C layers (suggested by Ross, 1969) and heat transport with a molecular conduction coefficient of 1.5mcal/cm sec-°C. Thus an average of at least 79 HFU is conveyed to the 56°C water. Another value of the interface thickness, 1 to 2m, may be estimated from light scattering measurements (Ryan, personal communication). A thickness of 1.5m requires a minimum flux of 166 HFU through the ocean bottom. As noted previously, the measured conductive heat flux through the sediment averaged 15–20 HFU with a local maximum of 60 HFU. We interpret these calculations to indicate that the heat flow into the 56°C water is not occurring exclusively by uniform molecular conduction through the sediment.

The Nature of the Heat Transfer Mechanism

Convection appears to be the only mechanism capable of providing the required amount of heat. It may take the form of (1) forced hydrothermal discharge of hot brine through localized vents or (2) thermal convection between a buried heat source and the 56°C water, either by convection of brine through the porous sediment or convection of the sediment itself. All combinations of these mechanisms and that of conductive transport are possible but the convective mechanisms appear quantatively to be more important.

Turner (1969) explained the origin of the thermally stratified layers in the brine pools on the basis of experiments involving uniform heating from below (Turner, 1964, 1965). Because the existence of the layered, isothermal water structure throughout the entire Atlantis II Deep has been well established, the heat flux through the bottom must be fairly uniform over an area of 60km². Localized heat sources capable of inducing widespread fluid motions in the 56°C water layer sufficient to cause the turbulent mixing necessary to maintain the observed isothermal conditions seem unlikely to us. In addition, the thermal conditions in the sediment should reflect the nature of the source. Localized hydrothermal discharge of hot water should cause a widespread negative gradient in the upper layers of sediment. Also, at locations removed from the source, the sediment temperature profiles should be relatively uncomplicated; thus, gradients within a given non-source area should be similar and exhibit little variation at an individual station. But none of these expected conditions is observed.

In support of a hypothesis of convection in the sediments we offer the following observations: (1) the temperature profiles show that the sediment becomes hotter rather than cooler with increasing depth, (2) the temperature profiles vary laterally and extreme positive and negative gradients are frequently observed within a single profile, (3) lastly, many of the thermal measurements were made in the deepest parts of the Atlantis II Deep. The absence of water warmer than the brine in the very area suggested as a possible source area by Bischoff (1969) seems to suggest that such water is not present.

Because data on the permeability and viscosity of the sediments does not exist, it is difficult to argue that convection is

theoretically possible, that is, that the critical Rayleigh number is exceeded. However, the results of experimental work by Elder (1965, 1967) on convective flow through porous media can be used to calculate a value for the average permeability which, if consistent with the permeability in other geothermal areas, would suggest that convection is at least possible. The following assumptions are implicit in this calculation:

(a) Convection is occurring by movement of the brine through the sediment.

(b) The ratio of the computed conductive heat flow (15 to 20 HFU) to the required heat flow (80 to 166 HFU) gives the upper and lower values of the Nusselt number (4 to 11).

(c) The following parameters which seem reasonable to us are used in our calculation: (1) coefficient of thermal expansion of the brine, $\alpha = 4 \times 10^{-4}(°C)^{-1}$, (2) thermal conductivity of the saturated sediment, $K_m = 1.8$mcal/°C-cm-sec, (3) acceleration of gravity, $g = 10^3$cm/sec², (4) thickness of convecting layer, $H = 1$km, (5) kinematic viscosity of the brine, $v = 10^{-2}$ poise, (6) permeability, K in cm², (7) temperature difference across the convecting layer, $\Delta T = 10^2$ °C.

We estimate the Rayleigh number R from Elder's relation between the Nusselt and Rayleigh numbers and solve the equation $R = K\alpha g\Delta T H / K_m v$ for the permeability; we obtain 2×10^{-9}cm². This value agrees satisfactorily with 10^{-10}cm² obtained by Wooding (1963) for the Wairakei geothermal system, 1 to 5×10^{-9}cm² by Elder (1965), also for the Wairakei system, and 3.5×10^{-10}cm² by Pálmason (1967) for geothermal areas in Iceland. It thus appears that by using geophysically reasonable values for the above parameters, convection of the brine through the sediments is at least theoretically possible.

The evidence cited above, we believe, demonstrates the necessity of some form of mass transfer but argues against transfer by localized hydrothermal discharge. Convective transfer through the sediment seems to be the most likely mechanism operating now. That convection in the sediments can proceed by either of two mechanisms was suggested above. The existence of sedimentary layers in the Atlantis II Deep, discussed by Bischoff (1969), allows us to choose between these mechanisms. Clearly, convection has proceeded by the migration of fluid through the porous sediment. Bulk convection of the sediment as a viscous fluid would result in a homogeneous sediment layer.

In addition to the convective transport of heat at the present time, the forceful expulsion of hot brine into the Atlantis II Deep must have occurred at one or more times in the past. Several pieces of evidence are suggestive: (1) the existence of sedimentary layers, (2) the chemical nature of these deposits, (3) hydrographic evidence for the water in the Discovery Deep having its source in the Atlantis II Deep.

Conclusions

The heat balance in the 56°C water in the Atlantis II Deep requires thermal transfer mechanisms other than molecular conduction for the transport of heat through the sediment. The existence of distinct, isothermal water layers throughout the entire Atlantis II Deep suggests that heat transfer through the bottom of the 56°C water is more uniform than would be expected if transfer occurs by hydrothermal discharge through a few localized vents. Uniform convection of the brine through the porous sediment is the most likely mechanism at the present time, although it was probably preceded by periods of hydrothermal discharge. Information on the mechanical properties of the sediment, additional measurements of the temperature structure in the sediment and further mineralogical and geochemical data will be required to define the nature and past history of the heat source and transfer mechanisms which exist beneath the hot brine area.

Acknowledgments

The authors are grateful to the Woods Hole Oceanographic Institution for the opportunity to participate in the Red Sea cruise. The cooperation of the scientists, crew and officers of the R/V CHAIN is gratefully acknowledged. Financial support was provided by the Office of Naval Research under Contract Nonr 1841(74).

References

Birch, F. S. and A. J. Halunen, Jr.: Heat-flow measurements in the Atlantic Ocean, Indian Ocean, Mediterranean Sea, and Red Sea. J. Geophys. Res., **71,** 583 (1966).

Bischoff, J.: Red Sea geothermal brine deposits: their mineralogy, chemistry, and genesis. *In: Hot brines and recent heavy metal deposits in the Red Sea,* E. T. Degens and D. A. Ross (eds.). Springer-Verlag New York Inc., 368–401 (1969).

Brewer, P. G., C. D. Densmore, R. Munns, and R. J. Stanley: Hydrography of the Red Sea brines. *In: Hot brines and recent heavy metal deposits in the Red Sea,* E. T. Degens and D. A. Ross (eds.). Springer-Verlag New York Inc., 138–147 (1969).

Elder, John W.: Physical processes in geothermal areas. *In: Terrestrial Heat Flow, Geophys. Monograph,* W. H. K. Lee (ed.), **8,** 211 (1965).

——: Steady free convection in a porous medium heated from below. J. Fluid Mech., **27,** 29 (1967).

Hunt, J. M., E. E. Hays, E. T. Degens, and D. A. Ross: Red Sea: detailed survey of hot brine areas. Science, **156,** 514 (1967).

Langseth, Marcus G.: Techniques of measuring heat flow through the ocean floor. *In: Terrestrial Heat Flow, Geophys. Monograph,* W. H. K. Lee (ed.), **8,** 63 (1965).

—— and P. T. Taylor: Recent heat flow measurements in the Indian Ocean. J. Geophys. Res., **72,** 6249 (1967).

Lee, W. H. K.: *Terrestrial Heat Flow, Geophys. Monograph,* W. H. K. Lee (ed.), 8, 276 (1965).

Munns, R. G., R. J. Stanley, and C. D. Densmore: Hydrographic observations of the Red Sea brines. Nature, **214,** 1215 (1967).

Pálmason, Gudmundur: On heat flow in Iceland in relation to the mid-Atlantic Ridge. Iceland and Mid-Ocean Ridges, "*RIT*," **38,** 111 (1967).

Pugh, David: Origin of hot brines in the Red Sea. Nature, **214,** 1003 (1967).

Ross, D. A.: Temperature structure of the Red Sea brines. *In: Hot brines and recent heavy metal deposits in the Red Sea,* E. T. Degens and D. A. Ross (eds.). Springer-Verlag New York Inc., 148–152 (1969).

—— and J. M. Hunt: Third brine pool in the Red Sea. Nature, **213,** 687 (1967).

Sclater, John: Heat flux through the ocean floor. *Ph.D. Thesis,* Cambridge University, Cambridge, England (1965).

Turner, J. S. and H. Stommel: A new case of convection in the presence of combined vertical salinity and temperature gradients. Proc. Natl. Acad. Sci., **52,** 49 (1964).

Turner, J. S.: The coupled turbulent transports of salt and heat across a sharp density interface. Int. J. Heat Mass Transfer, **8,** 759 (1965).

————: A physical interpretation of the observations of hot brine layers in the Red Sea. *In: Hot brines and recent heavy metal deposits in the Red Sea,* E. T. Degens and D. A. Ross (eds.). Springer-Verlag New York Inc., 164–173 (1969).

Von Herzen, R. P. and A. E. Maxwell: The measurement of thermal conductivity of deep-sea sediments by a needle-probe method. J. Geophys. Res., **64,** 1557 (1959).

Wooding, R. A.: Convection in a saturated porous medium at large Rayleigh number or Peclet number. J. Fluid Mech., **15,** 527 (1963).

Basalt from the Axial Trough of the Red Sea *

RICHARD L. CHASE †

Woods Hole Oceanographic Institution
Woods Hole, Massachusetts

Abstract

Fragments of volcanic glass and basalt raised in six cores between 19° and 23°N, two from Atlantis II Deep, are Recent or sub-Recent in age, have specific gravity between 2.5 and 2.7 and refractive index of $1.607 \pm .003$. One sample contained laths of plagioclase (An 60-82). The rocks contain no normative nepheline and are poor in K_2O (0.2 wt per cent). In these respects they resemble subalkaline oceanic basalts and the basalts of the island of Jebel Teir at the southern end of the axial trough. Their titania content is intermediate between that of alkaline basalts and oceanic subalkaline basalts and comparable with that of Ethiopian rift basalts. Their phosphorus and total iron contents are higher than that of oceanic subalkaline basalts and comparable to alkaline basalts of the African Rifts. Petrologically, as structurally, the axial trough of the Red Sea seems to be intermediate between rift valley and ocean.

Introduction

During the cruise of **R.V. CHAIN** to investigate the hot brine region in the Red Sea in 1966, fragments of volcanic rock were raised in six cores from the axial trough of the Red Sea between latitudes 19° and 23° North. Drake and Girdler (1964) postulated that strongly magnetized north-south dikes beneath the trough cause the large-amplitude magnetic anomalies there, but Girdler (1966) suggested that basalt flows of oceanic type would have sufficiently high remnant magnetism to cause the observed magnetic anomalies. Girdler postulated that the basalt is a component of the upper part of the crust. A layer characterized by a compressional wave velocity of about 4.5km/sec is present in two of the refraction profiles of Drake and Girdler (1964) from the axial trough. This velocity lies in the range characteristic of basalt from the ocean floor (Chase and Hersey, 1968). The fragments described here, with others from cores raised by **R.V. VEMA** (Rosenberg-Herman, 1965) constitute direct evidence of volcanic extrusives in the axial trough.

Location of Volcanic Fragments

The locations of cores from **CHAIN** cruise 61 containing volcanic fragments are listed in Table 1 and shown in Fig. 1. Two of the samples (84 and H-727-1) are from the Atlantis II Deep, two (164 and 165) are from smaller basins to the north and two (153 and 155) are from the largest basin in the Red Sea, south of the brine area. All of the samples were raised from depths greater than 1,800m. Two samples are from the surface of the seafloor. The others were buried beneath one to several meters of sediment, which was, according to Berggren *et al.* (1969), Recent or sub-Recent in age. Also listed in Table 1 are locations of cores from **VEMA** cruise 14 containing volcanic fragments. These cores, two from the southern deeps and one from the main deep of the Red Sea, were described by Rosenberg-Herman (1965).

* Woods Hole Oceanographic Institution Contribution No. 2183.
† Now at Department of Geology, University of British Columbia, Vancouver 8, B.C., Canada.

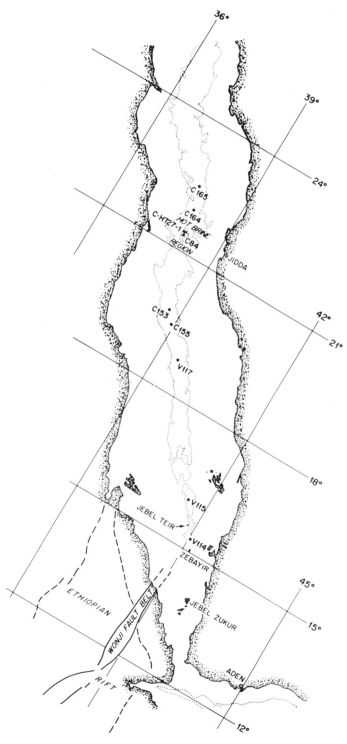

Fig. 1. Locations of bottom samples containing volcanic material in the Red Sea. Outer dotted line is contour at 962m (500 fathoms uncorrected) from Admiralty Chart C. 6359. Inner dotted line is contour at 1,931m (1,000 fathoms uncorrected), enclosing hot brine region. Dashed lines delineate margins of Ethiopian Rift. Numbers prefixed by V are **VEMA** cruise 14 cores, by C are **CHAIN** cruise 61 samples.

Richard L. Chase

Table 1

Station	Lat.	Long.	Depth (Corr. m)	Length of Core (cm)	Depth of volcanic in core (cm)
		CHAIN-61 Cores			
84P	21° 21′N	38° 03.7′E	1,984	870	870
H-727-1	21° 24.4′N	38° 03.6′E	1,895	0 *	0
153P	19° 43′N	38° 41′E	2,703	755	
155P	19° 23.5′N	38° 54.0′E	2,030	424	424
164	21° 59′N	37° 58.5′E	2,282	1,178	0
165P	22° 28.2′N	37° 46.5′E	2,167	81	81
		VEMA-14 Cores †			
114	15° 17′N	42° 02′E	768	611	throughout
115	16° 04.5′N	41° 28′E	1,194	590	90–110
117	18° 48′N	39° 31′E	1,372	190	40–60

* Fragments obtained during hydrographic lowering.
† From Rosenberg-Herman (1965).

Description of Samples

84P. From the bottom of 10-m piston core in Atlantis II Deep. The sample consisted of 5 angular fragments of black, glassy basalt, together weighing 5 grams, and 16 angular fragments of partly crystalline fine-grained basalt, together weighing 20 grams.

H-727-1. From the lead weight used in a hydrographic lowering in the Atlantis II Deep. Two angular fragments of black, glassy basalt, with spheroidal vesicles, less than a millimeter in diameter, comprising a few per cent of the total volume.

153P. From a piston core in the deepest depression of the axial trough south of the hot holes. About a dozen small angular fragments of black, basaltic glass ranging in largest dimension from 1cm to 1mm.

155P. From bottom of a 424-cm piston core raised from the deepest depression of the axial trough south of the hot holes. Several grains of friable, black, basaltic glass coated and laced with red-brown alteration product.

164. From a depression in the axial trough north of the hot holes. Angular chips of glassy basalt weighing one hundred grams. The core was only 5cm long. Thus, the flow appears to be so recent that it has negligible sediment cover.

165P. From bottom of an 81-cm piston core raised from a depression in the axial trough north of the hot holes. The corer was bent during the lowering, indicating that it struck a hard surface. One thin platelet of partially altered basaltic glass, from the surface of a flow, was recovered.

Details of the **VEMA**-14 cores, from Rosenberg-Herman (1965), are as follows:

114. A core 4m long, consisting of layers with from 20 to 98 per cent volcanic debris, including a pebble of basalt at 280cm and yellow, brown and black glass. Though the core was obtained in water 700m deep, the site is on the flank of a platform occupied by the volcanic island of Zebayir and other islets. Thus, the volcanic material may well be derived from the volcanic centers which formed the islands.

115. From the southern end of the axial trough, some 150km north of the volcanic island of Jebel Teir. Transparent and vesicular volcanic ash is a component of a section of the core interpreted to have been deposited during the last glaciation.

117. From the axial trough some 100km south of **CHAIN** samples 153 and 155. Dark brown glass is in the top 60cm of the core, reported as postdating the last glaciation.

Specific Gravity

The specific gravity of some fragments is listed in Table 2. The procedure was to clean glassy fragments of alteration products and adhering sediment, both with a steel probe and an ultrasonic vibrator, and to determine the specific gravity using a widemouthed pycnometer.

The values for the glasses are intermediate between those of basaltic glass and andesitic glass recorded by Daly *et al.* (1966), possibly because all the alteration products were not removed before weighing.

Petrography

84P. Thin-section showed crystals up to 0.7mm long of labradorite-bytownite [An 60-82, determined on the universal stage by measuring $\alpha'\Lambda$ (010) in section $\perp\times$] and a dusting of opaque oxides in a groundmass of featherlike aggregates of a rapidly quenched material of higher refractive index than the plagioclase, probably clinopyroxene. Other **CHAIN** samples were wholly glassy. The refractive indices of two of these are listed in Table 2.

114. A thin-section was made from a basalt pebble taken 2.8m from the top of this core. Fresh euhedral olivine up to 1.2mm long and smaller plagioclase laths lie in a groundmass of fibrous and featherlike pyroxene and plagioclase. Irregularly shaped vesicles up to 3mm across make up 20 per cent of the rock. Larger vesicles are lined with manganese oxides.

Table 2   Specific Gravity and Refractive Index of Basalt

Sample No.	S. G.	R. I.	Material
H-727-1	2.516	–	glass
153P	2.528	–	glass
155P	–	1.607 ± .002	glass
164	2.686	1.608 ± .002	glass
84P	2.713	–	partly crystalline

Chemical Composition

The chemical composition of two samples is shown in Table 3, columns 1 and 2. In preparing samples for analysis, all visible palagonitic alteration was removed from the fragments, which were then crushed to pass a 70 micron mesh and analysed by F. T. Manheim by the precision spectrometric (optical emission) method. Samples and standards were fluxed with lithium tetraborate according to Landergren *et al.* (1964) at 1000°C and excited in form of graphite pellets with a high voltage spark. SiO_2 was determined by difference of sum of cations plus P_2O_5 and ignition loss at 1000°C from 100.0. This value is generally more accurate than a direct value. H_2O^+ is taken to equal ignition loss at 500°C and is a minimum value. The values given for Fe_2O_3 are maximum values, being computed from the gain in mass when the samples were heated from 500°C to 1000°C. Because Fe_2O_3 is a maximum value, the CIPW norms tend to show greater silica saturation than they otherwise would. Hence, the high normative quartz content of sample 155 is suspect.

Discussion

The paleontological evidence from **CHAIN** and **VEMA** cores indicates that volcanism in the axial trough occurred in the Pleistocene and probably in the Holocene Epochs.

Basalts of Holocene or Pleistocene age are also known from islands in the southern end of the Red Sea (Gass, Mallick, and Cox, pers. comm.) from Southern Arabia and the Ethiopian Rift (Aden Volcanics) (Mohr, 1963; Gass *et al.*, 1965), and from various localities along the margins of the Red Sea and associated faults and rifts (U. S. G. S. — Aramco, 1963). An eruption of basalt was recorded near Medina in 1256 A.D., and thermal springs are known south of Jidda (Furon, 1963). Cenozoic basalts cover large areas of western Arabia. Average compositions of basalt from Jebel Zukur and nearby islands, and from Jebel Teir, in the southern part of the Red Sea (Gass, Mallick and Cox, pers. comm.), are listed in columns 4 and 6 respectively of Table 3.

The Red Sea connects through the Ethiopian Rift with the African rift valleys. The

Table 3 Chemical Analyses and CIPW Norms

Col. 1 — Station 164, Red Sea.
Col. 2 — Station 155, Red Sea.
Col. 3 — Average of 8 alkali olivine basalt analyses, Chyulu Range, Kenya.
Col. 4 — Average of 5 basalt analyses, Jebel Zukur and adjacent islands, south of axial trough, Red Sea.
Col. 5 — Average of 25 basalt analyses, Aden Series, Ethiopian Rift.
Col. 6 — Average of 6 basalt analyses, Jebel Teir, south end of axial trough, Red Sea.
Col. 7 — Average of 33 oceanic subalkaline basalt analyses.
Col. 8 — Standard deviations from the mean of Col. 7.

	1	2	3	4	5	6	7	8
SiO_2	47.86 *	48.71 *	44.28	49.28	47.3	50.04	49.51	0.75
TiO_2	2.10	1.70	2.64	3.05	2.0	1.93	1.33	0.30
Al_2O_3	13.10	12.80	12.20	16.41	14.1	16.56	16.26	1.55
Fe_2O_3	4.82	7.80⎫	13.81	2.97	6.7	3.31	1.77	0.70
FeO	9.16	7.46⎭		8.05	7.3	7.49	7.33	1.21
MnO	0.26	0.25	0.20	0.23	0.2	0.19	0.16	0.03
MgO	8.30	7.30	10.81	3.82	5.95	5.19	8.36	1.25
CaO	10.60	10.20	10.37	8.98	10.8	11.33	11.38	0.55
Na_2O	2.90	2.20	3.16	5.10	2.5	3.00	2.66	0.25
K_2O	0.20	0.19	1.15	1.30	1.3	0.36	0.23	0.15
H_2O^+	0.13	0.54⎫	1.19	0.43⎫	1.45	0.39	0.65	0.25
H_2O^-	0.05	0.34⎭		0.43⎭		0.26	0.32	0.26
P_2O_5	0.48	0.45	0.68	–	0.40	–	0.14	0.05
Cr_2O_3	0.04	0.06	–	–	–	–	–	–
	100.0 *	100.0 *	100.49	100.05	100.00	100.05	100.10	
q	–	6.54		–	0.7	2.22	–	
or	1.11	1.11		7.78	7.5	2.22	1.36	
ab	24.63	18.34		27.25	21.4	25.15	22.50	
an	21.96	24.46		18.07	22.5	30.86	31.74	
ne	–	–		8.52	–	–	–	
di	21.78	18.59		22.21	26.5	20.80	19.32	
hy {en	2.80	11.30		–⎫	11.6	6.70	7.81	
hy {ofs	1.32	3.17		–⎭		4.22	3.79	
ol	11.61	–		7.40	–	–	7.22	
mt	6.96	11.37		4.41	5.6	4.87	2.57	
il	3.95	3.19		5.78	3.3	3.65	2.53	
ap	1.34	1.01		–	0.9	–	0.31	

* SiO_2 determined by difference of sum of other oxides and ignition loss at 1000°C from 100.0. This value is generally more accurate than a direct determination. Actual direct values found are 51.3% and 49.9%.

younger volcanics of Ethiopia (Pliocene to Holocene) are known as the Aden Volcanics. The average of twenty-five basalt analyses of the Aden Volcanics, from Mohr (1963), is listed in column 5 of Table 3. The alkaline basalts of the rift valley region are represented in column 3 of Table 3 by an average of eight analyses of olivine basalts from the Chyulu Range in Kenya (Saggerson and Williams, 1964).

Several workers, most recently Girdler (1966), Vine (1966) and Laughton (1966), have postulated that the Red Sea is a locus of ocean-floor spreading. If this is true, the axial trough may be analogous to the central part of a mid-oceanic ridge, and the basalts therein similar to those of the deep ocean which originated as extrusions at the central part of a mid-oceanic ridge. Sub-

alkaline basalts * have been described from many sites on mid-oceanic ridges. Rather than compare the Red Sea basalts with each suite described, I have computed an average composition for subalkaline basalts from mid-oceanic ridges, with standard deviations (Table 3, columns 7 and 8). To compute the average composition, thirty-three analyses were selected, twenty-eight published, from papers by Muir and Tilley (1964, 1966), Chernysheva and Bezrukov (1966), Engel and Engel (1964a and b), Engel et al. (1965), Nichols et al. (1964) and five unpublished, from the Mid-Atlantic Ridge, by T. Asari for the author. The analyses were selected from those collected

* The terms "subalkaline basalt" and "alkaline basalt" were defined by Chayes (1966).

from mid-oceanic ridges and described as basalts or tholeiites. The shallowest depth for a selected sample was 1,450m. Analyses in which potassium content was greater than one percent or which had suffered alteration, shown by hydration (water content greater than two percent) or oxidation [Fe_2O_3/FeO (weight per cent greater than 0.5], were rejected.

The small quantities of basalt fragments recovered in the cores from the axial trough of the Red Sea, the fact that they came from the margins of basalt flows which are likely to have reacted with seawater during cooling and the coating of alteration or reaction products on most of them make it unlikely that the analyses obtained give a very accurate tally of the composition of the flows of which they once formed a part. Certain differences and similarities between them and the other basalts, whose compositions are posted in Table 3, are, however, probably real and noteworthy. Compared with the alkaline basalts of Kenya, the Red Sea basalts are high in silica, low in titania, have about equal amounts of iron and are much lower in potassium. Compared with the alkali olivine basalt from Jebel Zukur at the southern end of the Red Sea, they have low titania, high total iron, low sodium and much lower potassium. The basalts from Jebel Teir and the Ethiopian Rift resemble each other in content of titania, magnesium and calcium but differ in total iron and potassium. Their similarities are notable because Jebel Teir lies on the extension of the Wonji fault belt of the Ethiopian Rift (Fig. 1), along which fissure basalts have erupted in Recent times (Mohr, 1967). The island basalts, however, approach the Red Sea basalts in paucity of potassium.

The Red Sea and Ethiopian basalts have comparable content of titania, total iron and sodium, but the potassium of the Ethiopian basalts is six times higher. The low potassium content of the Red Sea basalt is typical of oceanic basalts as can be seen by comparison with the average oceanic subalkaline basalt. The main differences between the Red Sea basalts and the oceanic average are the higher total iron and phosphorus of the former.

The CIPW norms shown in Table 3 in-dicate that one of the Red Sea basalts is saturated in silica and has normative quartz. The other has hypersthene and olivine but no nepheline. The norms should be viewed with caution, however, because the method of analysis tends to yield a greater value than actual for the ratio Fe_2O_3/FeO. Thus the true normative quartz content of sample 155 (col. 2) is probably less than the indicated value of 6.54. The norm for the Kenyan basalts (col. 3) cannot be computed because the Fe_2O_3/FeO ratio is not known. The Jebel Zukur basalts (col. 4) are undersaturated alkali olivine basalts with normative nepheline in contrast to the basalts from the Ethiopian Rift and Jebel Teir (cols. 5 and 6), which are both saturated and have normative quartz. The oceanic basalts show neither quartz nor nepheline in the norm, in this way resembling one of the Red Sea basalt analyses.

Summary

The Red Sea basalts resemble oceanic basalts in their low potassium content and lack of normative nepheline, and the alkaline basalts of the African rifts in their high total iron and phosphorous. In titania, they are intermediate between oceanic and rift basalt. Thus the petrologic character of the axial trough of the Red Sea, judged from the as yet scant evidence of the chemical composition of its effusives, seems to reflect a structural character intermediate between rift valley and ocean.

Acknowledgments

Dr. David A. Ross supplied the rock fragments from the **CHAIN** 61 cores. Dr. Yvonne Rosenberg-Herman supplied the basalt fragment from the core obtained by Dr. Charles L. Drake on **VEMA** cruise 14. Dr. Frank Manheim analysed samples by emission spectrometry. Drs. I. G. Gass, I. J. Mallick and K. G. Cox supplied analyses from which the average compositions of basalt from Jebel Teir and Jebel Zukur were computed. Mr. Frank B. Wooding III determined values of specific gravity and made thin sections. Professors Arthur F. Buddington, Harry H. Hess and Dr. William G. Melson

criticized the manuscript, which was typed by
Miss Elaine Gurney.

This work was supported by the National
Science Foundation under grants GA 584 and
GP 5355 and by the Office of Naval Research
under contract Nonr 4029(00).

References

Chase, R. L., and J. B. Hersey: Geology of the north
slope of the Puerto Rico Trench. Deep Sea Research, (in press).

Chayes, F.: Alkaline and subalkaline basalts. Am.
Jour. Sci., **264**, 128 (1966).

Chernysheva, V. I., and P. L. Bezrukov: Serpentinites
from the crest of the Arabian-Indian (Carlsberg)
Ridge. Dok. Akad. Nauk SSSR, **166,** 961 (1966).

Daly, R. A., G. E. Manger, and S. P. Clark, Jr.: Density of Rocks. Mem. Geol. Soc. Amer.,**97,**20(1966).

Drake, C. L., and R. W. Girdler: A geophysical study
of the Red Sea. Geophys. J., **8,** 473 (1964).

Engel, A. E. J., and C. G. Engel: Composition of
basalts from the mid-Atlantic ridge. Science, **144,**
1330 (1964a).

———, ———: Igneous rocks of the East Pacific Rise.
Science, **146,** 477 (1964b).

Engel, C. G., R. L. Fisher, and A. E. J. Engel: Igneous
rocks of the Indian Ocean floor. Science, **150,** 605
(1965).

Furon, R.: *Geology of Africa.* Hafner, New York, 377
p. (1963).

Gass, I. G., I. J. Mallick, and K. G. Cox: Royal
Society Volcanological Expedition to the South
Arabian Federation and the Red Sea. Nature, **205,**
952 (1965).

Girdler, R. W.: The role of translational and rotational
movements in the formation of the Red Sea and

Gulf of Aden. Geol. Survey of Canada Paper 66-14,
65 (1966).

Landergren, S., W. Muld, and B. Rajandi: On the
geochemistry of deep-sea sediments. Appendix:
Analytical methods. Reports of the Swedish Deep-
Sea Expedition, **10,** special investigations no. 5,
148 (1964).

Laughton, A. S.: The Gulf of Aden, in relation to the
Red Sea and the Afar Depression of Ethiopia. Geol.
Survey of Canada Paper 66-14, 78 (1966).

Mohr, P. A.: The Ethiopian Cenozoic lavas; a preliminary study of some trends: spatial, temporal
and chemical. Bull. Geophys. Observatory of Haile
Selassie I, University, Addis Ababa, **6,** 103 (1963).

———: Major volcano-tectonic lineament in the Ethiopian Rift system. Nature, **213,** 664 (1967).

Muir, I. D., and C. E. Tilley: Basalts from the northern part of the rift zone of the Mid-Atlantic Ridge.
Jour. Petrology, **5,** 409 (1964).

———, ———: Basalts from the northern part of the
Mid-Atlantic Ridge II. The Atlantis collections near
30°N. Jour. Petrology, **7,** 193 (1966).

Nichols, G. D., A. J. Nalwalk, and E. E. Hays: The
nature and composition of rock samples dredged
from the Mid-Atlantic Ridge between 22°N and
52°N. Marine Geol., **1,** 333 (1964).

Rosenberg-Herman, Y.: Etudes des sediments quater-
naire de la Mer Rouge. In: *Thèses presentées a la
Faculté des Sciences de l' Université de Paris,*
Serie A, 1123. Masson & Cie, Paris (1965).

Saggerson, E. P., and L. A. J. Williams: Ngurumanite
from Southern Kenya and its bearing on the origin
of rocks in the Northern Tanganyika Alkaline district. Jour. Petrology, **5,** 40 (1964).

U. S. G. S.-Aramco: Geologic map of the Arabian
Peninsula, 1:2,000,000. U.S. Geol. Survey Misc.
Geol. Investigations Map I-270A (1963).

Vine, F. J.: Spreading of the ocean floor: new evidence. Science, **154,** 1405 (1966).

Water

General Circulation of Water Masses in the Red Sea *

G. SIEDLER

Woods Hole Oceanographic Institution
Woods Hole, Mass., and
Institut für Meereskunde der Universität Kiel
Kiel, Germany

Abstract

A summary is given of the main features of the Red Sea circulation caused by evaporation and wind stress. Average temperature and salinity distributions are discussed, and some estimates of the water budget and water renewal times are calculated.

The circulation of water masses in the Red Sea and the corresponding temperature-salinity distribution is determined by the usual climatic conditions found in this region. At present the two dominating factors are the evaporation processes at the Red Sea surface and the wind field effects over the Red Sea and the adjacent parts of the Arabian Sea.

The Red Sea is connected to other oceanic regions by three channels: the Strait of Bab el Mandeb in the south, which is divided into a large and a small channel by Perim Island, and the Suez Canal in the north. The average water exchange through the small channel of the Strait of Bab el Mandeb and through the Suez Canal is small compared to the flow through the main channel of the Strait of Bab el Mandeb. A sill to the north of Bab el Mandeb, somewhat more than 100m deep, controls the flow. As evaporation far exceeds precipitation in this region and as no river inflow occurs, the water budget is to a good approximation given by the water exchange through the large channel of Bab el Mandeb and by the amount of evaporation.

In prehistoric times changes of the circulation and of the associated budget must have occurred which were caused by additional river inflow and precipitation and also by variation of the sill depth near Bab el Mandeb due to sea level lowerings. Only the present water structure and circulation pattern will be discussed here.

Because of the loss of fresh water due to the arid climate, the surface salinity of Red Sea water is high. Density of sea-water increases with salinity and decreases with temperature. As the effect of salinity on density is not completely compensated by high temperatures, this will lead to a density which is large compared to the density on the same level surface in the upper layer of the adjacent Arabian Sea. Such a density distribution cannot be stationary without a circulation pattern of the following type: inflow of Arabian Sea water through Bab el Mandeb near the surface, vertical convection in the Red Sea, outflow of Red Sea water near the sill (Fig. 1). A typical section from the region of Bab el Mandeb with temperature-salinity data as well as data obtained from direct current measurements is given in Fig. 2.

The basic pattern of the Red Sea circulation was first described by Krümmel (1911) on the basis of the data obtained by Gedge (1898) on the British H.M.S. **STORK** and Luksch (1901a, 1901b) on the Austrian S.M. **POLA**. A few data from the Red Sea

* Woods Hole Oceanographic Institution Contribution No. 2184.

131

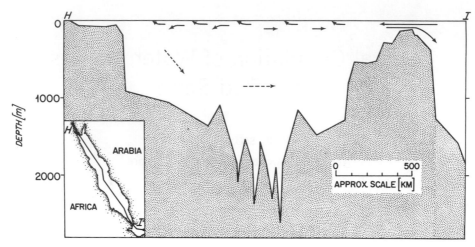

Fig. 1. Schematic circulation pattern in the Red Sea. Solid arrows indicate average direction of water motion, dotted arrows give direction of possible deep water motion.

were collected before these expeditions by Makaroff (1894) on the Russian ship **VITIAZ.** Krümmel made a first attempt to estimate the water budget of the Red Sea.

A major step forward in the knowledge of the spatial distribution of temperature and salinity and of variations with time of the currents at the southern entrance was achieved by the measurements of Vercelli (1927) and Picotti (1927) on the Italian **R.N. AMMIRAGLIO MAGNAGHI.** Further data were obtained by Matthews (1927, 1928) on the British H.M.S. **ORMONDE,** by Van Riel (1932, 1937) on the Dutch research vessel **WILLEBROD SNELLIUS,** on the British H.E.M.S. **MABAHISS** (Sewell, 1934a, 1934b, 1934c; Thompson, 1939a, 1939b; Mohamed, 1940a) and by Egyptian investigations on **MABAHISS** (Mohamed, 1940b). Defant (1930) derived a theoretical model of the circulation in straits where pressure gradient forces and frictional forces are supposed to balance each other and applied it to the data known at that time for the region of Bab el Mandeb.

Numerous measurements of the structure of water masses and currents in the Red Sea have been obtained in recent times to improve our knowledge of variations in space and time. Most data were collected on the British R.R.S. **DISCOVERY II** in 1950–51 and during French investigations aboard **NORSEL** (Crepon, 1965), aboard the American ships R.V. **ATLANTIS** and R.V. **VEMA** in 1958 (Neumann and Densmore, 1959; Neumann and Mc-

Gill, 1961), the S.S.S.R. ship **A.I. VOYEKOV** (Muromtsev, 1960), the French **COMMANDANT ROBERT GIRAUD** in 1962–1963, the British R.R.S. **DISCOVERY III** in 1964, the German F.S. **METEOR** in 1964–1965 (Dietrich *et al.*, 1965; Siedler, 1968) and the American R.V. **ATLANTIS II** in 1963 and 1965 (Miller and Risebrough, 1963; Miller, 1964).

A comparison of some of the data with a model of the circulation in the Red Sea was recently carried out by Phillips (1966), who studied similarity solutions for convective turbulent flows driven by a uniform flux of buoyancy and suggested a general function describing the vertical distribution of buoyancy in the Red Sea.

The basic structure and circulation of the Red Sea water masses is perturbed by effects of seasonal changes in evaporation and wind stress. Evaporation and cooling are greatest during the winter season (Privett, 1959). Therefore differences in the structure of the surface layers in summer and winter have to be expected.

Average distributions of temperature and salinity for the summer and winter season are given in Figs. 3 to 6. The average temperature distribution at the Red Sea surface in summer varies from below 26°C in the northern part to a maximum of more than 30°C between 16 and 20°N. Surface salinities vary from below 37 parts per thousand (per mill) near Bab el Mandeb to more than 40 per mill in the northern part of the Red Sea. Comparing these numbers

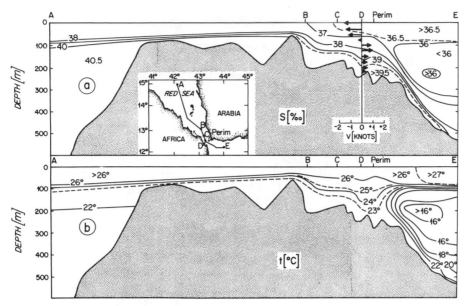

Fig. 2. Salinity (a) and temperature (b) distribution in the southern Red Sea and the inner Gulf of Aden, indicating outflow of Red Sea Water near the bottom and inflow of Gulf of Aden Water near the surface. Arrows give average current profile from direct measurements (Siedler, 1968).

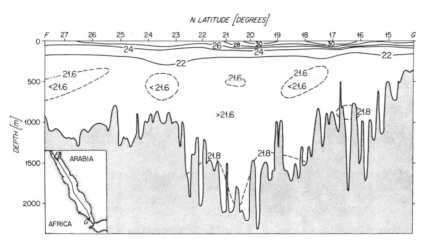

Fig. 3. Average temperature distribution in the Red Sea during summer. Numbers at isotherms indicate temperature in degrees C (Siedler, 1968).

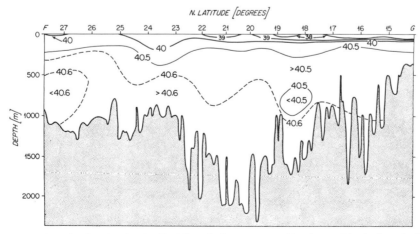

Fig. 4. Average salinity distribution in the Red Sea during summer. Numbers at isohalines indicate salinity in parts per thousand (Siedler, 1968).

133

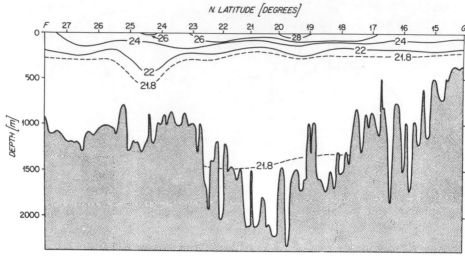

Fig. 5. Average temperature distribution in the Red Sea during winter. Numbers at isotherms indicate temperature in degrees C (Siedler, 1968).

with the sections for the winter, surface temperature is about 2°C lower in winter than it is in summer. However, no large changes in average salinity occur from summer to winter. Although it must be expected that the loss of fresh water due to evaporation is larger in winter, the salinity remains at about the same level. Advective processes apparently compensate for the increase in evaporation.

The transition zone from surface to deep water is generally between the 22°C and 24°C isotherms. Vertical profiles below the transition zone are very smooth and

constant over long time periods. Below the transition zone, temperature continues to decrease slightly, but increases in temperature and salinity can be observed in the deepest water. The differences in the structure of the deep water for summer and winter shown in these figures are probably not significant but are generated by a sampling pattern which, because of the use of data from different expeditions, is irregularly distributed.

Average vertical gradients of temperatures as well as of salinities in the surface layers are lowest in the north. As evapora-

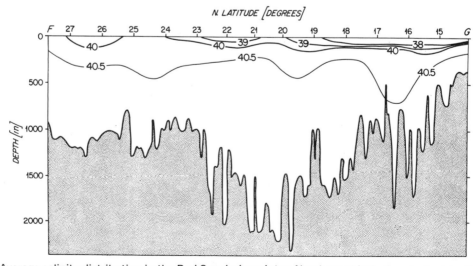

Fig. 6. Average salinity distribution in the Red Sea during winter. Numbers at isohalines indicate salinity in parts per thousand (Siedler, 1968).

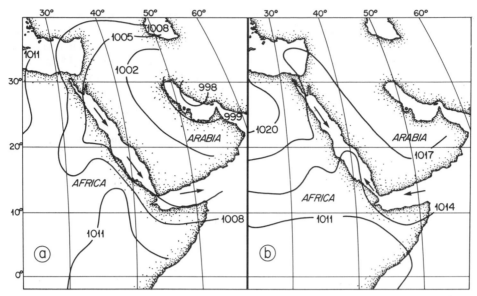

Fig. 7. Typical pattern of atmospheric pressure field over the Red Sea and the Gulf of Aden during summer (a) and winter (b). Arrows indicate wind directions, numbers at isobars indicate pressure in millibars (according to Preliminary Climatic Atlas of the World, Weather Information Branch, U.S. Army Air Force, Special Series No. 1, Chicago, 1943).

tion and cooling are at their maximum in winter, Neumann and McGill (1961) suggest that deep water may be formed at that time in the northern part of the Red Sea. Most of the main average circulation discussed earlier will occur in the surface layers, probably down to 150–200m.

The basic circulation of the Red Sea is affected by marked seasonal variations of the wind field which is controlled by the monsoon wind system. During the summer there usually is a large area of low atmospheric pressure over central Asia connected to the North African low-pressure trough and a high-pressure area over eastern Africa (Fig. 7a). Rather steady SW winds are found in the Arabian Sea, changing to NNW in the southern part of the Red Sea. During the winter this wind field pattern is reversed because high-pressure areas are generated over central Asia and northern Africa and a low-pressure field is formed over eastern Africa (Fig. 7b). The average winds in the northern part of the Red Sea are from NNW during the whole year.

The seasonal change in weather conditions thus leads to winds in the direction of the average surface water flow in the Strait of Bab el Mandeb during winter and in op-

posite direction to the average surface water flow during summer. Average wind speed is higher in winter than it is in summer. It can therefore be expected that the inflow of water from the Gulf of Aden will be stronger due to wind stress at the surface in winter than it is in summer. Thompson (1939a) suggested that the wind effect might be strong enough to generate an outflow at the surface in the strait during summer due to wind stress and an inflow at some intermediate depth. Neumann and McGill (1961) showed that this was not so in June, 1958. The effect of wind was not sufficiently large to change the two-layer current system, but a three-layer system may develop later in the summer. It has been stated before that evaporation is larger in winter than it is in summer and that it is expected that the advection of water from the Gulf of Aden be larger in winter. Thus, both the wind effects and the evaporation will cause an increase of the amount of inflow into the Red Sea to a level where it is balanced by the absolute pressure field in the Red Sea.

It should be possible to check the relevance of changes in the density field and the wind stress by direct current measurements near Bab el Mandeb. The data available up to now (Gedge, 1898; Vercelli, 1927;

Siedler, 1968) are representative only for short time periods. However, the recent measurements in the winter of 1964 (Siedler, 1968) indicate that a marked increase in the net inflow through Bab el Mandeb was found when the wind speed increased suddenly. It therefore seems probable that the wind generated flow will add an appreciable amount of water to the net inflow through the strait.

To determine the total water budget of the Red Sea, only indirect methods can be applied at the time being. During this century, no significant change of mean sea level and mean salinity could be detected. If the total volume of Red Sea water remains constant, the net inflow per year through Bab el Mandeb has to be equal to the total volume of the mean evaporation. Privett (1959) calculated a mean height of evaporated water of 183cm/yr, Neumann (1952) of 215cm/yr. If one assumes an evaporation rate of approximately 200cm/yr, the total volume V_E of evaporating water per year is given as the product of this number and the area of the Red Sea surface: $0.44 \times 10^{12} m^2$ (Kossinna, 1921). The result is:

$$0.88 \times 10^{12} m^3/yr$$

Estimates of the average transported volumes of inflowing and outflowing water volumes can be obtained from the continuity equation for constant values of the annual averages of sea level and salt content of the Red Sea from Knudsen's Hydrographical Theorem (Knudsen, 1900). The volume V_o of outflowing water is determined from the average salinities S_i of inflowing and S_o of outflowing water and the volume V_E of evaporated water by:

$$V_o = V_E S_i (S_o - S_i)^{-1}$$

Taking $S_i = 36.6\%_0$ and $S_o = 39.7\%_0$, one obtains:

$$V_o = 10.5 \times 10^{12} m^3/yr$$

or

$$V_o = 1.2 \times 10^9 m^3/hr \text{ or } .33 \times 10^6 m^3/sec$$

This number agrees well with the outflow rate of $V_{om} = 1.5 \times 10^9 m^3/hr$ obtained for a certain time period by direct measurements (Siedler, 1968).

From these numbers, it is possible to get some crude estimates for the renewal time of the Red Sea water. If it is assumed that only water in the upper 150m of the Red Sea takes part in the circulation process discussed earlier and that each water parcel travels on the largest loop possible, then this surface layer will be renewed when the volume of outflowing water equals the volume V_s of this surface layer. If the Red Sea were at least 150m deep everywhere, the volume of the surface layer would be approximately $V_S = 65 \times 10^{12} m^3$. With an outflow rate of $V_o = 10.5 \times 10^{12} m^3/yr$ the renewal time would be $t_S \approx 6$ years. If it is assumed that the whole Red Sea water with a volume of $V_D = 215 \times 10^{12} m^3/yr$ (Kossinna, 1921) takes part in the circulation process, the renewal time with the above outflow rate will be $t_D \approx 20$ years. As the major part of advection and vertical convection processes very probably occur in the surface layers, the renewal time for the surface layer down to 150m will be near t_S, while the renewal time for the deep water is expected to be much larger than t_D. A similar conclusion is indicated by C^{14} dated Red Sea deep water, which has an age of about 200 years (Craig, 1969).

Acknowledgments

Work for this publication was partly supported by the Office of Naval Research, Washington, D.C., under Contract N00014-66-C0241 and by the Deutsche Forchungsgemeinschaft, Bad Godesberg.

References

Craig, H.: Geochemistry and origin of the Red Sea brines. *In: Hot brines and recent heavy metal deposits in the Red Sea,* E. T. Degens and D. A. Ross (eds.). Springer-Verlag New York Inc., 208–242 (1969).

Crepon, M.: Circulation superficielle dans L'Océan Indien. Cahiers Océanographiques, **17,** 221 (1965).

Defant, A.: Die Bewegungen und der thermohaline Aufbau der Wassermassen in Meeresstrassen. Sitzungsberichte der Preussischen Akademie der Wissenschaften, Phys.-Math. Klasse, XIV (1930).

Dietrich, G., G. Krause, E. Seibold, and K. Vollbrecht: Reisebericht der Indischen Ozean Expedition mit dem Forschungsschiff METEOR 1964–1965. METEOR Forschungsergbnisse Reihe A, 1, 1 (1965).

Gedge, H. J.: Report on the undercurrents in the Straits of Bab el Mandeb from observations by Lieutenant and Commander H. J. Gedge, R.N., H.M.S. STORK London (1898).

Knudsen, M.: Ein hydrographischer Lehrsatz. Ann. d. Hydr. u. Marit. Meteorol., 28, 316 (1900).

Kossinna, E.: Die Tiefen des Weltmeeres. Veröff. Inst. f. Meeredkunde Berlin Reihe A, 9, 63 (1921).

Krümmel, O.: *Handbuch der Ozeanographie,* 2, Stuttgart (1911).

Luksch, J.: Expedition S.M. POLA in das Rote Meer, nördliche Hälfte (October 1895–Mai 1896). Wissenschaftliche Ergebnisse, physikalische Untersuchungen. Denkschrift d. kaiserl. Akad. d. Wissensch. in Wien, 65, 351 (1901a).

Luksch, J.: Expedition S.M. Schiff POLA in das Rote Meer, südliche Hälfte (Sept. 1897–April 1898). Wissenschaftliche Ergebnisse, physikalische Untersuchungen. Denkschrift d. kaiserl. Akad. d. Wissensch. in Wien, 69, 337 (1901b).

Makaroff, S. O.: Le VITIAZ et L'Océan Pacifique. St. Petersburg (1894).

Mathews, D. J.: Temperature and salinity observations in the Gulf of Aden. Nature, 120, 512 (1927).

————: Temperature and salinity observations in the Gulf of Aden. Nature, 121, 92 (1928).

Miller, A. R.: Highest salinity in the world ocean? Nature, 203, 590 (1964).

———— and R. W. Risebrough: Preliminary cruise report ATLANTIS II–Cruise 8, International Indian Ocean Expedition, July 5–Dec. 20, 1963. Woods Hole Oceanographic Institution, 64, 11, unpublished manuscript (1963).

Mohamed, A. F.: Chemical and physical investigations. The distribution of hydrogen-ion concentration in the north-western Indian Ocean and adjacent waters. John Murray Expedition 1933–1934, Scientific Reports, 2, 121 (1940a).

————: The Egyptian exploration of the Red Sea. Proc. Royal Soc. London, B, 128, 306 (1940b).

Muromtsev, A. M.: A contribution to the hydrology of the Red Sea, Dokl. Akad. Nauk. SSSR, 134, 1443 (1960).

Neumann, A. C. and D. Densmore: Oceanographic data from the Mediterranean Sea, Gulf of Aden and Indian Ocean. ATLANTIS Cruise 242, Woods Hole Oceanographic Institution, 60, 2, unpublished manuscript (1959).

Neumann, A. C. and D. A. McGill: Circulation of the Red Sea in early summer. Deep-Sea Research, 8, 223 (1961).

Neumann, J.: Evaporation from the Red Sea. Israel Exploration Journal, 2, 153 (1952).

Phillips, O. M.: On turbulent convection currents and the circulation of the Red Sea. Deep-Sea Res., 13, 1149 (1966).

Picotti, M.: Campagna idrografica nel Mar Rosso della regia nave AMMIRAGLIO MAGNAGHI 1923–24, Ricerche di oceanografia chimica. Annali Idrografici 11, Genova, Instituto Idrografico della Regia Marina (1927).

Privett, D. W.: Monthly charts of evaporation from the North Indian Ocean including the Red Sea and the Persian Gulf. Quart. J. Roy. Meteorol. Soc., 85, 424 (1959).

Sewell, R. B. S.: The John Murray Expedition to the Arabian Sea. Nature, 133, 86 (1934a).

————: The John Murray Expedition to the Arabian Sea. Nature, 133, 669 (1934b).

————: The John Murray Expedition to the Arabian Sea. Nature, 134, 685 (1934c).

Siedler, G.: Schichtungs- und Bewegungsverhältnisse am Südausgang des Roten Meeres. METEOR Forschungsergebnisse Reihe A, 4, 1 (1968).

Thompson, E. F.: Chemical and physical investigations. The general hydrography of the Red Sea. John Murray Expedition 1933–34, Scientific Reports, 2, 83 (1939a).

————: Chemical and physical investigations. The exchange of water between the Red Sea and the Gulf of Aden over the "Sill." John Murray Expedition 1933–34, Scientific Reports, 2, 105 (1939b).

Van Riel, P. M.: Einige ozeanographische Beobachtungen im Roten Meer, Golf von Aden und Indischen Ozean. Ann. Hydr. Marit. Meteor., 60, 401 (1932).

————: The Snellius Expedition. Leiden (1937).

Vercelli, P.: Campagna idrografica della R. N. AMMIRAGLIO MAGNAGHI in Mar Rosso 1923–24, Ricerche de oceanografia fisica. Annali Idrografici, 11, Genova, Instituto Idrografico della Regia Marina (1927).

Hydrography of the Red Sea Brines *

P. G. BREWER, C. D. DENSMORE, R. MUNNS, R. J. STANLEY

Woods Hole Oceanographic Institution
Woods Hole, Massachusetts

Abstract

The temperature and salinity characteristics of the Red Sea brines and their overlying waters are discussed, and data from the 1966 **CHAIN** cruise are presented. Most papers on this subject have reported "salinity" values derived from volumetric dilution of the sample with distilled water and conductometric measurement of the dilute sample. The errors in this procedure are described, and the true salinity of a number of samples has been measured. Layering in the brines is discussed. Some practical difficulties of sampling in the deeps are outlined.

Introduction

The Red Sea brines are a geochemical phenomenon, and a discussion of the hydrographic properties of the brines must necessarily be crude by normal oceanographic standards. The site was discovered through the interaction of the brines with their environment, producing small but significant changes in the Red Sea bottom water surrounding the area, and these anomalies were detected through the application of standard hydrographic techniques. Thus, a study of the hydrographic properties of the brines is essential to an understanding of their discovery and also their stability and physico-chemical properties.

The deep water of the Red Sea is contained in an enclosed basin; the only communication with the open ocean is in the south through the Straits of Bab el Mandeb. Here the sill depth is shallow, about 125m at its shallowest point. The surface circulation of the Red Sea, above 200m, is controlled mainly by evaporation (Neumann and McGill, 1962). Below this depth the entire basin is filled with warm saline water of remarkably constant temperature and salinity: the temperature is close to 21.7°C and the salinity 40.6 per mill. Wide deviations from these values do occur in the areas surrounding the water mass, in the Gulfs of Suez and Aqaba, and most certainly in the shallow pools bordering the sea. The dissolved oxygen content of the water mass is low (about 2.0 ml/l).

Thus, the hot brines are an anomaly in the midst of an extremely stable hydrographic environment. Mixing of the hot saline water with Red Sea bottom water, either by diffusion or a more active process, has been observed by several expeditions since the initial Albatross station in 1948. These data (Fig. 1) have been discussed by Miller *et al.* (1966) and Riley (1967).

Difficulties of Working in the Area

The total working area covers less than 150km², and to get an instrument cast into a particular deep calls for extremely precise ship handling and constant reference to buoys and echo sounder. Abruptness of the gradients at the brine interface makes placing of Nansen bottles extremely important. Without taking into account inaccuracies of wire metering and change of depth of the instruments due to wire angle, one may very easily miss the crucial mixing between normal sea water and the brine. This can be

* Woods Hole Oceanographic Institution Contribution No. 2185.

avoided by close bottle spacing and use of a pinger on the bottom of the hydrographic cast.

The surface currents in the central Red Sea are variable and unpredictable, though often setting from west to east or east to west. Frequently, when a cast is exactly placed for maximum effect, by the time the messengers have tripped the lower bottles, the cast has dragged out of the brine and is imperilled by a sharply rising bottom. The same hazards are encountered with coring gear and thermistor probes, cameras and pingers. A great deal of time can be expended in maneuvering ship and resetting casts.

Occasionally, in detailed examination of the brines, there have appeared data which give evidence of thermometer malfunction or poor salinity technique. It has become evident that this is not necessarily so, that because of sharp layering and mixing at interfaces (Table 1) it is possible to get true temperatures and salinities that lie wide of the TS curve in either direction. All these anomalies will be found to be at depths where turbulence rules. The rotation of a tripped Nansen bottle can radically change its environment in these layers.

Hydrography

In 1966 **CHAIN** made thirty-one hydrographic casts in the Discovery and Atlantis II Deep areas (Fig. 2). The high brine temperatures were measured with Richter and Wiese 60°C unprotected thermometers converted to protected thermometers for this purpose. These were the same thermometers used by **ATLANTIS II** in 1965. These were calibrated before, between and after the **ATLANTIS II** and **CHAIN** cruises. Calibration differences were of the order of 0.01°. Since only two of the thermometers were available, many casts were necessary to delineate the water structure.

Table 1   Salinity Increase Between Different Depths

Atlantis II Deep sta. 727		Discovery Deep sta. 717	
1,990m			
	6‰		
2,000m		1,999m	
			0‰
	18‰	2,004m	
			8‰
2,010m		2,009m	
	83‰		7‰
2,015m		2,014m	
	0‰		12‰
2,020m		2,019m	
	0‰		21‰
2,025m		2,024m	
	0‰		45‰
2,030m		2,029m	
	0‰		4‰
2,035m		2,034m	
	0‰		15‰
2,040m		2,039m	
	0‰		120‰
2,045m		2,044m	
	170‰		36‰
2,050m		2,051m	
	0‰		0‰
2,055m		2,053m	

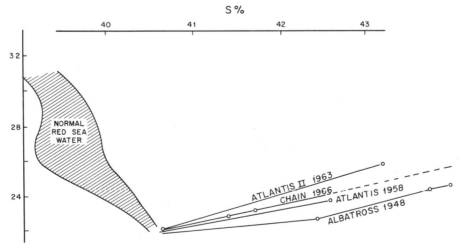

Fig. 1. Typical temperature-salinity envelope for the Red Sea with anomalies noted by some previous expeditions.

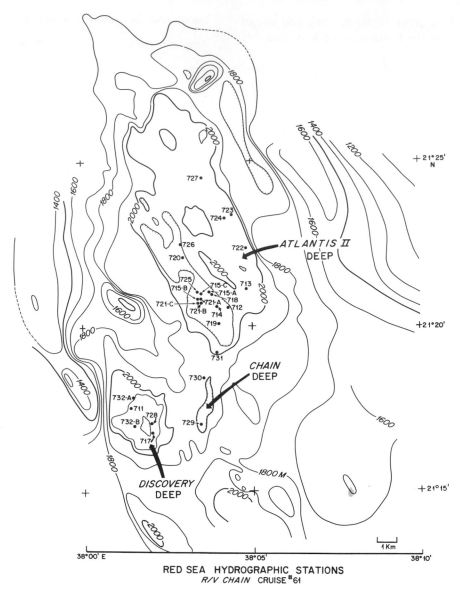

RED SEA HYDROGRAPHIC STATIONS
R/V CHAIN CRUISE #61

Fig. 2. Location of hydrographic casts made in the hot brine area during the 1966 **CHAIN** cruise. Contours in meters.

Depths of samples within the brines have been extrapolated from thermometric depths above the brine, taking wire angle (usually slight) into account. Unprotected thermometers of sufficient range to determine thermometric depth in the 56° brine are not available. It is estimated that the depths are accurate to within ±5m.

Conductive salinities were measured by the Schleicher-Bradshaw salinometer. Samples with salinities greater than 48 per mill were diluted with distilled water to bring them within the range of the instrument. The procedure was to pipet an aliquot of the sample (25ml) into an Erlenmeyer flask and to add an equal volume of distilled water. If the sample was still too saline to be measured, further similar dilutions were made: one part in eleven dilutions were made for the most saline samples. In spite of the inaccuracies introduced by shipboard operation the reproducibility of the procedure was about ±0.5 per mill.

Dissolved oxygen determinations were made by using the Winkler method. The values between normal sea water and the heaviest brines are probably suspect: the high concentrations of salts and heavy metals would probably interfere with both the scavenging action of the manganous precipitate and the subsequent titration. The dissolved oxygen content of the brines was very low or zero and below the useful working range of the Winkler method.

Gravimetric Salinities

The great majority of samples collected were diluted, and their salinity measured by conductive salinometer as described above. The virtue of this procedure is that results can be obtained quickly and compared with those of other workers who have used the same technique. Two large errors, however, are introduced by this procedure: dilution should be by weight and not by volume, and the ionic composition of the sample should be identical with the calibrating solution (standard sea water). For most oceanic samples this is so, but the Red Sea brines differ markedly from normal sea water in ionic composition, and a salinity value computed from a conductivity or a chlorinity determination will be in error.

The chemical composition of the Red Sea brines has been discussed by Brewer et al. (1965a and b) and Miller et al. (1966). Both papers have drawn attention to the great difference between these brines and the normal sea water with which they are surrounded. Table 2 lists data for the major ion composition of a sample from the Discovery Deep and for a sample of normal sea water; samples from the 56° water of the Atlantis II Deep have different minor element concentrations, but the gross composition is similar (Brewer and Spencer, 1969).

Swallow and Crease (1965) reported salinity data obtained by dilution of a sample by weight with distilled water and conductivity measurement, and they gave values of ca. 271 per mill for the Discovery Deep

brine. Munns et al. (1967) reported salinity data for a large number of samples from both the Atlantis II and the Discovery Deeps. These values were obtained by volumetric dilution with distilled water and conductometric determination. The data are given at the end of this chapter and show a salinity of ca. 320 per mill for the most saline samples (Tables 8 and 9).

Brewer et al. (1965a and b) made a complete chemical analysis of the Discovery Deep brine and reported a direct measurement of the salinity determined gravimetrically. The true salinity of a large volume sample was found to be ca. 255g/kg while summation of the analyses for individual ions gave a value of 256g/kg.

In this chapter the value obtained by volume dilution and conductivity measurement will be referred to as salinity, whereas the true value for the weight of dissolved solids will be referred to as the gravimetric salinity. On the TS plot (Fig. 6) salinities have been converted to approximate total dissolved salts by weight.

Gravimetric salinities were determined by the method of Morris and Riley (1964). An aliquot of the brine sample, varying from 1½ml to 10ml depending upon the salinity, was weighed into a tared platinum crucible containing sodium fluoride. The sample was evaporated carefully to small volume under an infrared lamp, and ethyl alcohol was added to precipitate small crystals of salt and thus prevent decrepitation on heating. The salts were heated to 650°C in a muffle furnace, dried in an evacuated desiccator and weighed. Corrections were made to reduce the weights to vacuo, and a small correction for the bromide present was made to comply with the Knudsen formula. This last correction has doubtful validity in these unusual brines. The correction, however, was applied so as not to introduce an inconsistency in the results as the samples tended towards true sea water. Chlorinity determinations were carried out by gravi-

Table 2   The Major Components of a Sample of the Discovery Deep Brine Compared to Normal Sea Water (g/kg)

	Na+	K+	Ca++	Mg++	Sr++	Cl-	Br-	SO4=
Discovery Brine	92.68	2.13	4.65	0.79	0.04	155.1	0.13	0.70
Normal Sea Water (S‰ = 35‰)	10.76	0.39	0.41	1.29	0.008	19.35	0.066	2.71

metric titration using potassium chromate indicator.

The gravimetric salinity data show that the values derived by conductivity measurements, or from chlorinity determinations, are too high (Tables 3 and 4). An exception to this was noted in three samples from the Discovery Deep between 1,995 and 2,004m. It was surmised that this might reflect an alteration of the ionic composition due to the dissolution of marine organisms floating on the strong density interface above the brine. Since chemical analysis failed to detect significant differences, the results are probably due to an error in the conductivity measurement.

Samples from the Atlantis II Deep formed a reddish brown precipitate on exposure to air, caused by the oxidation of ferrous iron to ferric iron. (Nothing is known of the quantity of particulate matter contained in the brine *in situ* and this will introduce an uncertainty). Gravimetric salinity determinations were made on the filtrate, hence true salinity values for the Atlantis II Deep are probably slightly higher than those reported by ca. 0.01 to 0.05 per mill in the 56° brine.

Morcos and Riley (1966) investigated the salinity, chlorinity and density relationships of waters from the Suez Canal region in the salinity range 42–48 per mill and proposed the use of the empirical equation S per mill $= 1.802$ Cl per mill for conversion of chlorinities to salinities in this limited range. No satisfactory relationship was found for water of high salinity that had been contaminated by dissolution of evaporite beds; similarly, in the Red Sea brines no simple relation holds. The brines progressively become proportionately richer in chloride with depth, the ratio to salinity

Table 3   Comparison of Gravimetric and Conductometric Salinities, and Chlorinity, in the Atlantis II Deep Station No. 721

Depth	Temperature	Chlorinity	Conductometric Salinity	Gravimetric Salinity
1,944	22.20	22.60	40.80	40.98
1,994	–	24.07	43.27	43.18
2,007	–	34.04	59.52	59.63
2,017	–	80.13	151.50	134.99
2,027	–	80.70	151.25	134.99
2,037	–	79.29	151.88	135.44
2,054	56.56	156.03	321.42	257.06
2,098	56.46	156.10	–	257.33

Table 4   Comparison of Gravimetric and Conductometric Salinities, and Chlorinity, in the Discovery Deep Station No. 717

Depth	Temperature	Chlorinity	Conductometric Salinity	Gravimetric Salinity
1,985	22.99	–	42.48	–
1,990	23.22	23.99	43.07	43.04
1,995	23.90	25.00	41.28	44.59
1,999	–	25.72	42.60	45.83
2,004	24.80	26.48	44.10	46.98
2,009	26.11	29.09	52.52	51.05
2,014	27.51	33.40	60.04	57.42
2,019	29.72	40.09	72.78	69.25
2,024	–	51.57	94.32	87.99
2,029	–	74.63	140.25	125.10
2,034	–	77.05	144.90	129.28
2,039	–	84.38	160.15	141.43
2,044	41.90	138.70	280.20	231.37
2,151	–	155.10	316.73	256.60
2,153	–	154.70	316.64	256.84
2,158	–	155.30	317.10	256.39
2,160	–	–	317.38	–

varying from ca. 1.80 for waters of salinity 42 per mill to ca. 1.65 for the most saline samples.

The plots of depth and temperature and depth and salinity (Figs. 3 and 4) were drawn from 119 temperature measurements greater than 22° and 188 salinity determinations greater than 41 per mill made during the 1966 **CHAIN** cruise. Some **ATLANTIS II** observations were also included. These show several noteworthy features. In the hotter brine of the Atlantis II Deep, there was an average temperature increase of 0.5° during an interval of twenty months, (Table 5), which suggests a tremendous input of heat, probably from below. No significant temperature change was found in the Discovery Deep. In neither of the Deeps was there any verifiable change of salinity. Averages of 24 samples of over 300 per mill from Discovery Deep and 41 samples from Atlantis II Deep, however, indicate that the latter has a higher salinity by 1.7 per mill (conductivity measurement).

A 44° layer about 30m thick occurs above the 56° brine in the Atlantis II Deep. This may be a transient feature. The implications of this layer are discussed by Turner (1969). Above the 44° layer in the Atlantis II Deep is a mixing layer of 120m between the 22° and 44° waters. That this is an active zone is shown by the varied tempera-

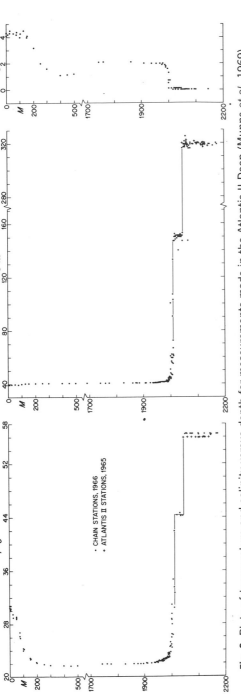

Fig. 3. Plots of temperature and salinity versus depth for measurements made in the Atlantis II Deep (Munns *et al.*, 1969).

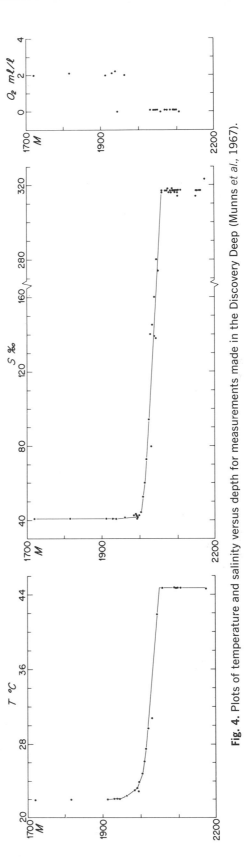

Fig. 4. Plots of temperature and salinity versus depth for measurements made in the Discovery Deep (Munns *et al.*, 1967).

Table 5   Temperature Changes 1964–1966

Location	Date	No. of Observations	Range	Temperature Spread	Average
Atlantis II Deep	1965	13	55.90–55.94	0.05	55.92
Atlantis II Deep	1966	15	56.31–56.56	0.25	56.48
Discovery Deep	1964	4	44.40–44.90	0.5	44.7
Discovery Deep	1966	8	44.65–44.75	0.10	44.72

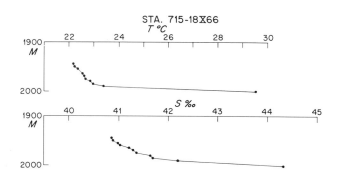

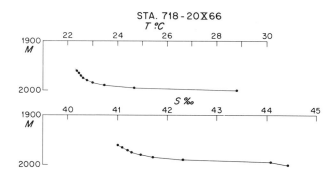

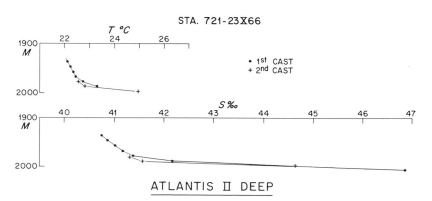

ATLANTIS II DEEP

Fig. 5. Temperature and salinity plots made on three different days in the Atlantis II Deep (Munns *et al.*, 1967).

Table 6   Depth of Salinity Interface

	Greatest Depth of Normal Water (m)	Minimum Depth of Highest Salinity (m)
R.R.S. DISCOVERY, 1964	1,945	2,030
R.V. ATLANTIS II, 1965	1,940	2,044
R.V. CHAIN, 1966	1,944	2,058

As **ATLANTIS II** had widely spaced bottles, and only one in the brine, these figures are not definitive.

ture and salinity plots of the transition on different days (Fig. 5). In each instance the shape of the temperature curve is similar to that of the salinity, showing that the variation is not caused by errors of salinometry or thermometry. The absolute depth of each cast may be in error but the relative depths of samples within a cast should be quite accurate. The variations in the plots suggest activity within this mixing layer.

A difference may exist in the depth of the Discovery Deep brine as found by the several expeditions (Table 6). This is the only indication of activity seen in this deep. A TS plot of the two deeps plus the two stations on the saddle between them is most illuminating (Fig. 6). One of the saddle stations, 731, closely follows the curve of the Atlantis II water while the other, 729, diverges widely to follow the Discovery water. Oddly enough, the water on the saddle below 2,000m (both stations) is con-

siderably cooler and less saline than that of either deep (Table 7). A reasonable assumption from these data is that vertical dissipation of heat from the various basins is primarily dependent on their depth and form of heating. Possibly the Atlantis II Deep, heated from below, periodically water spills across the saddle into the Discovery Deep, where it slowly cools and mixes with Red Sea bottom water. The saddle water, comparatively much greater in surface area to volume, would erode at an accelerated rate.

The Discovery brine is similar to that of the deeper Atlantis II brine with the exception of its 12° lower temperature. It has been suggested that the Discovery brine is cooling from the bottom, as **R.R.S. DISCOVERY** obtained a bottom temperature of 0.4° cooler than those above. Cooling has also been indicated by electronic instrumentation used by Pugh (1967) from **R.V. CHAIN.** It is, perhaps, more probable that

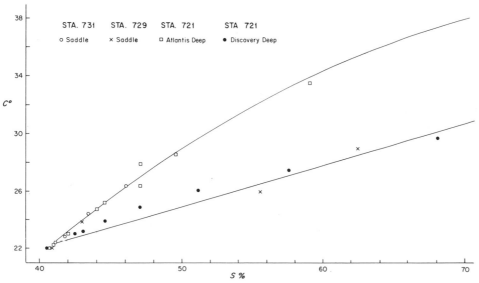

Fig. 6. TS diagrams of stations in the Atlantis II Deep, Discovery Deep and from the saddle area between the two deeps.

Table 7 Temperature and Salinity Measurements on the Saddle between the Deeps

	Atlantis II Deep Station 723		Saddle Station 731			Discovery Deep Station 717		Saddle Station 729	
	Temp.	Sal.	Temp.	Sal.		Temp.	Sal.	Temp.	Sal.
1,985	22.94	41.66	22.85	41.73					
2,000	28.62	50.76	24.41	43.67	2,000	~24	42.60	23.80	42.92
2,005	~33	61.98	26.38	46.96	2,005	24.80	44.10	—	—
					2,020	29.72	72.78	25.98	57.90

Underlined temperatures are extrapolated.

Table 8

Depth	Temperature	Salinity	Sigma-t	Sound Velocity m/s	Oxygen Content ml/l

CHAIN Station 725, October 30, 1966, Atlantis II Deep

Depth	Temperature	Salinity	Sigma-t	Sound Velocity m/s	Oxygen Content ml/l
1	30.77	39.213	24.64	1,551.9	4.30
10	30.79	39.197	24.62	1,552.1	4.18
25	29.01	39.469	25.44	1,549.0	4.34
75	26.20	39.302	26.24	1,543.6	4.15
99	24.36	39.87	27.24	1,540.5	3.97
149	22.82	40.27	28.01	1,538.0	2.97
198	22.17	40.41	28.30	1,537.4	3.22
298	21.78	40.53	28.50	1,538.2	1.58
397	21.71	40.58	28.56	1,539.7	1.12
497	21.70	40.63	28.61	1,541.4	1.20
597	21.71	40.56	28.55	1,543.1	1.42
797	21.73	40.62	28.59	1,546.7	1.67
997	21.76	40.64	28.60	1,549.9	1.91
1,169	21.79	40.61	28.56	1,552.8	2.01
1,366	21.84	40.60	28.54	1,556.2	2.05
1,562	21.89	40.60	28.53	1,559.6	2.09
1,759	21.95	40.62	28.52	1,563.1	2.07

CHAIN Station 727, October 31, 1966, Atlantis II Deep

Depth	Temperature	Salinity	Sigma-t	Sound Velocity m/s	Oxygen Content ml/l
1,980	22.74	41.40	28.89	1,569.8	2.01
1,990	23.02	41.72	29.05	1,571.0	1.80
2,000	26.43	47.99	32.77	1,586.4	0.54
2,010	—	66.7 *	—	—	0.11
2,015	—	150.4 *	—	—	0.00
2,020	—	150.4 *	—	—	0.06
2,025	—	149.2 *	—	—	0.06
2,030	—	150.1 *	—	—	0.05
2,035	—	150.1 *	—	—	0.05
2,040	44.26	150.5 *	—	—	0.06
2,045	—	147.0 *	—	—	0.05
2,050	—	318.6 *	—	—	0.00
2,055	—	317.8 *	—	—	—
2,065	—	318.6 *	—	—	—
2,075	56.49	318.6 *	—	—	0.00

* Shipboard processed values as described in text.

Table 9 CHAIN Station 732, A and B, November 8, 1966, Discovery Deep

Depth	Temper- ature	Salinity	Sigma-t	Sound Velocity m/s	Oxygen Content ml/l
1,720	21.92	40.60	28.52	1,562.3	2.00
1,817	22.00	40.61	28.50	1,564.2	2.14
1,913	22.03	40.67	28.54	1,565.9	2.04
1,938	22.10	40.70	28.54	1,566.5	2.19
1,944	22.03	40.67	28.54	1,566.4	0.00
1,962	22.38	41.07	28.74	1,568.1	1.96
1,964	22.87	42.10	29.39	1,571.1	—
2,043	—	138.4 *	—	—	0.06
2,058	44.70	316.1 *	—	—	0.00
2,093	44.73	316.4 *	—	—	0.08
2,106	44.73	317.0 *	—	—	0.00

* Shipboard processed values as described in text.

as the 44° water loses heat at the top and sides increased density causes it to pool in the bottom of the basin.

These brines far exceed the limits of oceanographic formulae for potential temperature and density. Crude measurement of density was made by means of thermal expansion of known weights of several samples with the following results: bottom water, 41.6 per mill, at 22°, Sigma-t = 28; Atlantis II 44° brine at 44.35°, Sigma-t = 89; Atlantis II 56° brine at 56.5°, Sigma-t = 178; Discovery 44° brine at 44.72°, Sigma-t = 183.

Table 8 shows two stations in Atlantis II Deep which are typical of the water column. Table 9 presents a single station in Discovery Deep, in the deep water only. There is some scatter of temperature within the brine layers (Table 5) which appears to be greater in 1966 than in 1965. The scatter in 1966 of 0.25° in Atlantis II Deep is an order of magnitude greater than the thermometric accuracy, again indicating present day activity within the brines of this deep.

References

Brewer, P. G., J. P. Riley, and F. Culkin: Chemical composition of the hot salty water from the bottom of the Red Sea. Nature, **206**, 1345 (1965a).

——, ——, ——: The chemical composition of the hot salty water from the bottom of the Red Sea. Deep-Sea Res., **12**, 497 (1965b).

Brewer, P. G., and D. W. Spencer: A note on the chemical composition of the Red Sea deep brines. *In: Hot brines and recent heavy metal deposits in the Red Sea*, E. T. Degens and D. A. Ross (eds.). Springer-Verlag New York Inc., 174–175 (1969).

Miller, A. R., C. D. Densmore, E. T. Degens, J. C. Hathaway, F. G. Manheim, P. F. McFarlin, R. Pocklington, and A. Jokela: Hot brines and recent iron deposits in deeps of the Red Sea. Geochim. et Cosmochim. Acta, **30**, 341 (1966).

Morcos, S. A., and J. P. Riley: Chlorinity, salinity, density, and conductivity of sea water from the Suez Canal region. Deep-Sea Res., **13**, 741 (1966).

Morris, A. W., and J. P. Riley: The direct gravimetric determination of the salinity of sea water. Deep-Sea Res., **11**, 899 (1964).

Munns, R. G., R. J. Stanley, and C. D. Densmore: Hydrographic observations of the Red Sea brines. Nature, **214**, 1215 (1967).

Neumann, A. C., and D. A. McGill: Circulation of the Red Sea in early summer. Deep-Sea Res., **8**, 223 (1962).

Pugh, D. T.: Origin of hot brines in the Red Sea. Nature, **214**, 1003 (1967).

Riley, J. P.: The hot waters of the Red Sea bottom and their related sediments. Oceanogr. Mar. Biol. Ann. Rev., **5**, 141 (1967).

Swallow, J. C., and J. Crease: Hot salty water at the bottom of the Red Sea. Nature, **205**, 165 (1965).

Turner, J. S.: A physical interpretation of the observations of hot brine layers in the Red Sea. *In: Hot brines and recent heavy metal deposits in the Red Sea*, E. T. Degens and D. A. Ross (eds.). Springer-Verlag New York Inc., 164–173 (1969).

Temperature Structure of the Red Sea Brines *

DAVID A. ROSS

Woods Hole Oceanographic Institution
Woods Hole, Massachusetts

Abstract

Temperature measurements obtained by a temperature telemetering pinger show a stratification in the hot brines of the Red Sea.

In the Atlantis II Deep a 44°C water, which starts at a depth of 2,009m and has a thickness of 28m, overlies the bottom 56°C water. The interface between these layers is sharp and only about 5m thick.

In the Discovery Deep a 4m thick 36°C water overlies a bottom 44°C water. The interface between these layers is not as sharp as that of the Atlantis II Deep.

The Chain Deep has water of at least 34°C. Layering may be present; however, the deepest part of this hole was not reached.

Introduction

Previous expeditions to the Red Sea had detected the existence of hot brines. Slightly anomalous temperatures, 3 or 4°C above the normal Red Sea bottom water, were measured during the Swedish **ALBA-TROSS** Expedition (Bruneau *et al.,* 1953), on an **ATLANTIS** cruise (Neumann and Densmore, 1959) and on an **ATLANTIS II** cruise (Miller, 1964). The British research vessel **R.R.S. DISCOVERY** in 1964 sampled 44°C water in what is now called the Discovery Deep (Swallow and Crease, 1965). Water temperatures as high as 56°C were observed during an **ATLANTIS II** cruise in an area that is now called the Atlantis II Deep (Miller *et al.,* 1966). Observations made aboard the German vessel **R.V. METEOR** confirmed these observations and also suggested that some internal temperature structure existed within the brines (Krause and Ziegenbein, 1966). The temperature data from the **METEOR** Expedition differed slightly from that of the other expeditions: they measured temperature with an electronic device which apparently did not function perfectly in the hot water (Ziegenbein, personal communication, 1967).

One of the objectives of the **CHAIN** cruise was to examine in detail the temperature structure of the known deeps and to explore for other possible deeps by looking for areas of anomalous temperature. Two techniques were used for this objective: conventional hydrographic equipment using Nansen bottles with high temperature recording reversing thermometers, and an electronic temperature telemetering pinger. The results of the former (Munns *et al.,* 1967; Brewer *et al.,* 1969) provided considerable data on the temperature structure. This data consisted only of discrete observations from each lowering. The electronic temperature telemetering pinger had the advantage of obtaining a continuous measurement of the temperature structure. This device was also valuable as a survey tool and was used to discover the Chain Deep (Ross and Hunt, 1967).

The objective of this paper is to present the results of the temperature telemetering pinger measurements in the Red Sea. A detailed discussion of the possible causes of the temperature stratification is presented in this volume by Turner (1969), Pugh (1969), and Erickson and Simmons (1969).

* Woods Hole Oceanographic Institution Contribution No. 2186.

Methods

A Benthos, Inc., Depth Telemetering Pinger was modified to measure temperature by replacing the depth-sensing element with a thermistor. The temperature telemetering pinger (TTP) emits two sound pulses within a one-second time interval. The first ping (P_1) is emitted at precisely a one-second time interval. The second ping (P_2) occurs after P_1 at a time interval that varies with temperature. The time difference between the pings is maintained by a thermistor (20–70°C range for the Red Sea work) that produces a resistance (varying with temperature) which controls the telemetering electronics. The pings from the TTP were received by the **CHAIN's** 12kc echo sounder and recorded on a Precision Graphic Recorder (PGR). The TTP was used independently as a temperature measuring device and as a regular pinger with other instruments to indicate height above the bottom, at the same time also measuring temperature.

There are three limitations to the data obtained by the TTP. The first is the accuracy of the temperature measurements which, because of the large temperature range of the thermistor, is ±1.25°C. In most instances, however, a more exact temperature value may be obtained by equating TTP results with those obtained by the reversing thermometers of the hydrocasts. Some temperature measurements of intermediate water layers were not sampled by hydrocasts, and therefore the above accuracy applies. One major disadvantage of the large range of temperature accuracy is that the instrument cannot be used to detect small, subtle temperature changes, such as those that apparently occur near the bottom of the Discovery Deep (Pugh, 1967, 1969) and in parts of the Atlantis II Deep (Erickson and Simmons, 1969).

The second limitation of the TTP measurements, and also those of conventional hydrocasts, is the difficulty in knowing exactly the position of the instrument. The determination of the depth of the instrument was made by subtracting its elevation above the bottom as observed on the PGR record from the water depth as indicated by the ship's echo sounder. A possible error in depth of the instrument and therefore of the different water layers may result if the pinger is not directly beneath the ship but in deeper or shallower water. Since numerous lowerings were made into the deeps, an average depth of the layers could be determined. The standard deviation of these depths (Table 1) is about 5 parts in 2,000 (or 0.25 per cent). Another measure of water depth was made by using an inverted echo sounder, lowered with the TTP, which measures the thickness of the overlying water. These results are relatively close to those determined by the TTP.

The thermistor had a response time of 0.7 seconds in which it would indicate 63 percent of a temperature change. A lowering rate of 40m/minute or less was used in the brine area. The thermistor response time would therefore produce a depth error of less than a meter for large temperature changes.

The third limitation to the data is that the depths and thicknesses of the layers are not absolute. A correction for the speed of sound in normal Red Sea water has been made (Matthews, 1939); however, no correction for the increase in sound velocity within the brines has been applied. The velocity of sound in the brines certainly is higher than that of normal Red Sea water; however, no corrections for such abnormal salinity and temperature values are available. The thicknesses and depths of the different temperature layers would be increased if the corrections were available. Pugh (1969) has extrapolated the formulae of Horton (in Vigoureux and Hersey, 1962) beyond the range of their established validity to the conditions of the Atlantis II and Discovery Deeps and obtained velocities of 1,910m/sec and 1,890m/sec respectively for the bottom waters. The velocity of sound in the normal Red Sea water is about 1,550m/sec.

Results

Atlantis II Deep

Lowerings into the Atlantis II Deep established that two distinct layers (Fig. 1a) of water existed: a 44°C water and a 56°C

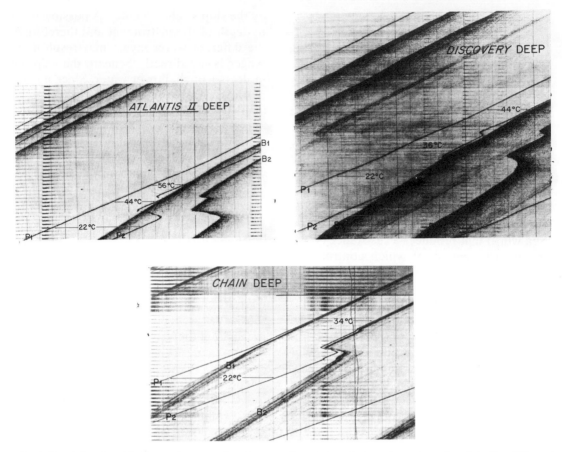

Fig. 1. Precision Graphic Recorder records of temperature telemetering pinger lowerings into the different deeps. The time difference between P_1 and P_2 indicates the temperature. B_1 and B_2 are the bottom returns from the two pulses P_1 and P_2. The distance between a pulse and its bottom return indicates the height of the instrument above the bottom. The jumps in P_2 in the last figure do not indicate a step-like pattern in the temperature distribution but are caused by stopping and starting of the winch.

water. More precise measurements by reversing thermometers are 44.2°C and 56.5°C (Munns *et al.*, 1967; Brewer *et al.*, 1969). The depths and thicknesses of these layers are shown in Table 1. The transition zone between the 44° and 56° water is very sharp and is probably 5m or less in thickness. Despite several attempts, no hydrographic samples were obtained from this zone. The significance of this sharp boundary is discussed in Turner (1969) and Erickson and Simmons (1969).

The depths reported here differ slightly from the hydrographic data of Munns *et al.* (1967) and Brewer *et al.* (1969). These discrepancies are probably due to the different depth determining techniques. The hydrographic depths were extrapolated from thermometric depths above the brine, tak-

ing wire angle into consideration. The depth technique used in the TTP observations has been previously described.

Discovery Deep

Two water layers were also found in the Discovery Deep (Fig. 1b). The upper 36°C layer, which was clearly observed on 7 of the 9 lowerings into this deep, was only about 4m thick. This layer was not sampled by hydrographic techniques. Its consistency in temperature and depth indicates that it is a real layer and is not due to an instrumental error. The transition zone between the 36°C layer and the underlying 44°C water (44.7°C by reversing thermometers) is more subtle than that of the 44°–56° boundary in the Atlantis II Deep.

Table 1 Depths to Different Layers (All Corrected Meters, but No Correction for the Increase of the Speed of Sound in the Brine)

	Average Depth TTP (m)	Standard Deviation (obs.)		Average Depth Inverted Echo Sounder	
Atlantis II Deep					
Start of hot water	1,984	5.3	(32)	1,977	(3)
Top of 44° water	2,009	4.4	(32)	2,010	(3)
Bottom of 44°	2,037	5.5	(28)		
Thickness of interface	5	2.4	(28)		
Top of 56° water	2,042	5.7	(28)	2,048	(1)
Discovery Deep					
Start of hot water	1,986	8.8	(9)	1,994	(2)
Top of 36° water	2,023	6.4	(9)	2,025	(2)
Bottom of 36°	2,027	2.6	(9)		
Thickness of interface	15	5.1	(9)		
Top of 44° water	2,042	8.4	(9)	2,041	(1)

Chain Deep

The Chain Deep (Fig. 1c) was discovered during the **CHAIN** 61 cruise with the TTP (Ross and Hunt, 1967). A maximum temperature of 34°C was detected at a depth of 2,042m. The temperature at this point, which was not in the deepest part of the deep, was still increasing. A higher temperature is therefore probable.

Discussion

The temperature structure in the Atlantis II, Discovery and Chain Deeps are similar in that the contact between the hot brine and the overlying normal Red Sea water is a gradual one. The Atlantis II and Discovery Deeps have two distinct layers: a 44° and 56°, and a 36° and a 44° respectively. Turner (1969) has suggested that the intermediate layers (the 44° in the Atlantis II, the 36° in the Discovery) form by mechanical mixing processes, such as internal waves, with some additional transfer of heat from the deeper, warmer layers. The deeper, 44° water of the Discovery Deep probably represents some past overflow of water from the Atlantis II Deep, which is the suspected source area of the hot water. The available data suggest that the brines of these two deeps are not connected at present (Fig. 2).

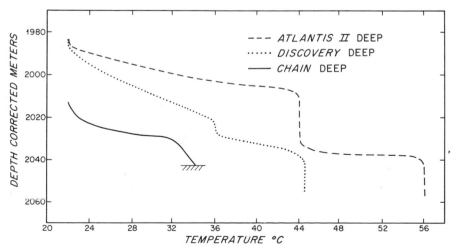

Fig. 2. General temperature distribution in the 3 Deeps (Ross and Hunt, 1967). Data obtained by temperature telemetering pinger lowerings during **CHAIN** cruise 61.

Table 2 Areas and Volumes of Different Layers in the Hot Brine Area (Assuming Slopes Are Straight Lines Between Layers)

	Area (at midpoint of layer)	Volume
Atlantis II Deep		
Start to top of 44°	59.9km²	1.49km³
44° to 56°	49.3km²	1.63km³
56° to bottom	–	2.31km³
Discovery Deep		
Start to top of 36°	10.9km²	0.40km³
36° to 44°	8.9km²	0.17km³
44° to bottom	–	0.74km³

The evidence of Turner (1969), Erickson and Simmons (1969), Bischoff (1969), Pugh (1967, 1969) and Cooper and Richards (1969) suggest that the Atlantis II Deep is the source of the hot brine. The recent increase in temperature of the 56° water (0.5°C increase in 20 months, Munns *et al.,* 1967) indicates that heat is presently being supplied to this deep. The possible mechanisms of heat addition are hydrothermal discharge and thermal convection from a buried heat source through the sediments (Erickson and Simmons, 1969). The same authors believe that the latter is occurring now but was probably preceded by hydrothermal discharge.

The areas and volumes of the different layers are given in Table 2. These values, combined with the chemical analyses of the brine, could be used to make a rough appraisal of the heavy metals in these waters. The variability of the heavy metal composition in the brine, however, is enough to make any appraisal questionable.

References

Bischoff, J. L.: Red Sea geothermal brine deposits: their mineralogy, chemistry, and genesis. *In: Hot brines and recent heavy metal deposits in the Red Sea,* E. T. Degens and D. A. Ross (eds.). Springer-Verlag New York Inc., 368–401 (1969).

Brewer, P. G., C. D. Densmore, R. Munns, and R. J. Stanley: Hydrography of the Red Sea brines. *In: Hot brines and recent heavy metal deposits in the Red Sea,* E. T. Degens and D. A. Ross (eds.). Springer-Verlag New York Inc., 138–147 (1969).

Bruneau, L., N. G. Jerlov, and F. Koczy: Rep. Swedish Deep-Sea Expedition, **3,** 4. Appendix, Table 2 (1953).

Cooper, J. A., and J. R. Richards: Lead isotope measurements on sediments from Atlantis II and Discovery Deep areas. *In: Hot brines and recent heavy metal deposits in the Red Sea,* E. T. Degens and D. A. Ross (eds.). Springer-Verlag New York Inc., 499–511 (1969).

Erickson, A. J., and G. Simmons: Thermal measurements in the Red Sea hot brine pools. *In: Hot brines and recent heavy metal deposits in the Red Sea,* E. T. Degens and D. A. Ross (eds.). Springer-Verlag New York Inc., 114–121 (1969).

Krause, G., and J. Ziegenbein: Die Struktur des heissen salzreichen Tiefenwassers im zentralen Roten Meer. METEOR Forschungsergebnisse, Reihe A, **1,** 53 (1966).

Matthews, D. J.: Tables of the velocity of sound in pure water and sea water for use in echo-sounding and sound ranging, 2nd ed. London Hydrographic Dept., Admiralty, 52 p. (1939).

Miller, A. R.: Highest salinity in the world ocean? Nature, **203,** 590 (1964).

———, C. D. Densmore, E. T. Degens, J. C. Hathaway, F. T. Manheim, P. F. McFarlin, R. Pocklington, and A. Jokela: Hot brines and recent iron deposits in deeps of the Red Sea. Geochim. et Cosmochim. Acta, **30,** 341 (1966).

Munns, R. G., R. J. Stanley, and C. D. Densmore: Hydrographic observations of the Red Sea brines. Nature, **214,** 1215 (1967).

Neumann, A. C., and C. D. Densmore: Oceanographic data from the Mediterranean Sea, Gulf of Aden and Indian Ocean. Woods Hole Ocean. Inst., Ref. 60-2, unpub. manuscript (1959).

Pugh, D. T.: Origin of hot brines in the Red Sea. Nature, **214,** 1003 (1967).

———: Temperature measurements in the bottom layers of the Red Sea brines. *In: Hot brines and recent heavy metal deposits in the Red Sea,* E. T. Degens and D. A. Ross (eds.). Springer-Verlag New York Inc., 158–163 (1969).

Ross, D. A., and J. M. Hunt: Third brine pool in the Red Sea. Nature, **213,** 687 (1967).

Swallow, J. C., and J. Crease: Hot salty water at the bottom of the Red Sea. Nature, **205,** 165 (1965).

Turner, J. S.: A physical interpretation of the observations of hot brine layers in the Red Sea. *In: Hot brines and recent heavy metal deposits in the Red Sea,* E. T. Degens and D. A. Ross (eds.). Springer-Verlag New York Inc., 164–173 (1969).

Vigoureux, P., and J. B. Hersey: Sound in the sea. *In: The Sea,* **1,** M. N. Hill (ed.). Interscience Publishers, New York (1962).

Suspended Matter in the Red Sea Brines and Its Detection by Light Scattering *

WILLIAM B. F. RYAN
EDWARD M. THORNDIKE
MAURICE EWING

Lamont Geological Observatory of Columbia University
Palisades, New York

DAVID A. ROSS

Woods Hole Oceanographic Institution
Woods Hole, Massachusetts

Abstract

A dense layer of suspended particulate matter exists below a depth of 1,900m in the Atlantis II Deep in the Red Sea. This layer was detected with a light scattering meter (nephelometer) at two locations within this deep and was found to conform generally to the zone of hot brines. The nepheloid layer scatters light with a constant intensity except for a few very thin internal layers of greater light scattering which correlate with the interface between the different temperature and salinity brines. The intensity of light scattering is found to increase gradually over an interval of eighty meters in the transition zone between the hot brine and the normal Red Sea deep water. The particles which produce the light scattering are interpreted as the colloidal suspensions and mineral precipitates created by the interaction of the reducing, acidic, and metal-containing hot brine with the oxidizing and alkaline overlying normal Red Sea deep water.

The Nephelometer

The Lamont Geological Observatory nephelometer (Thorndike and Ewing, 1967) consists of three basic units: a light source, a deep-sea camera and a baffle and attenuator. The arrangement of these units is shown in Fig. 1. The light source is a small tungsten filament lamp operated by nickel-cadmium cells. The baffle and attenuator unit prevents the direct beam of light from reaching the camera lens, except for a weakened part of the beam which has passed through the attenuator. The camera records on a continuous strip of film the direct-attenuated beam of light along with light that has been scattered by particles in the sea water. The record of the attenuated beam is used as a measure of any changes in the output of the light source or absorption of the light by the sea water.

In the instrument used aboard **CHAIN** on cruise 61 the nephelometer was mounted between the strobe light and the camera unit of a standard Ewing-Thorndike deep-sea multiple-exposure camera (Thorndike, 1959). The lowermost housing contained the strobe light for the bottom camera, followed by housings which contain the nephelometer camera, the nephelometer light source, and then the bottom camera. The baffle and attenuator unit is mounted midway between the nephelometer camera and light source. A compass is secured to an arm attached to the front of the unit per-

* Lamont Geological Observatory Contribution No. 1049. Woods Hole Oceanographic Institution Contribution No. 2187.

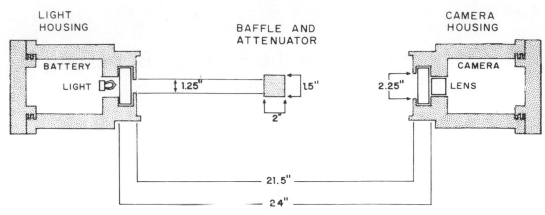

Fig. 1. Schematic sketch of the Lamont Geological Observatory nephelometer.

mitting the orientation of the obtained sea-floor photographs.

Operation at Sea

During both lowerings in the Atlantis II Deep (Table 1) the instrument was attached to standard hydrographic wire 50m below a temperature telemetering pinger (Ross, 1969). An internal watch puts one minute time marks on the nephelometer film. In order to determine the depth of the instrument while lowering and raising the package to and from the sea floor, an external clock is used to monitor the amount of wire out on the supporting cable. The two time pieces must, of course, be synchronized.

For the two lowerings in the Atlantis II Deep the bottom hit was not detected and the entire unit was submerged beneath the sediment-water interface. Since the exact moment of hit was to be used to synchronize the two clocks, there is a time uncertainty which is estimated at less than 30 seconds. The instrument was lowered and raised through the brines at a speed varying from 10 to 50m per minute. Thus there may be an error of about 25m in determining the height of the instrument off of the sea floor.

Table 1

Nephelometer Lowering	CHAIN 61 Station	Latitude	Longitude	Depth in Meters
39	86	21° 21′N	38° 03.5′E	2,167
40	133	21° 21′N	38° 04′E	2,164

Measuring the Film

The nephelometer film is measured with a recording optical densitometer. The densitometer recordings are plotted against depth in corrected meters and the values from 0 to 5 are arbitrary units that represent the range from the least scattering to the greatest scattering. Fig. 2 shows portions of the four film records obtained in the Atlantis II Deep.

The Nepheloid Layer in the Atlantis II Deep

Jerlov (1953) first reported the presence of a high concentration of suspended sediment in the deep water of the central Red Sea. At **ALBATROSS** station 254 a few miles southeast of the hot brine areas he detected a zone of increasing scattering from 1,900m to the bottom at 1,936m. This is the station where the anomalous high temperatures were noted for the first time (Bruneau *et al.*, 1953).

The densitometer records from the two **CHAIN** 61 stations confirm the presence of the high concentration of suspended sediment below 1,900m. The plot of light scattering vs. depth for lowering 39 (**CHAIN** station 86) is shown in Fig. 3. The high near surface readings are caused by ambient light in the surface waters rather than by scattered light.

The water column from 300 to 1,900m is remarkably free from detectable light scattering material. However, from 1,900m to the bottom at 2,167m a dense nepheloid layer is present. The concentration of light scattering material increases gradually from 1,900m to 1,980m and then remains rather constant to the bottom. Within this deep layer there are three or possibly more (see film records 40u and 40d in Fig. 2) discrete zones of significantly greater light scattering. Since the width of the dark bands on the film is exactly the width of the slit aperture of the nephelometer camera, it can be estimated that the bands have resulted from the instrument passing through a dense zone of suspended matter less than one or two meters in thickness.

The direct-attenuated beam recording is also plotted in Fig. 3 (marked a). This trace also shows these thin dark bands because in this instance the thin layer of extreme concentrations of suspended matter produces multiple scattering causing additional light to be brought into the path of the direct beam. From 2,040 to 2,167m there appears to be a slight genuine increase

in the absorption of the direct beam by the brine solution.

The amount of light recorded in the direct beam presumably by multiple scattering on passage through the thin zones is remarkable. Multiple scattering of this intensity is not ordinarily observed, even in the passage of nephelometers through dense mud clouds produced when the instruments strike soft mud bottoms.

In nearly all instances where the nephelometer has come into contact with a mud bottom, dark streaks are created on the film upon raising the instrument as the result of detecting loose sediment washing from the instrument package. It is surprising that in these two lowerings where the entire camera was submerged beneath the sediment-water interface, no such streaks were produced. However, a small mud cloud was detected on film strips 39*u* and 40*u* (i.e., the horizontal dark bands at the bottom of the film

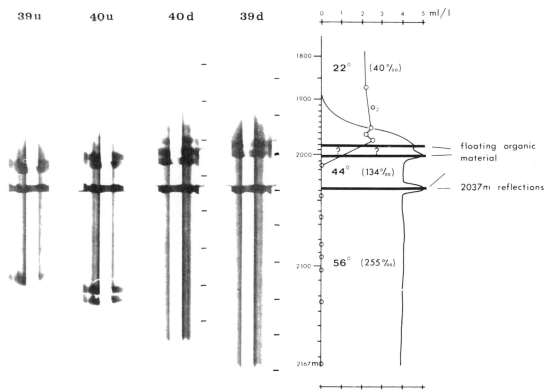

Fig. 2. Photographs of the film negatives for stations 39 and 40 in the Atlantis II Deep. The graph at the right shows the light scattering and oxygen content at depths corresponding to film negative 39 d (d = down, u = up). Oxygen is reported in ml/l (0–5 ml/l), light scattering on a scale from 0–5 (5 = greatest scattering). The two vertical bands in the middle of the film strips are the baffle and attentuated traces. The horizontal dark bands indicate the presence of very thin layers of extreme light scattering. The bands at the very bottom of film strips 40 u and 39 u indicate a cloud of mud stirred up when the camera unit hit bottom. The sudden disappearance of all the traces is caused by the submergence of the nephelometer camera below the sediment-water interface.

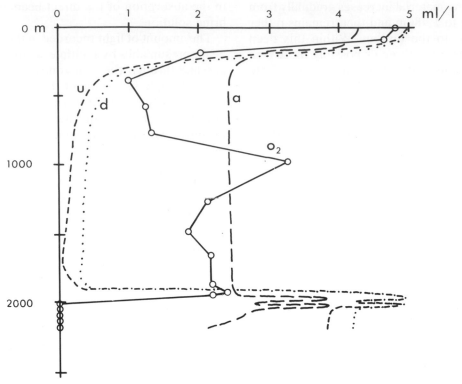

Fig. 3. Graph of light scattering and oxygen content *vs.* depth in corrected meters. The measurement of light scattering is in arbitrary units of 0 to 5 representing the range from least light scattering to the greatest. The values are for lowering 39. The raising is marked by u, lowering by d, and the attenuated beam by a. The oxygen values are from Miller *et al.* (1966).

strips which are found on the scattering trace only). Apparently one must assume that the brine sediments are non-adhesive.

Origin of the Nepheloid Layer

It is apparent from Fig. 2 that the nepheloid layer in the Atlantis II Deep is associated with the presence of the hot brines. No other layers of similar stratification have been found in the other ocean basins or trenches. The thin zones of intense scattering seem to correlate with the interfaces of various hot water masses in this deep. The depth at which the scattering becomes constant also corresponds to the point at which the dissolved oxygen concentration drops to less than 0.1ml/l (Brewer *et al.*, 1969). Ross *et al.*, 1969, report reflections of acoustical energy produced by a high energy sparker from the water column at a depth of about 2,037m. At this same depth one of the thin bands is detected.

Each of the thin bands of intense scatter-

ing is located near an interface between brines of contrasting temperatures and salinities. The zone at 2,037m separates the 56° bottom water from the overlying 44° water. The zone at 2,000m occurs in the transition zone between the 44° water and the overlying Red Sea deep water. Intermediate temperatures in this transition zone have been reported by Krause and Ziegenbein (1966) from hydrographic stations taken from **METEOR**.

The zone at a depth of 1,980m occurs at the base of the normal Red Sea water. Nephelometer lowering 40 indicates the presence of several very thin intermediate temperature water masses in the transition area.

The light scattering particles comprising the nepheloid layer may be the amorphous iron oxides and other heavy mineral deposits that are forming near the contact of the Red Sea deep water and the hot brine solution. It is known that the interaction of the slightly acid anoxic brines with the overlying normal Red Sea alkaline oxidizing deep water will promote the precipitation of metallic oxides and sulfides (see for example Bischoff, 1969). The precipitates settle into the anoxic brines where the lack of oxygen prevents further precipitation. This interpretation is supported by the evidence of constant light scattering in the oxygen depleted water masses. The constant light scattering is believed to indicate a constant flux of the metallic oxides and sulfides to the sea floor.

The discrete zones of intense scattering suggest accumulations of organic particulate matter which floats on the denser brine layers. The 56° water has a specific gravity of 1.205 and is completely free of bacteria with which to decompose the organic material (Watson and Waterbury, 1969). It is probable that even in the case of diatoms and foraminifera the soft parts are not separated from the hard shells through the process of decomposition, and relatively small quantities of tests are found in the metal-rich sediments. However, high rates of sedimentation may also result in smaller concentrations of biogenic material. Some of the material in these layers may also consist of trash discarded from surface vessels. Instruments lowered into the brine areas have returned to the sea surface coated with organic slime and draped with newspaper shreds and other assorted pieces of garbage.

Summary

A nepheloid layer in the Atlantis II Deep results from the precipitation of metallic oxides due to the interaction of the brine solution with the overlying normal Red Sea deep water. Discrete zones of intense light scattering within the brines may be very thin layers of organic matter which float on the various layers of brines of contrasting density. It is possible that in the thin layers of intense scattering the precipitate is very finely divided. An aggregation of these fine particles would increase the settling rate. Thus, for a constant flux of precipitate to the sea floor the light scattering would be decreased by such aggregation.

This nepheloid layer represents a rather unique steady state sedimentation phenomenon with no other known counterparts in any of the world's oceans.

Acknowledgments

This work was supported by the U.S. Navy, Office of Naval Research under contract N00014-67-A-0108-0004, and by the National Science Foundation under contract NSF-GA-550 and NSF-GA-584. The shipboard operation of the nephelometer was conducted by Chan Hilliard with the assistance of Peter Leonard.

References

Bischoff, J. L.: Red Sea geothermal brine deposits; their mineralogy, chemistry and genesis. *In: Hot brines and recent heavy metal deposits in the Red Sea*, E. T. Degens and D. A. Ross (eds.). Springer-Verlag New York Inc., 368–401 (1969).

Brewer, P. G., C. D. Densmore, R. G. Munns, and R. J. Stanley: Hydrography of the Red Sea brines. *In: Hot brines and recent heavy metal deposits in the Red Sea*, E. T. Degens and D. A. Ross (eds.). Springer-Verlag New York Inc., 138–147 (1969).

Bruneau, L., N. G. Jerlov, and F. F. Koczy: Physical and chemical methods. *Reports of the Swedish Deep-Sea Expedition*, **v. 3**, Physics and Chemistry, Fasc. ii, Appendix XXIX–XXX (1953).

Jerlov, N. G.: Particle distribution in the ocean. *Reports of the Swedish Deep-Sea Expedition*, **v. 3**, Physics and Chemistry, Fasc. ii, 73 (1953).

Krause, G., and J. Ziegenbein: Institut für Meereskunde der Universität Kiel, **53** (1966).

Miller, A. R., C. D. Densmore, E. T. Degens, J. C. Hathaway, F. T. Manheim, P. F. McFarlin, R. Pocklington, and A. Jokela: Hot brines and recent iron deposits in deeps of the Red Sea. Geochim. et Cosmochim. Acta, **30**, 341 (1966).

Ross, D. A.: Temperature structure of the Red Sea brines. *In: Hot brines and recent heavy metal deposits in the Red Sea*, E. T. Degens and D. A. Ross (eds.). Springer-Verlag New York Inc., 148–152 (1969).

——, E. E. Hays, and F. C. Allstrom: Bathymetry and continuous seismic profiles of the hot brine region of the Red Sea. *In: Hot brines and recent heavy metal deposits in the Red Sea.* E. T. Degens and D. A. Ross (eds.). Springer-Verlag New York Inc., 82–97 (1969).

Thorndike, E. M., and M. Ewing: Light Scattering in the Sea. Society of Photo-Optical Instrumentation Engineers, Underwater Photo Optics Seminar, October, 1966 (1967).

Thorndike, E. M. Deep-sea cameras of the Lamont Observatory. Deep-Sea Research, **5**, 234 (1959).

Watson, S. W., and J. B. Waterbury: The sterile hot brines of the Red Sea. *In: Hot brines and recent heavy metal deposits in the Red Sea*, E. T. Degens and D. A. Ross (eds.). Springer-Verlag New York Inc., 272–281 (1969).

Temperature Measurements in the Bottom Layers of the Red Sea Brines

DAVID T. PUGH *

Department of Geodesy & Geophysics
Cambridge, England

Abstract

Detailed measurements of temperatures in the water layers immediately above the sediments in Atlantis II and Discovery Deeps are presented. In the bottom 60m of Discovery Deep there is a cooling of 0.5°C as the sediment is approached; above 60m the temperature gradient is adiabatic except for a discrete temperature change at 80m. The water column at the bottom in Atlantis II Deep is stable. These results suggest periodic outflow of hot saline water from Atlantis II Deep into Discovery Deep.

Introduction

On the **R.V. CHAIN** 61 cruise in November, 1966, a number of temperature profiles were measured in the water layers immediately above the sediments in Atlantis II and Discovery Deeps. Profiles in the two deeps are remarkably different and are used here to delimit the area of origin of the waters and the way in which they acquire their thermal energy. Measurements made from **R.R.S. DISCOVERY** on passage from the Indian Ocean in May, 1967, are also included.

Measuring Techniques

Two parameters must be known before it is possible to construct temperature profiles. In addition to temperature measurements, it is necessary to know the elevation of the sensors above the sediment water interface at the instant of each measurement.

Temperature Measurements

Temperatures were measured using a modified form of the Cambridge Heat Flow Recorder (Lister, 1963). This apparatus was designed for working at temperatures near 0°C, such as are found in the deep parts of the main oceans. The electronics, however, functioned well in the 44°C water at the bottom of Discovery Deep. In the 56°C water of Atlantis II Deep the performance was not as good; however, because the electronics were not well connected thermally to the instrument pressure case, there was a delay of some eight minutes after entering the 56°C water before the temperature record deteriorated. During this period good temperature records were obtained. Two thermistors were used as the temperature sensing elements, and these were balanced against fixed internal resistances. Thermistor resistance variations were measured relative to a standard pair of balanced resistors within the recorder case. Relative accuracy of temperature measurements depends on the sensitivity setting of the recorder. In Discovery Deep the relative accuracy is ±.003°C; in Atlantis II Deep it is not less than 0.005°C. The absolute accuracy of the measurements is limited to about ±0.2°C.

* Present address, Admiralty Research Laboratory, Teddington, Middlesex, London.

Measurement of Elevation

The elevation of the recorder above the sediment surface was determined using pinger reflections. These reflections were generally numerous and diffuse, indicating rough bottom topography. The strongest reflection was used for measurements. In order to convert the time differences between the direct and reflected pinger pulse to elevation, it is necessary to know the velocity of sound in the bottom water. Direct measurements of sound velocity in these waters have not been made so that it is necessary to rely on extrapolation of formulae beyond the range of their established validity and on measurements made in saturated brines at atmospheric pressure and normal temperatures. Horton (Vigoureux and Hersey, 1962) has published a formula for the velocity of sound in sea water as a function of temperature, salinity and pressure, and this has been extrapolated to give a velocity of 1,910m per second for the bottom waters of Atlantis II Deep, and 1,890m per second for the bottom waters of Discovery Deep. For comparison with a measurement of sound velocity in 25.49 per cent brine at atmospheric pressure and 17°C, of 1,762m per second, Horton's formula was extrapolated to these conditions and gave a velocity of 1,790m per second. If a velocity of 1,900m per second is used for both deeps, errors are probably little more than 1 per cent, which is quite adequate.

Measurements and Results

Three stations were occupied in Atlantis II Deep, and one in Discovery Deep. The positions of these stations are shown in Fig. 1. Also shown is the position of a temperature gradient—heat flow station (No. 6277) made from **R.R.S. DISCOVERY** in May, 1967. Position fixing is based on bottom topography. Figs. 2 and 3 show the *in situ* temperature gradients observed in both deeps.

In Atlantis II Deep, there is a gradual uniform increase of temperature with elevation above the bottom. This gradient varied from station to station about a mean value of 0.30°C per km (measuring from the bottom upwards) with a standard deviation of 0.10°C per km. Temperatures were measured to within about 5m of the bottom at all three stations, and in none of these stations was there evidence for any change in the temperature gradient immediately above the sediment. In contrast, the temperature profile in the bottom water layer of Discovery Deep shows a number of interesting details. These were observed both on raising and lowering the recording equipment and with both thermistors. The water is colder in the bottom 60m than in the main body of the layer, which was 160m deep at this station. Swallow and Crease (1965) observed a similar decrease in temperature using pressure corrected unprotected thermometers, but the accuracy of measurement using this technique was not good.

From the sediment interface upwards, the temperature increases to an elevation of 60m. The temperature change is most rapid between 20m and 40m. Between 60m and 77m there is a gradual decrease of temperature with elevation. The sharp temperature change between 77m and 83m indicates microlayering within the 44°C water. Above the 83m level, the gradual decrease of temperature with elevation continues to the top of the water layer. The Discovery station 6277 showed a decrease of temperature with elevation, but because the measurements were made in a shallower part of the deep (by about 100m), the cooler bottom layer was not encountered. A temperature increase with depth into the sediment of 1°C per 3.2m was observed, which is about five times the normal sediment temperature gradient for oceanic heat flow. Because the corer struck hard carbonate layers, a core was not obtained for thermal conductivity determinations. Unfortunately no time was available for further measurements.

Analysis

Before proceeding to a discussion of the the implications of these temperature gradients for theories of the origin of the

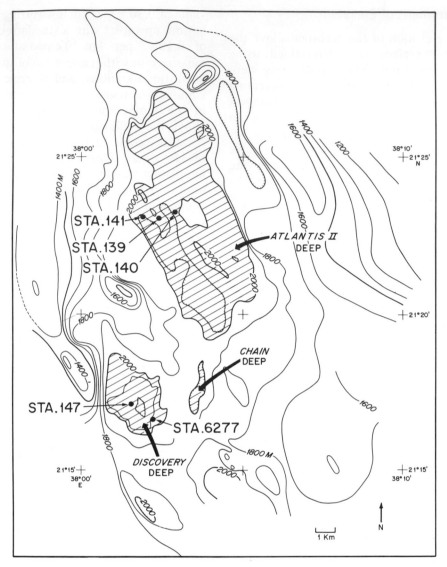

Fig. 1. Map of the Atlantis II and Discovery Deeps area showing station positions (Ross *et al.,* 1969).

waters, the physical aspects of three features of the profiles may be elaborated.

The Colder Water Layer in Discovery Deep

This bottom layer of cooler water suggests that cooling is taking place from below and that, initially, the water was in contact with slightly cooler sediment. The resulting downward flow of heat into the sediment would result in sediment temperature gradients markedly different from the temperature increase with depth into the sediment usually observed. (Erickson and Simmons, 1969). The temperature of the sediment, before the warmer water came

into contact with it, can only have been about 1°C cooler than that of the water because the final interface temperature should be approximately intermediate between the initial water and sediment temperatures.

If the process of heat transfer was solely one of thermal conduction, then the residence time of the bottom water could be calculated from the shape of the cooler part of the temperature profile. Heat transport by water motion, however, will occur and distort the picture. In particular, water which is cooling next to the sediment on the higher parts of the sloping interface will settle down into the deeper parts. This settling would reduce the temperature

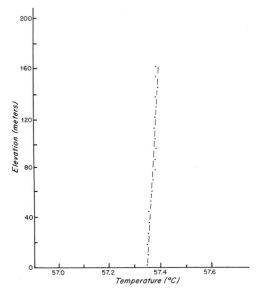

Fig. 2. *In situ* temperature profile for the bottom water layer in Atlantis II Deep (station 141). The temperature gradient is 0.25°C per km.

gradients in the water immediately above the sediment, which would otherwise be the maximum encountered along the temperature profile. The lower temperature gradient in the bottom 20m of Discovery Deep suggests that this process has occurred. The curve for cooling by thermal conduction of

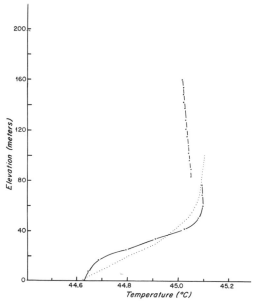

Fig. 3. *In situ* temperature profile for the bottom water layer in Discovery Deep (station 147). The dotted line shows the theoretical temperature profile for a water cooling of 0.5°C by conductive heat transfer into the sediment over a period of 100 years.

heat into the sediments over a period of 100 years is shown in Fig. 3. This indicates that the maximum residence time is of that order since water movements would make the heat transfer more efficient.

The Adiabatic Temperature Gradient

The consistent decrease of temperature with elevation between 80m and 160m in Discovery Deep suggests that this temperature decrease may be due to adiabatic compression of the water. The adiabatic temperature gradient must be estimated, for comparison, by extrapolation of formulae based on data for normal sea water since little is known about the thermodynamic properties of the bottom water in these deeps.

Polynomial expressions for the adiabatic temperature gradient as a function of temperature, salinity and pressure have been computed from data in the normal range of these parameters encountered in the oceans (e.g. Fofonoff and Froese, 1958). However, this kind of expansion is usually unreliable for gross extrapolations, and it is better to calculate the adiabatic temperature gradient from first principles.

The adiabatic temperature gradient is given by:

$$\left(\frac{\partial T}{\partial P}\right)_{ad} = \frac{T}{Cp}\left(\frac{\partial v}{\partial T}\right)_P$$

where T is the absolute temperature

Cp is the specific heat at constant pressure

$\left(\frac{\partial v}{\partial T}\right)_P$ is the rate of change of specific volume with respect to temperature at constant pressure (see for example Fofonoff, 1962).

$\left(\frac{\partial v}{\partial T}\right)_P$ at atmospheric pressures has been calculated by Manheim (1967) for 25 per cent brine solution, and a value of 0.000413cm³ per °C at 50°C was obtained. The data of Cox and Smith (1959) for Cp may be extrapolated to a value of 3.45 joules per g °C at 20°C and atmospheric pressure in a 25 per cent brine solution. After making small corrections for pressure in both cases, and for temperature in the case of Cp (Sverdrup, Johnson and Fleming, 1942), we obtain an adiabatic temperature gradient of 0.000379°C per decibar for Discovery Deep bottom water and 0.000394°C per decibar for the bottom water in Atlantis II Deep. The latter is slightly higher because the bottom water temperatures in Atlantis II Deep are higher. An error of less than 10 per cent may be estimated, and this is mainly due to uncertainty in the extrapolated value of Cp.

Temperature gradients in °C per m may be converted to °C per decibar if the value of gravitational acceleration and the *in situ* density of the water are known. The measured temperature gradient in Discovery Deep waters between elevations of 80m and 160m from the bottom is −0.483°C per km. This is 0.000403°C per decibar, assuming a density of 1.2 g per cm³ for the brines. Thus, within experimental error, this gradient is adiabatic, and the water should be well mixed. This gradient appears to continue for about 20m below the discontinuity, before the cooler layer is reached. In Atlantis II Deep, the potential temperature gradient becomes 0.80°C per km when the adiabatic correction is made, which means that the water column is relatively quite stable.

The Temperature Discontinuity

The sudden decrease in temperature of 0.05°C between 77m and 83m elevation is a smaller version of the larger discontinuities observed above the 44°C water layer. There are no intermediate temperature determinations so that discontinuity may be sharper than 0.05°C in 6m. If the layering has been in existence for more than a year, thermal conduction above would have smoothed out the boundary to a larger extent than that observed, and a mechanism such as that suggested by Turner (1969) is necessary to maintain the discontinuity. Turner's mechanism requires some heating from below, but this could be due to geothermal heat flow, which need only be negative in the deeper parts below the cooler water layer. Above a level corresponding to the 60m elevation in the profile, geothermal

heat flow is probably positive, maintaining the well mixed layer between 60m and 80m, and the sharp discontinuity at the 80m level. A salinity difference of about 0.05 per mill is necessary to maintain the vertical stability of the water column. Such a difference of salinity could not be detected by the methods of analysis used.

Discussion

Early theories suggesting non-local origin of the waters (Charnock, 1964) were no longer tenable when the high concentration of salt dissolved in the waters was realized as mixing during transport, must dilute the concentration considerably. It has also been suggested that the brine pools are relics of a period when the Red Sea was evaporated almost to dryness, which have subsequently acquired their heat by geothermal warming from below. This is not possible for Discovery Deep because of the colder water at the bottom, and in Atlantis II Deep such gradual heating would almost certainly have resulted in gradients much closer to the adiabatic than those observed. Heating of the Atlantis II Deep bottom water by conduction alone would require temperature gradients of the order of 1°C per 10cm over the whole floor of the deep to account for the heating of 0.6°C in 20 months that has been observed by Munns *et al.* (1967). Erickson and Simmons (1969) show that such gradients are not observed throughout the deep, thereby supporting the case against geothermal heating from below, though some of the heat must be supplied in this way.

The probable source of the waters is an injection of hot saline water from the underlying rocks into the bottom water. It is not possible to uniquely determine the source of injection of the waters using the evidence presented in this chapter. What follows is the simplest explanation consistent with the observations.

While the injection of hot saline water could occur independently in the two deeps, it is difficult to account for the layering in Discovery Deep in this way. This water is probably an overflow from Atlantis II Deep. The two layers may be due to overflow on separate occasions. During the overflow good mixing of the overflowing water mass occurred, and this uniformity has been maintained by geothermal heating from below, except in the deepest part where loss of heat to cooler sediments has occurred.

If periodic overflow occurs into Discovery Deep, then the injection of hot saline water into Atlantis II Deep is likely to be sudden rather than continuous. A sudden injection of warmer, lighter water at the bottom of Atlantis II Deep could rise rapidly to the top of the 56°C layer and then mix downwards from there, resulting in the stable stratification observed, but gradual infusion from below would be more likely to mix with the whole column of 56°C and reduce its vertical stability. Alternatively, injection could occur mainly at the top of the 56°C layer.

It is easy to estimate that if the rise of 0.6°C in 20 months is due to injection of water at 200°C, an increase in the thickness of the 56°C water layer of about 60cm is necessary to accommodate the new water. This would not be detectable from hydrographic measurements made on the two occasions.

Craig (1965) has shown by isotopic analysis that the water emitted is recirculated sea water. This water acquired its high temperature and salinity within the rocks and is probably injected periodically into Atlantis II Deep. Measurements of water temperatures in the bottom waters of Atlantis II Deep over a longer period are required to confirm this.

References

Charnock, H.: Anomalous bottom water in the Red Sea. Nature, **203**, 591 (1964).

Cox, R. A., and N. D. Smith: The specific heat of sea water. Proc. Roy. Soc. London, **A252,** 51 (1959).

Craig, H.: Isotopic composition and origin of the Red Sea and Salton Sea geothermal brines. Science, **154,** 1544 (1966).

Erickson, A. J., and G. Simmons: Thermal measurements in the Red Sea hot brine pools. *In: Hot brines and recent heavy metal deposits in the Red Sea,* E. T. Degens and D. A. Ross (eds.). Springer-Verlag New York Inc., 114–121 (1969).

Fofonoff, N. P.: Physical properties of sea water. *In: The Sea,* **1,** M. N. Hill (ed.). Interscience, New York, 3 (1962).

——, and C. Froese: Program for oceanographic computations. Fisheries Research Board of Canada, Manuscript Report Series, **27** (1958).

Horton, J. W., quoted by P. Vigoureux, and J. B. Hersey: *In: The Sea,* **1,** M. N. Hill (ed.). Interscience, New York, 476 (1962).

Lister, C. R. B.: Geothermal gradients using a deep sea corer. Geophys. Journal, **7,** 571 (1963).

Manheim, F. T.: Salt content of Atlantis II and Discovery Deep brine samples. Woods Hole internal memorandum (1967).

Munns, R. G., R. J. Stanley, and C. D. Densmore: Hydrographic observations of the Red Sea brines. Nature, **214,** 1215 (1967).

Ross, D. A., E. E. Hays and F. C. Allstrom: Bathymetry and continuous seismic profiles of the hot brine region of the Red Sea. *In: Hot brines and recent heavy metal deposits in the Red Sea,* E. T. Degens and D. A. Ross (eds.). Springer-Verlag New York Inc., 82–97 (1969).

Sverdrup, H. U., M. W. Johnson, and R. H. Fleming: *The Oceans.* Prentice-Hall, Englewood Cliffs, New Jersey (1942).

Swallow, J. C., and J. Crease: Hot salty water at the bottom of the Red Sea. Nature, **205,** 165 (1965).

Turner, J. S.: A physical interpretation of the observations of hot brine layers in the Red Sea. *In: Hot brines and recent heavy metal deposits in the Red Sea,* E. T. Degens and D. A. Ross (eds.). Springer-Verlag New York Inc., 164–173 (1969).

A Physical Interpretation of the Observations of Hot Brine Layers in the Red Sea

J. S. TURNER

Department of Applied Mathematics and Theoretical Physics
University of Cambridge, England

Abstract

The observed temperature structure of the Red Sea brines, which occur as well-mixed layers separated by sharp interfaces, is compared with other natural examples of layering and also with laboratory experiments designed to study the phenomenon. The existence of such layers is characteristic of liquids which are stabilized with salt but are made unstable by heating from below. Various possible mechanisms for their formation and maintenance are evaluated from a physical point of view. The available evidence supports the idea that the source of the brines is in the Atlantis II Deep, which overflowed into the other holes in recent times.

Introduction

A striking feature of the temperature and salt concentration profiles of the Red Sea brines is the existence of well-mixed uniform layers, alternating with thin regions where the gradients are large. In the Atlantis II Deep two layers and two interfaces are clearly marked below the normal Red Sea water, and the transition above the bottom layer is so sharp that no sample has been obtained from it despite many attempts. In Discovery Deep there is a bottom layer of maximum depth about 200m and probably also a thin intermediate layer, though the evidence here is weaker. The properties we wish to explain are summarized in Table 1, which has been assembled from Ross and Hunt (1967), Munns *et al.* (1967), Brewer *et al.* (1969)

and Ross (1969). Only passing reference will be made in this chapter to Chain Deep, the third deep hole where hot salty water has been detected but not adequately sampled.

The interface thicknesses recorded in Table 1 have been defined as the depths over which the largest gradients are observed. Above what has been called the upper interface in Table 1 there are regions about 100m thick where the gradients are much smaller, but the properties are still approaching those of normal Red Sea water. In Atlantis II Deep this top transition region has been observed to vary in detail from day to day, but similar evidence of "activity" has not been seen above Discovery Deep (Munns *et al.*, 1967; Brewer *et al.*, 1969). Another difference between the two deeps is that the water at the bottom of Atlantis II Deep has heated up during the period of observation (nearly two years), whereas a temperature decrease at the bottom of Discovery Deep implies cooling from the bottom (Pugh, 1967; 1969). All these features must be taken into account in any satisfactory physical interpretation of the layers.

An obvious first assumption is that the observed structure implies the separate formation of the several layers at different times in the past. This is unsatisfactory, however, since it does not explain the continuing sharpness of the interfaces, which should be smoothed out by diffusion. Instead, we shall develop the idea that layering can be produced and maintained by convection, which is driven by heating from below. Though the density steps are

Table 1 Summary of the Observations of Temperature and Gravimetric Salinity as Functions of Depth in Atlantis II and Discovery Deeps

	Atlantis II Deep				Discovery Deep			
	Depth Range (m)	Thick-ness (m)	Temp. (°C)	Salin. (‰)	Depth Range (m)	Thick-ness (m)	Temp. (°C)	Salin. (‰)
"Normal water"	Above 1,984		22	41	Above 1,986		22	41
Upper interface	{ 1,984 ⎰ 2,009	25			1,986 ⎰ 2,023	37		
Intermediate layer	{ 2,037	28	44	135	2,027	4	36	127
Lower interface	{ 2,042	5			2,042	15		
Bottom layer	Below 2,042		56	257	Below 2,042		45	256

dominated by the salinity differences and are very stable, each layer separately has an unstable temperature distribution across it and can thus be stirred by the heating. This not only makes the properties very uniform within a layer, but also serves to counteract the broadening effect of diffusion by eroding the edges of the layers and thereby sharpening the interfaces.

Considered in the above way, the layers in the Red Sea can be regarded as extreme examples of layers observed in other areas, about which something has already been learned, mainly through laboratory experiments. It will be helpful to summarize these related observations before applying the same ideas to the case of interest here.

Related Observations of Layering

The best documented example of layer formation in a closely comparable situation to that in the Red Sea has been reported by Hoare (1966). He studied an Antarctic lake which is fresh at the top but salty at the bottom, probably because of a single sudden intrusion of salt water in the past. The lake is also cold at the top and hot at the bottom, but the temperature and conductivity are not smooth functions of depth; many steps are observed, with uniform layers typically 1½m thick separated by regions of large gradient where the temperature changes by about 1°C in approx-

imately 15cm. Here one suspects that the salinity distribution was initially smooth and that layers have been produced later by the applied heating. It seems likely that a similar structure would be found in other lakes where salt intrusions are known to be present (such as those in Canada and Norway). The possibility of formation of convecting layers in this way should also be kept in mind in the discussion of "solar ponds," the artificially stratified lakes designed to trap solar heat in a saline layer underlying a fresh one (Tabor, 1966).

Many salinity-temperature-depth measurements obtained with continuously recording instruments in the open ocean show a similar stepped behavior. Layers are apparent on the records of many observers, though little of the data has yet been published with this aspect in mind. The clearest examples occur when one property (heat or salt) is stabilizing and the other destabilizing, while the net density gradient is of course stable. Most of these cases, however, differ from that of the Red Sea in one important respect: the net density difference is very small (D is only slightly larger than 1, in the notation of the next section) rather than being large as it is for the Red Sea brines. Layers are again observed when a stable salinity gradient is partially counteracted by heating from below; it is also worth mentioning that layering can occur in the opposite situation, where a stable temperature distribution is made unstable by the addition of salt at the top. Measurements of this latter kind have been reported by Tait and Howe (1968) and interpreted using laboratory experiments by Turner (1967).

Laboratory Experiments

The results of several different kinds of laboratory experiments are available. These shed light on both the mechanism of formation of layers from a gradient and on the transfer processes across the interfaces once they have formed. Any extrapolation to the larger scale and more extreme conditions of the Red Sea must be made cautiously, but they are at least of some qualitative value in isolating the physical ideas on which this chapter is based.

Formation of Layers from a Gradient

Turner and Stommel (1964) described the behavior of a stable salinity gradient when it is heated uniformly from below in a laboratory tank. Convective stirring initially produces a layer at the bottom that is well mixed in both heat and salt and grows by incorporating fluid from above. This layer does not continue to grow indefinitely, but, instead, another layer forms above it which behaves in the same way. In time, many such layers can form with sharp interfaces separating turbulent convecting regions. The system can be maintained in this state because the much larger molecular diffusivity for heat compared with salt allows heat to escape from a layer to cause a convective stirring above, while most of the salt is left behind to maintain a net stable density difference across the interface. A photograph of this experiment was reproduced by Swallow (1965) in an early discussion of the Red Sea results, and another experiment in which fewer, deeper layers were formed is shown in Fig. 1.

Recent quantitative experiments (Turner, 1968) have established a relationship between the maximum depth of the lowest layer before a second forms above it, the salinity gradient and the heating rate. The results are valid for relatively small gradients and large heating rates and can be

Fig. 1. A laboratory experiment showing layering in a plastic cylinder about 30cm in diameter. An initially stable salinity gradient has been heated from below to produce three well-mixed layers (which are marked with fluorescent dye). The strong stirring motions caused by the heating are apparent from the disturbances they have produced on the interfaces.

shown *not* to be appropriate for the Red Sea situation. Although the quantitative results cannot be used directly, these laboratory experiments do suggest that layers formed in the Red Sea entirely by heating from below could only be very thin, if they formed at all, and that some other process is needed to explain the formation of intermediate layers of the depths observed.

The Fluxes of Salt and Heat across Sharp Interfaces

In another series of experiments (Turner, 1965) measurements were made of the transports of heat and salt across a single interface separating two convecting layers, the motions being maintained by heating from below. Again the qualitative behavior will be of more interest here than the precise numerical values. The fluxes of both heat H and salt F_s from the lower to the upper layer were found to depend systematically on the ratio of the separate contributions of salinity and temperature to the density difference between the layers, i.e., on

$(\Delta\rho)_s/(\Delta\rho)_T = D$. As D was increased both fluxes were reduced, since molecular processes then became increasingly important compared to turbulent mixing near the interface. At the highest value of D used in these experiments ($D = 7$) molecular diffusion was shown to dominate. This is an important result for the present application since all the interfaces between layers in the Red Sea are more stable than those studied in the laboratory.

The appearance of the interface in these experiments is also of interest here. A convenient way to cover a large range of D in a single experiment is to remove the source of heating below and allow the system to approach its final equilibrium state of constant temperature but finite salinity difference. As D increases (and H decreases) the transition region is seen to become thicker, but provided the only heating is from below, there remains at the center of the interface a region of stable gradient where only molecular transfer can occur. On a few occasions during such transient experiments with decreasing heat flux an intermediate well-mixed layer was observed to form from an initially sharp interface; the effect could then be attributed to side wall heating or cooling, producing circulations which penetrated into the interior of the fluid. Boundary effects should clearly be kept in mind also in the case of interest here.

Immediate Deductions from the Red Sea Observations

Using these related observations and experimental results as a background, we can now discuss the Red Sea data in more detail. This section will state the deductions that can confidently be made about the present state of the brine layers; a more speculative discussion of their past history will follow.

The Relative Importance of Temperature and Salinity

An important parameter in the laboratory experiments, and also on the larger scale, is the density ratio D of the contributions of salinity and temperature to the density difference across an interface. The data of Table 1 have been used to calculate D for the two interfaces in each of the two deeps, and the results are shown in Table 2. There is some uncertainty about the conversion of salinity to density difference because of the change in composition between the layers, but this is unimportant for the present purpose.

All of the interfaces are very stable when compared with the laboratory experiments. The salinity steps are nearly the same in the two regions, but because of the lower temperatures the interfaces in Discovery Deep are the more stable. Because of the high stability of all the interfaces it is likely that molecular processes alone will determine the transfer across them, though their thickness will depend on the strength of the convection on each side (as discussed further below).

The Molecular Heat Fluxes between the Layers

If it is accepted that the transfer of heat is by molecular conduction, limits can be put on the fluxes by evaluating the temperature gradients. These gradients have been calculated (Table 2) from the thicknesses and temperature steps set down previously. The corresponding heat fluxes in Table 2 have been calculated assuming a thermometric conductivity of $1.5 \times 10^{-3}\mathrm{cm^2/sec^{-1}}$. The fluxes are considerably larger in the Atlantis II Deep, corresponding to the larger gradients there.

Profiles of comparable accuracy are not available for salinity, and so the salt fluxes cannot be calculated directly. Though laboratory results at lower values of D are known which relate F_s to H, the extrapolation to much larger D is uncertain and will not be used here. Qualitatively, however, it is safe to say that there will be a considerable net unstable buoyancy flux upwards, i.e., in terms of the density differences they produce the heat flux will greatly outweigh the salt flux. In all our subsequent calculations the diffusive salt flux will be neglected.

Table 2   The Relative Contributions of Salinity and Temperature to the Density Differences across the Four "Interfaces" in Atlantis II and Discovery Deeps, and the Calculated Temperature Gradients and Molecular Heat Fluxes

	Atlantis II Deep		Discovery Deep	
	Upper Interface	Lower Interface	Upper Interface	Lower Interface
Salinity difference (‰)	94	122	86	129
Density diff. $(\Delta\rho)_s$ (parts per thousand)	65	85	60	90
Temperature diff. (°C)	22	12	14	9
Density diff. $(\Delta\rho)_T$ (parts per thousand)	7.1	5.4	4.1	3.6
$D = (\Delta\rho)_s/(\Delta\rho)_T$	9	15	15	25
Temperature gradient (°C/cm)	9×10^{-3}	24×10^{-3}	4×10^{-3}	6×10^{-3}
Heat flux (μcal/cm² sec)	13	36	6	9

The Heat Flux through the Bottom

An important observation, which should be used cautiously until it is confirmed by measurements taken over a longer time interval, is that of Munns *et al.* (1967), who have deduced a temperature increase of 0.56°C in 20 months for the bottom layer in Atlantis II Deep. The mean depth of this layer, defined as the volume of 56° water divided by the area of the interface bounding its top (i.e. the 2,042m depth) contour, can be found from Table 2 given by Ross (1969), and is 54m. These two results taken together imply a mean net heat flux into the bottom layer of 55 microcalories per square centimeter per second (μcal/cm² sec). Adding the heat flux out of the top of this layer (Table 2) gives a value of about 90μcal/cm² sec for the heat flux through the sea floor, averaged over the area covered by the bottom brine layer in the Atlantis II Deep.

This is a very large value for the bottom of the ocean, but not unreasonable for geothermal areas on land (see for example Elder, 1965). It implies that Atlantis II Deep must be a localized area of current tectonic activity, in agreement with the geophysical observations. It does not mean, of course, that such large heat flows occur either continuously in time or uniformly over the whole area. Erickson and Simmons have measured temperature gradients in the sediments at a number of points on the bottom of Atlantis II Deep which are of both signs but imply a mean heat flux of 20μcal/cm² sec. The large variations in these measurements, combined with the

mean value deduced above (90μcal/cm² sec), make it likely that the heat is supplied not purely by conduction, but by some flow of water between a deep heat source and the sea floor. Convection in the sediments, as suggested by Elder (1965), is a possibility, but both the high salinity and the evidence from the sediments (Bischoff, 1969; Deuser and Degens, 1969) suggest the mechanism of direct injection of hot salty water through a vent or vents in the bottom. A few point measurements cannot be expected to sample the extreme values of the heat flux, but the existence of the stable saline layers in fact gives a much more accurate method of averaging over an area than is available elsewhere, even on land with a closely spaced network of gradient measurements.

This mechanism of inflow from localized sources is still consistent with the other observations. Horizontal uniformity can be maintained through most of the bottom brine layer if the injected water rises in the form of narrow plumes, which become diluted by mixing with their environment and then spread out across the whole layer, along the interface at its top. Continuing this process will fill the layer from the top with increasingly warm fluid and push previously affected layers downwards. Turner (1969) has shown in another context that this model can be used to reconcile a constant rate of heating throughout the depth of the layer with the slightly stable density gradient implied by the measurements of Pugh (1969).

Comparison between Atlantis II and Discovery Deeps

In strong contrast to the heating of Atlantis II Deep is the observation by Pugh (1967; 1969) that there is a cooler layer near the bottom of Discovery Deep, which implies a heat flux out through the bottom. It would be misleading to interpret this cooling entirely as a one-dimensional process and to base the calculation of heat flux on the thickness of the cooled layer at one place, since water cooled anywhere along a sloping bottom will tend to pour down and produce a deeper layer near the bottom of the depression. An order of magnitude estimate of the heat loss at the position of measurement may, however, be obtained by assuming that there is molecular conduction through the maximum gradient measured by Pugh. The heat loss calculated in this way is $0.3 \mu cal/cm^2$ sec, which is 30 times smaller than that deduced for the heat loss through the top of the layer, and for the purposes of the heat balance the transfer through the bottom can be neglected for Discovery Deep. The same measurements have been used by Pugh (1969) to put an *upper limit* on the residence time of the water now at the bottom of Discovery Deep. A cooled layer of the thickness observed (about 50m) could have been produced purely by molecular conduction in a time of the order 100 years.

The difference between the deduced heat fluxes for the two regions gives quantitative support to the conclusions which can be drawn by comparing the observed structure with the laboratory experiments. The lower interface in Atlantis II Deep is much sharper, and the intermediate layer is deeper and more uniformly mixed than in Discovery Deep, as would be expected with strong heating from below. The scatter of temperature measurements reported by Munns *et al.* (1967) is also larger in Atlantis II Deep, especially during the **CHAIN** cruise of 1966, and this is probably due to the larger fluctuations associated with convection over a non-uniform heat source on the bottom. The larger temperature fluctuations in the detailed hydrographic profiles between the 22° and 44°

waters above the Atlantis II Deep could also be a result of the stronger convection. Random internal waves, produced by the convective stirring, can propagate upwards into this region, though it is possible too that transient disturbances from above were affecting this transition layer during the period of observation.

Speculations about the History of the Layers

Arguing only from the present state and the changes over a few years, it is difficult to propose mechanisms for the formation and development of the layered structure that are not controversial. In this section various mechanisms are examined: some of these can definitely be rejected as being inconsistent with the evidence, but a plausible, self-consistent picture can be constructed.

The Origin of the Salty Water

There seems little doubt now that the hot salt water originates locally from a source or sources at the bottom of the Red Sea. The earliest suggestion, made by Charnock (1964) when only small increases of salinity and temperature had been detected, that this water might have formed round the shores and flowed down along the bottom, was shown to be untenable once the extreme conditions were observed. The chemical properties, which are quite different from those of normal Red Sea water, confirm this view. It is likely that the bottom water is, or has been in the recent past, in contact with a solid salt deposit.

The very similar salinity of the bottom layers of Atlantis II and Discovery Deeps suggests a common source. When considered with the heat flux data already discussed, this similarity gives weight to the suggestion that the brine has originated in the thermally active Atlantis II Deep, overflowed into Discovery Deep and also into Chain Deep, a smaller depression on the side of a saddle between the other two Deeps (Ross and Hunt, 1967). The difference in temperature between the bottom layers in the two holes does not contradict this hypothesis; it will be suggested later that the present temperatures could easily have been produced by subsequent heat transfer, after some catastrophic event had caused brine with the same salinity and temperature to overflow from Atlantis II Deep into Discovery Deep.

Formation of the Intermediate Layer

The possibility that the intermediate layer could have been formed purely by heating from below has already been rejected (during the discussion of the laboratory experiments) for the reason that in the Red Sea the heating rate is far too small in relation to the salinity gradient. It is likely, however, that during the processes of injection of brine through the sea floor and overflow into a second hole a smooth salinity gradient would be produced in the transition region rather than a stepped structure, and some other mechanism whereby a step could later form from a gradient must be considered.

The most likely process is mixing due to internal waves or seiches on the interface, which are induced by convection below or by external disturbances coming in from above. As such waves approach the boundaries of a basin, Thorpe (1966) has shown that they steepen and can break. The maximum steepening, and therefore the most vigorous overturning and mixing, occur at the level of steepest gradient near the center of the transition region. The mixed fluid of uniform density will then tend to spread back into the interior of the basin, where it can be maintained in a well-mixed state by heating from below, even though this heating was not vigorous enough to produce a layer in the first place. Within the accuracy of measurements to date, the salinities of the intermediate layers are nearly the same, although their thicknesses are different.

The next question is: can a mechanical mixing process, such as that produced by wave action, account entirely for the observed properties of the intermediate layers with respect to the undiluted bottom layer and the ordinary Red Sea water

above? If this were so, mixing of the two components in a certain proportion should explain both the intermediate salinity and temperature simultaneously. Using the gravimetric salinity values of Table 1 as the reference, the proportions of the mixture necessary to produce the right intermediate salinity are shown for the two holes in Table 3, together with the temperatures achieved at the same time. The last line of the table shows that the observed temperatures are considerably greater than those predicted: an extra transfer of heat relative to salt into the intermediate layer is implied. All our earlier discussion about molecular transports through the interfaces does indeed make such a heat flux seem very likely. The last three columns of Table 3 refer to a later calculation which takes this into account.

The Time Necessary to Achieve the Present State in Discovery Deep

It is impossible unambiguously to deduce a complete time history for a phenomenon which involves several different mechanisms, each of which could have modified the properties of the layers suddenly or more gradually. Nevertheless, it is instructive to carry out a calculation based on the following additional hypotheses. Suppose that the brine now in Discovery Deep all overflowed suddenly from Atlantis II Deep at some time in the past due to an event such as a volcanic eruption, and that at that time both bottom layers had the same salinity and temperature. Suppose further that over a short period, compared with the total time since then, some process such as wave mixing produced the observed layer and interface structure shown in Table 1 and that the present day heat fluxes (Table 2) can be regarded as typical mean values for the whole of this time.

The cooling rate of the bottom layer in Discovery Deep can be found from Table 2 if the mean depth is known. This depth is 90m, using again the table of areas and volumes given by Ross (1969), and the deduced cooling through the top of the layer is about 0.03°C per year. The heating rate in Atlantis II Deep obtained by direct

Table 3 The Comparison between the Temperatures Obtainable in the Intermediate Layer by Simple Mixing, and Those Which Are Observed in Atlantis II and Discovery Deeps. A Similar Calculation Is Shown for a Hypothetical "Original Brine," Supposed to Be Present in Both Holes Immediately after an Overflow into Discovery Deep.

Layer	Atlantis II Deep			Discovery Deep			"Original brine"		
	Upper	Intermediate	Bottom	Upper	Intermediate	Bottom	Upper	Intermediate	Bottom
Measured salinity (‰)	41	135	257	41	127	256	41	127	256
Fraction in mixture	0.57		0.43	0.60		0.40	0.60		0.40
Measured temp. (°C)	22	44	56	22	36	45	22	36	
Predicted temp. (°C)		36			31			32	(46)
Increase in temp. (°C)		8			5			4	

measurement (Munns *et al.*, 1967) is 0.33°C per year. When these rates and the temperatures observed at the present time are used with the above hypotheses, two equations are obtained from which one can deduce both the initial temperature of the "original" brine, 46°C, and the time since it overflowed, 30 years.

Another estimate of this time can be made from the properties of the intermediate layers. The last section in Table 3 refers to the temperature which would be attained by mixing the "original" brine at 46°C (the temperature just deduced) with 22°C water in the correct proportions to match the measured salinity. (The numbers recorded in Table 3 refer to Discovery Deep, but the intermediate layer temperature obtained for Atlantis II Deep will be little different, because of the very similar salinities.) The predicted temperature is 32°C, and in Discovery Deep there is therefore a 4°C rise in temperature to be explained. Using the difference between the heat fluxes across the top and bottom interfaces (Table 2) and the layer depth (4m) gives an age of 20 years. This estimate is less reliable than the preceding one based on the bottom layers because of the greater uncertainties in the measurements, especially in the depth of the intermediate layer, but the two values are consistent with one another. Both are considerably less than the upper limit imposed on the age by Pugh's measurements of the cooling at the bottom.

This same method can be applied to the intermediate layer in Atlantis II Deep. An initial temperature of 32°C is again predicted, and a 12°C rise in temperature has to be accounted for in a layer of depth 28m. This could have been produced by the observed net heat flux (Table 2) in 45 years. This time is significantly longer than the above estimates of the age of the saline water in Discovery Deep. It implies that the layers in Atlantis II Deep were already in existence, and were being heated from below, before the overflow into Discovery Deep took place. The sediments in the brine area (Ku *et al.*, 1969) do indicate that there have been many sudden injections of brine in the past, extending over much longer times than those deduced here. The calculations just performed suggest that one of the more recent injections of brine, occurring several decades ago, was sufficiently violent to cause an overflow into Discovery Deep. Other more complicated sequences of events are of course not ruled out by the data so far collected.

In conclusion, the strong dependence of such calculations on the assumptions made must be emphasized again. If, for example, the injection of brine and the consequent heating rate are so irregular that the change in temperature over two years is not a good indicator of longer term means, then the deductions will be invalid. The calculations have been intended to illustrate a method of approach rather than to produce final answers, and they should certainly be revised as further observational data become available.

Summary

The brine layers of the Red Sea have been considered as an extreme case of layering due to the combination of temperature and salinity gradients which have opposing effects on the density. The layers have been compared and contrasted with other observations in the ocean and lakes and with laboratory experiments that illustrate the important physical processes. All the interfaces between the well-mixed layers in the Red Sea are so stable that molecular diffusion alone can account for the flow of heat upwards; these heat fluxes have been calculated from the measured gradients. The nearly uniform layers are stirred by convection due to the heat transport, and the boundaries between them are maintained because of the dominant stable salinity distribution.

A tentative history of the layers has been suggested, based on a self consistent interpretation and the backward extrapolation of present measurements. A sudden overflow of brine from Atlantis II Deep into Discovery Deep is postulated, followed by the formation of an intermediate layer by stirring due to wave action. Atlantis II Deep appears to be heating at a rate of about 0.3°C/year, which implies a large value of the mean heat flux (90 μcal/cm^2 sec) through its bottom, probably produced by the injection of hot brine. Discovery Deep is cooling, partly from below but mainly by conduction to the water above. If the water in the two holes had the same properties at the time of overflow, which is suggested by the similarity of the salinities, the present temperatures of both the bottom and the intermediate layers are consistent with a very recent origin of the brine in Discovery Deep and an earlier formation of the layers in Atlantis II Deep.

Acknowledgments

This work has been supported by a grant from the British Admiralty. I am grateful to many people at Woods Hole Oceanographic Institution for discussing their observations with me, and especially to Dr. David A. Ross who stimulated my interest in this subject during the summer of 1967.

References

Bischoff, J. L.: Red Sea geothermal brine deposits: their mineralogy, chemistry, and genesis. *In: Hot brines and recent heavy metal deposits in the Red Sea*, E. T. Degens and D. A. Ross (eds.). Springer-Verlag New York Inc., 368–401 (1969).

Brewer, P. G., C. D. Densmore, R. G. Munns, and R. J. Stanley: Hydrography of the Red Sea brines. *In: Hot brines and recent heavy metal deposits in the Red Sea*. E. T. Degens and D. A. Ross (eds.). Springer-Verlag New York Inc., 138–147 (1969).

Charnock, H.: Anomalous bottom water in the Red Sea. Nature, **203**, 590 (1964).

Deuser, W. G. and E. T. Degens: O^{18}/O^{16} and C^{13}/C^{12} ratios of fossils from the hot brine deep area of the central Red Sea. *In: Hot brines and recent heavy metal deposits in the Red Sea*, E. T. Degens and D. A. Ross (eds.). Springer-Verlag New York Inc., 336–347 (1969).

Elder, J. W.: Physical processes in geothermal areas. Terrestrial Heat Flow, American Geophysical Union, Geophysical Monograph 8 (1965).

Hoare, R. A.: Problems of heat transfer in Lake Vanda, a density stratified Antarctic lake. Nature, **210**, 787 (1966).

Ku, T. L., D. L. Thurber, and G. Mathieu: Radiocarbon chronology of Red Sea sediments. *In: Hot brines and recent heavy metal deposits in the Red Sea*, E. T. Degens and D. A. Ross (eds.). Springer-Verlag New York Inc., 348–359 (1969).

Munns, R. G., R. J. Stanley, and C. D. Densmore: Hydrographic observations of the Red Sea brines. Nature, **214**, 1215 (1967).

Pugh, D. T.: Origin of hot brines in the Red Sea. Nature, **214**, 1003 (1967).

———: Temperature measurements in the bottom layers of the Red Sea brines. *In: Hot brines and recent heavy metal deposits in the Red Sea*, E. T. Degens and D. A. Ross (eds.). Springer-Verlag New York Inc., 158–163 (1969).

Ross, D. A. and J. M. Hunt: Third brine pool in the Red Sea. Nature, **213**, 687 (1967).

———: Temperature structure of the Red Sea brines. *In: Hot brines and recent heavy metal deposits in the Red Sea*, E. T. Degens and D. A. Ross (eds.). Springer-Verlag New York Inc., 148–152 (1969).

Swallow, J. C.: Hot salty water. Oceanus, **11**, 3, 3 (1965).

Tabor, H. Z.: Solar ponds. Science Journal, **2**, 6, 66 (1966).

Tait, R. I. and M. R. Howe: Some observations of thermo-haline stratification in the deep ocean. Deep Sea Research **15**, 275 (1968).

Thorpe, S. A.: Internal gravity waves. Ph.D. dissertation, University of Cambridge (1966).

Turner, J. S.: The coupled turbulent transports of salt and heat across sharp density interfaces. Int. J. Heat Mass Transfer, **8,** 759 (1965).

——: Salt fingers across a density interface. Deep Sea Research, **14,** 599 (1967).

——: The behavior of a stable salinity gradient heated from below. J. Fluid Mech., **33,** 183 (1968).

——: Buoyant plumes and thermals. *In: Annual Review of Fluid Mechanics,* (in preparation) (1969).

—— and H. Stommel: A new case of convection in the presence of combined vertical salinity and temperature gradients. Proc. U.S. Nat. Acad. Sci., **52,** 49 (1964).

A Note on the Chemical Composition of the Red Sea Brines *

PETER G. BREWER AND DEREK W. SPENCER

Woods Hole Oceanographic Institution
Woods Hole, Massachusetts

Abstract

Samples of brine from the Atlantis II Deep, the Discovery Deep, and the layer of 44°C brine overlying the Atlantis II Deep, have been analyzed for the major, and some minor, components. The data are compared with those of other workers and some significant differences are noted. In particular, the Atlantis II brine has been found to have different concentrations of chloride, bromide, sodium and potassium than previously reported. The major ion compositions of the Atlantis II brine and Discovery brine are similar.

The chemical composition of the 44°C brine in the Atlantis II Deep suggests that it may have formed from the mixing of the highly saline deeper brine and the overlying Red Sea Bottom Water. This brine is rich in manganese but low in iron, owing to precipitation of ferric hydroxide at the 56°/44° interface.

Introduction

This short paper on the chemical composition of the brines had not been originally planned as a contribution to this volume. However, plans made by the editors for a detailed chemical survey of the brines did not come to fruition, and so these few data are presented in order to avoid a gap in the geochemical picture.

Only four samples, representative of each brine mass, have been analyzed, and therefore it is not possible to discuss small variations in ionic composition within each body of water or to examine in detail the processes taking place at the brine interfaces. With this in mind, discussion has been limited to the analytical data alone.

Discussion of Previous Work

The first investigation of the chemical composition of the brines was carried out by Brewer et al. (1965). They reported the concentration of the major ions in a series of samples, varying with depth, in the Discovery Deep and also carried out analyses for minor components in a large volume sample taken from near the bottom of the cast (Table 4). The results showed significant differences in the ionic composition of the brine compared to normal sea water; specifically, the brine was found to be rich in sodium chloride and to have a high calcium to chloride ratio. Magnesium, sulfate and bromide were depleted relative to chloride, and certain trace metals, notably manganese and to a lesser extent zinc, were greatly enriched.

Miller et al. (1966) reinvestigated the site of the brine discovery and located the nearby Atlantis II Deep. The brine contained in the Atlantis II Deep was found to be significantly different from the Discovery brine, having a higher in situ temperature and forming a brownish precipitate shortly after sampling. A sample of this brine was analyzed for major and some minor components (Table 1). Unfortunately no great reliability can be placed on

* Woods Hole Oceanographic Institution Contribution No. 2188.

Table 1   The Ionic Composition of the
Atlantis II Deep 56°C Brine

Element	Miller *et al.* (1965) (g/kg)	This paper (g/kg)
Cl	163.0	156.03
Br	0.07	0.128
$SO_4^=$	1.0	0.840
Si	0.015	0.0276
Na	88.0	92.60
K	3.0	1.87
Ca	5.4	5.15
Mg	0.8	0.764
Sr	–	4.8×10^{-2}
Fe	6×10^{-2}	8.1×10^{-2}
Mn	7×10^{-2}	8.2×10^{-2}
Zn	2×10^{-3}	5.4×10^{-3}
Cu	8×10^{-4}	2.6×10^{-4}
Co	–	1.6×10^{-4}
Ni	–	–
Pb	–	6.3×10^{-4}

* Data from Miller *et al.* divided by 1.196 to convert to units of grams/kilogram.

these figures. Craig (1966) has pointed out that the reported chloride concentration is too high, and Riley (1967) has shown that the analyses appear to be only semiquantitative since the sum of cations and anions do not balance. In spite of these difficulties it is apparent that the two brines have a similar major ion composition. However, the Atlantis II brine gave every indication of being anoxic and rich in iron in contrast to the Discovery brine, which was reported to have a low but significant oxygen content and to be much lower in its concentration of dissolved iron.

Samples and Methods of Analysis

The samples of the Atlantis II 56° brine and the Discovery brine were collected by a Niskin sampler. The samples were of large volume, of the order of 2kg weight, and had been stored in screw top polyethylene bottles for 18 months prior to analysis. The bottles had not been opened or manipulated in any way.

The samples of brine from the 44° layer in the Atlantis II Deep had been stored in soft glass, 8-oz salinity bottles. The total volume of sample was about 200ml, and the bottles had been opened several times for withdrawal of aliquots for other measurements.

The only available sample that may represent the brine contained in the newly found Chain Deep was also stored in a soft glass salinity bottle. The bottle was only half full when it reached our laboratory, and, as before, aliquots of sample had been withdrawn previously. The final analysis was carried out on some 50ml of sample.

In view of the time factor, our lack of control over the samples and having regard for the facilities available, we used atomic absorption spectrophotometry for the determinations of the cations. For the major elements the samples were diluted gravimetrically with distilled water and matrix effects were minimized by using the method of standard additions. The large dilution factors and the extrapolation involved in this method limit our precision, and previous work has established this precision at about ±2 per cent. Better precision could be obtained with aqueous standards, but the accuracy of the data would be poor without careful work to eliminate chemical interferences in the flame. Thus the figures we report are not as precise as the major ion analysis reported for the Discovery Deep (Brewer *et al.*, 1965), but they are considerably more reliable than the analyses for the Atlantis II Deep reported by Miller *et al.* (1966).

Chloride was determined by gravimetric titration with silver nitrate, using potassium chromate indicator, and bromide was determined by the method of Morris and Riley (1966). Sulfate analyses were carried out gravimetrically using barium sulphate as described in Bather and Riley (1954). In these unusual waters the accuracy of this method must be open to doubt, and we did not have an opportunity to calibrate the method by standard additions of sulfate. Samples that had been stored in glass were not analyzed for silicate; the samples stored in polyethylene were analyzed colorimetrically using the method of Grasshoff (1964) and were calibrated by standard additions.

The transition metals were determined by atomic absorption spectrophotometry. Manganese in all samples, and iron in the sample from the 56°C brine, was present in sufficient quantity to permit direct determination in diluted aqueous solution. Other metals were concentrated by a solvent extraction technique using ammonium pyrollidine dithiocarbonate as a chelating agent and extracting the chelate into methyl isobutyl-ketone (Brewer and Spencer, 1968), using the method of standard additions.

An attempt was made to determine fluoride in the brines using a fluoride specific ion electrode; however, the fluorine content of the brines was found to be below the useful range of this device (ca. $<10^{-6}$M). The colorimetric method of Greenhalgh and Riley (1961) was found to be subject to severe interferences (Brewer, 1967) in samples of this type, and no attempt was made to determine fluoride colorimetrically.

Results

The Atlantis II 56°C Brine

The sample of brine taken for analysis had precipitated a thick reddish-brown sediment, and this precipitate was redissolved prior to analysis by the addition of HCl. Distilled water was also added to make sure that the addition of excess chloride did not cause any precipitation of salts. Corrections were applied to the results obtained to compensate for these additions. The results of the analyses are shown in Table 1 and are compared with those of Miller et al. (1966). It should be noted that the data of Miller et al. were originally presented in units of grams per liter. The data have been corrected to units of grams per kilogram by dividing by the density of the sample, reported as 1.196.

Our data supports the main conclusions reached by Miller et al. yet also shows some significant differences. The data for the anions reveals that the chloride content is almost 7g/kg lower than had been previously reported, and the bromide and silicate contents are higher by a factor of approximately two. The major cation analyses show the sodium content to be somewhat higher than previously had been supposed, and the potassium concentration

to be almost 40 per cent lower. A strontium analysis is reported for the first time in this brine, and the value lies close to that reported for the Discovery brine by Brewer et al. (1965). The total ionic strength of the brine is approximately 4.6M.

The transition metal determinations show that iron and manganese are the principal components, in agreement with past work. However, our analyses reveal that both these elements are present in significantly greater proportion than reported in Miller et al., as also is zinc. Our estimation of the copper content of the brine is only one third of that earlier reported. Values for cobalt and lead are given for the first time. Miller et al. noted "the indicated presence of H_2S"; however, the present work does not support the inference, and we found no evidence of sulfide in our samples.

The Atlantis II 44°C Brine

Analysis of the 44°C brine revealed some interesting new information (Table 2). The trace metal concentrations in particular are in marked contrast to those in 56°C brine immediately below. Over 99 per cent of the iron seems to have been lost between the two water masses due to the oxidation of divalent iron to the ferric state and its subsequent precipitation. The turbidity maximum noted by Krause and Ziegenbein (1966) and Ryan et al. (1969) may be at-

Table 2   The Ionic Composition of the 44°C Layer in Atlantis II Deep

Element	Concentration in g/kg
Cl	80.04
Br	0.101
$SO_4^=$	2.26
Si	—
Na	46.90
K	1.07
Ca	2.47
Sr	2.7×10^{-2}
Fe	2.00×10^{-4}
Mn	8.2×10^{-2}
Zn	1.52×10^{-4}
Cu	1.72×10^{-5}
Co	8.0×10^{-7}
Ni	1.2×10^{-6}
Pb	8.8×10^{-6}

tributed to this phenomenon. Much of the zinc, copper, cobalt and lead also seems to be removed at this interface.

The behavior of manganese is strikingly different. Our result indicates no change in the concentration of manganese across the interface in spite of the fact that the total salinity is about half that of the 56°C brine. This would seem to indicate another source of manganese, or, as Bischoff (1969) has suggested, that the dissolved manganese is in equilibrium with a solid phase. Certainly more work is needed to clarify this interesting point.

Intuitively the layer of 44°C water would seem to have formed from a mixing of Red Sea bottom water and the 56° brine. The relationships of the major ions in the respective water masses bear this out to some extent. Table 3 shows the composition of a hypothetical water formed by mixing 56°C brine and Red Sea bottom water in the proportion indicated by the chloride content of the water mass. We can see that to within ca. ±5 per cent we have an approximation to the 44°C water mass. However, the sulfate values are anomalous and do not admit to the simple mixing hypothesis.

Table 3 A Comparison of the Major Ion Compositions of a Hypothetical Mixed Water and the 44°C Water Found in the Atlantis II Deep

Element	Hypothetical Water (g/kg)	"44°" Water (g/kg)	% Difference
Na	47.8	46.9	−1.9%
Ca	2.45	2.47	+0.8%
K	1.13	1.07	−5.3%
Mg	1.21	1.19	−0.7%
Sr	0.025	0.027	+9.0%
SO₄	2.04	2.26	+10%

The Discovery Deep Brine

The results for the sample of Discovery Deep brine are presented in Table 4. The agreement with the data of Brewer *et al.* (1965) is generally good though the concentrations of calcium and sulfate differ by more than can be explained by analytical error. The major ion composition is very similar to that of the Atlantis II 56°C brine, but the minor element concentrations are

Table 4 The Ionic Composition of the Discovery Deep Brine

Element	Brewer *et al.* (1965) (g/kg)	This paper (g/kg)
Cl	155.2	155.3
Br	0.123	0.119
SO₄⁼	0.749	0.695
Si	0.0055	0.0035
Na	92.77	93.05
K	2.15	2.14
Ca	4.70	5.12
Mg	0.810	0.810
Sr	0.046	0.046
Fe	2.2×10^{-4}	2.7×10^{-4}
Mn	0.0540	0.0546
Zn	1.0×10^{-3}	7.7×10^{-4}
Cu	1.4×10^{-5}	7.5×10^{-5}
Co	–	1.29×10^{-4}
Ni	–	3.42×10^{-4}
Pb	–	1.65×10^{-4}

markedly different. Once again, manganese is the principal component though the concentration is lower than in the 56°C brine. With the exception of cobalt, other trace metals are less by an order of magnitude when compared to the Atlantis II Deep.

The Precipitate from the 56°C Brine

In an attempt to investigate the nature of the precipitate thrown down by the 56°C brine, samples were taken from four glass salinity bottles that had been stored for six months. The precipitate was filtered off on to a Millipore* HA filter (pore size 0.45μ) and washed with a little distilled water. The precipitate was dissolved in hydrochloric acid and the resulting solution analyzed by atomic absorption spectrophotometry. The data are presented in Table 5, reported as a ratio to iron which formed the vastly greater proportion of the precipitate. It can be seen that lead forms a larger fraction of the precipitate than does manganese. If this can be taken as indicative of the processes taking place *in situ*, then it would suggest that lead is being preferentially removed from the water mass but that little manganese deposition is occurring. This is in contrast to the observation of Krause and Ziegenbein (1966)

* Trade name.

Table 5   The Ratio of Some Minor Components to Iron in the Precipitate Formed from the Atlantis II Deep 56°C Brine

Depth of Sample	2,121m	2,057m	2,054m	2,065m
Cu × 10⁻⁴	3.73	2.3	2.55	5.08
Ni × 10⁻⁴	1.20	0.9	2.43	6.24
Co × 10⁻⁴	0.6	0.7	3.24	0.9
Mn × 10⁻⁴	0.9	1.5	1.97	7.7
Pb × 10⁻⁴	11.9	20.4	18.2	29.1

Table 6   The Chemical Composition of the Brine from the Chain Deep

Element	Concentration (g/kg)	Ratio to Chloride	Ratio to Chloride in Normal Sea Water
Cl	41.9	—	—
Br	0.0842	2.01×10^{-3}	3.48×10^{-3}
SO$_4^=$	2.81	0.067	0.1400
Na	24.0	0.574	0.5556
K	0.782	0.0187	0.0205
Ca	1.18	0.0282	0.0213
Mg	1.42	0.0339	0.0668
Sr	0.0125	0.298×10^{-3}	0.41×10^{-3}
Mn	0.005	—	—

that manganese forms a major part of the precipitate.

The Chain Deep Brine

The finding of the Chain Deep was a significant, new contribution from the 1966 CHAIN cruise. This third brine pool in the Red Sea lies on a saddle between the Atlantis II and Discovery Deeps and has a maximum depth of 2,066m (Ross and Hunt, 1967). The maximum temperature that has been recorded in the Deep is 34.0°C at a depth of 2,042m and the maximum salinity reported is 74.2 per mill, based on a conductivity determination on a sample from 2,024m depth. The ionic composition of the brine is different from that of normal sea water, and this value is certainly too high.

The narrow, steep-sided deep presents problems for adequate sampling, and attempts to obtain representative samples of the hottest, most saline brine contained there were unsuccessful. The only sample available for chemical analysis was taken from Station No. 729 at a depth of 2,024m; the *in situ* temperature was 29.07°C and the conductometric salinity was 74.2 per mill.

The chemical composition of this sample showed that although the brine was of much lower salinity than the brines from the Atlantis II and Discovery Deeps, it nevertheless had features in common. Table 6 shows the major ion composition of the brine and also the ratio of each element to chloride in comparison with normal sea water. It can be seen that the brine is enriched in sodium and calcium and depleted in magnesium and sulfate relative to normal sea water. These trends are similar to those found in the other brines but here are exhibited to a much lesser degree. The manganese content of the sample is high, and although it is only one tenth that of the Atlantis II and Discovery Deep brines, it is still enriched about one thousandfold in comparison to normal sea water.

The sample analyzed here is almost certainly not representative of the deepest brine from the Chain Deep. We have every reason to suppose that hotter brines from the Deep would be more saline and probably closer in ionic composition to the brines from the other two deeps.

Acknowledgments

The authors are grateful to Dr. H. Trüper and Mr. C. D. Densmore for providing samples and to Mr. N. Corwin for carrying out the silicate analysis.

This work was supported by Contract No. GA-584 from the National Science Foundation, and Contract No. AT(30-1)1918 from the Atomic Energy Commission Report No. NYO-1918-172.

References

Bather, J. M., and J. P. Riley: Chemistry of the Irish Sea. Part I, sulphate-chlorinity ratio. J. Cons. Int. Explor. Mer., **20,** 145 (1954).

Bischoff, J.: Red Sea geothermal brine deposits: their mineralogy, chemistry, and genesis. *In: Hot brines and recent heavy metal deposits in the Red Sea,* E. T. Degens and D. A. Ross (eds.). Springer-Verlag New York Inc., 368–401 (1969).

Brewer, P. G., J. P. Riley, and F. Culkin: The chemical composition of the hot salty water from the bottom of the Red Sea. Deep-Sea Res., **12,** 497 (1965).

Brewer, P. G.: Investigation of a Red Sea deep brine and certain micro-nutrients of sea water. Ph.D. thesis, University of Liverpool, 142 p. (1967).

——, and D. W. Spencer: Determination of trace metals in sea water using atomic absorption spectrometry. In press (1968).

Craig, H.: Isotopic composition and origin of the Red Sea and Salton Sea geothermal brines. Science, **134,** 1544 (1966).

Grasshoff, K.: On the determination of silica in sea water. Deep-Sea Res., **11,** 597 (1964).

Greenhalgh, R., and J. P. Riley: Determination of fluoride in natural waters with particular reference to sea water. Analytica chim. Acta, **25,** 179 (1961).

Krause, G., and J. Ziegenbein: Die Struktur des heissen salzeichen Tiefenwassers im zentralen Roten Meer. METEOR Forschungsergebnisse, A, **No. 1,** 53 (1966).

Miller, A. R., C. D. Densmore, E. T. Degens, J. C. Hathaway, F. G. Manheim, P. F. McFarlin, R. Pocklington, and A. Jokela: Hot brines and recent iron deposits in deeps of the Red Sea. Geochim. et Cosmochim. Acta, **30,** 341 (1966).

Morris, A. W., and J. P. Riley: The bromide/chlorinity and sulphate/chlorinity ratio in sea water. Deep-Sea Res., **13,** 699 (1966).

Riley, J. P.: The hot saline waters of the Red Sea bottom and their related sediments. Oceanogr. Mar. Biol. Ann. Rev., **5,** 141 (1967).

Ross, D. A., and J. M. Hunt: Third brine pool in the Red Sea. Nature, **213,** 687 (1967).

Ryan, W. B. F., E. M. Thorndike, M. Ewing, and D. A. Ross: Suspended matter in the Red Sea brines and its detection by light scattering. *In: Hot brines and recent heavy metal deposits in the Red Sea,* E. T. Degens and D. A. Ross (eds.). Springer-Verlag New York Inc., 153–157 (1969).

Trace Element Composition of Red Sea Geothermal Brine and Interstitial Water *

R. R. BROOKS,† I. R. KAPLAN,‡ AND M. N. A. PETERSON §

Abstract

Analytical data are presented for concentration of Na, K, Ca, Mg, Li, Sr, Mn, Fe, Pb, Cu, Zn, Cd, Co and Ni in the brine of Atlantis II and Discovery Deeps and also the interstitial water. Enrichment of alkali metals, Ca, Mn, Fe, Pb, Cu and Zn was found in the brine, and Mg was depleted relative to sea water. All metals were enriched in the interstitial water, with the exception of Mg and Fe. The base metals are much more enriched in the interstitial water than the brine.

On the basis that trace metal enrichment is higher in brine and interstitial water samples from the Atlantis II Deep than from the Discovery Deep, it is suggested that the source of the brine is derived from beneath the floor at Atlantis II Deep.

Introduction

Details on the chemical composition of the Red Sea brines have been given by Miller *et al.* (1966) and Riley (1967). It was recognized that the overall salinity of Discovery Deep appeared constant, although small differences were noticed due to minor stratifications. There also may have been changes occurring between the times of sampling. In the Atlantis II Deep evidence was produced for two layers underlying the normal Red Sea water.

This study considers the trace element constituents (Fe, Mn, Zn, Cu, Pb, Cd, Co, Ni, Li, Sr) and also Na, K, Ca and Mg of the brines and the interstitial waters squeezed from the sediment. It was hoped that the results would assist in (1) defining the origin of the brines in the deeps and (2) explaining the formation of the unusual sediment deposits.

Methods

Sampling

Hydrocast samples were collected by a Niskin sampler (coated with teflon) and were stored in polyethylene bottles at 4°C for six months prior to analysis. Hydrocast stations analysed are: 715, 718, 722(1), 722(2), 726, 727 (Atlantis II Deep); 717, 728(1), 728(3) (Discovery Deep); and 729 (Chain Deep) all shown in Fig. 1.

Interstitial water was analysed from eight cores: 84K, 120K, 128G, 128P (Atlantis II Deep); 81G, 119K (Discovery Deep), 95K (a rise within Atlantis II Deep), 118K (an area south of the deeps), the location of which are given in Fig. 1. Cores 84K, 120K and 118K were stored at −20°C. Core 95K was stored at 2 to 4°C, whereas cores 119K, 81G, 85K, 128G and 128P were left at ambient temperatures from time of collection until February, 1967. All samples were subsequently stored at 4°C until analysed (June, 1967).

* Publication No. 691 Institute of Geophysics and Planetary Physics, University of Southern California, Los Angeles, California.

† Present address: Department of Chemistry, Massey University, Palmerston North, New Zealand.

‡ Present address: Department of Geology and Institute of Geophysics and Planetary Physics, University of California, Los Angeles.

§ Present address: Scripps Institution of Oceanography, La Jolla, California.

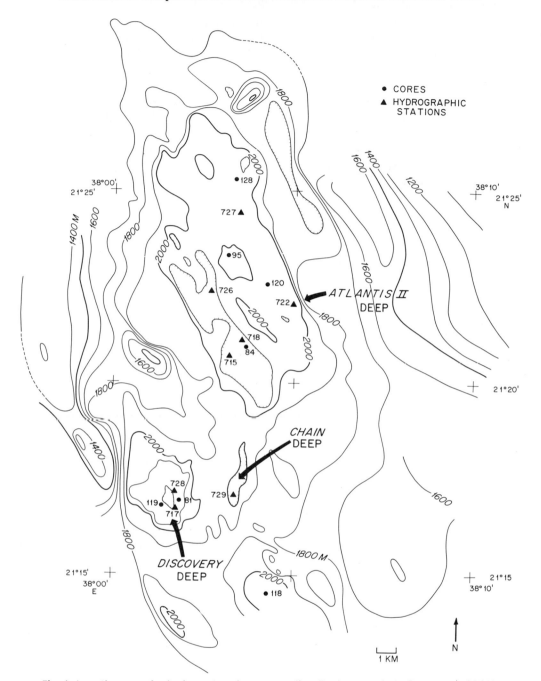

Fig. 1. Location map for hydrocast and core sampling (bathymetry from Ross *et al.*, 1969).

Hydrocast Analyses

Temperature and salinity measurements on the collected waters were made on board the R/V **CHAIN** following the description of Munns *et al.* (1967). On storage of the brines it was noted that some of the deeper samples from Atlantis II Deep had precipitated ferric hydroxide on the walls of the storage bottles. All samples from this brine body were therefore first treated with $^{1}/_{10}$ their volume of 6 N HCl and gently heated. All the brine samples were then diluted with distilled water to a total volume three times that of the original. Trace element analysis on Zn, Cu, Fe, Co and Pb was performed on 100ml of Atlantis II Deep bottom brine by additional dilution to 350ml, whereas the other samples were not further diluted. The pH was adjusted to 3.5 either by addition of purified 10 N ammonium hydroxide to the acid-treated samples or by addition of 6 N HCl to the untreated waters. Ammonium pyrrolidinedithiocarbamate (APDC) and methyl isobutyl ketone (MIBK) were then added to each sample and the metal complex was allowed to enter the organic solvent phase, which was then analysed on a Perkin-Elmer 303 Atomic absorption spectrophotometer. The procedure followed has been described in detail by Brooks *et al.* (1967).

The concentrated brines were further diluted eleven times and analysed directly for Mg, Ca and K by atomic absorption spectrometry and for Li by flame photometry. Na was analysed by an atomic absorption spectrometer on samples diluted 1,200-fold.

Interstitial Waters

Water was squeezed out of the sediment using a teflon-lined press described by Presley *et al.* (1967). As soon as possible after squeezing, pH and redox potential (Eh) were measured on the waters. The waters from all Atlantis II and Discovery Deep cores were diluted threefold. Trace metal analysis was made directly on the solutions without organic extraction as in the case of the hydrocasts. Further dilution was performed, as above, in order to analyse the alkali metals.

Results

General

The results are presented in Tables 1 and 3. Tables 2 and 4 show the relationship of the major ions to sodium, which is considered a conservative element. The relationship of alkali metals to sodium is graphically shown in Figs. 2 and 5, and the rela-

tionship between Fe and Zn in Atlantis II Deep is shown in Fig. 7.

Sulfate has been analysed separately from several of the hydrocast stations and interstitial waters. Values in the hydrocast samples varied from about 3.0g/l in the surface layers, 1.2–1.9g/l in the intermediate layer of Atlantis II Deep and 0.6g/l in the bottom water. In general the sulfate concentration showed an inverse relationship to sodium concentration (Kaplan *et al.,* 1969).

Brines from Hydrocasts

Analytical results for the alkali metals and alkaline earths (Na, K, Li, Ca and Mg) and six transition and base metals (Mn, Fe, Zn, Cu, Co and Pb) are given in Table 1. It is apparent that trends are present both for the major elements and for the trace elements. Table 2 displays the change in concentration relative to sodium, the major cation in the brines. This will enable an evaluation of whether concentration of other cations in the brine has been dominated by processes which have caused increase in sodium concentration or whether independent mechanisms were operational.

Sodium shows an eight- to ninefold increase in concentration over its concentration in normal Red Sea water. Three layers are distinctly evident in the Atlantis II Deep and a thin intermediate layer is evident in Discovery Deep (hydrocast No. 717).

Potassium shows a sixfold to sevenfold increase in the bottom Atlantis II and Discovery Deep brines over its concentration in normal Red Sea water. There is also good evidence for three distinct layers as seen in hydrocast No. 722 (1 and 2), 726 and 727. The Discovery Deep displays some layering as seen in 717, but only a thin layer is apparent.

There is a general relationship between potassium and sodium as seen in Fig. 2. Potassium does not, however, increase to the same extent as sodium. In Figs. 2, 4 and 5, the broken line, which intersects the origin and the value for normal sea water, is a locus which should be followed by

Table 1 Trace Elements in Hydrocasts from Red Sea Brines (ppm)

Code	Na (%)	K	Mn	Fe	Zn	Cu	Pb	Co	Li	Mg	Ca	Temp. (°C)	Salinity (‰)	Depth (m)
Atlantis—715														
10	1.66	470	<1	0.12	0.96	0.58	0.06	0.006	0.21	1,770		22.69		1,978
11	1.36	470	<1	0.08	0.90	0.16	0.02	0.007		1,530	460	23.56		1,993
12	2.30	700	19	0.12	3.60	1.03	0.14	0.008		1,450	910	>30.8		2,008
13	5.18	1,270	84	1.05	6.12	0.95	0.17	0.015	1.78	1,180	2,200	>30.8		2,023
14	5.12	1,270	81	1.50	4.68	0.35	0.31	0.013		1,180	2,300	44.61		2,038
15	11.20	2,620	90	92	8.82	0.56	0.56	0.006		830	5,200			2,053
16	11.20	2,560	90	97	7.92	0.16	0.37	0.009		810	5,100	56.52	315.92	2,068
17	11.20	2,620	93	97	7.92	0.23	0.12	0.017		800	5,200		318.78	2,083
18	11.20	2,680	93	93	7.65	0.32	0.17	0.005		800	5,200	56.55	320.38	2,098
19	11.20	2,640	90	95	8.82	0.16	0.22	0.002		810	5,100		318.78	2,113
20	10.90	2,620	90	93	7.92	0.13	0.08	<0.001		800	4,700		319.55	2,128
21	10.90	2,680	90	95	6.66	0.04	0.31	0.008		780	4,800		318.78	2,143
22	10.90	2,680	90	93	7.20	0.31	0.16	0.002		790	5,100		317.79	2,158
23	11.80	2,960	100	107	7.65	0.15	0.15	0.008	4.72	860	5,500		319.11	2,173
(core) 24	10.80	2,560	87	97	71.40	0.03	0.15	0.014		780	5,100			
50m below surface		420	1	0.30	0.05	0.02	0.01	0.001	0.21					
Atlantis—718														
14	1.31	400	<1	0.06	0.07	0.15	0.03	0.016	0.22	1,550	420	21.98	40.610	1,860
13	1.25	400	<1	0.06	0.16	0.04	0.01	0.003		1,500	420	22.37	41.000	1,960
12	1.25	400	<1	0.07	0.16	0.05	0.01	0.002		1,530	420	22.44	41.090	1,965
11	1.24	400	<1	0.12	0.17	0.04	0.00	0.002		1,500	420	22.51	41.190	1,970
10	1.24	400	<1	0.08	0.44	0.07	0.00	0.003		1,500	420	22.60	41.280	1,975
9	1.28	400	<1	0.36	0.22	0.07	0.01	<0.001		1,530	420	22.77	41.460	1,979
8	1.32	400	<1	0.12	0.20	0.03	0.01	<0.001		1,500	420	23.00	41.720	1,984
7	1.30	470	<1	0.07	0.13	0.08	0.01	0.003		1,510	420	23.49	42.310	1,989
6	1.40	470	<1	0.17	0.25	0.15	0.06	0.002		1,510	800	24.65	44.080	1,994
5	1.66	550	4.5	0.07	0.39	0.07	0.01	<0.001		1,470	1,300	28.80	44.440	1,999
4	2.00	640	15	0.13	1.26	0.07	0.04	0.005		1,360	1,400		56.870	2,004
3	3.48	990	45	0.58	1.92	0.11	0.11	0.005		1,300	1,500			2,009
2	5.16	1,270	82.5	1.20	3.12	0.26	0.05	0.008	1.78	1,210	2,100			2,014
1	4.94	1,270	80	1.09	3.12	0.50	0.08	0.009		1,160	2,100	44.32		2,019

Code	Na (%)	K	Mn	Fe	Zn	Cu	Pb	Co	Li	Mg	Ca	Temp. (°C)	Salinity (‰)	Depth (m)
Atlantis—726														
9	1.31	400	<1	0.04	0.15	0.02		0.001	0.26	1,520	430	22.98	41.680	1,990
10	1.49	500	<1	0.08	0.87	0.05		0.001		1,490	530	25.99	39.208	1,990
11	2.14	600	14	0.02	1.70	0.11		0.004		1,440	830	33.48	61.920	2,010
12	5.08	1,150	77	2.78	11.03	0.61		0.008		1,180	2,300		149.88	2,020
13	5.08	1,150	78	2.47	8.06	0.28		0.012	1.84	1,180	2,300		150.90	2,025
14	5.07	1,120	78	2.83	4.98	0.20		0.014		1,180	2,300		151.26	2,030
15	5.03	1,150	78	2.47	3.46	0.53		0.013		1,180	2,300		150.18	2,035
16	5.07	1,150	78	2.94	4.41	0.26		0.013		1,210	2,200		150.36	2,040
17	5.07	1,150	80	3.73	3.15	0.10		0.014		1,190	2,300	56.43	150.72	2,045
18	11.10	2,590	87	87	9.14	0.45		0.002		800	5,100		317.13	2,050
19	11.10	2,590	87	88	8.38	0.29		0.003		800	5,100		316.69	2,055
20	11.10	2,500	87	88	9.53	0.39		0.008		800	5,200		319.44	2,060
21	10.30	2,620	84	85	7.82	0.25		0.004		710	5,100		317.13	2,065
22	10.90	2,620	83	83	10.40	0.32		0.004		780	5,100		316.47	2,075
23	11.20	2,620	84	90	9.95	0.29		0.006	4.70	820	5,100		318.78	2,085
Atlantis—727														
9	1.28	450	<1	0.04	0.07	0.01	0.01	<0.001	0.23	1,510	430	22.74	41.40	1,980
10	1.31	400	<1	0.01	0.06	0.02	0.01	<0.001		1,480	430	23.02	41.72	1,990
11	1.56	470	2.1	0.04	0.41	0.02	0.01	<0.001		1,490	590	26.43	47.99	2,000
12	2.30	610	18	0.02	2.02	0.06	0.01	<0.001		1,430	940		66.72	2,010
13	5.06	1,200	78	0.48	3.02	0.12	0.01	0.010		1,180	2,300		150.42	2,015
14	5.00	1,200	78	0.76	2.33	0.11	0.05	0.011		1,160	2,300		130.36	2,020
15	5.01	1,200	78	0.75	1.95	0.13	0.06	0.010	1.56	1,180	2,400		149.16	2,025
16	5.04	1,230	77	0.85	2.02	0.07	0.07	0.013		1,180	2,300		150.06	2,030
17	5.00	1,230	78	0.75	1.89	0.04	0.01	0.012		1,210	2,400		150.12	2,035
18	4.98	1,270	80	1.18	2.83	0.23	0.05	0.013		1,150	2,300	44.26	150.54	2,040
19	5.13	1,270	82	0.98	2.02	0.06	0.08	0.013		1,210	2,400		146.96	2,045
20	10.10	2,620	87	88	5.10	0.07	0.14	<0.001		700	5,200		318.56	2,050
21	8.48	2,700	87	87	4.91	0.22	0.13	<0.001		650	5,000		317.79	2,055
22	11.90	2,700	90	97	5.90	0.15	0.20	0.007		880	5,100		318.56	2,065
23	11.20	2,620	86	90	5.54	0.11	0.29	0.003	4.16	800	5,000	56.49	318.56	2,075

Code	Na (%)	K	Mn	Fe	Zn	Cu	Pb	Co	Li	Mg	Ca	Temp. (°C)	Salinity (‰)	Depth (m)
Atlantis—722 No. 1														
9	1.55	500	3	0.44	0.82	0.20	0.03		0.37	1,500	660	27.93	48.00	2,001
10	1.92	580	11							1,470	780		55.62	2,006
11	2.58	700	23	0.75	4.66	0.34	0.09			1,160	1,100		72.30	2,011
12	5.02	1,270	80							1,160	2,400		150.36	2,016
13	5.02	1,270	80	0.84	3.21	0.56	0.23			1,160	2,300		151.02	2,020
14	4.94	1,250	80							1,140	2,300		150.96	2,025
15	4.94	1,250	80	0.800	2.33	0.38	0.24			1,140	2,300			2,030
16	4.94	1,250	78						1.56	1,160	2,300		150.72	2,035
20	5.03	1,250	80	1.54	2.33	0.30	0.06			1,160	2,300		151.26	2,040
17B	5.03	1,300	80								2,400		153.00	2,045
17A	11.10	2,620	83	87	4.66	0.32	0.26			790	5,100	44.10	318.45	2,055
1	10.90	2,560	87	90	4.78	0.39	0.19			790	5,200		320.87	2,060
22	10.90	2,560	84	87	7.06	0.52	0.39			760	5,100		319.00	2,065
NIS 1	10.80	2,540	86	90	5.04	0.03	0.12			770	5,100			2,070
23	11.00	2,640	84	90	6.55	0.42	0.72			770	5,100			2,075
NIS 2	11.00	2,680	84	93	4.80	0.03	0.12		4.14	800	5,200		319.77	2,081
Atlantis—722 No. 2														
10	1.38	470	<1	0.06	0.19	0.20	0.03	<0.001	0.24	1,550	470	22.44	41.10	1,970
5	1.32	470	<1							1,550	480	22.62	41.34	1,975
12	1.32	470	<1	0.04	0.25	0.20	0.01	<0.001		1,530	480	11.76	41.45	1,980
13	1.33	500	<1							1,540	500	22.94	41.66	1,984
14	1.40	500	<1	0.10	0.19	0.12	0.04	<0.001		1,510	510	23.64	42.51	1,989
15	1.48	530	<1							1,500	540	25.22	45.21	1,994
16	1.75	600	4.5	0.05	0.69	0.11	0.06	0.019		1,540	660	28.62	50.76	1,999
20	2.13	680	15							1,430	860		61.98	2,004
18	3.62	990	50	0.49	1.83	0.24	0.23	0.050		1,290	1,600		107.04	2,009
19	5.03	1,350	78							1,150	2,200		150.36	2,014
17	5.06	1,350	78	0.91	2.14	0.15	0.04	0.060		1,150	2,200		151.02	2,019
1	4.90	1,320	78							1,120	2,300		150.90	2,024
22	4.80	1,270	77	1.00	3.78	0.20	0.10	0.074	1.53	1,110	2,200		150.98	2,029
23	4.91	1,270	78							1,110	2,300	44.20	150.84	2,034
NIS 3	8.80	2,140	81	62.7	2.28	0.05	0.05	0.030		900	4,000			2,044
NIS 5	9.40	2,540	81	92.4	4.28	0.03	0.06		4.24	870	4,600			2,074
Chain—729														
20	1.28	400	<1	0.002	0.04	0.02	<0.001	<0.001	0.21	1,510	420	22.06	40.880	1,930–1,932
21	1.34	400	<1	0.002	0.12	0.05	0.01	<0.001		1,530	500	23.80	42.920	1,999–2,002
22	1.75	530	2.4	0.002	0.63	0.06	0.02	<0.001		1,420	680	25.98	57.900	2,019–2,022
23	2.44	700	6.6	0.002	0.66	0.01	0.02	<0.001	0.66	1,470	910	29.07	66.380	2,024–2,021

Code	Na (%)	K	Mn	Fe	Zn	Cu	Pb	Co	Li	Mg	Ca	Temp. (°C)	Salinity (‰)	Depth (m)
Discovery—717														
1	1.33	400	<1						0.26	1,530	480	22.99	42.48	1,985
2	1.38	450	<1	0.07	0.26	0.06	0.02	<0.001		1,530	480	23.22	43.07	1,990
3	1.38	450	<1							1,480	490	23.90	41.28	1,995
4	1.50	470	<1	0.03	0.42	0.08	0.01	<0.001		1,530	530		42.60	1,999
5	1.50	470	<1							1,500	530	24.80	44.10	2,004
6	1.58	470	<1	0.03	0.31	0.06	0.02	<0.001		1,410	550	26.11	52.52	2,009
7	1.92	600	3							1,460	710	27.51	60.04	2,014
8	2.42	700	6	0.05	0.42	0.07	0.14	<0.001		1,440	740	29.72	72.78	2,019
9	3.16	870	15							1,390	760		94.32	2,024
10	4.68	1,160	27	0.08	0.76	0.15	0.12	<0.001		1,260	1,900		140.25	2,029
11	4.91	1,320	29						1.66	1,240	2,000		149.1	2,034
12	5.26	1,370	33	0.09	0.51	0.33	0.13	<0.001		1,180	2,200		160.15	2,039
13	10.20	2,560	62							1,010	4,400	41.90	280.20	2,044
14	12.00	3,000	66	0.06	1.07	0.17	0.21	<0.001		970	5,200		316.73	2,151
15	10.90	2,880	62							860	4,700		316.84	2,153
16	11.90	3,050	66	0.05	1.83	0.34	0.18	<0.001		970	5,200		317.16	2,158
17	10.90	2,750	62						4.04	875	4,700		317.38	2,160

Code	Na (%)	K	Mn	Fe	Zn	Cu	Pb	Co	Li	Mg	Ca	Temp. (°C)	Salinity (‰)	Depth (m)
Discovery—728-1														
13	3.86	970	18	<0.06	<1.95	<0.31	<0.11	<0.001	1.20	1,360	2,100		138.60	2,038
9	4.88	1,150	30						1.60	1,230	3,800		273.79	2,048
10	9.98	2,230	51	0.04	1.04	0.22	0.07	<0.001		1,040	4,900		316.80	2,058
11	11.20	2,620	56							900	5,100		317.13	2,068
12	11.00	2,560	56	>0.06	>1.95	>0.31	>0.11	<0.001		900			317.02	2,071
14	11.20	2,700	60							890	5,000		317.57	2,074
15	11.00	2,500	56	0.06	0.70	0.23	0.10	<0.001		870	4,800		315.81	2,077
16	11.00	2,540	56							880	4,900		316.80	2,080
17	11.00	2,640	56	0.07	0.82	0.18	0.18	<0.001		880	5,100		316.47	2,083
18	11.00	2,620	54							880	5,000		318.12	2,086
19	11.20	2,620	54	0.15	0.76	0.22	0.10	<0.001		910	5,000	44.75	316.31	2,089
20	11.20	2,560	54							900	5,000		317.13	2,091
21	11.20	2,560	54	0.06	0.91	0.25	0.09	<0.001		890	4,900		316.91	2,094
22	11.50	2,560	53							925	4,900	44.74	316	2,098
23	11.00	2,560	56	0.04	2.77	0.16	0.22	<0.001	4.02	880	4,700		317	

Cd — 13: 0.005, 22: 0.000
Ni — 13: 0.003, 22: 0.002

Code	Na (%)	K	Mn	Fe	Zn	Cu	Pb	Co	Li	Mg	Ca	Temp. (°C)	Salinity (‰)	Depth (m)
Discovery—728-3														
9	1.26	400	<1	0.07	0.45	0.08	0.02	<0.001	0.22	1,510	450	22.05	40.73	1,930
13	2.54	710	7.5	0.16	1.42	0.05	0.08	<0.001	0.76	1,400	1,100	30.80	79.92	2,030
22	11.10	2,560	56							885	4,900	44.72	316.36	2,099
NIS 1	11.00	2,560	57							890	4,900			2,149
NIS 2	11.30	2,560	57	0.10	0.50	0.001	0.005	<0.001		885	4,800			2,159
NIS 3	10.90	2,560	56							865	4,800			2,179
NIS 4	10.80	2,560	57											
G. Core		2,230	62	0.42	0.40	0.002	<0.001	<0.001	4.06	880	4,700			2,199

Table 2 Concentrations of Constituents of Hydrocasts (ppm) Divided by the % Sodium Content of Each Sample

Code	Na (%)	Depth	K	Mn	Fe	Mg	Ca	Li
717								
2	1.38	1,990	326	0.72	0.05	1,108	347	0.33
4	1.50	1,999	314	0.66	0.02	1,019	353	
6	1.58	2,009	298	0.53	0.02	892	348	
8	2.42	2,019	290	2.47	0.02	595	305	0.61
10	4.68	2,029	246	5.76	0.01	269	405	
12	5.26	2,039	260	6.27	0.01	224	418	
14	12.00	2,151	249	5.49	<0.01	80	433	0.54
16	11.90	2,158	256	5.54	<0.01	81	436	
728-1								
13	3.86	2,038	251	4.66	<0.01	352	544	0.31
10	9.98	2,058	223	5.11	<0.01	104	490	
12	11.00	2,071	232	5.09	<0.00	81	446	
14	11.20	2,074	241	5.35	<0.01	79	445	
16	11.00	2,080	230	5.09	<0.01	79	454	
18	11.00	2,086	238	4.90	0.01	79	446	
20	11.20	2,091	228	4.82	<0.01	80	446	
22	11.50	2,098	222	4.60	<0.01	80	426	0.60
728-3								
9	1.26	1,930	316	0.79	0.05	1,158	357	0.30
22	11.10	2,099	230	5.04	0.01	79	441	0.51
N1S2	11.30	2,159	226	5.04	<0.01	78	424	
N1S4	10.80	2,199	237	5.27	0.03	81	435	0.64
715								
10	1.66	1,978	284	0.60	0.07	1,066	259	0.22
11	1.36	1,993	347	0.73	0.05	1,124	338	
12	2.30	2,008	305	8.26	0.05	630	395	
13	5.18	2,023	245	16.21	0.20	227	424	0.59
14	5.12	2,038	248	15.82	0.29	230	449	
15	11.20	2,053	231	7.96	8.14	73	460	
16	11.20	2,068	228	8.03	8.66	72	455	
17	11.20	2,083	233	8.30	8.66	71	464	
18	11.20	2,098	239	8.30	8.30	71	464	
19	11.20	2,113	235	8.03	8.48	72	455	
20	10.90	2,128	240	8.25	8.53	73	431	
21	10.90	2,143	245	8.25	8.71	71	440	
22	10.90	2,158	245	8.25	8.53	72	467	
23	11.80	2,173	250	8.47	9.06	72	466	0.73
24	10.80	2,194	237	8.05	8.98	72	472	

Code	Na (%)	Depth	K	Mn	Fe	Mg	Ca	Li
718								
14	1.31	1,860	304	0.76	0.04	1,183	320	0.29
13	1.25	1,960	319	0.79	0.04	1,199	335	
12	1.25	1,965	319	0.79	0.05	1,223	335	
11	1.24	1,970	321	0.80	0.09	1,209	338	
10	1.24	1,975	321	0.80	0.06	1,209	338	
9	1.28	1,979	311	0.78	0.28	1,195	328	
8	1.32	1,984	302	0.75	0.09	1,136	318	
7	1.30	1,989	363	0.76	0.05	1,161	323	
6	1.40	1,994	337	0.71	0.11	1,078	571	
5	1.66	1,999	328	2.71	0.04	885	783	
4	2.00	2,004	319	7.49	0.06	679	699	
3	3.48	2,009	283	12.93	0.16	373	431	
2	5.16	2,014	246	15.98	0.23	234	406	
1	4.94	2,019	257	16.19	0.22	234	425	0.62
726								
9	1.31	1,990	304	0.76	0.02	1,160	328	0.34
10	1.49	2,000	338	0.67	0.05	999	355	
11	2.14	2,010	279	6.54	0.01	672	387	
12	5.08	2,020	226	15.15	0.54	232	452	
13	5.08	2,025	220	15.35	0.48	232	452	
14	5.07	2,030	227	15.38	0.55	232	453	0.62
15	5.03	2,035	222	15.50	0.49	234	457	
16	5.07	2,040	227	15.38	0.57	238	433	
17	5.07	2,045	227	15.77	0.73	234	453	
18	11.10	2,050	233	7.83	7.83	72	459	
19	11.10	2,055	233	7.83	7.92	72	459	
20	11.10	2,060	225	7.83	7.92	72	468	
21	10.30	2,065	254	8.15	8.25	68	495	
22	10.30	2,075	240	7.61	7.61	71	467	0.71
23	11.20	2,085	233	7.49	8.03	73	455	
727								
9	1.28	1,980	352	0.78	0.02	1,179	335	0.32
10	1.31	1,990	304	0.76	0.00	1,129	328	
11	1.56	2,000	302	1.34	0.02	955	378	
12	2.30	2,010	264	7.82	0.00	621	408	
13	5.06	2,015	237	15.41	0.09	233	454	
14	5.00	2,020	239	15.59	0.46	231	459	
15	5.01	2,025	239	15.56	0.14	235	479	

Code	Na (%)	Depth	K	Mn	Fe	Mg	Ca	Li
16	5.04	2,030	244	15.27	0.16	234	456	0.53
17	5.00	2,035	245	15.59	0.14	241	479	
18	4.98	2,040	255	16.06	0.23	230	461	
19	5.13	2,045	247	15.98	0.19	235	467	
20	10.10	2,050	259	8.61	8.71	69	514	
21	8.48	2,055	318	10.25	10.25	76	589	
22	11.90	2,065	226	7.56	8.15	73	428	
23	11.20	2,075	233	7.67	8.03	71	446	0.63
722-1								
9	1.92	2,001	262	1.56	0.23	781	346	0.33
11	2.58	2,011	272	8.91	0.29	569	426	
13	5.02	2,020	252	15.93	0.16	231	458	
15	4.94	2,030	253	16.19	0.16	230	465	0.54
20	5.03	2,040	248	15.90	0.30	230	457	
17A	11.10	2,055	236	7.47	7.83	71	459	
1	10.90	2,060	234	7.98	8.25	72	477	
22	10.90	2,065	234	7.70	7.98	69	467	
N1S1	10.80	2,070	235	7.96	8.33	71	472	
23	11.00	2,075	239	7.63	8.18	69	463	
N1S2	11.00	2,081	243	7.63	8.45	72	472	0.64
722-2								
10	1.38	1,970	342	0.72	0.04	1,123	340	0.30
12	1.32	1,980	357	0.75	0.03	1,159	363	
14	1.40	1,989	359	0.71	0.07	1,078	364	
16	1.75	1,999	342	2.57	0.03	879	377	
18	3.62	2,009	304	15.33	0.15	395	490	
17	5.06	2,019	266	15.41	0.18	227	434	
22	4.80	2,029	264	16.04	0.20	231	458	0.54
N1S2	8.80	2,044	243	9.20	7.12	102	454	0.77
729								
20	1.28	1,932	311	0.78	<0.01	1,179	328	0.28
21	1.34	2,002	297	0.74	<0.01	1,141	373	
22	1.75	2,022	302	1.37	<0.01	811	388	
23	2.44	2,022	288	2.70	<0.01	602	372	0.46

Table 3   Constituents of Saline Waters from Red Sea Interstitial Waters
(All concentrations in ppm except sodium)

Depth (cm)	Na (%)	K	Li	Zn	Mn	Pb	Cu	Co	Ni	Cd	Sr	Ca	Mg	pH	Eh mv
Core 118															
0–20	1.27	470	0.66	0.4	2.9	0.3	0.1	0.3	0.3	<0.03	11	687	1,561	7.30	+370
20–40	1.21	480	0.46	0.3	2.6	0.2	<0.1	0.3	0.3		10	515	1,518	7.40	+370
50–70	1.15	500	0.46	0.2	2.3	0.2	<0.1	0.3	0.3		9	515	1,452	7.43	+335
70–90	1.15	450	0.46	0.1	2.3	0.2	<0.1	0.2	0.3		10	584	1,436	7.20	+320
90–100	1.04	420	0.40	0.1	2.3	0.2	<0.1	0.3	0.3		9	584	1,472	7.27	+310
100–110	1.15	450	0.46	0.1	3.3	0.2	<0.1	0.3	0.3		9	652	1,541	7.35	+290
110–120	1.15	440	0.46	0.2	3.4	0.2	<0.1	0.3	0.5		8	584	1,386	7.50	+265
130–150	1.16	500	0.46	0.1	5.1	0.2	<0.1	0.3	0.5		9	652	1,502	7.49	+225
155–160	1.07	420	0.40	0.1	3.8	0.2	<0.1	0.4	0.5		8	584	1,346	7.37	+232
170–190	1.12	450	0.46	<0.1	3.2	0.2	<0.1	0.4	0.5		8	584	1,445	7.58	+230
190–200	1.15	420	0.46	<0.1	2.9	0.2	<0.1	0.3	0.5		10	807	1,346	7.32	+238
200–205	1.17	470	0.46	<0.1	3.8	0.2	<0.1	0.4	0.3		10	687	1,386	7.49	+235
205–210	1.15	500	0.53	<0.1	4.4	0.2	<0.1	0.4	0.5		10	755	1,439	7.37	+220
210–218	1.25	450	0.53	<0.1	4.4	0.2	<0.1	0.4	0.3		11	807	1,403	7.07	+225
218–225	1.22	420	0.46	<0.1	4.7	0.1	<0.1	0.4	0.3		10	687	1,254	7.32	+303
225–240	1.25	440	0.53	<0.1	6.1	0.2	<0.1	0.4	0.3		10	755	1,386	7.13	+300
240–260	1.37	440	0.53	<0.1	5.0	0.1	<0.1	0.4	0.3		10	807	1,403	7.23	+280
260–280	1.15	420	0.53	<0.1	4.2	0.2	<0.1	0.4	0.5		10	807	1,360	7.20	+290
280–300	1.16	420	0.53	<0.1	4.1	0.2	<0.1	0.4	0.5		10	807	1,343	7.20	+250
300–307	1.12	420	0.46	<0.1	3.3	0.1	<0.1	0.4	0.5		10	807	1,294	7.28	+310
307–311	1.07	450	0.40	<0.1	3.3	0.2	<0.1	0.4	0.5		8	687	1,234	7.40	+280
311–321	1.15	510	0.53	<0.1	4.2	0.2	<0.1	0.4	0.6		8	652	1,403	7.54	+270
329–345	1.15	560	0.59	<0.1	4.4	0.3	<0.1	0.4	0.6		10	687	1,383	7.48	+285
345–365	1.15	540	0.59	<0.1	3.8	0.2	0.1	0.5	0.5		10	687	1,409	7.44	+280
365–385	1.12	560	0.66	<0.1	3.3	0.2	0.2	0.4	0.6		10	687	1,363	7.48	+275
385–400	1.12	530	0.66	<0.1	3.2	0.2	0.2	0.5	0.8		10	807	1,492	7.34	+275
Core 95K															
30–40	1.52	580	0.73	4.0	2.6	0.4	0.1	0.4	0.5	0.04	16	1,338	1,346	7.22	+280
40–50	1.52	600	0.73	0.5	2.8	0.3	0.1	0.6	0.6	0.00	16	1,270	1,475	7.28	+291
50–60	1.69	650	0.73	0.1	2.8	0.3	0.2	0.4	0.3	0.04	16	1,133	1,475	7.29	+290
60–70	1.76	700	0.82	0.2	3.5	0.4	0.2	0.6	0.5	0.04	18	1,201	1,601	7.37	+290
70–80	1.64	710	0.92	6.0	4.8	0.4	0.1	0.7	0.6	<0.03	18	1,201	1,551	7.24	+308
80–90	1.62	680	0.82	0.2	5.0	0.4	0.2	0.7	0.6	<0.03	18	1,270	1,518	7.23	+295
90–100	1.58	790	0.73	0.2	6.1	0.3	0.1	0.6	0.6	<0.03	17	1,270	1,554	7.34	+268
110–120	1.34	590	0.53	0.1	4.2	0.2	0.1	0.5	0.1	<0.03	13	807	1,469	7.38	+271

Depth (cm)	Na (%)	K	Li	Zn	Mn	Pb	Cu	Co	Ni	Cd	Sr	Ca	Mg	pH	Eh mv
140–150	1.31	450	0.46	1.2	2.4	0.2	0.1	0.3	0.3	<0.03	10	687	1,346	7.34	+290
175–180	1.24	480	0.40	0.1	7.1	0.2	0.1	0.3	0.3	<0.03	11	687	1,472	7.43	+265
180–185	1.31	600	0.82	0.8	2.5	0.4	0.2	0.7	0.6	<0.03	23	1,888	1,799	7.25	+262
200–230	1.12	410	0.81	0.1	6.3	0.1	0.2	0.3	0.5	<0.03	8	515	1,271	7.45	+310
230–240	1.12	410	0.46	0.1	6.1	0.1	0.1	0.3	0.3	<0.03	8	515	1,287	7.33	+318
241–250	1.12	360	0.46	0.1	7.1	0.1	0.1	0.3	0.3	<0.03	8	515	1,205	7.31	+305
250–260	1.07	380	0.46	0.1	5.9	0.1	0.1	0.3	0.3	0.03	8	515	1,254	7.25	+300
260–270	1.12	422	0.53	<0.1	5.0	0.2	0.1	0.3	0.3	<0.03	9	583	1,360	7.37	+280
270–280	1.12	422	0.36	<0.1	5.0	0.2	0.1	0.3	0.5	<0.03	8	515	1,317	7.52	+270
282–300	1.16	440	0.53	<0.1	4.1	0.2	0.1	0.4	0.5	<0.03	10	652	1,429	7.29	+280
320–330	1.15	440	0.53	<0.1	4.1	0.2	0.1	0.4	0.6	<0.03	10	687	1,429	7.37	+270
390–395	1.05	420	0.36	<0.1	8.5	0.2	0.1	0.3	0.5	<0.03	7	489	1,290	7.54	+270
Core 119															
0–30	10.20	2,024	5.28	8	88	3.3	0.4	4.0	2.4	0.6	65	5,905	990	6.28	+570
30–55	10.59	2,024	5.61	21	50	3.0	0.4	4.2	2.6	0.4	65	5,664	911	6.29	+550
55–60	10.42	2,200	4.62	8	38	2.7	0.4	4.0	2.7	0.5	62	5,664	911	6.21	+480
60–70	10.70	2,200	4.62	7	36	3.3	0.4	3.9	2.7	0.5	58	5,664	908	6.08	+450
70–95	10.42	2,024	4.62	15	36	—	0.4	3.9	2.1	0.5	62	5,836	914	6.43	+460
100–110	11.33	2,530	5.61	8	38	3.6	0.4	4.0	2.9	0.5	73	6,248	1,016	6.47	+430
110–120	10.54	2,024	4.62	20	36	3.3	0.4	3.6	2.4	0.5	65	5,767	901	6.50	+430
120–135	10.88	2,200	6.27	43	40	3.6	0.3	4.3	2.7	0.5	65	5,905	974	6.40	+430
135–140	11.05	2,365	5.94	31	35	3.6	0.3	4.2	2.7	0.5	65	6,007	993	6.38	+400
140–160	10.54	2,365	5.61	30	29	3.0	0.4	4.7	2.6	0.5	65	5,836	914	6.53	+410
160–180	10.42	2,200	5.61	89	36	3.6	0.3	5.4	2.4	0.6	67	5,836	944	6.12	+400
180–200	10.88	2,530	6.27	107	46	3.0	0.3	5.7	2.7	0.5	65	6,007	977	6.58	+390
202–205	10.03	2,200	5.00	58	31	3.6	0.3	4.8	2.1	0.5	58	5,321	884	6.66	+380
205–213	10.20	2,365	5.00	46	27	3.3	0.3	4.4	2.4	0.5	58	5,493	911	6.56	+380
213–230	10.88	2,365	3.96	33	27	3.6	0.4	4.5	2.6	0.5	65	5,767	1,013	6.46	+230
230–250	10.20	2,365	4.29	13	22	3.6	0.4	4.9	2.6	0.5	65	5,836	1,026	6.57	+320
253–260	10.59	2,365	4.62	33	26	3.0	0.5	5.0	2.9	0.6	70	6,179	1,076	6.50	+320
265–278	9.97	2,200	4.62	17	26	3.6	0.4	4.4	2.7	0.7	62	5,493	1,000	6.46	+360
278–298	10.42	2,365	4.95	38	27	3.0	0.3	4.8	2.9	0.6	65	5,836	1,066	6.58	+380
300–316	10.20	2,200	4.95	18	27	3.0	0.4	3.2	2.9	0.7	62	5,664	1,026	6.60	+370
316–350	10.42	2,200	5.28	69	33	3.0	0.4	4.4	2.6	0.5	65	6,008	1,063	6.63	+360
350–358	9.74	2,200	4.29	38	25	3.0	0.4	5.6	2.3	0.5	54	5,149	977	6.60	+360
360	10.20	2,200	4.62	23	22	3.0	0.4	5.8	2.4	0.5	58	5,218	974	6.46	+360
360–390	10.42	2,365	5.80	30	26	3.0	0.3	5.8	2.3	0.5	62	5,321	1,033	6.43	+350
390–400	10.42	2,200	4.95	168	31	3.3	0.3	4.8	2.0	0.7	65	5,596	1,086		

Depth (cm)	Na (%)	K	Li	Zn	Mn	Pb	Cu	Co	Ni	Cd	Sr	Ca	Mg	pH	Eh mv
Core 128P															
112–122	10.59	2,024	4.29	23	57	3.3	1.0	3.9	2.7	0.7	59	6,179	997	6.12	+471
142–152	10.20	2,024	4.95	26	69	3.3	0.6	4.2	3.3	0.7	59	6,591	1,304	6.20	+481
152–172	10.20	2,024	4.62	26	71	3.0	0.3	3.9	2.7	0.6	63	6,110	792	6.17	+450
182–192	10.20	1,890	4.62	28	66	3.0	0.6	4.1	3.3	0.7	59	6,522	828	6.16	+430
202–212	10.42	2,200	5.28	40	79	3.0	0.5	4.0	3.5	0.6	63	6,591	858	6.12	+370
232–242	10.59	2,024	4.62	53	75	4.2	0.5	4.1	3.3	0.7	59	6,591	908	6.16	+335
264–284	10.42	2,024	4.62	51	75	3.3	0.5	4.2	3.5	0.5	59	6,591	848	6.10	+325
304–314	11.05	2,024	4.62	54	73	3.6	0.4	4.3	3.3	0.6	59	6,591	851	6.15	+310
334–354	9.06	2,024	5.28	51	75	3.6	0.6	3.9	3.3	0.6	59	6,866	875	6.14	+318
364–379	8.88	2,200	5.28	43	71	3.6	0.4	3.8	3.3	0.4	55	6,591	851	6.13	+335
400–420	8.88	2,200	4.95	21	61	3.0	0.5	3.5	3.0	0.5	55	6,591	858	6.10	+328
420–440	8.88	2,200	5.61	43	75	3.0	0.6	3.9	3.5	0.7	63	6,179	835	6.10	+288
470–480	9.30	2,200	5.28	26	93	3.0	1.2	3.9	3.3	0.7	63	6,351	894	6.10	+330
486–490	8.88	2,024	4.62	20	84	3.0	1.1	3.9	2.9	0.8	55	6,248	842	6.12	+320
490–500	9.06	2,200	4.62	28	125	3.0	1.0	4.0	3.0	0.7	55	6,351	875	6.11	+325
510–530	9.15	2,200	4.62	40	146	3.3	0.4	3.8	2.7	0.6	59	6,351	891	6.16	+345
560–580	8.67	1,540	4.62	23	165	—	0.7	5.4	2.9	1.5	55	6,007	815	6.12	+351
595–615	9.91	1,870	5.61	18	191	—	0.7	3.9	2.9	0.6	63	6,522	891	6.12	+381
645–675	9.23	1,705	5.28	21	177	3.0	0.7	4.2	2.9	0.7	59	6,248	835	6.15	+335
718–748	9.54	1,870	5.94	35	125	3.0	1.1	3.9	3.0	0.5	59	6,522	809	6.15	+337
778–808	9.31	1,870	5.61	7	67	3.0	2.0	3.9	3.0	0.5	55	6,179	776	6.15	+341
838–858	9.74	1,870	6.27	140	63	5.4	2.0	4.9	3.3	0.7	59	6,591	818	7.32	+290
Core 128G															
15–30	10.88	2,365	5.28	84	59	3.3	0.5	4.3	2.3	1.0	62	6,591	795	6.29	
40–50	10.88	2,365	5.28	87	59	3.3	0.6	4.1	2.4	1.1	62	6,522	795	6.37	
50–60	11.04	2,530	5.28	63	59	3.3	0.6	4.2	2.4	0.8	62	6,591	802	6.29	
70–83	10.99	2,530	5.61	71	58	3.3	0.6	4.4	2.4	0.8	65	6,522	782	6.34	
90–102	10.88	2,530	4.62	74	58	3.0	0.7	4.2	3.2	1.0	65	6,351	776	6.46	
102–115	10.71	2,530	5.61	64	50	3.3	1.0	4.2	2.7	1.6	65	6,351	776	6.43	
128–130	10.71	2,530	6.27	53	59	3.0	3.5	4.4	2.7	1.2	65	6,351	776	6.23	
132–145	10.88	2,365	6.27	26	77	3.3	2.2	4.5	2.9	0.7	70	6,351	785	6.16	
Core 81G															
0–10	10.08	2,530	4.62	3.3	57	3.6	0.5	4.1	2.6	0.6	63	5,664	875	6.48	
10–20	10.59	2,530	5.28	1.3	59	3.3	0.4	4.0	2.6	0.6	67	6,111	917	6.46	
30–40	10.59	2,530	5.28	5.0	94	3.6	0.4	3.8	2.6	0.6	67	6,008	941	6.54	
39–43	10.59	2,530	5.28	5.6	97	3.0	0.6	4.1	2.7	0.7	63	5,905	941	6.50	
43–48	10.59	2,365	5.28	6.6	85	3.9	0.9	4.2	2.4	0.8	63	5,836	924	6.59	
50–62	10.42	2,365	5.28	1.3	75	3.6	0.6	4.4	2.4	0.6	67	5,836	921	6.46	

Depth (cm)	Na (%)	K	Li	Zn	Mn	Pb	Cu	Co	Ni	Cd	Sr	Ca	Mg	pH	Eh mv
62–70	10.42	2,365	5.28	1.0	69	3.3	0.6	3.9	2.6	0.5	63	5,767	921	6.54	
73–87	10.76	2,695	4.95	1.3	57	3.9	0.9	4.1	2.7	0.6	72	6,179	1,000	6.48	
87–100	10.76	2,530	4.95	6.6	49	3.9	0.6	4.1	2.9	0.5	72	6,008	980	6.47	
100–110	10.59	2,530	5.28	6.6	44	3.6	0.7	3.7	2.7	0.5	–	5,905	967	6.43	
110–120	10.59	2,695	5.28	5.0	40	3.6	0.4	4.5	3.0	0.5	76	5,836	974	6.40	
Core 120K															
100–110	10.20	2,024	6.27	139	54	3.9	0.5	4.2	3.0	1.7	63	6,007	832	6.44	
120–130	10.59	2,024	6.27	132	63	3.9	0.6	4.3	3.3	1.1	63	6,007	809	6.20	
130–140	10.59	2,365	6.93	104	71	4.2	0.6	5.4	3.5	1.0	67	6,351	825	6.21	
140–150	11.02	–	7.92	112	69	4.2	0.6	3.9	3.3	1.0	80	6,179	891	6.45	
170–180	10.76	2,530	6.66	198	63	19.5	3.2	4.0	3.3	2.3	67	6,248	858	6.66	
190–200	10.59	2,200	6.66	153	73	7.8	3.9	4.2	3.5	1.7	67	6,179	838	6.54	
200–210	10.08	1,870	5.94	59	58	3.9	3.5	3.5	3.0	0.7	63	5,664	815	6.58	
210–220	10.76	2,200	6.93	40	73	3.9	1.8	3.9	3.3	0.5	72	6,351	894	6.35	
220–230	10.33	2,200	5.94	15	67	3.0	1.2	3.6	3.0	0.5	67	5,905	891	5.90	
235–240	10.59	2,365	5.94	13	67	3.3	4.7	3.2	3.0	0.6	63	6,007	1,574	6.36	
240–250	10.19	2,024	5.94	10	74	3.6	6.0	3.9	3.3	0.5	72	5,836	782	6.29	
270–280	11.50	2,024	5.61	407	61	3.6	7.4	3.9	3.5	0.5	76	6,523	851	6.42	
300–310	11.05	2,200	5.61	73	61	3.3	7.2	4.1	3.3	0.5	72	6,351	871	6.44	
330–340	11.04	2,200	5.61	25	67	3.0	9.3	4.1	3.3	0.5	67	6,351	831	6.37	
340–360	10.59	2,200	5.94	12	69	3.6	7.2	4.1	3.3	0.6	72	6,248	822	6.38	
365	10.76	2,200	4.95	8	89	3.0	14.1	4.1	3.3	0.5	63	6,179	792	6.22	
380–385	10.59	2,200	5.28	5	63	3.0	5.7	4.1	3.0	0.5	63	6,179	769	6.35	
400–405	10.59	2,200	4.95	66	85	3.0	15.9	3.9	3.0	0.5	63	6,179	835	6.24	
425–430	10.19	2,200	4.62	30	65	3.6	15.6	4.2	3.0	0.4	63	5,905	749	6.31	
455–460	10.59	2,530	5.28	51	65	3.6	10.5	4.1	3.0	0.6	72	6,519	823	6.29	
475–480	10.59	2,530	4.95	31	63	3.9	20.4	4.2	3.3	0.7	72	6,519	815	6.37	
490–500	10.59	2,365	6.27	201	53	9.6	3.9	4.0	3.3	2.0	88	6,179	855	6.83	
Core 84K															
110–120	9.74	2,200	4.95	195	57	10.8	0.6	3.5	2.3	1.7	51	5,836	815	6.54	+310
150–160	10.25	2,200	5.61	191	69	3.9	0.6	3.2	2.7	1.1	59	6,008	875	6.51	+310
180–190	10.42	2,365	5.28	168	69	3.6	0.4	3.7	2.9	0.9	59	6,179	815	6.60	+311
200–210	10.42	2,365	4.95	224	69	9.6	1.2	3.5	2.7	3.9	63	5,664	987	6.33	+280
230–245	10.08	2,200	4.62	215	65	8.1	2.9	3.7	1.5	1.5	55	5,664	875	6.23	+292
260–270	10.59	2,530	5.28	175	82	13.8	2.4	3.7	1.8	1.1	63	6,248	845	6.77	+322
310–320	10.42	2,200	5.28	101	79	5.7	1.9	3.7	1.5	0.5	67	6,111	908	6.46	+360
340–350	10.59	2,365	5.61	109	69	4.2	2.0	3.9	2.1	0.6	72	–	–	6.31	+350
385–395	10.59	2,200	5.94	218	73	5.4	1.9	4.2	1.8	2.2	72	5,836	931	6.61	+340

Brooks *et al.*, Trace Element Composition of Red Sea Geothermal Brine and Interstitial Water.

Table 4 Concentrations of Constituents of Cores (ppm) Divided by the % Sodium Content of Each Sample

Concentration Ratios

Core	Depth (cm)	Na (%)	K	Li	Zn	Mn	Pb	Cu	Co	Ni	Cd	Sr	Ca	Mg
120K	110	10.19	198	0.61	13.62	5.29	0.38	0.04	0.41	0.29	0.16	6.2	589	82
	130	10.58	191	0.59	12.46	5.94	0.36	0.05	0.40	0.31	0.10	5.9	567	76
	140	10.58	223	0.65	9.82	6.70	0.39	0.05	0.51	0.33	0.09	6.3	600	78
	150	11.01	200	0.71	10.16	6.26	0.38	0.05	0.35	0.29	0.09	7.2	561	81
	180	10.75	235	0.61	18.40	5.85	1.81	0.30	0.37	0.30	0.20	6.2	581	80
	200	10.58	208	0.62	14.44	6.89	0.73	0.36	0.39	0.33	0.15	6.3	583	79
	210	10.07	186	0.58	5.85	5.75	0.38	0.34	0.34	0.29	0.07	6.3	562	81
	220	10.75	204	0.64	3.71	6.78	0.36	0.16	0.36	0.30	0.05	6.7	590	83
	230	10.32	213	0.57	1.45	6.48	0.29	0.11	0.34	0.29	0.04	6.5	572	86
	240	10.58	208	0.49	1.22	6.32	0.31	0.43	0.29	0.28	0.05	5.9	567	149
	250	10.18	232	0.58	0.98	7.26	0.35	0.58	0.38	0.32	0.04	7.0	573	77
	280	11.50	176	0.48	35.39	5.30	0.31	0.63	0.33	0.30	0.04	6.6	567	74
	310	11.04	199	0.50	6.60	5.52	0.29	0.65	0.37	0.29	0.04	6.5	575	79
	340	11.03	199	0.50	2.26	6.06	0.27	0.84	0.37	0.29	0.04	6.0	575	75
	320	10.58	208	0.56	1.13	6.51	0.33	0.67	0.38	0.31	0.05	6.8	590	78
	365	10.75	204	0.46	0.74	8.27	0.27	1.31	0.38	0.30	0.04	5.8	574	74
	385	10.58	208	0.49	0.47	5.94	0.28	0.53	0.38	0.28	0.05	5.9	583	73
	405	10.58	208	0.46	6.23	8.02	0.28	1.50	0.37	0.28	0.04	5.9	583	79
	430	10.18	216	0.45	2.94	6.37	0.35	1.53	0.41	0.29	0.04	6.2	579	74
	460	10.58	239	0.49	4.81	6.13	0.33	0.99	0.38	0.28	0.05	6.8	616	78
	480	10.58	239	0.46	2.92	5.94	0.36	1.92	0.39	0.31	0.06	6.8	616	77
	500	10.58	223	0.59	18.98	5.00	0.90	0.36	0.37	0.31	0.18	8.3	583	81
128G	30	10.87	218	0.48	7.72	5.42	0.30	0.04	0.39	0.21	0.09	5.7	606	73
	50	10.87	217	0.48	7.99	5.42	0.30	0.05	0.37	0.22	0.10	5.7	599	73
	60	11.03	229	0.47	5.70	5.34	0.29	0.05	0.38	0.21	0.07	5.6	597	73
	83	10.98	230	0.51	6.46	5.27	0.30	0.05	0.39	0.21	0.07	5.9	593	71
	102	10.87	233	0.42	6.80	5.33	0.27	0.06	0.38	0.29	0.09	6.0	584	71
	115	10.70	236	0.52	5.97	4.66	0.30	0.09	0.39	0.25	0.15	6.0	593	72
	130	10.70	236	0.58	4.94	5.50	0.28	0.32	0.40	0.25	0.11	6.0	593	72
	145	10.87	218	0.57	2.38	7.07	0.30	0.19	0.40	0.26	0.06	6.4	584	72
81G	10	10.07	250	0.45	0.32	5.65	0.35	0.04	0.40	0.25	0.05	6.3	562	87
	20	10.58	239	0.49	0.12	5.57	0.31	0.03	0.37	0.24	0.05	6.3	577	86
	40	10.58	239	0.49	0.47	8.87	0.33	0.03	0.35	0.24	0.05	6.3	567	89
	43	10.58	239	0.49	0.52	9.15	0.28	0.05	0.38	0.25	0.06	5.9	558	89
	48	10.58	223	0.49	0.62	8.02	0.36	0.08	0.40	0.22	0.07	5.9	551	87
	62	10.41	227	0.50	0.12	7.19	0.34	0.05	0.41	0.23	0.05	6.4	560	88
	70	10.41	227	0.50	0.09	6.62	0.31	0.05	0.37	0.24	0.05	6.0	553	88
	77	10.75	250	0.46	0.12	5.29	0.36	0.08	0.38	0.25	0.05	6.7	574	93

Concentration Ratios

Core	Depth (cm)	Na (%)	K	Li	Zn	Mn	Pb	Cu	Co	Ni	Cd	Sr	Ca	Mg
	100	10.75	235	0.46	0.61	4.55	0.36	0.05	0.38	0.26	0.04	6.7	558	91
	110	10.58	239	0.49	0.62	4.15	0.33	0.06	0.35	0.25	0.04	6.8	558	91
	120	10.58	254	0.49	0.47	3.77	0.33	0.03	0.42	0.28	0.04	7.2	551	91
84K	120	9.73	226	0.50	20.02	5.85	1.10	0.06	0.35	0.23	0.17	5.2	599	84
	160	10.25	215	0.54	18.63	6.73	0.38	0.05	0.30	0.26	0.11	5.8	586	85
	190	10.41	227	0.50	16.12	6.62	0.34	0.04	0.35	0.27	0.08	5.7	593	78
	210	10.41	227	0.47	21.49	6.62	0.92	0.11	0.33	0.25	0.37	6.0	544	95
	245	10.07	218	0.45	21.32	6.44	0.80	0.28	0.36	0.14	0.14	5.4	562	87
	270	10.58	239	0.49	16.52	7.74	1.30	0.22	0.35	0.16	0.10	5.9	590	80
	350	10.58	223	0.52	10.29	6.51	0.39	0.19	0.36	0.19	0.05	6.8	577	86
	395	10.58	208	0.56	20.58	6.89	0.50	0.17	0.39	0.16	0.20	6.8	551	88
119	30	10.19	198	0.51	0.78	8.62	0.32	0.04	0.39	0.23	0.05	6.4	579	97
	55	10.58	191	0.52	1.98	4.72	0.28	0.03	0.39	0.24	0.03	6.1	535	86
	60	10.41	211	0.44	0.76	3.64	0.25	0.04	0.38	0.25	0.04	6.0	544	87
	70	10.69	205	0.43	0.65	3.36	0.30	0.03	0.36	0.25	0.04	5.4	529	85
	95	10.41	194	0.44	1.43	3.45	0.25	0.04	0.37	0.20	0.04	6.0	560	88
	110	11.32	223	0.49	0.70	3.35	0.31	0.03	0.35	0.25	0.04	6.4	551	90
	120	10.53	192	0.43	1.89	3.41	0.31	0.03	0.34	0.22	0.04	6.1	547	85
	135	10.87	202	0.57	3.95	3.67	0.33	0.02	0.39	0.24	0.04	6.0	543	89
	140	11.04	214	0.53	2.80	3.16	0.32	0.02	0.38	0.24	0.04	5.9	543	90
	160	10.53	224	0.53	2.84	2.75	0.34	0.03	0.44	0.24	0.05	6.2	554	87
	180	10.41	211	0.53	8.54	3.45	0.28	0.02	0.51	0.23	0.04	6.4	560	91
	200	10.87	232	0.57	9.83	4.22	0.33	0.02	0.52	0.24	0.05	6.0	552	90
	205	10.02	219	0.49	5.78	3.09	0.29	0.02	0.47	0.20	0.04	5.8	530	88
	213	10.19	232	0.49	4.50	2.64	0.35	0.02	0.43	0.23	0.04	5.7	538	89
	230	10.87	217	0.36	3.03	2.48	0.30	0.03	0.41	0.23	0.04	6.0	530	93
	250	10.19	232	0.42	1.27	2.15	0.35	0.04	0.48	0.25	0.04	6.4	572	100
	260	10.58	223	0.43	3.11	2.45	0.33	0.04	0.47	0.27	0.04	6.6	583	102
	278	9.96	220	0.46	1.70	2.60	0.30	0.04	0.44	0.27	0.05	6.2	551	100
	298	10.41	226	0.47	3.64	2.59	0.34	0.02	0.46	0.27	0.06	6.2	560	102
	316	10.19	215	0.48	1.76	2.64	0.29	0.04	0.30	0.28	0.05	6.1	555	100
	350	10.41	211	0.50	6.62	3.16	0.28	0.04	0.42	0.24	0.06	6.2	577	102
	358	9.73	226	0.44	3.90	2.56	0.30	0.04	0.58	0.23	0.05	5.5	529	100
	360	10.19	216	0.45	2.25	2.15	0.29	0.04	0.57	0.23	0.05	5.7	512	95
	390	10.41	227	0.55	2.87	2.49	0.28	0.02	0.55	0.22	0.04	6.0	511	99
	400	10.39	211	0.47	16.15	2.98	0.31	0.02	0.46	0.19	0.06	6.3	538	104

Concentration Ratios

Core	Depth (cm)	Na (%)	K	Li	Zn	Mn	Pb	Cu	Co	Ni	Cd	Sr	Ca	Mg
118	20	1.27	372	0.51	0.3	2.28	0.3	<0.03	0.3	0.3	<0.03	8.7	543	1,233
	40	1.21	396	0.36	0.3	2.13	0.3		0.3	0.3		8.4	426	1,254
	70	1.15	435	0.39	0.2	1.98	0.3		0.3	0.3		7.8	450	1,266
	90	1.15	393	0.39	<0.1	1.98	0.3		0.3	0.3		8.7	510	1,251
	100	1.04	402	0.36	<0.1	2.19	0.3		0.3	0.3		8.7	561	1,410
	110	1.15	393	0.39	<0.1	2.85	0.3		0.3	0.3		8.1	570	1,344
	120	1.15	384	0.39	<0.1	2.94	0.3		0.3	0.3		6.9	510	1,209
	150	1.16	429	0.39	<0.1	4.38	0.3		0.3	0.3		7.8	561	1,290
	160	1.07	393	0.36	<0.1	3.54	0.3		0.3	0.6		7.5	546	1,257
	190	1.12	402	0.39	<0.1	2.85	0.3		0.3	0.3		7.2	522	1,293
	200	1.15	366	0.39	<0.1	2.52	0.3		0.3	0.3		8.7	705	1,173
	205	1.17	402	0.39	<0.1	3.24	0.3		0.3	0.3		8.7	588	1,188
	210	1.15	435	0.45	<0.1	3.81	0.3		0.3	0.3		8.7	657	1,254
	218	1.25	360	0.42	<0.1	3.51	0.3		0.3	0.3		8.7	648	1,125
	225	1.22	345	0.36	<0.1	3.84	0.0		0.3	0.3		8.1	564	1,032
	240	1.25	354	0.42	<0.1	4.89	0.3		0.3	0.3		8.1	606	1,113
	260	1.37	321	0.36	<0.1	3.63	0.0		0.3	0.3		7.2	588	1,023
	280	1.15	366	0.45	<0.1	3.66	0.3		0.3	0.3		8.7	705	1,185
	300	1.16	360	0.45	<0.1	3.51	0.3		0.3	0.3		8.7	693	1,155
	307	1.12	375	0.39	<0.1	2.94	0.0		0.3	0.3		8.7	723	1,158
	311	1.07	420	0.36	<0.1	3.06	0.3		0.3	0.6		7.5	642	1,152
	321	1.15	444	0.45	<0.1	3.66	0.3		0.3	0.6		6.9	570	1,224
	345	1.15	489	0.51	<0.1	3.81	0.3		0.3	0.6		8.7	600	1,206
	365	1.15	471	0.51	<0.1	3.30	0.3		0.3	0.3		8.7	600	1,230
	385	1.12	501	0.57	<0.1	2.94	0.3		0.3	0.6		9.0	612	1,224
	400	1.12	474	0.57	<0.1	2.85	0.3		0.6	0.6		9.0	720	1,332
95K	40	1.52	381	0.48	2.6	1.71	0.3	<0.06	0.3	0.3	<0.03	10.5	882	888
	50	1.52	396	0.48	0.3	1.83	0.3		0.3	0.3		10.5	837	972
	60	1.69	384	0.42	<0.1	1.65	0.3		0.3	0.3		9.6	669	873
	70	1.76	399	0.45	0.1	1.98	0.3		0.3	0.3		10.2	684	912
	80	1.64	432	0.54	3.7	2.91	0.3		0.3	0.3		11.1	732	945
	90	1.62	420	0.48	0.1	3.09	0.3		0.3	0.3		11.1	786	939
	100	1.58	498	0.45	0.1	3.84	0.3		0.3	0.3		10.8	801	981
	120	1.34	441	0.39	<0.1	3.12	0.0		0.3	1.2		9.6	603	1,098
	150	1.31	342	0.33	0.9	1.80	0.3		0.3	0.3		7.5	522	1,026

Concentration Ratios

Core	Depth (cm)	Na (%)	K	Li	Zn	Mn	Pb	Cu	Co	Ni	Cd	Sr	Ca	Mg
	180	1.24	387	0.30	<0.1	5.70	0.3		0.3	0.3		8.7	552	1,185
	185	1.31	456	0.60	0.7	1.89	0.3		0.6	0.6		17.7	1,437	1,371
	230	1.12	366	0.72	<0.1	5.64	0.0		0.3	0.3		7.2	462	1,137
	240	1.12	366	0.39	<0.1	5.46	0.0		0.3	0.3		7.2	462	1,152
	250	1.12	321	0.39	<0.1	6.33	0.0		0.3	0.3		7.2	462	1,080
	260	1.07	354	0.42	<0.1	5.49	0.3		0.3	0.3		8.1	522	1,218
	270	1.12	378	0.45	<0.1	4.47	0.3		0.3	0.3		8.1	522	1,218
	280	1.12	378	0.30	<0.1	4.47	0.3		0.3	0.3		7.2	462	1,179
	300	1.12	378	0.45	<0.1	3.51	0.3		0.3	0.3		8.7	561	1,227
	330	1.16	384	0.45	<0.1	3.57	0.3		0.3	0.6		8.7	600	1,245
	395	1.05	399	0.33	<0.1	8.07	0.3		0.3	0.6		6.6	465	1,230
128P	122	10.58	191	0.40	2.17	5.38	0.31	0.09	0.37	0.25	0.06	5.6	583	94
	152	10.19	198	0.48	2.54	6.76	0.32	0.05	0.41	0.32	0.06	5.8	646	128
	172	10.19	198	0.45	2.54	6.96	0.29	0.02	0.38	0.26	0.05	6.2	599	78
	192	10.19	185	0.45	2.74	6.47	0.29	0.05	0.40	0.32	0.06	5.8	639	81
	212	10.41	211	0.50	3.83	7.58	0.28	0.04	0.38	0.33	0.05	6.0	633	82
	242	10.58	191	0.43	5.00	7.08	0.39	0.04	0.38	0.31	0.06	5.6	622	86
	284	10.41	194	0.44	4.89	7.19	0.31	0.04	0.39	0.33	0.04	5.7	633	81
	314	11.04	183	0.41	4.88	6.60	0.32	0.03	0.38	0.29	0.05	5.3	596	77
	354	9.05	223	0.58	5.62	8.27	0.39	0.06	0.47	0.36	0.06	6.5	758	97
	379	8.87	248	0.59	4.84	7.99	0.40	0.04	0.43	0.37	0.04	6.2	742	96
	420	8.87	248	0.55	2.36	6.86	0.33	0.05	0.42	0.33	0.06	6.2	742	97
	440	8.87	248	0.63	4.84	8.44	0.33	0.06	0.39	0.37	0.06	7.1	696	94
	480	9.29	237	0.56	2.79	10.00	0.32	0.12	0.41	0.37	0.07	6.8	683	96
	490	8.87	228	0.52	2.25	9.45	0.33	0.12	0.43	0.37	0.08	6.1	704	95
	500	9.05	242	0.50	3.09	13.79	0.33	0.10	0.43	0.32	0.08	6.1	701	97
	530	9.14	240	0.50	4.37	15.95	0.36	0.04	0.43	0.32	0.08	6.4	694	97
	580	8.66	178	0.53	2.65	19.03	0.34	0.08	0.43	0.31	0.06	6.3	693	94
	615	9.90	189	0.56	1.81	19.27	0.30	0.06	0.54	0.29	0.14	6.3	658	90
	675	9.22	185	0.57	2.27	19.17	0.32	0.07	0.42	0.31	0.06	6.4	677	90
	748	9.53	196	0.62	3.66	13.10	0.31	0.11	0.44	0.30	0.07	6.2	684	85
	808	9.30	201	0.60	0.75	7.19	0.32	0.20	0.41	0.32	0.05	5.9	664	83
	858	9.73	192	0.64	14.37	6.46	0.55	0.20	0.50	0.33	0.07	6.0	677	84

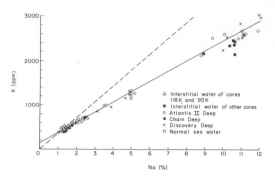

Fig. 2. Concentration of dissolved potassium relative to sodium in various samples. The broken line is the concentration path the brine should have followed if it were derived from evaporation of normal sea water.

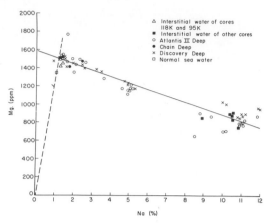

Fig. 4. Concentration of dissolved magnesium relative to sodium in various samples.

changes in ionic concentration if they were caused by a linear effect involving enrichment or dilution of normal sea water. It appears that K is not concentrating as rapidly as Na.

Calcium shows a more than tenfold increase in the bottom brines of both the Atlantis II and Discovery Deeps; intermediate concentrations of about 2,000ppm are also observed (Table 2, Fig. 3). Magnesium decreases in concentration by roughly a half relative to sea water (Table 2, Fig. 4).

Lithium shows almost a twentyfold increase and is consistently higher in the bottom waters of the Atlantis II Deep than in the bottom water of the Discovery Deep. It also demonstrates an intermediate concentration in the Atlantis II Deep as do the other alkali metals (Table 2, Fig. 5).

Manganese is highly concentrated in the enriched brines, with values up to 100ppm (compared with ~2ppb in normal ocean water). It reaches its maximum concentration quickly and does not display a distinct intermediate concentration zone as do the alkali metals or Fe (Fig. 6). Mn consistently shows a higher concentration in Atlantis II Deep than in Discovery Deep.

Iron shows a very large increase in the lower brine of Atlantis II Deep, but not in Discovery Deep. In one sample (No. 715, 2,158m) it has a concentration of 107ppm. Coincidentally, it appears to increase sharply where oxygen disappears as can be seen by comparing data for hydrocast No. 727 with the data published by Munns *et al.* (1967) and Brewer *et al.* (1969).

Zinc shows an increase with depth and

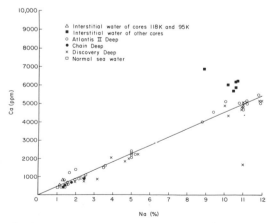

Fig. 3. Concentration of dissolved calcium relative to sodium in various samples.

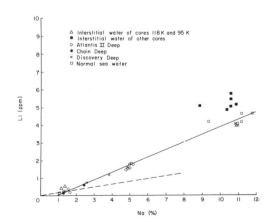

Fig. 5. Concentration of dissolved lithium relative to sodium.

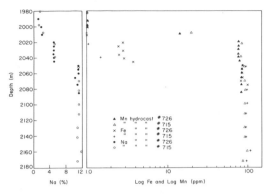

Fig. 6. Depth distribution of Na, Fe and Mn in two hydrocasts (715 and 726) from Atlantis II Deep, showing an intermediate brine layer represented by intermediate concentrations of Na and Fe but not Mn.

appears to follow manganese most closely. A plot of Zn against Fe in Fig. 7 shows that three distinct layers can be identified in Atlantis II Deep. Its concentration in Discovery Deep is lower.

Copper and lead showed increases in concentration over their concentration in sea water by one to two orders of magnitude. Their distribution was somewhat random, varying from layer to layer. In general the Atlantis II Deep had the highest concentrations.

Cobalt and nickel showed only slight enrichment over normal sea water. Discovery Deep appeared to show a lower content. We cannot explain the occasionally high reading (No. 722-2).

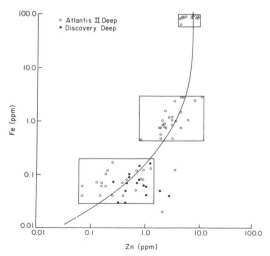

Fig. 7. Concentration of dissolved Zn relative to Fe in brine samples.

Interstitial Water

The analytical data from the interstitial waters show similarities in concentration for the alkali metals in Atlantis II and Discovery Deeps and the overlying brine.

Sodium is again high in both deeps and considerably lower in core 118K (outside the deeps) and core 95K (the rise within Atlantis II Deep). Calcium increases to over 6,500ppm (Table 3, core 128G) and generally averages between 5 to 6 $\times$ 10³ppm. There is also a slight increase in cores 118K and 95K over the concentration of Ca in sea water. Strontium shows a marked increase in the brine of the Atlantis II and Discovery Deep cores and correlates well with Ca. Magnesium decreases to values slightly below 800ppm. Potassium varies in concentration between 2,000 and 2,500ppm, and lithium is in the general range of 4.5 to 6ppm. For these alkali metals, the concentration follows the same general trend both in the brine and the interstitial water as can be seen by comparing Tables 2 and 4.

Manganese generally shows a slightly lower concentration than is present in the overlying brine, although some cores (core 128P, Table 4) appear to be exceptions. The Discovery Deep core (119K) displays lower concentrations than the Atlantis II Deep cores.

Copper shows slight increases in most cores over its concentrations in the brine, and in some cores (Table 4, core 120K) from ten- to hundredfold increase. Lead is consistently higher in the interstitial water by one to two orders of magnitude, generally falling in the range of 3–4ppm, with some notably higher exceptions. Even in cores 118K and 95K it showed a higher concentration. Zinc shows higher concentrations in the interstitial water than in the brine. It may rise in some layers to concentrations above 100ppm (core 120K, Table 3). Cadmium generally follows Zn. For example, both are in low concentration in cores 118K and 95K. Cd shows an increase by at least two orders of magnitude over its concentration in the brine.

Cobalt and nickel are also more consistently enriched in the interstitial water of the cores than in the brine. Cores 118K and 95K also show an enrichment. Average values for Co fall between 3 to 4ppm whereas Ni is between 2 to 3ppm.

Analysis for iron in interstitial water from various cores showed it to be depleted and in the ppb range. Since iron was enriched in the bottom brine of Atlantis II Deep, it appears that iron may have precipitated out of some of the cores during storage due to oxidation. This element was therefore not measured, since the results would probably be erroneous and insignificant.*

pH values average between 6.1 to 6.5 for the interstitial water from Atlantis II sediments. Discovery waters are ~6.5, and the waters from sediment of the rise above Atlantis II Deep (core 95K) and outside the Deeps (118K) show pH values above 7.0. They are somewhat lower at the top of core 95K and the bottom of core 118K, supporting other evidence for periodic interaction of these areas with brines caused by spill-over from Atlantis II Deep.

Eh measurements generally show high values ranging from +220 to +575mv. Most values appear to be in the range of +300 to +400mv, the lowest appearing in core 118K.

Discussion

We wish to propose an origin for the brines in the Atlantis II and Discovery Deeps and will then discuss some of the evidence supporting the hypothesis:

1. The origin of the water could be either Red Sea water or water released from the transition of gypsum to anhydrite. There may also be some contribution of "volcanic" water due to igneous heating of deep buried rocks.

---

* One should bear in mind that a precipitation of iron in the form of iron hydroxides may influence the distribution of some trace elements, which are known to coprecipitate with iron hydroxide. If iron has indeed precipitated out of solution, our data would give minimum values for the concentration of dissolved trace metals.

2. The water is heated by volcanic activity, allowing it to rise by lowering its density.
3. Circulation of the hot water through sedimentary sequences interbedded with evaporites underlying the Red Sea structures will permit enrichment of Na and Ca through leaching of halite, gypsum and sediment. K may be enriched by reaction of the brine with feldspars, whereas Mg is preferentially removed and converted to a magnesian silicate, *e.g.* chlorite. Dolomitization may be another process for enriching Ca and depleting Mg in the brine. Li, Sr and other trace metals are also leached out during this process. The base metals are transported as chloride complexes. Sulfate is removed by reduction with organic matter in the sediment.
4. In the immediate vicinity, the brine has only one major known exit into the Red Sea—the Atlantis II Deep.
5. Periodic eruptions cause brine to be introduced into the Atlantis II Deep. Overfilling of the Deep will cause water to spill over at the sill and enter the Discovery Deep.

Evidence Supporting Brine Origin Hypothesis

Past theories, reviewed by Riley (1967), have suggested that the brine has been formed:

(a) by evaporation of normal Red Sea water
(b) release of connate water (or water of crystallization) by geothermal heating
(c) sinking of Red Sea water near the Gulf of Aden and solution of evaporites (Craig, 1966)
(d) sinking of Red Sea water at flanks of the active zone, dissolution of halite from evaporites and then reaction of the brine with sediments under the driving force of geothermal heating which causes density gradients and allows the brine to rise (White, 1968). This last model is similar to

the mechanism proposed by White (1968) for the origin of the hypersaline brines of the Salton Sea areas, California and is the one favored by us. One possible modification of this hypothesis is that the source of the water may have partially come from the sediments (dehydration of gypsum).

Any model must explain several facts about the chemistry, such as abnormally low magnesium concentration, an increase in calcium concentration relative to sodium, depletion of dissolved sulfate in the brine and relatively high concentration of transition and base metals.

Evidence for the source of the water in the brine is still somewhat questionable. The strongest hypothesis, based on O^{18}/O^{16} and D/H experimental data, has been forwarded by Craig (1966). At this stage, however, we consider that the evidence for water coming entirely from overlying Red Sea water without any contribution from water of crystallization released during the transition phase of gypsum to anhydrite is inconclusive. As has been pointed out elsewhere (Kaplan et al., 1969), there is potentially several orders of magnitude more water available from this phase transition in the Red Sea sedimentary basin than is required to fill the presently known anomalous deeps.

Several lines of evidence point to this region being tectonically active so that a geothermal driving force is available to circulate the water through fissures. Na and Li are most probably enriched by solution of halite, and calcium by solution of gypsum at elevated temperatures, although all may be somewhat further enriched through leaching of the sediment. Mg may be depleted and K enriched, partially through reaction with clays and silicate minerals. Helgeson (1967) suggests that reactions of K-feldspar with sea water for example, would cause formation of chlorite and a lowering of pH according to the following equation:

$$2KAlSi_3O_8 + 8H_2O + 5Mg^{++}$$

$$\rightarrow Mg_5Al_2Si_3O_{10}(OH)_8 + 8H^+ + 2K^+$$

$$+ 3SiO_2 \quad (quartz)$$

This reaction cannot entirely account for the observed data since two molecules of K^+ are released for every five molecules of Mg^{++} removed, whereas in actual fact K increases by about sixfold and Mg decreases by only half. Therefore the reaction may account for removal of Mg, but additional leaching is required to dissolve K. There is some evidence that in situ exchange reactions occur within the sediment column, leading to enrichment of Ca and Li and depletion of K in interstitial water relative to the overlying brine (Figs. 2, 3 and 5).

The concentration of sulfate in the brine is about a third to a quarter of its concentration in sea water. Kaplan et al. (1969) have suggested that this may have resulted from the reduction of the sulfate by the organic matter in the shale at elevated temperatures causing the liberation of sulfide. The very narrow range of δS^{34} observed in the metal sulfides (+2 to +10‰) in the Atlantis II Deep (most of the values falling between +2 to +6‰) point to some large reservoir of sulfur.

The high metal concentration is due to the preferential solution of base metals by chloride-rich brine, which forms stable complexes with Pb, Ag, Cu, Cd and Zn, allowing these metals to remain in solution, even in the presence of low sulfide concentrations. The concentration of these metals was considerably higher in the interstitial water than in the overlying brine, probably for two reasons. When solid precipitates form (oxides or sulfides) these metals are scavenged and coprecipitated. The sediments show much higher concentrations of these metals than the brine (Bischoff, 1969; Kaplan et al., 1969). The brine in contact with the sediment will allow solution to occur following some equilibrium distribution between solid and liquid phases.

The stratified nature of the sediments in Atlantis II Deep and to a lesser extent in Discovery Deep indicate that cyclic processes caused them to precipitate, rather than a single continuous process. Thus, the slight differences in composition of the brine in the various hydrocast samples may represent incomplete mixing of somewhat different brines. Since the Red Sea area is tectonically rifted, brine may escape more or less easily depending on its proximity to a fissure. For this reason, brines are probably not circulated through the sediment in as many cycles as occurs in the Salton Sea deposits (Helgeson, 1968) and are not as enriched in trace metals or K.

We have proposed that the brine originates in the Atlantis II Deep. This is based on the fact that the brines in the Discovery Deep are not as enriched in trace metals as the Atlantis II Deep, and that there is a negative temperature gradient in the Discovery Deep, where the sediment surface is cooler than the brine (Pugh, 1967). In particular, reference has been made to lower Li, Mn and Fe in the Discovery brine. Since other alkali metals have a high concentration, it would appear that Li may be solubilized from a more restrictive source and the salts now present in the Discovery brine have had a lower contribution of Li. In particular, Fe is highly depleted in the Discovery brine, suggesting that during the period of transfer from Atlantis II Deep it became oxidized and precipitated. Mn is much more mobile and has been transferred more efficiently without being removed as an oxide, hence for instance, it is present in high concentration in the intermediate brine of the Atlantis II Deep, whereas Fe is depleted in this zone (Fig. 6). It is further possible that manganese is saturated in the lower brine and cannot attain a higher concentration. This may lead to a removal of Mn, perhaps as a carbonate (Bischoff, 1969), from the bottom brine due to supersaturation and a removal of Fe from the intermediate zone due to ferric hydroxide formation. Further, the S^{34}/S^{32} isotope data show a large difference in the composition of the sediments of the two deeps (Kaplan et al., 1969). In the Atlantis II Deep the isotope values show an origin of the sulfide from a nonbiogenic relatively homogeneous zone; the sulfide in Discovery Deep and other areas surrounding Atlantis II Deep (cores 118K and 95K) are derived from biogenic reduction of sulfate. There is some evidence, from the enrichment of trace metals in core 95K and δS^{34} measurements at the bottom of core 118K, that spill over has affected other areas besides Discovery Deep.

It is impossible to determine from our data if there is one or several vents and where they are located. Tentatively it may be stated that areas surrounding core station 84K and hydrocast station 715 may be the most active, but since the brine spreads over the bottom of the basin more data are needed to determine the exact location of the vents.

Acknowledgments

We acknowledge Robert Sweeney for assistance in analyzing the interstitial waters. This study was carried out under AEC contract AT (11-1) 34, Project 134.

References

Bischoff, J. L.: Red Sea geothermal deposits: their mineralogy, chemistry, and genesis. *In: Hot brines and recent heavy metal deposits in the Red Sea*, E. T. Degens and D. A. Ross (eds.). Springer-Verlag New York Inc., 368–401 (1969).

Brewer, P. G., C. D. Densmore, R. Munns, and R. J. Stanley: Hydrography of the Red Sea brines. *In: Hot brines and recent heavy metal deposits in the Red Sea*, E. T. Degens and D. A. Ross (eds.). Springer-Verlag New York Inc., 138–147 (1969).

Brooks, R. R., B. J. Presley, and I. R. Kaplan: The APDC-MIBK extraction system for the determination of trace elements in saline waters by atomic absorption spectrophotometry. Talanta, **14,** 809 (1967).

Craig, H.: Isotopic composition and origin of the Red Sea and Salton Sea geothermal brines. Science, **154,** 1544 (1966).

Helgeson, H. C.: Solution chemistry and metamorphism. *In: Researches in Geochemistry,* P. H. Abelson (ed.). J. Wiley and Sons New York, 362 (1967).

——: Geological and thermodynamic characteristics of the Salton Sea geothermal system. Amer. J. Sci., **266,** 129 (1968).

Kaplan, I. R., R. E. Sweeney, and A. Nissenbaum: Sulfur isotope studies on Red Sea geothermal brines and sediments. *In: Hot brines and recent heavy metal deposits in the Red Sea,* E. T. Degens and D. A. Ross (eds.). Springer-Verlag New York Inc., 474–498 (1969).

Miller, A. R., C. D. Densmore, E. T. Degens, J. C. Hathaway, F. T. Manheim, P. F. McFarlin, R. Pocklington, and A. Jokela: Hot brines and recent iron deposits in deeps of the Red Sea. Geochim. et Cosmochim. Acta, **30,** 341 (1966).

Munns, R. G., R. J. Stanley, and C. D. Densmore: Hydrographic observations of the Red Sea brines. Nature, **214,** 1215 (1967).

Presley, B. J., R. R. Brooks, and H. M. Kappel: A simple squeezer for removal of interstitial water from ocean sediments. J. Marine Res., **25,** 355 (1967).

Pugh, D. T.: Origin of hot brines in the Red Sea. Nature, **214,** 1003 (1967).

Riley, J. P.: The hot saline waters of the Red Sea bottom and their related sediments. *In: Oceanogr. Mar. Biol. Ann. Rev.,* M. Barnes (ed.)., **5,** 141 (1967).

Ross, D. A., E. E. Hays, and F. C. Allstrom: Bathymetry and continuous seismic profiles of the hot brine region of the Red Sea. *In: Hot brines and recent heavy metal deposits in the Red Sea,* E. T. Degens and D. A. Ross (eds.). Springer-Verlag New York Inc., 82–97 (1969).

White, D. E.: Ore forming fluids of diverse origins. Econ. Geol., **63,** 301 (1968).

Investigations of Atlantis II Deep Samples
Taken by the **FS METEOR**

MARTIN HARTMANN

Geologisch-Paläontologisches Institut der Universität Kiel
Kiel, Germany

Abstract

Information is presented on the geochemistry of the hot brines in Atlantis II Deep. Special emphasis is given to the origin of manganese and iron and to the source of the sulfur species in both sediments and brines. It is concluded that the fractionation of iron and manganese is principally a pH/Eh effect and that the sulfate ions in the brine waters are derived from the overlying Red Sea waters rather than supplied directly by the discharging brines.

Introduction

During the International Indian Ocean Expedition, the **FS METEOR** made two stations in the hot brine area. On November 22, 1964 at station 28 (21° 17′N, 38° 02′E) located in the Discovery Deep region, a bathysonde temperature measurement was run and a water sample taken. On April 27, 1965 continuous temperature and transparency measurements were run in the Atlantis II Deep (21° 23′N, 38° 03.8′E). In addition, two water samples from the brine and one surface sediment sample were obtained (Dietrich *et al.*, 1966).

The results of the temperature and transparency measurements have been published by Krause and Ziegenbein (1966). The investigation of the water and sediment samples from the Atlantis II Deep were reported by Hartmann and Nielsen (1966) and Hartmann and Lohmann (1968). This paper presents a short summary of these investigations.

The vertical temperature profiles of Krause and Ziegenbein (1966) indicate that in the Discovery Deep the temperature rises stepwise beginning at 1,988m. A maximum reading of 44.8°C is obtained at a water depth of 2,043m. Below this point the temperature remains constant until the sea bottom is reached at a depth of about 2,100m. In the Atlantis II Deep the temperature readings indicate three sharp steps, beginning at 1,962m depth of water. The corresponding temperatures are 41.2°C, 57.7°C and 58.4°C.* This last temperature was reached at a depth of 2,038m. A simultaneous decrease in transparency is observed with minimum values below the limit of detection.

Hartmann and Lohmann (1968) suggest the possibility of a relationship between the brine-temperature stratification and the very uneven relief of the bottom topography. In this way, progressively higher topographical levels govern the spillover points. This will result in stepwise enlargements of the surface areas of each successive layer. Such a widening results in larger contact zones along which a more effective heat transfer, and mixing with the next layer is possible. The very sharp boundary with normal Red Sea water may be regarded as the spillover

---

* Editors' note: — The differences in temperature values between the Krause and Ziegenbein results (measured electronically) and those of other workers are probably in part due to slight malfunctions of the formers' electronic equipment in the hot brine (Ziegenbein, personal communication to D. A. Ross, 1967).

Table 1

Sample Number	Origin of Sample	Remarks
1207/W	Lowest (?) salt water layer	Salt brine
1207/6	Upper salt water layer	Salt brine and sediment
1207/11	5m above sample 1207/6	Red Sea water
1207/121	121m above bottom	Red Sea water
1207/B	Middle salt water layer	Sediment sample
1207/PW	Middle (?) salt water layer	Pore water from sample 1207/B

level where the brine finally leaves the Atlantis II Deep. The catch basin for the overflow is probably the slightly deeper Discovery Deep. Table 1 lists the Atlantis II Deep samples which were available for investigation.

Investigation of the Water Samples

Table 2 shows the results of the chemical analyses of the water samples. Calculation of the $CaSO_4$ solubilities in highly concentrated NaCl solutions indicates that the concentrations of Ca^{2+} and SO_4^{2-} in the brine samples lie near the saturation point. The laboratory analyses indicate that sample 1207/6 must have been holding an *in situ* $CaSO_4$ precipitate in suspension.

It is known that the solubility of $CaSO_4$ decreases noticeably with rise in temperature and NaCl. This suggests that outflowing brine is not the exclusive source for the sulfate in the salt solution. Part of the sulfate has been introduced into the brine basin at a later time.

Mass-spectrometric investigations of the water sulfate support this inference (Hart-

mann and Nielsen, 1966). Sulfur isotope data in both brine and normal Red Sea water are identical. In case the sulfate had been derived predominantly from the solution of subsurface sulfate rocks, this identity in S-isotope ratios would be a great coincidence, for as Nielsen (1965) has pointed out, the isotope ratio of sulfate precipitates from the sea waters of the geologic past vary by several per mil. If the salt solution is completely sulfate-free as it flows from the underground, it cannot be determined from these investigations.

The addition of sulfate from the normal sea water probably takes place through simple mixing of brine and sea water as well as through precipitation of $CaSO_4$ in the mixing zone, with the resulting precipitate sinking down into the brine. The fact that sample 1207/6 must have held $CaSO_4$ in suspension indicates that this does take place. In the deeper zones of the brine the $CaSO_4$ precipitate will be redissolved as long as the saturation point has not been reached.

The iron and manganese content exceeds that of normal sea water by a factor of 10^4 or 10^5. At the observed pH such quantities

Table 2   Samples

	Red Sea (1207/121)	41.2°C (1207/6)	Interstitial (1207/PW)	57.7°C (1207/W)
Cl (g/l)	23.04	123.7	170.4	189.4
Ca (mg/l)	393	3,840	4,160	5,075
SO₄ (mg/l)	3,108	2,378	819	798
Mg (mg/l)	1,407	–	907	793
Fe (mg/l)	–	≪1 *	40	65
Mn (mg/l)	–	51	53	47
pH	8.17	6.50	6.26	5.61
δS³⁴(‰)	+20.3	–	+20.4	+20.3

* Because the separation of water and sediment by filtration was done after a storage period, it seems certain that the originally higher dissolved iron content has been reduced through oxidation and precipitation.

of iron could only be held in suspension in the bivalent form. In turn, the redox potential of the solution must be relatively low. Indeed, a measurement on the iron oxide mud of sample 1207/B gave an Eh value of about −100mV at a pH of 6.26. On the other hand, the presence of an iron oxide mud is proof that the redox potential in wide areas of the brine is not negative enough to completely redissolve the iron hydroxide. Hydroxides sinking down from the precipitation zone at the boundary with normal, oxygen-charged sea water and the feeding of Fe^{2+}, as well as other reducing agents (e.g., H_2S?) into the system from the hot brine source, maintain a dynamic balance in the solution. As a result of this process, a portion of the iron hydroxide is again reduced and taken back into solution. In this way the iron content of the salt brine can rise well above the concentration in the salt solution discharged from below.

Manganese, in the form of Mn hydroxide, must also precipitate upon contact with normal sea water. However, as the redox potential of Mn^{2+} lies in a more positive Eh range, the manganese hydroxide accumulation can start only when the iron hydroxide precipitation is nearly completed. The Mn hydroxides would also be almost completely redissolved as they sink into the salt brine. An enrichment process for Mn analogous to that for the iron can then take place; however, in this case it would be restricted to the uppermost layer of the brine body. The concentrations found in the analyses tend to substantiate this. Some manganese deposits have been found by Bischoff (1969). It is, however, questionable whether they account for all the Mn supplied by the brines. There is indication that most of the manganese has been flushed from the Atlantis II Deep and been deposited somewhere else.

Investigation of the Sediment Samples

A temperature measurement made on sediment sample 1207/B one half hour after collection, *i.e.,* after coming up through around 2,000m of ~20°C Red Sea water, gave a value of 65°C. The fact that the sample was not derived from the hot-

test points, as judged from the Cl^- content, would indicate that this high temperature was the result of a heat transfer from the underlying rock.

The sediment obtained in both bottom samples was a very soft, red-brown to dark brown mud with high water content, and in general conforms well to the description by Miller *et al.* (1966) of Atlantis II Deep sediment. Foraminifera tests are abundant. Quartz, some feldspar and montmorillonite-type clays could also be identified.

After the salt had been washed out with distilled water and the samples were dried at 110°C, a brown residue was obtained which in the case of sample 1207/B represented only 4.44 percent of the original wet sample weight. Table 3 shows the results of the chemical analyses of the two dried, salt-free sediment samples.

On the basis of x-ray analysis, most of the iron was of poorly crystallized or amorphous iron hydroxide. Pyrite could be identified by x-ray diffraction, however, only after the greater part of the sample was decomposed with hydrofluoric and dilute hydrochloric acids. The FeS_2 represented about 0.5 per cent of the total sample in sample 1207/B. For a full account on the mineralogical inventory see Bischoff (1969). Copper and zinc were analyzed by x-ray fluorescence. Miller *et al.* (1966) found a zinc content of up to 5.0 per cent Zn and were then able to identify sphalerite through x-ray diffraction. In samples 1207/B and 1207/6, however, the Cu and Zn seem to be only partially combined as sulfides, for after the subtraction of the pyrite sulfur far too little sulfur remains available for both elements to be totally combined as sulfides. Of particular significance is the lack of sufficient sulfur in sample 1207/6.

The sulfur isotope ratio determined in

Table 3   Chemical Analyses of Sediment

	1207/B	1207/6
Carbonate CO_2(%)	2.90	1.30
C_{org}(%)	0.5	−
Fe(%)	33.53	37.18
Mn(%)	0.215	0.367
Zn(%)	0.58	0.52
Cu(%)	0.57	0.37
S(%)	0.655	0.11
δS^{34}(‰)	+4.3	−

sediment sample 1207/B indicates that the sulfur in this case cannot be attributed to adsorption or to sulfate precipitated in the form of gypsum or anhydrite. Under such condition, δS^{34}-values similar to those for the water samples would be expected.

The isotope analyses also show that the combined sulfur in the sediment is not the result of a sulfate reduction by bacteria (Trüper, 1969) which took place after the sediment was deposited. If this were the case, sulfate of the interstitial water would show a relative enrichment of S^{34} in comparison to that of the free water. Therefore, it seems probable that sediment sulfur was brought in with the salt brine from the subsurface, most likely as dissolved sulfide. Later mixing with the iron hydroxide precipitate from the salt brine led to pyrite development.

References

Bischoff, J. L.: Red Sea geothermal brine deposits: their mineralogy, chemistry, and genesis. *In: Hot brines and recent heavy metal deposits in the Red Sea*, E. T. Degens and D. A. Ross (eds.). Springer-Verlag New York Inc., 368–401 (1969).

Dietrich, G., G. Krause, E. Seibold, and K. Vollbrecht: Reisebericht der Indischen Ozean Expedition mit dem Forschungsschiff METEOR 1964–1965. METEOR-Forschungsergebnisse, Reihe A, **1**, 1 (1966).

Hartmann, M. and L. Lohmann: Untersuchungen an der heissen Salzlauge und am Sediment des Atlantis II-Tiefs im Roten Meer. METEOR-Forschungsergebnisse Reihe C, **1**, 13 (1968).

—— and H. Nielsen: Sulfur isotopes in the hot brine and sediment of Atlantis II Deep (Red Sea). Marine Geology, **4**, 305 (1966).

Krause, G. and J. Ziegenbein: Die Struktur des heissen salzreichen Tiefenwassers im zentralen Roten Meer. METEOR Forschungsergebnisse, Reihe A, **1**, 53 (1966).

Miller, A. R., C. D. Densmore, E. T. Degens, J. C. Hathaway, F. T. Manheim, P. F. McFarlin, R. Pocklington, and A. Jokela: Hot brines and recent iron deposits in deeps of the Red Sea. Geochim. et Cosmochim. Acta, **30**, 341 (1966).

Nielsen, H.: Schwefelisotope im marinen Kreislauf und das δS^{34} der früheren Meere. Geologische Rundschau, **55**, 160 (1965).

Trüper, H. G.: Bacterial sulfate reduction in the Red Sea hot brines. *In: Hot brines and recent heavy metal deposits in the Red Sea*, E. T. Degens and D. A. Ross (eds.). Springer-Verlag New York Inc., 263–271 (1969).

Geochemistry and Origin of the Red Sea Brines

H. CRAIG

Scripps Institution of Oceanography
University of California at San Diego
La Jolla, California

Abstract

A steady-state model of the brine waters in Atlantis II Deep is presented. The deuterium and oxygen-18 concentrations in the water, and the dissolved argon content suggest a relative warm near surface Red Sea water as the source of the brine. In evaluating the overall environmental situation in the Red Sea in terms of temperature and salinity, the probable source lies about 800km to the south near the Strait of Bab el Mandeb. By integrating isotope data, and the trace and major element spectra of the brine, the origin, age and history of the brine waters become apparent.

Introduction

The deuterium and oxygen-18 concentrations in the water of the Red Sea geothermal brine show that the water is derived from the normal sea water of the Red Sea (Craig, 1966). Sea water of this isotopic composition is found in the south end of the Red Sea in the vicinity of the southern sill which controls the exchange of water with the Gulf of Aden. This area is underlain by a thick evaporite sequence of several thousand meters lying on the crystalline basement and emerging locally as salt domes. The brine is therefore believed to develop from downward circulating sea water in the region of the southern sill; addition of salt from the evaporite deposits increases its salinity and its temperature rises in accordance with the regional geothermal gradient. Driven by its increased density relative to normal sea water, the brine sinks and flows northward, emerging 2,000m lower and perhaps 1,000km farther north in the brine-filled deeps west of Mecca in the central rift of the Red Sea.

This picture of the origin and history of the brine was developed several years ago (Craig, op. cit.); since then new geochemical and chemical data have become available and a more accurate assessment of the chemical changes involved in the formation of the brine can be made. This paper summarizes the isotopic geochemical data, including recent radiocarbon measurements on the brine and overlying sea water (Craig and Lal, 1968), and presents a revised and more detailed discussion of the chemistry of the added salts.

Chemistry of the Brine

Eight reliable chemical analyses of the brines and associated waters are available. The Atlantis II Deep 44°C and 56°C brine layers, and the Chain Deep brine have been analyzed by Brewer and Spencer (1969). The Discovery brine (44.8°C water) has been analyzed by Brewer and Spencer (op. cit.) and by Brewer et al. (1965). In addition, Brewer et al. have analyzed a transitional water above the Discovery brine (69.2g/kg of Cl), and normal and slightly salt-enriched Red Sea Deep Water above this transitional water (22.54 and 22.68g/kg of Cl respectively); the 22.54 per mill water is essentially identical to

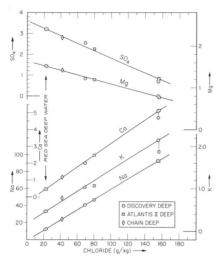

Fig. 1. Chloride variation diagram for major elements in the Red Sea brines.

normal ocean water of the same chlorinity.

In Fig. 1 the major element chemical data from these analyses are plotted on a chloride variation diagram. The linear relationships observed show that all intermediate salinity waters, including the Atlantis II 44°C water and the water in the Chain Deep, are formed by bulk mixing of the 156 per mill chloride brine and normal Red Sea Deep Water. The slight deviations observed for K in the Atlantis II brines are almost certainly analytical considering the conditions under which they were made (Brewer and Spencer, 1969). The Ca deviation of one Discovery brine sample represents the analysis by Brewer *et al.* (1965); the new Ca value by Brewer and Spencer (1969) is identical to their Ca value for Atlantis II brine.

The mixing lines for Mg and SO$_4$ in Fig. 1 have a special significance: they show that *the mixing relationship is linear between the brines and deep water of essentially the same salinity as that of present Red Sea Deep Water.* Concentration slopes for Ca, K, and Na *vs.* Cl in evaporation or dilution processes are close to those of the mixing curves in Fig. 1, so that mixing with evaporated Red Sea water would be difficult to detect with these components. However, the MgCl and SO$_4$Cl slopes for evaporation or dilution are approximately at right angles to the mixing curves observed, so that if the intermediate-salin-

ity brines had formed by dilution with an evaporated Red Sea water of higher salinity than present bottom water, a significant convex curvature in the mixing lines would be observed. Consideration of these two curves indicates that if the brine dilution occurred once, with no further mixing, the diluting Red Sea water could not have differed by more than about one per mill in chlorinity from present Red Sea Deep Water. Alternatively, if continuous mixing has occurred, it has been fast enough to remove any trace of an earlier sea-water component with a different chlorinity from that of the present.

Fig. 2 shows the same type of diagram for the minor elements Sr and Br. The Sr analyses of Faure and Jones (1969), made by isotope dilution, are also shown; one of their intermediate composition Atlantis II samples has the same values as the Chain Deep sample. These data also show a simple bulk mixing relationship between sea water and the 156 per mill brine.

The only other element for which a range of samples is available is Mn, which has the same concentration in both the 44°C and 56°C brines of the Atlantis II Deep (Brewer and Spencer, 1969). Bischoff (1969) believes that the Mn concentration in the 44°C water is buffered by the presence of a solid phase. Also, the concentration of iron in the Atlantis 44°C water and in the Discovery Deep is much lower than in the Atlantis 56°C brine, so much so that precipitation of iron as well as dilution must have occurred during the mixing process. Other trace metals may show similar effects, though the various data are not in good agreement.

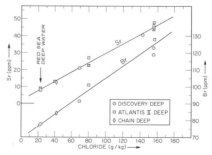

Fig. 2. Chloride variation diagram for strontium and bromine. Crossed symbols are Sr analyses made by isotope dilution (Faure and Jones, 1969).

The relationships in Figs. 1 and 2 show clearly that the Atlantis 44°C water, the Chain water, and the transitional Discovery Deep waters, are all formed by bulk mixing of the 156 per mill Cl brine with Red Sea Deep Water. In this process most of the brine components are conservative, but the concentrations of some trace metals are affected by chemical reactions and precipitation or solution.

Concentrations in the Original Brine

To what extent has the 155.5 per mill chloride brine in the Atlantis and Discovery Deeps been diluted with sea water mixing down from Red Sea Deep Water? One way to obtain an upper limit is by assuming that the original brine flowing into the depressions is saturated with e.g. NaCl. Addition of NaCl to the 155.5 per mill Cl brine in the laboratory resulted in saturation at 160.8g/kg of Cl at 25°C, corresponding to addition of 12.0g NaCl per kg of (unsaturated) brine (Craig, 1966). Solubility data indicate that NaCl is more soluble at 60°C than at 25°C by a factor of 1.023. If the observed brine is a mixture of sea water fluid and original brine fluid, the fraction of sea water (f_{SW}) is related to the concentrations of any element in the component fluids by:

$$f_{SW} = \frac{x_B - x_M}{x_B - x_{SW}} \qquad (1)$$

where x is the mass fraction of any element and the subscripts are B for original, undiluted brine, SW for Red Sea Deep Water, and M for the observed mixture now in the brine depressions (Cl = 155.5g/kg). Thus if we assume a saturation chloride concentration in original brine at 60°C, of 164.5g/kg, and a concentration of 22.5g/kg in RSDW, an upper limit of about 6 per cent is obtained for f_{SW}. (A more accurate estimate could be obtained by evaporating some brine to dryness, and saturating another aliquot of brine with its own salt at 60°C.)

A better estimate can be obtained by using equation (1) with an element whose concentration in the brine is very small compared to sea water concentration; in such a case an upper limit for f_{SW} is obtained as x_B goes to zero in (1). The best element for this calculation is F which has concentrations of 1.6ppm in sea water of 40.6 per mill salinity (RSDW), and 0.0512ppm in the Discovery brine according to Brewer et al. (1965), and a concentration less than 0.02ppm in Atlantis brine according to Brewer and Spencer (1969). Then assuming $x_B = 0$, that is, that the original brine has no fluoride, and $x_M = 0.0512$ppm, we find a maximum value of $f_{SW} = 3.2$ per cent. For any finite concentration of F in the original brine, the fraction of sea water would be less. A similar calculation can be made using the NO_3 concentrations measured by Brewer et al., and gives an upper limit of 5.5 per cent.

The fraction of any element in the observed brine, derived from admixture with sea water, is given by:

$$f_i(SW) = x_{SW} f_{SW}/x_M \qquad (2)$$

where x_M is the concentration of species i in the observed brine (i.e. mixture), and $f_i(SW)$ is the fraction of species i in the mixture which is derived from sea water. x_{SW} is the concentration of species i in Red Sea Deep Water of 40.6 per mill salinity. Using the maximum value of $f_{SW} = 3.2$ per cent calculated from the fluoride data, we obtain maximum values of $f_i(SW)$ for the individual species as shown in Table 1. In this calculation the sea water fraction for F is of course 100 per cent by definition.

The $f_i(SW)$ values are especially important in interpreting isotopic variations in the elements. For example Hartmann and Nielsen (1966) concluded from S^{34} measurements that the sulfate of the brine is derived from direct mixing with overlying Red Sea Deep Water. However, as shown in Table 1, only 12 per cent of the SO_4 in the observed brine can be so derived, and this is an upper limit which assumes that the original brine contains no fluoride at all. Therefore the RSDW is not a significant source of SO_4 by the process of direct mixing into the brine depressions.

Table 1 also shows that less than 4.4 per cent of the water and 3.0 per cent of the

Table 1 Maximum Values of $f_i(SW)$, the Fraction of Species i Which Can Be Derived from Red Sea Deep Water by Direct Mixing into the Brine Depressions, Calculated from Fluorine Mass Balance

Species	Brine Analyzed *	$f_i(SW)$ (%)
F	Disc.	100.0
SO$_4$	At.	12.0
I	Disc.	7.4
Mg	At.	6.3
H$_2$O	—	4.4
HCO$_3$	At.	3.0
Li	Disc.	2.7
B	Disc.	2.2
Br	At.	2.0
Cl	—	0.5

* Discovery brine: Brewer *et al.* (1965); Atlantis II brine: Brewer and Spencer (1969).

dissolved inorganic carbon of the brine can be derived from mixing with RSDW; these values are important for understanding the deuterium, oxygen-18, and C^{13} and C^{14} concentrations in the brine. The species "HCO$_3$" in Table 1 refers to total dissolved inorganic carbon, expressed as g/kg of bicarbonate ion; the concentrations used for HCO$_3$ are 0.143g/kg in the Atlantis brine (56°C water), measured by Weiss (1969), and an assumed concentration of 0.114g/kg in 19 per mill chlorinity sea water.

For all analyzed substances other than those listed in Table 1, $f_i(SW)$ is less than 1 per cent; the constituents in Table 1 are, of course, those with the highest concentration ratios x_{SW}/x_M. It should be emphasized that the tabulated values of $f_i(SW)$ are upper limits only. For example, the chloride concentrations (g/kg) which have been measured in the two high salinity brines are as follows:

Atlantis Deep	Discovery Deep	Reference
—	155.2	Brewer *et al.* (1965)
155.4 ± 0.1	155.5 ± 0.1	Craig (1966)
156.0	155.3	Brewer and Spencer (1969)

so that from the data of Brewer and Spencer we might conclude using equation (1), that Discovery brine has been diluted by addition of 0.5 per cent sea water to Atlantis brine. Because the Discovery brine is observed to be cooling at the bottom as well as the top (Pugh, 1967), it is widely believed that it represents an overflow from Atlantis Deep into an adjacent depression without direct influx of brine from below; the slightly oxidized nature of Discovery brine is consistent with this interpretation. Thus a slight dilution of the Discovery brine with sea water is plausible, and the value of 0.5 per cent is consistent with the results in Table 1. However, this effect should also be observable in sodium because of its high concentration in the brine, and the Na data measured to date (92.60g/kg in Atlantis brine, by Brewer and Spencer; 92.84 and 93.05g/kg in Discovery brine, by Brewer *et al.* and by Brewer and Spencer, respectively) do not agree with a chloride dilution in Discovery brine —in fact, they agree better with the chloride data reported by Craig (1966). Variations of the magnitude discussed here almost certainly reflect analytical and handling difficulties with such concentrated brines, and we must conclude that no significant differences in major element concentrations have yet been observed.

The upper limits for sea water contribution to the species listed in Table 1 depend on the F concentrations in the brines and RSDW—it would be important to analyze the Atlantis brine and the overlying deep water for fluorine in order to establish a more accurate upper limit for f_{SW}. Nevertheless, the present data show clearly that direct mixing with overlying sea water has not seriously affected the concentrations in the brines.

Origin of the Water

Deuterium-Oxygen-18 Salinity Relations

Fig. 3 shows the D-O^{18} isotopic relationships in waters of the Red Sea region (Craig, 1966). The δ values, in units of per mill are defined as

$$\delta = [(R/R_{SMOW}) - 1] \times 10^3$$

where R is the isotopic ratio HDO/H$_2$O or H$_2$O^{18}/H$_2$O^{16}, and *SMOW* is a defined

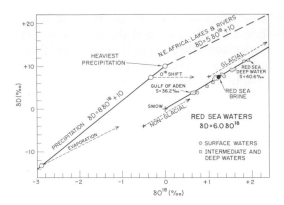

Fig. 3. Deuterium-oxygen-18 isotopic diagram for waters of the Red Sea region. SMOW is the standard Mean Ocean Water isotope standard; δ values are per mill variations relative to this standard.

Standard Mean Ocean Water with approximately the isotopic composition of average ocean water. The Red Sea waters whose compositions are plotted were collected on SIO Expedition Zephyrus in September, 1962; they extend over the salinity range from inflowing Gulf of Aden water at the south (S = 36.2 per mill) to Red Sea Deep Water (S = 40.6 per mill). Outside of the southern sill, in the Gulf of Aden and Indian Ocean, the waters are all less saline than the inflowing water of salinity 36.2 per mill, and have lower isotopic delta values. Within the Red Sea the D-O^{18} variations are linear with a slope of 6; this relationship is due to the evaporation-precipitation-inflow and outflow- and molecular exchange relationships (Craig and Gordon, 1965).

In glacial and non-glacial periods the isotopic composition and salinity of the ocean change slightly because of the formation and melting of continental ice sheets; the dashed lines in Fig. 3 show the maximum excursions of *SMOW* (marked by crosses) and corresponding isotopic trajectories assuming complete melting of continental ice during interglacial periods (Craig, *op. cit.*). Fig. 3 also shows the "precipitation locus" of slope 8 established for the isotopic variations in natural precipitation over the earth, and the trajectories of slope 5 followed by evaporating natural waters such as the North African

lakes and rivers shown in the diagram. A complete discussion of these effects is given by Craig and Gordon (1965) and they are briefly reviewed by Craig (1966). Together with the Red Sea water locus, these lines establish the isotopic relationships in possible source waters for the brines.

The isotopic and chloride data measured on the brines are given in Table 2 (Craig, *op. cit.*), and plotted in Fig. 3; they plot directly on the locus of present-day Red Sea waters, approximately in the middle of the salinity range. The isotopic compositions of the Discovery and Atlantis II brines are identical within experimental precision and fit the observed Red Sea D-O^{18} relationship exactly. An extraordinary set of coincidences would be required to derive such water from fresh waters by evaporation from an initial water on the precipitation locus (lower open circle on that locus, with $\delta O^{18} = -2.9$, $\delta D = -13.2$), or by O^{18} exchange of a fresh water with carbonate minerals — the "isotope shift" observed in continental geothermal areas, which does not affect the deuterium concentration (Craig, 1963). That is, either process would require an exactly specified initial isotopic composition of source water (because the slopes are fixed), coupled with an exactly specified process intensity (evaporation or oxygen isotope exchange), to terminate the isotopic trajectory exactly at its intersection with the Red Sea water locus. The probability of two such coincidences occurring simultaneously in either the evaporation or isotope exchange process is sufficiently remote to eliminate both possibilities; we

Table 2 Isotopic Data (Relative to SMOW) and Chloride Concentrations (by Mohr Titration) in Red Sea Brine Samples

Brine Sample	Cl(g/kg)	δD(‰)	δO^{18}(‰)
Discovery Deep	155.4	7.6	1.22
(Discovery Station 5580)	155.6	7.3	1.22
Atlantis II Deep	155.3	7.4	1.18
(Atlantis II Station 544)	155.5	7.6	
Average values	155.5	7.5	1.21

must conclude that the brine water is derived from the Red Sea water itself.

The specific source water can be identified from the isotopic-salinity relationships shown in Fig. 4. The deuterium and oxygen-18 values (Table 2) correspond to Red Sea water with an initial salinity of 38.2 ± 0.2 per mill for the water which formed the brine. Water of this salinity is found only in the southern end of the Red Sea, extending down to the southern sill at $13°41'N$ in the Hanish Islands.

The seasonal fluctuation in salinity profile over the southern sill is shown in Fig. 5, from Thompson (1939). The upper section (September) shows the situation during the summer when NNW winds prevail; there is a surface outflow of high salinity water and a sub-surface inflow of low salinity water, with water of about 36.4 per mill salinity lying on the sill; occasional outward flow of higher salinity water at depth may occur during the flood tide. At the end of September the flow changes to the winter pattern shown in the lower diagram for May, the last month of the winter season. The winds are SE, the surface flow is inward with outward flow of deep water averaging about 40.3 per mill salinity over the sill. Similar observations for June were made by Neumann and Densmore (Neumann and McGill, 1962)

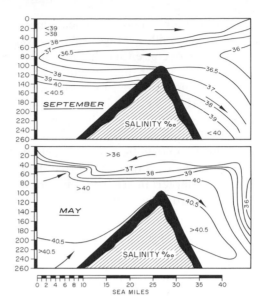

Fig. 5. Seasonal salinity variation with depth (meters) and distance (nautical miles) across the southern sill of the Red Sea (Thompson, 1939).

who found 40.2 per mill water on the sill, as compared with Thompson's observation of 40.4 per mill. The water lying on the sill, averaged over an annual cycle, will therefore have a mean salinity very close to 38.2 per mill.

Northward into the Red Sea, water of 38.2 per mill can be found as far north as 19.5° at the end of winter in May (Thompson, *op. cit.*); the 38.2 per mill isohaline shoals northward from the sill (Fig. 5) and intersects the surface at this latitude. Farther north the water is more saline at all depths. Neumann and Densmore (*op. cit.*) observed 38.2 per mill water extending north to about 18.5°N latitude in June. In summer the 38.2 per mill isohaline moves south and occurs at two depths enclosing a core of lower salinity water; in September, 1962, at 15.5°N, I found the 38.2 per mill isohaline at 96m and 38 per mill water at the surface, with lower salinity water in between. At 16.5°N the 38.2 per mill isohaline occurred at 35m and 80m depth, enclosing less saline water; northward the depths converged and the isohaline vanished at about 17.5°N (Zephyrus Expedition hydrographic data report, Scripps Institution of Oceanography).

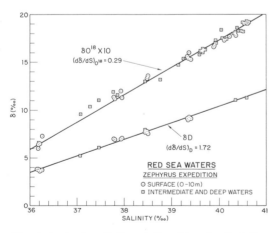

Fig. 4. Isotopic-salinity relationships for Red Sea waters. The deuterium and oxygen-18 isotopic delta values for the Red Sea Brine (+7.5 and +1.2) correspond to an initial salinity of 38.2 ± 0.2 per mill for the water which formed the brine.

Thus in order to find a significant amount of 38.2 per mill average salinity water over a year cycle, either directly or as an alternation between higher and lower salinities, one is restricted to latitudes south of about 18°N at best, and more likely south of about 17°N. That is, source water of this salinity must ultimately flow northward some 500 to 900km, under the bottom of the Red Sea, to reach the brine depressions, the larger figure applying to flow all the way from the southern sill.

Evidence from Dissolved Gases

Concentrations of N_2, Ar, total dissolved carbonate (ΣCO_2), and C^{13} in the Atlantis and Discovery brines have been measured by Weiss (1969). In Table 3, the N_2, Ar and ΣCO_2 concentrations are given as cc(STP) of gas per kg of 38.2 per mill salinity water; that is, the measured concentrations (cc/kg of brine) are multiplied by 1.295, the ratio of water content in 38.2 per mill salinity normal sea water to the water content of the brine. The effect of salt addition or subtraction to the presumed source water of the brine is thus eliminated. The equilibrium concentrations in 38.2 per mill sea water, relative to air at a total pressure of 1 atmosphere including the saturation vapor pressure of H_2O, are also shown; N_2 and Ar values are from the experimental data of Douglas as treated by Weiss (*op. cit.*), while the ΣCO_2 value is a rough estimate

from measurements on surface and deep waters in the Pacific, normalized for the salinity difference (Weiss and Craig, 1968).

The sea water salinity of 38.2 per mill used in Table 3 is the value obtained from the D-O^{18} isotopic data as the salinity of the source water of the brine. From the *T-S* diagram for Red Sea waters (Neumann and McGill, 1962) it is found that Red Sea water of 38.2 ± 0.1 per mill has a temperature of $28 \pm 1°C$, this temperature corresponding uniquely to that salinity value. Therefore, if the gases are conservative, their concentrations in the brine as predicted from the stable isotope data, should correspond to the saturation values in Red Sea water of $T = 28°$, $S = 38.2$ per mill. As shown in Table 3, the Ar values in such sea water and in Atlantis II brine are indeed found to be identical. (The temperature and salinity uncertainty ranges, $\pm 1°C$ and 0.1 per mill, correspond to variations of only ± 0.004 and 0.1 in Ar and N_2 concentrations, respectively.) Weiss (*op. cit.*) has developed this argument *ab initio* to derive minimum initial temperatures for sea water source fluid; he also shows that the Ar and N_2 concentrations in the brines are completely different from equilibrium solubilities in the brine. The Ar data thus provide striking confirmation of the D-O^{18} isotopic identification of source water for the Red Sea brine.

Discovery brine contains about 10 per cent excess Ar relative to Atlantis II brine;

Table 3   Dissolved Gases in Sea Water and the Red Sea Brines

Gas Concentrations Are Expressed as cc(STP)/kg of 38.2‰ Salinity Water, Calculated from Data of Weiss (1969). Δ Values Are Excess Concentrations Relative to Sea Water, δ Values Are Relative to PDB-1 Isotopic Standard

	Sea Water $S = 38.2‰$, $T = 28°C$	Atlantis II Brine	Discovery Brine
N_2	8.05	13.09	13.49
ΔN_2	—	5.04	5.44
Ar	0.2110	0.2110	0.2320
ΔAr	—	0.000	0.021
ΣCO_2 (ppm)	(125)	184.7	39.4
$\Delta\Sigma CO_2$ (ppm)	—	59.7	−85.6
δC^{13}(‰)	(−1)	−5.6	−16.8
$\delta C^{13}(\Delta CO_2)$(‰)	—	−15.2	+6.3

this is presumably excess radiogenic Ar^{40} and, if so, it is consistent with a model of Discovery brine as a stagnant overflow from Atlantis Deep as discussed by Weiss (1969). The excess N_2 of about 5cc/kg in each brine is attributed by Weiss to decomposition of organic matter in sediments, as observed in other ground waters. Oxidation of NH_3 in this amount would consume about 7.5cc/kg of O_2, while the solubility of O_2 in the initial sea water would be about 4.1cc/kg, so that at least half of the excess N_2 would probably have to result from denitrification of nitrate or nitrite in sediments. (The observed nitrate depletion in the brines relative to Red Sea normal water (Brewer *et al.*, 1965) could produce only 3 per cent of the excess N_2.) N_2 and Ar concentrations in the brines are greater than in overlying Red Sea deep water, so the brine concentration cannot have been increased by direct mixing— nitrogen isotope measurements should show whether the excess N_2 is indeed derived from sediments.

Table 3 also shows that Atlantis II brine contains about 32 per cent excess CO_2 relative to sea water, while Discovery brine has only about 20 per cent of the Atlantis brine concentration, or about 30 per cent of the original sea water concentration; the brines also differ in isotopic composition from each other and from the estimated sea water value. Mean values of δC^{13} for the CO_2 added to or subtracted from the sea water to give the brine concentrations are also shown. The mean δ value of -15.2 per mill for excess CO_2 added to Atlantis II water is quite characteristic of carbon derived from a mixture of heavy carbonate and light organic matter, as observed in ground waters, or as a mean isotopic composition of particulate carbon in the oceans (e.g. whole foraminifera), while the mean δ value of 6.3 per mill for CO_2 removed from Discovery water is much more similar to that of carbonate precipitated inorganically from sea water. These differences are discussed in a later section.

Dissolved gas studies on brine samples are continuing and a detailed study is necessary for confirmation of these conclusions; however, the present data show very clearly that inert gas concentrations are also a powerful tracer for geothermal waters, and give strong support to the source water origin deduced from the isotopic water data.

Origin of the Added Salts

Although most components of the brine are greatly enriched relative to their sea water concentrations, certain elements such as Mg are present in much lower concentrations. It is convenient for the material balance calculations to classify the brine components in terms of their enrichment factors relative to the presumed source water of 38.2 per mill salinity normal sea water, as defined by the process of addition or subtraction of salts to or from the source water. The enrichment factor thus defined is

$$E_i = \frac{(C_i/C_w)_{\text{Brine}}}{(C_i C_w)_{SW}} \qquad (3)$$

in which C_i and C_w are the concentrations of component i and of water (g/kg) and the subscript SW refers to normal sea water of 38.2 per mill salinity. In such a process E_w is of course unity; a value of $E > 1$ indicates a net addition of the component to the source water, while $E < 1$ indicates that the component has been removed from the original sea water. Enrichment factors for the various components are shown in Table 4, which also shows the concentration data used; these data refer to Atlantis II brine except when data are available only for Discovery brine.

In these and subsequent calculations the Na and Cl values given by Brewer and Spencer (1969) for Atlantis brine were adjusted slightly because their analysis shows an anion excess over neutrality by about 0.5 equivalent per cent. As noted earlier, their value of 156.03g/kg of Cl is higher than the value given by Craig (1966) for Atlantis brine and their own values for Discovery brine, and Craig (*op. cit.*) found both brines to have the same Cl content, 155.5g/kg. Also their Na value in Atlantis brine is *lower* than their value and that of Brewer *et al.* (1965) for Discovery brine;

Table 4 Enrichment Factors in the Red Sea Brine Relative to Normal Sea Water of 38.2‰ Salinity, Calculated for Addition or Subtraction of Salt

Component	Brine Analysis	Concentrations (g/kg)		Enrichment Factor
		Brine	Sea Water	
Pb	Atl. (1)	6.3×10^{-4}	3×10^{-8}	27,200
Mn	"	8.2×10^{-2}	4.2×10^{-6}	25,300
Fe	"	8.1×10^{-2}	$<2 \times 10^{-5}$	>5,200
Zn	"	5.4×10^{-3}	$<5 \times 10^{-6}$	>1,400
Ba	Atl. (2)	9×10^{-4}	16.7×10^{-6}	70
Cu	Atl. (1)	2.6×10^{-4}	5.5×10^{-6}	61
Ca	Atl. (1)	5.15	0.450	15
Si (3)	"	2.76×10^{-2}	2.4×10^{-3}	15
Na	"	92.85	11.75	10
Cl	"	155.5	21.13	9.5
Sr	"	4.8×10^{-2}	8.9×10^{-3}	7
K	"	1.87	0.423	5.7
Br	Atl. (1)	0.128	7.36×10^{-2}	2.2
B	Disc. (4)	7.8×10^{-3}	5.1×10^{-3}	2.0
Li	"	2.62×10^{-4}	2.07×10^{-4}	1.6
HCO_3^-	Atl. (5)	0.143	0.125	1.5
H_2O	Atl. (1)	742.50	961.65	1.0
Mg	Atl. (1)	0.764	1.413	0.70
I	Disc. (4)	3.0×10^{-5}	6.5×10^{-5}	0.60
SO_4	Atl. (1)	0.84	2.96	0.37
NO_3 (3)	Disc. (4)	4.4×10^{-5}	7.5×10^{-4}	0.08
F	"	5.12×10^{-5}	1.4×10^{-3}	0.05

Notes: (1) Brewer and Spencer (1969); values for Na and Cl slightly adjusted as described in text.
(2) Miller *et al.* (1966).
(3) Si and NO_3 sea water values are data for Red Sea Deep Water from reference (4), normalized to 38.2‰ salinity.
(4) Brewer *et al.* (1965). Miller *et al.* (1966) found $B = 11 \times 10^{-3}$ in Atlantis II brine, corresponding to an enrichment factor of 2.8.
(5) Brine value from Weiss (1969), sea water value as in Table 3, expressed as bicarbonate ion.

these differences account for the anionic imbalance. Accordingly the Cl and Na concentrations in Atlantis brine were taken as 155.5 and 92.85g/kg respectively as the most probable best values. This makes the total dissolved salt concentration in Atlantis brine 257.50g/kg as the sum of the analytical values (including total CO_2 calculated as HCO_3).

Table 4 shows that the observed enrichment factors fall roughly into four classes: (1) trace metals (Mn, Fe, etc.) with factors greater than 50, ranging up to 27,000; (2) normal components such as Na, K, Ca, and Sr with enrichments close to that of chloride—about 10 ± 5; (3) minor components only slightly enriched, with factors between 1 and 2; and (4) the components Mg, SO_4, F, and I which have factors less than one and have thus been subtracted from the original sea water. If Z_A and Z_S are the number of grams of salts respectively added to, and subtracted from, one kg of original sea water (38.2 per mill salinity, 38.5g/kg of total salts), then

$$(Z_A - Z_S) = 1,000 \left(\frac{C_{W\text{-}SW}}{C_{W\text{-}B}} - 1 \right) = 295.1 \text{g/kg}$$

is the net salt addition as given by the concentrations of water in the sea water and in the brine. The value of Z_S is given by

$$Z_S = \sum_s C_{i-SW} - 1.2951 \sum_s C_{i-B} \quad (4)$$

in which the summations are over the components with enrichment factors less than one (*subtracted* components: Mg, I, SO_4, F) in sea water and in brine. Thus the salt balance calculation gives

$Z_S = 2.30$g/kg, subtracted salts
$Z_A = 297.4$g/kg, added salts

as the amounts subtracted from, and added to one kg of 38.2 per mill salinity sea water, to form the brine.

Composition of the Added Salt

From the values of $(Z_A - Z_S)$ and Z_A for the brine, the mass fractions of the various components in the 297.4g of added salt per kg of initial sea water are given by

$$x_{i-A} = \frac{1.2951 C_{i-B} - C_{i-SW}}{297.4} \qquad (5)$$

in which x_{i-A} is the mass fraction of component i in the added salt and the C_i are the concentrations in brine and 38.2 per mill salinity sea water. The calculated composition of the added salt, with the major cations expressed as chlorides, is given in Table 5, which shows that 99.8 per cent of the added

salt consists of NaCl, KCl, and CaCl$_2$ with NaCl greatly predominant (93 per cent), and that the remaining components range from about 1 to 300ppm in concentration. The composition as a whole is quite normal for the halite-sylvite zones of evaporite deposits. Stewart (1963) has reviewed the chemistry of evaporite deposits in detail — trace element ranges in halite rocks of marine evaporites, given in his Table 24 (*op. cit.*) include: Sr, 52–180ppm; Si, 22–240ppm; Cu, 1–6.5ppm; Ni, 1.6–3.5ppm; Pb, 1–5ppm; and Ba, 1–3ppm. Bromine values in halite-sylvite rocks range from 140–3,300ppm (Table 28, *op. cit.*), and boron values, from two sets of data on halite rocks, range up to 50 and up to 379ppm (Table 27, *op. cit.*). All these values are quite comparable to the added salt composition in Table 5. Fe concentrations observed in halite-sylvite rocks range from 21–448ppm (Stewart, *op. cit.*, p. 37). Only Mn and Zn are considerably higher in the

Table 5   Composition of Salts Added to Normal Sea Water (S = 38.2‰) to Make the Red Sea Brine, and of Salts in Two Lousiana Brines from Salt Domes

	Red Sea Brine: Added Salts	Timbalier Bay, La. (1)	West Bay, La. (2)
Major Components (per cent)			
NaCl	92.7	89.9	81.2
KCl	1.3	0.3	0.8
CaCl$_2$	5.8	5.0	12.7
MgCl$_2$	—	2.85	2.09
SO$_4$	—	0.002	0.08
HCO$_3^-$	0.020	0.10	0.06
Minor Components (ppm)			
Sr	179	76	900
Br	310	591	1,964
B *	17	138	47
Si	112	83	37
Li *	0.45	38	95
F *	—	7.5	7.0
I *	—	131	90
Mn	357	7.6	150
Fe	353	59	550
Zn	23.5		25
Cu	1.11	0.14	2.0
Co	0.7		
Ni *	1.5	10.3	10.5
Pb	2.74		3.0
Ba	3.86	1,032	85

(1) Timbalier Bay oil field, Lafourche Parish, La.; chloride concentration 89.7g/kg of brine. Water analysis, from White *et al.* (1963, Table 12), has excess Cl equivalent to 1.7% of total salts over amount for neutrality.

(2) West Bay, Plaquemines Parish, La.; chloride concentration 124.0g/kg of brine. Water analysis, from White *et al.* (1963, Table 13), has excess Cl equivalent to 2.5% of total salts over amount for neutrality.

* Calculated from Discovery brine concentration data (*cf.* Table 3). Boron concentration would be 31ppm using Atlantis brine value of Miller *et al.* (1966).

added salt composition than in halite rocks, where observed values range up to 20ppm for Mn and about 3ppm for Zn; these two elements are concentrated in anhydrite and gypsum zones, and in salt clays.

A more direct comparison can be made with the salts found in brines collected around salt domes; Table 5 also shows the composition of the salts, recalculated to a water-free basis, in two such waters from Lousiana salt domes (White *et al.*, 1963). The Timbalier Bay sample occurs in Pliocene sands at a depth of 1,800m with a temperature of 70°C; the well is on the north flank of a salt dome which was encountered at a depth of 2,370m. The West Bay brine is found at 2,500m in Miocene sandstone, only 60m from salt in the crest of the West Bay salt dome, which is reported to lack an anhydrite cap; the sample and bottom hole temperatures were 42°C and 87.5°C respectively. These two samples are probably the most representative analyses available of waters which have been in contact with salt domes, and as seen in Table 5, with the exception of Mn, the concentrations of all other elements bracket those in the added salts of the Red Sea brine. Some elements in the added salt are, however, conspicuously low—Mg, Li, F, I, Ba, and possibly B; reasons for this are discussed later on.

White (1965) has discussed the Ca/Cl and K/Na ratios in natural waters with respect to origin and mineral interaction; in a Ca/Cl vs. Cl plot (Fig. 1, *op. cit.*) the salt dome waters in Table 5 are more similar than any others to the Red Sea brine. The K/Na ratio of the brine is characteristic of the lowest temperatures (<100°C) for mineral-water interaction and is also most similar to the salt dome waters; while the brine has probably never equilibrated Na and K with the solid phases, the ratio at least indicates that it has not been subjected to high temperatures in the presence of feldspars and micas for any prolonged period.

In general, the composition of the added salt agrees very well with what we know about the composition expected for solution of salts from the halite-sylvite zone of evaporite and salt dome deposits; the high concentrations of trace elements in the brine demand only ppm concentrations in the original salt deposits, and, with the possible exception of Mn and Zn, are not higher than expected. Even Mn and Zn are not greatly enriched with respect to the salt dome waters in Table 5.

Isotopic Data on the Added Salts

The isotopic composition of lead in the Atlantis II brine has been measured by Delevaux *et al.* (1967), together with that of a series of Cenozoic ore leads from Saudi Arabia and Egypt. Although considerable variation in the ore leads is observed, the brine lead is very similar in both 207/204 and 208/204 ratios to lead in a Tertiary vein in Rabigh, Saudi Arabia, and to a series of galena samples in Miocene gypsum and lime sediments in Egypt (Um Gheig, Um Ans, and Bir Ranga deposits, Table 2, *op. cit.*). All of these leads are normal Tertiary J-type leads with isochron and 208/204 ages close to zero, quite different from Precambrian ore leads; in particular their isotopic composition is precisely what would be expected for lead from a Tertiary evaporite sequence.

The isotopic composition of strontium in the brine has been measured by Faure and Jones (1969), and found to be similar to that of strontium in marine carbonates, again consistent with derivation from marine sediments or from shells of marine organisms.

Subtraction of Salts from the Brine

Components in the Red Sea brine with enrichment factors less than 1 include Mg, SO_4, I, NO_3, and F. The amount of a component subtracted from the original sea water of 38.2 per mill salinity, in grams removed per kg of original sea water, is given by

$$Z_{i-S} = C_{i-SW} - 1.2951 C_{i-B} \qquad (6)$$

The amounts subtracted, and the percentage of each component which has been removed, from the original sea water are:

Mg = 0.424g/kg   (30.0 per cent lost)

SO_4 = 1.872g/kg   (63.2 per cent lost)

$I = 2.61 \times 10^{-5} g/kg$ (40.2 per cent lost)

$NO_3 = 6.9 \times 10^{-4} g/kg$ (92.4 per cent lost)

$F = 1.33 \times 10^{-3} g/kg$ (95.0 per cent lost)

for a total of $Z_S = 2.30 g/kg$ removed. The SO_4/Mg ratio in the subtracted salt is 4.4, compared to the stoichiometric ratio in $MgSO_4$ of 4.0; the difference is within the analytical errors, so that the possibility that these components may actually have been removed as $MgSO_4$ must be considered. The F, I, and NO_3 brine concentrations were measured only in Discovery brine (Brewer *et al.*, 1965), which generally has lower trace metal concentrations than Atlantis brine, so that concentrations in Atlantis II brine may be significantly higher. However, Brewer and Spencer (1969) found that F in Atlantis brine was below their analytical limit of about $2 \times 10^{-5} g/kg$, while the concentration in Discovery brine was reported as $5 \times 10^{-5} g/kg$ (Brewer *et al.*, *op. cit.*), so F at least is not significantly higher in Atlantis brine.

It is worth noting that all these subtracted components occur in very low concentrations in halite-sylvite zones of evaporites (Stewart, 1963); fluorine in evaporites is found principally in sulfate deposits, while iodine is thought to be lost to the atmosphere during evaporation. Their low abundance, though not their subtraction from the source water, is thus consistent with the concentration pattern in the added salts. Fluorine and lithium occur in much higher concentrations in all so-called "volcanic" waters (White, 1965). The Li/Na ratio of 3×10^{-6}, and F/Cl of 3×10^{-7}, in Red Sea brine are the lowest ratios recorded for high salinity waters (cf. data in White *et al.*, 1963) and argue strongly against any significant magmatic or volcanic contributions to the brine.

Isotopic Composition of the Sulfate

The isotopic composition of sulfur in dissolved SO_4 ions has been measured by Hartmann and Nielsen (1966), and that of oxygen in the SO_4 ions by Longinelli and Craig (1967); the data can be summarized as follows:

	δS^{34} (per mill)	δO^{18} (per mill)
Red Sea Brine	20.3	7.3
Normal Red Sea Water	20.3	9.5

The S^{34} data refer to Atlantis II brine; O^{18} measurements were made on both Discovery and Atlantis II brines, with results of 7.5 and 7.2 respectively. O^{18} measurements on normal Red Sea water SO_4 gave $\delta = 9.3$ for Red Sea deep water, and $\delta = 9.6$ for SO_4 in surface inflow water from the Gulf of Aden. (Sulfur data refer to the meteoric sulfur standard, oxygen data to standard mean ocean water—*SMOW*.)

It has been shown in previous sections above that (1) the chloride and fluoride concentrations in the brine show that no significant fraction of the sulfate can be derived from direct mixing with overlying water (Table 1), and (2) that sulfate has actually been *removed* from the original water, 63 per cent of the original concentration having been lost. It is clear, therefore, that the δS^{34} values are simply those of the sulfate in the original sea water which formed the brine, and that SO_4 removal has taken place by a process which does not fractionate sulfur isotopes. The isotope data thus constitute additional evidence consistent with a brine origin from normal sea water.

The slight O^{18} decrease in the brines, relative to SO_4 in normal Red Sea water, probably represents an approach to isotopic equilibrium at the higher temperature of the brine, since the exchange time is probably less than 100 years at these temperatures (Longinelli and Craig, *op. cit.*); alternatively some oxygen isotope fractionation may have occurred in the removal process. Nevertheless, the S^{34} and O^{18} data together are more similar to values in sea water SO_4 than to anything else. Rafter (1967) has plotted S^{34} vs. O^{18} data from a wide variety of sulfate in natural barites, geothermal waters, and saline lakes; his diagram shows that the brine sulfate falls closer to sea water sulfate than to any other samples. Thus both the S and O isotopic labels indicate that the low sulfate concentration in the brine is a residual from the original sea water sulfate, most of which has been removed by a non-fractionating process for sulfur.

Kaplan *et al.* (1969) have studied sulfur isotope ratios in the brines and sediments in more detail. They find δS^{34} to be $+20 \pm 2$ per mill in sulfate in all the waters of both Atlantis and Discovery Deeps, with no significant differences between Atlantis 44° and 56° brines, Discovery brine, and Red Sea deep water overlying either brine; the isotopic data show no correlation with SO_4 concentration and individual variations appear to represent only analytical scatter. In Discovery Deep sediments, sulfides and elemental sulfur are very low in S^{34} ($\delta \sim -25$ per mill) and sulfate in interstitial water is about 7 per mill enriched in S^{34} relative to SO_4 in brine and sea water; this is clearly the result of *in situ* biogenic sulfate reduction in the sediments, as they point out. Biogenic sulfur and sulfides are also found in sediments outside the deeps and in the Atlantis II Deep central rise above the brine level. In all these areas the reduced sulfur is dominantly pyrite, with associated chalcopyrite, sphalerite, and elemental sulfur.

In Atlantis II Deep, sulfides and sulfur with $\delta S^{34} = +5$ per mill are found (range $+3$ to $+11$) (Hartmann and Nielsen, Kaplan *et al., op. cit.*); sulfur and sulfides are dominated by sphalerite, and $CaSO_4$ is also found. Sulfate in interstitial water is slightly depleted in S^{34} relative to brine and sea water ($\delta S^{34} = 15$ to 21 per mill, averaging about 18), and anhydrite is about $+23$ per mill, consistent with precipitation from brine or seawater (Kaplan *et al., op. cit.*). The origin of sulfur and sulfides in these sediments is obscure, especially as sulfur and sulfides of similar isotopic composition ($\delta S^{34} = +10$) were found by Kaplan *et al.* in Core 118K, 6km SE of the brine deeps and *below* normal biogenic sulfur and sulfides. In this core the dissolved sulfate in interstitial water is isotopically similar to that in Atlantis Deep cores ($\delta S^{34} = 16$ to 19 mill, averaging about 17.5), with the higher values (closest to sea water values) occurring in the upper region of light biogenic sulfur and sulfides.

The origin of the sulfide and sulfur in Atlantis Deep sediments, and the problem of their unique isotopic compositions, are obviously closely connected with the mechanism of sulfate depletion in the brine relative to its source water; these questions are considered in the next section.

Origin of Sulfide and Sulfur in Atlantis Deep Sediments

Two models have been proposed for sulfide origin and deposition in Atlantis Deep. Watson and Waterbury (1969) emphasize the difficulty of precipitating zinc sulfide and ferric hydroxide in the same environment, and postulate that precipitation of both takes place in the transition layer between the upper 44°C brine and normal Red Sea Bottom water, and that bacterial sulfate reduction is occurring in a shallow zone just above the 44°C brine. They suppose that just enough reduction takes place for ZnS precipitation so that ferrous iron diffuses on through this layer into the aerobic zone above where it is oxidized to the ferric state and precipitated. In order to account for the sulfide isotopic composition, δS^{34} about $+5$ per mill, they assume that a large enough fraction of the sulfate in the layer is reduced so that less than one enrichment stage is achieved. The sulfate in the reduction layer is assumed to be replaced continually by diffusion, so that it is never entirely used up; this is thus a steady-state model.

Later on, it is shown that a steady-state model of Atlantis Deep actually requires that essentially all of the Fe and Zn precipitation takes place within the 44°C convection layer. But one can readily show that the mass balance required by the W-W model cannot be achieved. Firstly, if we consider a closed system in which sulfate reduction is a Rayleigh process, with $\alpha = 1.040$ (the approximate isotopic separation factor for bacterial reduction), we find that starting with sea water sulfate ($\delta = +20$), 82 per cent of the sulfate has to be reduced in order for the *mean* δ value for all sulfide produced to be $+5$ per mill; the remaining 18 per cent of sulfate in the water then has $\delta S^{34} = +90$ per mill; this clearly has not occurred. In the proposed steady-state system, the sulfate is replaced by diffusion from both sides, and by the advective flux of new brine flowing into the depression. A simple diffusion-advection model (one-dimensional) with an advective flux of brine upward, and diffusion between the 44°C brine and sea water, gives an approximate solution for the mean value δ_J of sulfate at steady state:

$$\delta_J = \frac{\delta_{SW}[C_J(C_B - C_{SW}) - JC_{SW}] - \epsilon J(C_{SW} - C_J)}{[C_J(C_B - C_{SW}) - JC_{SW}] + 10^{-3}\epsilon J(C_{SW} - C_J)}$$

$$(7)$$

in which C is sulfate concentration and the subscripts SW, J, and B denote the values in Red Sea deep water, the thin sulfate-reduction layer (which can be assumed to be approximately uniform), and in the 56° brine. J is the sulfate reduction rate divided by the advective flux of brine into the deep, i.e. the removal rate in net brine-flow units. The factor ϵ is $10^3(\alpha - 1)/(1 + R)$, where R is the S^{34}/S^{32} absolute ratio $= 0.044$. In the steady-state replenishment model, the isotopic composition of precipitating sulfide at any time has to be given by:

$$\delta_S = [\delta_J - 10^3(\alpha - 1)](1/\alpha) \qquad (8)$$

The relationships in (7) can be fixed as follows: the value of J must not be significantly greater than the flux of Zn into the reduction layer, or most of the iron would precipitate as FeS, rather than predominantly as oxidized iron as observed. The net flux of Zn into the layer (in brine flow units as used for J) is simply its concentration in the 56° brine, $= 5.4 \times 10^{-3}$ when the concentration units are g/kg. The corresponding sulfate reduction rate $J = 7.9 \times 10^{-3}$ as a maximum value. With this value of J and $\alpha = 1.040$, the following values of sulfate concentration in the reduction layer, and δ of sulfate in the layer and of precipitating sulfide, are found:

Values for $C_J = 0.006$ are those for which $\delta_S = +5$ per mill as actually observed; the other values indicate the limits on C_J and the range of variation which could be encountered. It is seen that the sulfate in the reduction layer must be as heavy as +45 per mill, and the sulfate concentration as low as 6ppm, for this process to be operating at present. These values are quite impossible, as shown by the data of Kaplan *et al.* (1969) on sulfate concentration and δ values above Atlantis Deep. The essential point is that at steady-state the sulfate in the reduction layer has to be higher in S^{34} by one single-stage separation factor, as shown in the above values (this is true even if advection is omitted from the model). Thus the proposed steady-state process cannot account for the sulfide isotopic composition.

Kaplan *et al.* (1969) propose that the sulfate in the brine is derived from marine evaporites, some of which reacts with organic matter in shales to produce H_2S and perhaps elemental sulfur, removing sulfate from the brine. The metals are carried in as sulfides and precipitated in the Atlantis Deep, in some manner not described. The sulfide-sulfate fractionation of about 15 per mill is due either to a high-temperature isotopic equilibrium or to the difference in reduction rates of the two isotopes (α is observed to be about 1.022 for the relative rates at 18–50°C). They suppose that the similarly anomalous sulfides ($\delta = +10$) found at the base of core 118K, outside of the brine depressions, represent a thermal overflow of brine into that locality at some earlier time. Further, they conjecture that the water in the brine may originate from dehydration of gypsum to anhydrite (though such water would actually have a much different isotopic composition than is actually observed).

Arguments against the proposal of Kaplan *et al.* include the following:

1. The brine cannot have been in contact with shales at the high tempera-

C_J(g/kg) =	0	0.006	1.00	3.20(= C_{SW})
δ_J(per mill) =	59.4	45.2	20.6	20.3(= δ_{SW})
δ_S(per mill) =	18.6	5.0	−18.7	−18.9

tures (300–500°C) they require for equilibrium sulfur fractionation – an oxygen isotope shift in the water, due to exchange with silicates and carbonates, would surely occur (Craig, 1963, 1965).

2. Sulfate has been removed from the brine, not added from evaporites. If the removal had been accomplished by sulfate reduction, the remaining sulphate in the brine would now be enriched in S^{34} with respect to sea water sulfate. As shown previously, 63.2 per cent of the sulfate has been removed from the original sea water; the fraction of sulfate remaining is 0.368 which happens to be 1/e, so that in the Rayleigh process calculation, the delta value of the present brine sulfate should simply be:

$$\delta_{SO4} = \delta_{SW} + 10^3(\alpha - 1)$$

if the original δ value is that of sea water sulfate (+20.3). Thus for $\alpha =$ 1.020, the δ value of the present sulfate should be about +40, and for $\alpha =$ as low as 1.005 it should be +25. However, the sulfate in the brine is isotopically identical to that in sea water and has been removed by a non-fractionating process.

3. If reduction took place by means of organic carbon, δC^{13} of the dissolved carbon should be very low, about −30 per mill or less, whereas the Atlantis Brine dissolved carbonate has a δ value of −5.6 per mill (Weiss, 1969).

4. Sulfides and sulfur of about the same isotopic composition as Atlantis Deep sulfides, with $\delta S^{34} = +10$, occur outside the brine depressions, as found in core 118K (Kaplan et al., op. cit.); this sulfur is not biogenic and occurs as ZnS, FeS, and sulfur, at 390cm in the core, well below the biogenic sulfur ($\delta = -30$) at 110–175cm in the core. The existence of reduced sulfur so similar to that in Atlantis Deep sediments, some 6km outside of the area of the brines, shows that

sulfides of this isotopic composition are not uniquely products of hot brine introduction.

Kaplan et al. suppose that these sulfides and sulfur represent a thermal overflow of hot brine into the area of core 118K at some earlier time. However, this is demonstrably not so, as shown by the detailed study of Deuser and Degens (1969) on carbonates in this core. They show that pteropods at the 390cm level are similar to those of the same age in sediments 160km away, and that there is no isotopic shift in O^{18} and C^{13} in the carbonate overgrowths such as is observed in pteropods and foraminifera of the same age in Discovery Deep, and in fossils in Atlantis Deep. The C^{14} stratigraphy, from the measurements by Ku et al. (1969), show that the biogenic sulfides at 110–175cm in core 118K correspond to the time of rising sea level, about 11,000 years ago, when the Red Sea freshened again, as shown by the O^{18} data on the fossils. The 390cm level in core 118K corresponds to a time more than 20,000 years ago when Pleistocene sea level lowering had caused the Red Sea to become hypersaline. Two points concerning the sediments in core 118K should be emphasized: (1) the fossils are similar in O^{18} concentration to those of a core much farther away from the brine deeps, with no O^{18} shift such as is shown by Atlantis and Discovery Deep fossils, and (2) near the core bottom where the S^{34} enriched sulfides occur, casts of original shells and secondary overgrowths on other shells have the same δO^{18} values, indicating equilibrium with water which had surrounded them (Deuser and Degens, op. cit.). The isotopic evidence thus indicates very strongly that (1) the sediments of core 118K have not been exposed to thermal brines at any levels and (2), as Deuser and Degens point out, all available present evidence indicates that no brine was present in the deeps before about 10,000 years ago, a time much later than the deposition of the S^{34} enriched sulfides at the bottom of core 118K.

From the oxygen and carbon isotope data presently available, we have to conclude that deposition of S[34] enriched sulfides and sulfur at least as heavy as +10 per mill took place in Red Sea sediments as a normal, non-hydrothermal, process which could occur in the absence of any geothermal brines. The similarity of the sulfides and sulfur in Atlantis Deep and the bottom of core 118K ($\delta S^{34} = +3$ to +11 per mill), and of the dissolved interstitial sulfate (+15 to +20 per mill, averaging about +18 in both areas) implies very clearly that sulfide deposition in both areas occurred by similar processes which did not require, but also were not inhibited by, the presence of a geothermal brine. This in turn implies that sulfur introduction took place by an agency independent of the brine. The most plausible and simple explanation is that sulfur was introduced as H_2S from surrounding sediments into both these areas, precipitating sulfides and sulfur from geothermal brine in Atlantis Deep and from high salinity sea water in the Core 118K area.

The cyclic O[18] and C[13] correlation with Pleistocene glaciation and deglaciation, as measured by radiocarbon stratigraphy, is almost certainly a response to alternating hypersaline and freely circulating conditions in the Red Sea, as proposed by Deuser and Degens. Core 118K thus indicates the following sulfide cycle: during times of sea level lowering, corresponding to the bottom part of the core with large O[18] enrichment, the bottom waters of the Red Sea were so saline that biogenic sulfate reduction could not occur. H_2S introduction from sediments into the high salinity waters resulted in precipitation of sulfides and sulfur with δS^{34} about +10 per mill. When sea level rose again during glacial melting, and normal circulation was re-established about 11,000 years ago, micro-organisms could again live in the accumulating sediments and biogenic sulfate reduction produced sulfides with the characteristic δS^{34} of −30 per mill in the 110–175cm level of the core.

The sudden change in isotopic composition of elemental sulfur at the 175cm level of core 118K (Kaplan et al., op. cit., Fig. 3) is strikingly parallel to the O[18] and C[13]

depth profiles shown by Deuser and Degens (op. cit., Fig. 2):

Depth (cm)	δO^{18} (per mill)	δS^{34} (per mill)
90	0.3	−31
100	0.6	−15
175	6.5	+8
390	5.5	+10

(the O[18] values are those measured on the pteropod shells). Sulfides at the 175cm level are similar to those above, with $\delta S^{34} = -30$; at 390cm δS^{34} is +10, with no measurements in between. Deuser and Degens (op. cit.) point out that this transition layer consists of highly fractured and fragmented shells, indicative of turbulent sediment transport; a similar layer is observed at the same depth much farther south. The biogenic sulfides in this layer may therefore represent transported material; alternatively, biogenic reduction of sulfate and inorganic production of elemental sulfur may have persisted together for some time as the Red Sea freshened.

The slightly lower S[34] content of dissolved interstitial sulfate in both Atlantis Deep sediments and core 118K, relative to sea water and brine, indicate that brine and sea water sulfate have not been involved in the sulfide precipitation process at all, but that some oxidation of lighter sulfides and sulfur has taken place in situ, after sulfide deposition, as suggested by Kaplan et al. The somewhat lighter sulfides in Atlantis Deep sediments, averaging +5 per mill vs. +10 per mill in core 118K, may indicate that the introduced H_2S, probably generated by diagenesis in underlying sediments, was lower in S[34] than the sulfides formed from it, and that a larger fraction of the sulfur was precipitated in Atlantis Deep because of the much higher heavy metal content.

Still unexplained is the uniformly low dissolved sulfate concentration in the supposedly pre-brine levels of Discovery Deep sediments and in core 118K, of the order of 1g/kg, similar to that in the geothermal brines and Atlantis Deep sediments. The

Discovery Deep concentrations might be explained by diffusive interaction with overlying brine, but if the carbon and oxygen isotopic data correctly indicate that no brine has been present in core 118K, the low dissolved sulfate concentrations in this core are difficult to understand. Unfortunately no chlorinity determinations are available on the interstitial waters in Discovery Deep sediments and core 118K — without these data it is impossible to explain the sulfate concentrations and to be certain of the history of these sediments.

The Removal Process for the Subtracted Salts

The absence of S^{34} enrichment in the brine sulfate, and the absence also of a large C^{12} enrichment in the dissolved carbonate, indicate, as we have seen, that sulfate was not removed from the original water by reduction processes; nor do the sulfides in the sediments require any such origin. Similarly, it is difficult to account for the magnesium depletion in the brines by Mg fixation in dolomite or silicates such as chlorite: in such processes there would surely be oxygen isotope exchange between the water and the carbonates or silicates and an oxygen isotope shift would be observed in the water (Craig, 1963). Such exchange takes place in times of weeks for carbonates and feldspars at 100°C.

One process which could produce the observed depletion pattern of Mg, SO_4, I, and F, is that of membrane filtration, in which material is removed by selective filtration through clay minerals. White (1965) has recently reviewed the known and most probable characteristics of this process and has pointed out that Na^+, HCO_3^-, F^-, I, and B appear to be the most mobile components through such "membranes," followed by Cl^- and Mg^{++}; other components should be much less mobile. Neutral species and small, singly charged ions display the highest mobility, and White considers that I_2, H_3BO_3, and undissociated $NaHCO_3$ migration probably account for the mobility of these species.

Since it is known that an appreciable fraction of Mg and SO_4 exist as undissoci-

ated $MgSO_4$ in solutions as low in these components as sea water, selective loss of $MgSO_4$ by membrane filtration could account for the depletion of both these species in the stoichiometric proportions observed. About 10 per cent of the Mg ions in normal sea water are associated with SO_4 (Pytkowicz and Gates, 1968, and references therein) so that significant loss of uncharged $MgSO_4$ should occur in the filtration process. Sulfur isotope fractionation would probably be insignificant, because the sulfur atom is surrounded by oxygen atoms and probably has little interaction with the potential barriers involved in penetration of the membrane. The very modest boron enrichment in the brines can also be explained by a significant depletion in the filtration process subsequent to an initial higher enrichment.

The relative enrichment pattern in the bottom part of Table 4, is, with the exception of Li, therefore almost a classical pattern of expectation for the membrane filtration process. Lithium may actually be more highly enriched in the brine than indicated by the value of Brewer et al. (1965) used in Table 4; M. Peterson (personal communication, 1968) has measured concentrations of 4.5 and 4.0ppm Li in Atlantis II and Discovery brines respectively, and 1.7ppm in the 44°C brine layer of Atlantis Deep. These values are completely consistent with the mixing relations vs. chloride shown in Figs. 1 and 2, and, if correct, indicate that the enrichment factor for Li in the brines should be 28, rather than 1.6 as given in Table 4, making Li the most highly enriched element below the heavy metal group.

Salt and Water in Space and Time

Deuterium-O^{18} relationships in the waters of the Red Sea and the brines show, as we have seen, that water in the brines is isotopically identical to normal, present-day, Red Sea water of 38.2 per mill salinity. Although the data plotted in Fig. 3 show that the brine has about the median isotopic composition found in the Red Sea, it is important to note that this composition is by no means an *average* composition of Red Sea water; as discussed earlier, waters of equal or lower D and O^{18} concentrations ($S \leqslant 38.2$ mill) occur only in the southern part of the Red Sea, 400 to 800km south of the brine depressions in the region adjacent to and including the southern sill. There is no possibility that water of this composition can now be entering the recharge system on the flanks of the Red Sea adjacent to the brine depressions; all waters less than several hundred kilometers south of this latitude are already higher in D and O^{18} at all depths due to the progressive enrichment of inflowing water by evaporation and molecular exchange.

The southern sill at 13°41′N, is 850km south of the brine deeps, or just about 1,000km SE along the Red Sea axis. Water of the isotopic composition found in the brine corresponds closely to the mean water over the sill during an annual cycle (Fig. 5), and can be found in significant amounts up to about 430km SE of the brines, either as 38.2 per mill salinity water or as a virtual water which can be formed from direct mixing or annual cycling between more and less saline components. Such direct or seasonal mixtures are limited to depths of less than 200m within this region, as the water is always more saline than 40.2 per mill and higher in D and O^{18} below this depth. The sill depth itself is about 100m, and the 38.2 per mill isohaline lies about 40m above the sill in winter, and at levels of 60m above and 30m below the sill in summer, enclosing a core of lower salinity water. In summary, then, the brines occur about half-way along the length of the Red Sea, and water of similar isotopic composition can be found or formed in at most the southern one-third or one-fourth of the Red Sea, within a layer less than 200m below the surface.

Within this area Tertiary evaporite deposits are abundant both below the surface and as salt domes exposed on islands and along the shores (Girdler, 1958; Drake and Girdler, 1964; Heybroek, 1965). The seismic section measured on the southern sill itself by Drake and Girdler (*op. cit.*) shows some 2,500m of 3.7km/sec velocity material attributed to evaporites, lying be-

tween the crystalline basement below and some 250m of unconsolidated coral sand above—this section is shown in Fig. 6. Malone (discussion appended to paper of Girdler, 1958) has described the borehole on the island of Dahlak (15°N) which penetrated 2,000m of evaporites without ever leaving salt. Both the sill area and the flanks of the trough are extensively faulted and provide easy access for sea water to descend down into the salt.

In Fig. 6 some approximate isotherms have been marked in the section, assuming an average heat flow of 4×10^{-6} cal/cm^2 sec and conductivities as shown in the figure, and neglecting boundary effects. Since the sill depth falls off rapidly to the north and the brine deeps are some 2,000m below the surface, sea water can be heated to something like 100°C by the regional geothermal gradient which is more than sufficient to account for the observed brine temperatures, with no necessity for invoking any direct volcanic heating; these relatively low temperatures are consistent with the lack of an oxygen isotope shift in the water.

The minimum flow distance from source to brine depression is some 350km, for an origin at 18°N on the western flank of the Red Sea. The maximum distance is about

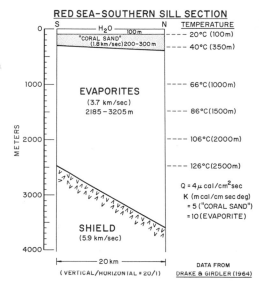

Fig. 6. Seismic section across the southern sill. Temperatures are calculated for the assumed heat flow and conductivity values shown, assuming no boundary effects.

1,000km northwest from the sill. The driving force for the flow is the density difference between 2,000m columns of essentially saturated brine and normal sea water, equivalent to about 40 atmospheres of pressure, so that the pressure gradient during the flow is of the order of 0.04 to 0.10 bars/km. At pressure gradients of 0.04 bars/km (corresponding to the 1,000km distance), petroleum flow rates of 75km/yr have been observed by Baker (1955) in fissure flow through rocks with fissures of about 0.25cm by 300m. Although viscosity and temperature effects are important, the flow velocity varied as the square of the fissure diameter, so that other things being equal, corresponding flow velocities of 1km/yr could be attained in fissures of only 0.03cm and a gradient of 0.04 bars/km. Although water is not petroleum, and such estimates are very crude, they indicate that even with fissure volumes of the order of 0.01 per cent of enclosing rock, there is no insuperable difficulty in carrying as much as 10^6 tons of brine per year even 1,000km through a section 10km by 300m of rock with a travel time of about 1,000 years. Possibly the distances and velocities are less and the times somewhat longer, but the fact remains that unless water at some earlier time had the present isotopic composition of the brine at considerably higher latitudes, the source is at least 350km to the south and more likely two to three times this distance, and the travel time must be significantly less than 10,000 years, according to the chronology of Ku et al. and Deuser and Degens. With a brine flux of this magnitude the salt addition to the Red Sea is only 10^{-5} of that added by the annual evaporation, so that even much greater fluxes would not be noticeable in the salt balance (Craig, 1966).

We must now ask about the history of the pattern of isotopic composition in Red Sea waters and in the ocean in general. A detailed consideration of the effects of the formation and melting of continental glaciers during the Pleistocene glaciation cycle, based on the new IGY information from Antarctica on ice thickness, indicates that average ocean water probably fluctuated between the limits marked for *SMOW* by crosses in Fig. 3, between times of maximum ice formation and complete melting of continental ice (if complete melting in ice actually occurred)—(Craig, 1966). The indicated maximum range of δO^{18} in *SMOW* is from about −0.5 per mill with complete melting of all present-day ice, to about +1.0 per mill during glacial maxima, which corresponds to a sea level lowering of about 200m below the present level. The dashed lines marked "glacial" and "interglacial" in Fig. 3 are D-O^{18} Red Sea slopes originating from these points of maximum oscillation.

Curray (1960), in reviewing all the sea level chronologies based on radiocarbon dating, concludes that during the past 20,000 years of emergence from a glacial period, sea level has been rising at an average rate of about 8.6m per 1,000 years. (This rate is equivalent to somewhat less than 1 per cent of the annual precipitation rate over the sea, so that the ocean is always about as well mixed isotopically as it is today.) Since the southern sill depth is about 100m, assuming the topography has not changed, the Red Sea should have been isolated when sea level fell 200m, and according to Curray's chronology, exchange of water with the Gulf of Aden should have been established again at 100/8.6 = approximately 11,600 years ago. This time is in striking agreement with the time of the 175cm transition from very high δO^{18} values in $CaCO_3$ in Red Sea sediments to normal values, established by the radiocarbon measurements of Ku et al. (1969) on the Deuser-Degens cores as about 12,000 years ago, and is surely not a coincidence (Deuser and Degens, 1969, Fig. 2).

The question of the isotopic composition of the Red Sea water during its isolation from the open sea by sea level depression below the sill is difficult to discuss theoretically. Craig and Gordon (1965) have thoroughly discussed the evaporation-exchange isotope effects for the ocean, for isolated water bodies, and open systems, but the Red Sea is an awkward size—intermediate between an ocean which controls its own environment, and a small water body which responds to the atmospheric vapor characteristics imposed by the ocean. However, the carbonate data of Deuser and Degens indicate that during the times of isolation of the Red Sea, δO^{18} of the water (at least *some* water) increased to about +7 per mill, and from this empirical value, certain interesting points ensue. Most importantly, the Red Sea did not evaporate down as a closed body of water with no water inflow at all; if it had, evaporation would have ceased when the activity of water equalled the mean relative humidity, and the continuing molecular exchange of water vapor would have brought the water to approximate isotopic equilibrium with atmospheric vapor, so that the water would have been isotopically similar to local precipitation (Craig, 1966b). Since this is not in accord with observations, there must have been a continual input of precipitation and runoff, and, as the activity of the water decreased due to increasing salinity, the evaporation rate decreased until a new steady state for input-evaporation and isotopic composition was attained for the isolation period. The relevant equation describing the situation is obtained approximately by replacing the humidity h by h/a, where a is the steady state activity of water, in equation (32) of Craig and Gordon for their open-system model, as discussed by them (*op. cit.*, p. 65). In this way one can correlate the average humidity, activity, and isotopic composition of fresh water inflow and of atmospheric vapor, and one can see that the empirical value of +7 per mill for δO^{18} is not unreasonable for the water although there are too many variables hanging free for a detailed model. Perhaps more indicative are the observations of Craig *et al.* (1963) that water evaporating into marine vapor reached a steady-state composition of $\delta O^{18} = +6$ to +7 per mill, of Longinelli and Craig (1967) that open systems such as the Dead Sea and W. Coast saline lakes have δO^{18} values of about +5 per mill, and that the present lakes and rivers of E. Africa evaporate to δO^{18} values of +6 per mill (Craig, 1963) — these values are, of course, independent of the initial isotopic composition of the evaporating water. The subject is admittedly difficult, tedious, and not well-advanced, but, to sum up, enough is known at present to indicate rather clearly that (a)

the Red Sea did not evaporate away, or reach a constant activity of water as an isolated body, and (b) the empirical value of +7 per mill for δO^{18} of the water, from the core carbonates, is a very reasonable figure for the value characteristic of periods when sea level fell below the sill and the Red Sea was cut off from the open ocean.

When sea level rose again due to melting of the ice sheets, the D/H and O^{18}/O^{16} ratios in sea water decreased because of the great depletion of these isotopes in precipitation stored in the ice sheets. At the time when sea level reached the sill, 100m below present level, *SMOW* would have had a δO^{18} value of about +0.5 per mill relative to the present zero value, due to melting of half the glacial increment of continental ice (Fig. 3). At this time, if oceanic circulation conditions were not drastically different from those of today, the entering water from the Gulf of Aden would have had a δO^{18} value of about +1.1 per mill, a salinity of about 37 per mill, and two opposing trends would have begun to affect the isotopic composition at the south end of the Red Sea: (1) entering water would have been slowly decreasing in δO^{18} from 1.1 to the present value of 0.6 per mill due to continued melting of ice, and (2) deep water to the north would have decreased in O^{18} and salinity due to the influx of sea water, and at some time later its density would have decreased to the point at which it could have begun to mix out over the sill at certain times. The Ku *et al.* chronology indicates that by 9,000 years ago, the water in the central section of the Red Sea had been essentially flushed to the isotopic composition and salinity it has today (Deuser and Degens, *op. cit.*, Figs. 2 and 5). Prior to this time, the proportion of deep water in the outflow over the sill would have been increasing as its density, salinity, and O^{18} concentration were decreasing by dilution so that it could mix with the near-surface waters. Thus the oxygen isotopic composition of the upper few hundred meters of water in the southern end of the Red Sea probably went through a highly damped, slight increase in δO^{18} relative to the surface inflow during the first few thousand years after the water flowed in, perhaps of the order of 1 or 2 per mill, and then declined to its present value of +1.2 per mill.

These considerations indicate that for a long period before the flow of brine into the deeps, and also since that emplacement began, there is no evidence that water in the immediate vicinity of the brine depressions could ever have been as low in deuterium and oxygen-18 as the water from which the brine has been made. The evidence from the oxygen-18 and radiocarbon stratigraphy in the Red Sea sediments therefore requires an origin of the brine far to the south, with the long flow trajectories of some 400 to 1,000km under the Red Sea floor before emergence into the deeps off Mecca.

The Glacial-Control Mechanism for Brine Flow

The isotopic evidence that the brine originates in near-surface waters at the south end of the Red Sea—on or adjacent to the southern sill—has the following consequence: during periods of continental glaciation when sea level falls below the sill and the level of the Red Sea drops, no water is available to form brine, and no flow will occur. As we have seen, the isotopic composition of the brine water indicates that it originates within the upper 200 —and probably upper 100—m of the southern waters; with present evaporation rates the Red Sea would drop this much in 50–100 years after sea level falls below sill depths, and considerably more before the decreased evaporation rate could be balanced by precipitation and runoff input. The oxygen isotope shift evidence from the work of Deuser and Degens on carbonates, indicating that brine was not present in the depressions before about 10,000 years ago, is thus completely consistent with the glacial-control mechanism proposed here: brine could not begin to flow *until* sea level rose and water covered the area where the flow trajectory begins— some 100m below the present level of the Red Sea. The correspondence between time of brine emplacement and overflow of sea water across the sill, so strikingly shown by the Ku chronology, is thereby explained as a natural consequence of the location of the point of origin of the brine.

Two further consequences of the glacial-control mechanism are interesting. Brine may have flowed during interglacial periods previous to the present one, so that the deep sediment record may show one or more layers of heavy metal deposition *below* the present deposits, each separated from the others by a hiatus marking a glacial period when no brine flow occurred. Secondly, when the level of the Red Sea falls after sea level drops, the flowing brine has a greater hydrostatic head and can overflow the depressions until the aquifer is drained to the new level of the Red Sea —such overflow may have contributed very significantly to the heavy metal deposits. Thus it will be important to obtain long sediment cores from the brine depressions to determine if sizeable metal deposits have been formed below the last glacial hiatus.

A Steady-state Model of Atlantis II Deep

The recent surveys by Hunt and his co-workers indicate that Atlantis II Deep may contain as much as 3×10^9 tons of brine, or 7.5×10^8 tons of salts, in a volume roughly 3km $\times$ 6km $\times$ 150m. From the isotopic data on the water and the glacial-control chronology discussed in the last section, we can estimate that the hold-up or residence time of brine in the depression might be as long as 3,000 years, but not much longer. With these values, the lower limit for the flux of new brine into Atlantis II Deep is of the order of 10^6 metric tons of brine/yr, corresponding to 2.5×10^5 tons of salt/yr. (For comparison, evaporation contributes about 3.6×10^{10} tons of salt per year.) This flux of new brine flows from the 56° layer in the bottom of the deep, to the lower salinity 44° convective layer above, and is then mixed out to the Red Sea. Assuming that a rough approximation to steady-state has been reached, the dynamics and chemical characteristics of the system resulting from this steady input will now be discussed.

The bottom of Atlantis II Deep is filled with 135m of 56°C brine with a chlorinity of 155.5g/kg; the temperature and salinity are approximately constant with depth (Pugh, 1967). Above this layer, from 2,040–2,010m, is a 30m thick layer of 44°C, chlorinity 80g/kg, brine. Both layers appear to be uniformly mixed by active convection cells, and the transition between them is so sharp that no adequate data are available on the shape of the gradient. Above the 44°C layer there is a transition zone of about 30m separating it from normal Red Sea deep water with a salinity of 40.6 per mill. The structure of this transition zone is shown in Fig. 7, in which measurements of Na concentration and temperature, made by M. Peterson on the WHOI **CHAIN** 61 survey, are shown. (The Na data, which were volumetric, have been converted to g/kg by an approximate correction factor from the sea water and brine densities, assumed to vary as the reciprocal of density between these limits—possible deviations are too small to affect the shape of the gradient shown.) Measurements made by Peterson on Ca, Mg, and K (all as g/kg directly) show identical gradients, as would be expected from the relationships in Fig. 1 which show that all these elements are conservative. Sodium data are used here because the analytical scatter is much smaller than for the less abundant elements. The data in Fig. 1 are a compilation of all measurements on six different casts (Atlantis 715, 718, 722-1, 722-2, 726, and 727) and some of the scatter is certainly due to the difficulty of correlating depths within such a narrow interval on many different casts.

Mixing at the Brine-Sea Water Interface

In principle, it might be expected that mixing between the 44°C brine and overlying sea water would be a simple one-dimensional diffusion + advection process, in which molecular or turbulent diffusion operates vertically, and the influx of new brine into the 56° bottom layer is compensated by a net upward advective flow from the 44°C layer into normal sea water, i.e.

$$\kappa C'' = w C' \qquad (9)$$

in which κ is the diffusion coefficient, w the advective velocity, and the primes denote

successive derivatives of concentration, C, with distance z (positive upwards). However, the Na gradient in Fig. 7 is exactly upside-down for such a process. The dashed line shown in the figure has been calculated from eqn. (9) with $\kappa/w = -4$m (advection down), for boundary conditions Na = 12.5g/kg at 1,980m and 46.9g/kg at 2,012m. That is, the linear gradient which would be observed for pure diffusion is displaced asymptotically downward from the upper boundary, so that in a one-dimensional model the advection must be downward. Since it is highly unlikely that sea water is being advected into the 44° brine from above we must conclude that the gradients in Fig. 7 reflect a two-dimensional mixing process in which a bottom current is flowing along Atlantic Deep above the 44° brine, removing the salts mixed into it by diffusion and (upward) advection. The profiles in Fig. 7 also characterize a two-dimensional process of this type in which horizontal flow has a quasi-parabolic velocity profile, falling off from the turbulent core of the flow towards the interface.

The temperature profile over Atlantis II Deep falls off more slowly than the salinity profile; Fig. 7 shows that temperatures at all depths in the transition region lie above the dashed line, which is calculated for the same one-dimensional mixing length used for the sodium-curve. If the salt diffusion is molecular, rather than turbulent, the thermal diffusivity in this system is about 100 times greater than the molecular diffusivity, so that a characteristic mixing length of the order of meters for salt would be of the order of hundreds of meters for temperature, and the temperature profile should be expected to be very close to linear unless

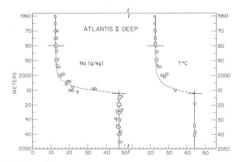

Fig. 7. Sodium and temperature profiles above the 44°C brine layer in Atlantis Deep, from data of M. Peterson (personal communication, 1968).

the horizontal advection rate is very great. More likely, the temperature profile represents an additional heat contribution by conduction from the surrounding sediments to the bottom current.

At the top of the transition zone (1,980m) salinity and temperature increase downward at a rate $dS/dT = 2$ per mill/°C, more than 5 times greater than the ratio of about 0.37 required for constant density; consequently the density increases rapidly, the stability is very high, and a convection cell does not develop above the 44° layer. However, above the hot sediments surrounding the brine depression there are probably rising convection currents in the normal bottom water, with downward flow of sea water above the brine itself and horizontal currents just above the interface. Diffusive mixing, probably turbulent but possibly molecular near the interface, takes place between the 44° layer and the horizontal bottom current, and in addition, in the steady-state model there is an advective flux out of the 44° layer equal to the influx of new brine into the 56°C bottom layer. A net brine influx of 10^6 tons/year would require an advective velocity of about 25×10^{-8} cm/sec out of the 56° layer (dimensions roughly 2×6km), and about 5×10^{-8} cm/sec out of the 44° layer (roughly 5×12km).

The 44°C brine layer is a mixture in almost equal proportions of 56° layer brine and normal Red Sea Bottom water, as demonstrated in Figs. 1 and 2, forming an independent convection cell which receives high-salinity brine from the 56° cell below, and mixes with the overlying sea water by diffusion. In this process there is a net advective flux of brine through the layer and the relative proportions of non-conservative brine salts are altered by chemical reactions, as discussed in the following section.

Steady-state Chemical Balance in the Atlantis II 44°C Brine Layer

The steady-state material balance for a component in the 44°C brine layer can be written as:

$$J_i = C_B - C_{44} - \phi_{SW}(C_{44} - C_{SW}) \quad (10)$$

in which C is the concentration of component i in 40.6 per mill salinity normal Red Sea bottom water (subscript SW), in 56°C Atlantis brine (subscript B), and in the 44° mixing cell (subscript 44). ϕ_{SW} is the total flux of sea water diffusing in from above, in units of the advection rate of new brine into the 56° layer, and J_i is the removal rate of component i from the 44° layer by any process other than diffusion and advection to the overlying sea water, also in 56°C brine-flux units. (The units of J_{Fe} are, e.g., (gFe/yr)/(kg 56° brine/yr)=g/kg when the concentration units are g/kg.) Thus, on the right-hand-side of (10) the first two terms are the advective flux of component i into, and out of, the 44° layer, respectively, and the last term is the net diffusive flux out to the sea water, all in units of the advective flux of new brine. The summation of the J_i is so small that the total advective flux from 44°C brine to sea water is essentially 1 in 56° brine-flux units, and ϕ_{SW} is also the total diffusive flux of material out to sea water.

For conservative elements (Figs. 1 and 2) $J_i = 0$, and ϕ_{SW} is obtained from the chloride concentrations (155.5, 80.0, and 22.5g/kg), as

$$\phi_{SW} = 1.3130 \quad (11)$$

in 56° brine-flux units. The steady-state chloride balance in the 44° cell (56° brine-flux units) is thus: 155.5g/kg advected in from below, 75.5g/kg as the net diffusion loss out to sea water, and 80.0g/kg advected out to sea water. For conservative elements ($J = 0$) we have

$$C_{44} = \frac{C_B + 1.313 \, C_{SW}}{2.313} \quad (12)$$

as the steady-state concentration, and for trace elements of very low concentration in sea water, the steady-state concentration is simply:

$$C_{44} \rightarrow 0.432C_B \quad (13)$$

as a criterion for $J = 0$. The principal elements for which J is clearly non-zero are Fe and Mn, and the various data on these elements are in good enough agreement to justify a detailed consideration.

Iron. The sediments in Atlantis II Deep contain precipitated iron as ferrous carbonate (siderite) and in the ferric form as hydrated iron oxides ranging from amorphous forms to goethite ($Fe_2O_3 \cdot H_2O$). Miller *et al.* (1966) proposed that iron oxides are precipitated by oxygen in the overlying sea water during pulses of brine, while Watson and Waterbury (1969), in their steady-state model for sulfide precipitation discussed in an earlier section, suppose that ferrous ions diffuse through their sulfate reduction layer and are continually oxidized and precipitated in the overlying oxygenated water as ferric hydroxide. On the other hand, Brewer and Spencer (1969) have proposed that iron is precipitated as $Fe(OH)_3$ at the 56°–44° interface.

Brewer and Spencer (*op. cit.*) report the Fe concentrations in Atlantis II brine as $C_B = 81$ ppm (56° brine), $C_{44} = 0.20$ ppm (44° brine). Peterson (personal communication) finds $C_B = 90$ ppm, $C_{44} \approx 0.9$ ppm (ranging from about 0.75 to 1.5); these measurements are similar enough that the calculations to follow are not seriously affected by the choice of data, and we shall use the Brewer and Spencer values. The concentration in normal sea water is less than 0.02 ppm (Brewer *et al.*, 1965), so that equation (13) shows that Fe is strongly non-conservative in the 44° layer. From equations (10) and (11) with the Brewer-Spencer data we find:

$$J_{Fe} = 80.5 \text{ppm/kg}$$

as the removal rate of iron by precipitation *within* the 44° layer (56° brine-flux units). The flux of iron from 44° layer out to sea water by mixing, that is Fe which can be precipitated above the 44° interface, is thus only 0.5 ppm/kg, so that essentially all the iron must be precipitated *within* the 44° layer.

Ferrous iron can be oxidized to the ferric state, and precipitated as ferric hydroxide, by the dissolved oxygen mixed in from overlying sea water:

$$Fe^{++} + \tfrac{1}{2}O_2 + 2H_2O +$$
$$e \rightarrow Fe(OH)_3 + H^+ \qquad (14)$$

We may assume that *all* the oxygen brought in by mixing is used to precipitate $Fe(OH)_3$. The dissolved O_2 concentration in Red Sea bottom water in this area is 2.5 cc/kg (Neumann and McGill, 1962), and C_B and C_{44}

will be zero, so from equation (10):

$$J_{O_2} = 3.28 \text{cc/kg}$$

is the removal rate for O_2 by precipitation in the 44°C layer. Equation (14) shows that 1 cc of dissolved O_2 corresponds to precipitation of 0.005 g Fe, so that the maximum precipitation flux for Fe by reaction with dissolved oxygen is:

$$J_{Fe}(O_2) = 16.4 \text{ppm/kg}$$

or only 20.4 per cent of the total removal rate of iron from the 44° layer. That is, 79.6 per cent of the precipitating iron must be removed without oxidation from dissolved O_2 mixed in from sea water. The total steady-state iron balance is as follows (fluxes in ppm/kg, fractions as percentage of the total Fe flux in and out of the 44°C layer):

In to 44° layer from 56° layer:
81.0 (100 per cent)
Out to sea water by diffusion-advection:
0.5 (0.6 per cent)
Precipitated as $Fe(OH)_3$ in 44° layer:
16.4 (20.3 per cent)
Other precipitation in 44° layer:
64.1 (79.1 per cent)

It is evident that another mechanism for Fe precipitation must be found, if the steady-state model is to work. Fe as well as Zn, Pb, and Cu, could be precipitated as sulfides in the 44° layer, by mixing in from sea water a small amount of reduced sulfur from biogenic sulfate reduction above the interface. But, as discussed in an earlier section, the sulfides should then have the biogenic δS^{34} values because only a small fraction of the sulfate can be reduced (since it does not show S^{34} enrichment); moreover, as Watson and Waterbury point out, FeS is not observed in the sediments, and it is highly unlikely that pyrite could precipitate directly in such a process.

At present, the most likely process seems to be precipitation of $FeCO_3$, since siderite is actually observed in the sediments, together with $MnCO_3$ (rhodochrosite). A precipitation flux of 64.1 ppm/kg of iron as $FeCO_3$ requires $J_{HCO_3} = 70.0$ ppm/kg as the removal flux of dissolved carbonate from the 44° layer. Using $C_B = 143$ ppm for 56° brine (Weiss, 1969), and assuming $C_{SW} =$

135ppm for Red Sea bottom water, as estimated in an earlier section, for the required value of J_{HCO_3} we find, from equation (10), $C_{44}(HCO_3) = 108$ppm. This is to be compared with the value for $J = 0$ (no precipitation of carbonate): $C_{44}(HCO_3) = 138.5$ppm. Therefore the $FeCO_3$ precipitation hypothesis can be tested directly, as a distinct minimum in the dissolved carbonate concentration should be observed in the 44° layer, relative to the brine and (actual) sea water concentrations, if this mechanism is responsible for the bulk of the iron precipitation.

Deuser and Degens (1969) measured a profile of C^{13} variations in dissolved carbonate through the brine; unfortunately the concentrations of carbonate were not measured. If we use their isotope data ($\delta_B = -4.2$, $\delta_{44} = -2$, $\delta_{SW} = +0.8$ per mill), then with $J_{HCO_3} = 70$, and with the isotopic analog to equation (10) we find ($\alpha - 1$) for the precipitation process $= 2.6$ per mill. Weiss (1969) has shown that their brine data are probably affected by loss of CO_2 during sampling, as he finds lighter values for the brine ($\delta_B = -5.6$) as would be expected if their samples have lost CO_2. Using his value for δ_B and $\delta_{44} = -3.4$ per mill as an approximate correction, we obtain ($\alpha - 1$) $= +6.1$ per mill. Thus the isotopic fractionation factor for carbon probably lies somewhere between 1.0006 and 1.006, according to the Deuser-Degens isotope data and the estimated value for HCO_3 concentration in Red Sea bottom water, a range which is easily consistent with our present rather scanty knowledge of the expected factors (Thode et al., 1965; Deuser and Degens, 1967; Wendt, 1968).

Let us assume $\alpha = 1.004$ for carbon fractionation in the precipitation process (C^{13} concentrating in $FeCO_3$). Using the Deuser-Degens data for δ_B and δ_{SW}, we calculate for $J_{HCO_3} = 0$, $C_{44} = 138.5$ppm, that $\delta_{44} = -1.4$ per mill. On the other hand, for $J_{HCO_3} = 70$, $C_{44} = 108$, we find the steady-state $\delta_{44} = -2.3$ per mill. That is, the difference is of the order of the scatter in their data, and one sees that the differences in carbonate *concentration* are a much more sensitive test of $FeCO_3$ precipitation than are the isotopic data. Of course the actual

concentration effects have to be compared with actual measurements on Red Sea bottom water.

Much of the iron in Atlantis II sediments is present as amorphous iron oxides of various oxidation states, which is probably transforming to goethite; in addition siderite and pyrite are present in considerable amounts. The observations of Kaplan et al. (1969), that oxidation of sulfur and sulfides is taking place in situ, as demonstrated by local increases in SO_4 concentration with concomitant S^{32} enrichment, have been described earlier. Possibly $FeCO_3$ is being oxidized to ferric oxides by in situ mechanisms such as use of carbonate ion for oxidation of sulfur and H_2S,

$$CO_3^= + H_2S + H_2O = SO_4^= + CH_4 \quad (15)$$

with oxidation of ferrous iron by SO_2 produced as an intermediate. Also $FeCO_3$ may react with H_2S to form FeS, and then FeS_2 by interaction with elemental sulfur. It is evident from Kaplan's work that very complicated sulfur chemistry is going on in the sediments, involving all sulfur valences, so that it is difficult at present to decipher the various oxidation-reduction reactions affecting the ferrous-ferric balance.

Iron Recycling. If $FeCO_3$ is being precipitated in the 44° layer, some of the precipitate may settle into the 56° brine, redissolve, and be transported back to the 44° layer. In this case the observed HCO_3 and Fe concentrations in the 56°C brine layer are actually higher than those in new brine entering the reservoir, rather than being equal to the entering brine concentrations as assumed until now. Let us assume that the entering brine actually has lost or gained no dissolved carbonate relative to the original 38.2 per mill sea water with its (estimated) concentration of 125ppm; then the entering brine concentration of HCO_3 is $125/1.295 = 96.5$ppm, and the recycle flux of HCO_3 into the 56° brine layer is $143 - 96.5 = 46.5$ppm/kg. With $J_{HCO_3} = 70$ppm/kg in the 44° layer, this means that $46.5/70$ or $^2/_3$ of the precipitating carbonate dissolves in the 56° brine and recycles, and $^1/_3$ is deposited directly in the sediments. Also $^2/_3$ of the Fe is recycled into the 56° brine, so that of the 79.1 per cent of the Fe flux listed above as precipitating by other means, 52.5 per cent of the total flux is recycling and 26.6 per cent is depositing in the sediments. For these values, the new brine entering the reservoir has an Fe concentration only 47.5 per cent of that observed in the 56° layer, i.e., 38.5ppm *vs.* 81ppm in the 56° layer itself, representing the enrichment due to recycling. These processes have to be considered when chemical reactions occur in stacked convection cells, and one is comparing the trace element concentrations with those of other water types as a criterion of origin.

Manganese. A Mn-Ca variation diagram is shown in Fig. 8 for both brines, using the recent data of Peterson from the WHOI **CHAIN** 61 Survey; both elements were measured directly as mg/kg. The Discovery brine data indicate considerable scatter, but show essentially a mixing relationship between the depleted Discovery brine and sea water. In Atlantis Deep, however, the 44° brine Mn concentrations are about twice as high as the value of 37ppm which would be expected for steady-state mixing between the 56° brine and sea water. Peterson's Mn data ($C_B = 86 \pm 4$, $C_{44} = 79 \pm 2$, and 56ppm in Discovery brine) agree very closely with those of Brewer and Spencer (1969), (82, 82, and 54.6ppm respectively), and the data for Atlantis II waters appear to be quite precise. The Ca data of Peterson on all brines are in exact agreement with those of Brewer and Spencer, but in Red Sea bottom water Peterson finds Ca = 432ppm *vs.* the value of 471ppm measured by Brewer *et al.* (1965) which is essentially the expected value from data on open ocean waters.

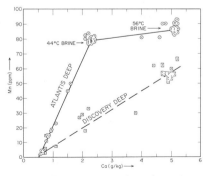

Fig. 8. Mn *vs.* Ca concentrations in waters of Atlantis and Discovery Deeps (data from M. Peterson, personal communication, 1968).

Using the Brewer-Spencer data in equation (10) we find $J_{Mn} = -108$ppm/kg (Peterson's data give −97), indicating an *addition* of Mn to the 44° layer, from sources other than sea water or influx from the 56° brine. The steady-state balance is as follows (56° brine-flux units and percentages of total flux out to sea water):

In to 44° layer from 56° layer:
$$82 \quad (43 \text{ per cent})$$
In to 44° layer from other source:
$$108 \quad (57 \text{ per cent})$$
Out to sea water by diffusion-advection:
$$190 \quad (100 \text{ per cent})$$

The input flux to the 44° layer can conceivably come from direct solution of manganese minerals in contact with the 44° layer, or from recycling of Mn, as in the discussion of iron recycling, but in this case by recycling of Mn precipitated in the overlying sea water back into the 44° layer. If such recycling is occurring, then Mn should appear non-conservative in the profile through the Atlantis transition zone above the 44° layer: this profile, from Peterson's **CHAIN** 61 data, is shown in detail in Fig. 9, with Mn plotted both against Ca and against the (corrected) Na data, with all points from Peterson's 6 Atlantis casts with Mn values between 44° brine and sea water concentrations plotted. (Three Ca points, at Ca = 0.61, 0.80, 0.95, were reported with Ca concentrations higher than shown by a factor of 2; they have been renormalized for Ca from a Ca-Na plot.) The unenclosed solid points at the low concentration extremes are those for which Peterson reported a manganese upper limit only, i.e., Mn < 1ppm. The lines in Fig. 9 are least squares fits to the enclosed points, and also include the mean 44° brine concentrations (Mn = 79ppm, Ca = 2.30g/kg, Na = 46.9g/kg) as one point. The regression lines are:

$$Mn = 45.50Ca - 24.44 \quad (S = 1.5)$$

$$Mn = 2.434Na - 34.587 \quad (S = 1.3)$$

with Mn in ppm, Ca and Na in g/kg, and standard errors of Mn regression on Ca and Na as shown. Thus the relationship of Mn to both conservative elements is exceedingly linear, and there is absolutely no in-

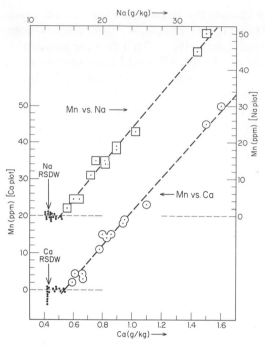

Fig. 9. Mn *vs.* Ca and Na concentrations in transition region above Atlantis Brine (data from M. Peterson, 1968). Na and Ca values in Red Sea Deep Water are shown by arrows; the unenclosed solid points show Na and Ca values measured for all waters with Mn reported as "less than one ppm." The lines are least squares fits, including the 44°C brine as one value, but excluding the Mn values for unenclosed points.

creases as the Mn concentration goes to zero cannot be discounted.

The question of the Mn offset at Red Sea deep water Ca and Na concentrations deserves serious attention because it bears so heavily on the possibility of Mn precipitation in the transition region and the possibility of manganese recycling from above. Nevertheless, it is extremely difficult to imagine a precipitation mechanism which would leave the Mn-Ca and Mn-Na relationships linear over the entire Mn range, and produce only a zero offset; moreover, the offset is probably not large enough to account for the amount of recycling required (108 ppm/kg, or 57 per cent of the Mn mixed out to sea water). Thus at present it seems more likely that Mn is dissolving directly into the 44° layer from previously precipitated manganese in contact with the layer on its flanks. It should be noted that the Mn concentration in the 56° brine is probably significantly increased above the original value in inflowing brine by the same mechanism, and in fact the similarity of the concentrations in the 44° and 56° layers may indicate simply that both are saturated with manganese. Therefore, the enrichment factors for Mn (as well as for Fe if iron is recycled) shown in Table 4 may not reflect the actual enrichment in the flowing brine at all. This may also be true of the high values for lead and zinc, which may simply reflect recycling or re-solution.

Zinc, Copper, Lead, Cobalt. The differences between the data of Brewer and Spencer (1969) and those of Peterson's data on concentrations of these elements are so large that at present no meaningful statement even as to the sign of *J* can be made without choosing one set of data over the other. For concentrations in the 56° and 44° brine respectively (all ppm), Brewer and Spencer find: Zn = 5.4 and 0.15; Cu = 0.26 and 0.017; Pb = 0.63 and 0.009; Co = 0.16 and 8 × 10⁻⁴, whereas Peterson's data indicate Zn = 7 (5 in Niskin bottles) and about 2.7; Cu = 0.3 (but 0.04 in Niskin bottles) and 0.3; Pb = 0.2 (0.10 in Niskin bottles) and 0.10; Co = 0.005 (but 0.03 in Niskin bottles) and 0.15. Similar discrepancies in lithium data have been noted earlier. The Niskin bottles contain no metal and may

dication of manganese precipitation above the 44° brine-sea water interface.

Both the Mn-Ca and Mn-Na lines go to zero Mn at indicated concentrations well above sea water values, at Ca = 0.537 g/kg and at Na = 14.20 g/kg. Peterson's average values for Red Sea deep water were obtained by averaging his 12 analyses from these casts with Na less than 13 g/kg. The values, with the standard error of the mean, are Ca = 0.432 ± 0.002 g/kg and Na = 12.54 ± 0.08 g/kg, so the offsets from the values at which Mn (= 0.004 ppm in normal sea water) vanishes, are quite significant. As pointed out above, Peterson's Ca value is about 0.04 g/kg lower than the most probable value for the deep water, but this difference is only a small fraction of the offset. The Mn offsets amount to 5 and 4 ppm less than zero at the sea water Ca and Na values, which are very large differences compared to the Mn scatter, but the possibility of a systematic error which in-

give better data for Cu and Pb, but other differences are difficult to explain. In any case, the Brewer and Spencer data indicate positive J values for all four metals with 93.5 per cent of the Zn flux from 56° brine to 44° brine, 85 per cent of the Cu flux, 97 per cent of the Pb flux, and 99 per cent of the Co flux, precipitated *within* the 44° layer. Peterson's data, however, show J approximately 0 for Zn (all removed by mixing), highly negative for Cu (input to 44° layer), and zero or negative for both Pb and Co. His data favor a conclusion that all these metals are conservative above the 56° brine, but any conclusion on the material balances must await a detailed study of the actual concentrations.

Discovery Deep

Element profiles for Discovery Deep, obtained by Peterson in his **CHAIN** 61 Survey, show a different pattern from that of Atlantis II Deep—from 1,980 to 2,010m they increase only slightly from sea water values, and then show an almost exactly linear slope to the brine interface at 2,065m. However, there are no samples between 2,040m, the depth at which the temperature has just increased to 36°C, and the brine interface, so that the question of the existence of a small 36° convection cell, implied by a few other measurements, cannot be answered. The temperature profile is non-linear and falls off more slowly from the brine interface than the concentrations. The profiles are roughly compatible with a model of a stagnant brine with no advective influx, but a more detailed set of measurements is required before this question can be answered. The intermediate brines in Discovery Deep, and the brine in Chain Deep, could not have formed by dilution with evaporated Red Sea water of much higher salinity than that of present Red Sea Deep Water, and remained stagnant since dilution, or the Mg and SO_4 mixing curves in Fig. 1 would be convex, as discussed in the first section of this paper. Thus mixing of these intermediate salinity brines with Red Sea Deep Water has probably been continuous and fairly rapid since these two depressions were filled.

Although the data plotted in Fig. 1 indicate a possibility that potassium concentrations might differ in the two deeps, Peterson's data show exactly similar values in Atlantis 56° brine and Discovery brine. All data show that Fe and Mn are much lower in Discovery brine than Atlantis 56° brine. Using the data of Brewer and Spencer (1969), for both brines, the Fe difference is 81ppm and the Mn difference is 27.4ppm; these differences correspond to removal of 88.5 and 30.4ppm of HCO_3 respectively, for a total of 119ppm of HCO_3 if the Fe–Mn depletion has occurred by precipitation of $FeCO_3$ and $MnCO_3$. The observed difference in HCO_3 concentration is 112ppm, in rather striking agreement with the carbonate precipitation model which thus provides a possible explanation for the carbonate difference.

Weiss (*op. cit.*) observed large differences in C^{13} concentration in the two brines; $\delta C^{13} = -5.6$ per mill in Atlantis 56° brine and -16.8 per mill in Discovery brine, and HCO_3 concentrations of 142.7 and 30.4ppm respectively. If a Rayleigh process has operated in Discovery Brine, with carbonate precipitation in a closed system, the fraction of carbonate remaining is 21.3 per cent, which requires a fractionation factor of 1.0073 for the process, roughly similar to the estimate for Atlantis II precipitation, and not an unreasonable value. The correspondence of depletion of both concentration of HCO_3 and C^{13} are difficult to explain by any mechanism requiring addition of light carbon, e.g., from organic matter, as the original carbonate has to be stripped out first. The closed system, carbonate precipitation model, is of course also consistent with a stagnant brine in Discovery Deep with no input of new brine.

Radiocarbon Measurements on Atlantis II Brine

Samples of Atlantis II brine and overlying sea water were collected, and the dissolved carbonate extracted for C^{14} measurements, by A. Jokela on the **CHAIN** 61 Expedition. The water samples were collected in 400 liter Niskin-type bag samplers, kindly made available by W. Broecker of Lamont Geological Observatory, and the CO_2 was extracted in 55 gallon drums using the Lamont technique with two plastic KOH bubblers on each drum. The extracted carbonate was measured at the Tata Institute for Fundamental Research; initial results have been reported briefly by Craig and Lal (1968). An unopened KOH bubbler was also returned from the ship as a blank and processed for CO_2 in the laboratory; it contained no CO_2 within a detection limit of 0.2cc. Samples were collected at 1,000m depth, at 160m above the brine interface (approximately 1,850 in depth), and in the 56° brine itself. C^{13} measurements were made on the CH_4 used for counting, in order to normalize for isotopic fractionation.

Table 6 shows the results of the radiocarbon measurements. The samples were counted as CH_4 in Oeschger counters at 90cm pressure; the net counting rate for NBS Oxalic acid was 11.413 ± .085 (one sigma) cpm. δC^{14} is the per mill difference in counting rate between the sample and $0.95 \times$ NBS Oxalic acid standard, and ΔC^{14} (per mill) is the difference corrected for the C^{13} variations (Lamont normalization):

$$\Delta C^{14} = \delta C^{14} - (2\delta C^{13} + 50)(1 + 10^{-3}\delta C^{14})$$

with all δ values in per mill. The errors shown are one standard deviation counting errors. Table 6 also lists the volumes of water from which CO_2 was extracted and the CO_2 yield expressed as ppm of HCO_3^-. The brine sample was extracted in two drums, and carbon from one first-stage KOH bubbler was measured separately in order to provide an independent check. Considerable fractionation occurs in the first bubbler stage, as shown by the δC^{13} value of this sample ($\alpha = 1.007$ from the C^{13} results and fractional yields), but the normalized C^{14} results are in excellent agreement.

The chlorinity of the collected brine sample was 152.34g/kg, *vs.* 155.5g/kg for the correct value, and it was observed that a small amount of sea water becomes trapped in the bag sampler when it is brought up unopened (A. Jokela, personal communication). The chlorinity difference indicates a 2 per cent contamination by sea water, and the mean ΔC^{14} value of the brine has been corrected for this, using ΔC^{14} of sea water $= -18$ per mill (average of the two measurements), an HCO_3 concentration in the brine of 143ppm (Weiss, 1969) and the previously assumed value of 135ppm for Red Sea deep water. The mean ΔC^{14} value of the brine, corrected in this way, is -866 per mill. The low yield of CO_2 from the brine (56 per cent, compared with the precise measurement by Weiss) reflects loss by gas evolution when the sample is brought to the surface, and the slight enrichment in C^{13} relative to a sealed sample (-4.8 per mill *vs.* -5.6 per mill) is in agreement with such an effect (Weiss, *op. cit.*).

Table 6   Radiocarbon Measurements on Red Sea Waters and Atlantis II Brine

TF*	Sample and Depth	Water Volume (liters)	ΣCO_2 as HCO_3 (ppm)	δC^{14}(‰)	δC^{13}(‰)	ΔC^{14}(‰)
690	Sea water—1,000m	120	104	39.3	−0.25	−12 ± 10
691	Sea water—160 m above brine, ~ 1,850m	197	98	28.1	+0.49	−24 ± 10
692	Brine: both KOH collectors on first drum, + second collector, second drum	(400)	(54)†	−842	−2.8	−849 ± 4
693	Brine: first KOH collector on second drum	(400)	(26)†	−846	−8.5	−851 ± 4
	Average brine, corrected for 2% sea water contamination	400	80	−	−4.8	−866 ± 4

* Tata Institute Radiocarbon Laboratory sample number.
†Yields of 9.6 and 4.7 liters STP of CO_2 obtained on samples 692 and 693 respectively.

A model for the interpretation of radio-carbon ages in flow systems with a reservoir has been developed previously (Craig, 1963) and has to be used for the Red Sea system in order to understand the measured values. In this model, there are two steady-state times: "t-time" and "τ-time," measuring the transit time for hydrodynamic or "pipe-line" flow from source, through aquifer, to the reservoir, and the residence time, or mean lifetime before removal, in the reservoir, respectively. Additionally, for short time periods of the order of the reservoir residence time, we must consider a third time, Δt, which measures the effect of the transient period before steady-state concentrations are attained. The general equation for radiocarbon concentration in the reservoir (i.e., the 56° brine) is:

$$\frac{R_T}{R_0} \frac{X}{X_0} = \frac{e^{-\lambda t}}{(1 + \lambda\tau)} (1 + T) \qquad (16)$$

in which R_T is the C^{14}/C^{12} ratio in the reservoir, measured at some time *after* the reservoir has filled with fluid, R_0 is the C^{14}/C^{12} ratio in the source fluid beginning the hydrodynamic flow to the reservoir, t is the flow time in the aquifer, and τ is the residence time in the reservoir. The terms X and X_0 are the C^{12}/H_2O ratios in the reservoir and in the source material; using these ratios, rather than concentrations as given previously (Craig, 1963) eliminates the dilution effect of the added salts. λ is the radioactive decay constant, $= (1/8,267)(\text{yr}^{-1})$.

In this model, "dead" carbon can be added to the fluid at any point in the aquifer or in the reservoir. With $t = O$, the aquifer transport is very fast, and the radiocarbon decrease is entirely due to the time spent in the mixed thermal reservoir and to dead carbon addition to the reservoir. With $\tau = O$, the effects are only those of radioactive decay during aquifer flow (as if pieces of wood were being transported), and addition of dead carbon during flow. With both t and $\tau = O$, the radiocarbon decrease is due to dilution only. The residence time τ is the ratio of the amount of carbon (or water) in the reservoir to the flux of carbon (or water) from the reservoir (assuming carbon and water leave in reservoir proportions).

The transient term T decreases to zero as the time since reservoir filling increases; when it is insignificant, R_T becomes R_S, the steady-state C^{14}/C^{12} ratio in the reservoir, which no longer varies with time. Therefore $(R_T/R_S) = 1 + T$, and the function T measures the deviation from the steady-state radiocarbon activity.

The form of T depends on the model assumed for the transient stage while the reservoir is filling, because it is necessary to know the radiocarbon concentration at the end-point of the filling process, when outflow from the reservoir begins, in order to integrate the differential equation from which (16) is derived. In the model used here, we assume the fluid flows in continually and mixes within the reservoir (by convection) with no outflow until the reservoir is filled to its steady-state capacity. The aquifer flow time from the source is t, and the filling time is τ; thus from the initiation of the first flow into the aquifer to the time it is filled and outflow begins, the time $(t + \tau)$ has elapsed. At this time the C^{14}/C^{12} ratio in the reservoir is given by:

$$\frac{R_{(t+\tau)}}{R_0} \frac{X}{X_0} = \frac{e^{-\lambda t}}{\lambda\tau} (1 - e^{-\lambda\tau}) \qquad (17)$$

and the reservoir begins to flush to steady-state (assuming constant input). Integrating from the C^{14}/C^{12} ratio in equation (17) to the value at some later time Δt, measured from the time when the reservoir is filled, the function of T in equation (16) becomes:

$$T = \frac{e^{-\frac{\Delta t}{\tau}(1+\lambda\tau)}[1 - e^{-\lambda\tau}(1 + \lambda\tau)]}{\lambda\tau} \qquad (18)$$

and R_T is the ratio at any time $(t + \tau + \Delta t)$ since flow through the aquifer began, or at time Δt since the reservoir was filled, and approaches R_S as $\Delta t \rightarrow$ infinity.

Equations (16) and (18) are solved for t as a function of given values of τ and Δt, for observed or assumed values of the left-hand-side, which is given by:

$$\frac{R_T}{R_0} \frac{X}{X_0} = \frac{X}{X_0} \left[1 - \frac{(\Delta_0 - \Delta_B)}{(1,000 + \Delta_0)} \right] \qquad (19)$$

(all delta values in per mill). Δ_B is the measured brine value in Table 6, $= -866$ per mill. The value of Δ_0 for the source water is assumed to be -50 per mill, according to present estimates of the activity in pre-nuclear, pre-industrial surface ocean waters. The values for Red Sea water in Table 6 have been increased by radiocarbon produced in nuclear weapon testing, as measurements made on Zephyrus Expedition (1962) gave values of -75 and -62 per mill in intermediate and deep waters (Craig, in preparation). The ratio (X/X_O) is the HCO_3^- enrichment factor, calculated in Table 4, assuming 125 ppm of HCO_3^- in the original 38.2 per mill salinity source water, $= 1.48$, and the value obtained in equation (19) is thus $1.48(0.414) = 0.209$.

Solutions to equation (16) are tabulated in Table 7, using 0.209 for the activity-dilution ratio. The final column shows the steady-state values of t as a function of τ, for $\Delta t =$ infinite, $T = 0$, and it is seen that in this system, in which the original carbon is assumed to have been increased by about 50 per cent by addition of dead carbon, the aquifer flow-time t is 12,940 years if the residence time in the reservoir goes to zero (this is the age which a block of wood with an original Δ value of -50 per mill would have for the present specific activity). For finite residence times, the time which can be allotted for aquifer flow decreases as the residence time increases, due to the holding period in the brine reservoir, and finally

$t = 0$ for a maximum residence time $\tau = 31,390$ years. All t-τ combinations between these values are possible steady-state solutions to equation (16). These values (as well as the transient values) are not strongly affected by the value assumed for (X/X_O); if the original HCO_3^- concentration is assumed to be 110 ppm, which is probably a minimum estimate for 38.2 per mill salinity water, the range of t is 0–11,890 years for $\tau = 26,560$ to zero. For $(X/X_O) = 1$ (no dilution), t varies from 0 to 16,200 years, and τ from 50,360 to zero.

For finite residence times, the time which can be allotted to aquifer flow decreases as the transient period Δt increases, since maximum C^{14} depletion in the reservoir takes place at steady-state concentrations which are essentially achieved at $\Delta t \approx 2\tau$; thus the values of t for a given τ decrease across Table 7 as Δt goes from 0 to infinity. (This system does not require a period of 4 to 5 τ to reach steady-state concentrations because the C^{14}/C^{12} ratio at the time the reservoir is filled is already close to the final steady-state value; it can be shown that the ratio $R_{(t+\tau)}/R_S$ varies from 1 to 1 as τ varies from zero to infinity, with an intermediate maximum value at $(\lambda\tau) = 1.793$ for all systems. For C^{14} the maximum value is 1.56 for $\tau = 14,825$ years.)

In columns 2, 3, and 4 of Table 7 for

Table 7   Solutions to Equations (16) and (18) for the Observed (R/R_O) $(X/X_O) = 0.209$: Values of t (Aquifer Flow Time), for Given Values of τ (Residence Time) and Δt (Period Since Filling), All in Years. Values Above the Dashed Line Represent Solutions Consistent with the Ku *et al.* Deuser-Degens Chronology

| τ | \multicolumn{4}{c}{Values of t, for} |
	$\Delta t = 0$	$\Delta t = 5,000$	$\Delta t = 12,000$	Steady-state
0	12,940	12,940	12,940	12,940
2,000	11,960	11,190	11,150	11,150
5,000	10,570	9,360	9,070	9,030
10,000	8,440	7,120	6,550	6,390
15,000	6,540	5,300	4,640	4,390
20,000	4,870	3,730	3,080	2,780
25,000	3,380	2,360	1,740	1,430
30,000	2,060	1,150	580	270

transient conditions, the total time involved since brine-flow began in the aquifer is $(t + \tau + \Delta t)$, time since brine began to flow into the depression is $(\tau + \Delta t)$, and the time since the depression filled to its steady-state level is Δt. It is evident that if the Ku *et al.*-Deuser-Degens chronology is correct, and brine-flow began during the past 12,000 years or so, solutions to the model are limited to the upper left quadrant of Table 7, from $\tau = 12,000$, $\Delta t = 0$, to $\tau = 0$, $\Delta t = 12,000$ years, above the dashed line in the table. However, these solutions require aquifer-flow times ranging from 8,400 to 12,900 years, and are thus generally not consistent with the isotopic composition of the water in the brine, which must date from not more than a few thousand years ago when the entering sea water approximately reached its present composition (say within 0.2 per mill for O^{18} and 1 per mill for D at most). Solutions for flow-times of a few thousand years are limited to residence times below the dashed line, greater than 15,000 years. We must therefore conclude that no solutions for the steady-state flow and dilution model are consistent with all the geochemical data now available.

It seems evident that in order to account for the extremely low radiocarbon activity in the brine, some isotope exchange with dead carbonate has to be assumed (carbonate, because of the relatively high C^{13}/C^{12} ratio in the brine). Exchange within the brine depression with the fossils deposited since brine flow apparently began, and recycling of carbonate would all add radiocarbon to the brine. On the other hand, exchange with very old carbonate during flow to the depressions would have the same effect in the model as increasing the numerical value of the C^{14} decay constant λ, by addition of a first-order rate constant to it. Thus in equation (16), if we set $(1/\lambda)$ equal to 4,133 years, twice the radioactive mean life, the flow times for $\tau = 0$ are all 6,470 years. For $\Delta t = 0$, $\tau = 10,000$, we obtain $t = 2,430$ years; for $\Delta t = 1,000$, $\tau = 5,000$, $t = 3,880$ years; for $\Delta t = 5,000$, $\tau = 2,000$, $t = 4,860$ years; etc. (The values of t become 50 per cent of those in Table 7 for $\tau = 0$, about 45 per cent for $\tau = 2,000$, about 38 per cent for $\tau = 5,000$, and about 27 per cent for $\tau = 10,000$ years.) An exchange-rate constant approximately equal to the radioactive decay constant therefore makes all the solutions above the dashed line in Table 7 consistent with the geochemical data, and requires that radiocarbon exchange out of the brine at an average rate about equal to the decay rate. This process is also consistent with the C^{13}/C^{12} ratio in the Atlantis II brine, which is about 6 per mill lower than that of sea water, in the direction in which dissolved carbon would shift if it approaches equilibrium with organically precipitated marine carbonates, especially with coralline material, which tends to be low in C^{13}.

Although more complicated processes can be designed to account for the low radiocarbon content of the brine, the exchange process with old carbonate is probably the simplest mechanism consistent with the present evidence on the origin of the brine. It is apparent that radiocarbon measurements on the Atlantis 44° brine and the Discovery brine would be extremely important in understanding the age and mixing relationships of these waters (e.g. the mixing rates in Atlantis Deep and the questions of the introduction time of, and possible influx into, Discovery brine). It is unfortunate that time could not be made available for sampling these two waters when Atlantis 56° brine was sampled, but hopefully future expeditions can afford the time.

Summary and Conclusions

The geochemical, chemical, and physical data on the Red Sea brines now available are consistent with a steady-state model of the Atlantis II Deep in which sea water circulates downward through evaporite sediments and flows northward, driven by the density difference between columns of brine and sea water, and heated by the local geothermal gradient which is quite sufficient to bring it to the observed temperature. The isotopic composition of the water, and the dissolved argon concentration, point to relatively warm, near surface waters, as the source of the brine. In particular, the deuterium and oxygen-18 concentrations indicate that the source water originates at least several hundred km south of the depressions, and probably about 800km south, near or on the southern sill, where water of 38.2 per mill salinity is abundant. The brine must flow at least 400km in a time of the order of a few thousand years, implying that fissure flow, probably in basalts along the central rift, is involved. The absence of an oxygen-18 shift in the water shows that temperatures are not much higher than 100°C at most.

The enrichment patterns of salts in the brine are consistent with an evaporitic origin, probably from the halite-sylvite zone, for most of the components. Certain constituents such as Mg, SO_4, I, and F have been depleted from the original sea water; this pattern and the almost stoichiometric loss of $MgSO_4$ may reflect loss by selective membrane filtration. Loss of NO_3 is probably due to nitrate reduction.

Flow of the present brine was probably initiated when sea level rose about 12,000 years ago to the level of the southern sill, and sea water became available on the sill and in the shallow regions adjacent to it. During glacial periods these areas were exposed and no supply of water was available at the faults which carry the water down. When the Red Sea is isolated during glacial periods, and its level falls, brine flow may temporarily increase because of the increased hydrostatic head, causing the brine to overflow its present level in the depression and possibly to precipitate large amounts of heavy metals by mixing with sea water. Several such cycles interrupted by periods of no flow during glacial times, may have occurred in the past.

At the present time, iron precipitation is occurring almost entirely in the 44°C layer of Atlantis Deep. Assuming a steady state, less than 1 per cent of the Fe entering this layer from the 56° layer is removed by mixing and the rest is precipitated. The supply of oxygen by mixing from overlying sea water is sufficient to precipitate only a small fraction of the iron being removed from this layer; most of the iron must be precipitating by another mechanism, possibly as ferrous carbonate. Magnanese is being supplied to the 44°C layer both from the 56° layer and from surrounding sediments, and there is no indication of Mn precipitation in this layer or in the overlying sea water. Iron recycling between the 44° and 56° layer may have increased the concentration of iron and other heavy metals in the 56° brine to steady-state levels considerably higher than in the inflowing new brine.

Sulfur and carbon isotope data show that the sulfides precipitating from the brine have not originated by sulfate reduction which has caused the observed sulfate depletion in the brine, nor are they precipitating at the 44° brine-sea water interface. In Atlantis Deep sulfides are probably precipitating because of introduction of H_2S from the surrounding sediments. A similar mechanism seems to have operated in the sediments outside of the brine area, during the period when the Red Sea was isolated by sea level lowering; at a later time when circulation was re-established, biogenic sulfate reduction began to produce sulfides and sulfur in these sediments.

At present there is no evidence requiring any contribution from volcanic or magmatic sources to the brines. The origin of Fe, Mn, and other trace metals, which show enrichment factors of the order of 10^2–10^4 (Table 4), remains unsettled. But proportions of these metals in the dissolved salts or added salts are quite comparable to those in the salts of salt-dome brines in non-volcanic areas (Table 5), and in most cases they are also similar to proportions in halite-sylvite rocks. Enrichments of these metals relative to major components can readily be accounted for by two processes:

(1) Water passing through evaporites will approach saturation with major components rather quickly, but will continue to extract trace metals which remain unsaturated.

(2) Recycling by steady-state precipitation in stacked convection cells can increase the concentration in the lowest layer by large factors, even orders of magnitude, relative to the actual fluid entering the

depressions. Thus the enrichment factors calculated for elements undergoing chemical reactions in such systems must be regarded as "apparent" enrichment factors only, until the original fluid can be sampled.

These and other concepts have been developed in considerable detail in order to provide a working framework for continued development and testing of our knowledge of the processes responsible for the origin and chemical history of the brine. Some of the present treatment is clearly speculative, some has the appearance of a tour-de-force treatment; but it is clearly important to determine whether or not a steady-state flow relationship has been reached or is being approached, most especially from the viewpoint of future exploitation of the brine. The arguments developed in the course of this work show that future expedition work should be aimed toward certain very detailed problems with important implications for understanding the processes operating in the brine areas:

1. Detailed temperature, salinity, dissolved oxygen, and trace element profiles across the brine-sea water transition regions should be made with great precision. Temperature-salinity profiles at various points above the brine can indicate the nature of the mixing processes involved, and oxygen, iron, manganese, SO_4 and other chemical profiles can delineate the precipitation processes, if any, occurring in this region.
2. Dissolved carbonate and C^{13} and C^{14} profiles should be measured from normal sea water continuously through the brines, because of the importance of carbonate chemistry and radiocarbon concentrations for many problems.
3. Chlorinity and sulfate measurements should be made simultaneously on interstitial waters both in the brine depressions and in normal sediments outside the depressions and correlated with S^{34}-C^{13}-O^{18}-C^{14} measurements on the sediments, in order to understand the very complicated history of sulfide-sulfate relationships in the area.

Acknowledgments

Many people helped immeasurably in the preparation of this paper; I am particularly grateful to A. Jokela for collecting the radiocarbon samples, to W. Broecker for providing the equipment, to D. Agrawal and S. Kusumgar of the Tata Institute for counting the samples, to John Hunt of Woods Hole for inviting the participation of A. Jokela and R. Weiss on the **CHAIN** 61 Expedition, and to E. Degens and D. Ross for their assistance. R. Weiss provided critical and helpful discussion of many points as the paper was written, H. Oeschger provided sanctuary for the writing, and Mrs. Patricia Renner performed the difficult job of typing and assembly with great skill during my absence. Professor M. Bass read the manuscript and clarified many points. SIO Expedition Zephyrus was supported by NSF grant G-24479; participation of A. Jokela and R. Weiss on **CHAIN** 61 was supported by ONR grant NONR-2216(23); preparation of the paper was supported by NSF grant GA-666.

References

Baker, W. J.: Flow in fissured formation. Proc. Fourth World Petroleum Cong., Section II/E, C. Columbo, Rome, 379–392 (1955).

Bischoff, J.: Red Sea geothermal brine deposits: their mineralogy, chemistry, and genesis. *In: Hot brines and recent heavy metal deposits in the Red Sea*, E. T. Degens and D. A. Ross (eds.). Springer-Verlag New York Inc., 368–401 (1969).

Brewer, P. G. and D. W. Spencer: A note on the chemical composition of the Red Sea brines. *In: Hot brines and recent heavy metal deposits in the Red Sea*. E. T. Degens and D. A. Ross (eds.). Springer-Verlag New York Inc., 174–179 (1969).

Brewer, P. G., J. P. Riley, and F. Culkin: The chemical composition of the hot salty water from the bottom of the Red Sea. Deep-Sea Res., **12**, 497 (1965).

Craig, H.: The isotopic geochemistry of water and carbon in geothermal areas. *In: Nuclear Geology on Geothermal Areas*, 1963 Spoleto Conference Proceedings, E. Tongiorgi (ed.). Consiglio Nazionale delle Richerche, Pisa, 17 (1963).

———: Isotopic composition and origin of the Red Sea and Salton Sea geothermal brines. Science, **134**, 1544 (1966).

———: Origin of the saline lakes in Victoria Land, Antarctica. Trans. Am. Geophys. Union, **47**, 112 (1966b).

——— and L. I. Gordon: Deuterium and oxygen-18 variations in the ocean and the marine atmosphere. *In: Stable Isotopes in Oceanographic Studies and Paleotemperatures,* 1965 Spoleto Conference Proceedings, E. Tongiorgi (ed.). Consiglio Nazionale delle Richerche, Pisa, 9 (1965).

Craig, H. and D. Lal: Radiocarbon age of the Red Sea Brine. Trans. Am. Geophys. Union, **49**, 193 (1968).

Craig, H., L. I. Gordon and Y. Horibe: Isotopic exchange effects in the evaporation of water. Jour. Geophys. Res., **68**, 5079 (1963).

Curray, J. R.: Sediments and history of Holocene transgression, continental shelf, northwest Gulf of Mexico. *In: Recent Sediments, Northwest Gulf of Mexico, 1951–1958,* Amer. Assoc. Petrol. Geologists, Tulsa, 221 (1960).

Delevaux, M. H., B. R. Doe, and G. F. Brown: Preliminary lead isotope investigations of brine from the Red Sea, galena from the kingdom of Saudi Arabia, and galena from United Arab Republic (Egypt). Earth and Planetary Science Letters, **3**, 139 (1967).

Deuser, W. G. and E. T. Degens: Carbon isotope fractionation in the system CO_2 (gas)-CO_2 (aqueous)-HCO_3^- (aqueous). Nature, **215**, 1033 (1967).

———, ———: O^{18}/O^{16} and C^{13}/C^{12} ratios of fossils from the hot brine deep area of the central Red Sea. *In: Hot brines and recent heavy metal deposits in the Red Sea,* E. T. Degens and D. A. Ross (eds.). Springer-Verlag New York Inc., 336–347 (1969).

Drake, C. L. and R. W. Girdler: A geophysical study of the Red Sea. Geophys. Jour., Roy. Astronom. Soc., **8**, 473 (1964).

Faure, G. and L. Jones: Anomalous strontium in the Red Sea Brines. *In: Hot brines and recent heavy metal deposits in the Red Sea,* E. T. Degens and D. A. Ross (eds.). Springer-Verlag New York Inc., 243–250 (1969).

Girdler, R. W.: The relationship of the Red Sea to the East African rift system. Quart. Jour. Geol. Soc., London, **114**, 79 (1958).

Hartmann, M. and H. Nielsen: Sulfur isotopes in the hot brine and sediment of Atlantis II Deep (Red Sea). Marine Geol., **4**, 305 (1966).

Heybroek, F.: The Red Sea Miocene evaporite basin. *In: Salt Basins around Africa,* Inst. Petroleum, London, 17 (1965).

Kaplan, I. R., R. E. Sweeney, and A. Nissenbaum: Sulfur isotope studies on Red Sea geothermal brines and sediments. *In: Hot brines and recent heavy metal deposits in the Red Sea,* E. T. Degens and D. A. Ross (eds.). Springer-Verlag New York Inc., 474–498 (1969).

Ku, T. L., D. L. Thurber, and G. G. Mathieu: Radio-carbon chronology of Red Sea sediments. *In: Hot brines and recent heavy metal deposits in the Red Sea,* E. T. Degens and D. A. Ross (eds.), Springer-Verlag New York Inc., 348–359 (1969).

Longinelli, A. and H. Craig: Oxygen-18 variations in sulfate ions in sea water and saline lakes. Science, **156,** 56 (1967).

Miller, A. R., C. D. Densmore, E. T. Degens, J. C. Hathaway, F. G. Manheim, P. F. McFarlin, R. Pocklington, and A. Jokela: Hot brines and recent iron deposits in deeps of the Red Sea. Geochim. et Cosmochim. Acta, **30**, 341 (1966).

Neumann, A. C. and D. A. McGill: Circulation of the Red Sea in early summer. Deep-Sea Res., **8**, 223 (1962).

Pugh, D. T.: Origin of hot brines in the Red Sea. Nature, **214**, 1003 (1967).

Pytkowicz, R. M. and R. Gates: Magnesium sulfate interactions in seawater from solubility measurements. Science, **161**, 690 (1968).

Rafter, T. A. and Y. Mizutani: Preliminary study of variations of oxygen and sulphur isotopes in natural sulphates. Nature, **216**, 1000 (1967).

Stewart, F. H.: Marine Evaporites. Chapter Y, Data of Geochemistry, U.S.G.S. Prof. Paper 440-Y, Washington, 1 (1963).

Thode, H. G., M. Shima, C. E. Rees, and K. V. Krishnamurty: Carbon-13 isotope effects in systems containing carbon dioxide, bicarbonate, carbonate, and metal ions. Canad. J. Chem., **43**, 582 (1965).

Thompson, E. F.: Chemical and physical investigations. The general hydrography of the Red Sea. John Murray Expedition 1933–34, Scientific Reports, **2**, 83; The exchange of water between the Red Sea and the Gulf of Aden over the "Sill." *ibid.,* **2**, 105 (1939).

Watson, S. W. and J. B. Waterbury: The sterile hot brines of the Red Sea. *In: Hot brines and recent heavy metal deposits in the Red Sea,* E. T. Degens and D. A. Ross (eds.). Springer-Verlag New York Inc., 272–281 (1969).

Weiss, R. F.: Dissolved argon, nitrogen and total carbonate in the Red Sea Brines. *In: Hot brines and recent heavy metal deposits in the Red Sea,* E. T. Degens and D. A. Ross (eds.). Springer-Verlag New York Inc., 254–260 (1969).

——— and H. Craig: Total carbonate and dissolved gases in equatorial Pacific waters. Trans. Am. Geophys. Union, **49**, 216 (1968).

Wendt, I.: Fractionation of carbon isotopes and its temperature dependence in the system CO_2-gas-CO_2 in solution and HCO_3-CO_2 in solution. Earth and Planetary Science Letters, **4**, 64 (1968).

White, D. E.: Saline waters of sedimentary rocks. Fluids in subsurface environments—a symposium. Am. Assoc. Petrol. Geol. Mem., **4**, 342 (1965).

White, D. E., J. D. Hem, and G. A. Waring: Chemical composition of subsurface waters. Chapter F, Data of Geochemistry, U.S.G.S. Prof. Paper 440-F, Washington, 1 (1963).

Anomalous Strontium in the Red Sea Brines *

GUNTER FAURE AND LOIS M. JONES

Department of Geology
The Ohio State University
Columbus, Ohio

Abstract

The hot brines of the Atlantis II and Discovery Deeps of the Red Sea contain strontium of very similar isotopic composition, characterized by an average Sr^{87}/Sr^{86} ratio of 0.7080 ± 0.0002. This strontium differs significantly in its isotopic composition from that of normal Red Sea water which was found to have a Sr^{87}/Sr^{86} ratio of 0.7097 ± 0.0004. The strontium in the brines is, in that sense, anomalous.

The concentrations of strontium in the 56°C brine of the Atlantis II Deep and in the 44°C brine of the Discovery Deep are very similar and average about 45ppm. The Sr/Chlorinity ratios of both brines are also similar and average 0.29. The concentrations of strontium, the Sr^{87}/Sr^{86} ratios and the chlorinities of all of the water samples from the Red Sea deeps are consistent with mixing of brine with normal sea water.

The Sr^{87}/Sr^{86} ratio of the authigenic minerals of sediment samples from the deeps changes with depth below the sea bottom. At a depth of several meters the Sr^{87}/Sr^{86} ratio of the sediment from both deeps has a value of 0.7112. This may represent the strontium present during the Pleistocene glaciation when the Red Sea was isolated from the open oceans.

Introduction

The isotopic composition of strontium in rocks and minerals of the earth is variable because of the continuing decay of naturally-occurring Rb^{87} to stable Sr^{87}. The ratio Sr^{87}/Sr^{86} of rock and mineral systems increases with time as a function of the Rb/Sr ratio of the system and the half-life of Rb^{87}, which is 50×10^9 years. Because of differences in their Rb/Sr ratios and geologic age, the Sr^{87}/Sr^{86} ratios of rocks and minerals vary widely.

According to a model proposed by Faure *et al.* (1965), the strontium in the oceans can be regarded as a mixture of three isotopic varieties of strontium which are derived from three different sources located on the continents and in the ocean basins. These sources and the Sr^{87}/Sr^{86} ratios of their strontium contributions are:

1. Old igneous and metamorphic rocks and detrital sedimentary rocks derived from them ($Sr^{87}/Sr^{86} = 0.715$);
2. Marine carbonate rocks ($Sr^{87}/Sr^{86} = 0.708$);
3. Young volcanic rocks containing "primary" strontium ($Sr^{87}/Sr^{86} = 0.704$).

Strontium from these sources is introduced into the oceans by rivers or by events and processes occurring in the ocean basins.

In contrast to the heterogeneity of the isotopic composition of strontium in the rocks and minerals of the crust, the strontium in the oceans has a uniform isotopic composition. Faure *et al.* (1965) reported an average Sr^{87}/Sr^{86} ratio of 0.7093 ± 0.0005 for ten samples of surface water from the North Atlantic Ocean. Later analyses by Faure *et al.* (1967), Burnett and Wasserburg (1967) and others have repeatedly confirmed this value and support the conclusion that the isotopic composition of strontium in the modern oceans is everywhere the same. Consequently, the

* Laboratory for Isotope Geology and Geochemistry Contribution No. 6.

Sr^{87}/Sr^{86} ratio may be used to differentiate strontium which may be of marine origin from strontium which is not marine.

The objective of this study is to determine the origin of the hot brines in the Atlantis II and the Discovery Deeps from measurements of the isotopic composition and concentration of strontium. We have also attempted to obtain information about the history of these brines by measuring the Sr^{87}/Sr^{86} ratios of authigenic minerals in sediment samples collected at varying depth below the bottom. The number of samples available to us was limited, but the data serve to demonstrate the usefulness of our approach and the need for a more extensive study of the strontium isotopic composition of these unusual brines.

No previous analyses of the isotope composition of strontium in the brines have been reported. Delevaux *et al.* (1967) measured the isotopic composition of lead from a "Red Sea brine" and found it to be similar to that of galena from vein deposits near Rabigh in Saudi Arabia.* Isotopic compositions of sulfur in brine and sediment of the Atlantis II Deep were reported by Hartman and Nielsen (1966). Miller *et al.* (1966) measured oxygen isotope ratios of brine and of normal Red Sea water and concluded (p. 356) that the "O^{18}/O^{16} ratios in the brines are similar to surface Red Sea waters or other sea waters which have been slightly concentrated by evaporation." This conclusion was confirmed and elaborated by Craig (1966, 1969) in a very elegant manner. On the basis of oxygen and hydrogen isotope compositions he showed that the brines are isotopically consistent with normal Red Sea water of post-Pleistocene age. Craig suggested that the salt content of the brines originated from leaching of evaporite rocks by sea water in the southern end of the Red Sea. The high temperature of the brines may be due to geothermal heating as the water flows northward through faults before it emerges 1,000km farther north in the brine pools. The results of our study are compatible with Craig's hypothesis.

* The temperature of this brine is stated as 74°F, which is almost certainly incorrect.

Analytical Procedures

The isotopic composition of strontium was measured on a 6 inch radius, 60° sector, single filament mass spectrometer (Nuclide Corp., Model 6-60-S). Strontium was separated from the water samples by passing them through ion exchange columns (length: 25cm, I.D. = 2cm) packed with Dowex 50 W-X8, 200–400 mesh, cation exchange resin. The eluant was 2 N vycor-distilled HCl. The strontium fraction was collected in several 15-ml polyethylene beakers and was identified by the radioactivity of carrier-free Sr^{89} tracer which was added to the water samples before they were placed on the resin columns. One or two 15-ml fractions were evaporated in vycor evaporating dishes for isotopic analysis of the strontium.

The strontium chloride separated from the water samples was dissolved in one drop of dilute vycor-distilled HNO_3 and was evaporated onto the tantalum filament of the source of the mass spectrometer. All filaments were precleaned in the mass spectrometer until no ion signal could be detected in the mass range of strontium at a temperature greater than that at which measurements are made. Stable ion beams were obtained by heating the strontium salt on the filament in the mass spectrometer at pressures of less than 1×10^{-7}mm of Hg. The ions were accelerated by an electric field of 2,000 volts, and the beam was swept by variation of the intensity of the magnetic field of the electromagnet. The signals were amplified by a Vibrating Reed Electrometer (Carey Model 31, input resistor 10^{12} ohms and critically damped response) and recorded on a 10-inch strip chart recorder (Honeywell). Each analysis consisted of 48 to 120 consecutive scans across the mass range 86 to 88. Peak heights were read directly off the ruled chart paper and were summed in sets of six. Sr^{87}/Sr^{86} and Sr^{86}/Sr^{88} ratios were calculated for each set of six scans and averaged for the complete run. A fractionation correction was applied to the measured Sr^{87}/Sr^{86} ratios, which is based on the assumption that the Sr^{86}/Sr^{88} ratio has a value of 0.1194 and that fractionation of strontium isotopes is proportional to mass differences. The reproducibility and accuracy of the analyses was monitored by repeated analyses of an isotope standard (Eimer and Amend, $SrCO_3$, lot 492327) by means of which our analyses can be related to those of other laboratories who have analyzed this standard.

The concentration of strontium in the water samples was measured by the method of isotope dilution using a "spike" enriched in Sr^{86} (97.64 atom per cent). The concentration of the spike solution was measured by separate isotope dilution runs using a "shelf solution" of normal isotope composition. Both the sea water samples and the spike solutions were weighed in closed containers to the closest 0.1mg. The strontium was separated as described above and the Sr^{86}/Sr^{88} ratios were measured on the mass spectrometer.

The sediment samples were first leached with double-distilled and demineralized water. The water-insoluble fraction was then treated for several hours with hot 2 N vycor-distilled HCl and HNO_3 in teflon dishes to dissolve authigenic carbonates, oxides and sulfides. The solutions were filtered into polyethylene beakers using S + S #576 filter paper and were then placed on the ion exchange columns. The strontium was separated as before and analyzed on the mass spectrometer.

Presentation and Discussion of the Results

The Sr^{87}/Sr^{86} Ratios of the Brines

The Sr^{87}/Sr^{86} ratios of brines from the Atlantis II and the Discovery Deeps are compiled in Table 1 and displayed in Fig. 1. All water samples were analyzed at least in duplicate, while one sample (Sta. 541A and B) was analyzed seven times. The range of Sr^{87}/Sr^{86} ratios for each sample is indicated in Fig. 1 by the length of the horizontal bar. The position of the mean is shown by a vertical bar. The Sr^{87}/Sr^{86} ratio of normal marine strontium (0.7093) is shown as a vertical dashed line in order to facilitate direct comparisons of the Sr^{87}/Sr^{86} ratios of the brines with that of normal marine strontium. The boundaries and transition zones between brines of different temperature are shown as horizontal lines.

The Atlantis II Deep contains two layers of brine separated by a thin transition zone. The lower brine (C61-18, temperature 56°C) was found to have a Sr^{87}/Sr^{86} ratio of 0.7082 ± 0.0005, based on two replicate determinations. The upper layer (C61-77, temperature 44°C) has a slightly higher Sr^{87}/Sr^{86} ratio of 0.7086 ± 0.0005, based on four replicate determinations. The 44°C brine of the Atlantis II Deep grades up-

ward into normal sea water represented by sample C61-64, which was found to have a Sr^{87}/Sr^{86} ratio of 0.7097 ± 0.0004. This value is in satisfactory agreement with strontium in other oceans.

The brine of the Discovery Deep has a temperature of about 44°C and is represented in our study by samples C61-63 and sta. 541 (A and B). The brine grades upward through an intermediate layer to normal Red Sea water. This intermediate layer is represented by samples C61-57 and C61-62.

The Sr^{87}/Sr^{86} ratio of the lower brine is 0.7080 ± 0.0002 which is the average of ten replicate analyses performed on samples C61-63 (Sr^{87}/Sr^{86} = 0.7077 ± 0.0002, 3 replicates) and Sta. 541A and B (Sr^{87}/Sr^{86} = 0.7081 ± 0.0003, seven replicates). Sample C61-62 comes from the lower portion of the mixed layer above the brine. Its Sr^{87}/Sr^{86} ratio is 0.7081 ± 0.0001 (2 replicates) and is indistinguishable from that of the underlying brine. The strontium concentration and Sr/Chlorinity ratio of this

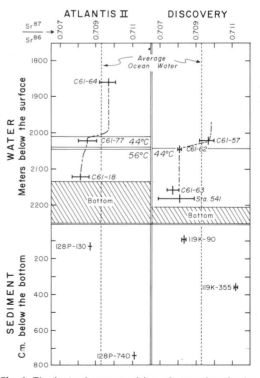

Fig. 1. The isotopic composition of strontium in the brines and sediment of the Atlantis II and the Discovery Deeps of the Red Sea.

Table 1   Isotopic Composition of Strontium in Brines from the Atlantis II and the Discovery Deeps of the Red Sea

Sample Number	Description	Depth (m)	Temp (°C)	$\dfrac{Sr^{86}}{Sr^{88}}$	$\left(\dfrac{Sr^{87}}{Sr^{86}}\right)$ corr.*
Atlantis II Deep					
C61-64	Normal	1,860	21.98	0.1184	0.7101
				0.1186	0.7092
	Average:			0.1185	0.7097 ± 0.0004
C61-77	Intermed.	2,019	44.32	0.1197	0.7083
				0.1195	0.7092
				0.1188	0.7082
				0.1180	0.7085
	Average:			0.1190	0.7086 ± 0.0005
C61-18	Brine	2,121	56.50	0.1193	0.7087
				0.1186	0.7076
	Average:			0.1190	0.7082 ± 0.0005
Discovery Deep					
C61-57	Intermed.	2,019	29.72	0.1186	0.7092
				0.1195	0.7098
				0.1183	0.7100
	Average:			0.1188	0.7097 ± 0.0003
C61-62	Intermed.	2,044	41.90	0.1188	0.7080
				0.1189	0.7082
	Average:			0.1189	0.7081 ± 0.0001
C61-63	Brine	2,158	44.2	0.1200	0.7075
				0.1181	0.7081
				0.1184	0.7074
	Average:			0.1188	0.7077 ± 0.0002
Sta. 541A †	Brine	2,180	44.0	0.1189	0.7083
				0.1185	0.7069
				0.1177	0.7073
Sta. 541B		"	"	0.1176	0.7086
				0.1197	0.7082
				0.1186	0.7083
				0.1166	0.7089
	Average:			0.1182	0.7081 ± 0.0003
Eimer and Amend SrCO₃, lot 492327 15 analyses				0.1180	0.7083 ± 0.0001

* Corrected for fractionation assuming $Sr^{86}/Sr^{88} = 0.1194$.

† The sample labelled Sta. 541A and B was given us by Mr. R. F. Weiss of the Department of Earth Sciences, University of California, La Jolla, California. It was originally obtained from Sir Edward Bullard and was collected on board the R.R.S. Discovery at station 5580 on September 11, 1964. A and B are duplicate samples taken from the same bottle. The salinity of this sample was reported by Swallow and Crease (1965). Craig (1966) determined its chlorinity as well as oxygen and hydrogen isotope ratios.

The errors assigned to the Sr^{87}/Sr^{86} ratios are one standard deviation of the mean.

water, to be discussed later, are also very similar to those of the brine below.

We conclude from the data in Table 1 that the brines of the Atlantis II and the Discovery Deeps contain strontium of very similar isotopic composition, having an average Sr^{87}/Sr^{86} ratio of 0.7080 ± 0.0002.

This value appears to be significantly lower than the Sr^{87}/Sr^{86} ratio of normal Red Sea water, which is 0.7097 ± 0.0004, according to our measurements. The strontium of the Red Sea brines is therefore isotopically anomalous. The relatively low Sr^{87}/Sr^{86} ratio of the brines may reflect the presence of primary strontium ($Sr^{87}/Sr^{86} = 0.704$) from magmatic sources at depth.

The Sr/Chlorinity Ratios of the Brines

The Sr/Chlorinity ratio of sea water has been determined repeatedly by numerous investigators and is believed to be a constant. This generalization, however, has been challenged by Angino et al. (1966) and by MacKenzie (1964). More recently, Riley and Tongudai (1967) repeated analyses of sea water published earlier by Cox and Culkin (1966), but they failed to confirm the conclusions of Angino et al. (1966). Riley and Tongudai (1967) obtained an average Sr/Chlorinity ratio (Sr in ppm, Chlorinity per mille) of 0.42 ± 0.02 for 38 samples taken from different depths. This value is in good agreement with that of Cox and Culkin (1966), who reported an average of 0.40 ± 0.02 for about 70 samples from the major oceans. These values are also in agreement with determinations by other investigators referred to by Riley and Tongudai (1967).

Table 2 shows the salinities, chlorinities and strontium concentrations of our water samples from the Red Sea. It is evident from the data that the 56°C brine of the Atlantis II Deep and the 44°C brine of the Discovery Deep have similar strontium concentrations of nearly 45ppm and chlor-

inities of about 156 per mill. The average Sr/Chlorinity ratio for both brines is 0.29. This value is significantly lower than the Sr/Chlorinity ratio of normal Red Sea water (C61-64) for which we have a value of 0.40. These results are in excellent agreement with those of Brewer et al. (1965) for brines from the Discovery Deep.

Craig (1969) has shown that the chemical composition of the water in the Red Sea deeps results from the mixing of brine with normal sea water. Fig. 2 shows that the strontium concentrations and chlorinities of our water samples follow this pattern closely. The isotopic composition of strontium in the water samples similarly exhibits the expected linear mixing relationship of the two end members, within the uncertainties of the measurements.

The Sr^{87}/Sr^{86} Ratios of the Sediments

Two samples of sediment from each of the two deeps were analyzed to determine the Sr^{87}/Sr^{86} ratios of the authigenic minerals. The location of the samples is shown in Ross and Degens (1969). The analytical data and brief mineralogical descriptions of the sediment samples are given in Table 3. The Sr^{87}/Sr^{86} ratios are plotted in Fig. 1 to facilitate their comparison to the strontium isotopic composition of the overlying brines.

The concentrations of strontium in the sediment samples were determined by X-

Table 2   Strontium Concentrations and Strontium/Chlorinity Ratios of the Red Sea Brines

Sample Number	Description	Conductometric Salinity (‰)	Chlorinity (‰)	Sr (ppm)	$\frac{Sr}{Chlor}$
Atlantis II Deep					
C61-64	Normal	40.61	22.53	9.0	0.40
C61-77	Intermed.	148.2	80.04	22.4, 22.7	0.28
C61-18	Brine	317.0	156.1	44.1	0.28
Discovery Deep					
C61-57	Intermed.	72.78	40.05	13.1	0.33
C61-62	Intermed.	280.2	142.2 *	40.5	0.29
C61-63	Brine	317.2	155.1	45.3	0.29
Sta. 541A	Brine	317	155.6 †	44.3	0.29
Sta. 541B		317	155.6	44.7	0.29

* The chlorinity was calculated by multiplying the salinity by 0.5075 which was obtained by interpolation on a plot of Chlor/Sal versus Salinity.

† The chlorinity was calculated by multiplying the salinity by 0.4907 which is the average Chlor/Sal ratio of C61-18 and C61-63.

The chlorinity and salinity determinations were made at the Woods Hole Oceanographic Institute.

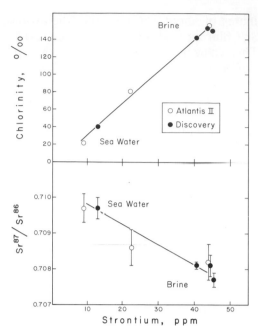

Fig. 2. These graphs illustrate the essentially linear relationship between chlorinity, Sr^{87}/Sr^{86} ratio and strontium concentration of water samples from the Red Sea Deeps.

ray emission spectrometry using the rock standard W-1 for comparison. The iron-rich sediment of the Atlantis II Deep contains slightly more than 100ppm of strontium. The sediment samples from the Discovery Deep contain $CaCO_3$ and have substantially higher strontium concentrations of about 140ppm (119K−90cm) and 640-ppm (119K−355cm), respectively. This

strontium is present in the water-insoluble minerals of the dry sediment.

The Sr^{87}/Sr^{86} ratios of the acid-soluble fraction of the sediments show an interesting variation with depth and, therefore, also with time in the past. In the Atlantis II Deep iron oxide sediment at a depth of 130cm below the bottom (core 128P) has a Sr^{87}/Sr^{86} ratio of 0.7087, while at a depth of 740cm the Sr^{87}/Sr^{86} ratio is 0.7112. A similar pattern of variations was observed in sediment from the Discovery Deep (core 119K). At a depth of 90cm the sediment consists mainly of iron oxide and has a Sr^{87}/Sr^{86} ratio of 0.7083, while at a depth of 355cm calcareous ooze has a Sr^{87}/Sr^{86} ratio of 0.7112.

The isotopes of strontium are not measurably fractionated by natural processes and the effects of fractionation due to selective volatilization and ionization of strontium isotopes in the source of the mass spectrometer are eliminated by the fractionation correction which is made routinely to all measured Sr^{87}/Sr^{86} ratios. Therefore, the Sr^{87}/Sr^{86} ratio of the authigenic minerals should be identical to that of the brine from which the minerals were precipitated. The variation of the Sr^{87}/Sr^{86} ratio of the sediment with depth in a core thus can be regarded as a record of the isotopic composition of the strontium in the water which occupied the deeps.

It is interesting to note that the Sr^{87}/Sr^{86}

Table 3   **Isotopic Composition and Concentration of Strontium in Sediment from the Atlantis II and the Discovery Deeps, Red Sea**

Depth (cm)	Description	Sr Conc. (ppm) *		$\frac{Sr^{86}}{Sr^{88}}$	$\left(\frac{Sr^{87}}{Sr^{86}}\right)$ corr.
		Dry	Leached		
Atlantis II Deep, Core 128P, 21°25.4′N, 38°03.4′E, depth 2,077m					
130	Iron oxide, siderite + rhodochrosite, quartz	117 ± 6	n.d.†	0.1195	0.7087
740	Iron oxide, siderite + rhodochrosite, sphalerite, quartz	102 ± 8	n.d.	0.1188	0.7112
Discovery Deep, Core 119K, 21°16.8′N, 38°01.7′E, 2,175m					
90	Iron oxide, calcite, quartz	142 ± 6	141 ± 4	0.1184	0.7084
				0.1193	0.7082
355	Calcite, quartz, siderite + rhodochrosite, feldspar?	636 ± 13	672 ± 10	0.1179	0.7112
				0.1181	0.7112

* Measured by X-ray emission spectrometry using W-1 (Sr = 190ppm) as comparison standard. The error is one standard deviation of the mean of four replicate determinations.
† Not determined.

ratio of the sediment in both depressions has the same value of 0.7112, at a depth of 740cm in the Atlantis II Deep and at 355cm in the Discovery Deep. This value is significantly higher than the Sr^{87}/Sr^{86} ratio of any of the present brines or that of present day sea water. Although we have no unique explanation of this observation, data by Ku *et al.* (1969) indicate that the deeper sediment layers were deposited during the Pleistocene at a time when the Red Sea represented a restricted environment. Alternatively, the high Sr^{87}/Sr^{86} ratio may indicate that these sediment layers were deposited by brines which had a Sr^{87}/Sr^{86} ratio of 0.7112 but which have since been displaced by the present brines.

The Sr^{87}/Sr^{86} ratio of the iron-rich sediment at a depth of 90cm in the Discovery Deep (core 119K−90cm) is 0.7083 and is thus identical to that of the overlying brine. We conclude therefrom that this sediment did precipitate from the brine which now occupies the Discovery Deep. The sediment at a depth of 130cm in the Atlantis II Deep (core 128P−130cm) has a Sr^{87}/Sr^{86} ratio of 0.7087. This value appears to be identical to the Sr^{87}/Sr^{86} ratio of the intermediate 44°C brine and may reflect the mixing of marine and brine strontium.

The data presented here indicate that the Sr^{87}/Sr^{86} ratio of the sediment in the deeps changes with depth and thus records historical information about the isotopic composition of strontium of the brines from which the sediment may have precipitated. A detailed investigation of these changes by analyses of a large number of sediment samples taken at increasing depth below the bottom should be undertaken.

Acknowledgments

This study could not have been undertaken without the help and cooperation of R. F. Weiss and Dr. D. A. Ross who provided the samples and much information. We are grateful to Drs. S. R. Hart and H. Craig who were instrumental in making the initial arrangements which got this project started. We also acknowledge with thanks the assistance of R. L. Hill and René Eastin who helped with the analyses. This paper was reviewed by Dr. H. Craig whose constructive criticism and suggestions we are pleased to acknowledge.

References

Angino, E. E., G. K. Billings and Neil Andersen: Observed variations in the strontium concentrations of sea water. Chem. Geol., **1**, 145 (1966).

Brewer, P. G., J. P. Riley and F. Culkin: The chemical composition of the hot salty water from the bottom of the Red Sea. Deep-Sea Res., **12**, 497 (1965).

Burnett, D. S. and G. J. Wasserburg: Evidence for the formation of an iron meteorite at 3.8×10^9 years. Earth Planetary Sci. Letters, **2**, 137 (1967).

Cox, R. and F. Culkin: Sodium, potassium, magnesium, calcium and strontium in sea water. Deep-Sea Res., **13**, 789 (1966).

Craig, H.: Isotopic composition and origin of the Red Sea and Salton Sea geothermal brines. Science, **154**, 1544 (1966).

———: Geochemistry and origin of the Red Sea brines. *In: Hot brines and recent heavy metal deposits in the Red Sea*, E. T. Degens and D. A. Ross (eds.). Springer-Verlag New York Inc., 208–242 (1969).

Delevaux, M. H., B. R. Doe, and G. F. Brown: Preliminary lead isotope investigations of brine from the Red Sea, galena from the Kingdom of Saudi Arabia, and galena from United Arab Republic (Egypt). Earth and Planetary Sci. Letters, **3**, 139 (1967).

Dixon, W. J. and F. J. Massey, Jr.: *Introduction to Statistical Analysis,* 2nd ed. McGraw-Hill Book Co. Inc., New York, 488 p. (1957).

Faure, G., P. M. Hurley, and J. L. Powell: The isotopic composition of strontium in surface water from the North Atlantic Ocean. Geochim. Cosmochim. Acta, **29**, 209 (1965).

Faure, G., J. H. Crocket, and P. M. Hurley: Some aspects of the geochemistry of strontium and calcium in the Hudson Bay and the Great Lakes. Geochim. Cosmochim. Acta, **31**, 451 (1967).

Hartman, M. and H. Nielsen: Sulfur isotopes in the hot brine and sediment of Atlantis II deep, Red Sea. Marine Geol., **4**, 305 (1966).

Ku, T. L., D. L. Thurber, and G. Mathieu: Radiocarbon chronology of Red Sea sediments. *In: Hot brines and recent heavy metal deposits in the Red Sea*, E. T. Degens and D. A. Ross (eds.). Springer-Verlag New York Inc., 348–359 (1969).

Mackenzie, F. T.: Strontium content and variable strontium/chlorinity relationship of Sargasso Sea water. Science, **146**, 517 (1964).

Miller, A. R., C. D. Densmore, E. T. Degens, J. C. Hathaway, F. G. Manheim, P. F. McFarlin, R. Pocklington, and A. Jokela: Hot brines and recent iron deposits in deeps of the Red Sea. Geochim. Cosmochim. Acta, **30**, 341 (1966).

Riley, J. P. and M. Tongudai: The major cation/chlorinity ratios in sea water. Chem. Geol., **2,** 263 (1967).

Ross, D. A. and E. T. Degens: Shipboard collection and preservation of sediment samples collected during CHAIN 61 from the Red Sea. *In: Hot brines and recent heavy metal deposits in the Red Sea,* Springer-Verlag New York Inc., 363–367 (1969).

Ross, D. A., E. E. Hays, and F. C. Allstrom: Bathym-etry and continuous seismic profiles of the hot brine region of the Red Sea. *In: Hot brines and recent heavy metal deposits in the Red Sea.* E. T. Degens and D. A. Ross (eds.). Springer-Verlag New York Inc., 82–97 (1969).

Swallow, J. C. and J. Crease: Hot salty water at the bottom of the Red Sea. Nature, **205,** 165 (1965).

Low Molecular Weight Hydrocarbon Analyses of Atlantis II Waters

J. W. SWINNERTON AND V. J. LINNENBOM

Ocean Sciences Division
U.S. Naval Research Laboratory
Washington, D.C.

Abstract

Preliminary data on the low molecular weight hydrocarbon content in Atlantis II waters are presented. Relative to normal sea waters the low molecular weight paraffins are enriched in the brines by a factor of about 10 to 1,000. The enrichment is most pronounced for methane and ethane.

During the past several years, a program of research on dissolved gases in sea water has been carried on at the Naval Research Laboratory, utilizing the technique of gas chromatography (Swinnerton et al., 1962a, b; Swinnerton and Linnenbom, 1967b). Because of the extreme sensitivity of this method, it has been possible to show for the first time the presence of low molecular weight hydrocarbons other than methane in the open ocean (Swinnerton and Linnenbom, 1967a). Since that time, the purpose of our work has been two-fold: (a) to obtain further data on the occurrence and distribution of the C_1 through C_4 hydrocarbons in sea water, and (b) to adapt the analytical equipment used in the laboratory for shipboard use, and to evaluate its performance under field conditions of use. Although certain trends in the distribution of hydrocarbons are beginning to appear, we feel at the present time that further exploratory sampling is necessary before any firm conclusions can be drawn regarding the mechanism of production of these gases in the sea.

Thus far, data have been obtained on the occurrence of these low molecular weight hydrocarbons in the open waters of the Atlantic and the Caribbean, in the anoxic waters of the Black Sea and the Cariaco Trench and in the Red Sea at the site of the Atlantis II hot spot. Sampling in all cases has been carried out by means of the usual Nansen casts. The precautions taken in handling samples were similar to those employed when making dissolved oxygen determinations. The water was transferred from the Nansen bottle to a one-liter glass reagent bottle fitted with a standard taper glass joint and returned to the Naval Research Laboratory for analysis. Some early work in the Gulf of Mexico indicated that this technique was effective in preventing contamination and loss of gas from solution. Samples analyzed within two hours of collection and compared with replicates analyzed several weeks later showed no significant difference in gas content. Use of preservatives (sodium azide and mercuric chloride) was found not to be necessary. In fact, the mercuric chloride was deleterious in that it removed all the unsaturates.

The method of analysis has been described (Swinnerton and Linnenbom, 1967b). Briefly, the method is based on purging of the dissolved gases from the sea water sample with a stream of pure helium, followed by trapping on a suitable adsorbent maintained at low temperature ($-80°C$). The adsorbent temperature is

Table 1   Low Molecular Weight Hydrocarbon Concentrations in Sea Water
($X \cdot 10^{-7}$ml/liter)

Location	Depth(m)	Methane	Ethane	Ethylene	Propane	Propylene	Iso-Butane	N-Butane	O_2(ml/l)
Red Sea — Atlantis II	50(a)	500	5.5	30.4	1.6	7.2	(c)	(c)	4.4
21°24′N	1,880(a)	400	9.4	3.1	2.0	2.0	(c)	(c)	2.0
38° 03′E	2,050	580,000	5,200	(b)	330	25	50	58	0.
Black Sea — 42°47′N	100(a)	680	7	7.8	7.0	22	(c)	(c)	0.5
30° 16′E	500	1,200,000	1,600	(b)	17	1.2	5.0	20	0.
	1,500	1,100,000	2,400	(b)	13	1.4	5.0	10	0.
	2,000	1,700,000	3,900	(b)	24	5.0	6.0	30	0.
Cariaco Trench — 10° 46′N	220(a)	560	8	(b)	4.2	3.1	0.9	(c)	0.5
65° 39′W	700	1,000,000	330	(b)	25	1.6	13	2.0	0.
	1,250	1,500,000	410	(b)	53	0.2	14	8.0	0.
Lake Nitinat (British Columbia)	10(a)	7,800	60	(b)	35	15	9.0	(c)	1.6
	90	13,000,000	220	(b)	390	8.0	50	4.0	0.
	200	16,000,000	250	(b)	440	10.0	58	7.0	0.

(a) Water from above anoxic zone, or in the case of Red Sea above hot brine.
(b) No separation of ethane and ethylene attempted in these cases, due to high concentrations of ethane.
(c) Trace amounts (10^{-8}ml/l).

then raised to 100°C, and the trapped gases are swept into the gas chromatograph by helium carrier, where they are separated on a suitable column and quantitatively measured with a hydrogen flame-ionization detector. The sensitivity of this method for the C_1–C_4 hydrocarbons is approximately 2×10^{-12} moles of gas, which corresponds to about 10^{-8}ml/liter based on a one-liter sample of sea water.

Table 1 presents some representative data on the occurrence of several hydro-carbons. This data was selected to illustrate the differences between anoxic bottom waters and the oxygenated layers directly above. For comparison, some data from Lake Nitinat, another anoxic body of water, are included.* Anoxic conditions develop in this lake below about 30m, in the Black Sea below about 200m, in the Cariaco Trench below about 350m and in the Atlantis II area in the Red Sea below about 2,000m.

With the exception of Lake Nitinat, the concentration of methane in the oxygenated waters is approximately 5×10^{-5}ml/l, a value which is fairly typical of the open ocean waters in the Atlantic as well. The higher value found near the surface of the lake is very likely due to diffusion of

* Samples from Lake Nitinat were obtained through the courtesy of Professor F. Richards of the University of Washington.

methane from the anoxic waters, which begin very close to the surface. Concentrations of methane greater by several orders of magnitude are found in the anoxic regions. This increase in concentration does not begin to appear until the level of zero oxygen concentration is reached. The concentrations of the higher saturated hydro-carbons in general show the same increase in the anoxic waters although in all cases their concentrations are much less than methane.

It seems clear at this point that, except for the Red Sea, the mechanism of produc-tion in the anoxic waters involves the anaerobic decomposition of organic matter. In the case of the Red Sea, the situation is less clear. Several facts may be noted: (a) in the hot brine from Atlantis II, the meth-ane/ethane and methane/propane ratios are considerably less than in the other anoxic waters. The methane/ethane ratio is, how-ever, greater than the values reported as being typical of gases of petroleum origin, i.e., 5–10. (b) In all the anoxic areas ex-cept the Red Sea hot spot, noticeable con-centrations of H_2S (in some cases, very marked) were observed. No H_2S could be detected in the Red Sea samples. From these facts, as well as some of the other observations reported, it is fair to say that the mechanism of hydrocarbon production in the case of the hot brine areas can hardly be the anaerobic decomposition of organic matter, as seems to be the case in the other anoxic waters. Our lack of adequate data at the present time, however, does not invite further speculation.

It may be of interest to note that in the upper layers of the open ocean an entirely different mechanism of hydrocarbon production appears to exist. We note marked concentration peaks in these waters, for example, which appear to exhibit a diurnal effect. This suggests a possible correlation with biological processes. Furthermore, the unsaturated hydrocarbons are noticeably higher in concentration than the saturated whereas in anoxic waters the reverse is true. The methane at the surface appears to be in equilibrium with the partial pressure of methane in the atmosphere.

One further point should be emphasized. Our oxygen determinations on the Atlantis II samples were obtained by Winkler methods and show zero oxygen concentration.

Elsewhere, however, traces of oxygen have been reported.

We were, unfortunately, rather restricted in our Red Sea sampling, and were unable to get as many samples as we would have liked. Certainly more information is needed before a clear picture of the situation can be formulated.

References

Swinnerton, J. W. and V. J. Linnenbom: Gaseous hydrocarbons in sea water: determination. Science, **156,** 1119 (1967a).

———, ———: Determination of the C_1 to C_4 hydrocarbons in sea water by gas chromatography. J. Gas Chromatography, **5,** 570 (1967b).

———, ———, and C. H. Cheek: Determination of dissolved gases in aqueous solutions by gas chromatography. Anal. Chem., **34,** 483 (1962a).

———, ———, ———: Revised sampling procedure for determination of dissolved gases in solution by gas chromatography. Anal. Chem., **34,** 1509 (1962b).

Dissolved Argon, Nitrogen and Total Carbonate in the Red Sea Brines

R. F. WEISS

Scripps Institution of Oceanography
La Jolla, California

Abstract

Concentrations of argon, nitrogen, and total carbonate in the Atlantis II and Discovery Red Sea brines have been measured by gas chromatography. The isotopic composition of the carbon in the total carbonate of both brines has also been measured. Nitrogen and argon are supersaturated by about 250 per cent relative to their solubilities in the brine in equilibrium with the atmosphere. The argon and nitrogen concentrations in the two brines differ by about 10 per cent and 3 per cent, respectively. Large differences were found in total carbonate concentration and in carbon isotopic composition. The possibility that the brines were formed by evaporation, either during some early stage or in coastal lagoons, is discounted by the dissolved gas results. The difference in argon content of the two brines is assumed due to radiogenic enrichment. This, and other observed differences between the brines, can be explained by a mechanism of independent origin or by two mechanisms of overflow from the Atlantis II Deep into the Discovery Deep. The argon and nitrogen data are consistent with an origin in which the brines were formed from high temperature Red Sea water with the addition of biogenic nitrogen. Water of sufficiently high temperature to be thus considered as a source of the brines can be found only in shallow coastal areas and in the surface and near-surface waters at the southern end of the Red Sea.

Introduction

Since the discovery of two remarkable pools of hot salty water at the bottom of the Red Sea (Miller, 1964; Swallow and Crease, 1965), relatively little attention has been paid to the use of dissolved gas concentration measurements to trace the origin of the brines. Brewer *et al.* (1965), struck by the effervescence of the Discovery brine when brought to the surface, collected a sample in a Squibbs type funnel for later laboratory analysis. Their results showed an argon to nitrogen ratio remarkably similar to that in air. The total carbonate content of 2.9cc CO_2 STP/kg brine was surprisingly low. When held at room temperature and atmospheric pressure, the sample evolved 21cc STP of gas per kilogram of brine. The authors did not speculate on the bearing of these very interesting results on possible mechanisms of brine origin. Prior to the present work there have been no direct measurements of dissolved gas concentrations in the Atlantis II brine, which also effervesces when brought to the surface. However, an estimate of the total carbonate content in the Atlantis II brine was obtained by weighing the carbonate precipitated for carbon-14 analysis from a large volume water sample collected on **R/V CHAIN** cruise 61 (A. Jokela and H. Craig, personal communication 1966). The number thus obtained was approximately 30cc CO_2 STP/kg, ten times greater than that observed in the Discovery brine. The present paper is intended to answer questions raised by earlier observations by presenting the first results of a continuing study of dissolved gas concentrations in both the Atlantis II and Discovery brines.

Sample Collection

Samples were collected on board the Woods Hole Oceanographic Institution **R/V CHAIN** cruise 61 during October, 1966. Samples of approximately 21cc were obtained using weight-operated Piggyback samplers (Weiss, 1968). This sampler, which attaches to the side of a Nansen bottle, obtains samples sealed in a length of copper tubing by two pinch-off clamps without exposure to atmospheric contamination. Sample C61-G1 was taken in the Atlantis II Deep (hydrographic station no. 712) at a depth of 2,115m, 51m above the bottom. The temperature was 56°C and the salinity 257g/kg (Brewer *et al.*, 1969). The total depth of the 56°C brine layer at this point was 126m. Sample C61-G9 was taken in the Discovery Deep (hydrographic station no. 717) at a depth of 2,151m, also 51m above the bottom. This is the approximate depth of the temperature maximum which lies above the bottom layer of decreasing temperature (Pugh, 1967). The temperature was 44°C and the salinity 256g/kg (Brewer *et al.*, 1969). The total depth of the 44°C brine layer at this point was 164m.

Extraction and Analysis

The dissolved gases were extracted in vacuum for gas chromatographic analysis and mass spectrometric determination of the C^{13}/C^{12} ratio in the total carbonate. The collection tube was oriented vertically and connected to a 125ml, flat-bottomed, pyrex extraction flask. A magnetic stirring bar and 0.5cc of concentrated H_3PO_4 to facilitate CO_2 extraction were added to the flask, which was connected to the vacuum line through a stopcock. Degassing of the acid was achieved by heating and stirring with the stopcock closed, followed by freezing with a dry ice bath and pumping away of the residual gas. This cycle was repeated until no residual gas was detected by a thermocouple vacuum gauge on the vacuum line.

With the stopcock closed, the lower clamp on the sample tube was removed and the pinch seal re-rounded. Both samples were clear and colorless as they flowed into the flask, a positive indication that they had been effectively sealed against contact with air oxygen which would produce the characteristic brown precipitate. In both brine samples evaporation in the evacuated flask caused a white precipitate to form. Each sample was degassed using the same method used to degas the acid, except that the gas was transferred by a Toepler pump through a U-tube chilled by a dry ice bath (to remove water vapor) and into a small U-tube system fitted with stopcocks for introduction into the gas chromatograph. The sample tube was also heated to drive droplets of waters into the chilled flask. Although no residual gas was observed after three boiling and freezing cycles, a total of five cycles was taken to ensure complete extraction.

Because of the possibility that some of the carbonate might have precipitated on the walls of the sample tube or reacted with the tube to form copper carbonate, it was necessary to perform a second extraction on each sample. With the stopcock closed, the assembly composed of the sample tube, flask and stopcock was removed from the vacuum line and turned upside down several times to flush the sample tube with acidified water. The assembly was then returned to the vacuum line and re-extracted. This gas was analyzed separately, showing a yield of approximately 23 per cent of the total carbonate in the Discovery sample and showing no measurable yield in the Atlantis II sample. A third extraction of the Discovery sample also gave no yield. The two CO_2 aliquots from the Discovery sample were later combined for the mass-spectrometric measurements.

Once the extractions were completed, the sample tubes were washed, dried and weighed together with their pinch-off clamps. Sample weight was determined to ±0.1 per cent by subtraction from the weights prior to extraction.

The gas chromatographic analysis of Ar, N_2 and CO_2 was performed in hydrogen carrier gas flowing at 30cc/min. Because of the difficulty of separating O_2 and Ar at other than cryogenic temperatures, any oxygen present was quantitatively converted to water in the presence of the hydrogen carrier by passage through a 7cm length of $1/8$ inch OD Teflon tubing packed with powdered palladium metal (Swinnerton *et al.*, 1964), followed by a buffer column of Porapak Q also packed in Teflon tubing. The separation of N_2 and Ar from CO_2, as well as the absorption of any water produced by oxygen consumption, was performed by a silica gel column and Ar was separated from N_2 by a Molecular Sieve 5A column. All columns were operated at 21°C. Because CO_2 is irreversibly absorbed by Molecular Sieve 5A at this temperature, the last column was fitted with a bypass valve to permit the CO_2 to pass into the detector after the analysis of Ar and N_2. The detector was a Carle Instruments thermistor micro-katharometer. A Disc integrator was used to determine peak areas.

In order to measure the C^{13}/C^{12} ratio in the total carbonate, the effluent CO_2 was trapped after chromatographic analysis by passing the carrier gas through a spiral glass trap immersed in liquid N_2. The trap was fitted with a fine, sintered-glass frit to catch suspended crystals of condensate and was connected to the vacuum line so that the CO_2 sample could be transferred to a glass sample tube for mass-spectrometric analysis. The efficiency of the trap was checked by cycling a test CO_2 sample through the chromatograph into the trap and then back into the chromatograph by means of the vacuum line and Toepler pump. After three cycles through the chromatograph and trap, sample loss was less than the 0.2 per cent reproducibility of the chromatographic measurement.

The gas chromatograph was calibrated for Ar and N_2 by analyzing measured amounts of dry air and for CO_2 by analyzing equal amounts of dry air and CO_2. A wide range of calibration sample sizes was analyzed to make certain that variations in sensitivity with sample size were not significant. Standard deviations for replicate analyses were 0.3 per cent of the absolute amount for Ar and 0.2 per cent for N_2 and CO_2.

Results and Discussion

Concentrations, C^{13}/C^{12} Ratios, and Solubilities

The analytical results are listed in Table 1. Argon and nitrogen concentrations in the Discovery brine are higher than in the Atlantis II brine by 10 per cent and 3 per cent, respectively. The Ar/N_2 ratios of 0.0161 in the Atlantis II brine and 0.0172 in the Discovery brine lie between the ratios of 0.0120 in air and approximately 0.028 in natural waters. The most marked differences between the two brines are in the total carbonate values; the Atlantis II brine is 4.8 times as concentrated as the Discovery brine and the Discovery brine C^{13}/C^{12} ratio is 11.2 per mill lower than in the Atlantis II brine.

The C^{13}/C^{12} ratios for the two brines reported here disagree with the results of Deuser and Degens (1969). In the Atlantis II brine, present results show a δ-value of −5.6 per mill (relative to the PDB-1 isotopic standard) while their results range between −3.1 and −4.7 per mill. Although their samples were drained from Nansen bottles into glass bottles with screw caps, and thereby may have lost CO_2 to the atmosphere, they conclude that there was no isotopic fractionation in this process because a sample taken in a Nisken sampler to which NaOH had been added prior to sampling gave the same isotopic results as the other samples. However, the 1.6 per mill spread in their results makes this test relatively insensitive to the 1.5 per mill difference between the present value of −5.6 per mill and their average value of −4.1 per mill. Even if the brines are undersaturated in CO_2 as Deuser and Degens suggest, the effervescence of other supersaturated gases would strip out dissolved CO_2. During the expedition, brine samples which had just been brought to the surface were always observed to effervesce when exposed to

Table 1   Gas Composition of the Brines

	N_2 cc STP/ kg Brine	Ar cc STP/ kg Brine	Total Carbonate cc CO_2 STP/ kg Brine	ppm as HCO_3^-	δ(C^{13}/C^{12})*
Atlantis II brine sample C61-G1	10.11	0.1629	52.40	142.7	−5.6‰
Discovery brine sample C61-G9	10.42	0.1791	11.17	30.41	−16.8‰

* Carbon isotopic composition of total carbonate reported in per mill δ-values relative to the PDB-1 isotopic standard.

atmospheric pressure. Craig (1963; personal communication) has found that similar removal of CO_2 in hot spring areas can increase the C^{13}/C^{12} ratio in the CO_2 left behind by up to 10 per mill. Unfortunately, it is impossible to determine if this mechanism is responsible for the observed differences and scatter since CO_2-concentration data is not available for their samples.

In the Discovery brine, the large difference between the present value of -16.8 per mill and Deuser and Degens' value of -5.8 per mill is also difficult to discuss without matching concentration data. Although according to present results the Atlantis II brine is 4.8 times as concentrated in CO_2 as the Discovery brine, Deuser and Degens (personal communication) did not notice any difference between the amount of CO_2 gas extracted from their Discovery brine sample and the amounts extracted from their Atlantis II brine samples. The observation of low CO_2 concentration in the Discovery brine by Brewer *et al.* (1965) is in qualitative agreement with the present results.

In Fig. 1, argon concentration is plotted against nitrogen concentration. All concentrations in this plot are expressed as cc STP/1000g H_2O, and are thus not affected by the dissolving or precipitation of salt. Analytical results were converted to these units using the value of 742.5 grams of H_2O per kg of brine.

Solubility values for fresh water and for 40.6 per mille salinity sea water (Red Sea Deep Water), in equilibrium with air at a total pressure of 1 atmosphere including the saturation vapor pressure of H_2O, are also plotted in Fig. 1. The solubility coefficients used are from Douglas (1964; 1965) after computer fitting of his data to polynomials in temperature and salinity. The computer work also showed that the Setchénow relationship is applicable to the salting out of argon and nitrogen in seawater over the salinity range 0 to 40 per mill; that is, that at constant temperature the logarithm of the solubility coefficient is a linear function of the salt concentration. Using this relationship the N_2 and Ar solubilities were extrapolated to the salinity

of the Red Sea brines. These values, which must be considered as only approximate, are plotted in the lower-left corner of Fig. 1. The Ar/N_2 ratio in air is also plotted.

Although the N_2 and Ar concentrations of the two brines are relatively close, it is important to note that the observed differences cannot be explained as due to either bulk loss from the Atlantis II sample or air contamination of the Discovery sample. In Fig. 1, bulk sample loss would move a point along a straight line toward the origin, while air addition would move a point away from the origin along a line with a slope equal to the Ar/N_2 ratio in air. If error limits are taken as twice the standard deviations and enough sample had been lost or air had been added to account for the observed difference in N_2 values, then the true Ar concentration in the Discovery brine would still have been higher than in the Atlantis II brine by at least 4.1 per cent and 5.5 per cent, respectively.

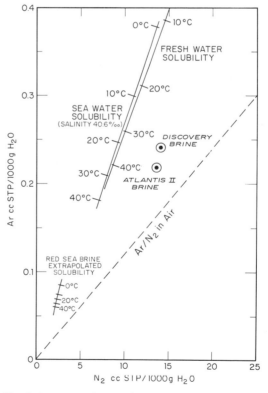

Fig. 1. Argon *vs.* nitrogen in the Red Sea brines and solubility values for various natural waters in equilibrium with the atmosphere.

Enrichment and Depletion of Nitrogen and Argon

Fig. 1 shows that the actual N_2 and Ar concentrations in the brines are supersaturated approximately 250 per cent with respect to the brines in equilibrium with air at 20°C. Although this difference could be produced by addition to the brine of radiogenic argon and large amounts of air, the existence of such a natural process is extremely improbable. Thus, the dissolved gas data add support to the isotopic evidence (Craig, 1966), that the brine was not formed by extensive evaporation during some early stage (Emery, 1965) or by density flow to depth after evaporation in coastal lagoons (Charnock, 1964).

Although a direct measurement of dissolved Ar and N_2 concentrations in normal Red Sea water has not yet been made, a number of workers (König et al., 1964; Linnenbom et al., 1965; Craig et al., 1967) have shown that the concentrations of both these gases in seawater are within 4 per cent of equilibrium saturation concentrations at the potential temperature of the water. Therefore, if the Red Sea brines are made from Red Sea water as indicated by the isotopic evidence, a mechanism of Ar depletion and/or N_2 enrichment is necessary. The fractionation between gas and liquid phases which has been used to explain observations in various geothermal areas (Hulston and McCabe, 1962; Mazor and Wasserburg, 1965; Gunter and Musgrave, 1966) cannot be used here because such a process cannot enrich one gas while depleting the other. In view of the hydrostatic pressure gradient and the geothermal gradient estimated by Craig (1966), the formation of a gas phase appears unlikely, and without the presence of a gas phase there remains no reasonable mechanism for the depletion of argon.

Production of nitrogen has been detected in flowing ground waters of Japan (Sugawara and Tochikubo, 1955). The measured nitrogen supersaturation of up to 70 per cent was associated with depleted oxygen and approximately saturated argon concentration. The source of the nitrogen was attributed to the decomposition of organic matter in the sediment through which the water had passed. Therefore it is reasonable to assume that nitrogen in the brines has been enriched by the decomposition of organic matter which was either present in the sediments through which the water flowed, or settled into the brine pools from the water above.

Differences between the Discovery and Atlantis II Brines

Because the isotopic evidence indicates that both brines originated from the same water, it appears likely that the difference in argon concentrations of the two brines is due to the addition of radiogenic argon. This difference and the observed differences in temperature, total carbonate concentration, carbon isotopic ratio and concentrations of trace metals, are not associated with the salinity difference which must be present if the Discovery brine is to be explained as a product of mixing between the Atlantis II brine and any of the surrounding waters. Thus, there remain three mechanisms or combinations thereof which could explain the observed differences between the two brines:

1. The two brines follow independent paths during the last stages of their formation, allowing additional radiogenic argon input to the Discovery brine and allowing the brines to achieve different temperatures and to undergo reactions to produce their chemical differences.

2. The Discovery brine was produced by an early episodic overflow of the Atlantis II brine (Pugh, 1967), thereby allowing the Discovery brine to represent an earlier discharge from the system while the Atlantis II brine is continually or periodically renewed and mixed into the overlying water column. Radiogenic argon which had accumulated in the sediments through which the brine passes would probably have been preferentially removed by the first waters to pass through the

system. Similarly, the first waters to be discharged by the system would be lower in temperature since they would lose considerably more heat to their as yet unheated surroundings. Thus one could reasonably expect an older discharge to have a higher radiogenic argon concentration and a lower temperature; a result which agrees extremely well with the proposed mechanism and with the observations. After its formation the Discovery brine would be allowed time to cool by conduction, thus producing the vertical temperature structure measured by Pugh (1967). The observed chemical differences could be explained either as the result of the effect of cooling on the equilibria involved or as being a function of the age of the discharge.

3. The Discovery brine is produced by repeated or continuous overflow in which *only part* of the brine which leaves the Atlantis II Deep flows into the Discovery Deep, the rest mixing into the overlying water column. This process would give results very similar to those of the previously described mechanism since, on the average, the Discovery brine would also represent an earlier discharge of the system than the Atlantis II brine.

Indications on the Source of the Brines

If it is assumed that the brines do not lose argon during formation, one can estimate directly from their Ar/H_2O ratios a range of possible temperatures and salinities for the initial seawater from which the brines were formed. Because the solubility of argon in seawater is a function of both temperature and salinity, any fixed argon solubility defines a temperature-salinity line along which the salting-out effect is compensated by the solubility change with temperature. Thus, the Ar/H_2O ratio of any seawater in equilibrium with the atmosphere uniquely defines a line of possible temperatures and salinities for this water. If the possibility that radiogenic argon has been

added after equilibration is allowed, then this line represents a *minimum* equilibrium temperature for any given salinity.

Since both brines are assumed to have formed from the same water and the difference in their argon content has been attributed to radiogenic enrichment, the minimum temperature-salinity line defined by the Atlantis II brines Ar/H_2O ratio must be taken as representing the initial water. This line runs from 29°C to 27°C over the 36 per mill to 41 per mill salinity range present in the Red Sea. Minimum temperatures determined from this line are subject to an error of ±3°C. Of this error, ±2.5° is due to the assumption that the initial water might deviate as much as 4 per cent from its equilibrium saturation argon concentration, and ±0.5° is due to analytical errors.

Although several factors in this estimate are tentative, it is interesting to note that the *only* areas of the Red Sea where waters warm enough and saline enough to lie on or above the minimum temperature-salinity line are found are in shallow coastal areas and in the surface or near-surface waters of the southern portion (Sverdrup *et al.*, 1942). When the minimum temperature-salinity line is superimposed on the temperature-salinity diagram for the Red Sea system (Neumann and McGill, 1962), the only waters which it crosses lie between 28.2°C, 37.8 per mill salinity, and 27.2°C, 39.7 per mill salinity, although a wider range of surface and near-surface waters must be included when error limits are taken into account. As is also shown by the isotopic data, the brines could clearly not have formed from Red Sea Deep Water (22°C, 40.6 per mill salinity).

On the basis of its D/H and O^{18}/O^{16} isotopic ratios Craig (1966) has fixed the salinity of the initial seawater forming the brines at 38.2 per mill. The intersection of this salinity value with the minimum temperature-salinity line gives a minimum temperature of 28°C ± 3°. It is most striking that when the latter temperature and salinity values are compared to the Red Sea system temperature-salinity diagram, the point they define falls squarely in the middle of that portion of the diagram which uniquely defines the surface and near-surface waters at the southern end of the Red Sea.

Future Measurements

Further work on the isotopic composition of the dissolved argon and nitrogen should be of great interest since the fraction of radiogenic argon can be calculated from the Ar^{40}/Ar^{36} ratio and the fraction of nitrogen of biogenic origin can be calculated from the N^{15}/N^{14} ratio (Richards and Benson, 1961; Benson and Parker, 1961). Once this is done, the assumption that no gas-liquid fractionation has occurred can be tested by the measurement of dissolved neon. The measurement of dissolved helium would also be of great interest because of the possibility of radiogenic enrichment.

Acknowledgments

I thank Professor H. Craig for his critical discussion and reading of the manuscript and for making the mass-spectrometric measurements; Dr. J. M. Hunt for his invitation to participate in the expedition; Dr. E. T. Degens and Dr. D. A. Ross for their invaluable assistance during the expedition; and C. D. Densmore, R. G. Munns and R. J. Stanley for their hydrographic work under less than comfortable conditions. This work was supported by NSF grant GA-666.

References

Benson, B. and P. D. M. Parker: Nitrogen/argon and nitrogen isotope ratios in aerobic seawater. Deep-Sea Res., 7, 237 (1961).

Brewer, P. G., C. D. Densmore, R. Munns, and R. J. Stanley: Hydrography of the Red Sea brines. In: Hot brines and recent heavy metal deposits in the Red Sea, E. T. Degens and D. A. Ross (eds.). Springer-Verlag New York Inc., 138–147 (1969).

Brewer, P. G., J. P. Riley, and F. Culkin: The chemical composition of the hot salty water from the bottom of the Red Sea. Deep-Sea Res., 12, 497 (1965).

Charnock, H.: Anomalous bottom water in the Red Sea. Nature, 203, 591 (1964).

Craig, H.: The isotopic geochemistry of water and carbon in geothermal areas. In: Nuclear Geology on Geothermal Areas, E. Tongiorgi (ed.). Consiglio Nazionale delle Ricerche Rome, 17 (1963).

——: Isotopic composition and origin of the Red Sea and Salton Sea geothermal brines. Science, 154, 1544 (1966).

——, R. F. Weiss, and W. B. Clarke: Dissolved

gases in the equatorial and south Pacific Ocean. J. Geophys. Res., 72, 6165 (1967).

Deuser, W. G. and E. T. Degens: O^{18}/O^{16} and C^{13}/C^{12} ratios of fossils from the hot brine deep area of the central Red Sea. In: Hot brines and recent heavy metal deposits in the Red Sea, E. T. Degens and D. A. Ross (eds.). Springer-Verlag New York Inc., 336–347 (1969).

Douglas, E.: Solubilities of oxygen, argon, and nitrogen in distilled water. J. Phys. Chem., 68, 169 (1964).

——: Solubilities of argon and nitrogen in sea water. J. Phys. Chem., 69, 2608 (1965).

Emery, K. O.: (1965) cited in Miller et al. (1966).

Gunter, B. D. and B. C. Musgrave: Gas chromatographic measurements of hydrothermal emanations at Yellowstone National Park. Geochim. Cosmochim. Acta, 30, 1175 (1966).

Hulston, J. R. and W. J. McCabe: Mass spectrometer measurements in the thermal areas of New Zealand. Part 1. Carbon dioxide and residual gas analysis. Geochim. Cosmochim. Acta, 26, 383 (1962).

König, H., H. Wänke, G. S. Bien, N. W. Rakestraw, and H. E. Suess: Helium, neon, and argon in the oceans. Deep-Sea Res., 11, 243 (1964).

Linnenbom, V. J., J. W. Swinnerton, and C. H. Cheek: Evaluation of gas chromatography for the determination of dissolved gases in sea water. Oceanog. Sci. Eng., Trans. Joint Conf. Marine Tech. Soc., Am. Soc. Oceanog., 1009 (1965).

Mazor, E. and G. J. Wasserburg: Helium, neon, argon, krypton and xenon in gas emanations from Yellowstone and Lassen Volcanic National Parks. Geochim. Cosmochim. Acta, 29, 443 (1965).

Miller, A. R.: Highest salinity in the world oceans. Nature, 203, 590 (1964).

——, C. D. Densmore, E. T. Degens, J. C. Hathaway, F. T. Manheim, P. F. McFarlin, R. Pocklington, and A. Jokela: Hot brines and recent iron deposits in deeps of the Red Sea. Geochim. Cosmochim. Acta, 30, 341 (1966).

Neumann, A. C. and D. A. McGill: Circulation of the Red Sea in early summer. Deep-Sea Res., 8, 223 (1962).

Pugh, D. T.: Origin of hot brines in the Red Sea. Nature, 214, 1003 (1967).

Richards, F. A. and B. B. Benson: Nitrogen/argon and nitrogen isotope ratios in two anaerobic environments. Deep-Sea Res., 7, 254 (1961).

Sugawara, K. and I. Tochikubo: The determination of argon in natural waters with special reference to the metabolisms of oxygen and nitrogen. Jour. Earth Sci. Nagoya Univ., 3, 77 (1955).

Sverdrup, H. U., M. W. Johnson, and R. H. Fleming: The Oceans, Prentice-Hall (1942).

Swallow, J. C. and J. Crease: Hot salty water at the bottom of the Red Sea. Nature, 205, 165 (1965).

Swinnerton, J. W., V. J. Linnenbom, and C. H. Cheek: Determination of argon and oxygen by gas chromatography. Anal. Chem., 36, 1669 (1964).

Weiss, R. F.: Piggyback water sampler for dissolved gas studies on sealed water samples. Deep-Sea Res., 15, 695 (1968).

Organisms

Bacterial Sulfate Reduction in the Red Sea Hot Brines *

HANS G. TRÜPER †

Woods Hole Oceanographic Institution
Woods Hole, Massachusetts

Abstract

Fresh brine samples and mud samples from cores were taken under sterile conditions. Enrichment media for sulfate-reducing and nitrate-reducing bacteria were inoculated with these samples and incubated at room temperature and the respective brine temperature. No positive development was found in cultures inoculated from brines or core muds from the brine area. Only one of five cores from outside the brine area contained sulfate-reducing bacteria, of which two strains were isolated. Water samples from the brine/seawater transition zone in the Atlantis II Deep also resulted in positive enrichments, of which three strains of sulfate-reducing bacteria were isolated. No nitrate-reducing bacteria could be enriched from any of the samples. The isolated *Desulfovibrio* strains were characterized. Their temperature optima were 35°–40°C. An experiment to train ten different strains of sulfate-reducing bacteria isolated from various marine environments to grow at elevated temperatures and salinities showed that only the sulfate-reducing bacteria from the Atlantis II Deep brine/seawater transition zone grew well at 10 per cent NaCl/44°C and ½ concentrated brine/35°C. The probable reasons responsible for the absence of bacterial sulfate reduction in the hot brines themselves and in the brine deep sediments are discussed.

Introduction

Dissimilatory sulfate-reducing bacteria are probably one of the oldest forms of bacterial life on earth. Their activities have been traced back for more than 3×10^9 years (Peck, 1966) by the rate of sulfur isotope fractionation in minerals and rocks. In contrast to most other bacteria and plants, which require sulfate only in amounts necessary for the formation of their amino acids and other sulfur containing cellular components (assimilatory sulfate reduction), dissimilatory sulfate-reducing bacteria utilize sulfate mainly as the terminal electron acceptor in their anaerobic respiration. They accumulate large amounts of sulfide, thus participating in the production and transformation of mineral deposits in nature. The basic enzymatic differences between assimilatory and dissimilatory sulfate reduction in bacteria have been clarified by Peck (1961).

Taxonomically, at present the dissimilatory sulfate-reducing bacteria comprise the two genera *Desulfovibrio* and *Desulfotomaculum* with altogether eight species (Campbell and Postgate, 1965, Postgate and Campbell, 1966). So far, only representatives of the five non-sporulating *Desulfovibrio* species have been found in marine or saline environments, while representatives of the three spore-forming *Desulfotomaculum* species seem to be freshwater bacteria.

Nutritional and Environmental Conditions Essential for the Development of Sulfate-reducing Bacteria

Sulfate is required as the terminal electron acceptor. Sulfite, sulfur or thiosulfate may replace sulfate.

* Woods Hole Oceanographic Institution Contribution No. 2189.

† Present address: Institut für Mikrobiologie, Universität Göttingen, Germany.

As the hydrogen (electron) donor, either molecular hydrogen or a variety of organic carbon compounds, such as organic acids (lactate, malate, etc.), alcohols (ethanol, iso-butanol, etc.), sugars or amino acids, are required. *Desulfovibrio* was widely believed to be facultatively autotrophic, *i.e.*, under special conditions (H_2 as the hydrogen donor) to be able to grow at the expense of carbon dioxide as the sole carbon source. Mechalas and Rittenberg (1960) and Postgate (1960) showed, however, that only 25 per cent of the cell carbon could be built up from CO_2 while the major part of cell carbon was derived from impurities in the usual laboratory chemicals. Thus, true autotrophy cannot be claimed for these micro-organisms.

Besides lactate as the favorable carbon source, usually yeast extract is added to the media for sulfate-reducing bacteria. The yeast extract provides a chelating effect on heavy metal ions and several amino acids and vitamins (Kadota and Miyoshi, 1960, 1963; Macpherson and Miller, 1963).

Although ammonium salts are normally the source of nitrogen for sulfate-reducing bacteria, some strains are able to fix atmospheric nitrogen (Sisler and ZoBell, 1951; Le Gall *et al.*, 1959). The following inorganic ions have been reported to be essential for the growth of marine *Desulfovibrio*: Na^+, K^+, Mg^{2+}, SO_4^{2-}, Cl^-, CO_3^{2-} and $H_2PO_4^-$ but not Ca^{2+} or Br^- (Hata, 1960). The most important cation is the absolutely required Fe^{2+} (Butlin *et al.*, 1949), the optimum concentration of 10–15 μg-atoms per liter (Postgate, 1956), *i.e.*, 0.56–0.84mg Fe^{2+}/liter. Since they are strictly anaerobic, sulfate-reducing bacteria need anoxic conditions with an Eh of $+100$ to -400mV (depending on the pH) with optimum development between -100 and -200mV (Baas Becking and Wood, 1955). They have been found and grown at pH values between 4.5 and 9.8, the optimum at 7.5.

The occurrence of sulfate-reducing bacteria has been reported from non-saline environments as well as from those with all salinity degrees up to saturation (Saslawsky, 1928; Hof, 1935; Rubentschick, 1946; Baas Becking and Wood, 1955; Baas Beck-

ing and Kaplan, 1956; ZoBell, 1957). Only the species *Desulfovibrio salexigens* and *Desulfovibrio desulfuricans* var. *aestuarii* are absolutely dependent on the presence of sodium chloride in their medium. While the former needs the chloride ion, the latter requires the sodium ion (Postgate and Campbell, 1966). According to ZoBell (1957) an extreme halophilic sulfate-reducing bacterium, *i.e.*, with optimum growth at salinities of 15 to 30 per cent, has yet to be found. The presence in natural environments with high salinities does not imply optimum growth at these conditions.

Only the spore-forming *Desulfotomaculum nigrificans* is an obligate thermophile between the so far isolated sulfate reducers. *Desulfovibrio* strains from marine sediments showed good growth between 20° and 45°C but not above 45°C (Rittenberg, 1941; Kimata *et al.*, 1955). Sulfate reducers from oil well waters grew at temperatures up to 50°C but not at 60°C (Gahl and Anderson, 1928). ZoBell (1957) found that sulfate-reducing bacteria isolated from oil and sulfur wells of 2,000–4,000m depth grew at 65°–85°C under a hydrostatic pressure of 200–400 atmospheres. Finally, growth was observed in a culture at 104°C under a pressure of 1,000 atmospheres (ZoBell and Johnson, 1949; Johnson, 1957). ZoBell and Morita (1957) isolated marine barophilic sulfate-reducing bacteria from 7,000 to 10,000m depth and were able to cultivate them for more than seven years at 3°–5°C under a pressure of 700 atmospheres.

Degree of Fulfillment of These Requirements by the Hot Brines

Normal seawater enriched with organic carbon (lactate plus yeast extract), phosphate, NH_4^+ and Fe^{2+} salts will support good growth of sulfate-reducing bacteria. Compared to normal seawater, the proportions of ion concentrations in the brine recovered from the Atlantis II Deep (Miller *et al.*, 1966) were increased for Na 9.74 times, K 9.26 times, Ca 15.67 times, Cl 10.0 times, Br 1.26 times, B 2.77 times, Fe 7.0×10^5 times, Mn 8.6×10^5 times, Zn 600 times, Cu 500 times and Ba 73.3 times. The sulfate concentration of the brine, however, is only 0.42 that of normal seawater. With 11.9mM/liter, the concentration of sulfate in the Atlantis II Deep brine comes close to Starkey's (1938) medium for sulfate reducers with 14.9mM SO_4^{2-}/liter. The lower Mg^{2+} content of the Atlantis II Deep brine (0.73 times, compared to normal seawater) would not have negative effects on sulfate-reducing bacteria (Baas Becking and Wood, 1955). The maximally tolerated amounts of Fe^{2+}, Zn^{2+} and Cu^{2+} were determined by Römer and Schwartz (1965) as 2.4, 0.085 and 0.0025g/liter, respectively. These values are higher than those found for the Red Sea hot brines by Miller *et al.* (1966). Therefore the concentrations of iron, zinc and copper in the brines cannot be considered harmful for sulfate-reducing bacteria.

The dissolved organic carbon material in the Atlantis II Deep brine amounts to 2mg/liter, which most probably has to be considered as refractory material (Watson and Waterbury, 1969). The high density of the brine seems to prevent smaller particles of organic material from sinking to the bottom. At the interface between brines and normal seawater, a layer of organic debris was observed. Since the shells of recent foraminifera and pteropods are found in the top layers of the sediments in the brine deeps, it is possible that organic matter attached to the shells is deposited. Theoretically, this could provide sufficient organic carbon for the development of sulfate-reducing bacteria. At present, no data on the amount of dissolved CO_2 and N_2 as well as on the amount of phosphate, nitrate, nitrite and ammonium ions in the brines are available.

The hot brines of Atlantis II and Discovery Deeps are free of molecular oxygen (Miller *et al.*, 1966; Munns *et al.*, 1967; Brewer *et al.*, 1969). The presence of free H_2S has been reported (Miller *et al.*, 1966) but has not been proven analytically. The high amount of Fe^{2+} could prevent the occurrence of free H_2S in the brines by immediate precipitation as iron sulfides. Direct Eh measurements have not been made. The amount of Fe in the brine from the Atlantis II Deep is about 100 times higher than the optimum concentration for the growth of sulfate-reducing bacteria. Measurements of the pH showed values of 5.0–5.5, thus coming close to acidities not favorable for the growth of sulfate-reducing bacteria. The salinity values of about 15 per cent in the Atlantis II Deep 44°C brine and of about 32 per cent in the Discovery Deep 44°C and the Atlantis II Deep 56°C brines would most probably allow the development of sulfate reducers if their other requirements are sufficiently met.

Methods

To obtain uncontaminated water and brine samples, the MJ-sampler (Jannasch and Maddux, 1967) was used. Since the sealing substance of this sampler, paraffin, was liquefied in the 56°C hot brine of the Atlantis II Deep, Niskin-samplers (Niskin, 1962) were used alternatively. Bottom samples were obtained from the different core types employed by the geologists. Samples were removed from top layers of the cores using sterilized vessels and instruments. The enrichment media were prepared on board **R/V CHAIN**. Baar's medium after Postgate (1965) was used for sulfate-reducing bacteria: 0.5g KH_2PO_4, 1.0g NH_4Cl, 1.0g $CaSO_4$, 2.0g $MgSO_4 \times 7H_2O$, 3.5g Na-lactate and 0.5g $Fe(NH_4)_2(SO_4)_2 \times 6H_2O$ were dissolved in 1,000ml Red Sea surface water or brine from the different localities. The solutions were adjusted to a pH of 7.5, sterilized by autoclaving and then, under sterile conditions, distributed to 60ml screw cap bottles. In addition, enrichment cultures for anaerobic nitrate-reducing bacteria were prepared using the following procedure (Claus, 1965): 3.0g beef extract, 5.0g peptone, 1,000ml seawater or brine, final pH 7.2. In the case of seawater, parallels with 1 per cent KNO_3 and 8 per cent KNO_3 were run; with brine, only medium with 1 per cent KNO_3 was used. Because of possible gas formation, nitrate reducer enrichments were done in glass-stoppered bottles.

The inoculated enrichment cultures were incubated in parallels at 25°C and, according to the origin (44°C or 56°C brine) of the respective samples, at 44°C or 56°C. Thus aliquots of each water or mud sample were used as inocula for the following enrichment culture types: (1) sulfate reducer medium — (a) seawater, 25°C; (b) seawater, 44°C/56°C; (c) brine, 25°C; (d) brine,

44°C/56°C; (2) nitrate reducer medium—(a) seawater, 1 per cent KNO_3, 25°C; (b) seawater, 8 per cent KNO_3, 25°C; (c) seawater, 1 per cent KNO_3, 44°C/56°C; (d) seawater, 8 per cent KNO_3, 44°C/56°C; (e) brine, 1 per cent KNO_3, 25°C; (f) brine, 1 per cent KNO_3, 44°C/56°C.

Besides the shipboard enrichments during **R/V CHAIN** cruise No. 61, October and November 1966, those for sulfate reducers were repeated at Woods Hole Oceanographic Institution. As inocula, aliquots of uncontaminated mud samples and of 10-liter enrichment cultures for anaerobic heterotrophic bacteria in trypticase medium (Watson and Waterbury, 1969) were used.

Pure cultures of sulfate-reducing bacteria were isolated by following the techniques recommended by Postgate (1965). Cell protein was determined with the phenol reagent method after Lowry *et al.* (1951). For the determination of sulfide, the methylene blue method (Pachmayr, 1960; Trüper and Schlegel, 1964) was used, basing on the reaction of S^{2-} with two molecules of dimethyl-para-phenylenediamin and subsequent oxidation with Fe^{3+} to methylene blue, which is measured spectrophotometrically. In preliminary experiments, this method had been proven to be useful also at a salinity of 30 per cent. All chemicals used were of analytical grade.

Experiments and Results

Enrichments of Bacteria

Enrichment cultures were inoculated with surface mud of the following cores from the brine deep area: numbers 124, 126, 127, 128, 129, 139A, 139B, 143, 158, 159 and 161 (Ross and Degens, 1969). Water and brine were sampled in connection with hydrocasts no. 724 (Atlantis II Deep, 5 samples between 2,075 and 2,101m depth, all from the 56°C brine), no. 728 (Discovery Deep, 7 samples between 2,177 and 2,237m depth, all from 44°C brine), and no. 721 (Atlantis II Deep, 5 Niskin-samples between 2,005 and 2,050m depth, within the 44°C brine/seawater transition zone). The no. 721 samples were combined in a carboy and trypticase-medium added (Watson and Waterbury, 1969). From outside the brine deep area, *i.e.*, from normal Red Sea water, 5ml samples of the Niskin-sampler cast no. 149

(20° 51.5′N, 38° 13.5′E) as well as 5-g samples from the following cores were used as inocula: cores no. 136, 152, 153, 156A and 165.

None of the Atlantis II Deep 56°C samples nor of the Discovery Deep 44°C samples gave positive enrichments under any of the employed enrichment conditions. No sulfate-reducing nor nitrate-reducing bacteria were found. From the combined samples of Niskin-sampler cast no. 721 from the 44°C brine/seawater transition zone above the Atlantis II Deep 56°C brine, three strains of sulfate-reducing bacteria (nos. 88-3, 88-4 and 88-5) were isolated. From outside the brine area, only mud samples from core no. 152 gave positive enrichments for sulfate reducers, and two strains were isolated from these enrichments (nos. 88-1 and 88-2). A mud sample of core no. 136 contained anaerobic protein decomposing bacteria that grew in Baar's medium, but no sulfate reducers could be isolated from this sample. All other samples gave negative enrichments. All enrichments for nitrate-reducing bacteria were negative.

Sulfide Determinations

Sulfide determinations in the hot brines directly after sampling with MJ- or Niskin-samplers were all negative. Despite former observation (Miller *et al.*, 1966), no free H_2S could be smelled. Obviously, the high iron content of the brines prevents the occurrence of free H_2S or HS^- ions.

Properties of the Isolated Sulfate-reducing Bacteria

The basic morphological and metabolic criteria for the taxonomical determination of the isolated strains are given in Table 1. The absolute NaCl requirement of all strains, together with the absence of a cholin metabolism would be reasonable facts to place the isolated strains in the species *Desulfovibrio salexigens* (Postgate and Campbell, 1966). The inability of the strains to metabolize malate, however, does not agree with the species description for *Desulfovibrio salexigens* but rather with that of the non-salt, requiring species *De-*

Table 1   Morphological and Basic Metabolic Properties of the Isolated Strains

The criteria were determined using the methods recommended by Postgate and Campbell (1966).

Strains	88-1	88-2	88-3	88-4	88-5
Cell morphology:					
(a) shape	vibrio	vibrio	vibrio	vibrio	vibrio
(b) width [μ]	0.4–0.6	0.4–0.5	0.5–0.7	0.5–0.7	0.3–0.7
(c) length [μ]	1.5–2.5	1.5–2.0	1.8–3.0	1.8–3.0	1.2–2.0
Absolute salt requirement	+	+	+	+	+
Metabolic properties:					
(a) growth with sulfate					
+ formate	+	+	+	+	+
+ acetate	–	–	–	–	–
+ lactate	+	+	+	+	+
+ malate	–	–	–	–	–
(b) growth without sulfate					
+ cholin	–	–	–	–	–

sulfovibrio vulgaris. Further studies are in progress in order to determine the taxonomical position of the Red Sea strains. There is no doubt that the isolated strains belong to the genus *Desulfovibrio.*

Using a special incubation block that allows parallel constant temperature settings over a range of 35°C (H. W. Jannasch, unpublished), the optimum growth temperatures as well as the temperature range allowing measurable growth of the strains were determined. Growth was measured as the amounts of cell protein and sulfide formed. As an example, Fig. 1 shows the relationships for *Desulfovibrio* (spec. strain 88-3). The temperature values for all five strains are given in Table 2.

All strains have relatively high optimum temperatures for growth. It is surprising that the optimum temperatures of the strains isolated from outside of the hot brine area are of the same magnitude as those of the strains from the Atlantis II Deep brine/seawater transition zone of about 44°C. Perhaps this type of sulfate-reducing bacteria is the predominant type of the central Red

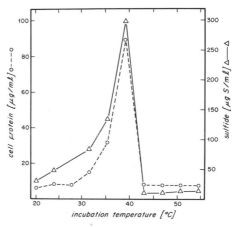

Fig. 1. Determination of the optimal growth temperature of *Desulfovibrio* spec. strain 88-3. Cells were grown in Starkey's medium after Postgate (1965), incubation time: 2 days. Protein was determined after Lowry *et al.* (1951), sulfide after Pachmayr (1960), Trüper and Schlegel (1964).

Sea bottom mud or the area where core no. 152 was taken once had a higher temperature or, finally, the organisms have been transported from the brine area to the location of core no. 152 (19° 48.5′ N, 38° 30.0′ E; 2,362m depth).

Adaptation of Sulfate-reducing Bacteria to Unusual High Salinities and Temperatures

By stepwise transfer to media with higher salinities and temperatures, it was attempted to adapt ten strains of sulfate-reducing bacteria to grow in Atlantis II Deep brine at 56°C. The strains had been

Table 2   Growth Temperatures of the Isolated Strains

Starkey's medium after Postgate (1965) was used.

Strain	Optimal Temperature	Temperature Tolerance
88-1	40.0°C	19°–48°C
88-2	35.0°C	20°–44°C
88-3	39.5°C	20°–42°C
88-4	40.0°C	19°–48°C
88-5	40.0°C	19°–48°C

isolated from various marine environments. In addition to Red Sea strains 88-2 and 88-4 (see above), the following strains were used (original habitats given in parentheses): *Desulfovibrio salexigens* strain 27-4 and *Desulfovibrio desulfuricans* strain 27-6 (both from Sippewissett, Massachusetts, salt marsh mud), *Desulfovibrio* spec. strains 44-1 and 44-2 (Lake Tiberias, Israel, mud from a 25m deep underwater salt spring), *Desulfovibrio desulfuricans* var. *aestuarii* strain 72-6 (continental shelf, U.S. East Coast, mud from 123m depth), *Desulfovibrio desulfuricans* var. *aestuarii* strain 99-2 (Pacific off Peru, mud from 1,900m depth), *Desulfotomaculum*(?) spec. strain 99-7 (Pacific off Peru, pasteurized mud from 4,200m depth), *Desulfovibrio* spec. strain CH-1 (bilge water, **R/V CHAIN**).

The cultures in 60ml screw cap bottles were started at 25°C and 3 per cent NaCl in Starkey's medium (Starkey, 1938; Postgate, 1965). As an indicator for growth, the production of sulfide was measured, which is possible in this kind of experiment since the basic organic composition of the medium is the same at the different salinities employed. Fig. 1 gives an example for the comparability of protein and sulfide. When the sulfide concentration reached constancy, usually after 2–7 days (depending on strain, salinity, and temperature), the strains were transferred into new medium with the next higher salinity and incubated at the same temperature or, vice versa, into medium with the same salinity and incubated at the next higher temperature. As Table 3 shows, different patterns were found for the ten strains. None of the strains grew after transfer to ¾ concentrated brine-medium or at 56°C. While some strains (72-6 and 99-2) show salt tolerance only, others (27-4 and 44-2)

Table 3   Adaptation of Ten Strains of Marine Sulfate-reducing Bacteria to High Salinities and High Temperatures

Starkey's medium after Postgate (1965) was used.

Salinity	25°C		35°C		44°C		56°C
	Strain no.	Sulfide [mM/1]	Strain no.	Sulfide [mM/1]	Strain no.	Sulfide [mM/1]	Strain no.
3% NaCl	27-4	10	27-4	10	44-1	9	
	27-6	9	27-6	8	88-2	2	
	44-1	10	44-1	9	88-4	6	
	44-2	9	44-2	9	CH-1	6	
	72-6	6	88-2	8			none
	88-2	10	88-4	9			
	88-4	10	99-7	5			
	99-2	10	CH-1	8			
	99-7	5					
	CH-1	10					
6% NaCl	27-6	9	44-1	9	88-2	2	
	44-1	8	88-2	8	88-4	6	
	72-6	6	88-4	9	CH-1	2	
	88-2	8	99-7	1			none
	88-4	8	CH-1	8			
	99-2	3					
	99-7	3					
	CH-1	8					
10% NaCl	44-1	8	88-2	5	88-4	8	
	72-6	8	88-4	8			
	88-2	7	CH-1	6			none
	88-4	7					
	CH-1	6					
½ conc. brine 17% NaCl	88-4	5	88-4	5	none	—	none
¾ conc. brine 25% NaCl	none	—	none	—	none	—	none

exclusively adapt to higher temperatures. Four out of ten strains were able to grow at 44°C, five to grow at 10 per cent NaCl. The number of strains able to grow simultaneously at high temperatures and at high salinities was small.

Only *Desulfovibrio* spec., strain 88-4, isolated from water of the Atlantis II Deep brine/seawater transition zone, was able to grow in ½ concentrated brine-medium at 25°C and 35°C as well as in medium with 10 per cent NaCl at 44°C. From the ten strains tested, this strain thus had the highest salt plus temperature tolerance. This fact demonstrates that in the brine/seawater transition zone, where the average salinity is 15 per cent and less and the temperature 44°C and less (Munns *et al.*, 1967; Brewer *et al.*, 1969), bacterial sulfate reduction can take place. The carbon source is most likely the organic debris floating on top of the brine. A quantitative importance of sulfate-reducing bacteria in the transition zone, however, remains questionable, as long as no measurements on their population density have been made.

Discussion

The brines of Atlantis II and Discovery Deeps, as well as the top sediment layers of eleven cores out of the hot brine area, were found to be free of sulfate-reducing bacteria. On first sight the brine deep area may have certain similarities with anoxic basins, such as the Black Sea or the Cariaco Trench (Richards and Vaccaro, 1956). An anoxic water mass is trapped in a basin and covered by normally oxygenated seawater of occasionally less density. The anoxic water contains dissolved hydrogen sulfide which, most probably, is due to bacterial sulfate reduction in the bottom mud, since frequently sulfate-reducing bacteria have been isolated from such areas (Skopintsev *et al.*, 1959). The Atlantis II Deep brine clearly does not fall in this category. It does not contain free measurable sulfide, even when the very sensitive methylene blue method was used for its detection.

Hartmann and Nielsen (1966), in their paper about sulfur isotope fractionation in the hot brine sediment of the Atlantis II Deep, state that "the difference in δ^{34}S between sediment sulfide and interstitial water sulfate might lead to the assumption of sulfide formation from bacterial reduction of sulfate. But this mechanism would lead to a pronounced enrichment in δ^{34}S of the interstitial water sulfate with respect to the overlying free brine. However, both these values were found identical within experimental error of ±0.1 per mill. Thus, sulfate reduction cannot take place in significant amounts in the sediment."

What are the reasons that no sulfate-reducing bacteria were found in the Red Sea hot brine deeps? Several answers are possible:

1. There is not enough organic material present to meet the rather complex requirements of sulfate-reducing bacteria. The amount of dissolved organic matter within the brine is very low (Watson and Waterbury, 1969). Most of the organic material floats on top of the brine.

2. Considering a possible inoculation or pollution of the brine area from outside, the polluting population of sulfate-reducing bacteria would need a rather high rate of simultaneous mutations to adapt to the high salinity, high temperature, low pH and other hostile environmental properties of the brines. Normal mutation rates of bacteria lie in the magnitude of 10^{-4} to 10^{-10} single mutations per generation (Schlegel and Jannasch, 1967). Nothing is known so far about the mutation rate of sulfate-reducing bacteria, nor about their generation times in natural environments.

3. The relatively high calcium content of the brine (6.44g/liter, Miller *et al.*, 1966) together with the high general salinity might act bacteriostatically upon sulfate-reducing bacteria. The calcium content of normal seawater is only 0.411g/liter; that of the enrichment medium used in this study is 0.416g/liter. Kuznetsova and Pantskhava (1962) found that the reason for

the absence of active bacterial sulfate reduction in certain stratal brines from oil wells was their high calcium content. Freshening, *i.e.*, dilution, allowed halophilic sulfate-reducing bacteria to develop.

4. The low number (one out of five) of positive enrichments for sulfate-reducing bacteria from Red Sea core muds outside the hot brine area indicates that the total number of these organisms in the investigated central area of the Red Sea is very low. Whether this is due to low primary production in the Red Sea cannot be said since pertinent data are not yet available.

The more or less combined effects of these four factors provide enough obstacles to hinder bacterial sulfate reduction in the Red Sea hot brines and the brine deep sediments, as well as to counteract a possible pollution of this area by sulfate-reducing bacteria from outside.

Acknowledgments

The author wishes to thank Dr. H. W. Jannasch for his interest and advice in this work. The technical assistance of Mrs. Grace C. Fraser is gratefully acknowledged.

This investigation was supported by the National Science Foundation (Grants GB-5199, GB-6314 and GA-584).

References

Baas Becking, L. G. M. and I. R. Kaplan: The microbiological origin of the sulphur nodules of Lake Eyre. Trans. Roy. Soc. S. Australia, **79**, 52 (1956).

Baas Becking, L. G. M. and E. J. F. Wood: Biological processes in the estuarine environment. I. Ecology of the sulphur cycle. Proc. Kon. Ned. Akad. v. Wet. Amsterdam, **58**, 160 (1955).

Brewer, P. G., C. D. Densmore, R. G. Munns, and R. J. Stanley: Hydrography of the Red Sea brines. *In: Hot brines and recent heavy metal deposits in the Red Sea*, E. T. Degens and D. A. Ross (eds.). Springer-Verlag New York Inc., 138–147 (1969).

Butlin, K. R., M. E. Adams, and M. Thomas: The isolation and cultivation of sulphate-reducing bacteria. J. Gen. Microbiol., **3**, 46 (1949).

Campbell, L. L. and J. R. Postgate: Classification of the spore-forming sulfate-reducing bacteria. Bacteriol. Rev., **29**, 359 (1965).

Claus, D.: Anreicherungen und Direktisolierungen aerober sporenbildender Bakterien. Zentr. Bakt. Parasitenk., I. Abt. Suppl., **1**, 337 (1965).

Gahl, R. and B. Anderson: Sulphate reducing bacteria in California oil waters. Zentr. Bakt. Parasitenk., II. Abt., **73**, 331 (1928).

Hartmann, M. and H. Nielsen: Sulfur isotopes in the hot brine and sediment of Atlantis II Deep (Red Sea). Marine Geol., **4**, 305 (1966).

Hata, Y.: Inorganic nutrition of marine sulfate-reducing bacteria. J. Shimonoseki Coll. Fisheries, **9**, 39 (1960).

Hof, T.: Investigations concerning bacterial life in strong brines. Rec. Trav. Botan. Neerlandais, **32**, 92 (1935).

Jannasch, H. W. and W. S. Maddux: A note on bacteriological sampling in seawater. J. Mar. Res., **25**, 185 (1967).

Johnson, F. H.: The action of pressures and temperature. *In: Microbial Ecology*, 7th Symp. Soc. Gen. Microbiol., R. E. O. Williams and C. C. Spicer (eds.). University Press, Cambridge, 134 (1957).

Kadota, H. and H. Miyoshi: A chemically defined medium for the growth of *Desulfovibrio*. Mem. Res. Inst. Food. Sci., Kyoto Univ., **22**, 20 (1960).

——, ——: Organic factors responsible for the stimulation of growth of *Desulfovibrio desulfuricans*. *In: Marine Microbiology*, C. H. Oppenheimer (ed.). C. C. Thomas, Springfield, Ill., 442 (1963).

Kimata, M., H. Kadota, Y. Hata, and T. Tajima: Studies on the marine sulfate-reducing bacteria. II. Influences of various environmental factors upon the sulfate-reducing activity of marine sulfate-reducing bacteria. Bull. Jap. Soc. Scient. Fish., **21**, 113 (1955).

Kuznetsova, V. A. and E. S. Pantskhava: Effect of freshening of stratal waters on development of halophilic sulfate reducing bacteria. Mikrobiologiya (Russ.), **31**, 129 (1962).

Le Gall, J., J. C. Senez, and F. Pichinoty: Fixation de l'azote par les bactéries sulfato-réductricies. Ann. Inst. Pasteur, **96**, 223 (1959).

Lowry, O. H., N. J. Rosebrough, A. L. Fair, and R. J. Randall: Protein measurement with the Folin phenol reagent. J. Biol. Chem., **193**, 265 (1951).

Macpherson, R. and J. D. A. Miller: Nutritional studies on *Desulfovibrio desulfuricans* using chemically defined media. J. Gen. Microbiol., **31**, 365 (1963).

Mechalas, B. J. and S. C. Rittenberg: Energy coupling in *Desulfovibrio desulfuricans*. J. Bacteriol., **80**, 501 (1960).

Miller, A. R., C. D. Densmore, E. T. Degens, J. C. Hathaway, F. T. Manheim, P. F. McFarlin, R. Pocklington, and A. Jokela: Hot brines and recent iron deposits in deeps of the Red Sea. Geochim. Cosmochim. Acta, **30**, 341 (1966).

Munns, R. G., R. J. Stanley, and C. D. Densmore: Hydrographic observations of the Red Sea brines. Nature, **214**, 1215 (1967).

Niskin, S. J.: A water sampler for microbiological studies. Deep-Sea Res., **9,** 501 (1962).

Pachmayr, F.: Vorkommen und Bestimmung von Schwefelverbindungen in Mineralwasser. Dissertation, Univ. of Munich (1960).

Peck, H. D., Jr.: Enzymatic basis for assimilatory and dissimilatory sulfate reduction. J. Bacteriol., **82,** 933 (1961).

———: Some evolutionary aspects of inorganic sulfur metabolism. Lecture Series on Theoretical and Applied Aspects of Modern Microbiology, Univ. of Maryland, 1 (1966).

Postgate, J. R.: Iron and the synthesis of cytochrome C₃. J. Gen. Microbiol., **15,** 186 (1956).

———: On the autotrophy of *Desulfovibrio desulfuricans.* Z. Allg. Mikrobiol., **1,** 53 (1960).

———: Enrichment and isolation of sulphate-reducing bacteria. Zentr. Bakt. Parasitenk., I. Abt. Suppl., **1,** 190 (1965).

——— and L. L. Campbell: Classification of *Desulfovibrio* species, the nonsporulating sulfate-reducing bacteria. Bacteriol. Rev., **30,** 723 (1966).

Richards, F. A. and R. F. Vaccaro: The Cariaco Trench, an anaerobic basin in the Caribbean Sea. Deep-Sea Res., **3,** 214 (1956).

Rittenberg, S. C.: Studies on marine sulfate-reducing bacteria. Dissertation, Univ. of Calif., Los Angeles, 115 p. (1941).

Römer, R. and W. Schwartz: Geomikrobiologische Untersuchungen. V. Verwertung von Sulfatmineralien und Schwermetall-Toleranz bei Desulfurizierern. Z. Allg. Mikrobiol., **5,** 122 (1965).

Ross, D. A. and E. T. Degens: Shipboard collection of sediment samples collected during **CHAIN** Cruise 61 from the Red Sea. *In: Hot brines and recent heavy metal deposits in the Red Sea,* E. T. Degens and D. A. Ross (eds.). Springer-Verlag New York Inc., 363–367 (1969).

Rubentschick, L. I.: Sulfate reducing bacteria. Mikrobiologiya (Russ.), **15,** 443 (1946).

Saslawsky, A. S.: Zur Frage der Wirkung hoher Salzkonzentrationen auf die biochemischen Prozesse im Limanschlamm. Zentr. Bakt. Parasitenk., II. Abt., **73,** 18 (1928).

Schlegel, H. G. and H. W. Jannasch: Enrichment cultures. Ann. Rev. Microbiol., **21,** 49 (1967).

Sisler, F. D. and C. E. ZoBell: Hydrogen utilization by some marine sulfate-reducing bacteria. J. Bacteriol., **62,** 117 (1951).

Skopintsev, B. A., A. V. Karpov, and O. A. Vershinina: Study of some sulfur compounds in the Black Sea under experimental conditions (in Russian). Trudy Marine Hydrophys. Inst. (MGI), Akad. Nauk SSSR, **16,** 89 (1959).

Starkey, R. L.: A study of spore formation and other morphological characteristics of *Vibrio desulfuricans.* Arch. Mikrobiol., **9,** 268 (1938).

Trüper, H. G. and H. G. Schlegel: Sulphur metabolism in Thiorhodaceae. I. Quantitative measurements on growing cells of *Chromatium okenii.* Antonie v. Leeuwenhoek J. Microbiol. Serol., **30,** 225 (1964).

Watson, S. W. and J. B. Waterbury: The sterile hot brines of the Red Sea. *In: Hot brines and recent heavy metal deposits in the Red Sea,* E. T. Degens and D. A. Ross (eds.). Springer-Verlag New York Inc., 272–281 (1969).

ZoBell, C. E.: Ecology of sulfate reducing bacteria. *In: Sulfate-reducing bacteria, their relation to secondary recovery of oil,* Science Symp. St. Bonaventure Univ., New York, 1 (1957).

——— and F. H. Johnson: The influence of hydrostatic pressure on the growth and viability of terrestrial and marine bacteria. J. Bacteriol., **57,** 179 (1949).

——— and R. Y. Morita: Barophilic bacteria in some deep sea sediments. J. Bacteriol., **73,** 563 (1957).

The Sterile Hot Brines of the Red Sea *

STANLEY W. WATSON AND JOHN B. WATERBURY

Woods Hole Oceanographic Institution
Woods Hole, Massachusetts

Abstract

The brines of the Atlantis II and Discovery Deeps, their sediments and the water overlying these Red Sea brines were examined for the presence or absence of bacteria. No bacteria were found in the Atlantis II sediments and brines, but they were present in the overlying transition and normal 22° waters. The sulfate reducing bacteria in this transition layer may account for the sulfur fractionation found in the metal sulfides in the sediments. In the Discovery Deep, bacteria were present in the sediment but were not demonstrated within the brines. The high temperature, salinity and metal concentrations in the Atlantis II brines are thought to act synergistically, preventing the growth of bacteria. The lower heavy metal concentration in Discovery Deep may make conditions slightly less hostile to life and permit the growth of bacteria in the sediments.

Introduction

The chemistry, geology and physical oceanography of the anaerobic hot brines of the Atlantis II and Discovery Deeps in the Red Sea has been previously described (Brewer et al., 1965, 1969; Charnock, 1964; Miller, 1964; Miller et al., 1966; Munns et al., 1967; Swallow and Crease, 1965). The aim of the present investigation is to determine if bacteria existed in the obdurate environment of these brines.

The maximum depth of the Atlantis II Deep is 2,175m. From this depth up to 2,040m the temperature is 56°, the salinity 258 per mill and the oxygen undetectable. The 56° brine is overlaid by 30m of 44° brine having a salinity of 135 per mill.

Traces of oxygen are present in the 44° water, but these amounts may have been introduced into the sample during analysis. The measured pH of the 56° water was 5.2 and the 44° water, 6.0. A transition layer, extending upward to 1,980m, overlaid the 44° water and separated it from the normal 22° Red Sea deep water. Within this transition zone the oxygen, salinity and temperature varied with depth and were intermediate between those of the 22° and 44° water layers. Because of the technical difficulties the measured pH may be quite different from the in situ pH.

Methods and Materials

Collection of Samples

All water samples used for bacteriological examination in these experiments were collected in 2-liter sterile, plastic sample bags by means of Niskin samplers. The depths and stations from which water was obtained are detailed in Tables 1, 2 and 3. The location of each station is shown in Fig. 1. Sediment samples were obtained by means of piston, free fall, gravity and Kasten coring devices. Only the upper 10cm of such cores were used in these experiments. The stations from which these samples were obtained are listed in Table 4 and the location of these stations is given in Fig. 1.

Treatment of Samples

After a few samples were inoculated into media and no growth of bacteria was observed, it was apparent that either no bacteria were

* Woods Hole Oceanographic Institution Contribution No. 2190.

Table 1 Water Samples from Atlantis II Deep

Depth (m)	Temperature °C	Hydrographic Station	Cultural Conditions (see key)	Type Cultural Conditions Yielding Positive Cultures	Smallest Sample Size in Which Bacteria Were Found (ml)
1,925	22	713	1	1	10^{-5}
1,946		713	1	no growth	—
1,956		713	1	no growth	—
1,958		713, 720	1, 3, 4, 6	1, 4	10^{-1}
1,967		713	1	1	10^{-1}
1,968	22	720	4, 6	4	250
1,977		713	1	1	10^{-1}
1,987		713	1	no growth	—
1,988	22	720	1, 3, 4, 6	1, 4	10^{-1}
1,994 *	24.65	715			
1,997			1	1	10^{-1}
1,998		720	1, 3, 4, 6	1	250
1,999 *	28.8	720			
2,010 *	33.43	713			
2,018		720	1, 3, 4, 6		
2,028		720	1, 3, 4, 6		
2,038	44	720	1, 3, 4, 6		
2,047		721	1, 3, 6		
2,048		720	1, 3, 4, 6		
2,057		721	1, 3, 6		
2,064	56	724	8, 9, 10		
2,065		714	5		
2,067		721	1, 3, 6		
2,070		714	6		
2,071		724	8, 9, 10		
2,075		712, 714	3, 5, 6		
2,077	56	721	1, 3, 6, 7		
2,078		724	8, 9, 10		
2,080		714	3, 5, 6		
2,084			7		
2,085		712, 724	1, 2, 3, 8, 9, 10		
2,087		721	1, 3, 6		
2,090	56	714	3, 6		
2,091		721	7		
2,095		712, 714	1, 2, 3, 5, 6		
2,097		721	7		
2,098		721	7		
2,100		714	3		
2,105		712			
		14	1, 2, 3, 5, 6		
		21			
2,107		721	1, 3, 6		
2,108		719	7		
2,110		714	3, 5, 6		
2,112		721	7		
2,115		714	3, 5, 6		
2,117		721	1, 3, 6		
2,119		721	7		
2,121		721	1, 3, 6		
2,126		721	7		
2,131		721	1, 3, 6		
2,133		721	7		
2,138		719, 721	7		
2,148		719	7		
2,158	56	719	7		

* Indicates sample taken with Nansen bottle; all other samples taken with Niskin sampler.

Cultural conditions:

1. 1ml sample inoculated into 9ml of 0.1% trypticase seawater broth in test tubes and serially diluted to 10^{-5}, incubation at 22°.
2. Same media as No. 1 with incubation at 44°.
3. Same media as No. 1 with incubation at 56°.
4. 250ml seawater sample dispensed to overflowing in 250ml Erlenmeyer screw cap flask and enriched with 1ml of 10% sterile trypticase and 2 drops of sodium sulfide and incubated at 22°.
5. Same as No. 4 with incubation at 44°.
6. Same as No. 4 with incubation at 56°.
7. 250ml seawater sample dispensed to overflow in screw cap Erlenmeyer flask and 1 gram of raw meat added to flask, sealed and incubated at 56°.
8. 1ml sample inoculated into 15ml cystine trypticase agar medium in test tube, sample overlaid with sterile vaseline, incubated at 56°. Agar made with brine from 56° layer of Atlantis II Deep.
9. 1ml of sample inoculated into 9ml of 0.1% trypticase medium made with brine from the 56° layer of Atlantis II Deep, incubation at 56°.
10. 1ml of sample inoculated into 9ml of trypticase soy agar plus 0.2% sodium sulfide, medium made with 56° brine from Atlantis II Deep, incubation at 56°.

Table 2 Water Samples from the Discovery Deep

Depth (m)	Temperature °C	Hydrographic Station	Cultural Conditions (see key)
* 1,985	22.99	717	
* 1,990	23.22	717	
* 2,019	29.72	717	
* 2,099	44.72	728	
2,149		728	1, 2, 11–20
2,159		728	1, 2, 11–20
2,179		728	1, 2, 11–20
2,199		728	1, 2, 11–20

* Indicates sample temperatures were obtained by means of Nansen bottle, no attempt was made to culture bacteria from these samples. All other samples collected with Niskin sampler.

Cultural conditions: One ml from each sample inoculated into 9cc of media in cotton-plugged test tubes. The media used and incubation temperatures are listed below.

1. 0.1% trypticase made with normal seawater, incubation at 44°.
2. 0.1% trypticase made with normal seawater, incubation at 22°.
11. 0.1% trypticase made with brine, incubation at 22°.
12. 0.1% trypticase made with brine, incubation at 44°.
13. Cystine trypticase agar made with normal seawater, incubation at 22°.
14. Cystine trypticase agar made with normal seawater, incubation at 44°.
15. Cystine trypticase agar made with brine, incubation at 22°.
16. Cystine trypticase agar made with brine, incubation at 44°.
17. Trypticase soy agar with 0.2% sodium sulfide, made with normal seawater, incubation at 22°.
18. Trypticase soy agar with 0.2% sodium sulfide, made with normal seawater, media overlain with vaseline after inoculation, incubation at 44°.
19. Trypticase soy agar with 0.2% sodium sulfide, made with brine, incubation at 22°.
20. Trypticase soy agar with 0.2% sodium sulfide, media made with brine, media overlain with vaseline after inoculation, incubation at 44°.

Note: No growth was observed in any of the media after 30 days incubation.

present or, if present, they were extremely difficult to culture. Because of the latter possibility a variety of media * and cultural conditions were used. Tables 1, 2 and 4 list the media and cultural conditions used for specific samples.

Water Samples from the Atlantis II Deep

Information concerning the water samples collected in the Atlantis II Deep is summarized in Tables 1 and 2. As shown in

* All commercial media used in these experiments were purchased from B.B.L. (Baltimore Biological Laboratory).

these tables bacteria were isolated from the 22° and transition layers but none were found in the 44° and 56° layers. In addition to the 1ml and 250ml samples inoculated into media listed in Tables 1 and 2, two 12-liter samples were collected from the 56° brines at station 721. These water samples were dispensed into 12-liter pyrex sterile carboys, enriched with 100ml of 10 per cent trypticase and overlain with sterile vaseline. After three months incubation at room temperature no bacterial growth was observed in either carboy.

A third 12-liter enrichment culture was made from water collected at this same station in the Atlantis II Deep. The water used in this third enrichment culture came from the transition, 44° and 56° layers. Water samples from each of the above layers were combined and emptied into a sterile 12-liter carboy. This water was then enriched with 100ml of 10 per cent trypticase and overlain with sterile vaseline to maintain anaerobic conditions. Within a week this water became turbid from bacterial growth and blackened from hydrogen sulfide production. Trüper (1969) isolated sulfate reducing bacteria from this enrichment culture.

The bacteria in this enrichment culture must have come from the transition zone since other experiments showed that the 44° and 56° layers were free of bacteria.

Water Samples from the Discovery Deep

Only four samples were collected from the hot brines of the Discovery Deep. The depths from which these samples were collected and the cultural procedures used in an attempt to isolate bacteria from them are listed in Table 2. All inoculated media remained sterile after a one-month incubation period.

Sediment Samples from the Atlantis II Deep

Thirteen sediment samples were collected in the Atlantis II Deep, and these samples were inoculated into a variety of media and incubated aerobically and anaerobically at 22°, 44° and 56°C as outlined in Table 4. With the exception of the sample

Table 3   Summary of Water Samples Collected in Atlantis II Deep

Depth (m)	Water Layer	Water Samples Collected	Samples from Which Bacteria Were Cultured
Surface–1,895	22° water	0	0
1,895–1,980	22° water	8	6
1,980–2,010	transition layer	4	3
2,010–2,038	44° brine	3	0
2,038–Bottom	56° brine	42	0

Table 4   Examination of Sediment Samples for Presence of Bacteria

Station	Type Media (see key)	Incubation Temperature °C	Highest Dilution Showing Bacterial Growth	Type Media Showing Bacterial Growth	Incubation Temperature Where Growth Occurred °C
Samples from Atlantis II Deep:					
71	1	22, 55			
89B	1, 3	22	10^{-4}	1, 3	22
89C	1, 3	22			
89D	1, 3	22			
94	1, 3	22			
106	1	55			
107	1	56			
108	1	56			
124	2, 4, 6	22, 56			
126	2, 4, 6	22, 55			
127	2, 4, 6	22, 56			
128	2, 4, 6	22, 55			
139A	1, 3, 5	22			
Samples from Discovery Deep:					
80	3, 5	22, 56			
81	3, 5	22, 56			
158	1, 2, 3, 4, 5, 6	22, 44	10^{-2}	1, 3, 5	22, 44
Samples from outside Atlantis II and Discovery Deeps:					
72	1, 3	22, 44, 56	10^{-3}	1	22, 44, 56
82B	3, 5	22, 56			
82D	3, 5	22, 56			
143	1, 3, 5	22	10^{-2}	1, 3, 5	22
159	1, 3, 5	22	10^{-2}	3	22
161	1, 3, 5	22	10^{-3}	1	22

Media and sample size: These sediment samples were inoculated into 9cc of media in cotton plugged test tubes. The composition of media 1–6 is listed below.

1. 0.1% trypticase made with normal seawater.
2. 0.1% trypticase made with Red Sea brine.
3. Trypticase soy agar plus 0.2% sodium sulfide, media after inoculation (overlain with sterile vaseline to maintain anaerobic conditions).
4. Trypticase soy agar plus 0.2% sodium sulfide, media made with brine from appropriate Deep, after inoculation media overlain with sterile vaseline.
5. Cystine trypticase agar, made with normal seawater, media after inoculation overlain with sterile vaseline.
6. Cystine trypticase agar, made with appropriate brine, media after inoculation overlain with sterile vaseline.

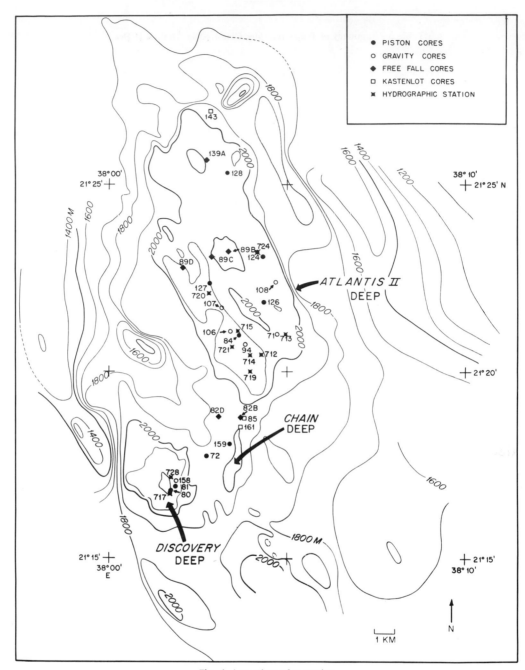

Fig. 1. Location of samples.

from 89B, we were unable to isolate bacteria from these samples.

Section 89B was in the general area referred to as the Atlantis II Deep. At this particular station, however, a ridge projected upward through the 44° brine into the normal 22° deep water. For this reason it was not surprising that both aerobic and anaerobic bacteria were isolated from the sediments of this station. As might be ex-

pected, the bacteria isolated from this sediment sample grew at 22° but failed to grow at 44° or 56°.

All of the above work on sediment samples was carried out on fresh samples aboard the ship. Samples were also frozen and shipped to Dr. Louis D. S. Smith at the Virginia Polytechnic Institute for additional studies. There they were examined for aerobic and anaerobic bacteria by inoculation into trypticase soy semi-solid agar and into a cooked meat medium with 0.3 per cent glucose. They were then incubated for three weeks at 44° and 56°. During this period bacteria did not grow in any of the media inoculated with sediment samples from the Atlantis II Deep.

Sediment Samples from the Discovery Deep

Three sediment samples from the Discovery Deep were inoculated into organic media aboard ship, and bacteria were cultured from one of these (Table 4). These bacteria grew at 22° and 44° both aerobically and anaerobically. Dr. Smith also cultured bacteria from a sediment sample collected in the Discovery Deep.

Sediment Samples Collected Outside of the Atlantis II and Discovery Deeps

Sediment samples collected in the areas adjacent to the Atlantis II and Discovery Deeps are listed in Table 4, and the location of the stations where these samples were collected is given in Fig. 1. Bacteria were cultured from four of these six samples. The samples 82B and 82D, from which we failed to isolate bacteria, were supposed to be from the edge of the ridge which separated the two Deeps but more likely came from the Atlantis II Deep. This would explain why bacteria were not isolated from them.

Dissolved and Particulate Organic Content of the Hot Brines and Sediments of the Atlantis II Deep

The dissolved organic carbon of the seawater above and in the hot brines of the Atlantis II Deep was determined by the method of Menzel and Vaccaro (1964). Water for these analyses was collected at station 721 at depths ranging from 1,965m to 2,115m (Table 5).

Prior to analyzing the dissolved organic content, the samples were filtered through a silver filter having a 2.0μ porosity to remove the iron which had precipitated in the

Table 5 Dissolved Organic Matter in the Water of the Atlantis II Deep Collected at Hydrographic Station 721

Depth (m)	mgC/l	Water Layer
1,965	1.83	22°
1,970	1.13	22°
1,975	1.55	22°
1,980	2.15	transition
1,995	1.16	transition
2,100	1.35	56°
2,105	1.50	56°
2,110	1.35	56°
2,115	1.03	56°

sample during storage. These filters were incinerated and the carbon content measured by a Perkin-Elmer carbon-nitrogen analyzer (Model 240). The carbon content of the filters was approximately one-third of the carbon content of the original water sample. Since the precipitated iron partially removed dissolved organic compounds from the water samples, the values given in Table 5 are probably one-third low.

Prior to organic analysis the sediment samples were evaporated to dryness overnight at 60°C and treated with hot 5 per cent trichloroacetic acid to remove carbonates. These samples were subsequently incinerated and the carbon content analyzed in a Perkin-Elmer carbon-nitrogen analyzer (Model 240).

Three sediment samples were analyzed for their carbon content (Table 6). Sample 84 came from the Atlantis II Deep, 85K from the ridge separating the two Deeps and 82D, as previously explained, was probably from the Atlantis II Deep.

Table 6 Organic Content of Red Sea Sediments

Station Number	% Organic Carbon	% Nitrogen	C/N Ratio
82D	0.26	0.0332	7.83
84	0.224	0.017	13.2
85K	0.664	0.0235	28.2

Discussion

We believe that the brines and sediments of the Atlantis II Deep are sterile but can only speculate why. Certainly the elevated temperatures and salinities are not solely responsible for the sterility of these environments since bacteria were found in the Discovery Deep sediments. Heavy metal toxicity was immediately suspected since copper, for example, in the Atlantis II brines had a concentration of one part per million, which would kill many bacteria.

Bacteria may exist in the brine of the Discovery Deep as the concentration of some of the heavy metals is an order of magnitude less than in the Atlantis II Deep. We failed to isolate bacteria from the Discovery Deep, but the results of our limited sampling are not adequate proof that this brine is sterile.

The reasons why bacteria were found in the interstitial waters of the sediments of the Discovery Deep but not the Atlantis II Deep are not clear. Both the temperature and heavy metal concentration are lower in the Discovery Deep than in the Atlantis II Deep. For example, the average copper concentration is about one part per million in the interstitial waters of the Atlantis II Deep and about 0.1 part per million in the Discovery Deep. Likewise, the temperature is 12° lower in the surface sediments of the Discovery Deep than in the Atlantis II Deep.

The heavy metal concentrations in the Atlantis II Deep would not inhibit all bacterial growth, but when all unfavorable conditions, including high salinity, high pressure and lack of decomposable organic matter and oxygen are combined, a synergistic effect might prevent the growth of all bacteria. Possibly the slightly lower temperature and heavy metal concentration in the Discovery Deep may be sufficient to permit bacterial growth.

For bacteria to grow indefinitely in any environment there must be a constant influx of oxidizable compounds to serve as an energy source. The high density of the brines probably prevents most particulate organic matter from settling into these waters. A small amount of organic detritus may settle into the brine, but it seems doubtful that this would provide the major source of organic carbon within the brines and sediments.

As iron was oxidized and precipitated in the samples of brines from the Atlantis II Deep, approximately 30 per cent of the dissolved organic matter was removed from the water. It is anticipated the same phenomenon should occur within the transition layer of the water column in the Discovery and Atlantis II Deeps. The organic matter brought down into the brines and sediments with the oxidation of iron may constitute the major fraction of organic matter within these environments.

At first glance the organic matter in the sediments would not appear higher than that found in normal deep sediments. In the Atlantis II Deep sediments the organic matter represented 0.2–0.3 per cent of the sample, which is normal for deep-sea sediments. When it is considered that 95 per cent of these sediments are composed of precipitated metals from the brines, the organic content is in reality higher than that found in normal sediments. If the weight of the brine precipitates were ignored, the organic carbon would comprise 4 per cent of the remaining sediments.

Since the brines of the Atlantis II Deep are sterile, all of the microbial activity must occur in the water column over the hot brines. Sulfate reduction is one of the microbial processes which probably occurs in these waters. Since sulfate-reducing bacteria are obligate anaerobes and since the brines are sterile, sulfate reduction could only take place in a thin layer of anaerobic water immediately over the 44° brine. Such an anaerobic zone in the transition layer has not been demonstrated but must exist since evidence of the presence of sulfate-reducing bacteria was found in these waters (Trüper, 1969).

Circumstantial evidence of the presence of sulfide was found when brass messengers sent through the transition layers returned blackened. The waters from the transition layers were also slightly blackish although the odor of hydrogen sulfide was not detected. This coloration was caused by a fine precipitate suspended in the water which may have been metal sulfide although experimental verification of this is lacking.

The best evidence that sulfate reduction occurs within the water column was found when a 12-liter carboy was filled with samples taken in the 44° brine, in the transition layer, and in the 22° water column at station 721. After enrichment with organic material and anaerobic incubation these waters blackened and the smell of hydrogen sulfide was immediately detected. Subsequent isolation of sulfide-reducing bacteria by Trüper (1969) confirmed that sulfate reduction must occur within this water column.

One of the samples dispensed into the above carboy came from the transition layer about 1 or 2m above the 44° brine. The sulfate-reducing bacteria had to come from this sample since the 44° brine was sterile and since the upper samples contained oxygen.

If sulfate reduction occurs in the transition layer, the sulfides would combine with zinc and precipitate into the sediments. This is confirmed by finding zinc sulfide in the sediments of the Atlantis II Deep. If these sulfides resulted from microbial reduction, sulfur isotope ratio should reflect this since sulfate-reducing bacteria are known to fractionate sulfur isotopes (Kaplan *et al.*, 1963).

Hartmann and Nielsen (1966) and Hartmann (1969) reported a δS^{34} of 4.3 for sediment sulfides and 20.3 and 20.4 for the sulfates of the brines and interstitial waters of the Atlantis II Deep. Since they found no enrichment of δS^{34} in the sulfates of the interstitial water, relative to those of the overlying brine, they concluded that the observed fractionation did not stem from microbial reduction of sulfate to sulfide. It was their opinion that the sulfide sulfur was brought through the sediment to the sea bottom in the brine.

Our studies have shown that since the brines and sediments are sterile, sulfate reduction could not occur in these environments. The fractionation observed by Hartmann and Nielsen (1966) in the sediments could be accounted for by sulfate reduction in the transition layer and the subsequent precipitation and sinking of these sulfides into the sediments.

The degree of fractionation of S^{32} over S^{34} in the Atlantis II Deep sediments reported by Hartmann and Nielsen (1966) indicates that the sulfate pool was relatively small, and this is substantiated by the theoretical considerations previously discussed concerning the necessity of sulfate reduction taking place in a thin layer of water overlying the 44° brine. The larger the pool the greater degree of fractionation would be expected since the bacteria preferentially reduce $S^{32}O_4$ over $S^{34}O_4$. Kaplan *et al.* (1963) reported a δS^{34} of -62.4 from a sediment sample from the San Diego Trough which approximates the maximum enrichment of S^{32} which could result from microbial sulfate reduction, indicating a large pool of sulfate.

If our hypothesis is correct that sulfate reduction occurred in a shallow layer of water overlying the 44° water, then complete exhaustion of sulfate might be expected if it did not diffuse into this thin layer of water. The fact that δS^{34} of the sediments was $+4.3$ indicates that the sulfates in this layer were not completely reduced and suggests that the pool was replenished continually by diffusion.

Since sulfides were precipitated principally as zinc sulfide and iron as iron hydroxide in the Atlantis II Deep, it seems paradoxical that both metals precipitated at the same time. Iron could have precipitated only when reduced iron came in contact with oxygen, and this had to occur in the aerobic part of the transition zone over the 44° water layer. If the anaerobic layer of the transition zone was extremely shallow, *i.e.*, 1 to 2m deep, then iron had to diffuse through this layer and come in contact with oxygen in the aerobic zone. If sulfides had been present in the anaerobic portion of the transition zone, this iron would have combined with them and precipitated as iron sulfides. This does not seem to be the case. It is likely that zinc, with a greater attraction for sulfides than iron, scavenged all the sulfides from this anaerobic zone, permitting the reduced iron to diffuse into the overlying aerobic zone where it precipitated as iron hydroxides. Although this theoretical sequence may account for the zinc sulfides and iron hydroxides in the sediments of the Atlantis II Deep, it cannot be responsible for the formation of iron sulfides found in the Discovery Deep sediments. Possibly in the

Discovery sediments sulfate reduction takes place within the sediments, but this does not seem likely since Trüper (1969) was unable to isolate sulfate-reducing bacteria from this environment.

The types of bacteria isolated from the sediment of the saddle (station 72), which separated the Atlantis II and Discovery Deeps, appear to reflect the history of this area. All evidence suggests that the 44° water of the Discovery Deep was formed by the flow of the higher salinity brines of the Atlantis II Deep into the Discovery Deep. For this event to have taken place, the level of the Atlantis II brines had to rise higher than the sill which separates the two deeps. Once the height of the brines exceeded sill depth, the hot water spread out laterally. This lateral movement was restricted as the bottom slopes upward except in the region of the Discovery Deep. Thus, the flow of the hot water must have been directed toward the Discovery Deep. Because of the density differences between these brines and the overlying 22° water, there should have been little mixing of the two water masses during transport of the brines. Some diffusion had to take place, and this is reflected by the salinity difference in the brines of the two deeps.

That the above sequence of events occurred seems to be verified by the isolation of bacteria which grew at 44° and 56° from the saddle separating the two Deeps. It will be noted that thermophilic bacteria were not found in sediments from other stations surrounding the Atlantis II Deep. The ancestors of the bacteria presently existing in the saddle area must have survived the high temperatures of the brines as they flowed over this saddle area. It is for this reason that many of the bacteria growing presently in these sediments can still grow at temperatures up to 56°.

Both facultative and anaerobic bacteria were cultured from the Discovery Deep sediments. Since this area has been anaerobic for a prolonged period, it is interesting that these bacteria have maintained their potential for aerobic growth. This potential suggests that the waters were previously aerobic.

Future cruises are planned to the Red Sea, and the authors would like to take this opportunity to mention some of the microbial problems which should be investigated. One of these problems concerns the decomposition of organic matter. In normal oceans little is known about the decomposition of particulate organic matter. Some investigators feel that a large fraction of this organic matter is decomposed within the water column while other workers believe that most of the decomposition occurs within the sediments. That Atlantis II Deep should be an excellent place to study the decomposition of particulate organic material since the unique density layering prevents most organic particles from sinking into the brine.

Certainly the brines of the Discovery Deep should be sampled extensively for bacteria on future cruises. We anticipate that if bacteria are present in these brines there are only 1–10 per liter which would require large enrichment cultures to detect their presence.

Samples should be taken at 1-m intervals in the transition layer in the Atlantis II Deep. These samples should be analyzed for oxygen and sulfur isotope ratios. Likewise, this layer should be carefully sampled for bacteria and the physiological types present should be documented. The possibility that the sulfides present in the sediments do arise from sulfate reduction in this area should be carefully examined, and a detailed study of the isotope fractionation within the sediments should be made.

A combined geological and microbiological study in this area was fruitful. It is hoped that in the near future similar collaborative studies can be renewed.

Acknowledgments

This investigation was supported by the National Science Foundation Grant GA 584 and by the Atomic Energy Commission Contract AT(30-1)-1918 (Ref. NYO-1918-163).

References

Brewer, P. G., J. P. Riley, and F. Culkin: The chemical composition of the hot salty water from the bottom of the Red Sea. Deep-Sea Res., **12,** 497 (1965).

——, C. D. Densmore, R. G. Munns, and R. J. Stanley: Hydrography of the Red Sea brines. *In: Hot brines and recent heavy metal deposits in the Red Sea,* E. T. Degens and D. A. Ross (eds.). Springer-Verlag New York Inc., 138–147 (1969).

Charnock, H.: Anomalous bottom water in the Red Sea. Nature, **203,** 591 (1964).

Hartmann, M.: Investigations of Atlantis II Deep samples taken by the FS METEOR. *In: Hot brines and recent heavy metal deposits in the Red Sea,* E. T. Degens and D. A. Ross (eds.). Springer-Verlag New York Inc., 204–207 (1969).

—— and H. Nielsen: Sulfur isotopes in the hot brine and sediment of Atlantis II Deep (Red Sea). Mar. Geol., **4,** 305 (1966).

Kaplan, I. R., K. O. Emery, and S. C. Rittenberg: The distribution and isotopic abundance of sulphur in recent marine sediments off southern California. Geochim. Cosmochim. Acta, **27,** 297 (1963).

Menzel, D. W. and R. F. Vaccaro: The measurement of dissolved organic and particulate carbon in seawater. Limnol. Oceanog., **9,** 138 (1964).

Miller, A. R.: Highest salinity in the world ocean? Nature, **203,** 590 (1964).

——, C. D. Densmore, E. T. Degens, J. C. Hathaway, F. T. Manheim, P. F. McFarlen, R. Pocklington, and A. Jokela. Hot brines and recent iron deposits in deeps of the Red Sea. Geochim. Cosmochim. Acta, **30,** 341 (1966).

Munns, R. G., R. J. Stanley, and C. D. Densmore. Hydrographic observations of the Red Sea brines. Nature, **214,** 1215 (1967).

Swallow, J. C. and J. Crease: Hot salty water at the bottom of the Red Sea. Nature, **205,** 165 (1965).

Trüper, H. G.: Bacterial sulfate reduction in the Red Sea hot brines. *In: Hot brines and recent heavy metal deposits in the Red Sea,* E. T. Degens and D. A. Ross (eds.). Springer-Verlag New York Inc., 263–271 (1969).

Late Pleistocene and Holocene Planktonic Foraminifera from the Red Sea *

W. A. BERGGREN

Woods Hole Oceanographic Institution
Woods Hole, Massachusetts

ANNE BOERSMA

Massachusetts Institute of Technology
Cambridge, Massachusetts

Abstract

Late Pleistocene and Holocene climatic fluctuations in the Red Sea are reflected in distributional patterns among planktonic foraminifera studied in 20 deep sea cores. Fluctuating percentages in two species, *Globigerinoides ruber* and *G. sacculifer,* have allowed the recognition of four biostratigraphic assemblage zones spanning the past 80,000 years. The youngest Zone, A, corresponds to the Holocene (base dated at *ca.* 11,000 years B.P.) is characterized by the dominance of *Globigerinoides sacculifer,* and represents the establishment of a normal Indian Ocean fauna following the Würm glaciation.

Fluctuations in the percentages of various species are related to salinity and temperature variations during this interval which are, in turn, related to glacio-eustatic movements of sea level and Late Pleistocene glaciation of the northern latitudes. Tectonic movements about 17,000 years B.P. may have played a role in relative changes of sea level during the latest phase of the Würm glaciation.

Globigerinoides ruber appears to be able to withstand extremely high salinity values better than any other tropical-subtropical species. The relationship between *G. sacculifer* and *G. ruber* may provide a useful tool in paleoecologic investigations in tropical regions which have undergone periodic isolation.

Introduction

During the 61st voyage of the **CHAIN** in the fall of 1966, approximately 60 deep sea cores were successfully recovered from the Red Sea. The majority of these cores were taken in the brine holes located between 21°15′ and 21°27′N and 38°00 to 38°5′E. Twenty of these cores were investigated for their micropaleontological contents, and particularly for their planktonic foraminiferal faunas. This investigation attempted to relate the distributional pattern of the planktonic foraminiferal populations to the climatic variations of the late Pleistocene in the Middle East.

Previous Research Concerning the Foraminifera of the Red Sea

Over 200 species of benthonic and planktonic foraminifera have been identified and reported from the Red Sea. Fichtel and Moll first described eight species of benthonic foraminifera from the Red Sea in 1798. Later, in 1826, d'Orbigny reported nine more species from Red Sea samples. In 1865 Parker and Jones described some 28 species of benthonic foraminifera and Heron-Allen and Earland reported 50 species from dredge hauls made in the Red Sea by the Cambridge Expedition. In 1941 Marie reported on the species of foraminifera which he found along the shores of Egypt.

In 1949 Said described 150 species, in-

* Woods Hole Oceanographic Institution Contribution No. 2191.

cluding 21 new species from the northern Red Sea. Said found that his benthonic species displayed a patchy distribution pattern; however distinctive faunas did appear to be characteristic of the three basins of the Red Sea. The majority of his species were Indo-Pacific, although two Mediterranean species, *Cibicides rodiensis* and *C. gibbosa,* were present. These latter two species he supposed corroborated the hypothesis of a Pliocene seaway between the Mediterranean and Red Seas. Surprisingly, a few western Atlantic forms were also present in his faunas.

A later paper by Said (1950) purported to "decipher the ecologic factors that determine the distribution of foraminifera in the northern Red Sea." Said proposed a depth zonation for the benthonic forams in the Red Sea proper and stated that temperature was not the regulating mechanism for their depth distribution. He suggested a correlation between the total number of benthonic foraminifera and the nitrogen content of the sediments. However, he admitted that other factors such as the median grain size must also be influencing the concentration of the foraminifera.

A supplement to this paper, contained in the same volume (1950), described thirty more species. Five of these were planktonic forms: *Globigerina bulloides, Globigerinoides sacculifer, Globigerinella aequilateralis, Orbulina universa,* and *Tretomphalus bulloides.*

Olausson (1960) reported foraminifera, coccoliths, pteropods, other gastropods, echinoderms and porifera remains from the Red Sea. He adopted the frequency curves of warm versus cold water species from Parker (1958) and applied them to the species of foraminifera which he discovered in the Red Sea. He designated *Hastigerina siphonifera, Globigerinoides rubra, Globigerinoides sacculifera,* and *Hastigerina pelagica* as warm water indicators and *Globigerina pachyderma* and *G. scitula* as cooler water forms. However, his descriptions of the cores dealt chiefly with the nature of the sediments and did not treat the foraminifera in any detail. His detailed analysis of the sediments includes descriptions of the lithified chunks which were present in his cores.

A recent study of Red Sea foraminifera was made by Herman (1965). She investigated fourteen long cores and six plankton tows and succeeded in relating the fluctuating distributional patterns of the planktonic foraminifera and two pteropod species to late Pleistocene temperature variations in the Middle East. In a comparison of Red Sea, Mediterranean and Indian Ocean planktonic forms she listed a fauna of 18 planktonic foraminiferal species in the Red Sea.

Methods of Core Analysis

The **CHAIN** 61 cores were sampled and prepared by the standard procedure outlined by Ericson *et al.* (1961).

Standard faunal counts were made on the +60 (250μ) fractions of all samples. Included in these counts were all species of planktonic foraminifera, benthonic foraminifera (counted as a group), pteropods (counted as a group), other fossils (counted as a group) and non-organic material. Counts were made of 200 specimens when possible. In the cases where 200 specimens did not result from the washing process, 100 or fewer individuals were counted and their relative distributions were figured on a percentage basis.

Material from the +120 fraction (125μ) and the +325 fraction (44μ) of the samples was also examined microscopically. It was found that these fractions contained predominantly young forms of the species found in the +60 fraction as well as five species: *Globigerinita glutinata, Globigerina quinqueloba, G. rubescens, Globigerinoides tenellus,* and *Globorotalia anfracta.* Since problems of consistent identification exist in material in the finer fraction, the species composition of the +60 fraction was considered a relatively accurate representation of the population structure as it reflected climatic conditions in the Red Sea basin.

General Nature of the Cores

Of the available cores, 20 were chosen for their greater lengths and their locations. It was hoped that the longer cores might

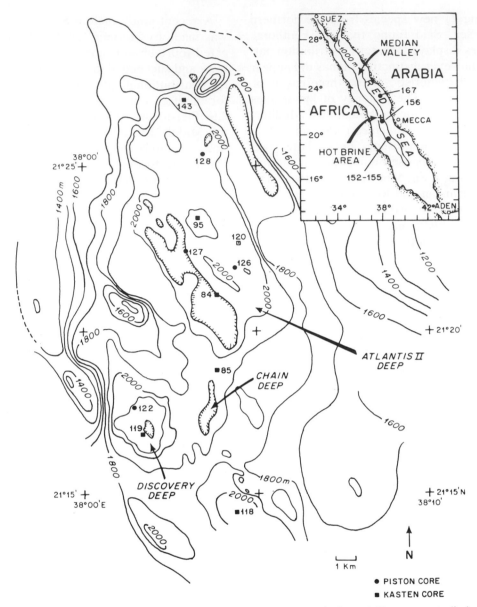

Fig. 1. Location of cores from the Red Sea in which planktonic foraminifera were studied.

exhibit a long stratigraphic section; however, all the cores appear to be late or post-Pleistocene in age. Radiocarbon dates made on cores 95K, 118K, 119K, 153P, and 154P (Ku *et al.* 1969) have been incorporated in the interpretation of results presented below (see Fig. 2).

Nine of the twenty cores studied were located within the area of deep brine holes. Ten other cores were located either on the periphery or well outside the hot hole area. One core, Station 5, was taken by the **CHAIN** on her forty-third voyage and comes from the southern part of the Red

Sea. The location, depth and length of all cores obtained on **CHAIN** cruise 61 in the Red Sea are given in Ross and Degens, 1969 (see Fig. 1).

The most abundant foraminiferal faunas washed out of brown or brownish-grey calcareous layers. The more brightly colored layers usually consisted of very fine clumps of precipitate material which washed entirely through the 325 mesh sieve. Certain shiny black, clayey layers contained a black grit and abundant pteropods, many of them oriented.

Only cores 167P, 152P, 153P, 154P,

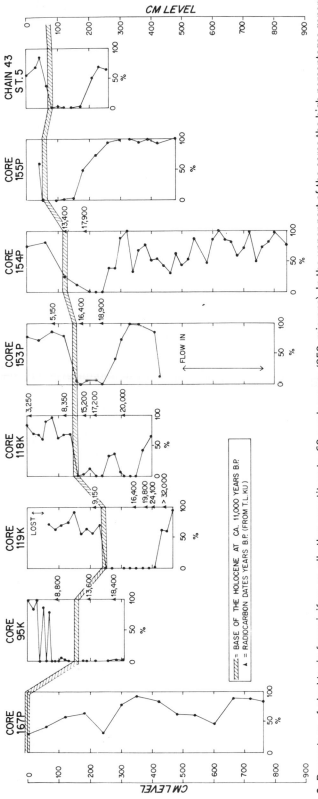

Fig. 2. Percentage of planktonic foraminifera *vs.* all other constituents, 60 mesh screen (250 microns). In the upper part of the cores the high percentages generally reflect the dominance of *G. sacculifer*; in the lower parts the dominance of *G. ruber*.

155P, 156A and B, and Station 5, **CHAIN** 43 from outside the hot hole area contained sediments which could be termed a true lutite or *Globigerina* ooze.

Planktonic Foraminiferal Fauna

The typical Indo-Pacific planktonic foraminiferal faunas inhabiting the Red Sea in other regions (Herman, 1965) are, in general, present in the Red Sea cores taken on **CHAIN** 61. A relatively restricted fauna of four or five warm water species occur in the coarse fraction (60 mesh = 250μ) of a given core. The following species are present, but not always abundant: *Orbulina universa, Hastigerina siphonifera + H. pelagica, Globigerinoides ruber, G. sacculifer, G. conglobatus* and *Globorotalia menardii.* The species *H. pelagica, G. conglobatus* and *G. menardii,* where present, rarely constitute more than 1 per cent of the fauna. In the finer fractions (230 mesh = 325μ) the following species have been identified (but do not form an appreciable part of the total fauna): *Globigerina quinqueloba, Globorotalia anfracta, Globigerinoides rubescens, G. tenellus* and *Globigerinita glutinata* (cf. Herman, 1965, p. 383, tab. E).

Several morphotypes of *Globigerinoides ruber* are present in the faunas. These include:

1. Small, compact white *ruber,* some with inflated final chambers and proportionately large apertures (Herman, 1965, variety A);
2. Small white *ruber,* thicker walled, with more compact tests and relatively smaller apertures (Herman, 1965, variety B);
3. Multi-chambered, elongate forms ascribed by some authors to *G. pyramidalis,* but included by most authors in *G. ruber;*
4. Multi-chambered, expanded forms with distinctive lateral chambers identified in Brady (1884) as *Globigerina helicina* (d'Orbigny) but included by Parker (1962) and others, as *G. ruber;*

5. Small, flattened, subquadrate specimens resembling juvenile *G. conglobatus,* also included by Parker (1962) as *G. ruber.*

In levels of supposed turbidites a Pliocene form, *Globorotalia miocenica,* was present. However, the origin of these specimens is unknown and obviously they are not representative of the stratigraphic sequence then being deposited.

Benthonic Foraminifera

The bottom conditions in the brine areas of the Red Sea should preclude many bottom-dwellers. In this area severe temperatures and salinities have been measured. The black muds suggest euxinic conditions at various times during the last glacial epoch.

In several instances benthonic foraminifera are present and compose approximately 1 per cent of the total fauna. The most abundant group of benthonics are the Miliolidae. Representatives of Buliminidae, Bolivinidae, Uvigerinidae, Lagenidae, Textulariidae, and Anomalinidae are also present at infrequent intervals.

It should be noted that the benthonic species reached their moderate abundance only in the cores coming from outside or on the periphery of the brine hole area.

Other Fossils

In addition to the other microfossil groups discussed in other chapters of this book (McIntyre, 1969; Goll, 1969; Wall and Warren, 1969; Chen, 1969), pelecypod shells, heteropods (primarily *Atlanta*), other marine gastropods, echinoderm and sponge remains, and otoliths are also to be found in the samples from these cores.

Discussion of the Cores

The following analyses are intended to give an idea of the distributional structure of the planktonic foraminifera in the cores. For this reason no detailed description of lithologies, mineral species, or other fossil groups is included. For more specific analysis of lithology and mineralogy, the reader should refer to Bischoff (1969), and for other fossil groups, to the papers of Chen (1969), Goll (1969), McIntyre (1969), and Wall and Warren (1969).

In nearly all samples the same four species of foraminifera are present in the +60 fraction. These are: *Orbulina universa, Hastigerina siphonifera, Globigerinoides ruber,* and *G. sacculifer.* A similar situation occurs in the −60 fraction where four different species, *Globigerina quinqueloba, Globorotalia anfracta, Globigerinoides rubescens,* and *Globigerinita glutinata,* occur along with numerous shell fragments and juveniles of the species in the +60 fraction. References to the "usual" or "typical" four species indicate the species cited above.

The most distinctive changes in distributional structure of the foraminifera occur in a zone of crusty carbonate layers. Although these layers are actually several distinct units, the sampling interval of 10, 20, and even 40cm does not allow an accurate delineation of each discrete layer. Therefore, the sequence containing these layers, shown to be the last to occur in the Pleistocene, is described as the "lithified" or "crusty carbonate" zone. When encountered in the cores, this zone appears much the same. For this reason it is not described in great detail with each occurrence.

cm Level	Description
	Core 95K
0–150cm	Samples contain clean foraminifera oozes composed of the typical four species of planktonic foraminifera in the +60 fraction; *Globigerinoides sacculifer* and *Hastigerina siphonifera* are the most abundant forms; both are small and tightly coiled; pteropods, heteropods, and pelecypods are also common; otoliths occur at 120cm; the four typical species of planktonics, plus juvenile forms, occur in the −60 fractions;
150–175cm	No foraminifera are present, only pteropods and carbonate crust;
175cm	Sample is fresh and clean, contains translucent pteropods and foraminifera; no carbonate material is present;

cm Level	Description
175–305cm	Same type of samples as in 150–175cm interval.
	Core 118K
0–150cm	Samples contain the usual four species of planktonic foraminifera in the +60 fraction, along with pteropods, heteropods, and an occasional *Hastigerina pelagica; Globigerinoides sacculifer* and *Hastigerina siphonifera* are the most abundant planktonics; juveniles, plus the typical four species of planktonic foraminifera, occur in the −60 fractions;
150–370cm	No foraminifera are present, only pteropods which increase to 33 per cent relative abundance; pteropods are crusty, and crusty carbonate chunks abound in the samples;
370–400cm	One planktonic species, *Globigerinoides ruber,* equals from 43 to 65 per cent of the total samples; pteropods and crusty carbonate also occur.
	Core 119K
0–150cm	Clean samples contain the four typical species of planktonic foramifera in the +60 and −60 fractions; specimens of *Globigerinoides sacculifer* and *Hastigerina siphonifera,* the most abundant species, are partially evolute; some individuals are iron-stained;
150–200cm	Samples are iron-stained; the sample at 250cm washed entirely through the 325 mesh sieve;
250–290cm	Clean samples, free of iron-staining, contain foraminifera and some pteropods; *Globigerinoides sacculifer* and *Hastigerina siphonifera* are the most abundant species;
290–410cm	Few, if any foraminifera, pteropods and crusty carbonate material comprise the samples; otoliths appear at 410cm;
410–455cm	One species of planktonic, *Globigerinoides ruber,* makes up from 63 to 97 per cent of the samples; pteropods and the crusty carbonate are also present.
	Core 120K
0–170cm	Three species of planktonic foraminifera occur in the +60 fraction: *Globigerinoides ruber, G. sacculifer,* and *Hastigerina siphonifera;* pteropods and carbonate crust are also present; the usual four species of planktonics, along with juvenile planktonics, occur in the −60 fractions; manganese micronodules lie embedded in the perforations in the foraminiferal tests;
170–400cm	Pteropods and foraminifera become rare and the crusty carbonate material predominates in all samples.

cm Level	Description

Core 143K

0–90cm Samples contain very large, translucent planktonic foraminifera; the typical four species are present in both the +60 and −60 fractions; specimens of *Hastigerina siphonifera* and *Globigerinoides sacculifer* are large and evolute, assuming exotic configurations as they uncoil;

90–120cm Samples contain 99 per cent foraminifera; pteropods are very scarce; crusty carbonate material occurs at 110cm;

120cm Pteropods reappear and constitute more than 30 per cent of the sample;

130–310cm All material washed through the 325 mesh sieve;

320–350cm Small, iron-stained residues contain few pteropods and no planktonic foraminifera.

Core 154P

0–110cm Samples contain clean, translucent foraminifera, the typical four species occurring in the +60 and −60 fractions; samples also contain pelecypods, pteropods, and some benthonic foraminifera; *Globigerinoides sacculifer* is the most abundant species and constitutes more than 40 per cent of the sample;

110–260cm Very abundant pteropods, crusty foraminifera, some otoliths, and the crusty carbonate material occur;

260–500cm One species, *G. ruber*, constitutes from 30 to 80 per cent of the samples; some benthonics and an occasional *Globorotalia menardii* occur; lithified fragments of the crusty carbonate occur at 260, 280, and 480cm;

520–792cm The typical four species of planktonic foraminifera occur in both fractions along with a few individuals of *Globigerinoides conglobatus* in the +60 fraction; *Globigerinoides sacculifer* and *G. ruber* are the most abundant forms; lithified carbonate material occurs at 520, 580, and 680cm.

Core 155P

0–110cm Samples contain clean foraminifera residues with the usual four species present in the −60 fractions; pteropods, heteropods, and otoliths also occur; planktonic species in the +60 fraction are: *Orbulina universa, Globigerinoides ruber, G. sacculifer, Hastigerina siphonifera,* and *Globorotalia menardii;* the planktonics are small and *Globigerinoides sacculifer* and *Hastigerina siphonifera* are tightly coiled;

75–160cm No foraminifera are present; pteropods equal 38 per cent of the sample; crusty carbonate material appears; pteropods are coated and filled with the carbonate crust;

180cm *Globigerinoides ruber* appears in the samples;

200–424cm Five species of planktonic foraminifera are again present; no pteropods appear; *Globigerinoides conglobatus, G. tenellus,* and some benthonic foraminifera occur rarely in the +60 fractions; *Globigerinoides ruber* remains the predominant species, equaling more than 50 per cent of the samples.

Core 167P

0–60cm Samples contain few planktonic foraminifera, but pteropods and the crusty carbonate material are abundant; pteropods are coated and filled with the carbonate crust; lithified fragments occur;

60–180cm The four typical species of planktonic foraminifera occur in the +60 and −60 fractions; *Globigerinoides ruber* is the most abundant form, constituting from 25–40 per cent of the samples;

180–757cm Samples are similar to those in the 60–180cm interval; *Globigerinoides sacculifer* is the most abundant species in nearly all samples; *Globigerinoides conglobatus, Globorotalia menardii,* and some benthonic foraminifera occur at some levels, always equalling less than 1 per cent of the sample; crusty carbonate material is present throughout.

CHAIN 43, St. 5 *

0–62cm Samples contain pteropods, heteropods, pelecypods, benthonic foraminifera, and five species of planktonics: *Globigerinoides ruber, G. sacculifer, Orbulina universa, Hastigerina siphonifera,* and *Globigerinita glutinata; Globigerinita glutinata* never equals more than 1 per cent of the +60 fraction sample; in the −60 fraction *Globigerinoides tenellus* occurs along with the usual four species;

62–80cm *Globorotalia miocenica,* a late Miocene-Pliocene form, occurs mixed in with Recent species; samples probably reflect turbidity current or slumping action;

* This core, located 17°39′N and 40°10′E, comes from much farther south than the **CHAIN 61** cores, but displays the same basic faunal and lithologic changes as the cores farther to the north. However, at many levels it does contain more species of planktonic foraminifera.

cm Level	Description
80–190cm	No foraminifera, only pteropods and crusty carbonate material occur; lithified chunks appear at 80, 100, 120, and 140cm;
190–255cm	*Globigerinoides ruber* is the only planktonic species present and equals from 50–70 per cent of the samples; pteropods and carbonate chunks are also present; *Globorotalia menardii, Globigerinoides conglobatus,* and few benthonic foraminifera occur sporadically.

Core 153P

0–160cm	The typical four species of planktonic foraminifera occur in both the +60 and −60 fractions; pteropods, heteropods, and pelecypods are present; *Globigerinoides sacculifer,* the most abundant species, comprises from 50 to 60 per cent of the fauna;
160–250cm	Foraminifera disappear and pteropods increase in abundance to 65 per cent of the samples; crusty carbonate material occurs;
250–330cm	*Globigerinoides ruber,* the only planktonic species present, equals 38 per cent of the samples, while pteropods make up the other 62 per cent; at 330cm *G. ruber* comprises 95 per cent of the sample;
330–440cm	*Globigerinoides ruber* decreases in abundance and iron-rich sediment predominates in the samples;
440–550cm	The typical four species of planktonics occur in both +60 and −60 fractions; *Globigerinoides sacculifer* is the most abundant species;
550–755cm	Contains disturbed and probably sucked-up material.

Ten other cores were also examined; however they were not studied in detail for the following reasons:

1. 84K, 85K, 126P, 127P, 128P, 122P were composed primarily of iron-rich mineral material, residues were small, and foraminifera were few, if present at all; 126, 127, and 128 did contain radiolaria (see Goll, 1969)
2. 156A, 156B, 154G were all less than 100-cm in length, penetrating only the upper part of the Holocene and thus were judged less important for stratigraphic purposes than other longer cores
3. 152P contained Pliocene species mixed with Pleistocene and Recent indicating disturbances, perhaps turbidity current or slumping.

Biostratigraphy

On the basis of the distribution pattern of planktonic foraminifera in cores examined from the Red Sea four assemblage zones have been recognized in the Late Pleistocene and Holocene. These zones, it should be emphasized, are based upon relative frequency patterns of two planktonic foraminiferal species and as such are of strictly local nature and validity. Insofar as they relate to general ecologic conditions within the Red Sea, they may eventually prove to be of local chronostratigraphic value; present data do not allow significant demonstration of this possibility.

In the following discussion "total planktonic foraminiferal fauna" refers to percentages calculated on the 60 mesh (250μ) fraction.

Zone A: This zone is characterized by relatively high percentages of *Globigerinoides sacculifer* (over 50 per cent) and the occurrence in varying numbers of *i. al., Orbulina universa, Hastigerina siphonifera + H. pelagica, Globigerinoides ruber* and *G. conglobatus.* In core 154P this zone extends from the surface down to about 120cm and corresponds to the Holocene.

Zone B: This zone is characterized by the rare or sporadic occurrence of planktonic foraminifera, generally *G. ruber* and *G. sacculifer,* or the total absence of planktonic foraminifera. In core 154P this interval extends from 120cm to about 260cm and in 118K from about 150cm to about 350cm.

Zone B corresponds to the upper part of dinoflagellate Zone D and most of Zone C. Dinoflagellate Zone C is defined on the absence or rarity of *Nematosphaeropsis* and *Hemicystodinium* and is characterized by a "sharp decline in all species" (Wall and Warren, 1969). The fact that biostratigraphic zonations of approximately the same interval based upon two different groups of microfossils are based upon the absence or rarity of generally common forms is probably related to the severe climatic conditions in the later part of the Pleistocene (see discussion under: Paleoecology).

Zone C: This zone is characterized by the dominance of *G. ruber* (30–80 per cent) over all other species of planktonic foraminifera. In the lower part of this zone *G. sacculifer* gradually decreases upwards from about 40 per cent of the fauna to 0 per cent (with sporadic fluctuations) at the expense of the steady increase in *G. ruber* (see Fig. 3, and Berggren, 1969, Fig. 2). The lower limit of this zone is placed at the level at which the divergence in percentage between these two species begins (*G. ruber* increases; *G. sacculifer* decreases). In core 154P this occurs at approximately 620cm.

Zone D: This zone is characterized by the joint occurrence in fluctuating percentages of *G. sacculifer* and *G. ruber*. The percentage of either of these two species generally varies between 10–45 per cent; together they account for between 50–100 per cent of the total planktonic foraminiferal fauna and usually over 75 per cent. In core 154P this zone extends from about 620cm to the bottom of the core (see Berggren, 1969, Fig. 2).

The relationship of this biostratigraphic zonation of Red Sea cores based on planktonic foraminifera to those based upon dinoflagellates and pteropods and to a chronostratigraphic scale of the Late Pleistocene and Holocene is shown in Berggren (1969, Fig. 2). The difference in boundaries between the dinoflagellate zonation and the planktonic foraminiferal zonation is due primarily to the lag effect in changes in paleogeographic conditions upon the dinoflagellate populations which flourished essentially in the more marginal parts of the Red Sea Basin. Correlation of CH 61-154P with another Red Sea core, V14-118 (Herman, 1965) is shown in Fig. 3.

The biostratigraphic zonation recognized here has been put into a time-stratigraphic frame of reference using a time-scale for the Late Pleistocene-Holocene adapted from Zagwijn (1963) and van der Hammen et al. (1967).

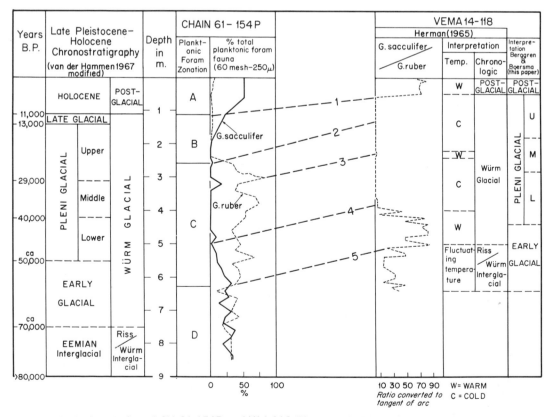

Fig. 3. Correlation of CH 61-154P and V14-118 (Herman, 1965). Explanation given in text.

Paleoecology

Factors influencing the regional distribution of living planktonic foraminifera have been discussed by several authors, in particular Bé (1959, 1960), Bé and Hamlin (1967), Parker (1962) and Bradshaw (1959). With regards to the present investigation data on two species, *Globigerinoides sacculifer* and *G. ruber*, may be pertinent to a general interpretation of the paleoecology of the Late Pleistocene and Holocene of the Red Sea.

Globigerinoides ruber is a eurythermal species with maximum abundance in warmer months (Bé, 1959). It is essentially a near surface water inhabitant (0–100m) and the presence of abundant zooxanthellae in tests restricts its distribution to the euphotic zone (Bé and Hamlin, 1967). The surface water temperature range for this species (including the pink variant) is 14°–30°C (highest concentrations are found between 21° and 29°C) and surface salinities either above 36.0 per cent or below 34.5 per cent (Bé, 1959; Bé and Tolderlund, 1968). The pink variant can tolerate temperatures up to 30°C.

Globigerinoides sacculifer is a warm stenothermal species with maximum abundance in warmer months. It is commonly found in the upper layers of the equatorial waters in which it is indigenous over a surface water range of 15°–30°C. The optimum surface water temperature range of *G. sacculifer* is 24°–30°C.

In a study of seasonal distributions of planktonic foraminifera in the Sargasso Sea off Bermuda (Bé, 1959), *G. ruber* was assigned to Group II (intermediate between Group I, temperate water species, and Group III, subtropical species) consisting of eurythermal species which are abundantly present throughout the year and reach maximum abundance during the warmer months. *G. sacculifer* was included in Group III—warm, stenothermal (subtropical) species which are generally restricted to the warmer months between June and December and reach peak abundance in October.

In a recent study Bé and Tolderlund (1969) have presented additional data on these two species pertinent to our discussion of Red Sea cores. *G. ruber* and *G. sacculifer* are shown to be the two most successful species in warm-water masses and may be "considered as competitive species vying to dominate the same ecological niche. Both occur in approximately the same geographic zones and water depths, and they differ mainly in the distributional patterns of their peak abundance areas."

These authors present supporting data which show that *G. ruber* reaches its maximum relative abundance in high salinity regions, such as the Caribbean Sea, the Gulf Stream system, Central and North Sargasso Sea, Canaries Current, the Brazil Current and central waters of the South Atlantic and Indian Ocean. In the eastern Mediterranean, where surface salinities exceed 39 per cent *G. sacculifer* is comparatively rare and exhibits an inverse relationship to *G. ruber* in the sediments. These authors suggest that average salinities of the tropical-subtropical Pacific and Atlantic Ocean may explain the generally inverse distributional relationships between *G. ruber* and *G. sacculifer*. Thus available data would suggest that *G. ruber* has a wider salinity range than *G. sacculifer*.

The general cooling trend which occurred during the Pleistocene between *ca.* 70,000 and 11,000 years ago is indirectly reflected in the distribution pattern of planktonic foraminifera in Red Sea cores. Glacio-eustatic changes in sea level were probably responsible for alternate periods of partial or complete isolation of the Red Sea from the Indian Ocean resulting in intervals of evaporation of surface sea water and, concomitantly, increase in salinity. These high salinity intervals have been recorded in the form of oxygen isotope measurements on pteropod and planktonic foraminiferal tests (Deuser and Degens, 1969). Four cycles were recorded and their relationship to biostratigraphic and chronostratigraphic data is shown in Berggren (1969, Fig. 2).

The four salinity-glacial cycles delineated by Deuser and Degens in core 154P correspond rather closely to the biostratigraphic subdivision of this core based upon planktonic foraminifera. This zonation, in turn, is based upon fluctuating percentages of the two dominant species in this core; these fluctuations are interpreted as being related to salinity and temperature changes.

In the lower part of core 154P (8m–*ca.* 6m), *G. sacculifer* and *G. ruber* exhibit moderate fluctuating percentages. Together they constitute between 50–100 per cent of the planktonic foraminiferal fauna over this interval (60 mesh screen = 250μ). Above this level *G. ruber* increases significantly whereas *G. sacculifer* occurs sporadically or is absent to a level between 3 and 2.5m where a sharp decrease occurs leading to the temporary disappearance of *G. ruber*. The relatively high percentages of *G. ruber* between 6 and 2.5m coincides approximately with Cycle II and the upper part of subjacent Cycle I. This interval corresponds, in turn, to the late Early Glacial and most of the Pleniglacial and approximately to dinoflagellate zones D and E, and the upper part of F (Wall and Warren, 1969). These authors have suggested that surface water temperatures may have varied between 13.5°–26.5°C in Zone E to 13°–18°C in Zone D to 22°-29°C in Zone F. This is compatible with data from the planktonic foraminifera. The estimates of surface water temperatures in Zone F are those which occurred, most likely, in shallow, near-shore areas. Surface waters in the central part of the Red Sea may be expected to have been somewhat cooler during this interval. The common occurrence together of *G. ruber* and *G. sacculifer* in planktonic foraminiferal Zone D (which corresponds approximately to dinoflagellate Zone F) suggests tropical-subtropical conditions with surface water temperatures within the range of their present day combined optimum, *i.e., ca.* 21°–30°C (Bé and Tolderlund, 1969). The continued flourishing of *G. ruber* (and low percentage of *G. sacculifer*) during planktonic foraminiferal Zone C may have been a result of its greater tolerance to in-

crease in salinity, whereas its relatively rapid disappearance at the top of Zone C and absence during a part of the suprajacent Zone B in this, and other, cores may have been a result of lowered surface water temperatures (average surface water temperatures below 14°–15°C). The interval encompassed by the upper part of foraminiferal Zone C and lower part of Zone B corresponds to the upper part of dinoflagellate Zone D and the lower part of Zone C and coincides with a period of extreme cold in northern Europe (the later phase of the Würm glaciation). Wall and Warren (1969) observe that dinoflagellate Zone C is characterized by a sharp decline in all species. There may be a direct relationship between the reduction in both planktonic foraminifera and dinoflagellates in the latest parts of the Würm glaciation, *i.e.,* between *ca.* 20,000–11,000 years B.P.

Herman (1969, Tab. 8) records, *i. al.,* *Globigerina bulloides* and a pteropod fauna dominated by *Creseis* spp. and characterized by *Limacina bulimoides, L. trochiformis,* and *Diacria trispinosa* in the latest glacial deposits (Stade, Unit III). *Creseis acicula,* the dominant form in the Last Glacial sediments in the Red Sea, has been found living in the Western Mediterranean during the winter in water temperatures as low as 13°C (Herman, 1969). *Styliola subula,* which does not occur in Holocene sediments in Red Sea cores, is present sparingly in sediments of the Last Glaciation, including the latest glacial phase (Stade, Unit III) and is known to tolerate temperatures lower than *Limacina trochiformis,* a common form in the Holocene.

The interval between *ca.* 20,000 years–11,000 years B. P. corresponds approximately to salinity-glacio-eustatic Cycle III (Berggren, 1969, Fig. 2). It would be expected that the relatively rapid shift in the δO^{18} ratios to lower values (which indicate restoration of more normal conditions through influx of Indian Ocean waters from the south) would also be reflected in a significant change in the planktonic foraminiferal fauna. In general, this does not occur. The microfauna at about 20,000 years B. P. is made up mostly by *G. ruber* and this species shows a sharp decline at about this time.

The general continuation through the interval of 20,000–11,000 years B.P. of a trend towards decrease in population which began somewhat earlier suggests the continuation of a change in the immediate environment which was responsible for this trend. That no significant faunal element from the south replaced the extant forms during this interval suggests that continued high salinity coupled with decrease in temperature as a result of glacio-eustatic lowering of sea level may have been the main factors involved in the maintenance of this trend towards decrease in populations. Tectonic movements about 17,000 years ago along the margins of the Red Sea may have also played a role in relative changes of sea level (Horowitz, 1967). In short, there was little in the northwestern Indian Ocean with which to replace or augment these populations at this time and cores in the northwestern part of the Indian Ocean may be expected to exhibit trends similar to those observed here in the interval between 20,000–11,000 years B.P.

However, the presence in some cores in the southern part of the Red Sea (e.g. V14-115, Herman, 1965; RC9-169, Herman, 1969) of *Globigerina bulloides* and pteropods with a lower temperature tolerance than Holocene forms, suggests that temperature is an important factor governing the distribution of species in the latest phase of the Last Glacial. As increased salinity and lowered temperature eliminated *G. ruber,* the sudden reestablishment of more normal connections with the Red Sea at about 20,000 years B.P., followed quickly by renewed evaporation and increase in salinity, may have allowed the development of these cooler forms in the southern part of the Red Sea at the expense of other less tolerant forms. Thus during the late Pleistocene (between *ca.* 80,000 and 20,000 years ago) surface water temperatures in the Red Sea may have dropped approximately 12°–13°C (from high values about 27°–28°C to below 15°C). This is in general agreement with the suggestion in Deuser and Degens (1969) that water temperature may have risen about 7°C during the Holocene (*i.e.,* over the past 10,000 years) on the assumption of a constant isotopic composition of the sea water during this time. Present day temperatures in the central part of the Red Sea average about 24°–25°C.

Most of the cores examined from the Red Sea terminate in Zone C (*G. ruber* being present to the almost total exclusion of other species). The boundary between zones B and C coincides approximately with the base of the youngest of several lithified carbonate layers. Fossil shells (mostly pteropods) are encrusted and filled with aragonite. Apparently there are several discrete layers of lithified, crusty material within this interval.

A distinct lithologic change occurs at the top of this lithified interval. The crusty material is absent, pteropods increase in abundance and four or five planktonic foraminifera make up the greater part of the microfauna in the coarse fraction. However, *G. sacculifer* is the dominant form (>50 per cent) in this interval—the Holocene. The top of the lithified sequence has been dated at about 11,000 years ago; the top of this layer may not be the same age in all cores, however, due to possible solution effects. This date corresponds essentially with the conclusion of the Würm glaciation in Europe. It has been suggested that at this time sea level was in the process of rising 90–100m in the Middle East causing a relatively rapid transgression of the Arabian Sea over the sill at the southern end of the Red Sea. This transgression restored the normal marine planktonic foraminiferal and pteropod populations in the Red Sea and can be expected to have a pronounced effect upon the indwelling populations.

Several species encountered in Red Sea cores are illustrated on Plate 1 and in Goll (1969).

Discussion

Herman (1965) recorded a relatively varied planktonic foraminiferal fauna from the Late Pleistocene-Holocene of the Red

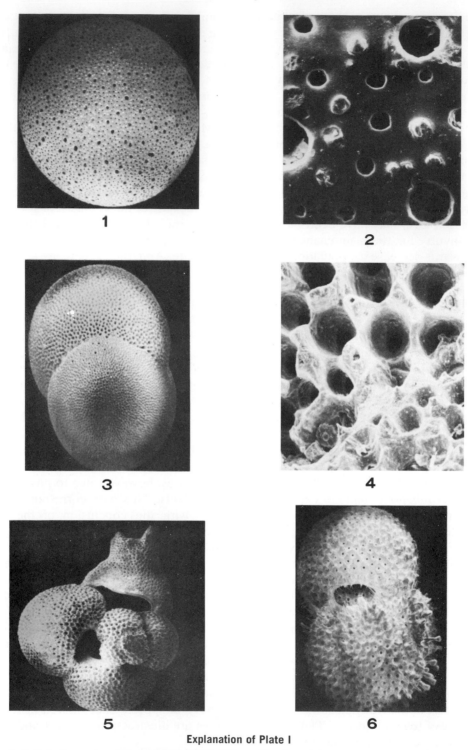

Explanation of Plate I

Figs. 1, 2. *Orbulina universa* from CH61-154P, 720cm; Fig. 1, *ca.* 70×; Fig. 2, *ca.* 900×. Note two distinct pore sizes, one roughly half the diameter of the other. Spine bases are also clearly visible between the pores.

Figs. 3, 4. Bilobate *Orbulina,* or *Biorbulina bilobata* from CH61-154P, 720cm; Fig. 3, *ca.* 70×; Fig. 4, *ca.* 900×. Fig. 4 is taken at the sutural junction between the two chambers in the central part of Fig. 3.

Figs. 5, 6. *Globigerinoides sacculifer* from CH61-154P, 720cm; Fig. 5 (spiral view), *ca.* 70×; Fig. 6 (edge view), *ca.* 70×. Note distally thickening spines and circular pores at base of reticulate "honey combed" ornament.

Sea. Eighteen species were recorded from 14 deep-sea cores and 6 plankton tows from the Red Sea region. Three additional cores were added to this total in a more recent study (Herman, 1969). We have examined several samples from cores studied by Herman (1965) (in particular V14-115) and have made a comparative analysis of her data. Because of the rather qualitative manner in which foraminiferal species data was presented (rare, frequent, common, abundant, and so forth) it is somewhat difficult to compare her data with those presented here (as per cent of total fauna). The following observations can be made:

1. In general the species distribution in cores studied by Herman (1965) are similar to those reported on here with the following exceptions: (a) *Globigerina bulloides* and *G. pachyderma,* reported from several levels in various cores (particularly V14-115, RC9-169) have not been found in any of the **CHAIN** Red Sea cores which we have studied. (b) *Globorotalia menardii,* reported as common in some intervals in V14-115 and RC9-169 (Herman, 1965, 1969) has been found only sporadically in our cores from the Red Sea. The absence of these forms in core CH61-154P may be due to differences in water mass circulation. Cores RC9-169 and V14-115 lie over 300km SE of CH 154P and V14-118.

The vertical changes in planktonic foraminiferal populations in Late Pleistocene Red Sea cores are due to temporary modifications in the hydrography of the Red Sea itself. This was occasioned by periodic interruption of the connection with the Indian Ocean to the south. It might be expected that fluctuations which could bring certain cold or warm species into the Red Sea from the south would be most evident in cores in the southern part of the Red Sea and that these species were unable to penetrate in significant numbers very far north into the Red Sea. This may explain the difference in latitudinal distribution of *G. bulloides, G. pachyderma* and *G. menardii-G. tumida* in Red Sea cores. Herman (1965, p. 385) observed that *G. bulloides* was common in Red Sea cores south of 18°N Lat. and suggested that the de-

crease in frequency to the north may be due more to salinity than to temperature.

2. Relatively high frequencies of *Globigerinoides ruber* were recorded in the lower part of Red Sea cores—*i.e.,* during Würm Glacial—followed by a short interval in which planktonic species were rare or absent (= Late Glacial), followed, in turn, by relatively high frequencies of *G. sacculifer* (Holocene). A marked similarity exists between the frequency distribution of *G. sacculifer* in Red Sea cores studied by Herman (1965, Fig. 15) and the **CHAIN** 61 cores which we have examined. By integrating data on the distribution of *Globigerinoides ruber* and *G. sacculifer* in V14-118 (Herman, 1965) and the nearby CH 61-154P an accurate correlation between the two cores is possible (see Fig. 3). An explanation of this figure follows.

1. The fluctuating curve (dashed line) which is drawn for V14-118 is taken from Herman (1965, Fig. 14). It represents a ratio of *G. sacculifer* to *G. ruber,* converted to tangent of arc. Thus relatively high frequencies of *G. sacculifer* in association with *G. ruber* will appear as increases to the right, whereas low frequencies of *G. sacculifer,* as well as its absence and the concomitant presence or absence of *G. ruber* will be recorded as low values or 0, respectively.

2. In the lower part of V14-118 (>6m– *ca.* 5m) a fluctuating ratio is seen which is reflected in the alternating percentages between >8m-*ca.* 6m in CH 61-154P. The boundary between planktonic foraminiferal zone C/D is drawn at *ca.* 630cm (*i.e.,* the level above which the percentage of *G. ruber* increases at the expense of *G. sacculifer*) and this boundary is extended to V14-118 (correlation line 5).

3. In CH 61-154P *G. sacculifer,* after exhibiting a persistent trend towards lower percentages, disappears at about 5m. This is reflected in V14-118 in the lower ratio values above 5m which decrease to 0 at about 4m. This level is correlated between the two cores (line 4).

4. The essentially flat level (*i.e.,* 0) of the ratio between 4m and 50cm in V14-118 is reflected in CH 61-154P by the absence or

sporadic occurrence of *G. sacculifer* between 5m and 120cm. Above 50cm (V14-118) and 120cm (CH 61-154P), respectively, the relatively high values of *G. sacculifer* are seen in both cores and heralds the repopulation of the Red Sea region by a normal Indian Ocean fauna. This level in CH 61-154P forms the boundary between planktonic foraminiferal zone B/C and is correlated with the corresponding level in V14-118 as line 1.

5. A level at about 3m in CH 61-154P, at which *G. ruber* reaches a peak value before showing a rapid decline and eventual disappearance is correlated, tentatively, with a weak positive value at about 230cm in V14-118 and which is interpreted by Herman (1965) to represent a brief warm interval during the Last Glacial. This level corresponds to about 29–30,000 years B.P. Another peak of *G. ruber* occurs in CH 154P at about 380cm which is approximately 37,000 years B.P. Both these levels fall within the Middle (or Interpleniglacial) interval (van der Hammen *et al.*, 1967). Zagwijn (1963, p. 177) observes that "slight improvements of climate seem to have recurred at times, for instance at about 37,000 years B.P. and 30,000 years B.P." The similarity in data is striking.

6. The boundary between planktonic foraminiferal zones B/C falls at about 260cm in CH 61-154P and is marked by the temporary disappearance of *G. ruber*. This level is correlated (line 2) with the level in V14-118 in which *G. ruber* shows a similar temporary disappearance (Herman, 1965, Tab. 6).

7. Herman's (1965) interpretation of the Late Pleistocene-Holocene chronostratigraphy and climatic variations are shown in the right side of Fig. 3. An alternate interpretation (which differs only slightly from her interpretation) is presented on the extreme right-hand column based upon the chronostratigraphic scale adapted in this investigation and which is shown on the left of Fig. 3. Herman (1965, p. 401) suggests that V14-115, 116 and 118 penetrate sediments deposited during the last interglacial (Riss/Würm). We would place this level somewhat higher, but the difference in interpretation is one of degree rather than kind.

Summary

Analysis of twenty cores, nine of them from within an area of deep brine holes in the Red Sea has shown that, in general, four species of planktonic foraminifera predominate in the +60 fraction. These are: *Orbulina universa, Hastigerina siphonifera, Globigerinoides ruber,* and *G. sacculifer.* Also present in some samples, but in some instances apparently inseparable from *H. siphonifera,* is *Hastigerina pelagica.* In the present study these two forms have been included together. *Globorotalia menardii* and *Globigerinoides conglobatus* are also occasionally present.

In the finer fractions the following species have been recorded: *Globigerina quinqueloba, Globigerinita glutinata, Globorotalia anfracta, Globigerinoides rubescens,* and *G. tenellus.*

On the basis of fluctuating percentages in *Globigerinoides ruber* and *G. sacculifer* a fourfold biostratigraphic zonation has been proposed for sediments spanning the past 80,000 years and correlation with other Red Sea cores taken on **VEMA** cruise 14 is made. Glacio-eustatic sea level changes are believed to have caused alternating periods of partial or total isolation of the Red Sea from the Indian Ocean resulting in increased evaporation and high salinities. Superimposed upon this is a trend of gradual but inexorable cooling of the surface waters of the sea. It is virtually impossible to separate the effects of salinity and temperature on the distributional patterns of planktonic foraminifera during the Late Pleistocene. However, a general explanation is attempted.

The occurrence together in moderate fluctuating percentages of *G. sacculifer* and *G. ruber* between 80,000 and 50,000 years ago is interpreted as indicating tropical or subtropical conditions in the Red Sea with surface temperatures ranging between 20°–30°C. From 50,000–25,000 years ago *G. ruber* continued to flourish in Red Sea sediments at the expense of *G. sacculifer* which exhibits a steady decline. This trend may be an indication of the adaptability (*i.e.,* tolerance) of *G. ruber* to survive under severe ecologic conditions, in this instance, a relatively strong increase in salinity. The relationship between *G. sacculifer* and *G. ruber* may serve as a useful tool in interpreting paleoecologic conditions in tropical-subtropical areas in which high salinities can be demonstrated.

Tectonic movements about 17,000 years ago may have also been involved in the modification of paleogeographic conditions of the Red Sea by glacial-eustatic sea level lowering causing high salinities during the latest part of the Würm glaciation (Horowitz, 1967). An interval between *ca.* 20,000 and 11,000 years ago is characterized by reduced populations or the complete absence of planktonic foraminifera and corresponds, at least in part, to a period of severe climatic conditions in northern Europe. The persistence of pteropods in this interval testifies to their ability to withstand greater extremes of climatic change — in this case extremely high salinity and relatively low surface water temperatures — than the planktonic foraminifera. This fact should also serve as a useful tool in future paleoecologic investigations in which pteropods and planktonic foraminifera are employed together.

With the gradual melting of the Würm glaciers and concomitant rise in sea level connection with the Indian Ocean was once again reestablished about 11,000 years ago and a normal, if somewhat, restricted, planktonic foraminiferal fauna was once again established in the Red Sea.

This investigation has led to the following interesting observations: under conditions of high temperature and salinity in tropical-subtropical areas *Globigerinoides sacculifer* and *G. ruber* appear to be the most resistant species. With further increase in salinity (and a probable decrease in temperature) *G. ruber* is the more resistant of the two. *Globigerinoides ruber* appears, then, to be the most tolerant species of planktonic foraminifera to increase in salinity. This agrees well with the most re-

cent data available on the present-day distribution of these two species. On the basis of our present investigation a sequence of elimination based on increase in salinity can be suggested (from the least resistant to the most resistant): *Orbulina universa* → *Hastigerina siphonifera* → *Globigerinoides sacculifer* → *Globigerinoides ruber.*

Acknowledgments

We should like to express our gratitude first to Dr. Robert Goll and to the following colleagues with whom we have had stimulating discussions on various aspects of Red Sea research: Drs. J. Bischoff, C. Chen, Y. Herman, T. L. Ku, J. Milliman, D. A. Ross, and D. Wall. We are particularly grateful to Dr. A. W. H. Bé, Lamont Geological Observatory, for allowing us access to unpublished data which has been of considerable value in our interpretative work on Red Sea cores. Miss Janice Oliphant helped with core sampling and sample preparation.

We would like to thank Mr. Gary Cogswell of JEOLCO in Medford, Massachusetts for allowing us to make the electron scanning microscope photographs on the JSM-2 instrument.

This work was supported by grant GA-676 and GA-584 from the National Science Foundation.

References

Bé, A. W. H.: Ecology of Recent planktonic foraminifera: Part 1 — Areal distribution in the western North Atlantic. Micropaleontology, **5,** 77 (1959).
———: Ecology of Recent planktonic foraminifera: Part 2 — Bathymetric and seasonal distributions in the Sargasso Sea off Bermuda. Micropaleontology, **6,** 373 (1960).
——— and W. H. Hamlin: Ecology of Recent planktonic foraminifera: Part 3 — Distribution in the North Atlantic during the summer of 1962. Micropaleontology, **13,** 87 (1967).
Bé, A. W. H. and D. S. Tolderlund: Distribution and ecology of living planktonic foraminifera in surface waters of the Atlantic and Indian Oceans. *In: S. C. O. R. Colloquium on Micropaleontology of Deep-Sea Sediments,* Cambridge, England (1969 in press).
Berggren, W. A.: Micropaleontologic investigations of Red Sea cores — summation and synthesis of results. *In: Hot brines and recent heavy metal deposits in the Red Sea,* E. T. Degens and D. A. Ross

(eds.). Springer-Verlag New York Inc., 329–335 (1969).

Bischoff, J. L.: Red Sea geothermal brine deposits: their mineralogy, chemistry and genesis. *In: Hot brines and recent heavy metal deposits in the Red Sea,* E. T. Degens and D. A. Ross (eds.). Springer-Verlag New York Inc., 368–401 (1969).

Bradshaw, J. S.: Ecology of living planktonic foraminifera in the north and equatorial Pacific Ocean. Contrib. Cush. Found. for Foram. Res. X, **2,** 25 (1959).

Brady, H. B.: Challenger Rept., **9,** 814 p. (1884).

Chen, C.: Pteropods in the hot brine sediments of the Red Sea. *In: Hot brines and recent heavy metal deposits in the Red Sea,* E. T. Degens and D. A. Ross (eds.). Springer-Verlag New York Inc., 313–316 (1969).

Deuser, W. G. and E. T. Degens: O^{18}/O^{16} and C^{13}/C^{12} ratios of fossils from the hot-brine deep area of the central Red Sea. *In: Hot brines and recent heavy metal deposits in the Red Sea,* E. T. Degens and D. A. Ross (eds.). Springer-Verlag New York Inc., 336–347 (1969).

Ericson, D. B., M. Ewing, G. Wollin and B. Heezen: Atlantic Deep-sea sediment cores. Geol. Soc. Amer. Bull., **72,** 193 (1961).

Fichtell, L. von and J. P. C. von Moll: *Testacea microscopica aliaque minuta e generibus Argonauta et Nautilus,* Wien, Osterreich, Camesina (1798), (1803 reprint).

Goll, R.; Radiolaria: History of a brief invasion. *In: Hot brines and recent heavy metal deposits in the Red Sea,* E. T. Degens and D. A. Ross (eds.). Springer-Verlag New York Inc., 306–312 (1969).

Herman, Y.: Etude des sédiments quaternaires de la Mer Rouge. *Doctoral thesis.* Univ. Paris, sér. A, 1123. Ann. Inst. Oceanogr., Monaco, Masson and Co., Paris, **2,** 341 (1965).

——: Vertical and horizontal distribution of pteropods in Quaternary sequences. *In: S.C.O.R. Colloquium on Micropaleontology of Deep-Sea Sediments,* Cambridge, England (1969 in press).

Horowitz, A.: The geology of Museri Island (Dahlak archipelago, southern Red Sea). Israel Jour. Earth Sci., **16,** 74 (1967).

Ku, T. L., D. L. Thurber and G. Mathieu: Radiocarbon chronology of Red Sea sediments. *In: Hot brines and recent heavy metal deposits in the Red Sea,* E. T. Degens and D. A. Ross (eds.).

Springer-Verlag New York Inc., 348–359 (1969).

Marie, P.: Sur la Familles des Foraminiféres des dépots litteraux actuels de la Mer Rouge et de Djibouti. Mém. Soc. Linn. Normandie n. sér. **1,** 53 (1941).

McIntyre, A.: The Coccolithophorida in Red Sea sediments. *In: Hot brines and recent heavy metal deposits in the Red Sea,* E. T. Degens and D. A. Ross (eds.). Springer-Verlag New York Inc., 299–305 (1969).

Olausson, E.: Description of sediment cores from the Red Sea and Mediterranean. Repts. Swed. Deep Sea Exped. VIII, **5,** 287 (1960).

d'Orbigny, A.: Tableau méthodique de la classe des Céphalopodes. Ann. Sci. Nat. **7,** 245 (1826).

Parker, F. L.: Eastern Mediterranean foraminifera. Repts. Swed. Deep Sea Exped. VIII, **4,** 219 (1958).

——: Planktonic foraminiferal species in Pacific sediments. Micropaleontology, **8,** 2, 219 (1962).

Parker, W. K. and T. R. Jones: On some foraminifera from the North Atlantic and Arctic Oceans, including Davis Straits and Baffin's Bay. Roy. Soc. London, Philos. Trans. **155,** 325 (1865).

Ross, D. A. and E. T. Degens: Shipboard collection and preservation of sediment samples collected during CHAIN 61 from the Red Sea. *In: Hot brines and recent heavy metal deposits in the Red Sea,* E. T. Degens and D. A. Ross (eds.). Springer-Verlag New York Inc., 363–367 (1969).

Said, R.: Foraminifera of the northern Red Sea. Cushman Lab. Foram Res., Spec. Pub. **26,** 1 (1949).

——: The distribution of foraminifera in the northern Red Sea. Cushman Found. Foram. Res. Contrib., **1,** 9 (1950).

——: Organic origin of some calcareous sediments from the Red Sea. Science, **113,** 517 (1951).

van der Hammen, T., G. C. Maarleveld, J. C. Vogel, and W. H. Zagwijn: Stratigraphy, climatic succession and radiocarbon dating of the Last Glacial in the Netherlands. Geol. en Mijnb., **46,** 3, 79 (1967).

Wall, D. and J. S. Warren: Dinoflagellates in Red Sea piston cores. *In: Hot brines and recent heavy metal deposits in the Red Sea,* E. T. Degens and D. A. Ross (eds.). Springer-Verlag New York Inc., 317–328 (1969).

Zagwijn, W. H.: Pleistocene stratigraphy in the Netherlands, based on changes in vegetation and climate. Verhand. K. Nederl. Geol. Mijnb., Gen. Geol., **21,** 173 (1963).

The Coccolithophorida in Red Sea Sediments *

ANDREW McINTYRE

Lamont Geological Observatory of Columbia University
and Queens College of the City University of New York

Abstract

Coccoliths delineate four and possibly five main zones within core C61-154P which are equivalent to the major glacial salinity cycles. The youngest and oldest zones contain oceanic warm water floras indicative of interglacial climates while the zones between represent restricted cooler waters with occasional influxes of oceanic species.

The presence in the lower portion of this core of two possible indicator species, *Gephyrocapsa ericsoni* cf. and *Coccolithus protohuxleyi*, allows correlation with Atlantic upper Pleistocene sediments (core V12-122) and substantiates the Eemian-Würm boundary date of approximately 70,000 years B.P.

Introduction

The Coccolithophorida are a group of unicellular, biflagellate, golden brown algae that secrete small, $1-25\mu$, calcite plates called coccoliths. Coccoliths are an important constituent of oceanic sediments and their fossil record extends into the Jurassic. Recent work indicates that they are excellent stratigraphic and climatic indicators for the Quaternary period (McIntyre, 1967; McIntyre *et al.*, 1967; Boudreaux and Hay, 1967).

In an attempt to apply recent work done by the author on coccolith stratigraphy of Atlantic cores (McIntyre, 1967; McIntyre and Bé, 1968) to the presumed extreme Red Sea climatic and oceanographic fluctuations of the upper Quaternary, core C61-154P was examined. The higher sedimentation rate combined with useful radiometric dates (Ku *et al.,* 1969) should allow

a greater discrimination of climatic effect on species as well as cross correlation between oceanic areas.

Methods

Samples were taken every 20cm from the top to 880cm and were the same as those used by Wall and Warren (1969).

Because of the extremely small size of coccoliths, only the electron microscope can be used for comprehensive counting. The limitations of EM work requires that the samples be cleaned of clay before viewing. Consequently all samples were treated by the method previously described by the author (McIntyre *et al.*, 1967). Counting unit quantities is impossible when a gram of sediment may contain over a million coccoliths. Consequently only relative percentage counts based on 200 individuals per sample were made. The generally corroded condition of much of the core material precluded the taking of micrographs except in isolated samples. Finally, the nearness of outcropping older sediments in the shores combined with the usual mixing effects and current distribution has produced some contamination of the flora by older species. Pre-Pleistocene contamination can be easily discriminated because of recrystallization effects as well as species type, e.g. discoasters; however, mixing within the Pleistocene samples tends to blur any sharp zonation.

Species List

The following 21 species have been identified in core C61-154P: *Acanthoica quattrospina* Lohmann, *Anthosphaera robusta* (Lohmann) Kamptner, *Calciopappus cau-*

* Lamont Geological Observatory Contribution No. 1293.

datus Gaarder & Ramsfjell, *Syracosphaera pirus* Halldal & Markali, *Syracosphaera pulchra* Lohmann, *Pontosphaera variabilis* Halldal & Markali, *Helicosphaera carteri* (Wallich) Kamptner, *Calciosolenia sinuosa* Schlauder, *Discosphaera tubifera* (Murray and Blackman) Lohmann, *Umbellosphaera irregularis* Kamptner, emend Paasche, *Umbellosphaera tenuis* (Kamptner) Paasche, *Rhabdosphaera stylifera* Lohmann, *Coccolithus huxleyi* (Lohmann) Kamptner, *Cyclococcolithus fragilis* (Lohmann) Deflandre, *Cyclococcolithus leptoporus* (Murray and Blackman) Kamptner, *Cyclolithella annulus* (Cohen) n. comb., *Gephyrocapsa ericsoni* McIntyre and Bé, *Gephyrocapsa protohuxleyi* McIntyre, *Gephyrocapsa oceanica* Kamptner, *Umbilicosphaera mirabilis* Lohmann, *Scyphosphaera elegans* (Ostenfeld) Deflandre.

In addition, a number of holococcoliths and a few incomplete unidentified heterococcoliths have been found. The total of all coccolith species in these samples numbers 33.

Coccolith Zonation

Four and possibly five discrete zones based on coccolith assemblages occur in C61-154P. Coccolith zone A, 0–120cm, contains a flora dominated by oceanic tropical to subtropical species. The coccoliths are well preserved, easing identification and counting as well as discrimination of older contamination which usually shows serious corrosion or secondary crystal overgrowth. Twenty-six species have been identified. 60 per cent of the flora are oceanic species of which *Umbellosphaera irregularis, Coccolithus huxleyi* and *Gephyrocapsa oceanica* are the most abundant (Fig. 1). A second component of the assemblage is made up of species indigenous to neritic shallow waters. Some of these species are known to have an alternate benthic generation which is a noncoccolith bearing filamentous stage (Boney and Burrows, 1966). Two species in this group, *Anthosphaera robusta* and *Scyphosphaera elegans* (Figs. 2C, 2D) average 30 per cent of total species number in Zone A. *S. elegans* is reported to be limited to Mediter-

ranean waters (Deflandre, 1942); however, a similar species has been found in the coastal waters of North America and the Indian Ocean by the author.

The sediments of Zone B, 120–280cm, are largely indurated and the preservation is exceedingly poor. The coccoliths when present are corroded often beyond recognition. The major coccolith structural types vary in resistivity to solution (e.g., holococcoliths are extremely weak; heterococcoliths are stronger, but within this group the subtypes vary from the generally weak caneoliths through scapholiths and cyrtoliths to the strongest forms, the placoliths). Thus under the solution conditions present in this zone there will be differential preservation and any counts made may be more representative of coccolith strength than the actual flora. The abnormally high numbers of the resistant coccolith *Helicosphaera carteri* in this zone is an example. Only the sediments between 220 and 260cm can be considered to contain even a partially representative coccolith assemblage. The outstanding facet of this assemblage is the dominance of the shallow water species *A. robusta* and *S. elegans* (Fig. 1). This is particularly significant since neither of these species is considered resistant to solution.

From 280–520cm the overall preservation is better except for the area around 480cm. The flora becomes increasingly cosmopolitan with depth. Zone C is thus typified by the reappearance of *Umbilicosphaera mirabilis, Cyclolithella annulus, Cyclococcolithus leptoporus* and *Cyclococcolithus fragilis*.

Zone D, 520–760cm, is dominated by a species identified as *Gephyrocapsa ericsoni cf.* (Figs. 1 and 2E). While nearly absent from younger sediments it constitutes over 30 per cent of total flora from 520cm to the base of the core. In addition, the return of oceanic, tropical and subtropical species in abundance gives the coccolith assemblages of this zone a strong similarity to that of Zone A. There is a minimum of corrosion within this zone as evidenced by the species diversity and presence in measurable quantities of the more fragile caneoliths.

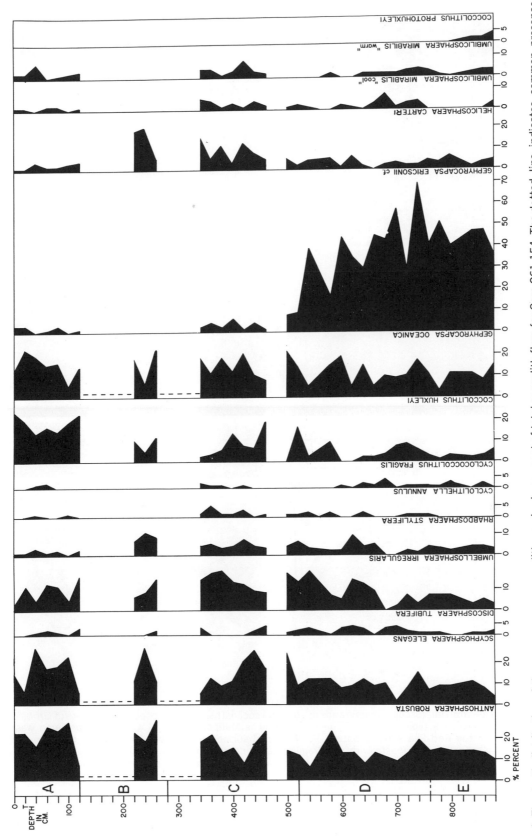

Fig. 1. The depth distribution of the more common coccolith species in per cent of total coccolith flora for Core C61-154. The dotted line indicates common presence in zones where counting was impossible due to corrosion. Since not all species are included values at any one depth will never equal 100 per cent.

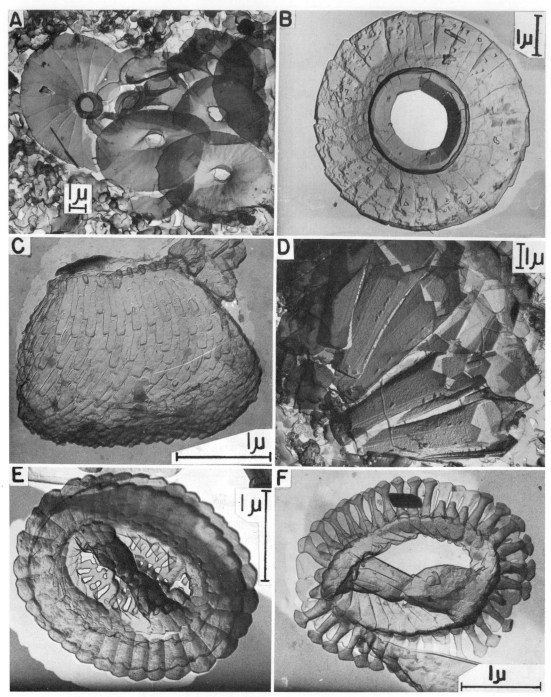

Fig. 2. Some electron microscope photographs of common coccoliths. A. A collapsed coccosphere of *Umbellosphaera irregularis,* a typical species. The background is Millipore filter on which the specimen was collected from the Red Sea surface waters. B. The distal surface of the cooler ecotype of *Cyclolithella annulus.* C. *Anthosphaera robusta* with the fragile basal ring missing as it is in most specimens from C61-154P. D. *Scyphosphaera elegans* collected live on a Millipore filter, only a portion of the coccosphere remains intact. E. The distal surface of *Gephyrocapsa ericsoni cf.,* which characterizes the flora of Zones D and E.   F. The distal surface of *Coccolithus protohuxleyi,* from Zone E.

Between 760 and 800cm a few minor changes occur in the flora and a new species, *Gephyrocapsa protohuxleyi*, appears (Fig. 2F). The near disappearance of *Cyclolithella annulus* and *Umbilicosphaera mirabilis* "cool" are probably climatic effects since they are present throughout the Pleistocene, but *G. protohuxleyi* is a form absent in today's sea. The presence of this species could be used to delineate a fifth zone E.

Climatic Fluctuations

Many species of Coccolithophorida are good temperature indicators, particularly in subtropical and transitional waters. The Red Sea, however, is in the tropical belt; even during the coldest periods of the glacial Pleistocene it is doubtful if cooler water floras would have been present. This would be particularly true if the sea were cut off from the Indian Ocean at these times by a drop in sea level. The lack of large numbers of *Cyclococcolithus leptoporus* and the complete absence of *Coccolithus pelagicus* is indicative of this.

The coccoliths species give the impression that a generally tropical flora with subtropical components persisted throughout the length of time represented by this core. Zones A and C through D contain tropical indicator species a good portion of the time. Only Zone B with its limited flora really gives any indication of a cooling trend in the Red Sea by the presence of cool ecotypes of subtropical species. The general rule that fewer species represent colder conditions is not true here; rather, restricted waters with higher salinities are in part responsible.

It has been shown that a number of species living today have coccolith ecotypes which can be assigned to cooler and warmer conditions within the normal range of an individual species (McIntyre and Bé, 1967). *Umbilicosphaera mirabilis* and *Cyclolithella annulus* (Fig. 2B) both show this phenomena. The "cooler" coccoliths in these species being found in waters around 18°C and 20°C respectively. In Zone B only the cooler forms are pres-

ent and restricted to a few samples. In Zones C and D both ecotypes are present for each species with dominance fluctuating in phase with the climatic changes.

Umbellosphaera irregularis (Fig. 2A) and *Umbellosphaera tenuis* are the best indicators of warm waters in the open ocean, the former being the higher temperature (tropical) species while the latter is subtropical. *U. tenuis,* while found in small numbers in the Red Sea today, was not identified from Zones C and D. Since *U. tenuis* is only distinguished from *U. irregularis* by the ribbed ornamentation on the distal coccolith surface, it is possible that due to corrosion this species may have been identified as *U. irregularis*. It is present in Zone B lending credence to the consideration of this zone as being cooler.

Deuser and Degens (1969) suggest that on the basis of salinity cycles erected from O^{18} values there were four major influxes of oceanic water from the Indian Ocean. These occurred at 11,000, 20–25,000, 40–45,000 and 65–200,000 years B.P. The latter three times of presumed influx correlate well with the periodic appearance of *Cyclococcolithus leptoporus* at 220–260cm, 380–440cm and 660–700cm and somewhat less perfectly with *C. fragilis*. These oceanic forms today are absent to rare in the Red Sea while common in the Indian Ocean between 10° and 40° latitude south.

Since they are not today a part of the flora of the Red Sea connected to the Indian Ocean why should they indicate influx of oceanic water in the past? The influx in question is the result of eustatic sea level rise occurring at warmer episodes within the generally colder glacial periods. During these colder times the subtropical floras were compressed into tropical areas (McIntyre, 1967). Thus *C. leptoporus* would have been common in equatorial Indian Ocean waters adjacent to the mouth of the Red Sea. Comparison of Deuser and Degens' O^{18} curves with *C. leptoporus* presence shows that they occur slightly later than the apex of the three "warm" peaks presumably in phase with the maximum melting effect and sea level rise. *C.*

fragilis, more tolerant than *C. leptoporus,* does show this effect for the two oldest peaks in the O^{18} curves.

Correlation with Atlantic Pleistocene Floras

While very little has been published on Quaternary coccolith stratigraphy in oceanic sediments, recent work by the author and his colleagues can be used for purposes of correlation (McIntyre and Bé. 1967). Because the Atlantic has a much lower average sedimentation rate than the Red Sea, roughly one-fourth, fine scale correlation based on climatic fluctuations within the Würm Glacial is not feasible at this time. However, correlation between major boundaries such as the pleniglacial (520cm) — early glacial and Würm — Eemian is possible.

Cores V12-122 (17°00′N, 74°24′W) from the Caribbean and V12-18 (28°42′S, 34°30′W) from the South Atlantic have been examined for upper Quaternary coccolith stratigraphy. V12-122 was chosen on the basis of well dated boundaries (Ku and Broecker, 1966) as well as previous paleontological studies (Ericson *et al.,* 1964). In this core, *Gephyrocapsa ericsoni cf.* has a distribution similar to C61-154P. While present throughout the length of V12-122 it is rare from 0 to about 180cm where it becomes common, analogous to the sharp increase at 500–520cm in C61-154P. *Gephyrocapsa protohuxleyi* is rare to absent above 210–220cm in V12-122 but forms a significant part of the flora below these zones. This distribution with the lag between *G. ericsoni cf.* and *C. protohuxleyi* is similar to that in C61-154P. The depth at which *G. protohuxleyi* last appears in abundance in V12-122 is considered the boundary between Ericson's X (interglacial) and Y (glacial) and is dated by Ku and Broecker using Th^{230} at 75,000 ± 8,000 years B. P. Sackett (1965) using Pa^{231} has dated the 210cm level at 76,000 years B. P. while Ericson *et al.* place the X-Y boundary at 68,000 ± 4,000 years B. P. The disappearance of *C. protohuxleyi* in C61-154P occurs between 800 and 760cm. Based on

extrapolation from C^{14} dates (Ku *et al.,* 1969) the base of Zone D at 760cm, the Eemian-Würm is dated at *ca.* 70,000 years B. P. The change in abundance of *G. ericsoni cf.* at approximately 180cm in V12-122 dated by extrapolation is roughly 59,000 years while this change occurs in C61-154P at *ca.* 50,000 years. V12-18 shows the same sequence of events but due to its much lower sedimentation rate the dates assigned have too broad a range.

Considering the difference in sedimentation rate between these cores as well as the usual difficulties inherent in radiometric dating, the correlation is very good. The disappearance in time of *C. protohuxleyi* in two widely separated areas at a major boundary would appear to make this species an excellent time-stratigraphic indicator, at least for tropical-subtropical waters. However, only after similar studies are completed on Pacific Quaternary cores can we really use *G. protohuxleyi* in this way.

Conclusions

1. The coccolith floras of Red Sea core C61-154P delineate 4 and possibly 5 zones that can be equated with glacial-salinity cycles.
2. The flora is divided into oceanic and shallow neritic species, the latter's high relative abundance in the glacial period indicating restricted, high salinity conditions.
3. The intermittent appearance of oceanic species, *Cyclococcolithus leptoporus* and *C. fragilis* at periods of presumed higher temperatures is indicative of oceanic influx due to eustatic rise of sea level.
4. The abrupt change in abundance of *Gephyrocapsa ericsoni cf.* and *Coccolithus protohuxleyi* respectively is correlatable with similar dated occurrences in Atlantic and Caribbean cores. The boundaries thus dated may be the early glacial-pleniglacial at 50–59,000 years B. P. and Eemian-Würm at approximately 70,000 years B. P.

Acknowledgment

This research was carried out under NSF Grant GA 1205.

References

Boney, A. D. and A. Burrows: Experimental studies on the benthic phases of Haptophyceae, I., effects of some experimental conditions on the release of coccolithophorids. J. Mar. Biol. Ass. U. K., **46,** 295 (1966).

Boudreaux, J. E. and W. W. Hay: Zonation of the latest Pliocene-Recent interval. Gulf Coast Assoc. Geol. Soc. Trans., **17,** 443 (1967).

Deflandre, G.: Coccolithophoridées fossiles D'Oranie. Genres Scyphosphaera Lohman et Thorosphaera Ostenfeld. Bull. Soc. Hist. Nat. Toulouse, **77,** 125 (1942).

Deuser, W. G. and E. T. Degens: O^{18}/O^{16} and C^{13}/C^{12} ratios of fossils from the hot brine deep area of the central Red Sea. *In: Hot brines and recent heavy metal deposits in the Red Sea,* E. T. Degens and D. A. Ross (eds.). Springer-Verlag New York Inc., 336–347 (1969).

Ericson, D. B., M. Ewing, and G. Wollin: Pleistocene epoch in deep sea sediments. Science, **146,** 723 (1964).

Ku, T. L., D. L. Thurber, and G. Mathieu. Radiocarbon chronology of Red Sea sediments. *In: Hot brines and recent heavy metal deposits in the Red Sea,* E. T. Degens and D. A. Ross (eds.). Springer-Verlag New York Inc., 348–359 (1969).

Ku, T. L. and W. S. Broecker: Atlantic deep-sea stratigraphy: extension of absolute chronology to 320,000 yrs. Science, **151,** 448 (1966).

McIntyre, A.: Coccoliths as paleoclimatic indicators of Pleistocene glaciation, Science, **158,** 1314 (1967).

———, and A. W. H. Bé: Modern coccolithophorida of the Atlantic Ocean, I, placoliths and cyrtoliths. Deep-Sea Research, **14,** 561 (1967).

——— and A. W. H. Bé: Coccolith Quaternary Stratigraphy of the Atlantic Basin. Abstract Geol. Soc. Amer. General Meetings Mexico City (1968).

———, ———, and R. Preikstas: Coccoliths and the Pliocene-Pleistocene boundary. Progress in Oceanography, **4,** 3 (1967).

Sackett, W. M.: Deposition rates by the protactinium method. Symp. on Marine Geochemistry, Grad. Sch. of Oceanography, Univ. of Rhode Island, Occ. Publ., **3** (1965).

Wall, D. and J. S. Warren: Dinoflagellates in Red Sea piston cores. *In: Hot brines and recent heavy metal deposits in the Red Sea,* E. T. Degens and D. A. Ross (eds.). Springer-Verlag New York Inc., 317–328 (1969).

Radiolaria: The History of a Brief Invasion *

ROBERT M. GOLL †

Woods Hole Oceanographic Institution
Woods Hole, Massachusetts

Abstract

Radiolaria occur in only three of the **CHAIN** 61 collection of sediment cores from the Red Sea: 126P, 127P, and 128P. These three cores were collected from the Atlantis II Deep and contain a total of 56 species of Radiolaria. The radiolarian sequences in these cores are believed to be synchronous, and the assemblage represents a restricted faunule derived from the Indian Ocean. The radiolarian episode began approximately 12,000 years ago and ended about 9,000 years ago. Sphalerite is associated with a portion of the radiolarian sediments of 127P and 128P.

Introduction

All the **CHAIN** 61 sediment cores from the Red Sea sampled for microfossils were examined carefully for Radiolaria, which were only found in three piston cores from the Atlantis Deep. Radiolaria are present between 576 and 632cm below the top of 126P, between 584 and 786cm below the top of 127P and between 845 and 874cm below the top of 128P. The concentration of specimens varies among the samples, but it is never as abundant as in samples from some abyssal regions of the open ocean. Sediment samples of approximately 15cc contain an average of 300 to 500 specimens, although six samples contained less than 100 specimens and three samples were barren. Presumably, highly variable sediment-accumulation rates have strongly influenced the total abundance of specimens in these samples.

The state of preservation of specimens was good in all the samples. In cores 127P and 128P, radiolarian specimens were comparable in size and appearance to other specimens of these species observed from other regions. Many radiolarian skeletons in 126P were unusually thin, although pitted surfaces characteristic of dissolution conditions were not observed.

The time relationships of these three cores has not been determined on the basis of calcareous microfossils. The concentration of biogenous calcite is generally low, and Foraminiferida are scarce and poorly preserved. Moreover, the pattern of foraminiferal-species frequency variation with core depth is not consistent among the three cores and cannot be correlated to other Red Sea cores.

Method of Study

In order to determine the time relationships of the radiolarian sections of 126P, 127P and 128P, it was necessary to compute the frequencies of each species in all the samples. Such statistical treatment is considered feasible because of the unusual depositional environment in which these cores were collected. The three cores are relatively closely spaced, the maximum horizontal separation being 5km, and they are all located in a basin in which deposition has almost certainly been continuous, although not uniform, for a long period of time. The restricted nature of the overlying water excludes the possibility of contami-

* Woods Hole Oceanographic Institution Contribution No. 2192.

† Present address: Lamont Geological Observatory of Columbia University, Palisades, New York.

nation of the burial assemblage by foreign radiolarian skeletons. Because the invasion of Radiolaria to this region of the Red Sea was apparently brief, it seems unlikely that large-scale sediment-redistribution phenomena influenced the burial assemblage. The intricate layering of sediment in these cores indicates that sediment reworking by burrowing organisms has not occurred. These considerations suggest that the burial assemblages in 126P, 127P and 128P represent a true, if only incomplete, history of radiolarian activity in the water column above the Atlantis II Deep.

Because of variability in the number of specimens present in the studied samples, uniform counts were not possible. In samples with sufficiently large assemblages, three hundred specimens were counted. All specimens were counted in samples of lower abundance. Strewn slides were made of the radiolarian residues of each sample, and no attempt was made to ensure random fractions of the residues. Repeated counts of a few of the samples were made to obtain an approximation of the range of standard deviations for this technique. Values for the percentage of each species in the total radiolarian assemblage were found to have a standard deviation of ±0.7 per cent for a count of 300 specimens. Counts of less than 200 specimens had an average standard deviation of ±2.3 per cent.

The Radiolarian Faunule

Frequency counts must be preceded by thorough taxonomic investigations. Unfortunately, studies of Radiolaria are severely hampered by the primitive state of their taxonomy, and the species relationships of all the Red Sea Radiolaria are not understood. 38 species have been identified, and they are listed below with reference to their description. Asterisks indicate species present in all 3 cores.

* *Acrosphaera inflata* HAECKEL, 1887, p. 101, Pl. 5, Fig. 7.
  Amphirrhopalum ypsilon HAECKEL, 1887, p. 522.
* *Anthocyrtis ophirensis* EHRENBERG, 1872a, p. 301, 1872b, Pl. 9, Fig. 13.
* *Astrosphaera hexagonalis* HAECKEL, 1887, p. 250, Pl. 19, Fig. 4.

* *Botryocystis scutum* (Harting) NIGRINI, 1967, pp. 52–54, Pl. 6, Figs. 1a–1c.
  Centrobotrys thermophila PETRUSHEVSKAYA, 1965, p. 115.
  Clathrocanium ornatum POPOFSKY, 1913, pp. 343–344, Pl. 7, Fig. 23.
  Clathrocyclas coscinodiscus HAECKEL, 1887, p. 1389, Pl. 58, Figs. 3–4.
* *Clathrocyclas ionis* HAECKEL, 1887, p. 1389, Pl. 59, Fig. 9.
  Dendrospyris binapertonis GOLL, 1968.
  Dendrospyris stabilis GOLL, 1968.
* *Dorcadospyris pentagona* (Ehrenberg) GOLL, 1968.
* *Euchitonia elegans* EHRENBERG, 1872a, p. 319, 1872b, Pl. 8, Fig. 3.
* *Eucyrtidium acuminatum* (Ehrenberg) POPOFSKY, 1913, p. 406, text–Fig. 127.
* *Eucyrtidium hertwigi* HAECKEL, 1887, p. 1491, Pl. 80, Fig. 12.
* *Giraffospyris angulata* (Haeckel) GOLL, 1969.
  Haliomma erinaceum POPOFSKY, 1912, p. 102, Pl. 4, Fig. 1.
* *Heliodiscus asteriscus* HAECKEL, 1887, p. 445, Pl. 33, Fig. 8.
  Heliodiscus echiniscus HAECKEL, 1887, p. 448, Pl. 34, Fig. 5.
  Liriospyris reticulata (Ehrenberg) GOLL, 1968.
* *Lithamphora furcaspiculata* POPOFSKY, 1908, p. 295, Pl. 36, Figs. 6–8.
* *Lithomelissa monoceras* POPOFSKY, 1913, p. 335, Pl. 32, Fig. 7.
* *Octopyle stenozona* HAECKEL, 1887, p. 652, Pl. 9, Fig. 11.
* *Panartus tetrathalamus* HAECKEL, 1887, p. 378, Pl. 40, Fig. 3.
* *Peridium spinipes* HAECKEL, 1887, p. 1154, Pl. 53, Fig. 9.
* *Psilomelissa tricuspidata abdominalis* POPOFSKY, 1908, p. 284, Pl. 33, Fig. 8.
* *Pterocanium praetextum* EHRENBERG, 1872a, p. 316, 1872b, Pl. 10, Fig. 2.
* *Sethophormis pentalactis* HAECKEL, 1887, p. 1244, Pl. 56, Fig. 5.
  Spirema giltschii DREYER, 1889, pp. 40–41, Pl. 6, Fig. 26.
  Spongaster tetras EHRENBERG, 1872b, p. 299, Pl. 6, Fig. 8.
  Spongoplegma antarcticum HAYS, 1965, p. 165, Pl. 1, Fig. 1.
* *Spongopyle stohrii* DREYER, 1889, p. 47, Pl. 5, Fig. 68.
* *Spongotrochus glacialis* POPOFSKY, 1908, p. 228, Pl. 26, Fig. 3.
* *Stylodictya polygona* POPOFSKY, 1912, pp. 131–132, Pl. 5, Fig. 3.

* *Tetrapyle quadriloba* HAECKEL, 1887, p.
    645.
  Tholospira cervicornis HAECKEL, 1887, p.
    700, Pl. 49, Fig. 6.
  Tholospira dendrophora HAECKEL, 1887,
    p. 700, Pl. 49, Fig. 6.
* *Tholospyris devexa* GOLL, 1969.
  Tholospyris procera GOLL, 1969.
* *Tholospyris scaphipes* (Haeckel) GOLL,
    1969.

I have divided the remainder of the fauna into 18 species that apparently have not been previously described. These species cannot be described properly until more specimens from other localities have been examined.

The radiolarian sequences in 126P, 127P and 128P are believed to be isochronous. Of the total 56 radiolarian species, only 36 are common to all three cores. However, 15 of the 20 species that are absent from one or two of the cores never represent more than 4 per cent of the total radiolarian fauna in any sample, and they are sporadically present in the samples from any one core. Because more specimens were counted in 127P, more rare species were encountered. Core 127P contains all 56 species whereas 43 species are present in 126P, and 44 species are present in 128P. These considerations suggest that the burial assemblages of these three cores were derived from the same radiolarian population that occupied the water mass above the Atlantis II Deep.

Origin and Age

In order to determine the direction of influx of Radiolaria into the Red Sea, Quaternary radiolarian burial assemblages in sediments from the Mediterranean Sea and Indian Ocean were compared with the Radiolaria in 126P, 127P and 128P. Only 21 species are common to both the piston cores from the Atlantis II Deep and **CHAIN** 61-56, a piston core from the eastern Mediterranean Sea. All the Radiolaria in the piston cores from the Atlantis II Deep are present as a part of the hundreds of species that constitute the radiolarian burial assemblage in **CHAIN** 43-7, a piston core

from the northwestern Indian Ocean. Therefore, it is concluded that the Radiolaria in 126P, 127P and 128P represent a restricted faunule that invaded the Red Sea through the Gulf of Aden.

On the basis of radiometric analyses, Ku *et al.* (1969) find a Holocene age for this radiolarian episode. Unfortunately, only one age determination is available for each of cores 127P and 128P, and there are no data for 126P. At 740cm in 128P, the age is *ca.* 7,500 years; at 785cm in 127P, the age is *ca.* 12,400 years. The latter date corresponds approximately with the Würm-Holocene boundary, and its position in 128P coincides with the first appearance of Radiolaria. It is significant that a marked increase in the occurrence of calcareous microfossils and dinoflagellates in other cores from the Red Sea has been found at horizons of comparable age in other sediment cores from the Red Sea (Berggren and Boersma, 1969). On the basis of approximate sediment-accumulation rates for 127P and 128P calculated from these two radiometric ages, the disappearance of Radiolaria in both cores occurred about 9,000 years ago.

It is not profitable to speculate on past climatic conditions of the Red Sea on the basis of the radiolarian faunule. There is no evidence that temperature is a controlling factor in the biogeographic distribution of Radiolaria. Moreover, there is no documentation for glacial-interglacial fluctuations of radiolarian-species frequencies such as have been found for Foraminiferida. Comparison of the radiolarian burial assemblage in the Atlantis II Deep with Table 1 of Nigrini (1967, p. 91) indicates that twelve of the species are restricted to "low latitudes" in the Indian Ocean based on studies of the tops of sediment cores.

Frequency Distribution of Dominant Species

Frequency curves for each species were constructed in order to facilitate correlation of the cores. Because 128P was allocated primarily for geochemical analysis, systematic micropaleontological samples were not available. Therefore, correlation with the other two cores was not possible, and only the species-frequency curves of 126P and 127P are discussed here. These two cores are only 2.5km apart, which enables exact faunal correlation. Most of the radiolarian species in 126P and 127P never constitute more than 6 per cent of the total assemblage in any sample, and their frequency curves show no significant variations with core depth. The radiolarian assemblages are markedly dominated by one or more of five species, which are illustrated in Fig. 1 and whose frequency curves are shown in Fig. 2. Correlation of 126P and 127P is based on fluctuations of these frequency curves.

Core 127P contains a much longer record of radiolarian activity than 126P, and certain generalizations can be drawn from the frequency curves of the dominate species in the former core. *Tholospira dendrophora* is more abundant in the upper and lower levels of the radiolarian sequence, whereas *Euchitonia elegans, Spongotrochus glacialis* and *Stylodictya polygona* are most abundant in the middle of the radiolarian sequence. These frequency variations seem to reflect changing hydrographic conditions of the water column overlying the Atlantis II Deep. During the periods of sediment deposition above and below the radiolarian sequence in 127P, conditions were not favorable for the presence of Radiolaria in the central region of the Red Sea. During the period of sediment deposition between 584 and 786cm in 127P, conditions slowly changed so that the water column in the central region of the Red Sea could support a radiolarian fauna. Eventually, conditions regressed to the original state. At the beginning and end of this period, species adapted to the most adverse conditions dominated the population. During the period of most favorable conditions, dominance shifted to species better adapted to this environment. Therefore, I believe that 127P contains the record of a complete episode of radiolarian activity. The species-frequency curves for the dominant species are much more erratic than would be expected for relatively stable conditions. Evidently, hydrographic factors underwent

significant fluctuations during this period.

The radiolarian sequence in 126P is only about one-fourth the length of the radiolarian sequence in 127P. The frequency curves for *Euchitonia elegans* are useful for correlation of the two cores. This species reaches a maximum abundance at 600–02cm in 126P, and the only levels of comparable abundance in 127P are: 598–600, 618–20 and 698–700cm. The level in 127P between 618–20cm is most like 600–02cm in 126P with regard to all five of the dominant-species frequency curves. These two levels are considered to be exactly time synchronous. Therefore, 126P records only a portion of the latter history of Quaternary radiolarian activity in the central region of the Red Sea.

Conclusions

During the Late Quaternary, a restricted faunule of Radiolaria briefly invaded the Red Sea into the water column above the Atlantis II Deep from the northwestern Indian Ocean. The history of this invasion has been observed in only three cores, 126P, 127P and 128P. Why Radiolaria were not found in cores south of the Atlantis II Deep is an unanswered question. Possibly, Radiolaria were not preserved in these cores. An alternative explanation is that the Radiolaria were not observed because of the masking effect of the large quantities of Foraminiferida present in these southern cores. Conditions were not generally favorable for calcite preservation in 126P, 127P and 128P; where they occur, Radiolaria are important constituents of the detrital fraction.

I believe that the concentration of dissolved silica in the central region of the Red Sea was an important factor relating to introduction and disappearance of Radiolaria. Gass and Mallick (1967) discuss the volcanic islands in the southern end of the axial trough of the Red Sea, some of which have been active in historical time. Herman (1965, Tables F–K) notes the sporadic presence of small quantities of Radiolaria in piston cores from the southern Red Sea region. In most cases, these Radiolaria are

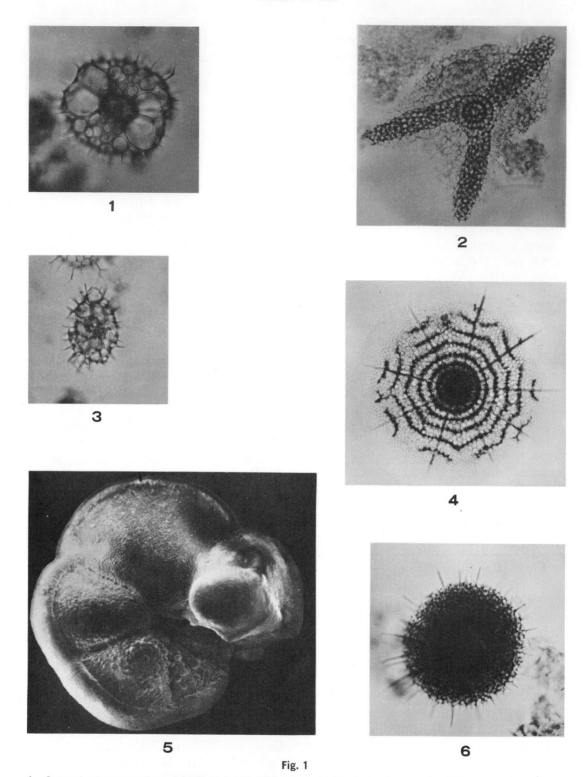

Fig. 1

1 *Octopyle stenozona* from **CHAIN** 61-127P, 594–95cm. Light photomicrograph ×189.
2 *Euchitonia elegans* from **CHAIN** 61-127P, 584–85cm. Light photomicrograph ×189.
3 *Tholospira dendrophora* from **CHAIN** 61-127P, 594–95cm. Light photomicrograph ×189.
4 *Stylodictya polygona* from **CHAIN** 61-127P, 764–65cm. Light photomicrograph ×189.
5 *Globorotalia menardii* from **CHAIN** 43, Sta. 5, 230cm; *ca.* 70×. Thickened calcite crust visible on early chambers of last whorl.
6 *Spongotrochus glacialis* from **CHAIN** 61-127P, 764–65cm. Light photomicrograph ×189.

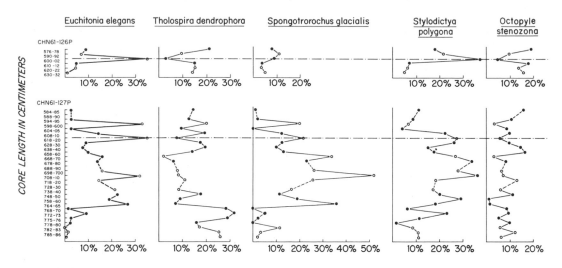

FREQUENCY OF EACH SPECIES MEASURED AS A PERCENTAGE OF
THE TOTAL RADIOLARIAN ASSEMBLAGE

Fig. 2. Frequency distribution with core length for five dominant-radiolarian species in **CHAIN** 61 cores 126P and 127P. Solid circles denote samples in which 300 radiolarian specimens were counted. Open circles denote samples in which 200 or less radiolarian specimens were counted. Dashed lines represent levels in these cores that are considered to be isochronous.

associated with volcanic glass. Evidently, these volcanic eruptions did not have a significant effect on the surface waters of the central region of the Red Sea during late Quaternary time. Only during one volcanic episode was sufficient silica liberated into the water column above the Atlantis II Deep to permit the existence of Radiolaria. Chase (1969) describes basaltic fragments from the sea floor of the central region of the Red Sea, but no siliceous fossils are associated with this material.

Zones of sphalerite-rich sediment are present in the radiolarian sequence of 127P, and sphalerite is also present in the bottom of 128P (Bischoff, 1969). Therefore, it appears that sphalerite deposition along the axis of the Atlantis II Deep was synchronous during late Quaternary time. The absence of sphalerite in the radiolarian sequence of 126P suggests that the sphalerite-depositional episode was not continuous throughout the entire area of the Atlantis II Deep. The radiolarian sequence in 126P is approximately 2,072m below present sea level, and the radiolarian sequence in 127P is approximately 2,113m below present sea level. This 51m vertical separation may have been a significant factor contributing to the absence of sphalerite

in 126P. Bischoff (1969) discusses the possible importance of the environment below the interface between the 56°C and 44°C waters of the hot holes in the formation of sphalerite. If the level of this interface varied in late Quaternary time, the radiolarian sequence in 126P could have been above the interface and outside of the sphalerite-depositional area, while the radiolarian sequence of 127P was below the thermal interface and received sphalerite deposition.

References

Berggren, W. A. and A. Boersma: Late Pleistocene and Holocene planktonic foraminifera from the Red Sea. *In: Hot brines and recent heavy metal deposits in the Red Sea*, E. T. Degens and D. A. Ross (eds.). Springer-Verlag New York Inc., 282–298 (1969).

Bischoff, J. L.: Red Sea geothermal brine deposits: their mineralogy, chemistry, and genesis. *In: Hot brines and recent heavy metal deposits in the Red Sea*, E. T. Degens and D. A. Ross (eds.). Springer-Verlag New York Inc., 368–401 (1969).

Chase, R. L.: Basalt from the axial trough of the Red Sea. *In: Hot brines and recent heavy metal deposits in the Red Sea*, E. T. Degens and D. A. Ross (eds.). Springer-Verlag New York Inc., 122–128 (1969).

312   Robert M. Goll

Dreyer, F.: Morphologische Radiolarienstudien I. Jenaischen Zeitschr. Naturw., 23, new ser. 16, 1 (1889).

Ehrenberg, C. G.: Mikrogeologische Studien als Zusammenfassung seiner Beobachtungen des kleinsten Lebens der Meeres-Tiefgrunde aller Zonen und dessen geologischen Einfluss. Monatsber. Kgl. Preuss. Akad. Wiss. Berlin, Jahrg. 1872, 265 (1872a).

————: Mikrogeologische Studien uber das kleinste Leben der Meeres-Tiefgrunde aller Zonen und dessen geologischen Einfluss. Abh. Kgl. Akad. Wiss. Berlin, Jahrg. 1872, 131 (1872b).

Gass, I. G. and D. I. J. Mallick: Royal Society Volcanological Expedition to the south Arabian Federation and the Red Sea. Nature, 205, 952 (1967).

Goll, R. M.: Classification and phylogeny of Cenozoic Trissocyclidae (Radiolaria) in the Pacific and Caribbean basins. I. Jour. Paleont. (in press) (1968).

————: Classification and phylogeny of Cenozoic Trissocyclidae (Radiolaria) in the Pacific and Caribbean basins. II. Jour. Paleont. (in press 1969).

Haeckel, E.: Report on the Radiolaria collected by H.M.S. CHALLENGER during the years 1873–76. Rept. Voyage CHALLENGER, Zool., 18, 1,803 p. (1887).

Hays, J. D.: Radiolaria and late Tertiary and Quaternary history of Antarctic seas. Biol. Antarctic Seas. 2, Antarctic Research Ser. 5 (Amer. Geophys. Union), 125 (1965).

Herman, Y. R.: *Etudes des sédiments quaternaires de la Mer Rouge.* Doctoral Thesis, Univ. Paris, Ser. A, 1,123, Masson and Co., Paris, 314 (1965).

Ku, T. L., D. L. Thurber and G. Mathieu. Radiocarbon chronology of Red Sea sediments. *In: Hot brines and recent heavy metal deposits in the Red Sea,* E. T. Degens and D. A. Ross (eds.). Springer-Verlag New York Inc., 348–359 (1969).

Nigrini, C. C.: Radiolaria in pelagic sediments from the Indian and Atlantic Oceans. Scripps Inst. of Oceanog., Bull., 11, 106 p. (1967).

Petrushevskaya, M. G.: Peculiarities of the construction of the skeleton of botryoid radiolarians (Order Nassellaria). Trudy Zoologicheskogo Inst. (Akad. Nauk SSSR), 35, 79 (1965).

Popofsky, A.: Die Radiolarien der Antarktis (mit Ausnahme der Tripyleen). Deutsche Südpolar-Exped. 1901–1903, 10 (Zool. 2), 3, 183 (1908).

————: Die Sphaerellarien des Warmwassergebietes. Deutsche Südpolar-Exped. 1901–1903, 13 (Zool. 5), 2, 73 (1912).

————: Die Nassellarien des Warmwassergebietes. Deutsche Südpolar-Exped. 1901–1903, 14 (Zool. 6), 217 (1913).

Pteropods in the Hot Brine Sediments of the Red Sea *

CHIN CHEN

Lamont Geological Observatory of Columbia University
Palisades, New York

Abstract

The distribution of pteropods in four cores from the hot brine area of the Red Sea and from one core 150 miles south of this area shows four stratigraphic zones. These zones can be correlated with eustatic sea level changes.

Introduction

Pteropods are an important group of planktonic organisms in the hot brine sediments of the Red Sea. Herman (1965, 1968) reported planktonic foraminifera and pteropods in Quaternary sediments of the Red Sea. The purpose of this study is to describe pteropod distribution in the hot brine sediments, and to correlate pteropod assemblage zones with lithology, planktonic foraminiferal zones, oxygen isotope analysis of planktonic organisms, and C^{14} dating.

Four Kasten cores from the hot brine area and one piston core outside of the brine area raised by the **CHAIN** cruise 61 of Woods Hole Oceanographic Institution have been available for pteropod study (Table 1). Cores CH61-143K and 95K are located in the Atlantis II Deep, CH 61-119K is in the Discovery Deep, and CH 61-118K is situated about 4 miles south of the Chain Deep. Core CH 61-154P is located in the southern central region of the Red Sea, approximately 150 miles south southeast of the Atlantis II Deep.

Results

Hot Brine Area

Three zones are recognized on the basis of correlation of vertical distribution of pteropod species with lithology, planktonic foraminifera species (Berggren and Boersma, 1969), oxygen isotope analysis (Deuser and Degens, 1969) and C^{14} age (Ku *et al.*, 1969). The vertical distribution of major pteropod species is shown in Fig. 1.

Zone A. The upper A zone is megascopically characterized by thick grey lutite layers which are interbedded with thin layers of light to dark brown and black sediments. In most cases pteropods are more abundant

* Lamont Geological Observatory Contribution No. 1288.

Table 1   Location, Depths and Lengths of Cores

Core	Latitude	Longitude	Depth (m)	Length (cm)
95K	21° 23.4′N	38° 03.2′E	1,910	335
118K	21° 14.4′N	38° 04.3′E	1,989	400
119K	21° 16.8′N	38° 01.4′E	2,175	400
143K	21° 27′N	38° 03′E	1,910	354
154P	19° 34′N	38° 59.5′E	1,276	792

313

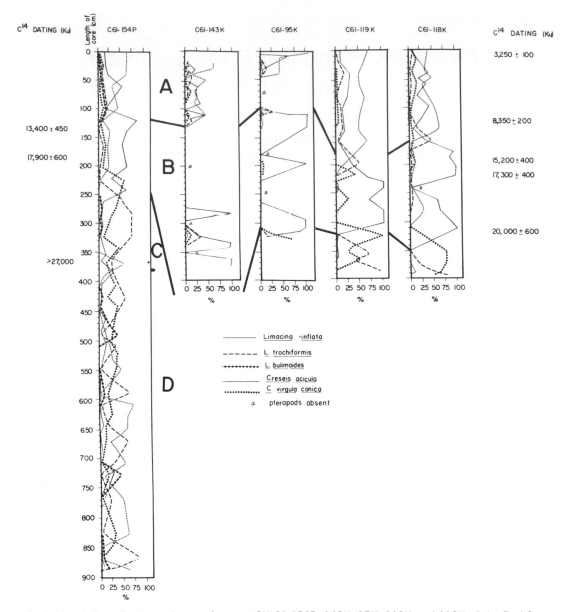

Fig. 1. Correlation of pteropod zones in cores CH 61-154P, 143K, 95K, 119K and 118K of the Red Sea.

in the grey and black layers than in the brown layers. A total of fourteen pteropod species is recorded in the upper zone. *Limacina inflata* ranks first of the total pteropod population in the upper zone, averaging about 40 per cent. *Creseis acicula* ranks second. Less variation of oxygen isotope ratio occurs in the upper zone of CH 61-118K.

Zone B. The iron-rich, reddish-brown layers are thicker in pteropod Zone B than in Zone A. As a result pteropods correspondingly decrease in number in the

brown layer. Thin, pocket-like lithified layers are distributed in the grey lutite. According to Gevirtz and Friedman (1966), "aragonite, contributed by pteropods, was precipitated in limestones, binding the grains together, and providing drusy fibrous infilling and syntaxial fibrous growth on grains. High salinity and temperature are necessary for the precipitation of aragonite." Thin black layers were found between the grey lutite.

Orientation of pteropods in the Zone B is megascopically observed. The C-axes (the longest dimension) of *Creseis acicula,* having pencil-like conical shells, are parallel to the lamination of thin black layers (Fig. 2), but random orientation appears in the thick, grey layers and porous limestones. The black layers contain a greater amount of H_2S, indicating a stagnant, reducing condition, and pteropod shells were deposited on the ocean floor without disturbance. The grey layers were deposited under more oxygenated conditions, and the light pteropod shells were disturbed by even very slight bottom currents or biological activity.

The number of pteropod species decreases from fourteen to five in the middle zone. The epipelagic species, *Creseis acicula,* constitutes an average of 75 per cent of the total pteropod population. Planktonic foraminifera are rare or absent in this zone. The oxygen isotope ratio increases abruptly within pteropod Zone B in CH 61-118K. The abrupt change may be a result of high salinities developed in the isolated Red Sea during periods of low sea level. Apparently, only epipelagic pteropod species are sufficiently tolerant

during a period of high evaporation and salinity.

In core 118K, C^{14} dating indicates ages of $8,350 \pm 200$ years B. P. at 120–125cm, and $15,200 \pm 400$ years at 180–185cm (Ku *et al.,* 1969). Thus, the boundary of 160cm between pteropod zones A and B is documented at about 11,000 years ago. This age corresponds to the start of the post-glacial time in Europe.

Zone C. The lower C Zone appears in the lowest part of the Kasten cores CH 61-95K, 118K and 119K. Core CH 61-143K does not penetrate the lower C Zone. In this zone the grey layer is thicker than the brown and black layers. The lower zone is characterized by a high percentage of *Creseis virgula conica* and *Limacina trochiformis* and an increase in the number of pteropod species.

Outside Hot Brine Area

Four pteropod zones can be recognized in core CH 61-154P (Fig. 1). The pteropod fauna assemblages in the upper three zones A, B and C of this core are correlated with

Fig. 2. Orientation of pteropods in the black layer of core CH 61-119K, 230–231cm.

those of the four Kasten cores in the hot brine area. The D Zone is typified by the alternation of *Limacina trochiformis* and *L. inflata* with an occasional abundance of *L. bulimoides*.

Discussion and Summary

The distribution of pteropods in the Red Sea cores can be compared according to the depth each major species inhabits in the water column. The pteropod fauna assemblage in Zone A is similar to the fauna in the water column (Herman 1968). The dominant species of Zone A is *Limacina inflata* which ranks first in the relative abundance of the total pteropod population in the upper 500m of the Sargasso Sea (Chen 1964).

Creseis acicula is the dominant form in Zone B. Leal-Rodriguez (1965) reported that *Creseis acicula* is the dominant species in shallow water at a depth of less than 20 fathoms (37m) off Veracruz coast, southwest Gulf of Mexico. It indicates that Zone B was deposited during a very low sea-level period. The orientation of *Creseis acicula* in the black layer indicates a stagnant, reducing condition of the depositional environment.

Zone C is characterized by *Limacina trochiformis* and *Creseis virgula conica* which are the dominant species in the Gulf Stream in the vicinity of continental shelf south of Cape Hatteras, North Carolina (Chen and Hillman, in press). McGowan (1960) stated that the highest concentration of *L. trochiformis* occurs in the upper 60m in the Pacific. This implies that an oceanic influx took place during the Zone C period.

Zone D shows the alternate dominance of two species, *Limacina inflata* and *L. trochiformis* with two peaks of *L. bulimoides* at 475 and 720cm in core 154P. Morton (1954) reported that *L. bulimoides* is the dominant pteropod species in the Benguela Current. He found that *Limacina bulimoides* is present in the upper 1,000m with its highest concentration occurring at a depth of about 250m. McGowan (1960) stated that *Limacina bulimoides* occurs in a greater depth than *L. inflata* in the Pacific. It can be inferred that the depth of the Red Sea during Zone D time was at least as deep as Zone A and perhaps deeper.

Acknowledgments

I thank Professors J. Hays and A. McIntyre for their help. The research was supported by a grant from the National Science Foundation.

References

Berggren, W. and A. Boersma: Late Pleistocene and Holocene planktonic foraminifera in the Red Sea. *In: Hot brines and recent heavy metal deposits in the Red Sea,* E. T. Degens and D. A. Ross (eds.). Springer-Verlag New York Inc.. 382–398 (1969).

Chen, C.: Pteropod ooze from Bermuda Pedestal. Science, **144,** 60 (1964).

——— and N. S. Hillman. Shell-bearing pteropods as indicators of water masses off Cape Hatteras, North Carolina. Bull of Marine Science (in press).

Deuser, W. and E. Degens: O^{18}/O^{16} and C^{13}/C^{12} ratios of fossils from the hot brine deep area of the Central Red Sea. *In: Hot brines and recent heavy metal deposits in the Red Sea,* E. T. Degens and D. A. Ross (eds.). Springer-Verlag New York Inc., 336–347 (1969).

Gevirtz, J. L. and G. M. Friedman: Deep-sea carbonate sediments of the Red Sea and their implications on marine lithification. Journal of Sedimentary Petrology, **36,** 143 (1966).

Herman, Y.: Etude des sédiments Quaternaires de la Mer Rouge. Ann. Inst. Oceanogr. Monaco, **42,** 339 (1965).

———: Evidence of climatic changes in the Red Sea cores. Means of Correlation of Quaternary Successions. Proceedings VIII INQA Congress (1968).

Ku, T. L., D. L. Thurber, and G. G. Mathieu: Radiocarbon chronology of Red Sea sediments. *In: Hot brines and recent heavy metal deposits in the Red Sea,* E. T. Degens and D. A. Ross (eds.). Springer-Verlag New York Inc., 348–359 (1969).

Leal-Rodriguez, D.: Distribucion de pteropodos en Veracruz, ver. Anales del Institito de Biologia, **36,** 249 (1965).

McGowan, J. A.: The systematics, distribution and abundance of Euthecosomata of the North Pacific. Unpublished Ph.D. diss., Univ. of California at San Diego, 197 p. (1960).

Morton, J. E.: The Pelagic Mollusca of the Benguela Current. Discovery Rept., **27,** 163 (1954).

Dinoflagellates in Red Sea Piston Cores *

DAVID WALL

Woods Hole Oceanographic Institution
Woods Hole, Massachusetts

JOHN S. WARREN

University of Cincinnati
Cincinnati, Ohio

Abstract

Dinoflagellates were studied in two piston cores (Ch 61-154P and 153P) from the axial trough of the Red Sea, approximately 150 miles south-southeast of the Atlantis II and Discovery Deeps. Some twenty species belonging to nine fossil genera were identified and on the basis of their vertical distribution (relative and semi-quantitative), 6 zones (A–F) were identified in core 154P and five (A–E) recognized in core 153P. Their distribution has been related tentatively to a Late Pleistocene-Holocene chronostratigraphic scale by means of radiocarbon dates, climatic data and extrapolation of rates of sedimentation. In these cores the distribution of dinoflagellates was determined by changing conditions of temperature, salinity and sea level and can be referred to four glacial cycles which have been recognized from oxygen isotope studies. A glacio-eustatic mechanism probably initiated these cycles and in this respect the geological histories of the Red Sea area and the Mediterranean provinces to the north are comparable.

Scope of the Investigation

Dinoflagellates have been studied in detail from piston cores 153P and 154P taken on R/V **CHAIN** cruise 61. Both cores were located in the southern central region of the Red Sea on the floor of the axial trough, approximately 150 miles south-southeast of the Atlantis II Deep (Table 1). Nevertheless, a thin, non-fossiliferous layer of brine-derived amorphous goethite was found in core 153P at 355cm, indicating that Late Pleistocene brines existed outside the larger, better known postglacial brine zones.

Table 1   Location, Depths and Lengths of Cores Studied

Core	Latitude	Longitude	Depth	Length
153P	19° 43′N.	38° 41′E.	2,704m.	755cm. (base disturbed)
154P	19° 34′N.	38° 59.5′E.	1,276m.	892cm.

Lithology of Cores

Core 153P. This core mainly comprises a series of brownish lutites, varying in color from yellow brown (10 YR 5/4) † to dusky yellowish-brown (10 YR 2/2). These are either homogeneous or diffusely mottled and only well laminated between 280–290cm and 320–450cm: in the latter horizon there are a few bands of ferruginous lutite (10 R 4/6). Lithified fragments are restricted to a short section between 170 and 204cm, where large pteropods are present. Basaltic fragments occur at two horizons, at 444cm and from 504 to 510cm. The base of this core is a homogeneous yellowish-brown lutite, and it may have been disturbed in coring.

Core 154P. This core consists of an alternating series of yellowish-brown (10 YR 4/2) and

---

* Woods Hole Oceanographic Institution Contribution No. 2193.

† Geological Society of America Rock Color Chart.

yellowish-gray (5 Y 6/2) or olive gray (5 Y 5/2) lutites with foraminifera and pteropods. There is a thick lithified section between 120 and 285cm and thin bands of similar material at 460 to 475cm and 555 to 575cm. Between all three horizons there are sporadic lithified fragments. There are thin bands of coarse, friable foraminifera-pteropod ooze at 555–575cm, 734–754cm and 882–884cm which grade upwards into lutites.

Samples and preparation. Twenty-four samples taken at intervals close to 25cm apart were studied to a depth of 575cm in core 153P. Forty-five samples from core 154P were examined; they were taken at 20cm intervals to a depth of 881cm. The samples were oven-dried at 100°C and weighed before treatment with cold dilute HCl and concentrated hot HF. Occasionally a second treatment of hot HCl was applied to remove a drusy precipitate that formed in HF. The organic matter obtained was given a brief ultrasonic vibration to disperse it before microscopic examination. No oxidation was applied. All the dinoflagellates in a measured, homogenized volumetric fraction of the organic material in aqueous suspension were counted to give a semi-quantitative estimate of the number of specimens in a gram of lutite in addition to percentage determinations. In core 153P, only dinoflagellates greater than 20μ overall size were counted, but this includes all the species identified in core 154P so that the results are comparable.

Systematic Details

The Nature of Fossil Dinoflagellates

Many living dinoflagellates produce morphologically distinctive resting spores (cysts) which possess a decay-resistant spore wall whose chemical composition is unknown, but it appears to be allied to the sporo-pollenin of higher plant spores and pollen and fossilizes in a similar manner. Living species of *Gonyaulax* and *Protoceratium* and allied genera in particular produce fossilizable resting spores which are common in many Quaternary marine sediments (Evitt and Davidson, 1964; Wall and Dale, 1967, 1968b; Rossignol, 1962, 1964; Wall, 1967). Since almost all fossil dinoflagellates originally were studied in older sediments (sometimes as "hystrichosphae-

rids") before relationships with living representatives were established, a separate system of classification and nomenclature was devised for them by paleontologists. This system is used here but it is hoped that in the future the modern and paleontological schemes will be integrated as better data become available.

Species List

The following are species identified in cores 153P and 154P: *Hystrichosphaera bentori* Ross., *H. bulloidea* Cooks. & Eis., *H. furcata* (Ehr.) Wetz., *H. membranacea* (Ross.) Wall, *H. mirabilis* Ross., *H. nodosa* Wall, *H. scabrata* Wall, *Nematosphaeropsis balcombiana* Defl. & Cooks., *Tectatodinium pellitum* Wall, *Leptodinium aculeatum* Wall, *L. dispertitum* Cooks. & Eis., *L. patulum* Wall, *L. paradoxum* Wall, *L. sphaericum* Wall, *L. strialatum* Wall, *Lingulodinium machaerophorum* (Defl. & Cooks.) Wall, *Operculodinium centrocarpum* (Defl. & Cooks.) Wall, *O. israelianum* (Ross.) Wall, *Hemicystodinium zoharyi* (Ross.) Wall, *Tuberculodinium vancampoae* (Ross.) Wall, various *Peridinium* cysts.

Vertical Distribution of Dinoflagellates

Species Distribution

Six assemblages of contrasting composition were recognized in core 154P and the latest five of these also were found in core 153P (Figs. 1, 2). These assemblages were labelled A through F in order of increasing age. They are species-associations characterized primarily by the relative abundances of four commonly represented species, namely, *Hystrichosphaera bulloidea*, *Hemicystodinium zoharyi*, *Nematosphaeropsis balcombiana* and *Leptodinium aculeatum* although occasionally these are outnumbered by other species such as *L. patulum* and *Operculodinium centrocarpum*. The stratigraphic intervals characterized by these species associations have been called zones although they are "assemblage zones" without phylogenetic significance. They appear to be useful how-

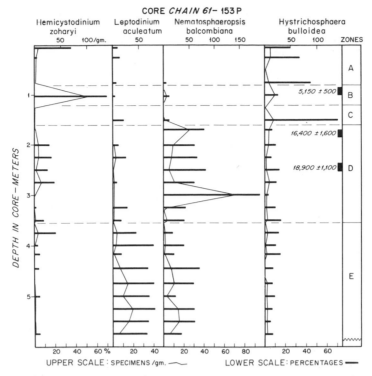

Fig. 1. Vertical distribution of common dinoflagellate species in core **CHAIN** 61-153P. (Only specimens larger than 20μ counted.) Dates from Ku *et al.*, 1969.

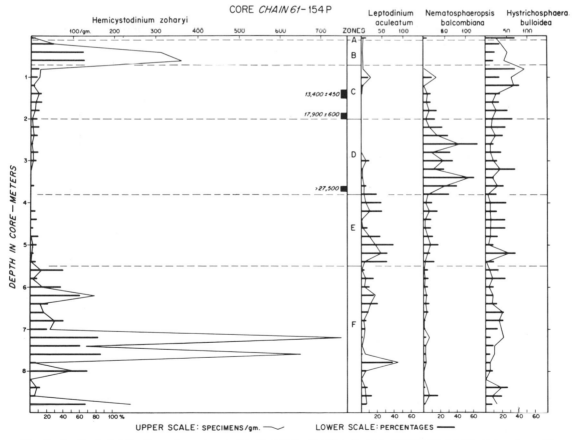

Fig. 2. Vertical distribution of common dinoflagellate species in core **CHAIN** 61-154P. Dates from Ku *et al.*, 1969.

ever for purposes of local stratigraphy and in this respect are comparable with many Quaternary pollen zones recognized in non-marine deposits of similar age.

Zone A. This zone is typified by a relative abundance of *H. bulloidea* and minor occurrences of *Hemicystodinium zoharyi*, but generally dinoflagellates are not abundant within it compared with the subjacent zone. A form of *H. bulloidea* whose spine endings lack the usual bifurcations occurred in the upper part of this zone in core 153P. Only the uppermost 10cm in core 154P had the characteristics of this zone compared with 80cm in core 153P.

Zone B. This zone is strongly dominated by *Hemicystodinium zoharyi* and is represented by *ca.* 40cm of sediment in core 154P but was restricted to one horizon in core 153P around 103cm, the two adjacent samples being barren.

Zone C. This zone is defined on the absence or rarity of *Hemicystodinium* and *Nematosphaeropsis* whose abundances typify zone B above and zone D below respectively. A sharp decline in all species can be found from the top of this zone to its base in both cores. In core 154P it is approximately 130cm thick and its upper section is characterized by *H. bulloidea*, *Leptodinium patulum* and *L. paradoxum*, but its base is almost barren. It is approximately 40cm thick in core 153P and is almost devoid of dinoflagellates, being recognizable only by virtue of the absence of *Hemicystodinium* and *Nematosphaeropsis* at this level.

Zone D. This zone is characterized by an abundance of *Nematosphaeropsis* and the absence or very small representation of *Leptodinium aculeatum*. It is well developed in both cores: in core 153P it occurs between 160 and 355cm and in core 154P between 200 and 380cm. In both cores there are two maxima of *Nematosphaeropsis;* the earliest maximum occurrence in both cores may be synchronous but the later abundances cannot be correlated from core to core. *Hemicystodinium* is rare or absent.

Zone E. In this zone, *Leptodinium aculeatum* replaces *Nematosphaeropsis* as the most common species, but the latter is well

represented. In core 153P this zone occurs from 355cm at least to 575cm; below this the core may be disturbed. In core 154P, zone E occurs between 380 and 550cm.

Zone F. From 550cm to its base, core 154P is typified by an abundance of *Hemicystodinium* associated with *L. aculeatum*. The former genus reaches values over 700/g and sometimes accounts for over 80 per cent of the assemblage. At the base of zone F, *L. aculeatum*, *H. bulloidea* and *O. centrocarpum* are common in restricted horizons, and this may mark a transition to an older zone which was not penetrated by the core.

Dinoflagellate Abundance

Semi-quantitative estimates of the abundance of dinoflagellates in a gram of lutite show certain similarities in the two cores (Fig. 3), but core 154P has a higher content throughout. This may be due to deposition of core 153P in deeper water, leading to a decreased supply of tests rather than dilution by increased sediment supply, since both cores have similar sedimentation rates and surface productivity probably did not vary considerably within the short distance of approximately 10 miles which marks the distance between the cores. Decreases in the number of specimens per gram seem to occur at times of transition from one species association to another marking the bases of our zones whereas specimens are quite abundant within the zones. They appear to be most abundant near the base of the zone and then decline towards its top.

Chronostratigraphy

General Statement

Fossil dinoflagellates have not been used previously to correlate cores from the sea floor. The first comprehensive account of their presence in deep-sea cores was published only as recently as 1967 (Wall, 1967). There also is a great lack of detailed information concerning the distribution of the resting spores of modern dinoflagellates on the sea floor although the work of Williams (1969) marks a good beginning to this task. Information concerning the marine distribution of living species is often inapplicable to geological problems because it refers only to the planktonic flagellated stage, and recent research reveals that the taxonomy of these species often seems inadequate when details of the organisms' encysted stages become available. They promise to offer the opportunity to subdivide some heterogeneous and polyphyletic groups (Wall and Dale, 1968b). There is therefore a lack of information to compare with the results of this investigation. The chronostratigraphy offered for cores 153P and 154P must be regarded as provisional in view of these difficulties. It has been compiled on the basis of radiocarbon dates (Ku *et al.* 1969), comparison with some other paleontological data described in this volume and with reference

to oxygen isotope work (Deuser and Degens, 1969) in addition to some extrapolation of rates of sedimentation.

The stratigraphic terminology and scale of reference adopted here is taken from van der Hammen *et al.* (1967) although it is directly applicable to the Netherlands and Denmark rather than southern provinces of the Red Sea. The reason it is used is that we feel future attempts to correlate marine dinoflagellate zones with land-based type sections will be successful only when epicontinental sequences, such as those found in the Netherlands, are used as standards. Here there are marine intercalations in basinal deposits whose nonmarine members have well documented stratigraphic positions based upon field relationships, carbon dates, and pollen zones. In contrast, reliable correlations between Alpine glaciated provinces and

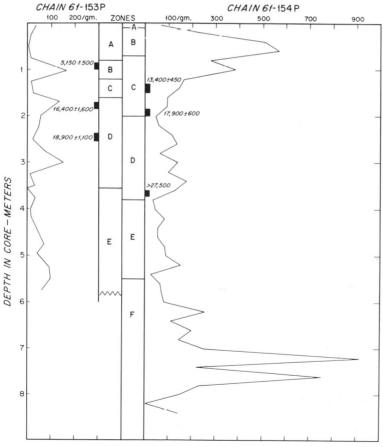

Fig. 3. Quantitative abundance of dinoflagellates relative to a gram of sediment in cores **CHAIN** 61-153P and 154P. Dates from Ku *et al.*, 1969.

marine sequences probably will always be very tenuous because of their geographic discontinuity.

Radiocarbon Dates

Five radiocarbon dates were obtained for cores 153P and 154P (Ku *et al.* 1969). The youngest was 5,150 ± 500 yrs. at 85–100cm in core 153P. This dates the maximum abundance of *Hemicystodinium* in this core in zone B as Late Atlantic. In core 154P this zone is thicker, and its age range probably is Atlantic to sub-Boreal, corresponding to the latter part of the postglacial climatic optimum or hypsithermal interval (Deevey and Flint, 1957, Table 1).

Four other radiocarbon dates fall within the dinoflagellate zones C and D. In core 154P, the middle of zone C (130–150cm) is dated at 13,400 ± 450 yrs. B.P., and a date of 17,900 ± 600 at 185–200cm has been obtained for the base of the same zone. In core 153P the top of zone D (170–185cm) has been dated at 16,400 ± 1,600 yrs. B.P., and the middle of the zone as 18,900 ± 1,100 at 237–259cm. Thus the boundary between dinoflagellate zones C and D appears to lie within the interval 15,000 to 18,500 yrs. B.P. and most probably lies very close to 17,500 yrs. The chronostratigraphic age of zone C is from Late Upper Pleniglacial to Early Holocene and includes the entire Late Glacial (Table 2).

The ages of the boundaries between zones D, E and F can be estimated only by extrapolation of rates of sedimentation. An average sedimentation rate of 10.7cm/-10^3 yrs. was estimated for core 154P: by extrapolation the base of zone D is *ca.* 35,580 yrs., and the top of zone F is *ca.* 51,500 yrs. The average rate of sedimentation for core 153P was estimated at 11.5cm/10^3 yrs. and the base of zone D as *ca.* 30,750 yrs. Zone F does not appear in this core. According to these estimates, the chronostratigraphic age of zone D is Upper Middle and Upper Pleniglacial, and zone E is Lower Glacial and Lower Middle Pleniglacial. Zone F in core 154P is believed to include the Early Glacial,

and at least the latter part of the Eemian Interglacial since extrapolation of rates of sedimentation suggest it is *ca.* 83,000 yrs. old at the base of the core.

The dinoflagellate zones D to F and the lowest part of zone C probably are lateral equivalents of the Entedebir Formation, a coralline limestone series which fringes much of the Red Sea area (Horowitz, 1967). This formation is overlain conformably by the Salt Plain evaporitic sediments and other lagoonal deposits which probably are of the same age as dinoflagellate zones A, B and the upper part of C.

Paleoecology

General Statement

Two sets of factors probably are of importance in accounting for the distribution of dinoflagellate spores in sediments, namely, those related to conditions surrounding the encystment and dinoflagellate species autecology (biological factors) and secondly, factors which influence the horizontal and vertical movements of resting spores after their formation towards and along the sea floor (dispersal factors) (Wall, 1969). The distribution of dinoflagellates in Red Sea cores probably was influenced quite strongly by biological and dispersal factors and seems to reflect not only changing productivity and communities but physiographic changes such as fluctuating sea level and concomitant sedimentation processes.

Dinoflagellates appear to indicate a cooling trend by a succession beginning with tropical species in the Late Eemian and Early Glacial, giving way to a mild temperate assemblage in the Late Weichselian (Würm) before reverting to tropical conditions in the Holocene. At the same time however, dinoflagellates in the Early Weichselian and Eemian and the Holocene are shallow water dinoflagellates which probably have been transported into the axial trough. They are rare throughout most of the Weichselian (zones D, E and C). This may indicate that coastal bays where these shallow water dinoflagellates proliferated were stranded by general lowering of the sea level in Pleniglacial and Late Glacial times.

Evidence for Climatic Change

Recent investigations by Wall and Dale (in prep.) have shown that the fossil species *Hemicystodinium zoharyi* is the resting spore of a modern tropical species, *Pyrodinium bahamense*. This species inhabits very warm, highly saline coastal waters in low latitudes around Bermuda, the Bahamas and Jamaica. Margalef (1961)

recorded temperatures greater than 29°C and salinities above 35.5 per mill in Phosphorescent Bay in Puerto Rico where *Pyrodinium* is abundant throughout the year.

Nematosphaeropsis balcombiana is one type of resting spore produced by *Gonyaulax spinifera* which is a temperate water species according to Kofoid (1911). Williams (1968) recently described a region in the North Atlantic between 45°N and 55°N where cysts of this type are abundant on the sea floor west of Ireland. The occurrence here of a biofacies rich in *Nemato-*

Table 2 Tentative Relationships Between Dinoflagellate Zones in Red Sea Cores CHAIN 61-153P and 154P and Several Selected Late Quaternary Chronostratigraphic Scales

Years B.P.	Estimate of mean temperatures-July 5°10°15°20°	Interstadials	Chronostratigraphy (van der HAMMEN et al. 1967)	Dinoflagellate Zones	Chronostratigraphy (EMILIANI, 1955; ROSHOLT et al.; 1961)	Oceanic isotopic stages	Chronostratigraphy (ERIKSSON 1967)
			Holocene	A	Postglacial	1	Postglacial
				B			
10,000		Allerød Bolling	Late Glacial	C	Late and Main Würm	2	Late Würm
20,000			Pleni- Glacial — Upper				Würm III
30,000		Denekamp	Pleni- Glacial — Middle	D		3	Würm II
40,000		Hengelo		E	Main and Early Würm		
50,000			Pleni- Glacial — Lower				
circa 60,000		Brørup	Early Glacial		Early Würm	4	Würm I
circa 70,000		Amersfoort		F			
			Eemian Interglacial		Riss-Würm Interglacial	5	Riss-Würm
circa 100,000							

sphaeropsis coincides with the position of the mild temperate climatic zone of Hall (1964). Late summer temperatures at the surface in this region are in the range of 13.5°C to 18°C, according to information in the U.S. Navy Marine Climatic Atlas of the World, vol. 1. These give an indirect estimate of the temperature conditions associated with the encystment of this dinoflagellate. Similarly, *Leptodinium aculeatum* (affinity unknown) was found in a biofacies along the coast of West Africa from around 35°N to 15°N. It has a lower latitude distribution pattern. Surface water temperature values for this region range from 22°C to 26.5°C in late summer and do not fall below 15°C. *Leptodinium aculeatum* thus appears to be a subtropical species from Williams' data supplemented by independent temperature data.

Applying these temperature estimates to the paleoecology of our cores, paleotemperatures for surface waters can be surmised during the deposition of zones B, D, E and F. Zones B and F were probably formed under tropical to subtropical conditions with shallow water temperatures in excess of 29°C at times and probably never below 22°C. The water probably also was highly saline. During formation of zone E, surface water temperatures may have been in the range 13.5 to 26.5°C while in zone D cooler summer temperatures probably prevailed and the annual temperature range was approximately 13 to 18°C. Thus we see a cooling trend during the Pleniglacial. The paucity of dinoflagellates in zone C prevents any reconstruction of climatic conditions but seems to represent a transition period between mild temperate conditions in the Late Pleniglacial and a return to tropical conditions in the Holocene.

Evidence for Sea Level and Salinity Changes

Evidence for fluctuating salinities and sea level changes during the Late Pleistocene and Holocene in the Red Sea region is provided by geochemical, sedimentological and paleontological data. Oxygen isotope studies on core 154P (which provides the most comprehensive results by virtue of its length) reveal fluctuations in O^{18} values (Fig. 4) which are too great in magnitude to be attributed to paleotemperature changes alone and indicate that the isotopic composition of Red Sea water did not remain constant in Quaternary times (Deuser and Degens, 1969). These authors suggest that there were significant salinity changes during the Late Pleistocene-Holocene period and that four salinity cycles can be seen in core 154P. Each cycle begins with a rapid decrease in salinity values to normal oceanic values relative to supersaturation levels which were established gradually following each influx. The Red Sea Basin during the Late Quaternary probably was an almost entirely landlocked sea subjected to greater than normal evaporation for long periods with resultant lowering of sea level. Four influxes of oceanic water from the Indian Ocean to the south probably occurred around 65,000–70,000 years B.P., 40–45,000 yrs., 20–25,000 years and 11,000 years ago. Deuser and Degens (1969) suggest these influxes were controlled by changes in the depth of a sill at the southern extremity of the Red Sea, either due to glacio-eustatic changes in sea level or to intermittent tectonic activity in the region of the straits of Bab el Mandeb.

Sedimentological investigations also suggest lowering of sea level occurred during a period in the Upper Pleniglacial when a lithified carbonate-rich layer was formed over much of the Red Sea area. In core 154P this layer is thickly developed between 120 to 285cm, but it is much thinner in core 153P where it is found from 170 to 204cm.

Paleontological evidence for a lower sea level during the Pleniglacial is derived mainly from the distribution of *Hemicystodinium*. This resting spore is the encysted phase of *Pyrodinium bahamense,* which is a dinoflagellate found in protected, shallow water coastal embayments. Its abundance in deep-water lutites in the Red Sea is incongruous with our knowledge of its present day distribution and strongly suggests that it was displaced from coastal environments into the axial trough by current action at times when it proliferated. This belief is strengthened by the presence of three turbidite units containing displaced reef-detritus in zone F of core 154P in association with *Hemicystodinium* (John Milliman, personal communication).

Horowitz (1967, Fig. 3) recently described a lagoonal facies surrounded by mangroves around Museri Island in the Dahlak Archipelago (southwestern Red Sea). He commented that here the Entedebir Formation is covered by abundant and spectacular mangrove lagoons where *Rhizophora* grows on intertidal flats or fringing hypersaline lagoons in which black, clayey sediments accumulate. This environmental situation closely compares with that found today in Phosphorescent Bay, Puerto Rico, where *Pyrodinium* (= *Hemicystodinium*) thrives in abundance. Transport of *Hemicystodinium* to the axial trough sediments during formation of zones F and B in particular probably occurred at times when direct communication existed from such lagoons or intertidal flats to the deep-sea. At other times these lagoons were stranded or destroyed temporarily by drying up.

In contrast, the dinoflagellate spores *Nematosphaeropsis balcombiana* and *Leptodinium aculeatum* probably were formed by species primarily found in the neritic zone but perhaps also capable of invading oceanic areas. The *Gonyaulax spinifera* group is well known in neritic situations and *Nematosphaeropsis* is the spore of a member of this group (Wall and Dale, 1967, 1968b). The latter has been found in some abundance in plankton hauls beyond the limits of the continental shelf at depths between the surface and 600m in the North Atlantic near Newfoundland, off the Shetlands, and in areas of mixing between the North Atlantic Drift and Irminger Sea under the synonyms *Pterococcus* and *Pterosperma labyrinthus* Ost. (Lohmann 1904, 1910; Gaarder, 1954). It is uncertain whether the organism inhabits these areas during its entire life history or the spores are drifted into them, but these records do show *Nematosphaeropsis* occurs in the phytoplankton many miles from the coast and outside neritic environments.

Applying this information to the Red Sea cores, we postulate that a gradual fall in sea level towards the end of the Early Gla-

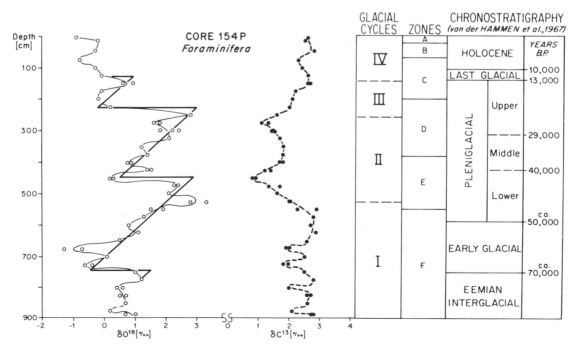

Fig. 4. Red Sea dinoflagellate zones related to isotopic data, postulated glacio-eustatic cycles and a chronostratigraphic scale. Isotopic data from Deuser and Degens, 1969.

cial stranded the coastal bays formerly occupied by *Hemicystodinium* (= *Pyrodinium*) until they were reestablished in the Holocene by the Middle Flandrian transgression. During the Pleniglacial only dinoflagellates capable of inhabiting neritic or oceanic areas survived. Modifications of dinoflagellate communities during the Pleniglacial may have been the result of a diminution in the extent of the neritic zone as well as a climatic cooling which is indicated by their temperature-encystment relationships discussed above.

Synopsis

Correlations can be drawn between the evidence for climatic, geochemical and physiographic change outlined above in an attempt to reconstruct the Late Pleistocene-Holocene history of the southern Red Sea. The common factor in all these distributions is the strong suggestion of glacio-eustatic influences which are typical of the Quaternary, and the history of the area seems best explained in these terms. There appear to have been four glacio-eustatic cycles since the Eemian Interglacial in the Red Sea, each beginning with an influx of water from the Indian Ocean followed by a period of high evaporation and falling sea level. The influxes can be considered as "transgressive" phases and the periods of high evaporation as "regressive" phases, although piston core studies by themselves do not provide direct evidence of shoreline migrations normally indicative of sedimentary cycles. However, the initiation of the "transgressive" phases of two older cycles in the Red Sea appear to be more or less contemporaneous with transgressions in the Mediterranean and may reflect worldwide sea level fluctuations in the Late Pleistocene (Bonifay and Mars, 1959). Each of these four cycles appears in part to have been characterized by different paleontological assemblages on the basis of which some local zonation of Red Sea sediments has been attempted. Dinoflagellates appear to have responded to these changes both in their transport to the axial trough (and perhaps productivity) and by

community changes leading to temporary depletions and extinctions of certain species.

Cycle 1 appears to have begun *ca.* 70,000 yrs. ago towards the end of the Eemian, with an ingressive phase in the lower Early Glacial followed by a period of high evaporation during the remainder of Early Glacial times. Its early phase may have been more or less contemporaneous with the Eutyrrhenian transgression in the Mediterranean (Bonifay and Mars, 1959; Eriksson, 1967, Table 4) and the Amersfoort Interstadial in the Netherlands and Denmark (Fig. 4). A tropical to subtropical climate prevailed during this cycle, and the Red Sea probably had a relatively wide continental shelf surrounded by reefs and shallow water, embayments not unlike those found around the shores of the Red Sea, Puerto Rico and Jamaica today.

Cycle 2 was initiated around 50,000 yrs. ago in Lower Pleniglacial times and was followed by an evaporative or regressive phase which lasted until Upper Pleniglacial times, around 20–25,000 yrs. ago. It may have been partly synchronous with the Neotyrrhenian transgression in the Mediterranean (Bonifay and Mars, 1959; Eriksson, 1967, Table 4). The early part of this cycle is characterized by an increased abundance of *Nematosphaeropsis* and *L. aculeatum* and the absence of *Hemicystodinium* which declined in abundance during the latter part of cycle 1. In the latter part of cycle 2 *Leptodinium aculeatum* declined to leave *Nematosphaeropsis* as the dominant species. Dinoflagellate zone E thus coincides closely with the transgressive phase of this cycle and zone D with its regressive phase. Climatic cooling and a lowering of sea level probably occurred during the latter phases of its development with some decrease in the area of submerged continental shelf.

Cycle 3 began with an influx of oceanic water around 20–25,000 years ago. It appears to be coincident with a decline in abundance of *Nematosphaeropsis* (top of zone D) but there was no introduction of any other species to take its place. Similar reductions in the abundance of dinoflagellates can be seen to coincide with the earlier influxes and this may be due to temporary interruptions in the supply of cells to the axial trough as the circulation pattern and sedimentary processes were modified. The influx of oceanic water at the beginning of cycle 3 also appears to be coincident with a climatic period of extreme cold in Europe, but there is no evidence of severe cooling in cores 153P and 154P. Cycle 3 was of relatively short duration and its "regressive" phase was brief; it probably lasted only until the early part of the Last Glacial. Extensive lithification of sea floor deposits occurred during the latter part of cycle 2 and during cycle 3 and ceased at the beginning of cycle 4.

Cycle 4 includes the Flandrian transgression and the postglacial climatic optimum. It began around 11,000 yrs. ago and was accompanied by climatic amelioration leading to the return of tropical to subtropical conditions in the Red Sea. At the same time the area of continental shelf below water probably increased and coastal bays around archipelagos were reestablished. Dinoflagellates show increases during the latter part of this cycle up until sub-Atlantic times which may represent the beginning of a final evaporative stage represented by present day conditions. The extensive brine-derived, iron-rich deposits of the Red Sea were formed throughout this final cycle.

Finally, from the viewpoint of Quaternary dinoflagellate studies it is interesting to note that in the Red Sea dinoflagellates apparently responded to a succession of glacio-eustatic events. In previous investigations both Rossignol (1962) and Wall and Dale (1968a) have observed correlations between glacio-eustatic events and dinoflagellate species-associations in Early Pleistocene sequences in Israel and England. However, these studies involved littoral and epineritic facies in epicontinental basinal regions; the present study concerns deep-water lutites. This indicates that dinoflagellates are worthy of study in a wide variety of marine terrains and can provide a new marine micropaleontological method in Quaternary research in future years.

Acknowledgments

This investigation was supported by grants GA-584 and GB-5200 from the National Science Foundation.

References

Bonifay, E. and P. Mars: Le Tyrrhénien dans le cadre de la chronologie Quarternaire méditerranée. Bull. Geol. Soc. France, **7, I,** 62 (1959).

Deevey, E. S. and R. F. Flint: Postglacial hypsithermal interval. Science, **125,** 182 (1957).

Deuser, W. G. and E. T. Degens: O^{18}/O^{16} and C^{13}/C^{12} ratios of fossils from the hot-brine deep area of the Central Red Sea. *In: Hot brines and recent heavy metal deposits in the Red Sea,* E. T. Degens and D. A. Ross (eds.). Springer-Verlag New York Inc., 336–347 (1969).

Emiliani, C.: Pleistocene temperatures. Jour. Geology, **63,** 538 (1955).

Eriksson, K. Gösta: Some deep-sea sediments in the western Mediterranean Sea. Progress in Oceanography, **4,** 267 (1967).

Evitt, W. R. and S. E. Davidson: Dinoflagellate studies: I. Dinoflagellate cysts and thecae. Stanford Univ. Publs., Geol. Sci., **10,** 1 (1964).

Gaarder, K. R.: Report on the scientific results of the MICHAEL SARS North Atlantic Deep-Sea Expedition, 1910. Coccolithineae, Silicoflagellatae, Pterospermataceae and other forms, **2,** 1 (1954).

Hall, C. A.: Shallow-water marine climates and molluscan provinces. Ecology, **45,** 226 (1964).

Horowitz, A.: The geology of Museri Island (Dahlak Archipelago, Southern Red Sea). Israel J. Earth-Sci., **16,** 74 (1967).

Kofoid, C. A.: Dinoflagellata of the San Diego Region: IV. The genus *Gonyaulax,* with notes on its skeletal morphology and a discussion of its generic and specific characters. Univ. Calif. Publ. Zool., **8,** 187 (1911).

Ku, T. L., D. L. Thurber, and G. Mathieu: Radiocarbon chronology of Red Sea sediments. *In: Hot brines and recent heavy metal deposits in the Red Sea,* E. T. Degens and D. A. Ross (eds.). Springer-Verlag New York Inc., 348–359 (1969).

Lohmann, H.: Eier und sogenannte Cysten der Plankton-Expedition. Ergebnisse der Plankton-Expedition der Humbolt-Stiftung, **4N,** 1 (1904).

——: Eier und sogenannte Cysten des nordischen Planktons. Nordisches Plankton, Zoologischer Teil, **1, 11,** 1 (1910).

Margalef, R.: Hidrografia y fitoplancton de un area marine de la costa meridional de Puerto Rico. Investigacion Pesquera, **18,** 33 (1961).

Rosholt, J. N., C. Emiliani, J. Geiss, F. F. Koczy, and P. J. Wangersky: Absolute dating of deep-sea cores by the Pa^{231}/Th^{230} method. J. Geol., **69,** 162 (1961).

Rossignol, M.: Analyse pollinique de sédiments marins quaternaires en Israel. II. Sediments pléistocènes. Pollen et Spores, **4,** 121 (1962).

———: Hystrichosphères du Quaternaire en Méditerranée oriental dans les sediments pléistocènes et les boues marines actuelles. Rev. Micropaleont., **7,** 83 (1964).

van der Hammen, T., G. C. Maarleveld, J. C. Vogel, and W. H. Zagwijn: Stratigraphy, climatic succession and radiocarbon dating of the Last Glacial in the Netherlands. Geol. en Mijnb., **46,** 79 (1967).

Wall, D.: Fossil microplankton in deep-sea cores from the Caribbean Sea. Palaeontolgy, **10,** 95 (1967).

———: The lateral and vertical distribution of dinoflagellates in quaternary sediments. Proc. Symposium Micropal. Bottom Sediments, Cambridge, 1967 (1969, in press).

——— and B. Dale: The resting cysts of modern marine dinoflagellates and their palaeontological significance. Rev. Palaeobotan. Palynol., **2,** 349 (1967).

———, ———: Early Pleistocene dinoflagellates from the Royal Society Borehole at Ludham, Norfolk, New Phytol. 67, 315 (1968a).

———, ———: Modern dinoflagellate cysts and evolution of the Peridiniales. Micropaleontology, **14** (1968b).

Williams, D. B.: The occurrence of dinoflagellates in marine sediments. Proc. Symposium Micropal. Bottom Sediments, Cambridge, 1967 (1969, in press).

Micropaleontologic Investigations of Red Sea Cores — Summation and Synthesis of Results *

W. A. BERGGREN

Woods Hole Oceanographic Institution
Woods Hole, Massachusetts

Abstract

The Red Sea is a natural laboratory for investigating relatively rapid changes in a restricted marine environment during the Late Pleistocene and Holocene. An investigation which integrates the combined data of micropaleontologic, radiocarbon dating, and oxygen and carbon investigations has yielded an interpretation of the paleoecology and paleoclimatology in the Red Sea during the past 80,000 years.

The distribution in the Late Pleistocene and Holocene of several planktonic microfossil groups (dinoflagellates, foraminifera, nannofossils, pteropods and radiolarians) has been investigated in cores from the Red Sea. Tentative biostratigraphic zonations of sediments spanning the last 80,000 years are based upon percentage fluctuations in these groups (with the exception of the radiolarians whose appearance is too brief) and have been correlated with a chronostratigraphic scale of the Late Pleistocene and Holocene.

Glacio-eustatic control of sea-level (resulting in alternating intervals of lowered sea-level and high salinity followed by relatively rapid invasions of normal marine water from the Indian Ocean) and the process of glaciation itself (resulting in a gradual decrease in water temperature) are believed to have been the primary factors governing the distribution of the fossil populations. The biostratigraphic zonations are closely related to, and reflect the influence of, the glacial-salinity cycles which have been recognized by means of oxygen isotope measurements.

Introduction

During **CHAIN** cruise 61 to the Red Sea approximately 60 deep-sea sediment cores were successfully recovered. The majority of these cores were taken in an area of deep-brine holes located between 21°15′ and 21°27′ North latitude and 38°00′ to 38°10′ East longitude.

Very little is known of the distribution of the microfauna in fossil sediments of the Red Sea. In an attempt to extend our knowledge in this respect, material from some of these cores was made available to several specialists. In this paper an attempt is made to integrate and synthesize the data presented by these different studies in order to formulate some generalized interpretations about the paleoecology and paleoclimatology of the Red Sea during the Late Pleistocene and Holocene.

Synopsis of Results

General

A chronostratigraphic scale for the Late Pleistocene and Holocene has been adapted from van der Hammen *et al.* (1967) for use in the interpretation and comparative analy-

* Woods Hole Oceanographic Institution Contribution No. 2194.

sis of data obtained by the various specialists whose work is presented in the papers above. Oxygen and carbon isotope studies have been made on several cores by Deuser and Degens (1969). Radiocarbon dates (Ku *et al.*, 1969) allow the distribution data and resulting biostratigraphic zonations of various microfossil groups to be placed within a proper time-framework and allow interpretation of Late Pleistocene and Holocene paleoclimatology and paleoecology in the Red Sea in terms of known paleoclimatic and paleoecologic conditions elsewhere, particularly northern Europe.

Most of the cores penetrate into the Middle or Upper Pleniglacial (*ca.* 25–35,000 years ago). In the hot-brine area, sedimentation rates were relatively high and penetration below the Holocene/Pleistocene level at *ca.* 11,000 years was minimal. The longest cores studied outside the hot-brine area penetrated into the Last (Eemian) Interglacial (CH 61-154P and 167P). A fourfold biostratigraphic (assemblage) zonation was recognized in the planktonic foraminifera, pteropods and coccoliths. A six-fold zonation was determined using dinoflagellates and elements of a fifth zone were discerned in coccoliths in the lowermost part of CH 61-154P (McIntyre, 1969). Though differing somewhat in their boundaries, the different zonations are intimately linked to and reflect the changing paleooceanographic conditions in the Red Sea as determined by oxygen-isotope measurements (Deuser and Degens, 1969).

Dinoflagellates

Wall and Warren (1969) use dinoflagellates for the first time in the correlation of marine sediments in deep-sea cores. The authors provide an interesting example of the manner in which information on reproduction and distribution of living marine organisms can be applied to an interpretation of local paleoecology and paleogeography. The abundant distribution in the Red Sea lutites of *Hemicystodinium* (the en-

cysted phase of *Pyrodinium bahamense*)—which lives in protected, shallow-water, coastal embayments—is viewed as incongruous and suggests displacement. On the other hand, dinoflagellate spores of *Nematosphaeropsis balcombiana* and *Leptodinium aculeatum* were formed by species which probably inhabited the neritic realm and perhaps, also, the oceanic environment. Thus the authors suggest that towards the end of the Early Glacial (*ca.* 50,000 years ago) a gradual fall in sea-level stranded the marginal, coastal environment in which *Hemicystodinium* (= *Pyrodinium*) lived until it was reestablished within the Holocene by the Middle Flandrian transgression. During the Pleniglacial only dinoflagellates capable of living in the neritic or oceanic areas survived. Examination of Fig. 2 reveals a striking similarity in the distributional patterns of *Nematosphaeropsis balcombiana* (a dinoflagellate) and *Globigerinoides ruber* (a planktonic foraminifer) supporting the suggestion by Wall and Warren. This similarity also suggests that a common set of factors (high salinity and low temperatures) may have been responsible for the sharp decrease and sporadic occurrence of these two species during the late phase of the Pleniglacial. The difference in the boundaries of the dinoflagellate zones compared to those based on other groups of microfossils may be explained by the fact that the zonation is based upon dinoflagellate species of widely differing ecology which responded in a different manner to changes in the paleooceanographic conditions, e.g., time lag compared to zonation based upon planktonic foraminifera and pteropods.

The dinoflagellates indicate a cooling trend through the Late Pleistocene beginning with tropical assemblages in the Late Eemian and Early Glacial, a mild temperaate assemblage in the Late Weichselian (Würm), followed by a tropical assemblage which became reestablished during the Holocene.

Surface water temperatures were suggested to have varied from 22°–29°C in dinoflagellate Zone F (shallow-water, coastal environment) to 13.5°–26.5°C in

Zone E and 13°–18°C in Zone D (open-sea environment).

Planktonic Foraminifera

The general cooling trends which occurred during the Pleistocene between *ca.* 70,000–11,000 years ago is indirectly reflected in the distributional pattern of planktonic foraminifera in Red Sea cores (Berggren and Boersma, 1969). Relatively few planktonic species are found in Red Sea cores but the distributional pattern of two, *Globigerinoides ruber* and *G. sacculifer,* were found useful in paleoecologic and paleoclimatic interpretations.

A fourfold zonation of the Late Pleistocene-Holocene sequence in the Red Sea was based upon fluctuating percentages in *G. ruber* and *G. sacculifer*. High salinity was suggested as the controlling factor causing the temporary disappearance of *G. sacculifer* in the Early Pleniglacial (*ca.* 45,000 years ago), whereas the effect of lowered temperature superimposed upon high salinities, was suggested as having caused the sharp decrease and, ultimately, temporary disappearance of *G. ruber* (*ca.* 22,000 years ago).

Paleoclimatic conditions varied probably from tropical-subtropical with surface water temperatures between 21°–30°C during the Late Eemian-Early Glacial to temperate with surface water temperatures ranging from below 14°–15°C to about 20°C during the Late Pleniglacial.

Pteropods

Four assemblage zones have been recognized based upon percentage frequency of several diagnostic species (Chen, 1969). The upper three zones, A–C, correspond to the planktonic foraminiferal zones A–C (Figs. 1 and 2).

Pteropod Zone A is characterized by high percentages of *Limacina inflata* (*ca.* 40 per cent and *Creseis acicula* and corresponds to the Holocene sequence in Red Sea cores.

The author suggests that the rather restricted nature of the pteropod faunas in Zone B may be due to high salinities in the Red Sea during a period of low sea level.

Pteropod Zone C is characterized by a high percentage of *Creseis virgula, C. conica* and *Limacina trochiformis*. These three zones are found in the hot-brine area but only outside this area was an additional lower zone found. Pteropod Zone D is characterized by the alternation of *L. trochiformis* and *L. inflata* with an occasional abundance of *L. bulimoides*.

Chen suggests that pteropod Zone B was a time of very low sea-level as *Creseis acicula,* the dominant form in this zone, is

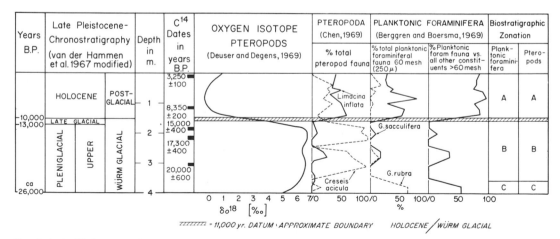

Fig. 1. Late Pleistocene-Holocene biostratigraphy, chronostratigraphy and oxygen isotope data of CH 61-118K.

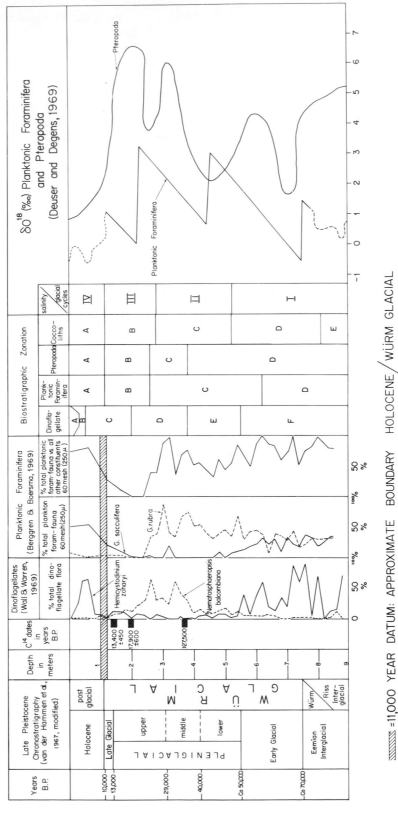

Fig. 2. Late Pleistocene-Holocene chronostratigraphy, comparative biostratigraphy and oxygen isotope data of CH 61-154P.

at present the dominant species in shallow-water (less than 20 fathoms = 37m) off the coast of Veracruz, Mexico. Zone C was a time of moderately deep seas and during Zone D the seas fluctuated between moderately deep and deep.

The epipelagic species, *Creseis acicula,* constitutes an average of 75 per cent of the total pteropod population of Red Sea cores. The C-axis (longest dimension) of *C. acicula,* is parallel to the laminations in the thin-bedded black mud layers in Zone B, whereas random orientation occurs in the more thickly bedded gray layers and calcarenites of Zone B. The high level of H_2S in the black layers suggests that pteropods were deposited on the ocean floor in a tranquil environment and did not suffer post-depositional disturbances by bottom currents; biological activity is exhibited by pteropods in the lutites and gray muds.

Coccoliths

The four or possibly five main coccolith zones described from Red Sea core CH 61-154P can be considered in the main the result of climatic control. There is very good correlation between the salinity/glacial cycles and floral change.

Zone A, 0–120cm, is typified by a modern subtropical to tropical oceanic coccolith assemblage with an additional small component of indigenous, shallow water species. These latter forms are present throughout the core and occasionally dominate the flora. This youngest zone is equivalent to the post-glacial. Zones B, 120–280cm, and C, 280–520cm, represent the Pleniglacial Period. Because of poor preservation over most of B and in portions of C, floral determinations are not always quantitative; however it is clear that a cooler, restricted, shallow water flora is present in B whereas with increasing depth in C the flora becomes cosmopolitan and warmer with the return of oceanic species. That this is still a period of cool conditions is shown by the presence of "cold" ecotypes of such species as *Umbilicosphaera mirabilis* and *Cyclolithella annulus.* Zone D, 520–760cm, is dominated by a species identified as *Ge-*

phyrocapsa cf. *ericsonii.* While nearly absent from younger sediments it constitutes over 30 per cent of total flora from 520cm to the base of the core. In addition the return of open ocean tropical and subtropical forms in relative abundance correlates well with the placement of this section in the late glacial Early Interglacial.

Below 760cm a fifth zone, E, may be present due to the appearance of a species intermediate in structure between *Coccolithus huxleyi* and the *Gephyrocapsa* complex as well as some minor floral changes. This particular species, *Coccolithus protohuxleyi,* has been found in Upper Pleistocene cores from the Atlantic Ocean. The final disappearance of this species in the Atlantic occurs at approximately 75,000 years B. P. in core V12-122. By extrapolation of sedimentation rates in CH 61-154P the disappearance of this species in the Red Sea is equivalent in time. In fact the correlation between oceanic species fluctuations in V12-122 and CH 61-154P is quite good particularly for *G.* cf. *ericsonii* and *C. protohuxleyi.*

Radiolarians

A total of 56 radiolarian species were found within intervals ranging from 30cm to about 2m in three cores from Atlantis II Deep and their occurrences are believed to be isochronous (Goll, 1969). All 56 species were encountered in CH 61-127P, 43 were recorded in 126P and 44 species were found in 128P, suggesting to the author that the thanatocoenoses in the three cores were derived from the same radiolarian population which inhabited the water column above the Atlantis II Deep.

The relationship of the Red Sea radiolarian fauna to the Mediterranean Sea and Indian Ocean faunas was investigated. Only 21 species are common to the Atlantis II Deep cores and a core from the Eastern Mediterranean. All the radiolarian species found in the Red Sea cores were present as "part of the hundreds of species that constitute the radiolarian burial assemblage in CH 43-7," a piston core from the northwestern Indian Ocean. The author suggests that the Red Sea radiolarians represent a restricted faunule which invaded the Red Sea through the Gulf of Aden. Radiocarbon dates have shown that this invasion occurred about 12,000 years ago and on paleontologic evidence — ended about 85,000 years ago. The appearance of radiolarians in the Red Sea is thus linked with the transgression at the Holocene/Pleistocene boundary which has been dated elsewhere in this volume at between 11–12,000 years ago.

The author suggests that dissolved silica, rather than temperature, may have been the controlling factor in the introduction and subsequent disappearance of radiolarians in the Red Sea sediments. The occurrence of sphalerite-rich sediments in the radiolarian sequence of core 127P and at the bottom of 128P suggests that sphalerite deposition was synchronous along the axis of the Atlantis II Deep. The absence of sphalerite in core 126P suggests that sphalerite deposition was discontinuous in this region; however, the author suggests that if the interface between the 56°C and 44°C waters in the hot holes varied, the radiolarian sequences in 126P and 127P, which have approximately 51m vertical separation between them, might have been above and below this sphalerite deposition, respectively.

Conclusions

A synthesis of micropaleontologic, radiocarbon dating and oxygen isotope data presented by various authors in this volume allows a generalized interpretation of Late Pleistocene and Holocene paleoecology and paleoclimatology in the Red Sea. Oxygen isotope data indicate that four distinct glacial/salinity cycles occurred during the past 70,000 years in the Red Sea. These cycles were characterized by gradual increases in salinity over an interval of approximately 20,000 years terminated abruptly by relatively rapid intervals during which the Red Sea and Indian Ocean reestablished connection resulting in more normal salinities. Although the climate during the Eemian Interglacial-Early Glacial was tropical-subtropical in this region, a gradual deterioration in climate occurred during the Pleniglacial (Würm Glacial). Thus the effects of lowered temperature and increased salinities may be expected to have played a significant role in the distribution of the various planktonic organisms which lived in the Red Sea during this time.

The paleoclimatology-paleoecology of the Red Sea during the Late Pleistocene-Holocene may be summarized in the following manner. A tropical-subtropical climate existed in the Red Sea region from at least 80,000 to about 50–60,000 years ago and corresponds approximately to Cycle I (see Fig. 2). During this time surface water temperatures probably varied between 21°–30°C similar to present day values in the area. The second evaporative cycle began about 50,000 years ago and lasted until about 25,000 years ago. Climatic cooling and lowering of sea level are suggested by micropaleontologic data and oxygen isotopic measurements. Cycle III (ca. 23,000–13,000 years ago) corresponds to the coolest part of the Late Pleistocene. The general impoverishment of the microfauna during this interval is probably directly related to lowered water temperatures, in addition to the effects of pronouncedly high salinities. Surface water temperatures may have reached values as low as 13°–14°C during this time. Cycle IV (corresponding to the Holocene) witnessed the reestablishment in the Red Sea of a normal marine microfauna from the Indian Ocean and a gradual rise in temperature to present day values.

The Red Sea has shown itself to be an excellent "laboratory" for the study of relatively rapid changes of short duration in the marine environment. By integrating data on the distribution of fossil microfaunas, radiocarbon dating and oxygen-carbon isotope measurements a picture of the paleoclimatic and paleoecologic conditions in the Red Sea over the past 80,000 years has emerged.

Acknowledgments

This investigation has been supported by grants GA-676 and GA-584 from the National Science Foundation.

References

Berggren, W. A. and A. Boersma: Late Pleistocene and Holocene planktonic foraminifera in the Red Sea. *In: Hot brines and recent heavy metal deposits in the Red Sea,* E. T. Degens and D. A. Ross (eds.). Springer-Verlag New York Inc., 282–298 (1969).

Chen, C.: Pteropods in the hot brine sediments of the Red Sea. *In: Hot brines and recent heavy metal deposits in the Red Sea,* E. T. Degens and D. A. Ross (eds.). Springer-Verlag New York Inc., 313–316 (1969).

Deuser, W. G. and E. T. Degens: O^{18}/O^{16} and C^{13}/C^{12} ratios of fossils from the hot-brine deep area of the central Red Sea. *In: Hot brines and recent heavy metal deposits in the Red Sea,* E. T. Degens and D. A. Ross (eds.). Springer-Verlag New York Inc., 336–347 (1969).

Goll, R.: Radiolaria: The history of a brief invasion. *In: Hot brines and recent heavy metal deposits in the Red Sea,* E. T. Degens and D. A. Ross (eds.). Springer-Verlag New York Inc., 306–312 (1969).

Ku, T. L., D. L. Thurber, and G. Mathieu: Radiocarbon chronology of Red Sea sediments. *In: Hot brines and recent heavy metal deposits in the Red Sea,* E. T. Degens and D. A. Ross (eds.). Springer-Verlag New York Inc., 348–359 (1969).

McIntyre, A.: The Coccolithophorida in Red Sea sediments. *In: Hot brines and recent heavy metal deposits in the Red Sea,* E. T. Degens and D. A. Ross (eds.). Springer-Verlag New York Inc., 299–305 (1969).

van der Hammen, T., G. C. Maarleveld, J. C. Vogel, and W. H. Zagwijn: Stratigraphy, climatic succession and radiocarbon dating of the Last Glacial in the Netherlands. Geol. en Mijnb., **46,** 79 (1967).

Wall, D. and J. S. Warren: Dinoflagellates in Red Sea piston cores. *In: Hot brines and recent heavy metal deposits in the Red Sea,* E. T. Degens and D. A. Ross (eds.). Springer-Verlag New York Inc., 317–328 (1969).

O^{18}/O^{16} and C^{13}/C^{12} Ratios of Fossils from the Hot-Brine Deep Area of the Central Red Sea *

WERNER G. DEUSER AND EGON T. DEGENS

Woods Hole Oceanographic Institution
Woods Hole, Massachusetts

Abstract

Carbon and oxygen isotope ratios of foraminifera and pteropod shells show considerable fluctuations with depth in all cores investigated. Distinct changes in the isotope ratios at certain depths can be correlated among cores. Foraminifera tests deposited in the Discovery and Atlantis II Deeps during the last 10,000 years show evidence of a beginning isotopic reequilibration with the hot brine. Older deposits show no such effect, suggesting a maximum age of the brine of about 10,000 years. The isotopic data strongly suggest the repeated occurrence of periods of evaporation in the Red Sea during the last 80,000 to 100,000 years. It seems likely that these periods coincided with those of lowered sea level which severely restricted the exchange of water between the Red Sea and the Indian Ocean. Alternatively the reduced exchange could have been caused by tectonic activity near the Strait of Bab el Mandeb.

Introduction

Shell carbonates from sediment cores collected on **R/V CHAIN** cruise 61 in and around the hot-brine deeps of the central Red Sea were analyzed for stable oxygen and carbon isotope ratios. It was hoped that such studies might yield information on the source and history of the hot brines. This report concentrates on foraminifera and pteropod shells because they were the most abundant fossil groups. The study of the oxygen isotope content of coexisting minerals might reveal additional clues to the history of the deeps. Such work is presently in progress.

Three cores, 118K, 119K and 154P, were investigated systematically from top to bottom. In addition, a few samples were analyzed from cores 84K, 126P and 127P. The sampling locations of these cores are shown in Fig. 1. The cores designated K are large-volume, or Kasten, cores, four meters long. Those designated P are piston cores, seven to nine meters long.

Experimental Procedure

The cores were sampled shortly after opening of the cases in which they had been stored aboard the ship, most of them in refrigeration. All surfaces were scraped to eliminate contamination and any material which might have been displaced by drag along the core barrel. The box cores were sampled in 5-cm sections so that a set of 80 samples (in the case of a complete 4-m core) represented the entire sediment column of a core. Only spot samples were taken from the piston core 154P, every 10cm in the top half meter and every 25cm in the remainder, each sample covering 10cm of core length. The sediment samples were kept in tightly sealed jars until further processing.

For the separation of the fossil material the samples were washed with cold tap water through a set of sieves of 1,000μ, 250μ and 125μ mesh size. In almost all instances the most useful material was contained in the 250–1,000μ fraction, except for the larger pteropod

* Woods Hole Oceanographic Institution Contribution No. 2195.

specimens. The smaller fractions contained mostly fossil fragments and inorganic material. After drying of the fractions, the foraminifera were picked by hand from the $250–1,000\mu$ fraction and the pteropods from the $250–1,000\mu$ and $>1,000\mu$ fractions.

In the case of the foraminifera no further purification steps were required before dissolving the tests in 100 per cent phosphoric acid to extract carbon dioxide for mass spectrometric analysis (McCrea, 1950). Preliminary analyses had shown that roasting of the tests in a helium atmosphere, a method developed by Epstein

et al. (1953) to remove volatiles which might exchange oxygen with the carbonate, did not produce results different from those obtained on untreated material. Emiliani (1966) reported similar observations. This feature can be attributed to the low content of organic matter in the foraminifera tests which amounts to 0.06 to 0.10 per cent. Similarly, the pteropod shells contain about 0.1 per cent organic matter, largely in the form of proteins which once acted as mineralization templates. In the case of pteropod shells, however, it was sometimes necessary to manually clean individual shells or select only

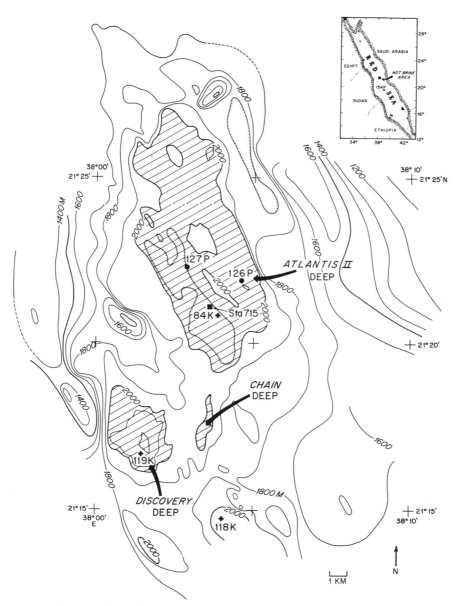

Fig. 1. Bathymetric map of hot brine area showing position of cores.

fractions of shells so as to avoid contaminating the samples with foreign material stuck to or contained in the shells.

In certain depth zones of the cores the pteropod shells were overgrown and partly or wholly filled with aragonite crystals, radially grown outward and inward, which could not be separated from the original delicate shells. Similar specimens have been described by Gevirtz and Friedman (1966). The secondary overgrowth usually far outweighs the original shell material so that the isotope ratios determined on such specimens are representative of the aragonite overgrowth but not necessarily of the original shells. The amino acid spectra are identical for fresh and incrusted shells. The total yield of organic matter in the incrusted specimens, however, amounts to less than 0.01 per cent, suggesting that this is all contained in the original shell.

The isotope ratio analyses were carried out on a mass spectrometer (Model RMS, Nuclide Corporation) with dual sample inlet and double collector of the general type described by Mc-Kinney *et al.* (1950). The results were corrected for contributions by the less abundant isotopic species and for instrumental characteristics and are reported in the δ-notation where

$$\delta\ (\%_0) = \left(\frac{R_{sample}}{R_{standard}} - 1\right) \times 1{,}000$$

and $R_{standard}$ is the O^{18}/O^{16} or C^{13}/C^{12} ratio, respectively, of the PDB-1 standard carbon dioxide (Craig, 1957). The analytical precision achieved, as evidenced by duplicate preparations and analyses of aliquots from the same carbonate sample, is usually ±0.1 per mill or better for both oxygen and carbon.

Results

Core 118K (6km SE of Discovery Deep)

Pteropods. Core 118K was studied in considerable detail. The measurements of δC^{13} and δO^{18} on the fossil pteropods are plotted in Fig. 2 against core depth. 5 radiocarbon dates, taken from Ku *et al.* (1969), are included to provide a chronological reference frame.

The most striking feature of the carbon and oxygen isotope data is the pronounced change in values observed at a core depth of around 150cm. The δC^{13} values exhibit an abrupt change of about 3 per mill within just a few centimeters. The δO^{18} values undergo a net change of about 5.5 per mill but with a transition zone of some 30cm. A

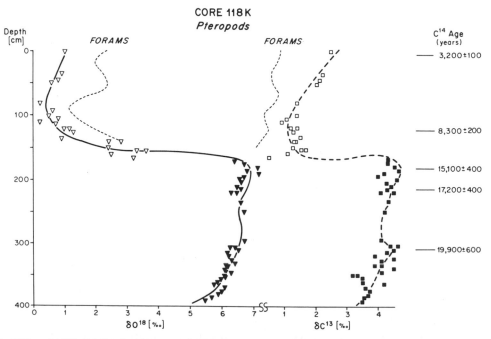

Fig. 2. δO^{18} and δC^{13} distribution in pteropod shells of core 118K. Open triangles and squares = unaltered shells; filled triangles and squares = aragonite incrusted shells.

good parallelism is observed between δC^{13} and δO^{18}, except in the transition zone between 140 and 165cm where carbon becomes isotopically somewhat lighter with depth while the opposite is the case for oxygen.

The transition zone is distinguished from the material above and below it not only by its isotopic content but also by the physical appearance of its fossils, which are in a highly fractured state. Above 140cm the appearance of the shells was very fresh and did not change significantly to the very top of the core. Below 170cm the shells exhibited aragonite filling and overgrowth to a varying extent. Near the bottom of the core, layers occurred in which only casts of the original shells remained, everything else apparently having been dissolved by secondary action. In those instances the casts were analyzed isotopically and the results proved identical to those obtained on overgrown shell material from nearby layers, thus suggesting that the casts were in isotopic equilibrium with the material which had once surrounded them.

Foraminifera. The variations with core depths of δO^{18} and δC^{13} in foraminifera tests are shown in Fig. 3. Foraminifera, which constitute a substantial fraction of the total sediment in the upper 1½m of the core, disappear rather abruptly below 150cm. Only four species were identified in significant numbers, *Globigerinoides rubra, Globigerinoides sacculifera, Hastigerina siphonifera* and *Orbulina universa* (Berggren, 1969). In a number of samples the species were separated into two groups for isotopic analyses, *i.e., O. universa* and the other three. No consistent differences were found, however, between samples of the two groups from the same sediment layers, implying that, on the average at least, they inhabited the same water depths. Consequently, no further separation of the foraminifera was carried out. Below 150cm foraminifera were rare and only two samples yielded enough material for an analysis, those at 230 and 250cm. Even in these instances it was necessary to wash 10 to 20 times the usual amount of sediment through the sieves in order to obtain enough tests.

The δO^{18} values of the foraminifera fluctuate between -0.9 and $+0.7$ per cent in a rather orderly manner, which probably reflects fluctuations in the water temperature over the last 10,000 years. Assuming constant isotopic composition of the water during this period of time, these fluctuations correspond to a temperature change of 7°C. The δC^{13} values show smaller fluctuations which in general are sympathetic with the oxygen values. Both of the isolated, deeper samples are significantly enriched in O^{18} and the lower one (250cm), also in C^{13}, compared to the samples from the upper part of the core.

Core 119K (Discovery Deep)

Pteropods. Only sixteen samples of pteropod shells from this core (Fig. 4) were analyzed because of its strong similarity with core 118K. The ranges of δO^{18}

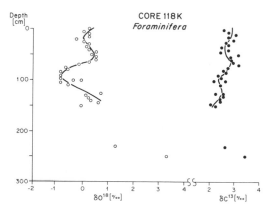

Fig. 3. δO^{18} and δC^{13} distribution in foraminifera tests of core 118K.

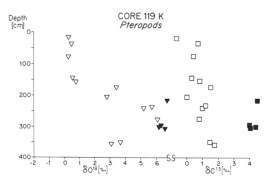

Fig. 4. δO^{18} and δC^{13} distribution in pteropod shells of core 119K. Open triangles and squares = unaltered shells; filled triangles and squares = aragonite incrusted shells.

and δC^{13} values are the same as in that core, and there is also a transition zone between 150 and 200cm separating the light values at the top of the core from the heavy values below. The appearance of the shells, both fresh and overgrown, was generally similar to that in core 118K. The occurrence of fresh looking shells below 340cm appears significant. The macroscopic appearance of these shells is similar to that of shells from the upper 1.5m; however, the δ-values fall in the range of the transition zone. Furthermore, within the thinned zone (relative to 118K) of aragonite-overgrown shells, no such consistency in either degree of overgrowth or in O^{18} and C^{13} enrichment was found as was manifest in 118K.

Foraminifera. Foraminifera tests were analyzed from all core depths where they occurred in sufficient abundance (Fig. 5). Analogous to the pteropods, the forams found below 340cm are similar in their appearance to those at the top of the core. The δO^{18} values below 340cm, however, cluster around 2 per mill, *i.e.*, significantly heavier than at the top, and the δC^{13} values are around 1 per mill, which is lighter than most values above 170cm. The δO^{18} fluctuations of the upper 1.5m correlate well with those of core 118K, but their amplitude is smaller and their mean value is lower by 0.9 per mill. There is also a slight indication of a similar shift towards lower values of δC^{13}.

Cores 84K, 126P and 127P (Atlantis II Deep)

Foraminifera. Core 84K consisted predominantly of inorganic precipitates and contained extremely little shell detritus. Only the samples between 145 and 245cm contained a sufficient number of foraminifera for analysis. The scarcity of the data does not allow a correlation with cores 118K and 119K, but because of the apparently high sedimentation rate in the Atlantis II Deep, the fossils at the 145 to 245cm interval probably correspond to some in the top meter of the two cores from outside the brine area. The δ values are comparable to those of the upper 1.5m of core 119K, δO^{18} varying from -0.7 to -1.4 per mill and δC^{13} from $+2.2$ to $+2.8$ per mill.

Two foraminifera samples from near the bottom of core 126P, at 734 and 744cm, had δO^{18} values of -0.6 per mill and δC^{13} values of 2.1 and 2.4 per mill, respectively. From inspection of the samples and cores, these are correlated with foraminifera from between 125 and 150cm in core 118K and thus may show a similar extent of oxygen isotopic shift as the samples from 119K and 84K.

Pteropods. At the bottom of core 127P (783cm) aragonite encrusted pteropods were found. Their δO^{18} was $+6.7$ per mill and δC^{13} was $+3.9$ per mill. These values are identical to those of 118K and show no indication of a shift toward lighter values.

Core 154P (About 160km SSE of Atlantis II Deep)

This core has special significance since it was one of the longest cores available and was collected about 100 miles south of the hot brine area. It could thus be expected to be free of any influence of events which took place in those deeps.

Pteropods. Fig. 6 shows the pteropod analyses of core 154P and, for comparison, a timetable of the glacial periods in Europe (van der Hammen *et al.*, 1967). The correlation is based on C^{14} age determinations on this core by Ku *et al.* (1969) given in

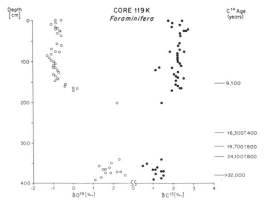

Fig. 5. δO^{18} and δC^{13} distribution in foraminifera tests of core 119K.

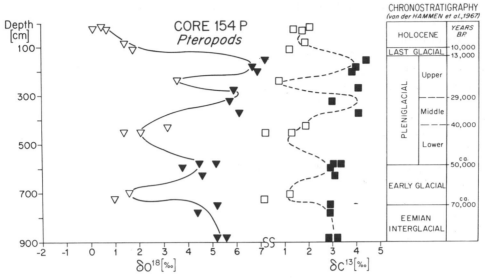

Fig. 6. δO¹⁸ and δC¹³ distribution in pteropod shells of core 154P. Open triangles and squares = unaltered shells; filled triangles and squares = aragonite incrusted shells.

Fig. 7. The volume of sample material available for isotopic analysis was quite small, and, as a consequence, not all horizons yielded sufficient pteropod shell material for analysis although there was almost always an abundance of foraminifera. When possible, separate materials of different appearance were analyzed separately, as, for example, at 275cm where fresh looking shells occurred with shells having an aragonitic overgrowth. Thus, two points plotted for the same depth are not to be taken as duplicate runs of identical material. The fact that such different materials were found together in some samples merely reflects the mode of sampling the core in 10cm strips. The alternating occurrence of fresh looking and overgrown shells throughout the core appears to be significant. In all instances the overgrown shells have higher O¹⁸ and C¹³ contents than the unaltered shells.

Foraminifera. A complete set of foraminifera samples from this core was analyzed and, where possible, in duplicate (Fig. 7). Extrapolation from C¹⁴ data (Ku et al., 1969), and paleontological evidence (Berggren et al., 1969) indicate that this core probably represents the sediment deposited in the last 80,000 to 100,000 years.

The C¹⁴ age confirms the correlation with 118K on the basis of δO¹⁸ of the uppermost layer of "heavy" pteropods. The oxygen values of the foraminifera reveal several cycles of gradually increasing and then relatively rapidly decreasing δO¹⁸ values, with the abrupt changes occurring between 750 and 725, 475 and 450, and 250 and 225cm. The trends are not always smooth. This may be partly due to the discontinuous 25cm interval sampling of the core. Many of the tests were extremely small and in such instances it was very difficult to ex-

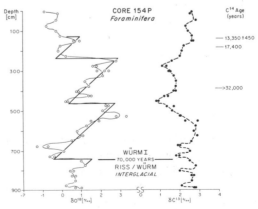

Fig. 7. δO¹⁸ and δC¹³ distribution in foraminifera tests of core 154P.

clude all foreign material adhering to or included in the tests. This may be the cause for the lower than average level of analytical precision as indicated in some of the duplicate analyses. The precision of the δC^{13} determinations is better, and the δC^{13} values form a rather smooth and continuous curve. The sudden changes in the δO^{18} values have no counterparts in the δC^{13} values, which show gentle increases and decreases leading to pronounced minima that generally coincide with the δO^{18} minima.

Discussion

The Red Sea is a long and narrow basin, having a limited water exchange with the Indian Ocean at its southern end where a sill rises to about 100m below present sea level. It is, therefore, not possible to interpret the oxygen isotope data from its sediments simply in terms of paleotemperatures, as is commonly done with deep sea sediments. One must assume that the isotopic composition of the water was not constant throughout the history of the Red Sea and, moreover, probably also changed with distance from the Gulf of Aden as the influence of the exchange with the open ocean diminished. Such a geographic effect is indeed observed in the Red Sea today (Craig, 1966). Using Craig's data for present-day Red Sea water, it is possible to calculate temperature extremes for the fossils found in the uppermost sediment layers, above the first horizon of incrusted pteropods, which represent the last 10,000 years. Below this level it is impossible to make such a calculation because salinity effects mask the temperature effect.

The Hot-Brine Deep Area

One prominent feature of cores from this area is the transition zone which occurs at a depth of 135 to 170cm in 118K and about 15cm deeper in 119K. This zone is marked by a rapid change from heavy δO^{18} (+6 to 7 per mill) and δC^{13} (+4 to 4.5 per mill) values for pteropods to very much lighter values and by the sudden appearance of an abundance of foraminifera tests. Forams are virtually absent below this zone in 118K and were only found below 340cm in 119K. Another characteristic of the transition zone is the extremely crushed state of its fossil shells. This zone can also be identified at 150cm in core 154P, 100 miles to the south. We interpret this zone as being the record of the abrupt change in climate which took place about 11,000 years ago and has been observed in many parts of the world (Broecker et al., 1960). This change led to the general type of climate that persisted to the present time. Such a rapid change may have caused a turbulent sediment transport from the south (due to overflow at the sill) to the north of the Red Sea and produced the distinct layer of fragmented shells.

Since the climatic changes after this event were relatively small, it may be correct to assume that the present range of O^{18}/O^{16} ratios of Red Sea surface waters is representative of the conditions of the last 10,000 years and to calculate temperature extremes for the waters in which the foraminifera and pteropods formed their shells. According to Craig (1966), the δO^{18} range for surface waters is 0.9 to 1.6 per mill on the PDB scale. Applying this range to the foraminifera data of 118K yields a maximum temperature range of 17 to 28°C, which is reasonable for this area. Foraminifera and pteropods from the same horizons generally differ slightly in their carbon and oxygen isotope contents. The δO^{18} values for pteropods are usually somewhat higher, and the δC^{13} values are slightly lower. These observations suggest a deeper depth habitat for the pteropods, where the water is colder and the C^{13}/C^{12} ratio of the bicarbonate may be lower due to the contribution of oxidized biogenic carbon of lower C^{13} content.

The isotopic shift observed in the upper 150cm of core 119K relative to core 118K, between foraminifera of the same age, has been discussed in detail in an earlier paper (Deuser, 1968). Foraminifera and, apparently, also pteropods have undergone an average shift of 0.9 per mill in the direction of O^{18} depletion in the Discovery Deep (Figs.

3 and 5). This is probably due to post-depositional isotope exchange with the hot brine. A similar shift is observed in foraminifera from the upper detrital layers in the Atlantis II Deep. Aragonite incrusted pteropods from below the transition zone, however, exhibit no isotopic shift in either deep. This suggests that the hot brine has not passed the sediment in broad fronts from below and that it was not present before about 10,000 years ago in either hole. If the brine rose in one or several discrete vents, these were not in the immediate vicinity of cores 126P and 127P. Core 84K was unfortunately too short to enable one to draw conclusions on this point based on the isotopic evidence.

Evaporation and Carbonate Precipitation

The almost complete absence of foraminifera from large segments of cores 118K and 119K and the high O^{18} and C^{13} contents of the pteropods, as well as their secondary aragonite overgrowth can best be explained as due to highly saline conditions during times of lowered sea level. At such times the relative rise of the sill at Bab el Mandeb diminished, and perhaps even occasionally stopped, the exchange of water between the Red Sea and the Indian Ocean. In the absence of a significant influx from the Indian Ocean, evaporation resulted in higher salinities and concurrently higher O^{18} content of the water (Epstein and Mayeda, 1953). The effect was probably the stronger the greater the distance from the Gulf of Aden. This may be the reason why foraminifera are practically absent from large parts of cores 118K and 119K while they are abundant in the time equivalent zones of core 154P, 100 miles farther south.*

The enrichment of C^{13} requires a different explanation because the isotopic composition of the carbon pool in the water

does not change by evaporation in the same manner as the isotopic composition of oxygen. Carbonates from highly saline environments have δC^{13} values around +4 per mill (Lowenstam and Epstein, 1957; Degens and Epstein, 1964). Recent analyses of a series of 20 oolites, grapestones and calcareous algae (Table 1), all from highly saline environments, show a δC^{13} range from +3.5 to +5.5 per mill. Two different processes appear to be responsible for the observed δC^{13} values. One set of samples is isotopically similar to the aqueous bicarbonate; the other set exhibits δC^{13} values which are several per mill higher than δC^{13} of marine bicarbonate. Organisms, such as molluscs, and foraminifera, which in the building of their calcareous shells rely exclusively on organic template phenomena (Degens et al., 1967), have shells with a δC^{13} value identical to or close to that of the utilized sea water bicarbonate. Other organisms, like certain calcareous algae, which precipitate carbonate in the process of photosynthesis, deposit $CaCO_3$ enriched in C^{13} by several per mill relative to the dissolved carbon supply. Similarly, calcium carbonate formed inorganically upon slow evaporation of sea water shows δC^{13} values around +4 to 5 per mill.

We suggest the following explanation for the observed differences in the C^{13}/C^{12} ratio of carbonates. The process of carbonate formation may be described by the following simplified equation:

$$Ca^{++} + 2HCO_3^- \rightleftharpoons CaCO_3 + H_2O + CO_2 \quad [1]$$

The deposition on organic templates precludes carbon-isotopic equilibrium in this process as indeed it precludes chemical equilibrium in all those instances where the deposition occurs in an undersaturated environment. All three carbon-containing molecular species, therefore, will have identical or very similar isotopic compositions. In contrast, the carbonate formation will be an equilibrium process with isotopic fractionation in an environment which slowly reaches saturation through evaporation. The calculated carbon isotope fractionation at equilibrium between aque-

---

* It is also conceivable that not only the distance from the Gulf of Aden but also the distance from the shore line has a marked influence on the salinity profile and in turn on the biological population. Berggren et al. (1969) present paleontological evidence in support of this inference.

Table 1 Oxygen and Carbon Isotope Ratios of Oolites, Grapestones, and Calcareous Algae

No.	Age	Type	Locality	δO^{18}	δC^{13}	Comments
1				−0.6	4.4	Samples 1 to 10 represent
2				−0.5	5.1	consecutive leachings of a
3				−0.6	5.2	0.5mm oolite fraction,
4				−0.5	4.5	starting from the outside
5				−0.5	4.6	(No. 1) towards the core
6	Recent	Oolite	Bahamas	−0.4	4.6	of the oolite (No. 10).
7				−0.3	5.0	
8				−0.4	5.2	
9				−0.7	5.2	
10				−0.3	5.5	
11				0.0	4.9	0.125mm fraction
12			Off Florida	1.9	4.1	
13			Off Georgia	1.8	3.5	
14	Pleistocene	Oolite	Off Georgia	2.0	4.8	
15			Off Georgia	2.0	4.7	Center layers of No. 14
16			Great Salt Lake	−4.0	3.3	
17	Recent	Grapestone	Bahamas	−0.3	4.9	
18				−0.3	4.6	
19		*Halimeda* sp.		−1.2	4.2	
20				−0.6	4.2	
21				−1.3	2.0	
22	Recent	*Udotea* sp.	Bahamas	−0.6	5.8	
23				−1.8	5.9	
24				−2.3	5.8	
25		*Rhipocephalus* sp.		−1.4	4.4	
26				−0.9	4.4	

Samples 20 to 26 after Lowenstam and Epstein (1957).

ous carbon dioxide and the carbonate ion is about 7 per mill at 25°C. Experiments indicate an even greater fractionation (Thode *et al.,* 1965). Thus, if during carbonate formation, as described by equation [1], the carbonate and the CO_2 are in isotopic equilibrium, their C^{13}/C^{12} ratios will differ by about 8 per mill, the carbonate having the higher ratio. If we assume a δC^{13} of 0 per mill for the bicarbonate, the value for the carbonate will accordingly be +4 per mill and that of the CO_2 −4 per mill because they form in a 1:1 molar ratio. While the carbonate will remain fixed, the CO_2 will quickly reequilibrate with the dissolved carbon pool and, in the case of continued evaporation and precipitation in a restricted environment, will lead to a gradual decrease of the C^{13}/C^{12} ratio in that pool.

Climatic Conditions of the Last 80,000 to 100,000 Years

The cyclic changes in both the O^{18}/O^{16} and C^{13}/C^{12} ratios as they are observed in core 154P can best be interpreted as being due to climatic changes which, in the special situation of the Red Sea, could strongly influence the properties of the marine environment. Lowering of sea level in response to continental glaciations reduced the water exchange at the southern end of the Red Sea. Even today the salinity of Red Sea waters is several per mill above normal ocean waters and evaporation would quickly raise it if exchange at the southern end were reduced. There are no major watersheds draining into the Red Sea which could offset this evaporation, even during periods of a cooler and less arid climate (pluvial times) than exists in the area today.

An alternate explanation for the observed cyclic changes, which must be considered in a tectonically active region like the Red Sea, is the reduction of water exchange with the Indian Ocean due to tectonic lifting of the southern sill or narrowing of the Strait of Bab el Mandeb. This would have the same effect on the isotopic composition of Red Sea waters as a sea level drop, except for the climatic effects which are associated with periods of lowered sea level. Tectonic activity might also produce catastrophic events, possibly leaving such records as the layer of intensely fragmented shells observed above the most recent layer of aragonite-incrusted pteropods.

Judging from the oxygen isotope measurements, there were at least three major cycles of slowly increasing salinity followed by much faster decreasing salinity during the time span covered by core 154P. As is indicated by the occurrence of the most recent aragonite-overgrown pteropods, the cycle between 200 and 125cm is correlated to that zone observed in core 118K. This conclusion is confirmed by the radiocarbon dates (Ku *et al.*, 1969). On the basis of these correlations and the preliminary radiocarbon dates available, we extrapolate an age of 80,000 to 100,000 years for the bottom of core 154P. Above the last change at 150cm the foraminifera data show a pattern similar to that observed in greater detail in core 118K that corresponds to the events of the last 10,000 years.

The oxygen data from the peteropods of core 154P (Fig. 6) also show the alternation of periods of evaporation and supersaturation, on the one hand, and "normal" conditions, on the other. Isotopically light pteropod shells without secondary overgrowth are found in the upper meter of the core and at or near those depths where the foraminifera data show pronounced changes toward lighter values. Where the foraminifera exhibit high δO^{18} values, overgrown pteropods occur with δ-values between 4 and 7 per mill.

The carbon-isotope data from the forams (Fig. 7) are an interesting complement to the oxygen data. At the beginning of each cycle, *i.e.*, when the O^{18}/O^{16} ratios begin to increase, the C^{13}/C^{12} ratios show a similar increase until a point is reached where carbonate supersaturation occurs. This point is characterized by the appearance of incrusted pteropods of high C^{13} content (Fig. 6) and by the beginning of a gradual return to lower C^{13}/C^{12} ratios in the forams while the O^{18}/O^{16} ratios continue to in-

crease. The initial rise in C^{13}/C^{12} ratio can best be interpreted as due to temperature effects.

In view of our brief discussion of carbonate precipitation, a likely explanation for this phenomenon is the following: during periods of continuing evaporation more saline waters sank to the bottom. Carbonate saturation was reached and isotopically heavy aragonite was precipitated, aragonitic pteropod shells serving as substrates. (No aragonite incrustation of foraminifera tests has been observed. This is most certainly related to their calcitic nature which interferes with the epitaxial growth of aragonite.) The influx of supersaturated waters may have varied from place to place, as suggested by varying degrees of overgrowth between shells from corresponding layers of cores 118K and 119K. The massive carbonate precipitation lead to a gradual decrease of the C^{13}/C^{12} ratio in the remaining dissolved carbon. This is reflected in the lower C^{13}/C^{12} ratios of the foraminifera tests formed in the upper layers of the water. When a rise in sea level or some other mechanism reestablished a significant influx of ocean water, supersaturation ceased, sinking pteropod shells were no longer overgrown, the O^{18}/O^{16} ratio in the water, and thereby in shells formed in equilibrium with the water, decreased, and, consequently, conditions were back to "normal" until a renewed drop in sea level lead to a repetition of the cycle.

Summary

The following conclusions are drawn from the detailed measurement of oxygen and carbon isotope ratios of the fossil content in three cores 118K, 119K and 154P:

1. All cores show pronounced fluctuations in both O^{18}/O^{16} and C^{13}/C^{12} ratios with depth, and these fluctuations can be used to correlate sediment horizons among cores from different locations.
2. The hot brine has reacted with fo-

raminifera tests and has produced an isotopic shift in the tests.

3. Hot brine appears to have been present, at least intermittently, in the Discovery Deep for about 10,000 years but does not seem to have originated in that deep.

4. If the brine entered the Atlantis II Deep from below, it must have risen along discrete vents rather than in a broad front.

5. Core 154P shows several cycles of isotope ratio variations produced by evaporation periods. These were probably due to glacial sea-level lowering and consequently reduced water exchange between the Red Sea and the Indian Ocean.

6. Aragonite incrusting of pteropod shells occurred in the latter part of all evaporation cycles and ceased after the beginning of the relatively rapid return to more normal conditions.

7. The isotopic composition of the dissolved carbon of an isolated basin, such as the Red Sea, experiences a significant shift towards lower C^{13}/C^{12} ratios during periods of massive carbonate precipitation.

Acknowledgments

This work was supported by the National Science Foundation under Grants GA-970 and GA 584. We thank Mrs. Edith Ross and Mrs. Charlotte Lawson for valuable laboratory assistance.

Appendix

C^{13}/C^{12} Ratios in the Waters of the Atlantis II Deep

C^{13}/C^{12} ratios in the dissolved carbon of a suite of water samples from the Atlantis II Deep (**CHAIN** 61, Hydro station 715) were measured in connection with the studies described in this chapter. $CuCl_2$ was added to the water samples immediately upon their being drained from Nansen bottles into glass bottles with tightly sealing

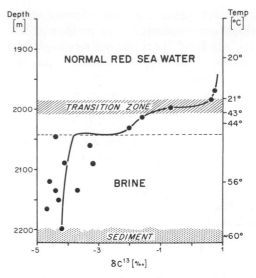

Fig. 8. δC^{13} in dissolved inorganic carbon of Atlantis II Deep, Station 715.

screw caps. One sample was collected in a Niskin sampler to which KOH was added prior to collection to secure complete recovery of the dissolved carbon. No isotope difference was found between waters taken from the Nansen bottles and the KOH-spiked Niskin sampler. This indicates that no loss in CO_2 has occurred during collection and in turn that the brines are undersaturated with respect to CO_2 *in loco*. CO_2 was extracted in the laboratory by acidifying 125cc of the water with H_3PO_4. The results are shown in Fig. 8. δC^{13} in the 56° water is about −4.5 per mill, and above this layer, through the 44° layer, there appears to be a more or less continuous transition to lighter values, approaching +1 per mill in the normal Red Sea bottom water. These data are thus consistent with the hypothesis that the 44° layer of the Atlantis II Deep is nothing but a mixing layer between the 56° brine and Red Sea bottom water (Turner, 1969). The δC^{13} of the dissolved carbon in the 44°C brine of the Discovery Deep appears to be similar to that of the 56°C Atlantis brine samples. Analysis of a single sample yielded a value of −5.8 per mill. It is tempting to speculate on a juvenile origin for this exceptionally light, dissolved carbon, as similarly light carbon has been reported for late-stage volcanic exhalations of apparently juvenile carbon dioxide (Taylor *et al.*, 1967). Alternatively, some light δC^{13} values may arise from the preferential removal of the C^{13} from the dissolved carbon pool in the course of carbonate precipitation.

In view of the high brine temperature and the low C^{13}/C^{12} ratio, it is surprising that no evidence for isotopic reequilibration can be detected with the isotopically very heavy deposits older than about 10,000 years, at least not in cores 126P and 127P. It appears that, if the brine did come from below in this hole, it did not rise in the immediate vicinity of either of these two core locations.

References

Berggren, W. A. Micropaleontologic investigations of Red Sea cores — Summation and synthesis of results. *In: Hot brines and recent heavy metal deposits in the Red Sea*, E. T. Degens and D. A. Ross (eds.). Springer-Verlag New York Inc., 329–335 (1969).

Broecker, W. S., M. Ewing, and B. C. Heezen: Evidence for an abrupt change in climate close to 11,000 years ago. Am. J. Sci., 258, 429 (1960).

Craig, H.: Isotopic standards for carbon and oxygen and correction factors for mass-spectrometric analysis of carbon dioxide. Geochim. et Cosmochim. Acta, 12, 133 (1957).

———: Isotopic composition and origin of the Red Sea and Salton Sea geothermal brines. Science, 154, 1544 (1966).

Degens, E. T. and S. Epstein: Oxygen and carbon isotope ratios in coexisting calcites and dolomites from recent and ancient sediments. Geochim. et Cosmochim. Acta, 28, 23 (1964).

Degens, E. T., D. W. Spencer, and R. H. Parker: Paleobiochemistry of molluscan shell proteins. Comp. Biochem. Physiol., 20, 553 (1967).

Deuser, W. G.: Postdepositional changes in the oxygen isotope ratios of Pleistocene foraminifera tests in the Red Sea. J. Geophys. Res., 73, 3311 (1968).

Emiliani, C.: Paleotemperature analyses of Caribbean cores P6304-8 and P6304-9 and a generalized temperature curve for the past 425,000 years. J. Geol., 74, 109 (1966).

Epstein, S., R. Buchsbaum, H. A. Lowenstam, and H. C. Urey: Revised carbonate-water isotopic temperature scale. Geol. Soc. Am. Bull., 64, 1315 (1953).

Epstein, S. and T. Mayeda: Variation of O^{18} content of waters from natural sources. Geochim. et Cosmochim. Acta, 4, 213 (1953).

Gevirtz, J. L. and G. M. Friedman: Deep-sea carbonate sediments of the Red Sea and their implications on marine lithification. J. Sed. Pet., 36, 143 (1966).

Ku, T. L., D. L. Thurber, and G. Mathieu: Radiocarbon chronology of Red Sea sediments. *In: Hot brines and recent heavy metal deposits in the Red Sea*, E. T. Degens and D. A. Ross (eds.). Springer-Verlag New York Inc., 348–359 (1969).

Lowenstam, H. A. and S. Epstein: On the origin of sedimentary aragonite needles of the Great Bahama Bank. J. Geol., 65, 364 (1957).

McCrea, J. M.: On the isotopic chemistry of carbonates and a paleotemperature scale. J. Chem. Phys., 18, 849 (1950).

McKinney, C. R., J. M. McCrea, S. Epstein, H. A. Allen, and H. C. Urey: Improvements in mass spectrometers for the measurement of small differences in isotope abundance ratios. Rev. Sci. Instr., 21, 724 (1950).

Taylor, H. P., Jr., J. Frechen, and E. T. Degens: Oxygen and carbon isotope studies of carbonatites from the Laacher See District, West Germany, and the Alnö District, Sweden. Geochim. et Cosmochim. Acta, 31, 407 (1967).

Thode, H. G., M. Shima, C. E. Rees, and K. V. Krishnamurty: Carbon-13 isotope effects in systems containing carbon dioxide, bicarbonate, carbonate, and metal ions. Canad. J. Chem., 43, 582 (1965).

Turner, J. S.: A physical interpretation of the observations of hot brine layers in the Red Sea. *In: Hot brines and recent heavy metal deposits in the Red Sea*, E. T. Degens and D. A. Ross (eds.). Springer-Verlag New York Inc., 164–173 (1969).

van der Hammen, T., G. C. Maarleveld, J. C. Vogel, and W. H. Zagwijn: Stratigraphy, climatic succession and radiocarbon dating of the last Glacial in the Netherlands. Geol. en Mijnb., 46, 79 (1967).

Radiocarbon Chronology of Red Sea Sediments *

TEH-LUNG KU

Woods Hole Oceanographic Institution
Woods Hole, Massachusetts

DAVID L. THURBER AND GUY G. MATHIEU

Lamont Geological Observatory of Columbia University
Palisades, New York

Abstract

Radiocarbon dates have been obtained on seven cores taken from the central part of the Red Sea in the axial trough. The observed sedimentation rates (over the past 20,000 years) range from about 5cm/1,000 yr to more than 60cm/1,000 yr. Variations are primarily due to the episodic precipitation of minerals associated with the geothermal activity of the hot-brine deeps. The brine-derived materials precipitate at a rate of more than 40cm/1,000 yr, whereas accumulation rates of the detrital silicates and calcareous shells are on the order of 2cm/1,000 yr and 8cm/1,000 yr respectively, for the area studied.

The abrupt world climatic changes at 11,000 years ago which mark the close of the Wisconsin glacial period (Broecker *et al.*, 1960) are found in these cores to be reflected by the following features of transition: (1) Lithified carbonate sediments, while common below the transition, are no longer formed. (2) Both planktonic foraminifera and pteropods become much more abundant in the recent sediments. The relative abundance of species also changes; species with less tolerance to high salinity now flourish at the expense of the more salinity-tolerant ones. (3) Shallow water dinoflagellates appear to be transported to the axial trough area at the present, but are absent below the transition zone. (4) δO^{18} of foram and pteropod tests decreases from values as high as +6 per mill (on PDB scale) in older sediments to near 0 per mill in the recent. These features imply that before the warming of world-wide climate at about 11,000 years B. P., the Red Sea environmental condition must have been one of high salinity

and regression of the sea. The sill located near the Strait of Bab el Mandeb must have been effective in limiting the water exchanges between the Red Sea and the Indian Ocean during the low sea stand of the Wisconsin. Such a condition may have prevailed with short-term exceptions back to about 70,000 years ago.

Intrusions of hot brine into the Atlantis II Deep may have occurred over the last 10,000 years. Records of geothermal events older than 20,000 years have been observed elsewhere.

Introduction

This paper presents radiocarbon ages obtained on seven sediment cores from the central part of the Red Sea. The samples were collected by **R/V CHAIN** during her cruise No. 61 in 1966. Three cores (95K, 127P, 119K) were from the hot brine deeps. Cores 143K and 118K were located outside but very close to the brine area, whereas cores 153P and 154P were from the axial trough approximately 100 miles to the south of the deeps. The sample locations and water depths are given in Table 1 and Fig. 1.

The results permit estimates of sedimentation rates over the past 30,000 years for these sedimentary columns. Such a time scale, combined with information gathered from paleontological, mineralogical and

* Woods Hole Oceanographic Institution Contribution No. 2196 and Lamont Geological Observatory Contribution No. 1292.

Table 1 Source of Samples

Core No.* (CHAIN 61)	Latitude (N)	Longitude (E)	Depth (m)	Core Length (cm)
95K	21° 23.4′	38° 03.2′	1,910	305
118K	21° 14.4′	38° 04.3′	1,989	400
119K	21° 16.8′	38° 01.4′	2,175	400
127P	21° 22.4′	38° 02.8′	2,106	836
143K	21° 27.0′	38° 02.9′	1,910	354
153P	19° 43.0′	38° 41.0′	2,704	755
154P	19° 34.0′	38° 59.5′	1,276	792

* K and P refer to Kasten and piston coring device respectively. (See Ross and Degens, 1969.)

stable isotope studies, forms a basis for the reconstruction of the late Pleistocene history of the Red Sea. It also allows correlations to be made between events that have taken place in the Red Sea and those in other geographic areas.

Nature of Core Samples

The cores are different in texture and lithology from those of the normal pelagic type. In general, the sediments are finely

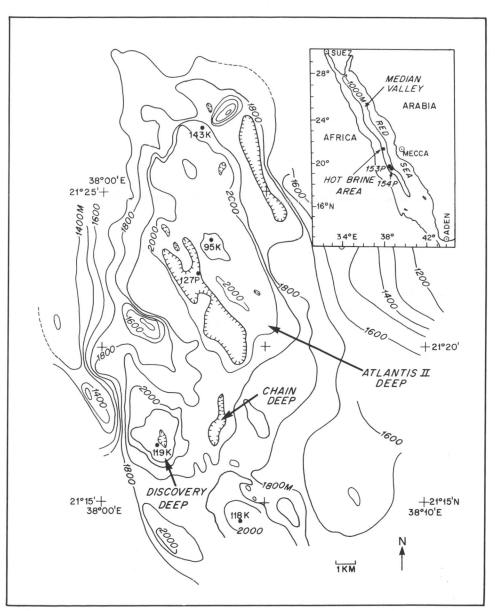

Fig. 1. Bathymetric map of hot-brine area showing location of cores. Note that cores 153P and 154P are shown in the index map.

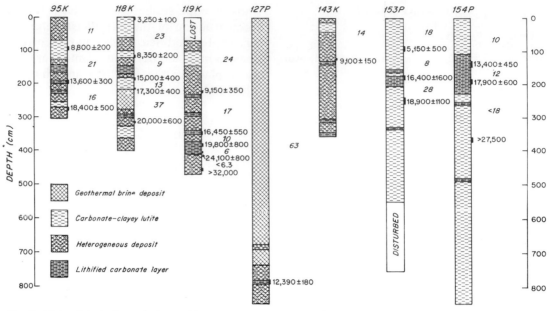

Fig. 2. Major lithologic variations with depth in the cores. Radiocarbon dates for individual segments of the cores are given in years before present. The italic numbers indicate the average sedimentation rates (cm/1,000 yr) over the interval between the two C14-dated sections.

bedded or diffusely mottled. They are characterized by layers exhibiting a great variety of color shades which vary in thickness from millimeters to tens of centimeters. Lamination is especially prominent for those cores in or near the deeps (e.g., 95K, 118K, 119K, 127P and 143K) where materials associated with the geothermal brine extrusions produce deposits of various compositions. In Fig. 2 we show, schematically, the major lithologic units, representing sediment types formed by one dominant process of deposition over another within the interval designated.

Geothermal brine deposits are characterized by materials precipitated out of the subterranean brine as it discharges into the bottom of the Red Sea. Bischoff (1969a) has separated these precipitates into three main facies: Fe-montmorillonite (dark brown), amorphous-goethite (orange to yellow), and sulfides (black). Generally homogeneous and finely-bedded, this unit contains a relatively small amount (<5 per cent) of detrital silicates and calcareous material. The latter may occasionally be present as thin laminae or lenses less than a centimeter in thickness.

Carbonate-clayey lutite is pelagic sedimentary material of varying color, from yellowish brown (10 YR 4/2)* to dark brown (10 YR 5/4)* and from yellowish gray (5 Y 6/2)* to olive gray (5 Y 5/2).* The major components are tests of planktonic foraminifera and pteropods, coccoliths, and clays. Quartz and feldspar are present in minor quantities. These are materials common to the normal deep sea calcareous oozes, but differ from the latter in having higher iron content and a less oxidized state, giving a darker appearance to certain layers of this unit.

Heterogeneous deposits are sections showing interbedding and greater degree of intermixing of the above-mentioned two units. The thickness of the interlayering bands is arbitrarily set as thinner than 10cm for easier illustration on the graph shown in Fig. 2. This unit is probably due to intermittent discharging of brine precipitates mixed and intercalated with the lutite material deposited in normal marine environments. One distinctive feature is that a large number of pteropod shells is often

* G.S.A. rock-color chart.

associated with black sulfide-rich layers which were presumably laid down under anaerobic conditions.

Lithified carbonate layers are indurated hard layers of aragonitic composition serving as cementing matrix for lutites containing pteropods and forams. Although these layers do not constitute a major lithologic facies in the sedimentary columns (rarely exceeding 1cm thick as an uninterrupted layer), their presence is indicative of inorganic precipitation of aragonite under hypersaline marine conditions (Gevirtz and Friedman, 1966; Milliman *et al.*, in press).

Methods

Sample Preparation

The cores were sampled shortly after their storage trays were opened; they were still moist. Because of possible contamination, the outer parts in contact with the core barrel were discarded. To obtain samples containing more than 0.5 grams of carbon for the C^{14} assay, we sampled sections of relatively high calcium carbonate content. Hence the choice of depth interval along a core was irregular because of the inhomogeneous lithology. A few samples were chosen from horizons indicating changes in foraminiferal assemblages.

The bulk core materials were wet-sieved (using distilled water) through sieves of 74μ-mesh size. We made C^{14} measurements only on the $>74\mu$ size fractions. These are composed dominantly of planktonic forams and pteropods, with shells of benthonic forams, pelecypods, heteropods, and other marine gastropods occasionally present in minute amounts. In certain sections containing the lithified carbonate layers, aragonite serves as cementing agent for the fossil shells and imparts overgrowths and infillings to the imbedded pteropod shells. We made no attempt to separate the cementing material from the original shells. In two cases, however, we have made separate runs on the lithified fragments (a mixture of cement and imbedded

aragonite-filled shells) and on the foraminifera and pteropod unfilled shells from essentially the same depth.

Radiocarbon Ages

Radiocarbon measurements were made at the Lamont Geological Observatory radiocarbon laboratory with the exception of one (core 127P, 783–790cm) which was made by Isotopes, Inc., Westwood, N.J. Details of the Lamont apparatus and procedures have been previously described (Broecker *et al.*, 1959).

The ages are reported in years before 1950, using the 5,568 yr half-life. The precision of measurement is given as the standard error on the 66 per cent confidence level, determined from counter characteristics, sample activity, and estimated errors in dilution, where necessary.

The ages are calculated on the assumptions that the samples obtained their CO_2 from normal surface sea water, 5 per cent deficient in radiocarbon relative to the atmosphere, and that this deficiency is counteracted by concentrating the C^{14} by 5 per cent relative to standard wood. Thus the ages are calculated without correction from the 0.95 NBS oxalic acid secondary standard.

The first assumption is supported by the general oceanographic situation in the Red Sea and unpublished (Lamont Geological Observatory) data on Red Sea and Indian Ocean water.

The δC^{13} (PDB) on some of the fauna ranges only from +3 to +4.5 per mill (Deuser and Degens, 1969) and is dependent on the presumed eustatic change of sea-level. Uncertainties due to isotope fractionations should be slight (about 100 years). Those due to variations in radiocarbon may be somewhat greater but should be less than 1,000 years, and would affect all samples of the same apparent age equally.

Other factors affecting the ages might include the following:

1. Mixing of the core during coring. This usually affects the top or bottom sec-

tions of the core and can usually be identified. The samples come from sections where the lithology is intact.

2. Penecontemporaneous mixing of the core by currents, organisms, or slumping. This is probably ubiquitous in deep-sea sediments but easily identifiable where lithology is clear as it is in these cores. The thinness and distinctness of the laminations here suggests that precision on age determination is limited more by the thickness of the samples sectioned than by mixing processes.

3. The presence of reworked material is a possible contribution of error. Windblown carbonate dust is excluded by dating only the coarse fraction (Rubin and Suess, 1955; Broecker and Kulp, 1957; Olsson and Eriksson, 1965). In the Red Sea one must also be cautious in excluding carbonate sands which may be largely derived from pre-Wisconsin coral reefs. The samples were examined megascopically and freed from any material possibly derived from this source. Generally the cores were devoid of such material.

4. Post-depositional exchange or cementation. We assume that sample treatment contributes insignificant error except to the oldest samples which are, at any rate, beyond the limits where C^{14} could be significantly detected.

However, two interesting possibilities for natural contamination may exist.

Craig (1968, personal communication) has suggested that radiocarbon in the deep brines is derived by exchange between HCO_3^- in solution and the sediments. Back exchange of dead CO_2 would make the sediments too old in apparent age, but probably the contribution has an insignificant effect on the sediments.

The cemented layers are also somewhat of a mystery. Agreement in age between cemented layers and uncemented fossils immediately adjacent suggest penecontemporaneous cementation and a similar source for the carbon.

We conclude that the radiocarbon ages reported here are a reasonably accurate representation of the age of the depth in the core from which they were taken. This may be further suggested by the fact that the C^{14} age sequence in a core is in accord with the stratigraphic positions.

Results and Discussion

The results are summarized in Table 2. They are also plotted in Fig. 2 as a function of depth in the cores, together with schematic descriptions of the major lithologic variations. In Table 2 and Fig. 2 we also give our estimates of the total sedimentation rate in centimeters per thousand years for the interval between each pair of radiocarbon dates. These were obtained by dividing the age difference by the depth (midpoint) difference.

Rates of Sedimentation

Variations in the rates of deposition from core to core and from one interval to another within a core are large. This is not unexpected in view of the inhomogeneity in lithology and sedimentation of these deposits. The bulk sedimentation rates given in Table 2 and Fig. 2 merely indicate an approximate range of variability as well as an order of magnitude estimate. The average water- and salt-free bulk density for the geothermal brine deposits (brine content ~80 per cent by weight) is about $0.3 gr/cm^3$, whereas that of lutite layers is about $0.7 gr/cm^3$. Hence it may be more desirable to express the rate values in terms of $gr/cm^2/1,000$ yr rather than $cm/1,000$ yr. However, the large lithology changes within an interval between two age measurements have precluded such an attempt.

The total sedimentation rates vary from about 5cm to more than 60cm per 1,000 years. We may broadly divide the sedimentary components into three categories: brine-derived material, carbonate, and detrital silicates. The precipitation rates of the brine-derived material can be estimated from core 127P (0–740cm) and core 118K

(210–285cm). They are in general larger than 40cm/1,000 yr. For estimation of the carbonate and silicate deposition rates, we use the data on 153P and 154P. These cores are located at about 100 miles to the south of the hot brine deeps and are relatively free of the brine precipitates. The average total sedimentation rate of the two lutite cores over the last 20,000 years is about 10cm/1,000 yr. These two cores have an average carbonate content of about 80 per cent. The corresponding accumulation rates for the carbonate and the detrital silicates are thus on the order of

Table 2 Age and Sedimentation Rate Data for Red Sea Cores

Core No.	C¹⁴ Lab No.	Depth Interval (cm)	Midpoint (cm)	C¹⁴ Age (yr)	Sed. Rates § (cm/10³ yr)
95K	L1168C	*90–100	95	8,800 ± 200	11
	L1168A	*190–200	195	13,600 ± 300	21
	L1168B	*270–275	272.5	18,400 ± 500	16
118K	L1165A	*0–10	5	3,250 ± 100	
	L1165C	*120–125	122.5	8,350 ± 200	23
	L1165D	†180–185	182.5	15,200 ± 400	9
	L1165Da	*180–185	182.5	14,800 ± 400	
	L1165E	†210–215	212.5	17,300 ± 400	13
	L1165F	†310–315	312.5	20,000 ± 600	37
119K	L1171A	*220–225	222.5	9,150 ± 350	24
	L1171CI	†340–350	345	16,400 ± 500	17
	L1171CII	‡340–350	345	16,500 ± 650	
	L1171DII	‡375–380	377.5	19,800 ± 800	10
	L1171B	*400–405	402.5	24,100 ± 800	6
	L1165G	*450–455	452.5	>32,000	<6.3
127P	I-3465	‡783–790	786.5	12,390 ± 180	63
143K	L1211B	‡120–125	122.5	9,100 ± 150	14
153P	L1209A	*85–100	92.5	5,150 ± 500	18
	L1209B	‡170–185	177.5	16,400 ± 1,600	8
	L1209C	*237–259	248	18,900 ± 1,100	28
154P	L1210A	‡130–150	140	13,400 ± 450	10
	L1210B	‡185–200	192.5	17,900 ± 600	12
	L1210C	*360–375	367.5	>27,500	<18

* Forams + pteropods.
† Pteropods.
‡ Forams + pteropods + aragonitic cement.
§ Averaged over the interval between two midpoints.

8cm/1,000 yr and 2cm/1,000 yr, respectively.*

As a general comparison, we may state that the accumulation rates of detrital silicates in the deep Red Sea are about five to ten times those in the open ocean pelagic area and are comparable to or slightly higher than those in the Caribbean Sea (see summary in Ku *et al.,* 1968). On the other hand, the Red Sea clay accumulation rates are lower than those of other land-bordered ocean basins like the Mediterranean (~15cm/1,000 yr for a core from the West Mediterranean, see Eriksson, 1967). Presumably this is due to the lack of major water-sheds draining into the Red Sea basin.

Late Pleistocene and Holocene Chronology

Evidence for changes in climatic and environmental conditions of the Red Sea in the geological past has been provided by studies of oxygen and carbon isotopes (Deuser and Degens, 1969) and of microfossils (Berggren, 1969) in the cores. It is clearly of importance to determine when these changes took place.

Planktonic foraminifera and pteropods are the two major groups of larger organisms in the sediments. A characteristic distribution pattern observed by Berggren and Boersma (1969) and by Chen (1969) has shown that both the foraminifera and pteropod abundances exhibit an abrupt decrease at a depth of about one to two meters down the cores. Immediately below this transition zone, the relative abundance of the two fossil assemblages is such that forams are rarely found, whereas pteropods continue, although with less specific diversity. Lithologically the transi-

* A number of cores (**VEMA** 14 cruise) covering a much larger area of the Red Sea studied by Herman (1965) indicate that the rates of lutite sedimentation may vary by two-to-threefold. The physiographic conditions in the Red Sea basin will certainly justify these variations. As the microfossil assemblages in these cores, especially the ones from the southern part of the Red Sea, show some differences from those of 153P and 154P (Berggren, 1969), oceanic circulation may also be influential in the sedimentation processes.

tion zone often marks the appearance of the youngest lithified layers in the sedimentary columns.

The tops of the lithified layer are correlated from core to core (Fig. 3). Based on our C^{14} dating, the high salinity condition which induced inorganic precipitation of aragonite and also perhaps greatly suppressed the presence of foraminifera (also shown in Fig. 3) in the Red Sea ended 10,000 to 13,000 years ago. This change is further reflected by the oxygen and carbon isotopic compositions of the pteropod and planktonic foraminiferal tests in three cores (118K, 119K, 154P) studied by Deuser and Degens (1969). As shown in Fig. 4 for core 118K, a 6 per mill difference in the δO^{18} values of pteropods is found between depths at 140cm and 165cm (the lithified material and the aragonite-filled and -encrusted pteropod shells start to appear at its lower boundary); values close to +1 per mill (on PDB-standard scale) are observed above the transition zone and values of around +6 per mill are found below. The magnitude of change is too great to be attributed to temperature fluctuations, and instead, the influence of evaporation is called upon for this observed change (Deuser and Degens, 1969). By interpolating the ages for sections above and below the transition zone, our estimate of the age for the midpoint of the transition (152cm) is 11,650 ± 300 years (assuming a constant sedimentation rate in the interpolated interval). This age corresponds closely to the hypothesized rapid warming of world-wide climate, which occurred at the end of Wisconsin (Würm) glaciation (Broecker *et al.,* 1960).

The presence of lithified aragonitic layers and the large δO^{18} values of microfossils below the transition in the core imply that conditions in the Red Sea about 11,000 years ago must have been highly saline. The situation can be visualized by considering that at depth the Red Sea is a closed basin; its water exchange with the Gulf of Aden is controlled by the sill located near the Strait of Bab el Mandeb. Since the sill depth is about 125m, sea-level lowering during the Wisconsin glacial times could have greatly reduced the water exchange,

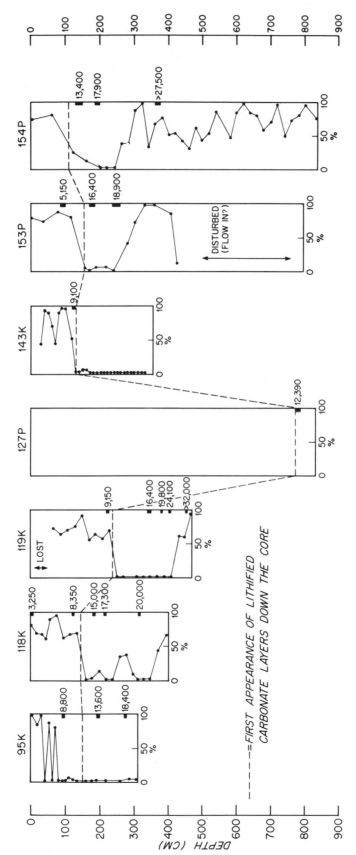

Fig. 3. Relative abundance of planktonic foraminifera in the cores (Berggren and Boersma, 1969). They are expressed as percentages of forams versus all the other constituents in the coarse fractions (>60 mesh). Note sudden decrease across the boundaries where the lithified aragonitic layers appear. Ages for the tops of the lithified zone are correlative from core to core and are about 10,000–13,000 years B. P., corresponding approximately to the time of the end of Wisconsin glaciation.

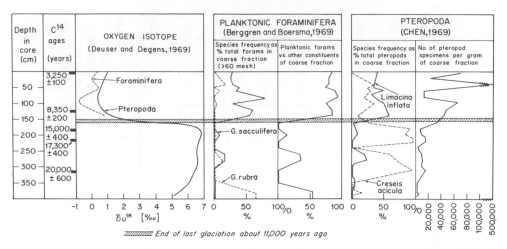

Fig. 4. Isotopic and paleontological data for core 118K showing changes occurring close to 11,000 years ago in the Red Sea. The environmental conditions before 11,000 years—lower temperature and higher salinity than those of the present—are reflected by the higher δO^{18} for surface water and by the suppression of organism activities. Changes in the species distribution across the transition are also distinctive.

resulting in higher net loss of water vapor through evaporation, hence higher salinity. Broecker *et al.* (1960) have demonstrated that the climatic change about 11,000 years ago was rather abrupt; a time span of less than 2,000 years from full glacial to the non-glacial conditions that have persisted throughout the Holocene. From the interpolated sedimentation rate and thickness of the transition zone, we also estimate the time encompassing the transition to be less than 2,800 years. This may be a maximum value, since the intervals sampled for the δO^{18} measurements and the possibility of sediment mixing all tend to increase the apparent range. With rapid transgression of the Arabian Sea over the sill, a pronounced effect upon the indwelling population and assemblages of planktonic foraminifera and pteropod would be expected (Fig. 4). It should be mentioned that the relatively small magnitude of fluctuations shown by the foraminifera δO^{18} data enable Deuser and Degens (1969) to estimate the Red Sea surface water temperature to vary, at most, over a range of 17° to 28°C in the post-glacial time.

An attempt to extend the chronology beyond the C^{14} age range is made by extrapolation of sedimentation rates. This is done for 154P, one of the longest cores available for the area. The core is rela-

tively free of hot brine influence, and judging from its lithology, would be expected to exhibit a more uniform sedimentation rate. As shown in Fig. 5, extrapolation of the average sedimentation rate (11cm/1,000 yr) obtained for the top 2m gives a time span of about 80.000 years covered by the total length of the core. It should be pointed out, however, that in view of the insufficient C^{14} data and dating by Th^{230} and Pa^{231} being unsuccessful for this core (see Ku, 1969), the extrapolated chronology beyond 20,000 years must be regarded as being tentative.

A distinctive feature is the cyclic change in the O^{18}/O^{16} of microfossils beginning at least 70,000 years ago. The 11,000 year datum in the core again marks the shifting of δO^{18} to values close to those observed for the present-day Red Sea surface waters. Before this time the δO^{18} increases gradually several times, these increases being separated by abrupt decreases. This pattern suggests long periods (15,000 to 30,000 years of duration) of lowered sea level and evaporation, each followed by short terms (2,000 years) of influx of water from the Indian Ocean.

The last cold (glacial) period in world climate began prior to 70,000 years ago (Emiliani and Geiss, 1967; Broecker *et al.*, 1958; Anderson *et al.*, 1960; and others). If our extrapolated ages are valid, the oxygen isotopic record shown here in 154P suggests an eustatic lowering of sea-level during the interval between 11,000 and 68,000 years B. P., and the resultant saline environments in the Red Sea. Aside from the isotope data, this situation is further reflected by the ecological consideration of the distribution of the dinoflagellate species *Hemicystodinium zoharyi* and the foraminifera species *Globigerinoides sacculifera* and *Globigerinoides rubra* (Fig. 5). According to Wall and Dale (personal communication), *Hemicystodinium* is the encysted phase of *Pyrodinium bahamense*

which has only been found in shallow water coastal embayments. A lower sea level would have stranded the coastal bays in which *Hemicystodinium* lived. Studies of Bé and Tolderlund (1968) have shown that *G. sacculifera* and *G. rubra* are contrasted in their tolerance with respect to the salinity conditions. The former favors the intermediate surface salinities (34.5 to 36.0 per mill) whereas the latter is an euryhaline species whose maximum frequencies often occur at the more extreme salinities, *i.e.*, either >36.0 per mill or <34.5 per mill. The distributional patterns of these phyto- and

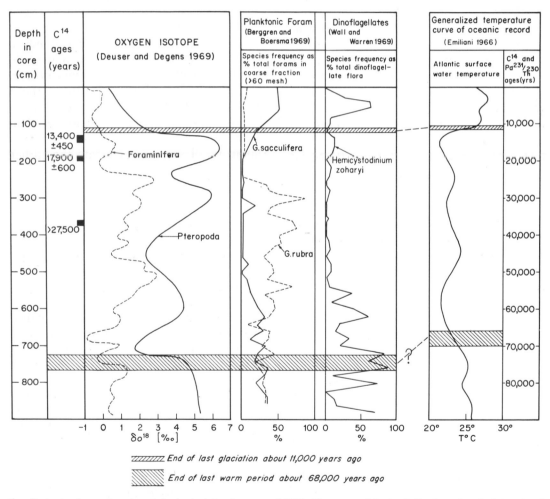

Fig. 5. Isotopic and paleontological data for core 154P. The record is tentatively extended to about 70,000 years B. P. by extrapolation of deposition rates for the top 2m of the core. Emphasis here is to show that during the world-wide cold climate period (as exemplified by Emiliani's oxygen isotope data on the Atlantic cores shown at the right of the figure) between about 11,000 and 68,000 years ago, condition in the Red Sea was one of lowered sea-level and high salinity. The cyclic fluctuations in the oxygen isotope results are characteristic; they are better represented by the foram data as measurements were made on much closer intervals. The higher δO^{18} values for pteropods are due to overgrowths and infillings of aragonite which might be precipitated from the more saline waters in deeper parts of the water column.

zooplanktons shown in Fig. 5 also serve to indicate the lowered sea-level and high salinity environments. Although the influxes were responded to by the surface salinity, hence by the oxygen isotopic compositions of the shells of the then-existing forams (*G. rubra*) and pteropods, they were apparently too brief for *Hemicystodinium* and *G. sacculifera* to proliferate. Reestablishment of these species did not occur until the transgression marking the end of Wisconsin-Würm glaciation about 11,000 years ago.

Wall and Warren (1969) pointed out that the cyclic nature of salinity changes below the 11,000 years level observed in core 154P may reflect the glacio-eustatic sea-level fluctuations recorded by the marine terraces in the Mediterranean (Bonifay and Mars, 1959; Eriksson, 1967, Table 4). On the other hand, Deuser and Degens (1969) further suggest the possibility that the influxes of normal oceanic water from the Indian Ocean before 11,000 years ago may also be controlled by events such as tectonic activities in the region of the Strait of Bab el Mandeb, which caused changes in the sill depth of the Red Sea. At present, a fuller account of the events occurring in this part of late Pleistocene history of the Red Sea is still prevented by paucity of data, and, as we have stressed, our interpretation of the chronology beyond 20,000 years still awaits further investigations.

History of Geothermal Brine Activity in the Atlantis II Deep

The Red Sea hot brine area appears to represent a geothermal system in which deposition of heavy metals is still active (Miller *et al.*, 1966; White, 1968). An assessment of the ages of these deposits will give information as to the geothermal activities in the past. Work of Bischoff (1969a, 1969b) suggests that the main locale of brine discharge is in the Atlantis II Deep. Cores taken from this Deep all contain geothermal brine deposits throughout their total lengths. Only 127P has reached a layer of lithified carbonate lutite at the bottom of the core. The C^{14} age of this lutite layer is dated at 12,390 ± 180 years B. P. Lithological correlations (Bischoff, 1969a, Fig. 2) and the uranium series disequilibrium ages (Ku, 1969) show that the brine deposits in all other Atlantis II Deep cores are younger than this age. Hence, if core 127P represents a continuous record of deposition, the most recent phase of brine activities in the Atlantis II Deep should have begun and have proceeded more or less uninterruptedly since 10,000 years ago. Brine-discharging activities older than 12,000 years in the area are only observable outside the Deep in cores of lower sedimentation rates. An 80cm thick layer of amorphous-goethite bed present in core 118K has an age between 17,000 and 20,000 years. Thin layers (up to 10cm thick) of geothermal deposits are further found in cores 118K and 119K at depths older than 20,000 years of age. As pointed out by Bischoff (1969a), these layers may represent "spill-overs" from the Atlantis II Deep. Cores longer than those presently available from the Deep should be of value to elucidate the earlier brine discharge history.

Acknowledgments

We would like to acknowledge the many discussions with our colleagues, especially Drs. W. Berggren, E. T. Degens, and D. Wall. Drs. D. A. Ross, W. Berggren, and W. Deuser, and Miss A. Boersma provided us with some of their unpublished results. Marylou Zickl, Paul Smith, and Jan van Donk aided in the laboratory. We thank Drs. W. S. Broecker and K. O. Emery for criticizing the manuscript.

This work was supported by the National Science Foundation through grants GA-584 to Woods Hole Oceanographic Institution and GA-1346 to Lamont Geological Observatory.

References

Anderson, S. T., H. deVries, and W. H. Zaguin: Climate change and radiocarbon dating in the Weichselian Glacial of Denmark and the Netherlands. Geol. Mijnbouw, **39**, 38 (1960).

Bé, A. W. H. and D. S. Tolderlund: Distribution and ecology of living planktonic foraminifera in surface waters of the Atlantic and Indian Oceans. Preprint (1968).

Berggren, W. A.: Micropaleontologic investigations of Red Sea cores—Summation and synthesis of results. *In: Hot brines and recent heavy metal deposits in the Red Sea,* E. T. Degens and D. A. Ross (eds.). Springer-Verlag New York Inc., 329–335 (1969).

Berggren, W. A. and A. Boersma: Late Pleistocene and Holocene planktonic foraminifera from the Red Sea. *In: Hot brines and recent heavy metal deposits in the Red Sea,* E. T. Degens and D. A. Ross (eds.). Springer-Verlag New York Inc., 282–298 (1969).

Bischoff, J. L.: Red Sea geothermal brine deposits: their mineralogy, chemistry and genesis. *In: Hot brines and recent heavy metal deposits in the Red Sea,* E. T. Degens and D. A. Ross (eds.). Springer-Verlag New York Inc., 368–401 (1969a).

———: Goethite-hematite stability relations with relevance to sea water and the Red Sea brine environment. *In: Hot brines and recent heavy metal deposits in the Red Sea,* E. T. Degens and D. A. Ross (eds.). Springer-Verlag New York Inc., 402–406 (1969b).

Bonifay, E. and P. Mars: Le Tyrrhénian dans le cadre de la chronologie Quaternaire mediterranéenne. Bull. Geol. Soc. France (7), I., 62 (1959).

Broecker, W. S. and J. L. Kulp: Lamont natural radiocarbon measurements, IV. Science, **126,** 1324 (1957).

Broecker, W. S., K. K. Turekian, and B. C. Heezen: The relationship of deep sea sedimentation rates to variations in climate. Am. Jour. Sci., **256,** 503 (1958).

Broecker, W. S., C. S. Tucek, and E. A. Olson: Radiocarbon analysis of oceanic CO_2. Internat. Jour. Applied Radiation and Isotopes, **7,** 1 (1959).

Broecker, W. S., M. Ewing, and B. C. Heezen: Evidence for an abrupt change in climate close to 11,000 years ago. Am. Jour. Sci., **258,** 429 (1960).

Broecker, W. S. and E. A. Olson: Lamont radiocarbon measurements, VIII, Radiocarbon, **3,** 176 (1961).

Chen, C.: Pteropods in the hot-brine sediments of the Red Sea. *In: Hot brines and recent heavy metal deposits in the Red Sea,* E. T. Degens and D. A. Ross (eds.). Springer-Verlag New York Inc., 313–316 (1969).

Deuser, W. G. and E. T. Degens: O^{18}/O^{16} and C^{13}/C^{12} ratios of fossils from the hot-brine deep area of the central Red Sea. *In: Hot brines and recent heavy metal deposits in the Red Sea,* E. T. Degens and D. A. Ross (eds.). Springer-Verlag New York Inc., 336–347 (1969).

Emiliani, C. and J. Geiss: On glaciation and their causes. Geol. Rundschau, **46,** 576 (1957).

Emiliani, C.: Paleotemperature analysis of the Caribbean cores P6304-8 and P6304-9, and a generalized temperature curve for the past 425,000 years. J. Geology, **74,** 109 (1966).

Eriksson, K. G.: Some deep-sea sediments in the western Mediterranean Sea. *In: Progress in Oceanography,* **4,** 267, M. Sears (ed.). Pergamon Press (1967).

Gevirtz, J. L. and G. M. Friedman. Deep-sea carbonate sediments of the Red Sea and their implications on marine lithification. Jour. Sedimentary Petrology, **36,** 143 (1966).

Herman, Y.: êtude des sediments Quaternaires de la Mer Rouge. *In: Ph.D. thesis,* University of Paris (1965).

Ku, T. L., W. S. Broecker, and N. Opdyke: Comparison of sedimentation rates measured by paleomagnetic and the ionium methods of age determination. Earth Planet. Sci. Letters, **4,** 1 (1968).

Ku, T. L.: Uranium series isotopes in sediments from the Red Sea hot-brine area. *In: Hot brines and recent heavy metal deposits in the Red Sea,* E. T. Degens and D. A. Ross (eds.). Springer-Verlag New York Inc., 512–524 (1969).

Miller, A. R., C. D. Densmore, A. Jokela, E. T. Degens, J. C. Hathaway, F. T. Manheim, P. F. McFarlin, R. Pocklington: Hot brines and recent iron deposits in deeps of the Red Sea. Geochim. et Cosmochim. Acta, **30,** 341 (1966).

Milliman, J., D. A. Ross, and T. L. Ku: Lithified deep-sea carbonates in the Red Sea. Jour. Sedimentary Petrology (in press).

Olsson, I. U. and K. G. Eriksson: Remarks on C^{14} dating of shell material in sea sediments. Progress in Oceanography, **3,** 253 (1965).

Ross, D. A. and E. T. Degens: Shipboard collection and preservation of sediment samples collected during CHAIN 61 from the Red Sea. *In: Hot brines and recent heavy metal deposits in the Red Sea,* E. T. Degens and D. A. Ross (eds.). Springer-Verlag New York Inc., 363–367 (1969).

Rubin, M. and H. E. Suess: U.S. Geological Survey radiocarbon dates II. Science, **121,** 481 (1955).

Wall, D. and J. S. Warren: Dinoflagellates in Red Sea piston cores. *In: Hot brines and recent heavy metal deposits in the Red Sea,* E. T. Degens and D. A. Ross (eds.). Springer-Verlag New York Inc., 317–328 (1969).

White, D. E.: Environments of generation of some base-metal ore deposits. Econ. Geol., **63,** 301 (1968).

Sediments

Shipboard Collection and Preservation of Sediment Samples Collected During **CHAIN** 61 from the Red Sea *

DAVID A. ROSS AND EGON T. DEGENS
Woods Hole Oceanographic Institution
Woods Hole, Massachusetts

Abstract

This paper contains a brief description of the different coring techniques used during the **CHAIN** 61 cruise to the Red Sea. Locations, core length, type of sampler used and water depth are given for the Red Sea stations.

Introduction

A systematic sampling of the sediments beneath the brine area was made during **CHAIN** 61. The principal objectives of this sampling program were:

(a) Determination of depth and extent of the heavy metal deposits.

(b) Collection of samples for shipboard chemical determinations.

(c) Application of heat probe to determine temperature changes in both water and sediment.

(d) Sampling of long cores for stratigraphical purposes. Samples were collected from outside the brine area to aid in understanding the regional geologic setting and history of the Red Sea.

The location of samples collected from the brine area is shown in Fig. 1. General information about these samples is given in Table 1. Data on samples collected outside the brine area, in other parts of the Red Sea, are presented in Table 2. Depths are reported in meters, corrected for changes in sound velocity according to Matthews (1939).

Collection of Samples

Samples were taken with five different sampling devices: piston corers, gravity corers, free-fall corers, Kasten or box corers and dredges. A brief description of the characteristics and techniques of these instruments follows.

Piston Cores

Piston cores were taken with a standard piston corer, using a free fall of about 7m. A short gravity corer was used to trip the piston corer. The position of the piston corer relative to the bottom was controlled by a pinger. All piston corers were equipped with thermistors to take temperature measurements for heat flow determinations.

Ten piston cores were raised from the brine area and seven from outside the brine area. An attempt for 30-foot cores (about 10m) was made at all stations, except for station 84 where a 40-foot corer was lowered. Although care was taken during the coring operation, cores 81P, 84P, 124P, 129P and 159P have some sediment disturbance.

The surface sediment in some of the brine area is a very porous, homogenous black ooze that may not have been adequately sampled by the piston corer. Comparison of the piston core and the trip

* Woods Hole Oceanographic Institution Contribution No. 2197.

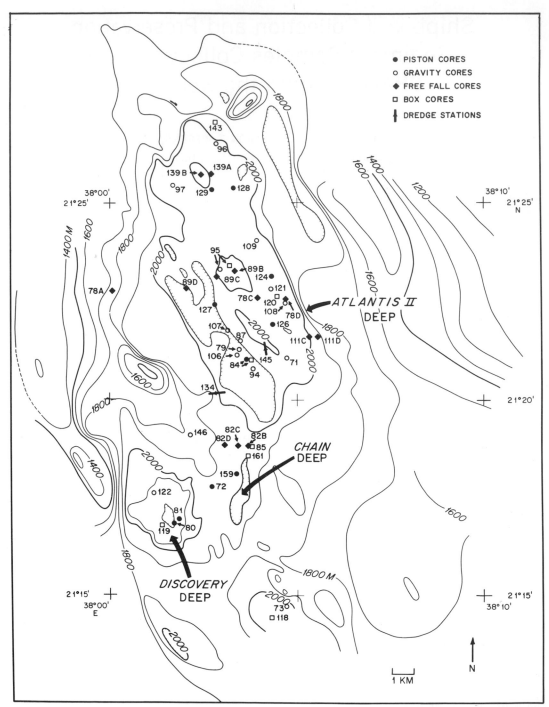

Fig. 1. Location of sediment samples obtained from the hot brine region of the Red Sea.

Table 1   Samples Collected from the Hot Brine Area of the Red Sea
During CHAIN 61

Station	Type of Sampling Gear	Total Length cms (approx.)	Latitude N	Longitude E	Depth meters (corr.)
71	Gravity	74	21°21'	38°04.7'	2,095
72	Piston	208	21°17.8'	38°02.7'	1,930
73	Gravity	50	21°14.7'	38°04.7'	2,027
78A	Free Fall	40	22°22.8'	38°00.5'	1,757
78C	Free Fall	120	21°22.5'	38°03.9'	2,056
78D	Free Fall	120	21°22.5'	38°04.7'	2,086
79	Gravity	208	21°21.3'	38°03.5'	2,151
80	Piston	870	21°16.8'	38°01.7'	2,199
81	Piston	880	21°16.9'	38°01.9'	2,164
82B	Free Fall	113	21°18.8'	38°03.7'	1,989
82C	Free Fall	101	21°18.8'	38°03.5'	1,943
82D	Free Fall	117	21°18.8'	38°03.1'	1,965
84	Piston	870	21°21'	38°03.7'	2,164
84	Kasten	400	21°21'	38°03.8'	2,164
85	Kasten	400	21°18.8'	38°03.8'	1,969
89B	Free Fall	111	21°23.2'	38°03.4'	1,914
89C	Free Fall	117	21°23.1'	38°02.9'	2,102
89D	Free Fall	121	21°22.8'	38°02.1'	1,058
94	Gravity	260	21°20.7'	38°03.8'	2,170
95	Kasten	335	21°23.4'	38°03.2'	1,910
95	Gravity	197	21°23.2'	38°03'	1,939
96	Gravity	60	21°26.4'	38°02.9'	2,047
97	Gravity	5	21°25.4'	38°01.7'	2,042
106	Gravity	84	21°21.1'	38°03.4'	2,164
107	Gravity	130	21°21.7'	38°03.2'	2,151
108	Gravity	243	21°22.4'	38°04.7'	2,071
109	Gravity	283	21°24'	38°03.9'	2,077
111C	Free Fall	122	21°21.5'	38°05.3'	2,016
111D	Free Fall	122	21°21.5'	38°05.6'	1,965
118	Kasten	400	21°14.4'	38°04.3'	1,989
119	Kasten	400	21°16.8'	38°01.4'	2,175
120	Kasten	400	21°22.6'	38°04.5'	2,060
121	Gravity	290	21°22.8'	38°04.3'	2,064
122	Gravity	296	21°17.6'	38°01.2'	2,126
124	Piston	139	21°23.1'	38°04.3'	2,056
126	Piston	810	21°21.9'	38°04.3'	2,066
127	Piston	836	21°22.4'	38°02.8'	2,106
128	Piston	874	21°25.3'	38°03.3'	2,077
129	Piston	205	21°25.3'	38°02.8'	2,027
134	Dredge	—	21°20.1'	38°02.9'	2,164
139A	Free Fall	122	21°25.7'	38°02.8'	1,972
139B	Free Fall	122	21°25.7'	38°02.5'	2,007
143	Kasten	354	21°27'	38°02.9'	1,910
145	Dredge	—	21°21.2'	38°04.2'	2,126
146	Gravity	5	21°19.1'	38°02.2'	1,815
158*	Gravity	162			2,204
159	Piston	294	21°18.1'	38°03.4'	1,980
161	Kasten	324	21°18.5'	38°03.7'	1,949

* Position uncertain, presumably in the Discovery Deep.

Table 2 Samples Collected in the Red Sea Outside of the Hot Brine Area During CHAIN 61

Station	Type of Sampling Gear	Total Length cm (approx.)	Latitude N	Longitude E	Depth meters (corr.)
69	Gravity	70	25°22.5′	36°08.9′	2,245
135	Gravity	23	21°30.0′	38°16.0′	753
136	Gravity	38	21°28.0′	38°13.0′	1,259
151	Gravity	298	20°42.0′	38°15.0′	2,301
152	Piston	711	19°48.5′	38°30.0′	2,362
153	Piston	755	19°43.0′	38°41.0′	2,704
154	Piston	792	19°34.0′	38°59.5′	1,276
155	Piston	424	19°23.5′	38°54.0′	2,027
156A	Free Fall	48	20°59.0′	38°19.0′	1,738
156B	Free Fall	63	20°58.5′	38°20.5′	1,329
164	Piston	5	21°59.0′	37°58.5′	2,281
165	Piston	81	22°28.2′	37°46.5′	2,164
167	Piston	757	23°20.0′	37°20.0′	826
168	Dredge	–	25°26.8′	36°13.0′	2,164

gravity core usually indicates a thicker sequence of this black material in the trip gravity core.* Cores taken from outside the brine area frequently must have hit relatively hard material, such as volcanic rock or consolidated calcareous sediment, indicated by the fact that at stations 155, 164 and 165 the corer noses were bent and a relatively short core was obtained.†

Gravity Cores

Gravity cores were obtained using a Benthos corer usually with a 10-foot (about 3m) plastic liner as the coring barrel; under certain conditions, a metal barrel with the plastic liner inside was used. Most of the gravity cores were equipped with thermistors for heat flow studies. A pinger was used to position the corer relative to the bottom sediments.

Fifteen gravity cores were collected from the brine area and four from outside the brine area. The cores were generally less than 3m in length and had little indication of disturbance.

Free Fall Cores

Free fall or boomerang cores (Sachs and Raymond, 1965) were taken in several localities in the Red Sea. The recovery

ratio of the free falls dropped into the brine area was relatively poor, i.e., 19 were dropped and 13 were recovered. The corer, which weighs about 40kg in water, accelerates quickly to its terminal velocity of 450m/min. The impact of this instrument into the soft sediment is apparently sufficient in some instances to bury the entire corer below the sediment surface and prevent its release and subsequent floating to the surface. Many of the recovered float portions were covered with mud, clearly indicating such a "super"-penetration. In addition, most of the free falls from the brine area were completely filled up upon recovery, also suggesting the possibility of "super"-penetration. Two free fall attempts outside of the brine area obtained relatively short cores.

Kasten Cores

A German made ‡ Kasten or box sampler was used at several stations. The sampler consists of a thin-walled case (15cm × 15cm × 4m), a core catcher, and a "push weight" of 1,000kg with a self closing valve (Fig. 2). Samples are obtained by lowering the device to the bottom and letting it sink slowly into the sediments by its own weight. The cores so obtained show no significant disturbance. Thin lamina sometimes less than a millimeter in thick-

---

* A comparison of the box cores with nearby trip gravity cores also reveals a thicker black ooze sequence for the latter ones.

† Fragments of volcanic rocks (i.e., volcanic glass and basalt) were found in the core catcher of some of the bent noses.

‡ Hydrowerkstätten, Feinmechanik und Apparatebau G.M.B.H., 23 Kiel-Hassee-Uhlenkrog 38, Germany.

LEGEND

A   Detrital (foraminifera tests; high-magnesium calcite)
B   Iron Sulfides
C   Goethite-Amorphous
D   Detrital (semilithified pteropod shells; secondary aragonite)
E   Iron Montmorillonite
F   Anhydrite
G   Sphalerite
H   Mangano-Siderite
I   Sulfides

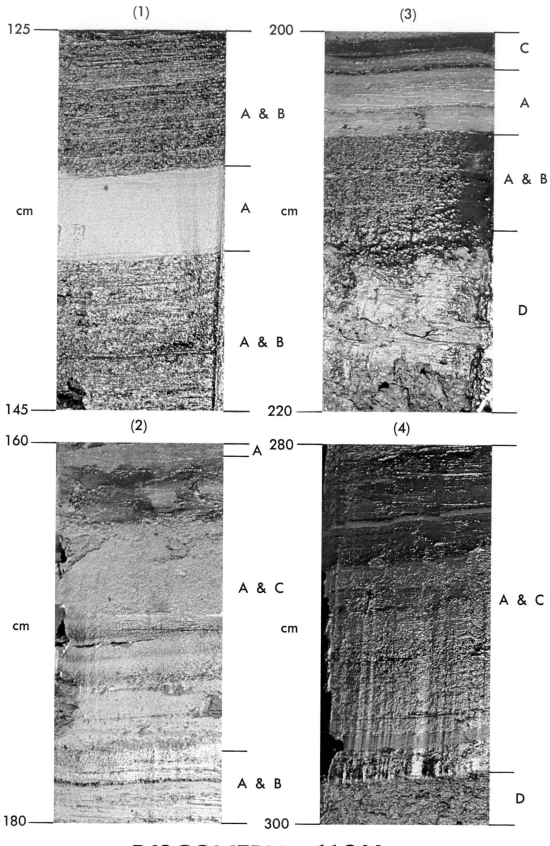

DISCOVERY 119K

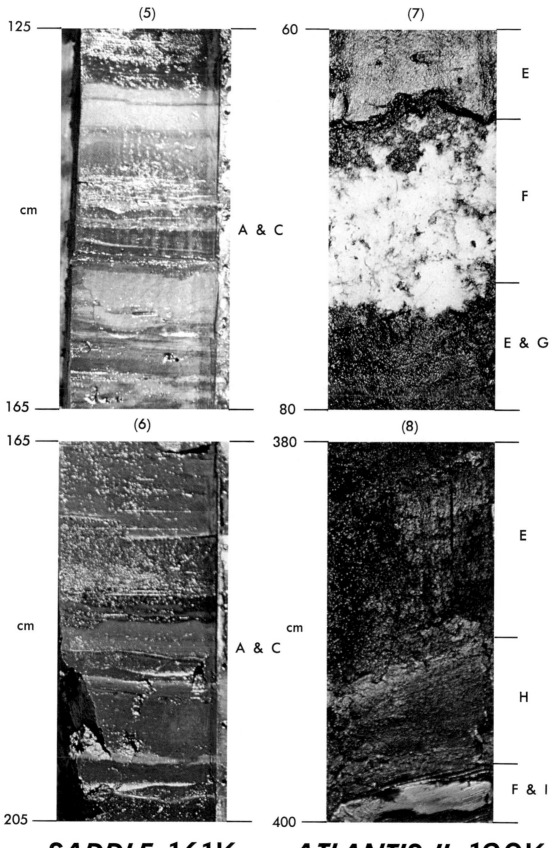

SADDLE 161K **ATLANTIS II 120K**

Fig. 2. Kasten corer on the fantail of the **CHAIN.** The "push weight" and self-closing valve is in the foreground (photograph by R. G. Munns).

ness are undisturbed (Color plate). Another advantage of the Kasten core is the large volume of material it collects. Eight Kasten cores were raised from the brine area and one about 7 miles south of the Atlantis II Deep.

Dredges

Dredging operations used a chain bag dredge with a pinger 100m above it to insure that the dredge was on the bottom. The dredge was towed upslope; usually, no significant increases in wire tension occurred during the dredging operation. Dredges 134 and 145, from the brine area, obtained a small piece of anhydrite and semi-consolidated clay respectively. Dredge 168, from outside the brine area, recovered carbonate rock.

Preservation of Samples

All the piston, trip gravity, gravity, and free fall cores were cut (when necessary) into 5-foot sections and stored in refrigerators at 2°C. Most of the Kasten cores were placed in 1-m long containers and either refrigerated at 2°C or frozen at −16°C. Kasten cores 84, 119 and 161 were stored unrefrigerated in their original containers. Some shipboard samples were taken for mineralogy, interstitial water and chemical studies.

References

Matthews, D. J.: Tables of the velocity of sound in pure water and sea water for use in echo-sounding and sound-ranging, 2nd Ed. London Hydrographic Dept., Admiralty, 52 p. (1939).

Sachs, P. L. and S. O. Raymond: A new unattached sediment sampler. Jour. Marine Res., **23,** 44 (1965).

Red Sea Geothermal Brine Deposits:
Their Mineralogy, Chemistry, and Genesis *

JAMES L. BISCHOFF †

Woods Hole Oceanographic Institution
Woods Hole, Massachusetts

Abstract

Chemical and mineralogical analyses were performed on samples from ten specially selected cores from the Red Sea geothermal deposit. The deposit was divided into seven bedded and laterally correlative facies as follows:

(1) detrital
(2) iron-montmorillonite
(3) goethite-amorphous
(4) sulfide
(5) manganosiderite
(6) anhydrite
(7) manganite

Distribution of the facies, their unconsolidated nature and age relations indicate that the solids were precipitated out of the overlying brine column, that the area of brine discharge is very local within the Atlantis II Deep, and that the chemistry of the brine has changed considerably with time.

Mechanisms of precipitation include simple cooling of subterranean brine as it discharges into the bottom of the Atlantis II Deep and mixing of the brine with the overlying sea water.

Introduction

The subjects of fundamental geochemical interest in the Red Sea geothermal system are the nature and source of the brine, the nature and distribution of the brine-derived deposits and the processes of precipitation of the deposits. The system appears to be an ore body in the process of formation (White, 1968), and the above subjects represent the major gaps in the understanding of general ore-forming processes. The Red Sea geothermal system is one of three remarkable similar systems: the others are the Salton Sea geothermal system (White et al., 1963) and the recently discovered Cheleken geothermal system in the Soviet Union (Lebedev, 1967a).

This chapter is concerned with the Red Sea brine deposits themselves, their distribution, mineralogy, chemistry and mode of precipitation. Conclusions regarding the mode of precipitation hopefully will have application to the origin of ancient ore bodies. Preliminary investigations on the mineralogy and chemistry of these deposits are reported in Miller et al. (1966).

Methods

Sampling

Ten cores were chosen for detailed study as representative of the various areas in the geothermal region (Fig. 1). Cores 84K, 120K, 126P, 127P, 128P (K refers to Kasten cores, and P to piston cores) are from the floor of the Atlantis II Deep below the brine-sea water interface. Core 95K is from within the Atlantis II Deep but atop a knoll above the brine-sea water interface. 85K and 161K are from near the brine-sea water interface in the sill area separating the Atlantis II and Discovery Deeps. 119K is from the floor of the Discovery Deep

* Woods Hole Oceanographic Institution Contribution No. 2198.

† Present address: Dept. of Geological Sciences, University of Southern California, Los Angeles, California.

and 118K from about 4 miles southeast of the Discovery Deep.

The sediments in these cores are well bedded on a scale ranging from a few millimeters to several meters. The sediments from below the brine-sea water interface are characterized by striking colors of black, buff, ochre, orange, blue, green and brown, while those above or outside the Atlantis II Deep, *i.e.,* 95K, 118K

and 119K, have more subdued colors of grey, buff and orange.

Samples were taken immediately upon opening the refrigerated cores. Samples were collected from each lithologic unit in each core, and in case of massive units, were collected every 50cm or so to insure proper representation. The samples were preserved at room temperature in air-tight, plastic-capped glass jars and were

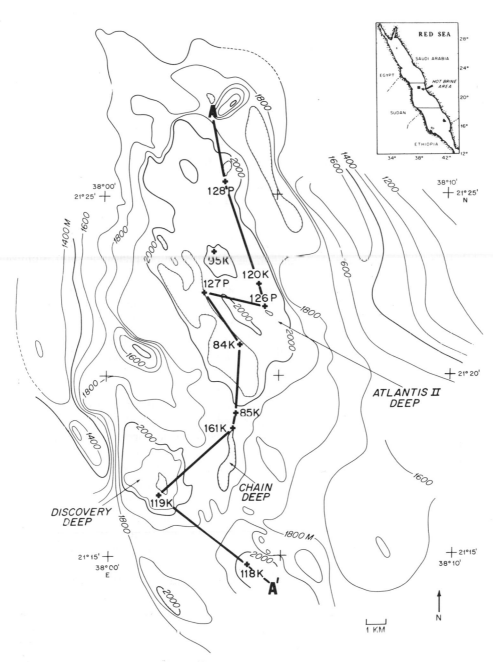

Fig. 1. Bathymetric map of geothermal area (Ross and Degens, 1969) showing core locations. Brine pools are limited to approximately below the 2,000m contour. Line A-A′ refers to a section shown in Fig. 2.

processed within one week to minimize oxidation and dehydration. Control studies on immediately processed samples indicated no such changes for the stored samples. In all, 133 samples were studied.

Analytical Techniques

Each sample was separated into three size fractions, $>62\mu$, 2-62μ and $<2\mu$, by first wet sieving through a 62μ screen with distilled water and then separating the $<2\mu$ fraction from the remainder by centrifuging, a process developed by Hathaway (1956). Each size fraction, now completely leached with distilled water, was dried at 50°C and weighed. The weight of the original wet sample and the individual dry separations were used for calculation of the weight per cent interstitial brine and the dry weight per cent of each size fraction (Table 1). Most of the brine-derived material consists of amorphous to microcrystalline spherules of various size, and it is the sizes of these spherules and spherule aggregates that largely constitute the size fractions (see Strangway *et al.*, 1969 for a discussion of crystal size in this material).

X-ray diffraction patterns were obtained for the $<2\mu$ and 2-62μ fractions, and grain mount slides for petrographic microscope examination were prepared from the 2-62μ and $>62\mu$ fractions (Table 1). A cured crystal monochromater was used on the x-ray unit to discriminate against the high levels of iron radiation.

Fifty-five of the original samples were selected for chemical analysis, using the 2-62μ fractions for which x-ray patterns had been obtained. This fraction was chosen because it generally accounts for more than 90 per cent of the dry material and because the $>62\mu$ fraction consists predominantly of detrital foraminiferal debris. Total carbon, analyzed as CO_2, was determined by an induction furnace-gasometric technique (Laboratory Equipment Corporation, Saint Joseph, Michigan). Total sulfur, including sulfate and sulfide, was volatilized to SO_2 in an induction furnace, and determined by an iodometric titration (Laboratory Equipment Corporation; see also Conrad *et al.*, 1959). Ferrous iron was determined by the classic dichromate titration (Kolthoff and Sandell, 1952). Zn, K and Mn were determined by atomic absorption spectroscopy. K and Mn were additionally determined by spark emission spectrometry as was Si, Ti, Al, Fe (total), Mg, Ca, Sr, Ba, Na, Cu, Pb and P. The emission spectrometer technique is that of Landergren *et al.* (1964) as modified by Manheim (in preparation) and consists of

fusing samples in a borate flux with internal standards of In and Be. The total composition of the samples were expressed as oxides and summarized (Table 2).

There appears to be some influence on Pb values by an Fe line on the spectrochemical analyses. This makes some of the low-heavy-metal, high Fe samples have erroneously high Pb (*e.g.* 118K-260, Table 2). It appears that about 50 per cent Fe_2O_3 gives about 0.03–0.04 per cent PbO. This is not a significant effect in the high heavy-metal samples (*e.g.* 85K-85, Table 2). Replicate analyses of the same sample by different analysts and laboratories are compared in Bischoff and Manheim (1969) and indicate the general reliability of the data. Analyses of most of the samples, moreover, sum close to 100 per cent, indicating that the above list of elements includes all major components.

Analytical Data: Facies Description

General Statement

Several facies, apparently correlative from core to core, were defined on the basis of physical appearance and mineralogy. Correlation was subsequently confirmed by age relations (Ku *et al.*, 1969). Distribution and lateral extent are shown in the cross section in Fig. 2, and averaged chemical composition in Table 3. Well defined bedding, inmixing of detrital material and age sequences leave little doubt that solids precipitated out of the overlaying water column and settled to their present position. As will be discussed later, various minerals appear to be forming at differing levels of the water column and settling to the same horizon on the sediment surface, resulting in a certain amount of mixing of the various facies; thus, considerable generalization has been necessary in defining and discussing the facies. In general, however, each facies represents a dominance of one process of mineral precipitation over another, and a description of each facies follows.

Detrital Facies. Buff-colored pelagic sedimentary material occurs throughout the geothermal area, both as discrete beds and as inmixtures with other facies. This material, essentially the same as described by

Table 1   Size Fraction, Brine Content and Mineralogy Data

Size fractions in terms of dry salt-free sediment and brine content in terms of weight per cent wet sediment. X's for mineral content refers to relative intensity of total x-ray pattern compared to standard of pure mineral. 4 X's indicates pattern is of equal intensity to pure standard, 2 X's half as strong and so on. Amorphous material content estimated by difference. For example, if an x-ray pattern of a given sample had sufficient peaks for a 2 X designation of one mineral and no other peaks were present, the 2 X's were assigned for amorphous material.

The cores are grouped by location in this table and in Table 2 and numerically within each group. For example, 95K, 118K and 119K are located above or outside the Atlantis II brine, 85K and 161K are located in the sill area, and the remaining, in numerical order, within the Atlantis II Deep.

SAMPLE (depth, cm)	interstitial brine content (%)	size distribution, μ	percent in size	amorphous	goethite	hematite	lepidocrocite	Fe montmorillonite	manganosiderite	anhydrite	sphalerite	chalcopyrite	pyrite	manganite	todorokite	detritals	miscellaneous and remarks
95K																	
-85	78.2	>62	21.7													XXXX	
		2-62	77.3	XX												XX	
		<2	1.0	XX												XX	
-110	71.1	>62	9.1													XXXX	
		2-62	89.8	X									tr.			XXX	
		<2	1.1													XXXX	
-150	72.8	>62	0.2													XXXX	
		2-62	99.2		XXXX												
		<2	0.6	X	XX											X	
-180	63.5	>62	5.2									X				XXX	
		2-62	92.0										XX			XX	
		<2	2.8	XX	X											X	
-185	56.2	>62	5.5		X							X				XX	
		2-62	89.5	X	X					tr.		tr.				XX	
		<2	5.0										X			XXX	
-190	---	>62	11.5													XXXX	tr. gypsum
		2-62	85.2		X											XXX	
		<2	3.3	X												XXX	
-340	80.8	>62	2.3													XXXX	
		2-62	95.6	XX	X								tr.			X	
		<2	2.1	XXX												X	
-395	23.5	>62	91.5													XXXX	
		2-62	8.1	X												XXX	
		<2	0.4	XX												XX	
118K																	
-40	55.5	>62	5.1	X												XXXX	
		2-62	93.7													XXX	
		<2	1.2													XXXX	
-90	79.5	>62	3.3	X	X											XX	
		2-62	93.9	XX	XX												
		<2	2.8		XX											XX	
-120	80.0	>62	19.8													XXXX	
		2-62	74.4	X				X								XX	
		<2	5.8						XX							XX	
-160	50.8	>62	17.7													XXXX	
		2-62	78.9										tr.			XXXX	
		<2	3.4						XX							XX	
-200	78.1	>62	1.7	X	XX											X	
		2-62	93.5	XX	XX												
		<2	4.8	XX	XX												
-260	80.8	>62	2.2	X	XX											X	
		2-62	85.5	XX	XX												
		<2	12.3	XX	XX												
-320	75.1	>62	1.8				XXX									X	
		2-62	93.0				XXX									X	
		<2	5.2				XXXX										
-385	49.1	>62	2.8													X	pyrrhotite XXX
		2-62	87.6	X									X			XX	
		<2	9.6					XX		tr.						XX	

Table 1 (Continued)

SAMPLE (depth, cm)	interstitial brine content (%)	size distribution, μ	percent in size	amorphous	goethite	hematite	lepidocrocite	Fe montmorillonite	manganosiderite	anhydrite	sphalerite	chalcopyrite	pyrite	manganite	todorokite	detritals	miscellaneous and remarks
119K -50	94.3	>62	5.2													XXXX	
		2-62	80.9	X				XXX									
		<2	13.9	X				XXX									
-90	68.2	>62	12.4													XX	Greigite-XX
		2-62	80.2	X	X											XX	
		<2	7.4	X												XXX	
-110	71.2	>62	11.2													XXX	Greigite-X
		2-62	85.2	X	X											XX	
		<2	3.6													XXXX	
-130	67.5	>62	20.6	X	tr.											XXXX	
		2-62	64.1													XXX	
		<2	15.3	X												XXX	
-139	55.8	>62	18.7													XXXX	
		2-62	67.4	XX												XX	
		<2	13.9					XX								XX	
-151	70.5	>62	28.2													XXXX	
		2-62	66.3		X											X XX	
		<2	5.5	X				XXX									
-160	68.3	>62	4.9						XXXX								
		2-62	84.2	XXX	X												
		<2	10.4		X			XX	tr.							X	
-205	67.3	>62	6.2													XXXX	
		2-62	79.3	X	X											XX	
		<2	14.5		X			XX								X	
-231	55.4	>62	9.7													XXXX	Greigite tr.
		2-62	84.5	XX	X											X	
		<2	5.8													XXXX	
-239	62.3	>62	36.2													XXXX	
		2-62	56.3	X												XXX	
		<2	7.5													XXX	
-262	84.9	>62	1.0		XX											XX	
		2-62	97.3	XX	XX												
		<2	1.7	XXXX													
-270	89.5	>62	0.7													XXXX	
		2-62	96.0				XXX									X	
		<2	3.3	X												XXX	
-273	58.2	>62	2.6													XXXX	
		2-62	82.3													XXXX	
		<2	15.1													XXXX	
-282	52.3	>62	1.8													XXXX	
		2-62	95.6	X									XX				
		<2	2.6													XXXX	
-295	46.6	>62	54.6													XXXX	
		2-62	45.1													XXXX	
		<2	0.3													XXXX	
-310	52.6	>62	13.00													XXXX	
		2-62	45.6	XX												XX	
		<2	41.4													X	

Table 1 (Continued)

SAMPLE (depth, cm)	interstitial brine content (%)	size distribution, μ	percent in size	amorphous	goethite	hematite	lepidocrocite	Fe montmorillonite	manganosiderite	anhydrite	sphalerite	chalcopyrite	pyrite	manganite	todorokite	detritals	miscellaneous and remarks
119K Con't.																	
-324	80.3	>62	1.2	X	X											XX	
		2-62	81.2	XXX					tr.							X	
		<2	12.6					XXXX									
-329	58.1	>62	68.2					XXX	X								
		2-62	11.5		XX			XX									
		<2	20.3		X			XXX									
-355	58.7	>62	1.4													XXXX	
		2-62	88.4		X											XXX	
		<2	10.2	XXXX													
-380	52.3	>62	26.1	X												XXX	
		2-62	72.6	X												XXX	
		<2	1.3													XXXX	
-425	60.2	>62	12.6													XXXX	
		2-62	82.1													XXXX	
		<2	5.3													XXXX	
-450	55.8	>62	15.6													XXXX	
		2-62	80.3	X												XXX	
		<2	4.1	X												XXX	
161K																	
-135	84.8	>62	1.5													XXXX	
		2-62	50.5	XXX	X												
		<2	48.0	XXXX				tr.									
-200	87.0	>62	1.2													XXXX	
		2-62	50.0	XXX	X												
		<2	48.8	XXX	X												
-221	80.2	>62	1.3													XXXX	dolomite XX
		2-62	87.8	X	XXX											tr.	dolomite tr.
		<2	10.9	XXXX	tr.												
-231	82.9	>62	30.8													XXXX	dolomite XX
		2-62	57.1	XXX	X												
		<2	12.1	XXX	X												
-237	84.3	>62	3.6		X									X	XX	
		2-62	78.8	XXX													woodruffite? X
		<2	17.6	XXX													woodruffite? X
-265	86.2	>62	0.8													XXXX	
		2-62	89.4	XX	XX												
		<2	9.8	XXX	X												
-280	85.8	>62	2.1													XXXX	
		2-62	60.6	XX	XX												
		<2	37.3	XXXX													
-325	88.6	>62	5.3													XXXX	
		2-62	69.2	XXX	X												
		<2	25.5		XXXX												
-370	84.8	>62	1.6													XXXX	
		2-62	79.8	X	XXX												
		<2	18.6	X	XXX												
-389	83.7	>62	4.1													XXXX	
		2-62	88.3	X	XXX												
		<2	7.6	XX	XX												

Table 1 (Continued)

SAMPLE (depth, cm)	interstitial brine content (%)	size distribution, μ	percent in size	amorphous	goethite	hematite	lepidocrocite	Fe montmorillonite	manganosiderite	anhydrite	sphalerite	chalcopyrite	pyrite	manganite	todorokite	detritals	miscellaneous and remarks
161K Con't.																	
-392	88.2	>62	1.5													XXXX	
		2-62	60.3	X	XX											X	
		<2	38.2	XX												XX	
85K																	
-30	91.7	>62	0.3													XXXX	
		2-62	93.4	XX	XX												
		<2	6.3	XX	XX												
-85	92.9	>62	0.4													XXXX	
		2-62	95.7	X	XXX												
		<2	3.9	XXX	X												
-120	81.3	>62	5.3	XXXX													
		2-62	85.9	XXXX													
		<2	8.8	X				XXX									
-160	74.8	>62	37.5	XX	X											X	
		2-62	57.8	XXX	X												
		<2	4.7	X	XXX												
-180	88.0	>62	0.5	XX	X											X	
		2-62	80.8	XX	XX												
		<2	18.7		XXXX												
-210	89.2	>62	0.5	XX	X											X	
		2-62	72.3	X	XXX												
		<2	27.2	X	XXX												
-240	87.3	>62	3.9	XX	X											X	
		2-62	95.1	XXX	X												
		<2	1.0	XXXX													
-270	87.4	>62	5.4	X	XX											v	
		2-62	84.1	X	XXX												
		<2	10.5	X	XXX												
-300	88.0	>62	1.1	XX	X											X	
		2-62	95.6	XX	XX												
		<2	3.3	XXXX	tr.												
-320	77.7	>62	0.7	XX	X											X	
		2-62	82.9	XX	XX												
		<2	16.4	XXXX	tr.												
-340	80.9	>62	5.3	X	X				X							XX	
		2-62	85.4	X	X											X	
		<2	9.3	X	X											XX	
-380	58.6	>62	17.1								XXX	X					
		2-62	82.5	X							XX	X					
		<2	0.4	X							.XX	X					
120K																		
-50	93.8	>62	9.6					XXX	tr.							XXXX	
		2-62	78.5	X				XXXX								tr.	
		<2	11.9														
-80	67.4	>62	28.9						XXXX								
		2-62	60.1						XXXX								
		<2	11.0						XXXX								
-100	68.9	>62	1.2	X				X				tr.				XX	
		2-62	85.3	X				XXX				tr.					
		<2	13.5	X				XXX			tr.	tr.					

Table 1 (Continued)

SAMPLE (depth, cm)	interstitial brine content (%)	size distribution, μ	percent in size	amorphous	goethite	hematite	lepidocrocite	Fe montmorillonite	maganosiderite	anhydrite	sphalerite	chalcopyrite	pyrite	manganite	todorokite	detritals	miscellaneous and remarks
120K Con't.																	
-125	79.6	>62	13.5	X					X							XX	
		2-62	83.2	XX					XX								
		<2	3.3	X	X											XX	
-140	93.2	>62	33.1													XXXX	
		2-62	48.0	X				X	XX								
		<2	18.9	X	X			XX	tr.								
-167	81.1	>62	18.2					X	X							XX	
		2-62	76.5					XX	X							X	
		<2	5.3					XX	X							X	
-200	93.4	>62	2.2													XXXX	
		2-62	73.9	X				XX	X								
		<2	23.9	XXX												X	
-282	95.2	>62	5.7							XX						XX	
		2-62	36.7	XXX				X								X	
		<2	57.6	XX				X								X	
-302	70.8	>62	39.8						XXXX								
		2-62	59.0						XXXX								
		<2	1.2	XXX				X									
-335	93.3	>62	1.6							XX						XX	
		2-62	39.3	XXXX													
		<2	59.1	XXX				X	tr								
-360	74.0	>62	90.2						XXXX								
		2-62	6.5	XX				X	X								
		<2	3.3	XXXX													
-396	91.8	>62	6.7					XX								XX	
		2-62	61.4	XX				XX									
		<2	31.9	XX				XX									
-420	50.4	>62	48.2						XXXX								
		2-62	51.4						XX		X		X				
		<2	0.4	XXX							X						
84K																	
-120	91.6	>62	3.6							X	X		X			XX	
		2-62	85.9					XX			X					tr.	
		<2	10.5					XXX			X					tr.	
-190	93.0	>62	2.9													XXXX	
		2-62	85.8	X				XX			X						
		<2	11.3					XXX			X						
-235	---	>62	78.8							XX	XX	tr.					
		2-62	21.2							XX	XX	tr.					
		<2	0.0														
-260	94.3	>62	4.5							XXXX							
		2-62	46.8			XX		XX			tr.						
		<2	48.7			XX		XX			tr.						
-328	67.3	>62	45.7						X	XXX							
		2-62	49.5			X		XX		X							
		<2	4.8			X		XX		X							
-350	77.8	>62	2.0													XXXX	
		2-62	91.8			XXXX											
		<2	6.2			XXXX											

James L. Bischoff

Table 1 (Continued)

SAMPLE (depth, cm)	interstitial brine content (%)	size distribution, μ	percent in size	amorphous	goethite	hematite	lepidocrocite	Fe montmorillonite	manganosiderite	anhydrite	sphalerite	chalcopyrite	pyrite	manganite	todorokite	detritals	miscellaneous and remarks
126P																	
-90	94.2	>62	2.7													XXXX	
		2-62	45.6	XX				XX									
		<2	51.7	X				XXX									
-280	93.7	>62	2.0													XXXX	
		2-62	85.3	XX				XX									
		<2	12.7	X				XXX									
-380	93.1	>62	3.0	X	X			X								X	
		2-62	81.7	X	X			X			X						
		<2	15.3					XXX			X						
-450	94.3	>62	1.5													XXXX	
		2-62	84.3	X				XXX			tr.						
		<2	14.2	X				XXX			tr.						
-650	90.2	>62	8.9	XXX												X	
		2-62	91.1	XXXX													
		<2	0.0														
-670	81.8	>62	47.8											XXX			Groutite X
		2-62	52.2	XXXX													
		<2	0.0														
-730	90.0	>62	1.2	X	XX											X	
		2-62	98.8	XX	XX												
		<2	0.0														
-744	55.6	>62	8.0													XXXX	
		2-62	61.8													XXXX	
		<2	30.2													XXXX	
-760	62.3	>62	9.2													XXXX	
		2-62	40.6	XX	X											X	
		<2	50.2	XX	X											X	
-800	89.8	>62	2.0	XXX												X	
		2-62	84.9	XXXX	tr.												
		<2	13.1		XX			XX									
127P																	
-80	95.7	>62	2.3													XXXX	
		2-62	86.4	XXXX													
		<2	11.3	XX				XX									
-150	94.9	>62	2.5													XXXX	
		2-62	84.2	X				XXX									
		<2	13.3					XXXX									
-237	97.3	>62	1.6													XXXX	
		2-62	91.8					XXX			X						
		<2	6.6					XXXX									
-260	96.0	>62	5.4													XXXX	
		2-62	81.7	XXX				X									
		<2	12.9	XX				XX									
-340	95.6	>62	9.8	X				X								XX	
		2-62	54.1	XX				XX									
		<2	36.1	X				XXX									
-380	96.9	>62	3.9													XXXX	
		2-62	86.3	X				XXX									
		<2	9.8	X				XXX									

Table 1 (Continued)

SAMPLE (depth, cm)	interstitial brine content (%)	size distribution, μ	percent in size	amorphous	goethite	hematite	lepidocrocite	Fe montmorillonite	manganosiderite	anhydrite	sphalerite	chalcopyrite	pyrite	manganite	todorokite	detritals	miscellaneous and remarks
127P Con't.																	halite XXXX
-427	92.6	>62	8.9														
		2-62	81.8	XX							X		X				
		<2	9.3	XXXX													
-455	82.6	>62	37.3		XXXX												
		2-62	58.2	X	XXX												
		<2	4.5	XX	XX												
-460	79.5	>62	33.6		XX				XX								
		2-62	63.6	X	XXX												
		<2	2.8	XX	XX												
-464	80.3	>62	3.8	XX	X											X	
		2-62	80.9	XX	XX												
		<2	15.3	XXX	X												
-495	92.6	>62	5.1						XXXX								
		2-62	92.8	XXXX	tr.			tr.									
		<2	2.1	XXXX													
-514	78.9	>62	9.4													XXXX	
		2-62	81.8	XX					XX								
		<2	8.8	X				XXX									
-527	93.4	>62	6.0	XX					XX								
		2-62	82.5	XX					XX								
		<2	11.5	XX												XX	
-550	80.6	>62	1.6						XX	XX							
		2-62	78.6	XX	X					X							
		<2	19.9	XXXX													
-610	59.3	>62	34.6						XX							XX	
		2-62	52.8						tr.		XXXX						
		<2	12.6								XXXX						
-645	89.7	>62	4.3						XXXX								
		2-62	79.4	XXX					X								
		<2	16.3	XXXX													
-710	85.1	>62	12.7													X	barite XXX
		2-62	86.4	X							X	X	X				
		<2	0.9	X							X	X	X				
-750	55.4	>62	70.4						XXX							X	
		2-62	28.5						XX							XX	
		<2	1.1	XX												XX	
-757	50.2	>62	2.3													XXXX	
		2-62	96.0	X					tr.							XXX	
		<2	1.7													XXXX	
-762	80.4	>62	1.4													XXXX	
		2-62	85.9	XX			XX									X	
		<2	12.7				XXXX										
-783	55.6	>62	24.1													XXXX	
		2-62	73.8	X												XXX	
		<2	2.1	XX												XX	

Table 1 (Continued)

SAMPLE (depth, cm)	interstitial brine content (%)	size distribution, μ	percent in size	amorphous	goethite	hematite	lepidocrocite	Fe montmorillonite	manganosiderite	anhydrite	sphalerite	chalcopyrite	pyrite	manganite	todorokite	detritals	miscellaneous and remarks
128P																	
-130	94.7	>62	6.5														
		2-62	80.9	XXX				X								XXXX	
		<2	12.6	XXX				X									
-162	95.7	>62	4.0														
		2-62	85.8	XXX				X								XXXX	
		<2	10.2	X				XXX									
-200	94.4	>62	5.5													XXXX	
		2-62	85.2					XXXX			tr.						
		<2	12.0					XXXX									
-226	91.7	>62	4.7													XXXX	
		2-62	90.6	XX				X			X						
		<2	4.7					XXX			X						
-244	95.2	>62	0.7													XXXX	
		2-62	69.2	X	X			XX									
		<2	30.1	X	X			XX									
-280	93.2	>62	1.4													XXXX	
		2-62	91.4		X			XXX			tr.						
		<2	7.2	X	X			XX	tr.								
-320	94.1	>62	2.1													XXXX	
		2-62	92.3	X				XXX			tr.						
		<2	5.2					XXX			X						
-357	95.0	>62	1.4													XXXX	
		2-62	86.1	XXX				X									
		<2	12.5	X				XXX			tr.						
-376	88.2	>62	6.7													XXXX	
		2-62	92.2	X				XX	X								
		<2	1.1	X				XXX									
-377	---	>62	100.0						XXXX								
		2-62	0.0						---								
		<2	0.0						---								
-420	95.2	>62	5.5													XXXX	
		2-62	93.6	XXX	X												
		<2	0.8	XX	XX												
-440	95.5	>62	3.8													XXXX	
		2-62	93.7	XXX	X												
		<2	2.5	XX	XX												
-470	72.6	>62	7.7													XXXX	XX
		2-62	92.0	X	X												
		<2	0.3	XXXX													
-500	79.8	>62	60.6											XXXX			
		2-62	39.0												XXXX		
		<2	0.4												XXXX		
-510	81.0	>62	53.9											XXX	X		
		2-62	45.9												XXXX		
		<2	0.2												XXXX		
-540	81.4	>62	14.3													XXXX	
		2-62	83.7	XX	XX												
		<2	2.0	XXXX													

Table 1 (Continued)

SAMPLE (depth, cm)	interstitial brine content (%)	size distribution, μ	percent in size	amorphous	goethite	hematite	lepidocrocite	Fe montmorillonite	manganosiderite	anhydrite	sphalerite	chalcopyrite	pyrite	manganite	todorokite	detritals	miscellaneous and remarks
128P Con't.																	
-560	86.0	>62	9.4													XXXX	
		2-62	85.8	XX	XX												
		<2	4.8	XXXX													
-600	85.2	>62	10.6													XXXX	
		2-62	82.7	X				X								XX	
		<2	6.7					XXXX				tr.					
-640	92.0	>62	4.9													XXXX	
		2-62	88.9	XXXX													
		<2	6.2	XX	X			X									
-700	91.7	>62	1.0						XXXX								
		2-62	93.8	XXX					X								
		<2	5.2	XXXX					tr.								
-740	77.8	>62	8.5													XXXX	
		2-62	90.2	XX				XX									
		<2	1.3	XX				XX									
-795	89.2	>62	22.9						XXXX								
		2-62	70.9	XX							X	tr.				X	
		<2	6.2								X	X				XX	

Gevirtz and Friedman (1966) for the normal deep-sea sediments of the Red Sea, consists primarily of aragonitic pteropod shells, calcitic foraminiferal tests, coccoliths, clastic quartz, feldspar and clays. Detrital material predominates in cores 95K and 118K and occurs largely as separate beds while within the remaining cores it is largely mixed with the other facies. Detritus rarely exceeds 5 per cent in the Atlantis II deposits, whereas detrital material accounts for approximately 50 per cent of the Discovery sediments (core 119K).

Two samples from the sill area (161K-221 and 161K-231, Table 1) contain euhedral crystals of well-ordered dolomite (~70μ), which on the basis of its association is believed to be of detrital origin. No dolomite was found within the Atlantis II Deep.

The dominant chemical components are CaO, SiO_2, Al_2O_3, MgO and CO_2. Anomalous, but minor amounts of Fe_2O_3 are also present (see Table 3).

Iron Montmorillonite Facies. Dark brown, finely bedded, montmorillonitic mud comprises the uppermost facies throughout the Atlantis II Deep. Although approximately 70cm of this material occurs at the top of core 119K from the Discovery Deep, it is elsewhere primarily restricted to below the brine-sea water interface within the Atlantis II Deep where its thickness ranges from 4 to 6m. Discrepancies in thickness of this unit between piston and Kasten cores may be due to superpenetration of the latter. If a correction were possible, the correlation diagram (Fig. 2) could be simplified. Moreover, one cannot be sure that the iron montmorillonite facies is absent in the sill area since the cores from that area are of Kasten type (85K and 161K).

The iron montmorillonite facies, representing the most recent and perhaps presently occurring precipitation from the brine, is very liquid, generally containing from 90 to 96 weight per cent interstitial brine and consisting of various proportions of montmorillonite and amorphous material. The montmorillonite appears to contain ferrous and ferric iron in the octahedral position and is believed to be of inter-

Table 2 Chemical analyses of Red Sea brine deposits (Sequence of cores as in Table 1.)

CHEMICAL ANALYSES OF RED SEA BRINE DEPOSITS[a]

	95K-110[b]	95K-150	95K-185	95K-395	118K-90	118K-160	118K-260	118K-320
SiO_2	19.	<2.	30.	20.	8.0	34.	6.	8.3
TiO_2	<.01	<.01	0.81	0.46	<.01	0.96	0.01	0.17
Al_2O_3	7.8	<.2	10.2	7.6	0.94	9.4	<.2	2.9
Fe_2O_3 (tot.)	38.	85.	8.3	6.7	61.	6.5	78.	41.
(FeO)	(.84)	(.45)	(2.26)	(.62)	(.42)	(1.46)	(.30)	(.58)
Mn_3O_4	0.40	0.40	0.94	0.59	0.15	0.28	0.47	0.33
MgO	3.0	0.36	4.0	4.0	1.3	3.9	0.9	1.7
CaO	10.1	1.9	20.5	28.4	8.7	20.1	1.9	15.7
SrO	0.04	<.01	0.10	0.13	0.02	0.11	<.01	0.05
BaO	<.03	<.03	<.03	<.03	<.03	<.03	<.03	<.03
Na_2O	0.79	<.1	1.3	0.92	0.3	1.5	<.1	0.5
K_2O	0.56	0.08	0.93	0.62	0.14	0.68	0.11	0.35
ZnO	0.16	0.21	0.04	0.09	0.13	0.10	0.19	0.13
CuO	0.10	<.01	<.01	<.01	0.09	<.01	0.01	0.02
PbO	0.03	0.08	<.02	<.02	0.07	<.02	0.07	0.02
P_2O_5	<.4	<.4	<.4	<.4	<.4	<.4	<.4	<.4
Ig. loss[c]	19.1	13.4	23.6	29.3	18.4	21.6	15.2	21.8
sum	99.	101.	101.	99.	99.	99.	103.	101.
CO_2	11.9	1.98	20.5	26.69	8.29	20.09	2.49	14.23
S	0.10	0.06	0.21	0.11	0.15	0.60	0.07	0.09
Ag	<.001	<.001	<.001	<.001	<.001	<.001	<.001	<.001

	119K-90	119K-151	119K-262	119K-355	161K-200	161K-200 <2μ
SiO_2	18.	15.	7.2	25.	12.5	11.0
TiO_2	0.05	0.23	0.02	0.48	0.03	0.04
Al_2O_3	2.2	4.7	1.5	6.3	<.2	<.2
Fe_2O_3 (tot.)	38.	8.3	57.	4.5	63.	62.
(FeO)	(3.72)	(3.59)	(2.82)	(1.04)	(3.57)	-
Mn_3O_4	0.74	0.31	0.43	0.39	0.32	0.30
MgO	2.1	3.0	1.1	4.4	0.56	0.40
CaO	8.7	37.6	6.2	25.5	1.6	3.0
SrO	0.03	0.12	<.03	0.10	<.01	<.01
BaO	<.03	<.03	<.03	<.03	<.03	<.03
Na_2O	3.8	2.5	3.2	2.4	<.1	.1
K_2O	0.40	0.69	0.18	0.74	0.13	0.76
ZnO	0.36	0.17	0.14	0.09	2.25	0.9
CuO	0.03	0.02	0.17	0.02	0.60	0.69
PbO	0.04	0.04	0.04	<.02	0.10	0.11
P_2O_5	<.4	<.4	<.4	<.4	<.4	0.9
Ig. loss	24.5	27.6	24.2	28.6	18.8	-
sum	99.	100.	101.	99.	100.	-
CO_2	9.13	32.89	6.01	24.86	1.83	2.79
S	0.18	2.98	0.34	0.14	0.50	0.56
Ag	<.001	<.001	<.001	<.001	.0043	.007

a. CO_2 and S determined by LECO combustometric tecnique, FeO by dichromate titration, ZnO by atomic absorption spectroscopy, K_2O and Mn_3O_4 by both atomic absorption and emission spectroscopy, remainder done by direct reading spark emission spectroscopy.

b. Refers to core number and depth of sample in core in cm.

c. Includes loss of volatiles such as H_2O, CO_2, and S, and gain of oxygen. Elements are in oxide form during spark emission analysis.

	161K-221	161K-237	161K-280	161K-325	85K-85	85K-120	85K-160	85K-270
SiO_2	8.5	7.9	14.	13.5	7.0	16.	16.	5.5
TiO_2	0.16	<.01	0.07	0.05	<.01	<.01	0.08	<.01
Al_2O_3	2.1	<.2	1.0	1.5	<.2	1.0	1.7	<.2
Fe_2O_3 (tot.)	58.	36.	61.	53.	70.	42.	39.	70.
(FeO)	(.25)	(<.001)	(.87)	(3.42)	(.15)	(14.0)	(24.07)	(.46)
Mn_3O_4	5.55	26.	1.39	2.4	0.70	2.1	0.87	0.63
MgO	1.5	1.5	1.1	0.95	0.51	2.4	2.2	0.3
CaO	3.9	3.1	2.4	5.0	2.35	6.6	7.2	1.0
SrO	0.03	0.04	0.02	0.05	0.02	0.4	0.36	0.01
BaO	0.03	0.03	<.03	0.04	0.03	0.11	0.09	0.03
Na_2O	1.7	3.6	2.9	2.7	1.9	1.6	1.1	2.0
K_2O	0.14	0.16	0.10	0.19	0.10	0.17	0.14	0.08
ZnO	0.67	1.74	1.66	2.22	0.70	4.2	8.67	0.15
CuO	0.38	0.12	0.55	0.71	0.67	3.1	2.4	0.05
PbO	0.07	0.13	0.09	0.08	0.15	0.33	0.47	0.15
P_2O_5	<.4	<.4	<.4	<.4	<.4	0.5	<.4	<.4
Ig. loss	19.4	20.2	13.3	16.4	14.8	18.8	19.1	22.9
sum	102.	101.	100.	99.	99.	99.	99.	103.
CO_2	4.32	2.27	2.93	5.06	2.35	7.85	8.07	1.25
S	0.27	<.02	0.28	0.62	0.15	3.06	5.55	0.11
Ag	<.001	<.001	0.0049	0.005	0.0031	0.024	0.024	<.001

	85K-320	85K-340	85K-380	120K-50	120K-100	120K-125	120K-167	120K-282
SiO_2	14.	8.4	3.9	32.	21.	21.	24.	32.
TiO_2	0.30	0.05	<.01	0.17	0.09	0.14	0.14	0.03
Al_2O_3	5.1	1.1	<.2	1.2	1.5	3.4	3.3	<.2
Fe_2O_3 (tot.)	57.	63.	32.	38.	32.	42.	39.	46.
(FeO)	(.71)	(6.44)	(19.1)	(8.18)	(24.1)	(10.18)	(14.52)	(9.07)
Mn_3O_4	0.76	1.95	2.5	1.23	5.58	5.40	5.74	2.13
MgO	2.8	0.86	1.0	1.6	1.3	1.3	1.4	1.2
CaO	2.2	8.5	1.8	6.1	4.0	2.3	2.3	2.1
SrO	0.08	<.01	0.29	0.31	0.31	<.01	<.01	0.02
BaO	<.03	<.03	0.12	0.04	0.07	<.03	<.03	<.03
Na_2O	0.9	1.0	<.1	1.6	2.0	2.1	2.9	2.1
K_2O	0.08	0.16	0.10	0.35	0.46	0.60	1.1	0.73
ZnO	0.64	0.72	13.27	1.66	3.88	0.42	0.37	0.09
CuO	0.05	0.07	8.8	0.64	0.76	0.04	0.04	0.42
PbO	0.04	0.10	0.16	0.08	0.20	0.05	0.05	0.08
P_2O_5	<.4	<.4	<.4	<.4	<.4	<.4	<.4	<.4
Ig. loss	16.5	15.3	24.6	13.0	25.8	20.3	20.7	15.0
sum	100.	101.	98.	98.	99.	99.	101.	102.
CO_2	1.87	7.59	-	7.22	-	14.26	18.63	8.58
S	0.25	0.11	19.50	-	-	0.14	0.19	-
Ag	<.001	<.001	0.02	0.0048	0.0054	<.001	<.001	0.0022

382 James L. Bischoff

Table 2 (Continued)

	120K-396	84K-120	84K-190	84K-260	84K-260 <2μ	84K-328	84K-350	127P-80
SiO_2	24.	26	18.	28.	27.	14.	5.0	25.
TiO_2	<.01	0.09	<.01	<.01	0.13	0.09	<.01	0.09
Al_2O_3	<.2	1.4	2.1	2.1	2.2	2.3	<.2	1.2
Fe_2O_3 (tot.)	31.	43.	28.	46.	47.	60.	85.	45.
(FeO)	(1.51)	(7.0)	(20.2)	(12.8)	(10.8)	(10.16)	(3.75)	(7.34)
Mn_3O_4	0.10	0.15	2.26	0.57	0.1	0.63	<.3	1.07
MgO	0.55	1.3	3.4	1.0	0.63	0.87	0.5	0.75
CaO	1.8	2.6	4.2	2.8	1.8	7.2	1.4	3.2
SrO	<.01	0.01	<.01	0.01	<.01	0.01	<.01	0.01
BaO	<.03	<.03	<.03	0.03	<.03	<.03	0.03	<.03
Na_2O	4.0	0.8	3.2	1.8	1.9	1.7	1.7	2.8
K_2O	0.73	0.17	0.16	0.17	0.14	0.18	0.10	0.40
ZnO	0.02	2.56	16.2	1.12	1.61	0.17	0.08	6.73
CuO	0.06	0.88	1.29	1.35	1.30	0.45	0.45	0.82
PbO	<.03	0.17	0.21	0.18	0.14	0.06	0.07	0.14
P_2O_5	<.4	<.4	<.4	<.4	<.4	<.4	<.4	<.4
Ig. loss	34.4	19.4	22.6	13.4	14.8	12.8	5.0	14.3
sum	97.	99.	102.	99.	99.	100.	99.	102.
CO_2	2.57	2.64	8.91	3.78	3.47	2.38	1.69	5.72
S	0.08	5.8	11.78	6.33	–	3.74	0.58	2.02
Ag	<.001	0.007	0.02	0.006	0.006	<.001	<.001	0.007

	127P-237	127P-427	127P-514	127P-610	127P-710	127P-762	128P-162	128P-200
SiO_2	22.	29.	28.	26.	40.	22.	28.	28.
TiO_2	0.04	0.11	0.40	<.01	0.06	0.23	0.11	0.11
Al_2O_3	1.0	1.5	6.7	2.7	1.5	3.8	1.3	1.7
Fe_2O_3 (tot.)	28.	27.	34.	17.	21.	44.	40.	38.
(FeO)	(8.24)	(14.6)	(22.7)	(8.4)	(11.6)	(1.06)	(6.02)	(7.43)
Mn_3O_4	0.57	0.43	2.77	0.56	0.83	1.55	0.89	0.70
MgO	1.0	0.21	1.9	0.4	0.39	1.2	1.6	1.2
CaO	6.9	5.1	3.3	1.2	1.7	7.0	7.8	7.8
SrO	0.02	0.05	0.01	<.01	0.02	0.02	0.03	0.03
BaO	0.03	0.06	<.03	<.03	0.04	<.03	<.03	<.03
Na_2O	4.8	4.0	1.0	2.9	1.7	2.0	3.5	3.5
K_2O	0.67	0.51	0.21	0.51	0.20	0.83	0.64	0.52
ZnO	2.54	6.59	0.48	21.0	7.85	0.39	0.86	2.20
CuO	0.33	1.5	0.36	3.7	4.0	0.15	0.61	0.38
PbO	0.04	0.21	<.4	0.27	0.31	0.06	0.06	0.06
P_2O_5	<.4	<.4	<.4	<.4	<.4	<.4	<.4	<.4
Ig. loss	31.6	21.0	21.7	25.6	19.5	16.9	16.4	16.4
sum	100.	97.	101.	102.	99.	100.	102.	101.
CO_2	6.27	4.77	21.5	5.39	6.86	6.78	8.91	7.48
S	–	13.1	–	18.2	16.3	0.75	0.50	1.10
Ag	0.0016	0.010	0.0031	0.016	0.017	<.001	0.0027	0.0027

	128P-226	128P-280	128P-320	128P-357	128P-376	128P-420	128P-470	128P-510	128P-795
SiO_2	25.	26.	20.	14.	31.	14.	9.0	7.1	17.
TiO_2	0.15	<.01	0.07	<.01	.08	0.05	0.05	0.10	0.09
Al_2O_3	1.6	1.2	<.2	<.2	1.3	<.2	1.7	1.3	2.3
Fe_2O_3 (tot.)	30.	42.	52.	17.	27.	51.	57.	25.	48.
(FeO)	(10.68)	(8.4)	(19.6)	(3.74)	(1.98)	(6.74)	(1.94)	(.70)	(12.08)
Mn_3O_4	0.95	0.91	0.91	2.5	0.44	0.81	1.23	45	0.66
MgO	1.2	0.8	1.5	0.65	1.4	0.63	0.72	1.6	0.25
CaO	8.2	5.15	7.0	4.15	10.4	5.4	7.4	2.7	1.6
SrO	0.03	0.01	0.02	0.02	0.04	<.01	0.01	<.01	0.03
BaO	<.03	<.03	<.03	<.03	<.03	<.03	<.03	0.05	0.04
Na_2O	4.4	3.4	0.7	6.1	3.0	3.5	2.3	3.3	3.0
K_2O	0.90	1.01	0.45	1.2	0.81	0.42	0.33	0.35	0.84
ZnO	4.65	2.41	3.49	2.64	0.53	0.42	0.10	1.11	5.52
CuO	0.34	0.58	0.31	0.12	0.42	0.16	0.03	0.07	1.2
PbO	0.07	0.11	0.06	0.03	0.07	0.06	0.05	0.07	0.07
P_2O_5	<.4	<.4	0.5	<.4	<.4	<.4	<.4	<.4	<.4
Ig. loss	22.2	15.5	14.0	50.6	20.4	22.2	18.2	14.0	20.8
sum	100.	99.	101.	99.	99.	99.	98.	102.	101.
CO_2	7.77	4.95	-	4.14	9.75	5.06	6.64	2.20	14.04
S	2.50	1.79	-	1.63	.34	1.59	-	1.14	6.79
Ag	0.0023	0.0058	0.0012	<.001	0.0018	<.001	<.001	0.0013	0.0012

Table 3   Averaged Chemical Analyses for Facies. Based on 43 "Selected" Typical Analyses from Table 2

	Detrital	Fe Mont-morillonite	Goethite-Amorphous	Sulfide	Manganite *
SiO_2	27.3	24.4	8.7	24.7	7.5
Al_2O_3	8.4	1.7	1.1	1.5	0.7
Fe_2O_3 (total)	6.5	37.1	64.2	24.3	30.5
FeO	1.4	11.7	2.7	13.4	0.4
Mn_3O_4	0.6	2.1	1.1	1.1	35.5
CaO	23.6	4.8	3.4	2.5	2.9
ZnO	0.08	3.2	0.7	12.2	1.4
CuO	<.01	0.8	0.3	4.5	0.1
CO_2	23.1	8.6	3.6	5.7	2.2
S	0.3	3.9	0.6	16.8	0.6

* Based on only 2 analyses.

mediate composition within the series between nontronite (dioctahedral) and the yet undescribed ferrous bearing montmorillonite (trioctahedral). Details of x-ray, chemical and Mössbauer analysis are now in process and will be presented later. The occurrence of montmorillonite in the Atlantis II Deep was noted in Miller *et al.* (1966).

Sphalerite, as a minor constituent, is usually present up to several per cent (Fig. 3), and probably accounts for the majority of the ZnO and S in the analyses (ZnO/S for sphalerite = 2.5). Other minor constituents include goethite and manganosiderite.

Averaged analyses of typical samples of this facies (Table 3) indicate the major components, FeO, Fe_2O_3 and SiO_2. Relatively high Na_2O values (1–5 per cent) are attributable to incomplete leaching of the interstitial brine and to adsorbed Na^+ at interlayer positions of the montmorillonite lattice.

Goethite-Amorphous Facies. Orange to yellow beds of the goethite-amorphous facies immediately underlie the iron montmorillonite facies within the Atlantis II Deep, usually with a rather abrupt transition zone of approximately 20cm, and extend considerably beyond the immediate area (Fig. 2). Similar material is found as a 1cm bed in core 153P located approximately 100 miles south of the Atlantis II

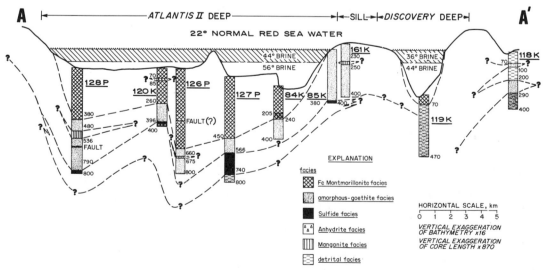

Fig. 2. Generalized cross-section through geothermal area projected along A-A' in Fig. 1 showing facies distribution. Small numbers beside cores refer to depth in cm. Note superposition of facies patterns to denote mixtures.

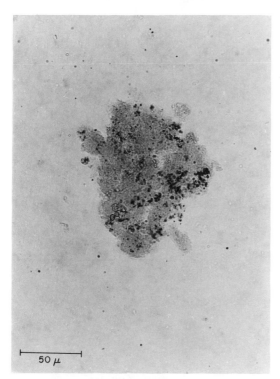

Fig. 3. Photomicrograph of iron-montmorillonite material. Dark spots are sphalerite. Refractive index of medium = 1.55.

Deep, but it is not clear whether its presence is due to the Atlantis II source or another, as yet undiscovered, "hot hole."

Facies thickness within the Atlantis II Deep is approximately one meter, with somewhat greater thickness in core 128P due to a clearly visible slump fault. The unit apparently thickens toward the sill area, where approximately four meters were penetrated by core 161K. Except for the upper 70cm of iron montmorillonite, cored sediments of the Discovery Deep are comprised entirely of this facies, although much diluted with detrital material. Core 118K, south of the deeps, contains several beds of goethite-amorphous material, relatively more crystalline than that in the Atlantis II Deep, and radiometric dating (Ku *et al.*, 1969) indicates that these beds are much older. Core 95K, from the "island" protruding above the brine in the Atlantis II Deep, contains similarly well crystallized material mixed with detrital material.

Interstitial brine contents in this facies are lower than in the iron montmorillonite muds, averaging approximately 80 per cent by wet weight. Poorly crystalline goethite

and amorphous "limonite" of varying proportions are the major mineral components, again with minor proportions of other minerals. Microscopically, the samples are composed of 1-30μ spherules (Fig. 4) with the goethite-rich samples showing faint birefringence. Refractive index ranges from less than 1.55 to slightly greater than 2.0, indicating the extreme range of hydration. Pyrite, pyrrhotite and greigite are found associated with goethite in cores 95K, 118K and 119K.

Well crystallized lepidocrocite, a polymorph of goethite, occurs in samples 118K-320, 119K-270 and 127P-762 (Table 1) and may indicate a correlative bed from inside the Atlantis II Deep through the Discovery Deep and outside to the adjacent area. The lepidocrocite samples are physically indistinguishable from the goethite samples and may have been overlooked in the other cores.

The solids of this facies in core 84K from the deepest part of the Atlantis II Deep have transformed evidently to well crystallized hematite, indicating a temperature anomaly and the possibility of the proximity

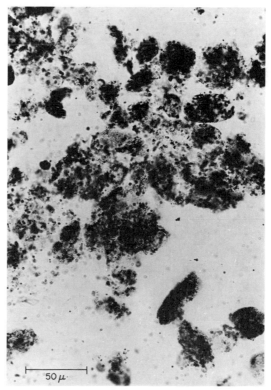

Fig. 4. Photomicrograph of goethite-amorphous material. Refractive index of medium = 1.55.

of the brine vent. Implications of this transformation are discussed elsewhere (Bischoff, 1969). Averaged analyses for this unit (Table 3) indicate the predominance of Fe_2O_3 and SiO_2. SiO_2 is somewhat lower than in the iron montmorillonite unit and evidently is amorphous, and FeO is markedly low. The somewhat high CaO values balance well with CO_2 and are attributable to detrital debris.

Sulfide Facies. The facies of most interest to those concerned with the genesis of syngenetic base-metal sulfide ores is the sulfide facies that occurs in an apparently continuous, homogeneously black bed throughout the bottom of the Atlantis II Deep (Fig. 2). It is the deepest horizon encountered and was penetrated by only four cores, 127P, 128P, 120K and 85K (Table 1). This unit was not observed beyond the brine-sea water interface of the Atlantis II Deep. 127P penetrated approximately a one-meter section, bottoming in detrital material, but it is not clear whether this detrital material represents the basement of the deposits, or a pause of brine activity such as is seen higher up in cores 126P and 128P. The remaining three cores barely penetrated the top 20cm of the sulfide unit, leaving a large uncertainty about its continuity and thickness. The absence of this facies in 126P may be that the unit pinched out locally, or that the core was not deep enough to penetrate it. Sphalerite, with lesser amounts of associate chalcopyrite and pyrite accounts for the bulk of the mineralogy (Plate 3). High FeO contents, excess S over requirements for Zn and Cu sulfides, and absence of the sensitive 32° x-ray peak of pyrite indicate that much x-ray amorphous iron-monosulfide must also be present. Coarsely crystalline barite is associated with this facies in one sample (127P-710, Table 1), being present in the 62μ fraction. Anhydrite is locally associated with this unit (120K-420 and 84K-235, Table 1). ZnO, FeO, CuO, SiO_2 and S are the major components (Table 3). The presence of radiolarian tests in the sulfide zones of 127P and 128P evidently account for a large part of the SiO_2.

Manganosiderite Facies. Thin, semi-lithified, buff-colored beds (1mm–2cm) of man-

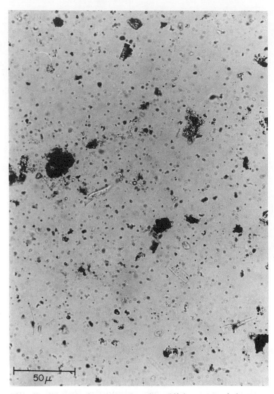

Fig. 5. Photomicrograph of sulfide material, predominantly sphalerite. Refractive index of medium = 1.55.

ganosiderite occur locally within the iron montmorillonite and goethite-amorphous facies within the bounds of the Atlantis II Deep, particularly in core 120K. The individual beds appear to be discontinuous and were too thin to show on the correlation diagram (Fig. 2). The mineral is well-crystallized manganosiderite (Fig. 6) and is of intermediate composition within the siderite-rhodochrosite isomorphous series. The beds also contain minor amounts of halite, and it is not clear whether the halite formed *in situ* at the same time as the manganosiderite or resulted from desiccation during storage. Halite was found only with manganosiderite, which argues for its formation *in situ*. The variation of the major diffraction peak position of manganosiderite for various samples indicates considerable range in the Fe:Mn ratio, and, as pointed out in Miller *et al.* (1966), some Ca may be present in the lattice.

Anhydrite Facies. White, massive beds (up to 20cm) of anhydrite were found locally in the Atlantis II Deep (cores 120K, 84K

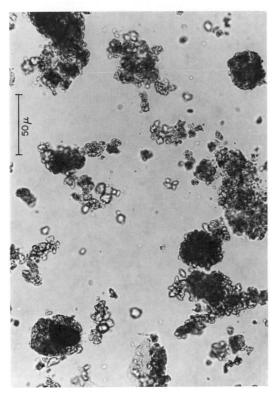

Fig. 8. Photomicrograph of manganite, reflected light.

Fig. 6. Photomicrograph of manganosiderite. Refractive index of medium = 1.55.

and 127P, Table 1 and Fig. 2) The beds are composed of very pure, well-crystallized anhydrite (Fig. 7).

Manganite Facies. The manganite facies, also of local occurrence within the Atlantis II Deep, is found in cores 161K, 126P and 128P, in black 15- to 50-cm beds, not unlike the sulfide facies in physical appearance. Well-crystallized manganite is the characteristic mineral (Fig. 8), with locally associated todorokite, (Mn, Mg, Ca, Ba, K, Na)$_2$Mn$_5$O$_{12}$ · 3H$_2$O (sample 128P-500), and groutite, MnOOH (126P-670), while woodruffite, (Zn, Mn)$_2$Mn$_5$O$_{12}$ · 4H$_2$O, represents this unit in sample 161K-237 (Tables 1 and 2). Mn$_3$O$_4$ and Fe$_2$O$_3$ are the major components (128P-510 and 161K-237, Table 2), with iron probably present mainly substituting for manganese and partly as amorphous limonite, and FeO markedly low.

Geochemical Relations

Chemical variations within individual facies, even within the same core, are very large. Plots of the concentrations of SiO$_2$, Fe$_2$O$_3$, FeO, Mn$_3$O$_4$, ZnO, CuO, Ag, BaO

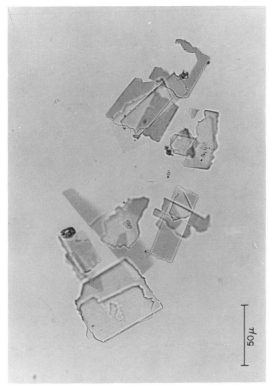

Fig. 7. Photomicrograph of anhydrite, crossed polarizers. Refractive index of medium = 1.55.

James L. Bischoff

and S against depth for each of the cores reveal these variations and some interesting covariations (Figures 9a–9f). ZnO, CuO, Fe, Ag and BaO covary markedly with S, throughout the various facies within the Atlantis II Deep, suggesting these elements (excepting Ba) to be present predominantly as sulfides. Ba does not fit in a sulfide lattice, and because the barium sulfate mineral, barite, is commonly asso-

ciated with Mississippi Valley type sphalerite deposits, Ba is most likely present as a sulfate. Outside the Atlantis II Deep only FeO varies with S while CuO and ZnO do not, a fact borne out mineralogically (cores 95K, 119K and 118K, Figs. 9a and b), and suggesting a process of sulfide precipitation different from that within the Atlantis II Deep. S and the metals of sulfide affinity, with the exception of Zn, which is present in

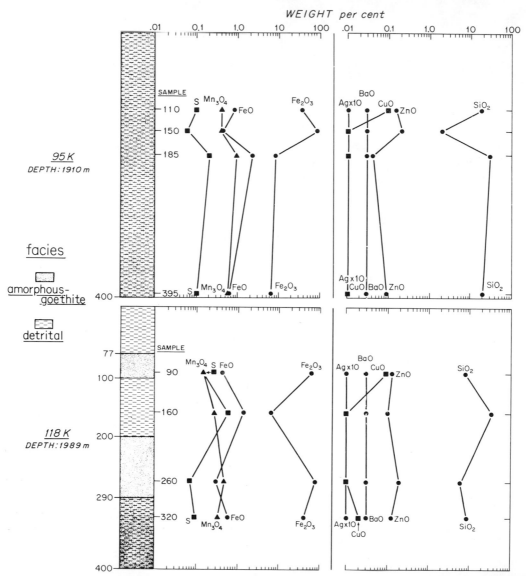

Fig. 9. Charts of cores showing facies and chemical variations with depth. Numbers to right of cores refer to sample locations in cm from top of core, numbers to left refer to facies boundaries. Chemical data on logarithmic scale.
a. Cores 95K * and 118K

* Discrepancy in core length and sample numbers of 95K between this paper and others in this book is because the core device did not completely penetrate. In this paper, samples were measured from top of core barrel, not top of sediment.

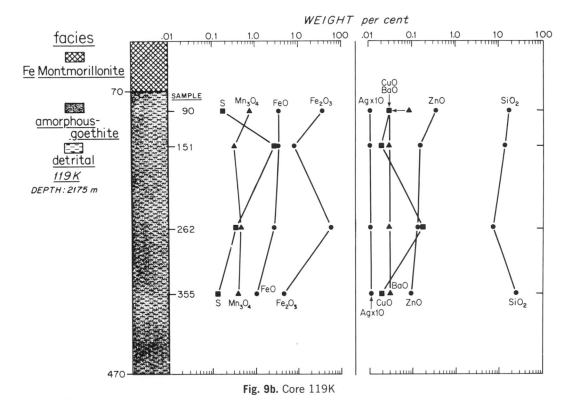

Fig. 9b. Core 119K

the lattice of woodruffite, are depleted in the manganite facies. CuO:ZnO ratios are rather low and constant in the sulfide facies but increase in the goethite-amorphous facies, suggesting co-precipitation of some Cu with colloidal ferric hydroxide. Such co-precipitation is expected at near neutral pH and appears to catalyze the oxidation process of ferrous iron (Hem and Skougstad, 1960). Within the iron montmorillonite facies (core 128P, Fig. 9f) ZnO is often in excess over that necessary to balance S. Electron probe work on the iron montmorillonite (G. Parks, personal communication) indicates occasional small areas of high Zn and no S. Thus, excess Zn is evidently neither in the montmorillonite lattice, nor in a sulfide.

CaO, MgO, Al_2O_3 and TiO_2 display strong correlation with the detrital facies, as expected, and were left off the plots in Fig. 9 to prevent undue complication. Mg:Ca ratios are very high throughout all facies (averaging about 1:5) and reflect the detrital component composition. There is insufficient dolomite or magnesian calcite present to account for this high ratio, so

detrital chlorite may be the source of the Mg. High K values in the detrital facies are probably due to detrital illite. In the brine-derived facies, however, there is insufficient detrital material to account for the K, and it must be co-precipitated with the iron montmorillonite, amorphous or perhaps sulfate phases. Fe_2O_3 and SiO_2 are major components throughout all the brine-derived facies and show remarkably little variation. As iron montmorillonite is the only recognized SiO_2 bearing mineral, large amounts of amorphous SiO_2 must therefore be present throughout the other facies. The surface of colloidal ferric hydroxide evidently catalyzes the polymerization and precipitation of SiO_2 (Parks, 1967), explaining the large amounts of SiO_2 usually associated with natural limonite.

Although the brine contains somewhat more dissolved Mn than Fe, the brine-derived deposits rarely contain more than one per cent Mn_3O_4, except in the very local beds of the manganite facies. This is undoubtedly explained by the higher oxidation potential of divalent manganese with respect to divalent iron, such that at the

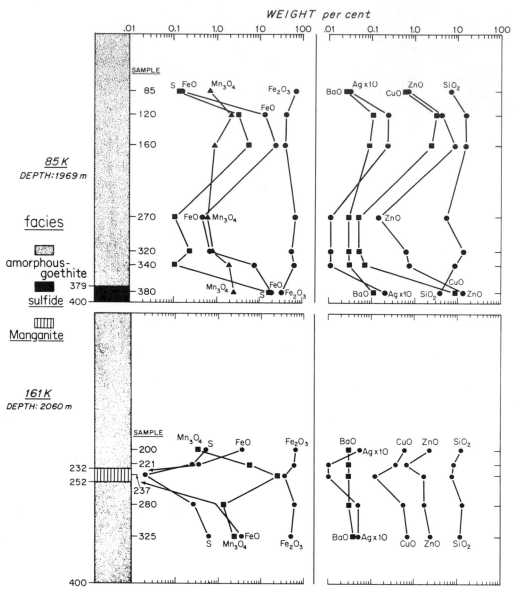

Fig. 9c. Cores 85K and 161K

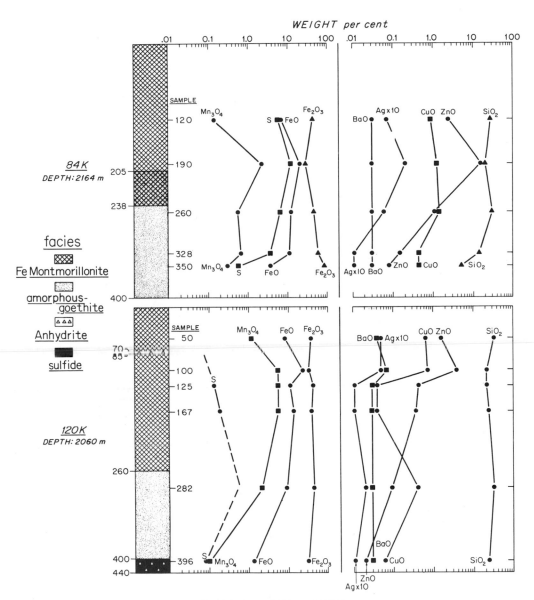

Fig. 9d. Cores 84K and 120K

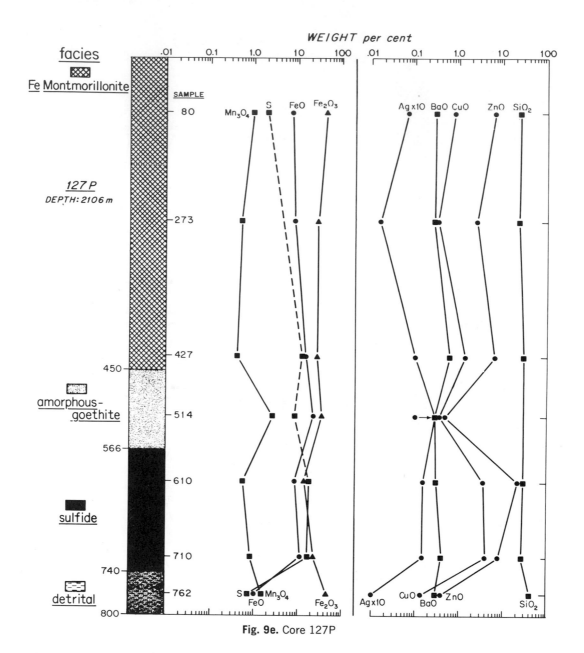

Fig. 9e. Core 127P

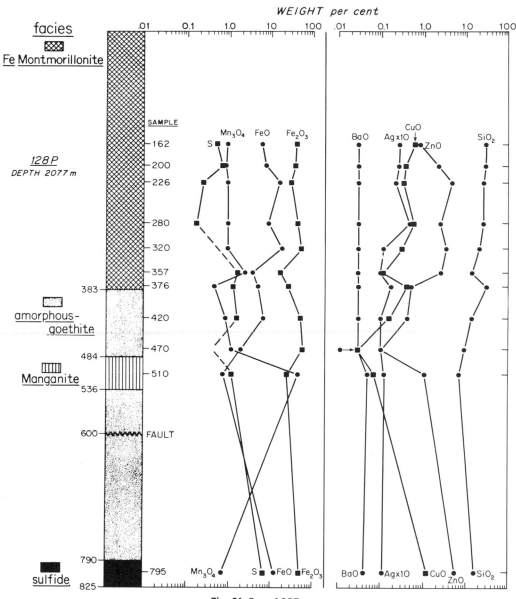

Fig. 9f. Core 128P

brine-sea water interface, the iron consumes the available oxygen and the manganese diffuses out of the system. The chemical data on the brine supports this model, the 56° brine being rich in both iron and manganese, while the 44° brine and the Discovery Deep brines are rich in manganese but markedly depleted in iron (Brewer and Spencer, 1969). Statistical treatment and further interpretations of the geochemical data will be discussed in Spencer and Bischoff (in preparation).

Processes

General Statement

The subject of most interest to those concerned with the genesis of ore bodies is the mode of concentration of the metals, that is, the mechanism of supersaturation and subsequent precipitation of the minerals from the brine. Although a detailed discussion of the origin of the brine is beyond the scope of this chapter, and would require additional data, the author prefers a model whereby Red Sea water circulates through the subterranean sediments of the Red Sea basin, exchanging various portions of original SO_4 for H_2S by early diagenetic bacterial activity, accumulating chloride salts from evaporite beds and heavy metals via chloride complexes from more shaley sequences, presumably fairly deep along channels of subsurface circulation. Encountering a high heat source while moving toward the central part of the Red Sea rift zone, the brine then percolates upward through fractures to the sea floor (see White, 1968, for a more complete discussion of possible origins of the Red Sea and Salton Sea geothermal brine systems). That the deposits of the Atlantis II Deep were precipitated from the overlying brine column and sedimented into their present position is amply demonstrated by the age sequences (Ku, 1969; Ku et al., 1969), bedding relations and gel-like nature of the deposits. The point of discharge of the brine, moreover, must be within the Atlantis II Deep judging from the distribution of the deposits and on the basis of the brine chemistry and temperature relations (Brewer et al., 1969).

From the occurrence of hematite, the point of discharge is probably near station 84 (Bischoff, 1969). Thus, a picture emerges of a hot subterranean brine rich in chloride salts and heavy metals, slowly percolating into the bottom of the Atlantis II Deep and precipitating solids which then deposit at a rate of about 40cm per thousand years.

An order of magnitude approximation of the rate of brine discharge based on the amount of silica in the Atlantis II sediments can be made. The assumptions and approximations are as follows:

1. All SiO_2 in the Atlantis II sediments was precipitated from the brine (this ignores detrital contributions).
2. Origin of dissolved SiO_2 in brine is from subterranean equilibration with quartz, with reasonable equilibration temperatures taken as 110°C and 170°C, with respective solubilities of 90 and 226ppm SiO_2 (from the data of Morey et al., 1962).
3. The amount of SiO_2 precipitated is represented by the difference between the high temperature quartz solubility and the present SiO_2 concentration in the Atlantis II 56° brine, 59ppm (Brewer and Spencer, 1969). This assumes no loss of SiO_2 during ascent of brine to discharge point and that no further precipitation below 59ppm takes place.

The mass of the dry salt-free sediment of the top 10m is taken as 19.4×10^7kg (Bischoff and Manheim, 1969). SiO_2 accounts for approximately 20 weight per cent of this material so that total precipitated SiO_2 in the top 10m is approximately 3.9×10^7kg.

Using the 110° quartz equilibrium value, each kilogram of brine precipitated 0.031 g of SiO_2, and 1.25×10^{12}kg of such brine was required to account for the deposit. Dividing by sediment age at the bottom of the 10m section (12,000 years, Ku et al., 1969) yields a brine discharge rate of 10^8kg of brine per year. Similar calculations for the 170° quartz equilibrium value yields a discharge rate of 1.9×10^7kg of brine per year. Comparing these calculated rates with the amount of brine at present within the Atlantis II Deep (ca. 7.2×10^9kg) indicates that between 0.26 and 1.4 per cent of the total brine pool is being added yearly.

Although the most recent precipitation from the brine is dominantly iron-montmorillonite, small amounts of the other minerals appear to be precipitating concurrently, although contributing very little to the bulk deposits. Variation during the past 10,000 years in volume of mineral precipitates has given rise to the succession of facies as seen in the cores. A small part of this variation could, perhaps, be explained by fluctuating physical conditions such as temperature and level of brine-sea interface, but the major changes must have been due to a variation in the chemistry of the brine. Today the brine is an iron hydroxide-silicate producer as evidenced by the iron montmorillonite facies; in the past it has been an iron hydroxide, iron-manganese carbonate and base-metal sulfide producer as evidenced by the succession of facies. The length of time which the brine was a base-metal sulfide producer, *i.e.*, how deep the sulfide bed extends, is an interesting and economically pertinent question (see Bischoff and Manheim, 1969).

What causes this chemical variation? Has the brine chemistry changed cyclically, or has it changed gradually with time? At present these questions must remain open.

Precipitation of Goethite and Amorphous Iron Hydroxide Phases

Precipitation of the limonitic material requires oxidation of dissolved ferrous iron and subsequent precipitation of ferric hydroxide. This process appears to be occurring within the 44° brine of the Atlantis II Deep as indicated by the marked depletion of dissolved iron in this layer noted by Brewer and Spencer (1969). Further evidence is that initially clear samples of the 56° brine rapidly precipitated ferric hydroxide upon exposure to air (Miller *et al.*, 1966).

The pertinent half reactions for this process are,

$$Fe^{+2} + 3H_2O \leftrightarrows Fe(OH)_3 + 3H^+ + e^-$$

$$O_2 + 2H_2O + 4e^- \leftrightarrows 4OH^-$$

Combining,

$$2Fe^{+2} + \tfrac{1}{2}O_2 + 5H_2O \leftrightarrows 2Fe(OH)_3 + 4H^+$$

with oxygen diffusing into the 44° brine from the overlying sea water. Note that this reaction releases hydrogen ions, lowers the pH and suggests that measured pH values from these brines are likely to be too low unless oxygen contamination was avoided. SiO_2 associated with the ferric hydroxide in the deposits was probably in a supersaturated state after cooling of the brine and simply polymerized on the catalyzing surfaces of the nascent ferric hydroxide colloids. Anomalous CuO concentrations associated with the goethite-amorphous facies probably resulted from similar surface adsorption during this period.

Such colloids, when first precipitated, are extremely small and strongly hydrated (Hem and Cropper, 1959) and would be expected to move laterally and up in the water column as well as down. Such mobility is offered as the explanation for the widespread occurrence of the goethite-amorphous facies outside the Atlantis II Deep. Enlargement of the brine pools and resultant elevation of the brine-sea water interface would also result in such a distribution, but this is unlikely, as the other facies are not so widely distributed.

After deposition, transformation of the amorphous ferric hydroxide to goethite, as a trend toward thermodynamic stability, would take place,

$$Fe(OH)_{3\,(limonite)} = FeOOH_{(goethite)} + H_2O$$

with resultant expulsion of adsorbed Cu and SiO_2. The goethite-amorphous facies in cores 95K and 118K is more goethitic than that of the Atlantis II Deep, and is evidently older (Ku *et al.*, 1969); CuO values are distinctly lower.

Transformation of goethite to hematite according to the reaction,

$$FeOOH_{(goethite)} = Fe_2O_{3\,(hematite)} + H_2O$$

a further dehydration, has evidently taken place in the sediments at station 84K, as mentioned earlier, and is discussed in Bischoff (1969).

The reasons for the occurrence of lepidocrocite instead of goethite in the several samples mentioned are not understood at present.

Precipitation of Iron-Montmorillonite

The somewhat limited distribution and high ferrous iron content of the iron-montmorillonite suggest precipitation in a zone of decreased oxygen availability, probably lower in the brine column than for the ferric hydroxide. Aside of the ferrous iron content, the iron-montmorillonite facies differs from the goethite-amorphous facies in having higher SiO_2, indicating a greater availability of SiO_2 from the brine at the time of precipitation. Thus, enough SiO_2 was available that when a portion of the iron was oxidized (below the 44° brine?), the montmorillonite became supersaturated and precipitated. Having an almost colloidal nature, portions moved over the sill and deposited in the Discovery Deep.

Precipitation of Anhydrite

Calcium sulfate solubility increases with decreasing temperature, and the brine, therefore, must become increasingly undersaturated with respect to anhydrite as it cools. Thus, the most likely mechanism of anhydrite precipitation is by mixing of SO_4^{--} from the overlying sea water with Ca^{++} of the brine. Sulfur isotope data of Hartmann and Nielsen (1966) and Kaplan et $al.$ (1969) indicate the identity of the anhydrite SO_4 with that of normal Red Sea SO_4 and support the mixing conclusion. Recent studies on the stability fields of gypsum and anhydrite (Zen, 1965; Hardie, 1967) show anhydrite to be the stable phase at the temperature and salinity of the Atlantis II Deep.

It is puzzling why anhydrite has not precipitated continuously and more abundantly in the system, unless replenishment of SO_4 in the overlying sea water is extremely slow. Perhaps SO_4 has been depressed in the sea water immediately overlying the 44° brine by anaerobic bacteria. Watson and Waterbury (1969) and Trüper (1969) have shown that particulate organic matter is collected in this zone due to the density differential and that sulfate reducing bacteria are present.

Precipitation of Manganosiderite

Formation of the iron-manganese carbonate as a direct precipitate from the brine is difficult to explain for the same reason as for anhydrite precipitation; carbonates become increasingly soluble with decreasing temperature. An attractive possibility is that the manganosiderite forms by replacement of infalling detrital calcium carbonate according to the reaction,

$$CaCO_3 + Me^{++} = MeCO_3 + Ca^{++}$$

where Me^{++} refers to divalent iron and manganese in the brine. At equilibrium,

$$\frac{aCa^{++}}{aMe^{++}} = K_r$$

where K_r is the equilibrium constant of the above reaction and is calculated by dividing the solubility product constant of calcite by that of either siderite or rhodochrosite. For 25°C and one atmosphere pressure, the values of K_r are $10^{+2.35}$ and $10^{+1.73}$, respectively (K_{sp} values from Latimer, 1952).

If the reaction were rapid, replacement should take place until either the aCa^{++}/aMe^{++} ratio reaches the equilibrium value or all the calcium carbonate is consumed. Although it is not possible to evaluate the activities of Ca^{++}, Fe^{++} and Mn^{++} in the brine, molal ratios will give a qualitative indication of the nearness of the brine to equilibrium with both phases. Molal ratios of Ca^{++}/Fe^{++} and Ca^{++}/Mn^{++} calculated from the data in Miller et $al.$ (1966) for the 56° brine yields $10^{+2.1}$ and $10^{+2.01}$, respectively, suggesting the brine at the present time is close to equilibrium.

It is suggested that periodic influx of brine was interrupted by periods of quiescence. At the beginning of such an influx, the brine was sufficiently concentrated in iron and manganese (and perhaps at a higher initial temperature) such that replacement of infalling detrital calcium carbonate took place rapidly; then, as the iron and manganese levels were reduced, additional infalling detrital calcium carbonate was unaffected. However, foraminiferal debris and manganosiderite were not found coexisting in the deposits, and petrographical evidence of partial replacement of foraminifera debris is lacking.

Another possibility is that CO_2 is produced from organic matter by bacterial activity in the Red Sea water above the 44° brine. Diffusion of such CO_2 into the brine could supply carbonate ion for manganosiderite precipitation.

Precipitation of Manganite

The occurrence of coarsely crystalline manganese oxide minerals presents an enigma. Certainly oxidation of divalent manganese is required, but if the process were the same as for the ferric hydroxide material, one would likewise expect a very poorly crystalline product. Marine manganese nodules are generally characterized by very poor crystallinity (F. T. Manheim, personal communication), and they generally have formation rates in terms of millions of years (Bender et al., 1966). Bricker (1965) has pointed out, however, that coarsely crystalline manganese oxide minerals can be produced by oxidation of minerals of lower oxidation state such as rhodochrosite. Evidence of extreme oxidizing conditions in this facies is found in the extreme low level of FeO (Table 3). Thus, the Red Sea manganese oxides are possibly local "weathering" products of preexisting manganosiderite beds. Such a process could have occurred during a particularly long quiescent period of brine discharge during which the brine had diffused away and normal sea water was in contact with the surface deposits in the Atlantis II Deep.

Precipitation of Sulfide

The mechanism of base-metal sulfide precipitation in general continues to be a subject of controversy. In the present case, several alternative mechanisms have potential applicability.

The base-metals clearly came with the brine (Miller et al., 1966) so that the central question concerns the source of sulfide. H_2S was not detected in the 56° brine (Brewer et al., 1969) although small amounts must be present; the Atlantis II

Deep brines and sediments are completely free of sulfate-reducing bacteria (Watson and Waterbury, 1969; Trüper, 1969). Sulfate-reducing bacteria are, however, abundant in the sea water immediately above the 44° brine of the Atlantis II Deep, evidently feeding on the organic matter which is collecting there (Watson and Waterbury, 1969). It is an attractive possibility that sufficient sulfide is produced in this area to precipitate the base-metals. Kaplan et al. (1969) have found that the δS^{34} content of the Atlantis II sulfides have a very narrow range, averaging about +6.0. Although bacterial reduction of sulfate in interstitial waters of marine sediments results in a wide range of sulfide δS^{34} values (Kaplan and Rittenberg, 1962), bacterial sulfate reduction in a water column with an infinite reservoir of available sulfate might be expected to have a narrow range of sulfide δS^{34}. Thus, the sulfur isotope data is not inconsistent with such a suggested mixing process for the precipitation of the base-metal sulfides. On the other hand, one might expect precipitation of the sulfides so high in the water column would result in a much wider distribution of these minerals than observed, and also more iron-sulfide.

Another possible mechanism is the mixing of two brines, one rich in H_2S, the other in base metals. This process appears to account for the precipitation of sphalerite in the storage tanks of the Chelekin geothermal brines (Lebedev, 1967b), where brines from several wells, each with a slightly different chemistry, are mixed. The chemical differences of the various Chelekin brines seem to be a function of depth, and overall they are remarkably similar to the Red Sea brines, particularly regarding temperature, total salinity, Ca/Mg ratio. The sulfides of the Atlantis II Deep may, therefore, have precipitated at a coalescing front between a metal-rich, sulfide-poor brine and a sulfide-rich, metal-poor brine, both coming along the same subterranean channel. Such a process, in fact, has been suggested by Beales and Jackson (1966) for the Pine Point deposits in Canada. This mechanism, however, fails to explain that sphalerite is forming at present within the iron-montmorillonite.

Barnes and Czamanske (1967) have suggested an oxidation process by which sphalerite might precipitate. Noting the common association of barite with sphalerite, they reason that, if the Zn is held in HS^- complexes, partial oxidation of sulfide results in barite precipitation and release of H^+ changing the sulfide distribution, and thereby releasing the Zn from HS^- and precipitation as sulfide. There is a strong correlation of Ba with the sulfides of the Atlantis II deposits, and even one occurrence of barite. However, the amount of Ba is very small compared to the heavy metals, and if HS^- complexes were responsible for metal transport, certainly H_2S would have been detected in the 56° brine. This process may operate in some geologic environments, but it appears unlikely for the Red Sea system.

The most simple and most probable process to the author is that of cooling. The association of epigenetic ore minerals with chloride brines has been known for a long time from fluid inclusion data (Sorby, 1858). Metal transport as chloride complexes in such solutions, first quantized by Garrels (1941), and later given a rigorous basis by Helgeson (1964), seems now, judging from the current literature, to be fairly well accepted. The formation of these metal chloride complexes involves an entropy increase and consequent increase in stability (and hence possible maximum concentration) with increasing temperature. If the brine approached equilibrium with sulfide minerals at depth and high temperature, most of the metals in solution would be as chloride complexes. During cooling, continuous release of metals from complexes would take place, with consequent supersaturation of the sulfide minerals. White (1968) has suggested a similar process for sulfides in the Salton Sea brine. Such a process for the Red Sea system would explain the narrow range of δS^{34} values, and the limited areal distribution of the sulfides as most precipitation would occur nearest the point of discharge in the Atlantis II Deep.

Barnes (1967) has pointed out that adequate Zn can be transported near 100°C as the $Zn(HS)_2$ complex (S >> Zn) and as $ZnCl_2$ (Zn >> S), but that it is not possible to transport just enough Zn and S to precipitate ZnS without a large excess of either.

The deposits as a whole and metals in the brine indicate that concentration of Zn and Cu is only slightly greater than S, making the bacterial reduction of SO_4 an attractive hypothesis. On the other hand, Hemley et al. (1967) have shown that ZnS is very soluble in silicate buffered chloride solutions at considerably elevated temperatures.

The iron sulfides within the goethite-amorphous facies outside the Atlantis II Deep (cores 118K, 119K) or above the brine-sea water interface (core 95K) require a different explanation. δS^{34} values scatter considerably around an approximate −25.0 average (Kaplan et al., 1969) and strongly suggest a diagenetic origin from bacterial activity. A diagrammatic summary of some of the above discussed precipitation processes is shown in Figs. 10 and 11.

The Red Sea Deposit as an Ore Body

Besides being of scientific interest as an ore-forming system, the Red Sea brine deposits would certainly be economically exploited were they on land (Bischoff and Manheim, 1969). In its present form the deposit bears little resemblance to any ore body with which the author is familiar, particularly those of the sedimentary iron ores. Sedimentary iron ores are not generally associated with base metals, and, on the other hand, base metal deposits most commonly considered syngenetic seldom contain much iron as oxides or clay minerals. It is tempting to speculate about the appearance of the deposit after tectonism and metamorphism have operated, but there are so many variables such as oxidation, reduction, addition or subtraction of components, etc., that the attempt would be pointless.

That portion of the deposit which has been sampled by coring appears to fall into the "syngenetic" category of ore deposits, i.e., bedded deposits forming at the same time as the surrounding sediments. It is, however, interesting to speculate about the

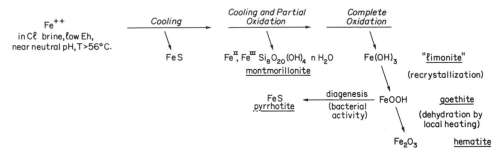

Fig. 10. Sequence of some suggested processes of formation of iron-bearing minerals from the Red Sea brine.

process in the country rock through which the brine must pass immediately prior to discharge into the Atlantis II Deep. If the model of base-metal sulfide precipitation by cooling and releasing of metal from chloride complexes is correct for the Atlantis II deposits, then such sulfides must be precipitating along channels through which the brine must pass, assuming continuously decreasing temperatures. The pathway of the brine must include fissures, faults, brecciated zones and bedding planes

of the carbonate rocks underlying the present sea floor. Garrels (1941) has shown that in such a system, if Pb and Zn have comparable concentrations, PbS will precipitate earlier (at higher temperatures) because of the weaker affinity of Pb for chloride complexing. Little Pb was found in the Atlantis II deposits (Table 2), and this may be due to such a preferential removal process at depth.

Summary of fluid inclusion data (Yermakov, 1965) for desposits of this type indi-

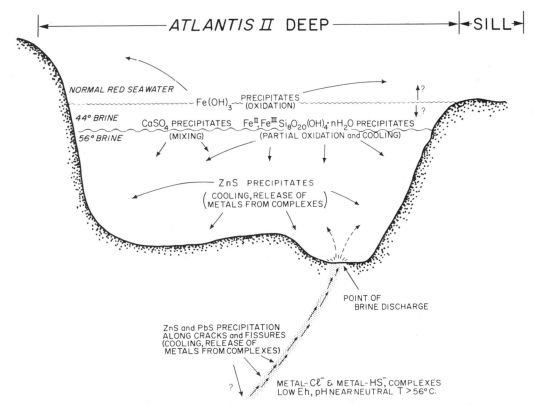

Fig. 11. Schematic representation of some of the suggested processes of mineral precipitation within the Atlantis II Deep.

cates chloride brines to be the mineralizing solutions and temperatures of deposition of around 150°C, conditions which one might expect deeper within the Red Sea geothermal system.

Thus, a model is suggested of epigenetic mineralization at depth within the Red Sea geothermal system with similarities to many of the Mississippi Valley type Pb-Zn deposits.

Acknowledgments

The author is indebted to K. O. Emery and the staff of the USGS-WHOI Continental Margin Program for generously providing laboratory space and equipment and to Mrs. S. McLeod and Mrs. H. Richards for assistance with the analytical work.

Sincere appreciation is due J. C. Hathaway and F. T. Manheim of the U.S. Geological Survey for advice and instruction on analytic techniques and for many hours of fruitful discussion. Completion of the project would have been impossible without their help.

D. E. White of the U.S. Geological Survey deserves special thanks for helpful suggestions as to organization and interpretation of the data.

This work was supported by National Science Foundation grant No. GA-584.

D. E. White, F. T. Manheim and K. O. Emery kindly read and improved the manuscript.

References

Barnes, H. L.: Sphalerite solubility in ore solutions of the Illinois-Wisconsin Districts. *In: Genesis of Strataform Lead-Zinc-Barite-Fluorite Deposits (Mississippi Valley type deposits), a Symposium.* J. S. Brown (ed.). Economic Geology Pub. Co., Monograph 3, New York, p. 326 (1967).

Barnes, H. L. and G. K. Czamanske: Solubilities and transport of ore minerals. *In: Geochemistry of Hydrothermal Ore Deposits*, H. L. Barnes (ed.). Holt, Rinehart and Winston, Inc., New York, 334–378 (1967).

Beales, F. W. and S. A. Jackson: Precipitation of lead-zinc ores in carbonate reservoirs as illustrated by Pine Point Ore Field, Canada. Trans. Canadian Inst. Mining and Metall. Appl. Earth Sci., **76,** B278 (1966).

Bender, M. L., T. L. Ku, and W. S. Broecker: Manganese nodules, their evolution. Science, **151,** 325 (1966).

Bischoff, J. L.: Goethite-hematite stability relations with relevance to sea water and the Red Sea brine

system. *In: Hot brines and recent heavy metal deposits in the Red Sea,* E. T. Degens and D. A. Ross (eds.). Springer-Verlag New York Inc., 402–406 (1969).

——— and F. T. Manheim: Economic potential of the Red Sea heavy metal deposits. *In: Hot brines and recent heavy metal deposits in the Red Sea,* E. T. Degens and D. A. Ross (eds.). Springer-Verlag New York Inc., 535–541 (1969).

Brewer, P. G., C. D. Densmore, R. Munns, and R. J. Stanley: Hydrography of the Red Sea Brines. *In: Hot brines and recent heavy metal deposits in the Red Sea,* E. T. Degens and D. A. Ross (eds.). Springer-Verlag New York Inc., 138–147 (1969).

Brewer, P. G. and D. W. Spencer: A note on the chemical composition of the Red Sea brines. *In: Hot brines and recent heavy metal deposits in the Red Sea,* E. T. Degens and D. A. Ross (eds.). Springer-Verlag New York Inc., 174–179 (1969).

Bricker, Owen P.: Some stability relations in the system $Mn-O_2-H_2O$ at 25°C and one atmosphere total pressure. American Mineralogist, **50,** 1296 (1965).

Conrad, A. L., J. K. Evans, and V. F. Gaylor: Rapid determination of fluorine, sulfur, chlorine and bromine in catalysts with an induction furnace. Analytical Chemistry, **31,** 422 (1959).

Garrels, R. M.: The Mississippi Valley type Pb-Zn deposits and the problems of mineral zoning. Econ. Geology, **36,** 729 (1941).

Gevirtz, J. L. and G. M. Friedman: Deep-sea carbonate sediments of the Red Sea and their implications on marine lithification. Jour. Sed. Petrol., **36,** 143 (1966).

Hardie, L. A.: The gypsum-anhydrite equilibrium at one atmosphere pressure. The American Mineralogist, **52,** 171 (1967).

Hartmann, M. and H. Nielsen: Sulfur isotopes in the hot brine and sediment of Atlantis II Deep (Red Sea). Marine Geol., **4,** 305 (1966).

Hathaway, J. C.: Procedures for clay mineral analyses used in the sedimentary petrology laboratory of the U.S. Geological Survey. Clay Minerals Bull., **3,** 8 (1956).

Helgeson, H. C.: *Complexing and Hydrothermal Ore Deposition.* Pergamon Press, New York-London, 128 p. (1964).

Hem, J. D. and W. H. Cropper: Survey of ferrous-ferric chemical equilibria and redox potentials. U.S. Geol. Survey Water-Supply Paper, **1459-A** (1959).

Hem, J. D. and M. W. Skougstad: Coprecipitation effects in solutions containing ferrous, ferric, and cupric ions. Geological Survey Water-Supply Paper, **1459-E** (1960).

Hemley, J. J., C. Meyer, C. J. Hodgson, and A. B. Thatcher: Sulfide solubilities in alteration controlled systems. Science, **158,** 1580 (1967).

Kaplan, I. R. and S. C. Rittenberg: The microbiological fractionation of sulfur isotopes. *In: Biogeochemistry of Sulfur Isotopes,* M. L. Jensen (ed.). Proceedings of a National Science Foundation Symposium, Yale University, April (1962).

Kaplan, I. R., R. Sweeney, and A. Nissenbaum: Sulfur isotope studies on Red Sea brines and sediments.

In: Hot brines and recent heavy metal deposits in the Red Sea, E. T. Degens and D. A. Ross (eds.). Springer-Verlag New York Inc., 474–498 (1969).

Kolthoff, I. M. and E. B. Sandell: *Textbook of Quantitative Inorganic Analysis,* 3rd ed. The Macmillan Company, New York, 759 p. (1952).

Ku, T. L.: Uranium series isotopes in sediments from the Red Sea hot brine area. *In: Hot brines and recent heavy metal deposits in the Red Sea,* E. T. Degens and D. A. Ross (eds.). Springer-Verlag New York Inc., 512–524 (1969).

——, D. L. Thurber, and G. Mathieu: Radiocarbon chronology of Red Sea sediments. *In: Hot brines and recent heavy metal deposits in the Red Sea,* E. T. Degens and D. A. Ross (eds.). Springer-Verlag New York Inc., 348–359 (1969).

Landergren, S., M. William, and B. Rajandi. Analytical Methods. *In: S. Landergren, On the Geochemistry of Deep Sea Sediments,* Reports of the Swedish Deep Sea Expedition, **10,** Special Investigation No. 5 (1964).

Latimer, W. H.: *Oxidation Potentials,* 2nd ed. Prentice-Hall, Englewood Cliffs, N.J., 392 p. (1952).

Lebedev, L. M.: O sovremennom otlozhenii samarodnovo svintza iz termal'nykh rassolov Chelekena (On contemporary deposits of native lead from the thermal brines of Cheleken). Dokl. Akad. Nauk SSSR, **174,** 197 (1967a).

——: Sovreminnoye obrazovaniye sfalerita v proizvodstvennykh sooruzheniyakh Chelekenskovo mestorozhdeniya. Dokl. Akad. Nauk SSSR, **175,** 920 (1967b).

Miller, A. R., C. D. Densmore, E. T. Degens, J. C. Hathaway, F. T. Manheim, P. F. McFarlin, R. Pocklington, and A. Jokela: Hot brines and recent iron deposits in deeps of the Red Sea. Geochim. et Cosmochim. Acta, **30,** 341 (1966).

Morey, G. W., R. O. Fourmer, and J. J. Rowe: The solubility of quartz in water in the temperature interval from 25°C to 300°C. Geochim. Cosmochim. Acta, **26,** 1029 (1962).

Parks, G. A.: Aqueous surface chemistry of oxides and complex oxide minerals, isoelectric point and zero point of charge. *In:* W. Stumm, Symposium chairman (chapter 6), *Equilibrium Concepts in Natural Water Systems,* American Chemical Society, Washington, D.C. (1967).

Ross, D. A. and E. T. Degens: Shipboard collection and preservation of sediment samples collected during CHAIN 61 from the Red Sea. *In: Hot brines and recent heavy metal deposits in the Red Sea,* E. T. Degens and D. A. Ross (eds.). Springer-Verlag New York Inc., 363–367 (1969).

Sorby, H. C.: On the microscopical structure of crystals, indicating the origin of minerals and rocks. Quart. Jour. Geol. Soc. London, **14,** 443 (1858).

Strangway, D. W., B. E. McMahon, and J. L. Bischoff: Magnetic properties of minerals from the Red Sea thermal brines. *In: Hot brines and recent heavy metal deposits in the Red Sea,* E. T. Degens and D. A. Ross (eds.). Springer-Verlag New York Inc., 460–473 (1969).

Trüper, H. G.: Bacterial sulfate reduction in the Red Sea hot brines. *In: Hot brines and recent heavy metal deposits in the Red Sea,* E. T. Degens and D. A. Ross (eds.). Springer-Verlag New York Inc., 263–271 (1969).

Watson, S. W. and J. B. Waterbury: Sterile hot brines of the Red Sea. *In: Hot brines and recent heavy metal deposits in the Red Sea,* E. T. Degens and D. A. Ross (eds.). Springer-Verlag New York Inc., 272–281 (1969).

White, D. E.: Environments of generation of some base-metal ore deposits. Econ. Geol., **63,** 301 (1968).

White, D. E., E. T. Anderson, and D. K. Grubbs: Geothermal brine well: mile deep drill hole may tap ore-bearing magmatic water and rocks undergoing metamorphism. Science, **139,** No. 3558, 919 (1963).

Yermakov, N. P. *Research on the Nature of Mineral Forming Solutions.* Trans. by V. P. Sokoloff. Edwin Roedeer (ed.). Pergamon Press, N.Y. (1965).

Zen, E-An: Solubility measurements in the system $CaSO_4$-$NaCl$-H_2O at 35°, 50°, and 70°C and one atmosphere pressure. Jour. Petrology, **6,** 124 (1965).

Goethite-Hematite Stability Relations with Relevance to Sea Water and the Red Sea Brine System *

JAMES L. BISCHOFF †

Woods Hole Oceanographic Institution
Woods Hole, Massachusetts

Abstract

Goethite is the dominant Fe^{+3} bearing mineral in all cores from the Red Sea brine deposits with the exception of core station 84 within the Atlantis II Deep where hematite has been transformed from original goethite and limonite.

Review of existent thermochemical data allows construction of an approximate stability diagram relating fields of goethite and hematite $+ H_2O$ to T and a_{H_2O} at 1 to 200 atmospheres total pressure, conditions relevant to sea water and the Red Sea brines.

All conceivable "normal" submarine environments and the Atlantis II brines plot well within the goethite-H_2O stability field. Transformation of goethite to hematite at station 84 appears to result from a local temperature anomaly, suggesting a close proximity to a zone of brine discharge.

Introduction

The goethite-amorphous facies of the Red Sea brine deposits is the most geographically widespread of the facies observed, occurring throughout the Atlantis II and Discovery Deeps, and also several miles outside the immediate geothermal area. The facies evidently forms as a result of discharging Fe^{++} rich, oxygen-poor brine mixing with the overlying oxygen-rich sea water. As oxidation of Fe^{++} to Fe^{+++} occurs, goethite and amorphous limonite are precipitated and settle to the bottom. Goethite associated with limonite is the dominant Fe^{+++} bearing mineral throughout the "goethite" facies, except for core station 84 where goethite and limonite have evidently completely transformed to well-crystallized hematite (Bischoff, 1969). Thus, a study of the stability conditions of goethite and hematite in saline environments was prompted in order to understand bottom conditions leading to transformation.

Thermochemical Data

The transformation of goethite to hematite in water can be written as

$$2FeOOH \text{ (goethite)} = Fe_2O_3 \text{ (hematite)}$$
$$+ H_2O \text{ (liquid)} \quad (1)$$

and all data and discussion to follow will refer to this reaction.

The phase relations of hematite-goethite have been studied for many years and are still apparently unresolved, a consequence of the extremely sparing solubility of the two minerals and, hence, their reluctance to react. Fish (1966) has recently reviewed the subject in detail.

The relevant variables for goethite-hematite stability for sea water and Red

* Woods Hole Oceanographic Institution Contribution No. 2199.

† Present address: Dept. Geological Sciences, University of Southern California, Los Angeles, California.

Sea brine are temperature and activity of water at pressures of 1 to 200 atmospheres. The activity of water (a_{H_2O}) is important as H_2O appears in the reaction, and its activity is decreased by dissolved salts, as in sea water and the Red Sea brine. 200 atmospheres approximates the pressure at 2,000m in sea water, the approximate depths of the Atlantis II Deep brine layers.

Fish (1966) summarizes the existent data into four cases, each with a combination of three measured thermochemical quantities sufficient to calculate the free energy of reaction (1) at variable pressure and temperature. In this work Fish's latter two categories are considered as most valid and will be used, along with a third category using an analogy to the corundum-diaspore transformation.

Category A (ΔH°_{593}, $\Delta C_{p(a)}$, T_{eq})

ΔH°_{593} is taken from the work of Sabatier (1954) who, using DTA (differential thermal analysis), determined the enthalpy of the reaction

$$2FeOOH \rightarrow Fe_2O_3 + H_2O \text{ (gas)} \quad (2)$$

to be +16,000 calories per mole hematite.

Hüttig and Garside (1929) determined the heat capacity of goethite to be 18.2 cal deg^{-1} $mole^{-1}$. Combining this with heat capacities of water and hematite (Rossini et al., 1952) yields ΔC_p for the reaction as +6.6 cal deg^{-1} $mole^{-1}$.

ΔH°_{298} for reaction (1) is then calculated, using the heat of vaporization of water, and ΔC_p by:

$$\Delta H_T = \Delta H^\circ_{298} + \Delta C_p (T\text{-}298) \quad (3)$$

and assuming constant ΔC_p,

$$\Delta H^\circ_{298} = 6,500 \text{ cal mole}^{-1}$$

Other values of ΔH°_{298} quoted by Fish range from 3,500 to 7,400 cal $mole^{-1}$ and show the uncertainty of the above value.

The value of T_{eq}, the temperature of which goethite and hematite in pure water are at equilibrium, is also uncertain. A value at 130°C is taken from the work of Posjnack and Merwin (1922) who found that goethite tended to precipitate from solutions below 130°C and hematite above. Smith and Kidd (1949) report the transformation of synthetic hematite in H_2O to goethite at 108°C, indicating a lower limit of T_{eq}. They also report the transformation of synthetic goethite in pure water to hematite at 141°C – giving an upper limit to T_{eq}. Thus, Posjnack and Merwin's temperature of 130°C is taken as a valid approximation.

Category B (ΔS°_{403}, $\Delta C_{p(b)}$, T_{eq})

ΔS°_{403} is taken from Schmalz (1959) who studied the transformation of amorphous $Fe(OH)_3$ in water to hematite and to goethite at temperatures of 138 to 179°C and pressures of 800–974 bars. The resulting boundary between hematite and goethite at these temperatures and pressures projected convincingly to the 130°C temperature of Posjnack and Merwin (1922). Using the slope of this projected boundary line, ΔV of reaction and the Clapeyron-Clausius equation, Schmalz estimated ΔS°_{403} to be +3.7 cal deg^{-1} $mole^{-1}$.

It must be pointed out that the boundary Schmalz determined may be merely a kinetic boundary, as the transformation of amorphous $Fe(OH)_3$ to both hematite and goethite is characterized by a decrease in free energy at all the temperatures and pressures considered. Either mineral could, thus, form metastably. The good projection of the boundary to the equilibrium T of Posjnack and Merwin (1922) and Smith and Kidd (1949) suggest, however, the possible validity of Schmalz's boundary as a true univariant line.

As in category A, T_{eq} is set at 130°C. ΔC_p is taken as +5.6 cal deg^{-1} $mole^{-1}$ estimated by Schmalz using an analogy to the corundum-diaspore pair.

Category C (ΔH°_{593}, ΔC_p, ΔS°_{298})

ΔH°_{593} is the same as in category A and ΔC_p as in category B. Since the transformation of diaspore to corundum is structurally analogous to the goethite to hematite, the entropy changes are probably almost the same. Thus, ΔS°_{298} of the diaspore-corundum transformation, +12 cal/deg mole, is used (Rossini et al., 1952).

Construction of T-a_{H_2O} Stability Diagram

Standard Pressure

The diagram is first constructed for 1 atmosphere. T_{eq} is chosen as 130°C for cases A and B, and for these only the equilibrium a_{H_2O} need be calculated.

With the solid phases at unit activity and under conditions of variable a_{H_2O}, the following equation holds for reaction (1):

$$\Delta G_{a_{H_2O}} = \Delta G_{298}^\circ - RT \ln a_{H_2O} \quad (4)$$

The equilibrium constant K for reaction (1) is simply the activity of water at equilibrium with both hematite and goethite at 25°C, 1 atm pressure. Thus, at equilibrium,

$$\Delta G_{a_{H_2O}} = 0$$

and

$$\Delta G_{298} = RT \ln a_{H_2O} = RT \ln K \quad (5)$$

The first step then is to evaluate ΔG_{298} for both sets of data.

Category A

ΔS_{298}° is needed to calculate ΔG_{298}° from the equation:

$$\Delta G_{298}^\circ = \Delta H_{298}^\circ - T\Delta S_{298}^\circ \quad (6)$$

At T_{eq},

$$\Delta G_{403}^\circ = 0$$

and

$$\Delta H_{403}^\circ = 403 \ (\Delta S_{403}^\circ) \quad (7)$$

using equation (3) to solve for ΔH_{403}° and substituting into (7),

$$\Delta S_{403}^\circ = +17.75 \text{ cal deg}^{-1} \text{ mole}^{-1}$$

ΔS_{298}° from ΔS_{403}° is calculated from the relations,

$$\Delta S_T^\circ = \Delta S_{298}^\circ - \Delta C_p \ln \frac{298}{T}$$

$$\Delta S_{298}^\circ = +15.73 \text{ cal deg}^{-1} \text{ mole}^{-1} \quad (8)$$

Substituting into (6) and then (5)

$$a_{H_2O} \text{ (equil)} = 0.04$$

Now such a low activity of water is out of the range of natural aqueous solutions and can be visualized only in terms of relative humidity. Although saturated NaCl brines at 25°C have a_{H_2O} at 0.75, 0.04 is plotted as a hypothetical point on the a_{H_2O} coordinate. A line drawn between T_{eq} at 130°C and a_{H_2O} of 0.04 then is an estimate of the univariant line at 25°C according to data in category A (Fig. 1).

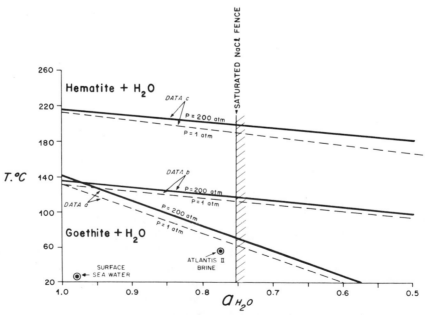

Fig. 1. Calculated stability relations of the system; goethite-hematite-H$_2$O as a function of temperature and activity of water.

Category B

Substituting ΔS°_{403} and ΔC_p into (8) yields

$$\Delta S^\circ_{298} = +2.0 \text{ cal deg}^{-1} \text{ mole}^{-1}$$

solving for H°_{403} from (7) and substituting into (3)

$$\Delta H^\circ_{298} = +904 \text{ cal mole}^{-1}$$

and substituting into (6) and then (5)

$$a_{H_2O} \text{ (equil)} = 0.6$$

As in category A, T_{eq} and a_{H_2O} are connected by a straight line (Fig. 1).

Category C

Equilibrium temperature is calculated by (6), and the equation,

$$\left(\frac{\partial \Delta G}{\partial T}\right)_P = -\Delta S \qquad (9)$$

expressing ΔS as in (8). Substitution into (9), integration, and solution for T yields, $T_{eq} = 214°C$.

Substitution into (5) yields

$$a_{H_2O} \text{ (equil)} = 0.007$$

Again, T_{eq} and a_{H_2O} are connected by a straight line (Fig. 1).

Pressure Effect on T_{eq}

The value of T_{eq} at $P = 200$ atm is calculated from the Clapeyron-Clausius equation,

$$\frac{dP}{dT} = \frac{\Delta S'}{\Delta V'} \qquad (10)$$

Assuming the univariant line to be linear for the pressure range concerned and using ΔS°_{403} and ΔV_{403} for $\Delta S'$ and $\Delta V'$,

$$\Delta T = \frac{(196)\Delta V_{403}}{\Delta S_{403}}$$

ΔV_{403}, taken as $+8.23$cc is calculated using molar volumes of hematite and goethite at *STP* and the molar volume of water at 130°C and 4 atm pressure. Then, for category A,

$$\Delta T = +2.25°C$$

$$T_{eq} \text{ 200 atm} \cong 135°C$$

for category B,

$$\Delta T = 10.7°C$$

$$T_{eq} \text{ 200 atm} \cong 141°C$$

for category C,

$$\Delta T = 2.5°$$

$$T_{eq} \text{ 200 atm} \cong 217°C$$

Pressure Effect on a_{H_2O}

An estimate of equilibrium a_{H_2O} at $P = 200$ atm requires the relations

$$\left(\frac{\partial \Delta G}{\partial P}\right)_T = \Delta V \qquad (11)$$

which becomes, assuming constant ΔV for the pressure range considered;

$$\Delta\Delta G_P = \Delta G^\circ_{298} + \Delta V(P\text{-}1) \qquad (12)$$

In this case ΔV is taken as $+7.05$cc using the molar volume of water at 25°.

Category A

Substituting into (12) and then (5)

$$a_{H_2O} \text{ (200 atm)} = 0.037$$

Category B

Doing the same as above

$$a_{H_2O} \text{ (200 atm)} = 0.575$$

Category C

$$a_{H_2O} \text{ (200 atm)} = 0.0007$$

These new values are then plotted as before, and the approximate univariant lines constructed (Fig. 1).

Discussion

If the thermochemical data used can be considered approximately correct, particularly T_{eq}, several conclusions can be drawn from Fig. 1. Increase in pressure expands the stability field of goethite, and, consequently, normal submarine environments will always plot within the goethite field, particularly at depth and lower temperatures.

The 56°C Atlantis II brine plots close to the NaCl saturation fence and closer to the estimated univariant lines than normal sea water, but it is still well within the goethite field by either set of data. The preponderance of goethite and total lack of hematite (except at 84K) throughout the Atlantis II Deep tends to support the conclusion that goethite is the stable phase under the general conditions.

The implication, therefore, of the transformed hematite in core 84K is that local temperature anomaly exists, and is sufficiently high to be in the hematite field, and for transformation to occur in less than 5,000 years (the oldest probable age for material at the bottom of core 84K (Ku, 1969).

This conclusion prompted some simple experiments. Three glass ampules were prepared with 200mg charges, each of Red Sea goethite (core 127P), Red Sea hematite (core 84K) and Red Sea amorphous limonite. Each was fuse-sealed with 10ml of Atlantis II brine. The hematite run was kept at 50°C while the goethite and the amorphous runs were kept at 115°C. After several days the amorphous material had transformed to hematite. The goethite and hematite runs, however, remained dormant, even after several months.

The crystallization of the amorphous limonite to hematite in such a short time means that the "goethite" facies of the Atlantis II Deep has never been at temperatures approaching 115°C; otherwise, the amorphous material would have transformed to hematite—an irreversible reaction.

Conclusions

A temperature anomaly is implied to exist in the vicinity of station 84, and the remaining areas have remained relatively cool, considerably below 115°C.

Station 84 is located in the deepest part of the Atlantis II Deep, and may be near the discharge source (or one of the sources) of the brine. The subterranean brine must be considerably hotter at depth judging from abundant brine-derived amorphous silica in the Atlantis II deposits (Bischoff, 1969), and one would expect that the point of discharge be an area of anomalous temperature.

Acknowledgments

The author is grateful for helpful advice and encouragement from F. T. Manheim of the U.S. Geological Survey. Thanks are due to W. Fyfe of Manchester University and I. Kaplan of the University of California for criticizing the manuscript. Financial support was provided by NSF Grant GA-584.

References

Bischoff, J. L.: Red Sea goethermal brine deposits, their mineralogy, chemistry and genesis. *In: Hot brines and recent heavy metal deposits in the Red Sea,* E. T. Degens and D. A. Ross (eds.). Springer-Verlag New York Inc., 368–401 (1969).

Fish, F. F., Jr.: The stability of goethite on Mars. Jour. Geophys. Res., **71**, 3063 (1966).

Hüttig, G. F. and H. Garside: Zur Kenntnis des Systems Eisenoxyd-Wasser. Z. Anorg. Allgem. Chem., **179,** 49 (1929).

Ku, T. L.: Uranium series isotopes in sediments from the Red Sea hot brine area. *In: Hot brines and recent heavy metal deposits in the Red Sea,* E. T. Degens and D. A. Ross (eds.). Springer-Verlag New York Inc., 512–524 (1969).

Posjnack, F. D. and H. E. Merwin: The system Fe_2O_3—SO_3—H_2O. J. Am. Chem. Soc., **44,** 1965 (1922).

Rossini, F. D., D. D. Wagman, W. H. Evans, S. Levine, and I. Jaffe: Selected values of chemical thermodynamic properties. Nat. Bur. Stand. U.S. Circ. 500 (1952).

Sabatier, G.: La mesure des chaleurs de transformation à l'analyse thermique différentielle. Bull. Soc. Franc. Mineral. Crist., **77,** 1077 (1954).

Schmalz, R. F.: A note on the system Fe_2O_3—H_2O. Jour. Geophys. Res., **64,** 575 (1959).

Smith, F. G. and K. J. Kidd: Hematite-goethite relations in neutral and alkaline solutions under pressure. Amer. Min., **34,** 403 (1949).

Chemical Composition of Sediments and Interstitial Brines from the Atlantis II, Discovery and Chain Deeps

RUTH L. HENDRICKS, FREDRIC B. REISBICK, EDWIN J. MAHAFFEY
AND D. BLAIR ROBERTS

Kennecott Copper Corporation, Exploration Services
Salt Lake City, Utah

MELVIN N. A. PETERSON

Scripps Institution of Oceanography
LaJolla, California

Abstract

A total of 110 sediment samples and 58 interstitial water samples have been analyzed by wet chemical and optical spectrographic methods. Sample handling and storage procedures included the use of inert atmosphere and refrigeration to inhibit any changes in the samples prior to analysis. Most of the interstitial waters are essentially saturated with respect to sodium chloride and are enriched in iron, copper, zinc and other heavy metals, as reported by other workers. The interstitial waters obtained from sediments around the edges of the deeps show more similarity to normal sea water.

The sediment samples show a wide variation in physical appearance and in chemical composition, from oxidized red muds to dark sulfide-rich muds and a few light-colored carbonate-rich samples. Extreme variations in composition are found between adjacent sample sites, and considerable variation is found even within different splits from the same core. Vertical variations in the cores are also extreme, sometimes within a few centimeters of depth. The moisture content of the sediment samples, as received, varied widely with an average of about 55 per cent by weight. The *in situ* water content was higher. The water soluble salt portion of the sediment was not determined separately, but it made up more than half of the solid sediment in many cases. Some of the sulfide-rich parts of this deposit contain high concentrations of zinc, copper and other heavy metals, but the average metal content of the deposits is much lower than would be indicated by these few high values.

Introduction

Samples received at Kennecott Exploration Services from the Atlantis II, Discovery, and Chain Deeps were catalogued into 110 sediment samples and 58 interstitial water samples. All were stored at −20°C until the time of analysis. The purpose of the analytical work was twofold:

1. to determine the concentrations of economically important metals in the deposits.

2. to provide data for study of the chemistry of the deposits, with a view toward understanding the origin of comparable occurrences in other parts of the world.

Sediment samples were in plastic bags which had been placed in one-quart glass jars at the time of collection, under a non-oxidizing atmosphere. Where samples had been squeezed and the interstitial water preserved, the water portion was in a polyethylene bottle in the jar containing the squeezed sediment. With the exception of a few small samples, a portion of each water and sediment sample remains in storage in a nitrogen atmosphere at −20°C for future reference.

In order to obtain a maximum of reliable data, all samples plus selected duplicates were analyzed both by spectrographic and wet analytical methods. This also provided basis for a determination of the accuracy of procedures used. Spectrographic analyses of the waters were made on evaporated residue of one milliliter for each sample, with results calculated back to the water sample. These analyses included determination of 24 elements per sample. In addition, 23 elements were specifically sought but not detected in any sample. Wet chemical analyses of the water included 19 determinations on each sample, plus 2 additional (cobalt and nickel) on selected samples. The spectrographic analyses of the sediments include 36 determinations on each sample, and, in addition, 13 elements were sought but not detected in each sample. The wet chemical analyses include 28 determinations on each sample, plus determination of moisture loss to 50°C on all samples and four other determinations (fluorine, lithium, borate and carbonate) on selected samples. Both brine and sediment samples contained a higher percentage of light metal salts than normally encountered in geochemical analyses. This necessitated considerable research to adapt procedures normally used on rock and water samples. New procedures were developed where necessary. The location of the cores is given in Ross and Degens (1969).

Spectrographic Analysis

Red Sea brine and sediment samples were analyzed by semiquantitative, DC arc emission spectrographic procedures. A 3-meter modified Eagle mount concave grating spectrograph was used in the analyses. This spectrograph, employing a 30,000 lines per inch grating, provides a desirable high reciprocal linear dispersion of 2.7Å/mm in the first order. Spectra were recorded photographically and interpreted densitometrically, using working curves established by standards of known composition. Analytical methods for the brine samples and sediment samples are described separately.

Interstitial Brine Samples

Frozen volumes of interstitial brines were thawed and sampled. Sample preparation and analytical standards are described and the method of analysis is outlined.

Sample Preparation. Milliliter aliquots of the brine samples were evaporated under heat lamps to residue salts. Evaporation was performed in crucibles which were lined with a commercial plastic wrapping material to prevent adherence of the residue to the crucible. The brine residues were then transferred to glass vials and dried overnight at 110°C. The dried residues were weighed, and the weights recorded as grams residue on evaporation (R.O.E.). Prior to analysis, the dry residues were ground to a fine powder in an agate mortar.

Analytical Standards. Analytical standards were prepared by adding weighed amounts of elements, as oxides or carbonate salts, to a pure sodium chloride base. The standards were prepared in such a way that no standard contained more than four chemical elements, and the elements combined in a given standard were selected for their relative purity to minimize the possibility of trace element contamination. The initial standards contained 1.0 per cent by weight of each element, and successive dilutions were made to provide a graded series of concentrations varying by a factor of 1/3. A commercial mixing device was used to insure homogeneity of the standard dilutions.

Method of Analysis. Twenty milligram portions of the standards were mixed with 10 milligrams of SP-1 grade graphite. The mixture was tamped into a shallow-cupped, undercut graphite electrode (Ultra Carbon #1988). Using a 1/8-inch graphite cathode, the standards were arced to completion in two minutes at 12 amps DC. An arc atmosphere of 70 per cent argon, 30 per cent oxygen, delivered at 4.8 l/m, was used to reduce cyanogen band interference and improve sensitivity. A rotating step sector was used in recording the spectra on Kodak SA-1 plates. Spectrographic operating conditions are given in Table 1.

Analytical line intensities were determined densitometrically. Analytical working curves were established by plotting optical density against log concentration, after replicate arcings of the standards. Analysis line wavelengths and limits of detection for elements determined in the brine residues are given in Table 2.

The brine residue samples were analyzed in the same manner as the standards. Analysis line intensities were determined and element concentrations, as parts per million in residue, were obtained from the standard curves.

Analytical Results. The element concentrations in parts per million in residue were multiplied by the weight, in grams, of residue on evaporation in order to convert analytical values to mg/l in the brines. Results of the brine analyses are given in Table 5.

In addition to those elements reported in Table 5, the following elements were sought but not detected: As, Be, Bi, Co, Ga, Hf, In, La, Li, Nb, P, Pd, Pt, Re, Sc, Ta, Te, Tl, Th, W, U, Yb and Y. The elements Sb, Cd, Ce and Zr, detected in several samples, are reported only as trace amounts because of their extremely low

concentration. Vanadium is not reported because of calcium interference.

Sediment Samples

Representative sample splits were taken from previously dried sediments. The sediment samples were analyzed by two semi-quantitative spectrographic procedures and determinations were made for 49 elements. Sample preparation, the standards used and analytical methods are described below.

Sample Preparation. Sediment samples, previously dried at 50°C and pulverized, were mixed by shaking, and representative sample splits were obtained by means of a vibrating splitting device. Residual brine salts were not leached from the samples. Prior to analysis, samples were ground to −325 mesh and dried for an additional 15 hours at 110°C. Additional sample weight loss in drying from 50° to 110°C was, on the average, only about 2 per cent, and can be considered insignificant for all practical purposes.

Sediment sample colors varied from green to red, indicating wide differences in sample composition. Because of this variation, and the presence of brine residues, the samples were diluted with purified silica in the ratio of one part sample to one part silica in order to obtain a more uniform matrix.

Analytical Standards. Because of interest in determining as many elements as

Table 1 Spectrographic Operating Conditions

	Interstitial Brines	Sediments
Electrodes:		
Sample (anode)	Ultra Carbon #1988	Ultra Carbon #1988 or National #L4018
Counter	Ultra Carbon #7259	Ultra Carbon #7259
Analytical Gap, mm	2.5	2.5
Current, DC amps	12	12 or 10
Slit Width, microns	25	25
Sector Transmission, 3 step	50%:12.5%:3.125%	50%:12.5%:3.125% or 100%:25%:6.25%
Atmosphere	70% A:30% O_2 @ 4.8 l/m	70% A:30% O_2 @ 4.8 l/m or 70% A:30% O_2 @ 1 1/m
Wavelength Range (Å)	to include 2200–3600	to include 2200–3600
Photographic Emulsion	Kodak SA-1	Kodak SA-1
Development	Kodak D19, 4 min @ 23°C	Kodak D19, 4 min @ 23°C

Table 2   Spectrographic Wavelength and Limits of Detection

	Interstitial Brine Residues		Sediments	
Element	Wavelength (Å)	Detection Limit (wt. % in residue)	Wavelength (Å)	Detection Limit (wt. % in sediment)
Al	2378.4	0.03	2378.4	0.4
	3082.2	0.01		
Sb	*	*	2877.9	0.002
As			2288.1	0.004
Ba	2335.3	0.01	2335.3	0.06
Be			3130.4	0.0002
Bi			3067.7	0.0002
B	2496.8	0.01	2496.8	0.004
Cd			2288.0	0.00004
Ca	2994.9	0.01	3181.3	0.2
Ce			3201.7	0.1
Cr	2843.2	0.001	2843.2	0.002
Co			3453.5	0.0004
Cu	3247.5	0.0001	3274.0	0.001
			2824.4	0.04
Ga			2943.6	0.0004
Ge			3039.1	0.00002
Au	2675.9		2675.9	0.00008
Hf			3134.7	0.03
In			3256.1	0.00004
			3039.3	0.00004
Fe	3021.1	0.01	2843.6	0.2
	2843.6	0.03		
La			3337.5	0.006
Pb			2833.1	0.0002
			2663.2	0.08
Li			3232.6	0.04
Mg	2776.7	0.01	2776.7	0.02
	3329.9	0.3		
Mn	2799.8	0.001	2798.3	0.002
	3070.3	0.01	3070.3	0.1
Mo	3170.3	0.0001	3170.3	0.001
			3194.0	0.001
Ni	3050.8	0.0001	3050.8	0.0006
	3414.8	0.0001		
Nb			3195.0	0.01
P			2554.9	0.06
Pd			3404.6	0.0002
Pt			2659.4	0.003
K	3446.7	0.3	3446.7	0.2
Re			3460.5	0.001
Sc			3353.7	0.001
Si	2519.2	0.01	2568.6	
Ag	3382.9	0.00001	3280.6	0.00003
Na			3302.3	0.02
Sr	3380.7	0.03	3464.5	0.04
	3464.5	0.01		
Ta			3311.2	0.2
Te			2385.7	0.001
Tl			2767.9	0.0002
Th			2870.4	0.1
Sn	3175.0	0.001	3175.0	0.00006
Ti	3168.5	0.001	3168.5	0.01
W			2947.0	0.0002
U			2860.4	0.2
V			3184.0	0.002
Yb			3464.4	0.003
Y			3327.9	0.004
Zn	3345.0	0.003	3345.0	0.002
			3345.6	0.6
Zr			3273.0	0.004

* No detection limits established for elements sought but not found.

possible in the sediment samples, and the inherent difficulty in preparing reliable synthetic standards, the analytical curves for most elements were established by natural geologic standards. The standards used include various dilutions of: the U.S. Geological Survey standard granite G-1 and standard diabase W-l; the several standard ores and rocks obtainable from the National Bureau of Standards; and several natural rock and synthetic standards developed by this laboratory.

Method of Analysis. Two methods of analysis were used: a total energy method by which the sample is completely volatilized; and, a second method employing a larger quantity of sample and selective volatilization to obtain increased analytical sensitivity for the elements Sb, As, Bi, Cd, Ge, Au, In, Pb, Re, Te, Tl, Sn, W and Zn. Analysis line wavelengths and limits of detection for the elements determined are given in Table 2.

The total energy method was used to determine 48 elements. A 5mg sample was intimately mixed with 15mg of a buffering agent comprised of SP-1 grade graphite containing 12.5 per cent GeO_2. The mixture was tamped into a shallow-cupped, undercut electrode (Ultra Carbon #1988) and arced to completion at 12 amps DC in 1.5 minutes. An arc atmosphere of 70 per cent A : 30 per cent O_2, delivered at 4.8 1/m, was used to reduce background interference and stabilize the arc. Spectrographic operating conditions are given in Table 1.

Increased analytical sensitivity was achieved for several elements by a method which records only the early period of arcing when certain elements are selectively volatilized from a large quantity of sample. 90mg of sample were mixed with 15mg of a carrier buffer comprised of equal amounts of NaCl, Na_2CO_3 and elemental sulfur. The mixture was loaded into a deep-cupped electrode (National #L4018), and arced for one minute at 10 amps DC in an atmosphere of 70 per cent A : 30 per cent O_2 delivered at 1 l/m.

Spectra were recorded photographically and analytical line intensities were determined with a densitometer. Analytical curves for both methods were established by plotting optical density against log concentration for multiple arcings of the standards. Internal standards and emulsion calibration were not used. Photographic plates from the same emulsion lot were used for all sample analyses, and developer temperature was controlled to ± .3°C.

Two standards were included with the 12 samples on each photographic plate. Element determinations were adjusted by a correction factor whenever deviation of the standards indicated variation in the spectrographic operating conditions.

Analytical Results. Dilution of the sediments with silica was designed to duplicate, as nearly as possible, the matrix of the standards used for analysis. This dilution, coupled with the high buffer to sample ratio used in the total energy method, should have essentially eliminated the so-called matrix effects caused by variations in the sample composition.

Results of the semiquantitative sediment analyses are given in Table 6. Data are corrected for sample dilution and reported as weight percent in sediment.

Chemical Analysis

Interstitial Brine Samples

Eleven of the 58 brine (interstitial water) samples were comparable to normal sea water in terms of salinity. These were frozen when in storage at −20°C. The remaining 47 samples were nearly saturated with sodium chloride. Ten contained a ferric oxide precipitate; 13 contained a crystalline precipitate which dissolved upon warming to room temperature and agitating.

Splitting was done in a glove-bag filled with nitrogen. Thirty milliliters of each sample was placed in a polyethylene bottle and returned to cold storage. Aliquots were removed at the same time for ferrous iron and cuprous copper determinations. Complexing agents were added at once in the nitrogen atmosphere to prevent valence changes.

Ferrous iron was determined spectrophotometrically as the 1,10-phenanthroline complex. This complexed both ferrous and ferric iron; the former was red-orange, the latter, yellow. The red-orange complex was read at 508mμ. A Beckman model DU spectrophotometer was used for all colorimetric determinations.

Cuprous copper was complexed with neo-cuproine (2,9-dimethyl-1,10-phenanthroline hemihydrate) and extracted into iso-amyl alcohol. Since concentrations were below the range of colorimetric methods, the copper concentration in the iso-amyl alcohol was determined by atomic absorption at 3247.5Å. Neo-cuproine complexes only that copper present as cuprous ion.

Analyses for elements present in an oxidized ionic state were made on undiluted brines. Sulfur present as sulfate ion was determined gravimetrically by precipitation as barium sulfate. Some sulfur was present as sulfide ion, which oxidized to sulfate ion upon standing at room temperature and exposure to air.

Boron present as borate ion was determined spectrophotometrically with carminic acid reagent. The range of the procedure is 1 to 15 micrograms of boron. A wavelength of 585mμ was used.

Phosphorus present as phosphate ion was determined by a molybdovanadophosphoric acid method modified for low ranges. The yellow complex was read at 400mμ. The procedure has an analytical range of 1 to 60 micrograms of phosphorus.

Dissolved silica present in the monomeric state was determined spectrophotometrically by a heteropoly blue method modified for use in a range of 2 to 50 micrograms of SiO_2. A wavelength of 650mμ was used. It was found from spectrographic analysis of dried solids and atomic absorption analysis that monomeric silica comprised a small portion of the total silica present in the brines.

Chloride content was determined by titration with a standardized silver nitrate solution. An Orion specific ion electrode for chloride with a scale expander and recorder was used as an end point indicator. For determination of bromide, interfering iron and manganese were removed by precipitation with calcium hydroxide. Bromide was then oxidized with sodium hypochlorite, potassium iodide added, and the liberated iodine titrated with a standardized sodium thiosulfate solution. The method is accurate to 10ppm bromide with a lower limit of 50ppm.

For the remaining analyses, the brines were acidified to prevent precipitation of metal salts. A one to one dilution with 10 per cent hydrochloric acid was used on the concentrated brines and one part 20 per cent HCl to 10 parts brine for the less saline samples. For atomic absorption analysis, it was necessary to dilute the brines to a point at which the content of solids in solution was less than 6 per cent. A higher concentration of solids quickly clogs the aspiration tube and burner. It also physically blocks and scatters light from the hollow cathode lamp, thus giving an erroneously high absorption reading.

Table 3 lists elements determined by atomic absorption spectrophotometry, wavelength used, analytical range in solution, fuel used, and reagents added to decrease interference or to increase sensitivity. A Techtron model AA-3 atomic absorption instrument was used for all atomic absorption analyses.

For those elements present in high concentrations, appropriate dilutions were made to put the concentration in analytical range. For the determination of silver, an extraction into triisooctyl thiophosphate (TOTP) and benzene was made. This eliminates interference and increases sensitivity. The TOTP benzene extract was diluted 60 per cent with reagent grade ethanol to improve combustion.

All standards were in a solution of sodium chloride comparable to the sodium chloride concentration of the brine samples as diluted. On about 13 samples, determinations requiring a large amount of sample were not made due to an inadequate amount of sample. The usual amount of sample available was 50ml; a smaller amount was available on many samples. These data are shown in Table 7.

Sediment Samples

The sediment samples were brought to room temperature in the jars before splitting in a nitrogen atmosphere. Any brine which

Table 3   Operating Conditions for Atomic Absorption Determinations—Interstitial Waters

Element	Wavelength (Å)	Fuel	Sol'n. Range (ppm)	Additive	
Fe	2483	Air-Acetylene	.1–40		
Cu	3248	Air-Acetylene	.05–3		
Na	5890	Air-Acetylene	1–5		
K	7665	Air-Acetylene	1–5		
Ca	4227	Air-Acetylene	1–5	1% La	To eliminate interference from P, Al, and Si
Mg	2852	Air-Acetylene	.1–5	1% La	
Sr	4607	Nitrous oxide-Acetylene	1–5	1% La	
Mn	2795	Air-Acetylene	.05–30		
Zn	2138	Air-Acetylene	.05–10	40 to 50% ethyl alcohol	To improve sensitivity and noise suppression
Al	3093	Nitrous oxide-Acetylene	1–20		
Co	2407	Air-Acetylene	.1–5		
Ni	2320	Air-Acetylene	.1–5		
Ag	3281	Air-Acetylene	.01–5		

had leaked out of the plastic bag and into the jar was discarded. A number of the jars were not tightly sealed, and contamination or oxidation was suspected.

Each sample was mixed as thoroughly as possible in the plastic bag before splitting. In many cases this was difficult because of the glutinous consistency of the muds. Those with a high water content settled rapidly, thus necessitating constant mixing throughout the splitting process. Some muds were visibly streaked and heterogeneous. It must be realized that splits of these samples are not representative in every case.

Splits were made in nitrogen and placed in preweighed containers for cuprous copper, ferrous and total HCl soluble iron, water soluble sulfate, carbonate and percent water determinations. A sample of approximately one hundred grams was taken to be dried for total determinations. With the exception of the hundred-gram portion, splits were removed in closed containers and weighed. They were then returned to the nitrogen atmosphere for addition of reagents. Percent water was determined by weight loss upon drying at 50°C. The hundred-gram split was dried at 50°C. When analyses were made on wet samples, results were calculated and reported on the basis of dry sample weight.

Samples numbered 84K 400cm, catcher, (1) and (2) were from two jars with the same label. As can be seen from the analyses, there is some difference in the samples. Samples numbered 97G, bottom of core, (A) and (B) were packaged in the same jar but in separate plastic bags. Since they were obviously different sediments, they were dried and analyzed separately.

After all sediment samples had been split, four were picked to resplit in an effort to check the adequacy of splitting techniques and procedures, and the stability of oxidation conditions. Some surprising trends appeared in the oxidation states of these samples. In all four, the total HCl soluble iron checked well with the original split. The ferrous iron content, however, was significantly higher in all four samples. Whether this reduction reaction had been taking place during the time the samples were stored or during the time they were warmed for splitting is not known.

In all four samples, the water soluble sulfate values were lower in the repeat splits. This may indicate either another reduction reaction or formation of an insoluble sulfate. Cuprous copper showed no significant change. Percent water checked quite well, indicating reasonably good splitting techniques. These muds were all more than 65 per cent water. Total analyses, however, showed some difference in composition when compared with the original splits.

Oxidation state analyses were more difficult in the case of the sediment samples than the brines since it was necessary either to dissolve the desired species selectively or to effect dissolution without producing a valence change. In the case of iron, a cold, 50 per cent HCl dissolution was used. This dissolved iron minerals, with the exception of some sulfides, without oxidation or reduction of the iron. 1,10-phenanthroline spectrophotometric determinations were made for both ferrous and total HCl soluble iron at 508mμ.

The most successful dissolution for the determination of cuprous copper was a leach with neo-cuproine. This took all cuprous copper into solution as the cuprous neo-cuproine complex. Native copper, if present, was dissolved; cupric copper was unaffected. Approximately 24 hours were necessary for a total leach. In cases of low cuprous content, the complex was extracted into iso-amyl alcohol and determined by atomic absorption at 3247.5Å. In those sediments of high cuprous content (leach solution of orange color), the complex was extracted into chloroform and determined spectrophotometrically at 457mμ.

For determination of water soluble sulfate, the samples were water washed, agitated and filtered. This was done in a nitrogen atmosphere. Sulfate in the filtrate was determined gravimetrically by precipitation as barium sulfate.

Carbonate was determined gasometrically by a modified Orsat method involving the evolution of carbon dioxide upon addition of HCl. A number of the sediments were above the range of the procedure. Some evolved sufficient H$_2$S to interfere with the determination. Both conditions are designated in the table of analytical results.

All remaining analyses were done on dried sediment. Drying was slow on many samples. Those with high water and sodium chloride content formed a crust which interfered with drying. Dried sediments were ground by hand with a porcelain mortar and pestle.

Total sulfur was determined gravimetrically by precipitation as barium sulfate. A Leco sulfur analyzing apparatus was unsatisfactory for this type of sample. Seven samples were analyzed for total HCl soluble sulfate. The values were found to be considerably higher than water-soluble sulfate, principally due to the presence of anhydrite.

Total silicon was determined on the sediments by the heteropoly blue spectrophotometric method at 650mμ wavelength. Results were reported as SiO$_2$. A sodium hydroxide fusion was used to decompose silicates; the melt was taken up in water and adjusted to the proper pH (1.3) with HCl. The analytical range of the procedure is 25 to 400 micrograms of SiO$_2$. Barium was determined by atomic absorption spectrophotometry in the same fusion solution at 5535.5Å. Lanthanum solution was added to a final concentration of one percent to decrease interference.

A perchloric acid digestion was used for determination of phosphorus. This digestion converts all phosphorus present to phosphate ion which can then be measured spectrophotometrically. Because of interference by the high iron content, it was necessary to modify a heteropoly blue method for phosphorus in steel. The range of the procedure is 20 to 200 micrograms of phosphorus.

Boron present in the sediments as borate was dissolved in concentrated HCl. It was determined spectrophotometrically by the carminic acid method described in the section on brine analysis.

Determination for gold was made using a digestion of HF and perchloric, then *aqua regia*. Gold was precipitated as the telluride and finally extracted into methyl isobutyl ketone for atomic absorption determination at 2428Å.

Determinations of gold, platinum, and palladium were also made on four composites made from 51 sediment samples collected and dried on board ship by M. N. A. Peterson. These were prepared by *aqua regia* digestion, chemical preconcentration and emission spectrographic analysis of the concentrate. The average gold value for the four composites was 0.27ppm Au. Platinum and palladium were not detected (detection limits were 0.01ppm Pd and 0.1ppm Pt).

Determination of water soluble chloride was as described in the section on brine analysis. Determination on a few samples for total chloride showed chloride to be present as water soluble salts. Determination of bromide on a selected few sediments showed any present to be below the range of the procedure.

About one-tenth of the sediments were analyzed for fluoride. Preparation was made with a sodium carbonate fusion. The melt was taken up in water and the pH of the solution adjusted to 7.8 with H_2SO_4. An Orion specific ion electrode was used to determine fluoride concentration. All of the chosen samples were below the optimum analytical range of the procedure, but values are reported for nine samples.

The remaining analyses were done by atomic absorption spectrophotometry. A four acid digestion (HCl, HNO_3, HF and $HClO_4$) taken to perchloric fumes was used. The same instrument (Techtron model AA-3) was used, as with brine samples, but with the addition of a Techtron electronic digital readout. The digital readout, considered far superior to the meter with which the instrument is originally equipped, greatly enhanced sensitivity and precision of all analyses. Table 4 lists elements determined by atomic absorption spectrophotometry and wavelength, fuel, analytical range and additives used.

Suppression was encountered in determination of sodium at ranges above five ppm in solution. This difficulty was eliminated by dilution to the one to five ppm range.

A similar problem was encountered in calcium determination. Even with the addition of lanthanum solution to eliminate interference from P, Al and Si, suppression was severe at solution concentrations above 2ppm Ca. Varying magnesium concentrations affected calcium sensitivity and necessitated matching sample and standard Ca/Mg ratios. High sulfur concentrations in many samples appeared to cause additional suppression, presumably due to the formation of sulfate during digestion.

Duplicate silver analyses were made on 14 sediment samples for comparison with spectrographic and aqueous phase atomic absorption analyses. The lower analytical limit of aqueous atomic absorption analysis is 5ppm in dried sediments. The 14 samples covered the full range of silver values. They were prepared with a hydrofluoric and nitric acid digestion and extracted and determined as described in the section on brine analysis.

Duplicate digestions were made on about fifteen percent of the sediment samples. Results are listed with original analyses in Table 8. In general, duplicates checked well, indicating adequate sample grinding and mixing as well as suitable digestion methods.

Table 4   Operating Conditions for Atomic Absorption Determinations — Sediments

Element	Wavelength (Å)	Fuel	Sol'n Range (ppm)	Additive	
Na	5890	Air-Acetylene	1–5		
K	7665	Air-Acetylene	1–5		
Li	6708	Air-Acetylene	1–5		
Ca	4227	Air-Acetylene	.5–3	1% La	To eliminate
Mg	2852	Air-Acetylene	1–5	1% La	Al, P, and Si
Sr	4607	Air-Acetylene	1–5	1% La	interference
Cu	3248	Air-Acetylene	.05–5		
Pb	2170	Air-Acetylene	1–5		
Zn	2139	Air-Acetylene	1–5		
Cd	2288	Air-Acetylene	1–5		
Mn	2795	Air-Acetylene	1–50		
Fe	2483	Air-Acetylene	1–50		
Co	2407	Air-Acetylene	1–5		
Ni	2320	Air-Acetylene	1–5		
Cr	3579	N_2O-Acetylene	1–5		To improve
Mo	3133	N_2O-Acetylene	2–30	Alcohol	sensitivity and
Al	3093	N_2O-Acetylene	10–50	Alcohol	noise control
Ag	3281	Air-Acetylene	.05–5		

Table 5 Spectrographic Analysis of Interstitial Brines

Sample Location	Residue on Evaporation (g/ml)	Al	Ba	B	Cd	Ca *	Cr	Cu	Au	Fe	Pb	Mg *
71 15–25cm	0.304	<30	–	40	–	5,000	–	<0.3	–	100	0.3	9oo
71(b) 59–74cm	0.310	<30	–	30	–	4,000	–	0.2	–	40	2	7oo
78C 15–25cm	0.307	<31	–	30	–	6,000	–	1	–	100	<0.3	8oo
78D	0.309	<31	–	30	–	7,000	–	0.3	–	200	0.3	9oo
80 25–15cm	0.297	<30	–	30	–	6,000	–	1	–	<30	–	1,000
84K 206cm	0.249	<25	–	30	–	4,000	<3	3	–	24 est.	2	7oo
84K 365–370cm	0.304	<30	–	30	–	6,000	–	<0.3	–	17 est.	<0.3	9oo
†84K 400cm	0.301	<30	–	50	–	5,000	–	–	–	18 est.	<0.9	9oo
80P 120–130cm	0.311	<31	–	30	Tr.	7,000	–	<0.3	–	30	0.3	1,000
80 310–320cm	0.299	<30	–	50	–	4,000	–	0.3	–	15 est.	0.5	2,000
80 580–595cm	0.304	<30	–	50	–	5,000	–	<0.3	–	9 est.	0.3	1,000
80 840–850cm	0.299	<30	50	90	–	4,000	<3	–	–	12 est.	<0.9	1,000
81P 15–25cm	0.294	<30	–	40	–	4,000	–	–	–	21 est.	<0.9	1,000
81P 315–325cm	0.293	<30	–	40	–	6,000	–	–	–	40	<0.9	1,000
81P 605–615cm	0.304	<30	–	60	–	6,000	–	–	–	70	<0.9	1,000
81P 855–865cm	0.297	<30	30	60	–	4,000	–	–	–	24 est.	<0.9	1,000
82B 15–25cm	0.047	<5	–	10	–	6oo	–	0.1	–	5	0.05 est.	1,000
84K 143cm	0.316	<32	–	30	–	8,000	–	0.3	–	2 est.	<0.3	1,000
84K 263cm	0.317	<31	–	–	Tr.	6,000	–	<0.3	–	50	1	9oo
84K 0–30cm	0.313	<31	–	–	–	6,000	–	<0.3	–	90	0.3	8oo
84G 89–99cm	0.316	<31	–	30	–	5,000	–	<0.3	–	60	0.6	8oo
84P 290–315cm	0.315	80	–	30	Tr.	5,000	<3	0.6	–	100	0.9	9oo
84P 595–605cm	0.322	<32	–	30	–	2,000	–	0.3	–	60	0.6	1,000
84P 865–875cm	0.259	<30	–	40	–	7,000	–	–	–	30	<0.9	9oo
85K 30cm	0.037	<4	–	4	–	4oo	–	0.2	–	3 est.	<0.04 est.	1,000
85K 190cm	0.046	5	10	20	–	5oo	–	0.1	–	5	0.07 est.	1,000
85K 258cm	0.044	<4	–	10	–	6oo	–	0.1	–	2 est.	0.04 est.	1,000
85K 300cm	0.053	<5	–	10	–	5oo	–	0.2	–	3 est.	0.08 est.	9oo
85K 340cm	0.053	<5	–	10	–	6oo	–	0.3	–	5 est.	0.05 est.	1,000
†85K 400cm	0.05	<5	–	10	–	8oo	–	0.1	–	4 est.	0.05 est.	1,000
89C 0–10cm	0.313	<31	–	30	–	8,000	–	<0.3	–	150	0.3	8oo
89C 10–30cm	0.314	3	–	30	–	4,000	–	–	–	100	0.3	8oo
89C 20–30cm	0.318	<32	–	30	–	10,000	–	<0.3	–	100	0.3	1,000
95K 0–10cm	0.043	7	–	10	–	5oo	0.4	0.1	–	5	0.06 est.	1,000
‡95K 270–305cm	0.035	<4	–	9	–	4oo	0.7	0.3	–	3	0.04 est.	7oo
106G 0–30cm	0.312	<31	–	30	–	7,000	–	0.3	–	100	0.6	9oo
107G 0–30cm	0.306	<31	–	50	–	6,000	–	<0.3	–	30	2	9oo
108G 0–30cm	0.312	<31	–	–	–	6,000	–	<0.3	–	100	0.3	8oo
109G 5–35cm	0.312	3	–	<31	–	5,000	–	–	–	50	0.3	8oo
109G 140–170cm	0.303	<31	–	–	–	4,000	–	0.5	–	40	0.3	8oo
117K 200cm	0.046	<5	–	20	–	6oo	–	0.05	–	5	0.05 est.	1,000
117K 255cm	0.038	<4	–	10	–	6oo	–	0.1	–	6	<0.04 est.	1,000
119K 30–40cm	0.307	<31	–	50	–	5,000	–	<0.3	–	70	0.9	9oo
119K 50–60cm	0.313	<31	–	50	–	8,000	–	0.3	–	90	0.9	9oo
119K 90cm	0.313	<31	50	60	–	5,000	–	<0.3	–	50	3	1,000
119K 110cm	0.316	<32	–	30	–	7,000	–	0.3	–	50	3	1,000
119K 240cm	0.319	<32	–	50	–	6,000	–	0.3	–	16 est.	0.6	1,000
119K 345–360cm	0.318	<32	–	30	–	4,000	–	0.05	–	6 est.	0.5	9oo
119K 400–430cm	0.306	<31	–	30	–	6,000	–	0.3	–	6 est.	0.5	9oo
121G 0–20cm	0.314	9	90	90	–	7,000	–	8	Tr.	200	0.3	8oo
121G 280–290cm	0.308	3	–	30	–	6,000	–	–	–	40	5	1,000
§126P 33–63cm	0.312	5	–	30	–	6,000	–	–	–	100	20	9oo
126P 284–314cm	0.309	3	–	<31	–	5,000	–	–	–	100	0.3	9oo
161K 0–70cm	0.248	<25	–	40	–	3,000	–	<0.3	–	12 est.	0.3	8oo
161K 70–130cm	0.285	<29	–	40	–	5,000	–	0.3	–	14 est.	0.6	9oo
161K 130–180cm	0.282	<28	–	30	–	5,000	–	<0.3	–	60	0.4	8oo
161K 180–275cm	0.291	<3	–	<29	–	4,000	–	–	–	12 est.	0.3	8oo
161K 275–320cm	0.295	<3	–	<30	–	5,000	–	–	–	12 est.	0.9	7oo

* Non-significant zeros are shown in lower case.
† Catcher.
‡ From bottom above catcher.

(All values in milligrams per liter of brine)

Mn	Mo	Ni	K *	Si	Ag	Ce	Sr	Sn	Ti	Zn	Zr	Sb
100	—	—	5,000	600	0.03	—	90	—	—	<3	—	—
60	—	—	2,000	30	0.03	—	90	—	—	—	—	—
70	—	0.3	2,000	90	—	—	90	—	—	—	—	—
60	—	0.3	3,000	200	—	—	70	—	<3	—	—	—
100	—	0.3	3,000	60	0.03	—	90	—	—	—	—	—
60	—	0.4	2,000	100	0.04	—	80	2	<3	—	—	—
60	—	—	3,000	60	—	—	80	—	—	—	—	—
70	—	—	2,000	200	—	—	80	—	—	—	—	—
50	—	0.3	3,000	30	<0.03	—	100	<3	<3	—	—	—
30	—	—	4,000	90	<0.03	—	80	—	—	—	—	—
30	—	—	3,000	60	0.03	—	90	—	—	—	—	—
30	—	0.4	2,000	600	0.03	—	90	—	—	—	—	—
30	—	—	2,000	30	—	—	80	—	—	—	—	—
15	—	—	2,000	30	—	—	90	—	—	—	—	—
10	—	—	2,000	200	—	—	90	—	—	—	—	—
9	—	—	1,000	90	—	—	100	—	—	—	—	—
4	0.07	0.05	800	30	0.009	Tr.	10	—	<0.5	—	—	—
60	—	0.3	3,000	200	—	—	70	—	—	—	—	—
90	1	—	3,000	50	—	—	100	—	—	—	—	—
90	0.03	—	3,000	80	—	—	90	—	—	—	—	—
90	<0.3	—	4,000	100	—	—	100	—	—	<95	—	—
80	0.3	2	3,000	200	0.06	—	100	<3	10	100	—	—
100	0.3	—	6,000	60	0.03	—	90	—	—	<97	—	—
80	—	—	1,000	90	—	—	80	—	—	—	—	—
5	0.1	<0.04	800	20	0.007	—	10	—	<0.4	—	—	—
5	0.07	<0.05	700	600	0.01	—	10	—	<0.5	—	—	—
5	0.07	<0.04	900	40	0.008	—	10	—	<0.4	—	—	—
3	0.2	<0.05	800	50	0.01	—	20	—	<0.5	—	—	—
5	0.05	0.05	800	20	0.01	—	20	—	—	—	—	—
6	2	0.2	2,000	50	0.008	—	10	—	<0.5	—	—	Tr.
100	—	—	3,000	50	—	—	90	—	—	—	—	—
90	0.3	—	2,000	30	0.03	—	100	1	—	—	—	—
90	—	—	10,000	<32	0.3	—	100	—	—	<96	—	—
3	0.09	0.04	900	40	0.006	—	10	—	1	—	—	—
3	0.1	0.1	600	20	0.005	—	10	—	<0.4	—	—	—
100	—	—	6,000	30	0.03	—	90	—	—	<94	—	—
100	—	—	5,000	<31	0.06	—	100	—	—	<92	—	—
100	—	0.3	3,000	<31	—	—	90	—	—	—	—	—
100	—	—	3,000	60	0.03	—	90	—	—	—	—	—
80	—	<0.3	3,000	<30	0.05	—	90	—	<3	—	—	—
4	0.07	<0.05	1,000	20	0.005	—	10	—	<0.5	—	—	—
4	0.06	<0.04	1,000	30	0.004	—	10	—	—	—	—	—
90	—	—	3,000	90	—	—	100	—	—	<92	—	—
100	0.3 est.	—	7,000	50	0.09	—	90	—	—	<94	—	—
90	—	—	3,000	300	0.09	—	100	—	—	<94	Tr.	—
100	—	—	6,000	30	0.03	—	100	—	—	<95	—	—
100	—	—	5,000	30	0.03	—	100	<3	—	<96	—	—
70	—	—	2,000 est.	<32	0.03	—	100	—	—	<96	—	—
60	0.5	—	6,000 est.	<31	0.06	—	100	—	—	<92	—	—
100	—	0.5	3,000	600	3	—	90	—	—	—	Tr.	—
100	—	—	4,000	90	0.2	—	90	—	—	—	—	—
100	—	0.6	4,000	100	0.3	—	90	1.6 est.	—	—	—	—
100	0.3 est.	—	3,000	200	0.03	—	100	—	—	—	—	—
100	<0.3	0.3	2,000	400	0.05	—	70	10	<3	—	—	—
200	—	—	3,000	30	0.03	Tr.	100	<3	—	<85	—	—
200	—	—	3,000	40	0.04	—	80	—	—	<85	—	—
200	—	—	2,000	40	0.03 est.	—	90	—	—	—	—	—
200	—	—	3,000	60	0.03 est.	—	90	1.2 est.	—	—	—	—

§ This sample is suspect because it was precipitated by heating.
— = Sought but not detected.
Tr. = Trace.
All spectrographic analyses by F. B. Reisbick.

Table 6   Semiquantitative Analysis of Red Sea Sediments

Sample Location	Al	Sb	As	Ba	Be	B	Cd
71 0–15	0.4	—	—	—	—	0.004	0.001
71 15–25	0.6	0.002	0.01	0.06	—	0.1	0.02
71 44–59	0.4	—	0.004	—	—	0.004	0.005
71(B) 59–74, H5.0	0.6	—	0.008	0.06	—	0.006	0.008
78C 0–15	—	—	0.004	—	—	0.004	0.0006
† 78C 15–25	0.4	—	0.006	—	—	0.1	0.001
78D 0–15	0.4	—	0.006	0.06	—	0.008	0.002
† 78D 15–25	—	—	0.006	—	—	0.006	0.005
80P 0–15	1.	—	0.006	0.06	—	0.03	0.00015
80 15–25	3.	—	0.01	—	—	0.02	0.0006
80P 100–120	3.	—	0.004	—	—	0.006	0.00004
80 290–310	4.	—	0.004	—	—	0.008	0.00004
80 310–320	4.	—	0.004	—	—	0.01	0.00004
80 580–595	3.	—	0.004	—	—	0.004	0.00004
80 595–615	3.	—	0.004	—	—	0.008	—
80 840–850	3.	—	0.004	—	—	0.006	0.00004
80 850–870	3.	—	0.004	—	—	0.006	—
81P 0–15	4.	—	0.004	—	—	0.006	—
81P 15–25	4.	—	0.004	—	—	0.006	—
81P 290–305	4.	—	0.006	—	—	0.006	0.00004
81P 315–325	4.	—	0.006	—	—	0.006	0.00004
81P 590–605	3.	—	0.004	—	—	0.008	0.00004
81P 605–615	3.	—	0.004	—	—	0.006	0.00004
81P 855–865	3.	—	0.004	—	—	0.004	0.00004
81P 865–880	3.	—	0.004	—	—	0.006	0.00004
82B 15–25	0.4	—	0.05	0.1	—	0.015	—
82B 15–25 (f.f.)	0.6	—	0.01	0.15	—	0.006	0.015
82D Near catcher	—	0.002	0.06	—	—	0.004	0.0002
84G 89–99	0.4	—	0.05	0.2	—	0.02	—
84G 99–114	—	—	0.005	—	—	—	0.002
84K 0–30	0.6	—	0.008	—	—	0.004	0.006
84K 143	—	—	—	—	—	0.004	0.002
84K 206	0.6	—	0.008	0.06	—	0.004	0.002
84K 263	1.	0.009	0.09	—	—	—	0.04
84K 300	0.6	—	0.02	0.3	—	0.06	0.015
84K 304	0.4	—	0.02	—	—	—	0.015
84K 310–325	0.6	—	0.006	—	—	—	0.015
84K 325–348	0.6	—	0.008	0.06	—	—	0.01
84K 325–348 L. P.	1.	—	0.02	—	—	—	0.02
84K 365–370	—	—	0.02	—	—	—	0.00015
84K 365–370 (352–400 same)	0.4	—	0.008	0.2	—	—	0.00006
84K 400, Catcher	0.6	—	0.004	>0.5	—	—	0.00006
80P 120–130	1.5	—	0.004	—	—	0.004	0.00004
84P 290–315	0.6	0.002	0.008	0.3	—	0.006	0.006
84P 315–345	0.4	—	—	0.4	—	0.004	0.008
84P 580–595	0.4	—	0.01	>0.5	—	0.004	0.009
84P 595–605	0.6	—	0.008	0.2	—	—	0.009
84P 850–865	0.6	—	0.01	—	—	—	0.00006
84P 865–875	—	—	0.01	—	—	—	0.00004
85K 30, Unit 6	0.8	0.002	0.04	>0.5	—	0.03	0.0005
85K 135	0.6	0.002	0.015	>0.5	—	0.006	0.04
85K 170, Clinton type	0.4	0.004	0.08	>0.5	—	0.02	0.003
85K 70, Unit 6	—	—	0.03	0.3	—	0.02	0.0001
85K 190, Unit 5	—	—	0.05	0.3	—	0.02	0.0002
85K 200, Unit 5	—	—	0.03	0.4	—	0.02	0.0002
85K 280	—	0.003	0.03	0.2	—	0.01	0.0004
85K 300, Unit 3	—	0.003	0.015	0.4	—	0.01	0.00015
85K 340, Unit 2	0.8	0.006	0.02	>0.5	—	0.01	0.002
85K 360, Unit 2	0.8	0.002	0.01	>0.5	0.0008	0.008	0.0001
85K 258, Unit 4	—	0.003	0.03	>0.5	—	0.02	0.0004
85K Unit 1, Core catcher	0.4	0.01	0.03	>0.5	—	—	0.07
89C 0–10	0.6	—	0.004	0.06	—	0.004	0.005
89C 10–30	0.8	—	0.015	0.1	—	0.006	0.01
89C 0–20	—	—	—	>0.5	—	—	0.0004
89C 20–30	0.6	—	0.01	>0.5	—	0.006	0.003
94C 130–145	0.4	—	—	—	—	—	0.0002

* Not reported because of iron interference.

† Squeeze and remainder.

— = Sought but not detected.

(All values in weight %, oven dried sample)

Ca	Cr	Co	Cu	Ga	Ge	Au	In	Fe	La	Pb
`3.	0.003	0.006	0.06	0.0004	—	—	—	6.	—	0.015
3.	*	0.02	0.4	0.0015	0.00015	—	—	>10.	—	0.04
3.	0.003	0.006	0.1	0.0006	—	—	—	6.	—	0.02
3.	0.004	0.008	0.2	0.001	0.0001	—	—	>10.	—	0.04
4.	0.003	0.003	0.04	0.0004	—	—	—	8.	—	0.009
3.	*	0.004	0.08	0.0006	0.00003	—	—	>10.	—	0.02
3.	0.004	0.003	0.04	0.0006	—	—	—	10.	—	0.02
3.	0.003	0.003	0.08	0.0006	0.0001	—	—	>10.	—	0.02
8.	0.006	0.002	0.02	0.0004	—	—	—	8.	—	0.003
>10.	0.006	0.0015	0.01	0.0004	0.00004	—	—	8.	—	0.006
>10.	0.006	0.002	0.008	0.0006	—	—	—	6.	—	0.0009
>10.	0.006	0.002	0.008	0.0006	—	—	—	6.	—	0.0015
>10.	0.007	0.003	0.01	0.0006	—	—	—	8.	—	0.009
>10.	0.006	0.0015	0.006	0.0004	0.00002	—	—	6.	—	0.009
>10.	0.006	0.002	0.008	0.0006	—	—	—	6.	—	0.001
>10.	0.006	0.0015	0.008	0.0004	0.00002	—	—	6.	—	0.001
>10.	0.006	0.002	0.008	0.0006	—	—	—	6.	—	0.0009
>10.	0.008	0.003	0.006	0.0006	—	—	—	6.	—	0.0008
>10.	0.006	0.001	0.006	0.0006	—	—	—	6.	—	0.0009
>10.	0.006	0.002	0.008	0.0006	—	—	—	6.	—	0.0015
>10.	0.006	0.0015	0.008	0.0006	—	—	—	6.	—	0.0015
>10.	0.006	0.002	0.008	0.0006	—	—	—	6.	—	0.0015
>10.	0.006	0.0015	0.008	0.0004	—	—	—	6.	—	0.0015
>10.	0.006	0.001	0.0C8	0.0004	0.00002	—	—	6.	—	0.001
>10.	0.006	0.002	0.008	0.0006	—	—	—	6.	—	0.0015
2.	*	0.0006	0.03	0.0004	0.00006	—	—	>10.	—	0.015
3.	*	0.01	0.4	0.002	0.00015	—	—	>10.	—	0.04
2.	*	0.001	0.15	0.0004	0.00015	—	—	>10.	—	0.015
0.8	*	—	0.15	0.0006	0.00015	—	—	<10.	—	0.03
3.	0.003	0.003	0.06	0.0004	0.00002	—	—	6.	—	0.02
3.	0.004	0.008	0.2	0.0008	0.00006	—	—	>10.	—	0.03
3.	0.003	0.006	0.1	0.0004	—	—	—	6.	—	0.015
3.	0.004	0.008	0.2	0.001	0.00004	—	—	8.	—	0.02
>10.	0.004	0.04	1.	0.004	0.0002	0.00015	—	10.	—	0.06
10.	0.004	0.02	1.	0.004	0.00004	0.0001	—	10.	—	0.03
10.	0.003	0.02	0.6	0.002	0.00002	0.0001	—	8.	—	0.03
>10.	0.004	0.015	0.8	0.002	0.00003	0.00015	—	10.	—	0.02
>10.	*	0.01	0.4	0.0015	0.00003	0.00008	—	>10.	—	0.009
10.	0.004	0.02	0.8	0.003	0.00006	0.00008	—	>10.	—	0.02
2.	*	0.0008	0.2	0.0006	0.00004	—	—	>10.	—	0.003
4.	*	0.003	0.2	0.001	—	—	—	>10.	—	0.0009
>10.	*	0.001	0.2	0.001	0.00003	—	—	>10.	—	0.0015
>10.	0.006	0.002	0.008	0.0004	—	—	—	4.	—	0.001
2.	0.004	0.015	0.3	0.002	0.0004	—	—	>10.	—	0.06
3.	0.003	0.006	0.2	0.0006	0.00002	—	—	6.	—	0.02
10.	0.004	0.01	0.4	0.002	0.00006	0.00008	—	8.	—	0.04
>10.	0.004	0.01	0.6	0.002	—	0.0001	—	8.	—	0.03
2.	*	0.003	0.2	0.001	0.00004	—	—	>10.	—	0.002
2.	*	0.0015	0.2	0.001	0.00004	0.00008	—	>10.	—	0.002
1.	*	—	0.2	0.001	0.0002	0.0001	—	>10.	—	0.04
1.5	*	0.0004	0.2	0.002	0.00007	0.00015	—	>10.	—	0.15
0.8	*	—	0.4	0.001	0.0004	0.0001	—	>10.	—	0.04
1.5	*	—	0.2	0.0006	0.00008	—	—	>10.	—	0.03
0.6	*	0.0006	0.03	0.001	0.0002	—	—	>10.	—	0.03
1.5	*	0.0004	0.04	0.0008	0.0001	—	—	>10.	—	0.03
1.	*	—	0.02	0.0006	0.00015	—	—	>10.	—	0.02
0.8	*	—	0.15	0.0008	0.0001	—	—	>10.	—	0.02
2.	*	0.0006	0.2	0.001	0.0006	—	—	>10.	—	0.03
2.	*	0.0006	0.2	0.001	0.0001	—	—	>10.	—	0.009
0.6	*	—	0.04	0.0006	0.0003	—	—	>10.	—	0.04
2.	*	0.006	4.	0.003	0.00015	0.0001	0.00015	>10.	0.006	0.03
3.	0.004	0.008	0.15	0.0008	0.00003	—	—	8.	—	0.02
4.	*	0.02	0.2	0.0015	0.0001	—	—	>10.	—	0.03
>10.	0.002	0.0015	0.03	—	—	—	—	4.	—	0.005
>10.	0.004	0.01	0.15	0.0006	0.00003	—	—	>10.	—	0.02
>10.	—	0.002	0.02	—	—	0.00008	—	0.6	—	0.003

Table 6   Semiquantitative Analysis

Sample Location	Mg	Mn	Mo	Ni	P	K	Sc
71 0–15	0.6	0.1	0.003	—	—	0.6	—
71 15–25	0.6	0.015	0.01	0.001	—	—	—
71 44–59	0.6	0.1	0.004	0.0008	—	0.8	—
71(B) 59–74, H5.0	0.8	0.1	0.006	0.0015	—	0.4	—
78C 0–15	0.6	0.15	0.006	0.0006	—	0.6	—
† 78C 15–25	0.6	0.3	0.008	0.0006	—	—	—
78D 0–15	0.6	0.2	0.006	0.0006	—	0.6	—
† 78D 15–25	0.6	0.15	0.006	—	—	0.3	—
80P 0–15	2.	>10.	0.006	0.003	—	0.4	—
80 15–25	2.	8.	0.004	0.003	—	0.6	—
80P 100–120	3.	0.8	0.0015	0.004	—	1.	0.001
80 290–310	3.	0.8	0.0015	0.004	—	1.	0.001
80 310–320	3.	0.8	0.0015	0.006	—	1.	0.001
80 580–595	2.	0.8	0.001	0.004	—	1.	0.001
80 595–615	3.	0.8	0.0015	0.004	—	1.	0.001
80 840–850	2.	0.6	0.001	0.004	—	1.	0.001
80 850–870	2.	0.6	0.001	0.004	—	0.6	0.0001
81P 0–15	3.	1.	0.001	0.006	—	1.	0.001
81P 15–25	2.	0.3	0.001	0.004	—	0.2	0.001
81P 290–305	3.	0.6	0.002	0.004	—	1.	0.001
81P 315–325	3.	0.3	0.0015	0.004	—	—	0.001
81P 590–605	3.	0.6	0.0015	0.004	—	0.8	0.001
81P 605–615	2.	0.3	0.001	0.004	—	—	0.001
81P 855–865	2.	0.2	0.001	0.003	—	0.2	—
81P 865–880	3.	0.6	0.0015	0.004	—	0.8	0.001
82B 15–25	2.	10.	0.006	0.0006	0.15	—	—
82B 15–25 (f.f.)	1.	0.2	0.01	0.002	—	—	—
82D Near catcher	2.	>10.	0.004	0.0015	0.15	—	—
84G 89–99	1.	8.	0.004	0.001	0.2	—	—
84G 99–114	0.6	0.1	0.004	—	—	0.8	—
84K 0–30	0.8	0.1	0.008	0.0008	—	0.3	—
84K 143	0.8	0.15	0.0015	0.0008	—	0.8	—
84K 206	2.	0.2	0.008	0.001	—	0.6	—
84K 263	3.	0.04	0.2 (est.)	0.004	—	—	—
84K 300	4.	0.09	0.06	0.006	—	0.3	—
84K 304	2.	0.07	0.06	0.003	—	0.3	—
84K 310–325	3.	0.15	0.03	0.006	—	—	—
84K 325–348	1.	1.	0.015	0.002	—	0.2	—
84K 325–348 L.P.	2.	0.3	0.03	0.004	—	—	—
84K 365–370	0.8	0.15	0.02	—	—	0.2	0.001
84K 365–370 (352–400 same)	0.8	0.6	0.03	0.0008	—	0.2	—
84K 400, Catcher	1.	0.2	0.01	0.001	—	0.3	—
80P 120–130	2.	0.3	0.001	0.004	—	0.8	—
84P 290–315	1.5	0.15	0.008	0.002	—	1.	—
84P 315–345	0.6	0.15	0.003	0.0008	—	0.8	—
84P 580–595	0.8	0.6	0.01	0.003	—	0.3	—
84P 595–605	1.	0.2	0.01	0.004	—	—	—
84P 850–865	0.8	0.2	0.03	0.0006	—	—	—
84P 865–875	0.4	0.1	0.02	—	0.08	—	—
85K 30, Unit 6	1.	0.4	0.006	—	0.08	—	—
85K 135	2.	2.	0.002	0.0006	—	0.2	—
85K 170, Clinton type	0.8	1.5	0.01	0.0006	0.09	—	—
85K 70, Unit 6	2.	0.6	0.008	—	—	0.3	—
85K 190, Unit 5	0.8	4.	0.015	0.0006	0.15	—	—
85K 200, Unit 5	2.	4.	0.006	—	0.08	0.3	—
85K 280	1.5	1.	0.008	—	0.08	0.2	—
85K 300, Unit 3	1.	0.2	0.003	—	0.15	0.2	—
85K 340, Unit 2	1.	4.	0.003	0.001	0.2	—	—
85K 360, Unit 2	2.	3.	0.002	0.001	—	—	—
85K 258, Unit 4	0.8	0.4	0.008	—	0.2	—	—
85K Unit 1, Core catcher	1.	3.	0.02	0.004	—	—	—
89C 0–10	0.6	0.1	0.005	0.001	—	0.6	—
89C 10–30	0.8	0.2	0.01	0.0015	0.06	0.4	—
89C 0–20	0.3	0.2	0.002	—	—	0.4	—
89C 20–30	0.4	0.2	0.006	0.0008	—	0.2	—
94C 130–145	0.15	0.015	0.0015	—	—	0.3	—
95K 0–10	2.	6.	0.002	0.006	0.06	0.3	0.001

of Red Sea Sediments (continued)

Ag	Na	Sr	Tl	Sn	Ti	W	V	Y	Zn	Zr
0.001	>4.	—	0.0002	—	—	—	0.004	—	0.3	0.06
0.006	>4.	—	0.001	0.00008	0.015	0.0003	0.01	—	4.	0.01
0.002	>4.	—	0.0006	—	—	—	0.004	—	1.	0.006
0.004	>4.	0.04	0.0006	—	—	—	0.008	—	2. (est.)	0.02
0.0008	>4.	0.04	0.0002	—	—	—	0.004	—	0.15	0.02
0.0015	>4.	—	0.0004	—	0.01	—	0.006	—	0.3	0.02
0.0008	>4.	—	0.0002	—	—	—	0.004	—	0.3	—
0.001	>4.	—	—	—	—	—	0.02	—	1.	0.006
0.0001	>4.	0.06	—	0.00006	0.1	—	0.015	0.004	0.08	0.04
0.00015	>4.	0.08	—	0.00006	0.15	—	0.01	—	0.2	0.02
—	>4.	0.1	—	0.00006	0.2	—	0.01	0.004	0.05	0.02
—	>4.	0.1	—	0.00006	0.2	—	0.01	0.004	0.05	0.02
—	>4.	0.15	—	0.00006	0.3	—	0.015	0.004	0.04	0.02
—	>4.	0.08	—	0.00006	0.2	—	0.01	0.004	0.04	0.02
—	>4.	0.1	—	0.00006	0.2	—	0.01	—	0.05	0.02
—	>4.	0.1	—	0.00006	0.2	—	0.01	0.004	0.05	0.02
—	>4.	0.1	—	0.00006	0.2	—	0.01	—	0.05	0.03
—	>4.	0.1	—	0.00006	0.3	—	0.015	0.004	0.02	0.04
—	3.	0.1	—	0.00006	0.3	—	0.01	0.004	0.02	0.03
—	>4.	0.15	—	0.00006	0.2	—	0.02	0.004	0.03	0.02
—	>4.	0.2	—	0.00006	0.2	—	0.015	0.004	0.05	0.02
—	>4.	0.15	—	0.00006	0.2	—	0.015	—	0.05	0.02
—	4.	0.15	—	0.00006	0.2	—	0.01	0.006	0.04	0.04
—	3.	0.1	—	0.00006	0.2	—	0.01	0.004	0.05	0.02
—	>4.	0.15	—	0.00008	0.2	—	0.015	—	0.05	0.06
0.0002	>4.	—	—	0.00008	—	—	0.02	—	0.4	0.02
0.008	>4.	—	0.0006	—	0.03	0.0002	0.01	—	4.	0.02
0.0004	2. (est.)	—	—	0.00006	—	—	0.02	—	0.3	0.08
—	2.	0.04	—	—	0.01	—	0.015	—	1.	0.004
0.001	>4.	—	—	—	—	—	0.003	—	0.3	0.006
0.003	>4.	—	0.0003	—	0.01	—	0.008	—	2.	0.006
0.002	>4.	—	—	—	—	—	0.004	—	0.5	0.08
0.003	>4.	—	0.0006	0.00006	—	—	0.006	—	0.5	—
0.02	4.	0.06	0.004	0.0001	0.015	—	0.01	—	8. (est.)	0.06
0.004	>4.	0.04	0.0008	0.00006	0.02	—	0.01	—	1.	0.01
0.004	>4.	0.04	0.0001	0.00006	—	—	0.006	—	5.	0.08
0.004	>4.	0.06	0.0006	0.00006	0.015	—	0.008	—	4.	0.02
0.003	>4.	0.1	0.0003	—	0.02	—	0.006	—	1.5	0.01
0.008	>4.	0.06	0.0008	0.00008	0.06	—	0.008	—	6.	0.01
0.0008	>4.	—	0.0006	—	—	0.0002	0.006	—	0.04	0.01
0.0006	>4.	—	—	0.00006	—	—	0.006	—	0.03	0.08
0.001	>4.	0.1	—	0.00006	0.03	0.0002	0.008	—	0.05	0.006
—	2.	0.08	—	0.00006	0.15	—	0.01	—	0.03	0.02
0.008	>4.	0.06	0.0008	—	—	0.0003	0.008	—	2.	—
0.004	>4.	0.08	—	—	—	—	0.006	—	1.	0.004
0.006	>4.	0.15	0.0008	0.00006	—	—	0.008	—	0.6	0.01
0.006	>4.	0.08	0.001	0.00006	0.02	—	0.008	—	3.	0.02
0.0008	>4.	—	—	—	—	—	0.006	—	0.03	0.04
0.0008	2.	—	0.0002	0.00006	—	0.0002	0.006	—	0.02	0.01
0.008	2.	0.08	—	—	0.015	0.0003	0.01	—	0.8	0.01
0.02	2.	0.3	0.0003	—	0.01	0.0004	0.01	—	3.	0.01
0.003	2.	0.2	—	—	0.01	0.0004	0.006	—	0.8	0.02
0.002	>4.	0.06	—	—	—	0.0002	0.008	—	0.5	0.015
0.00008	2.	0.04	—	—	—	0.0003	0.02	—	1.	0.015
0.0002	>4.	0.08	—	—	—	—	0.015	—	0.5	0.06
0.0003	>4.	—	—	—	—	0.0004	0.006	—	0.3	0.008
0.002	>4.	0.06	—	0.0001	—	0.0004	0.02	—	0.3	—
0.004	2.	0.06	—	0.00008	0.04	0.0002	0.02	—	1.	0.03
0.0008	3.	0.06	—	0.00006	0.04	—	0.01	—	0.3	0.02
0.0003	2.	0.08	—	—	—	0.0003	0.008	—	0.6	0.01
0.01	<1.	0.04	0.0015	0.0004	—	—	0.01	—	>10.	0.02
0.002	>4.	—	0.0003	—	0.01	—	0.006	—	1.5	0.006
0.006	>4.	—	0.0006	0.00006	0.06	0.0003	0.01	—	2. (est.)	0.02
0.0006	>4.	0.2	—	—	—	—	0.003	—	0.07	0.01
0.002	>4.	0.3	0.0004	—	0.06	—	0.006	—	0.5	—
0.0004	>4.	0.06	—	—	—	—	0.004	—	0.08	0.01
—	1.5	0.15	—	—	0.15	—	0.02	—	0.2	0.02

Table 6   Semiquantitative Analysis

Sample Location	Al	Sb	As	Ba	Be	B	Cd
95K 0–10	3.	—	0.01	—	—	0.008	—
95K 10–20	3.	—	0.008	0.06	—	0.006	0.00004
95K 40–45	3.	—	0.006	—	—	0.006	—
95K 45–55	4.	—	0.004	—	—	0.006	0.00008
95K 55–60	3.	—	0.008	—	—	0.015	—
95K 270–305 From bottom	0.6	—	0.01	—	—	0.02	0.0002
95K From above catcher	0.4	—	0.01	0.06	—	0.015	0.00015
97G From bottom of core	—	—	0.004	>0.5	—	0.004	0.00008
106G 0–30	—	—	0.004	0.06	—	—	0.002
107G 0–30	—	—	0.004	—	—	0.004	0.00004
108G 0–30	—	—	0.004	—	—	0.004	0.002
108G 173–193	—	—	0.004	—	—	0.004	0.005
109G 5–35	—	—	0.004	—	—	0.006	0.00015
109G 140–170	—	—	—	—	—	0.004	0.0003
111C 122, Core catcher	—	—	0.004	—	—	0.004	0.002
111D 122, Core catcher	4.	0.002	0.03	0.2	—	0.006	0.009
117K 200	0.6	—	0.006	—	—	0.02	—
117K 255	0.6	—	0.007	—	—	0.03	0.00004
117K 322	—	—	0.009	—	—	0.02	—
117K 335	2.	—	0.004	—	—	0.004	—
117K 290–400	4.	—	—	—	—	0.004	—
118K 75	3.	—	0.004	—	—	0.008	0.00008
118K 220	1.	—	0.006	—	—	0.004	0.00006
118K 310	3.	—	0.006	—	—	0.004	0.00004
118K 360	1.	—	0.008	—	—	0.015	0.00006
119K 30–40	0.6	—	0.006	0.2	—	0.004	0.01
119K 50–60	0.4	—	—	0.2	—	0.006	0.0008
119K 90	0.6	—	0.006	>0.5	—	0.006	0.007
119K 100	—	—	—	0.3	—	0.006	0.002
119K 110	0.8	0.002	0.04	>0.5	—	0.004	0.01
119K 115	—	—	—	0.1	—	0.004	0.00004
119K 220	—	—	0.004	0.06	—	0.004	—
119K 240	0.6	—	0.01	0.2	—	0.006	0.0003
119K 325–345	—	—	0.006	0.06	—	0.004	0.00004
119K 345–360	0.4	—	0.01	—	—	0.006	0.00004
119K 400–430	0.4	—	0.004	0.2	—	0.004	0.0001
121G 0–30	—	—	0.004	—	—	0.004	0.0002
121G 270–280	—	—	0.004	—	—	0.004	0.0001
121G 270–290	0.4	—	0.006	—	—	0.006	0.00006
126P 284–314	0.4	—	0.008	—	—	0.006	0.007
126P 33–63, Top	—	—	0.006	—	—	—	0.0006
161K 0–70	0.6	—	0.01	0.2	—	0.02	0.005
161K 70–130	0.4	—	0.01	0.05	—	0.01	0.008
161K 130–180	0.3	—	0.03	0.3	—	0.03	0.0003

of Red Sea Sediments (continued)

Ca	Cr	Co	Cu	Ga	Ge	Au	In	Fe	La	Pb
>10.	0.006	0.0015	0.02	0.0004	—	—	—	8.	—	0.002
>10.	0.004	0.002	0.01	0.0004	—	—	—	6.	—	0.0015
>10.	0.006	0.0008	0.006	0.0006	0.00003	—	—	8.	—	0.0009
>10.	0.004	0.001	0.006	0.0006	0.00002	—	—	6.	—	0.001
10.	*	0.0008	0.004	0.0008	—	—	—	>10.	—	0.0015
3.	*	0.003	0.04	0.0006	—	—	—	>10.	0.01	0.002
2.	*	0.004	0.04	0.0006	—	—	—	>10.	0.01	0.002
3.	0.004	—	0.02	0.0004	0.00008	—	—	>10.	—	0.002
4.	0.002	0.006	0.08	0.0004	—	—	—	6.	—	0.015
4.	0.003	0.004	0.04	0.0004	—	—	—	6.	—	0.01
4.	0.003	0.002	0.06	0.0004	0.00002	—	—	6.	—	0.01
3.	0.002	0.004	0.1	0.0006	0.00002	—	—	6.	—	0.02
3.	0.003	0.0006	0.02	0.0004	0.00002	—	—	6.	—	0.006
3.	0.002	0.001	0.03	0.0004	—	—	—	6.	—	0.006
2.	0.003	0.001	0.04	0.0004	0.00002	—	—	8.	—	0.02
>10.	0.006	0.008	0.4	0.002	0.0002	0.0001	—	8.	—	0.04
4.	*	0.002	0.003	0.0006	—	—	—	>10.	—	0.0009
3.	*	0.004	0.004	0.0006	—	—	—	>10.	—	0.0008
2.	*	0.001	0.002	0.0006	—	—	—	>10.	—	0.0007
>10.	0.004	0.0006	0.008	—	0.00002	—	—	2.	—	0.0008
>10.	0.02	0.002	0.02	0.0006	0.00004	—	—	6.	—	0.002
>10.	0.003	0.001	0.01	0.0004	—	—	—	8.	—	0.002
>10.	0.004	0.001	0.006	0.0004	0.00003	—	—	6.	—	0.0015
>10.	0.004	0.003	0.003	0.0006	0.00002	—	—	6.	—	0.001
8.	*	0.004	0.008	0.0004	—	—	—	>10.	0.008	0.003
6.	*	0.006	0.3	0.0015	0.0002	—	—	>10.	—	0.03
4.	0.004	0.002	0.06	0.0004	—	—	—	8.	—	0.006
4.	*	0.02	0.2	0.002	0.0002	—	—	>10.	—	0.09
3.	0.003	0.006	0.06	0.0004	0.00002	—	—	8.	—	0.03
3.	0.004	0.01	0.4	0.002	0.001	—	—	10.	—	0.08
3.	0.002	—	0.08	0.0004	—	—	—	3.	—	0.005
3.	0.002	0.0006	0.03	—	0.00002	—	—	6.	—	0.004
2.	*	0.004	0.2	0.0008	0.0001	—	—	>10.	—	0.015
3.	*	0.0008	0.06	—	0.00004	—	—	8.	—	0.005
2.	*	0.001	0.1	0.0006	0.0002	—	—	>10.	—	0.009
3.	*	0.0008	0.1	0.0004	0.00003	—	—	8.	—	0.008
3.	0.003	0.001	0.03	—	—	—	—	6.	—	0.01
3.	*	0.0008	0.06	0.0004	0.00004	—	—	8.	—	0.008
3.	*	0.0008	0.06	0.0004	0.00003	—	—	>10.	—	0.004
3.	*	0.006	0.15	0.001	0.00006	—	—	>10.	—	0.03
2.	0.002	0.0006	0.02	—	—	—	—	6.	—	0.009
4.	*	0.0015	0.2	0.0008	0.00004	—	—	>10.	—	0.03
2.	*	0.001	0.2	0.0008	0.00006	—	—	>10.	—	0.02
3.	*	0.002	0.15	0.0008	0.00008	—	—	>10.	—	0.06

Table 6   Semiquantitative Analysis

Sample Location	Mg	Mn	Mo	Ni	P	K	Sc
95K 10–20	2.	6.	0.002	0.006	—	0.4	—
95K 40–45	2.	1.	0.002	0.004	0.06	0.8	0.001
95K 45–55	3.	0.3	—	0.006	0.05	1.	0.001
95K 55–60	4.	0.6	0.01	0.003	0.1	0.3	0.001
95K 270–305 From bottom	1.	0.15	0.015	0.0008	0.7	—	—
95K From above catcher	1.5	0.15	0.01	0.002	0.4	—	—
97G From bottom of core	1.	4.	0.002	—	—	0.3	—
106G 0–30	0.6	0.1	0.004	0.0008	—	0.6	—
107G 0–30	0.6	0.1	0.003	—	—	0.8	—
108G 0–30	0.4	0.1	0.003	—	—	0.6	—
108G 173–193	0.6	0.8	0.004	0.0008	—	0.7	—
109G 5–35	0.8	0.1	0.004	—	—	1.	—
109G 140–170	0.6	0.2	0.001	—	—	1.	—
111C 122, Core catcher	1.	0.1	0.003	—	—	1.	—
111D 122, Core catcher	3.	2.	0.006	0.008	0.07	0.8	0.001
117K 200	2.	0.2	0.01	0.003	—	—	—
117K 255	2.	1.5	0.006	0.002	0.08	—	—
117K 322	2.	0.2	0.004	0.002	0.2	—	—
117K 335	2.	0.2	—	0.002	—	0.6	0.001
117K 290–400	3.	0.2	—	0.004	0.1	0.8	0.002
118K 75	2.	1.	0.002	0.003	—	0.6	—
118K 220	2.	0.2	0.0015	0.003	—	1.	—
118K 310	2.	0.15	0.003	0.004	—	0.8	0.001
118K 360	2.	0.2	0.003	0.004	—	0.2	—
119K 30–40	1.5	1.	0.002	0.0008	—	0.3	—
119K 50–60	0.8	0.2	0.002	—	—	0.6	—
119K 90	1.	0.4	0.004	0.002	—	0.6	—
119K 100	0.6	0.1	0.002	—	—	1.	—
119K 110	0.8	2.	0.008	0.006	—	0.2	—
119K 115	1.	0.1	0.001	0.0006	—	1.	—
119K 220	0.6	0.3	0.003	0.0006	—	0.7	—
119K 240	1.	1.	0.008	0.002	—	0.8	—
119K 325–345	0.6	0.15	0.006	0.0008	—	0.6	—
119K 345–360	0.8	0.15	0.01	0.0006	—	0.8	—
119K 400–430	0.6	0.3	0.003	0.001	—	0.6	—
121G 0–30	0.3	0.1	0.003	—	—	0.6	—
121G 270–280	0.8	0.2	0.003	—	—	0.6	—
121G 270–290	1.	0.8	0.006	0.0008	—	0.6	—
126P 284–314	0.6	0.1	0.008	0.0008	—	0.6	—
126P 33–63, Top	0.4	0.1	0.003	—	—	0.6	—
161K 0–70	0.8	0.1	0.008	0.0006	—	0.2	—
161K 70–130	0.6	0.3	0.006	—	—	0.3	—
161K 130–180	0.8	>10.	0.008	0.0008	0.09	—	—

Elements sought but not found: Bi, Ce, Hf, Li, Nb, Pd, Pt, Re, Ta, Te, Th, U, and Yb.
Spectrographic analyses by F. B. Reisbick.

of Red Sea Sediments (continued)

Ag	Na	Sr	Tl	Sn	Ti	W	V	Y	Zn	Zr
—	2.	0.2	—	—	0.1	—	0.015	—	0.15	0.01
—	2.	0.2	—	0.00006	0.2	—	0.01	0.004	0.04	0.03
—	1.5	0.2	—	0.0001	0.4	—	0.01	0.004	0.05	0.04
—	>4.	0.08	—	—	0.2	—	0.01	0.004	0.05	0.01
0.00006	1.	0.04	—	—	0.02	0.0002	0.03	0.008	0.15	0.01
0.00006	2.	0.04	—	—	0.01	—	0.03	0.008	0.09	—
—	>4.	0.6	—	—	0.01	—	0.004	—	0.08	0.08
0.0015	>4.	—	0.0003	—	—	—	0.004	—	0.6	—
0.0008	>4.	—	0.0002	—	—	—	0.003	—	0.007	—
0.0008	>4.	—	0.0002	—	—	—	0.004	—	0.4	0.004
0.0015	>4.	—	0.0006	—	—	0.0002	0.006	—	1.	0.02
0.0006	>4.	—	—	—	—	—	0.004	—	0.02	0.008
0.0006	>4.	—	—	—	—	0.0002	0.006	—	0.07	0.02
0.0006	>4.	—	0.0002	—	0.06	—	0.004	—	0.1	0.06
0.004	3.	0.06	0.002	0.00015	0.4	—	0.01	—	0.6	0.04
—	>4.	0.08	—	—	0.03	—	0.008	—	0.02	0.04
—	>4.	—	—	—	0.04	—	0.006	—	0.05	0.04
—	3.	—	—	—	—	—	0.01	—	0.01	—
—	1.	0.6	—	0.00006	0.2	—	0.008	—	0.015	0.02
—	2.	0.1	—	0.00015	0.4	—	0.01	0.004	0.04	0.01
0.0001	>4.	0.1	—	—	0.15	—	0.008	—	0.06	0.02
—	>4.	0.2	—	—	0.08	—	0.01	—	0.05	0.01
—	>4.	0.1	—	0.00006	0.6	—	0.01	0.004	0.03	0.015
—	>4.	0.1	—	0.00006	0.1	—	0.015	0.006	0.02	0.01
0.004	>4.	0.04	0.0002	—	0.04	0.0002	0.01	—	2.	—
0.001	>4.	0.06	—	—	—	—	0.006	—	0.1	0.03
0.006	>4.	0.2	0.0008	—	0.03	—	0.008	—	2.	0.006
0.0008	>4.	0.08	0.0002	—	—	—	0.006	—	0.1	0.02
0.004	>4.	0.4	0.002	—	0.06	0.001	0.015	—	1.5	0.015
0.0003	>4.	0.04	—	—	—	—	0.008	—	0.007	—
0.0006	>4.	—	—	—	—	—	0.002	—	0.008	0.01
0.002	>4.	0.04	0.0003	—	0.04	—	0.006	—	0.02	0.01
0.0006	>4.	—	—	—	—	—	0.003	—	0.01	0.015
0.0008	>4.	—	—	—	0.02	—	0.008	—	0.015	0.03
0.0006	>4.	—	0.0003	—	—	0.0002	0.01	—	0.03	0.02
0.0006	>4.	—	0.0003	—	—	—	0.003	—	0.05	0.015
0.001	>4.	—	—	—	—	—	0.003	—	0.01	0.01
0.001	>4.	—	—	—	—	—	0.004	—	0.006	0.03
0.003	>4.	—	0.0006	—	0.01	—	0.008	—	1.5	0.01
0.0004	>4.	—	0.0002	—	—	—	0.002	—	0.07	—
0.003	>4.	0.06	—	—	0.04	—	0.01	—	0.6	—
0.003	>4.	—	—	—	0.015	—	0.01	—	1.	0.01
0.0008	4.	0.08	—	—	0.04	—	0.01	—	1.	0.015

Table 7   Chemistry

Sample Location	Fe^{++} (mg/liter)	Fe Total (mg/liter)	Cu^{+} (mg/liter)	Cu Total (mg/liter)	Cl (g/100 ml)	Br (mg/liter)	Na (g/100 ml)	K (g/100 ml)	Ca (g/100 ml)
71 15–25cm	85	138	0.03	0.8	18.5	110	11.1	0.28	0.61
71(b) 59–74cm	51	66	0.02	0.4	19.0	140	11.0	0.35	0.50
78C 15–25cm	120	162	ND	0.3	18.5	180	10.3	0.33	0.71
78D	117	174	0.02	0.4	18.8	180	11.0	0.35	0.66
80 25–15cm	<1	2	ND	0.4	18.3	150	10.2	0.35	0.54
84K 206cm	21	21	ND	0.5	19.1	150	11.1	0.33	0.64
84K 365–370cm	16	19	0.30	0.4	19.1	170	10.3	0.34	0.69
* 84K 400cm	9	15	0.02	0.8	18.8	60	11.7	0.35	0.50
80P 120–130cm	7	10	ND	1.2	18.7	140	10.7	0.38	0.43
80 310–320cm	10	14	0.24	0.3	18.5	170	11.0	0.33	0.55
80 580–595cm	4	8	0.18	0.6	18.4	160	10.8	0.35	0.52
80 840–850cm	10	13	0.10	0.7	18.4	110	11.5	0.31	0.53
81P 15–25cm	21	26	ND	0.7	17.7	110	10.4	0.34	0.44
81P 315–325cm	20	30	0.03	1.0	18.4	110·	10.5	0.35	0.57
81P 605–615cm	39	57	ND	0.5	18.6	85	11.0	0.35	0.42
81P 855–865cm	16	25	ND	0.7	18.5	140	10.8	0.34	0.42
82B 15–25cm	<1	1	ND	ND	2.6	–	1.2	0.04	0.05
84K 143cm	24	25	0.35	0.6	18.9	170	11.3	0.35	0.64
84K 263cm	35	39	ND	1.6	19.3	·160	11.4	0.40	0.61
84K 0–30cm	50	60	ND	0.6	19.1	170	11.3	0.40	0.59
84G 89–99cm	34	41	ND	1.2	18.8	160	11.1	0.43	0.60
84P 290–315cm	19	25	ND	0.7	18.5	160	11.1	0.43	0.61
84P 595–605cm	25	53	0.10	0.4	18.7	50	11.3	0.42	0.65
84P 865–875cm	21	29	0.02	0.6	18.4	140	11.1	0.38	0.56
85K 30cm	<1	0.1	0.41	0.5	2.3	–	1.1	0.034	0.046
85K 190cm	<1	1.8	0.14	0.2	2.7	–	1.4	0.040	0.059
85K 258cm	<1	0.2	0.10	0.1	3.0	–	1.1	0.039	0.049
85K 300cm	<1	0.3	–	0.2	3.0	–	1.6	0.044	0.063
85K 340cm	<1	0.4	0.47	0.6	3.0	–	1.5	0.048	0.069
* 85K 400cm	<1	2	–	0.2	3.0	–	1.5	0.045	0.070
89C 0–10cm	93	123	ND	1.1	18.8	170	11.2	0.39	0.62
89C 10–30cm	94	113	ND	1.3	18.8	–	12.7	0.39	0.59
89C 20–30cm	98	108	0.14	0.3	18.6	140	10.8	0.42	0.65
95K 0–10cm	<1	0.3 est.	–	0.1 est.	3.0	–	1.2	0.034	0.045
† 95K 270–305cm	<1	0.5	–	0.1 est.	2.3	–	1.1	0.035	0.047
106G 0–30cm	102	110	0.10	0.1 est.	18.7	160	10.2	0.40	0.63
107G 0–30cm	20	38	ND	0.3	18.6	150	10.2	0.40	0.59
108G 0–30cm	97	119	ND	0.5	18.6	170	11.4	0.42	0.62
109G 5–35cm	28	49	ND	1.3	18.7	130	10.6	0.32	0.53
109G 140–170cm	24	39	ND	1.1	18.7	160	10.9	0.37	0.61
117K 200cm	<1	3.6	–	0.2	2.4	–	1.1	0.040	0.049
117K 255cm	<1	5.4	–	0.1	2.0	–	1.0	0.031	0.045
119K 30–40cm	49	62	ND	0.1 est.	18.5	120	10.8	0.42	0.61
119K 50–60cm	60	73	ND	0.2	18.8	120	10.2	0.41	0.63
119K 90cm	32	35	ND	0.4	18.7	150	10.3	0.40	0.60
119K 110cm	33	33	0.01	0.2	18.6	120	10.0	0.40	0.60
119K 240cm	8	10	0.03	0.4	18.9	130	10.3	0.41	0.62
119K 345–360cm	<1	3	0.05	0.5	18.9	120	10.5	0.40	0.61
119K 400–430cm	ND	2	ND	0.2	18.4	120	10.2	0.40	0.60
121G 0–20cm	121	136	ND	1.2	18.7	130	10.9	0.36	0.58
121G 280–290cm	19	25	ND	1.5	18.3	130	10.2	0.34	0.55
‡ 126P 33–63cm	89	126	ND	1.7	18.5	130	10.5	0.36	0.58
126P 284–314cm	74	83	ND	1.5	19.1	140	10.5	0.39	0.58
161K 0–70cm	1	1	ND	0.7	15.8	180	9.2	0.23	0.52
161K 70–130cm	7	10	0.10	0.2	17.1	160	9.4	0.36	0.51
161K 130–180cm	–	2	0.27	0.3	17.1	120	9.0	0.36	0.52
161K 180–275cm	2	4.1	ND	2.1	17.5	130	9.7	0.39	0.52
161K 275–320cm	2	10	ND	1.5	17.9	130	10.3	0.43	0.55

* Catcher.
† From bottom above catcher.
‡ This sample is suspect because it was precipitated by heating.

of Interstitial Brines

Mg (mg/liter)	Sr (mg/liter)	Mn (mg/liter)	Zn (mg/liter)	Al (mg/liter)	Co (mg/liter)	Ni (mg/liter)	Ag (mg/liter)	B (present as borate) (mg/liter)	$SO_4^=$ (mg/liter)	SiO_2 (monomeric) (mg/liter)	PO_4 (mg/liter)
825	51	99	15	ND	—	—	0.04	10	400	4	19
740	64	71	2.3	5	—	—	ND	12	330	13	13
750	65	109	16.7	2 est.	—	—	0.05	11	200	14	17
740	57	93	20.5	7	—	—	0.04	11	300	13	13
975	75	151	1.9	2 est.	—	—	0.08	9	260	3	6
815	61	93	14	2 est.	—	—	0.05	11	80	25	7
850	63	80	2.3	3 est.	—	—	0.38	10	310	18	4
825	58	75	0.7	3 est.	—	—	ND	10	360	14	4
1,315	73	29	0.8	6	0.9	0.7	—	13	—	5	4
1,300	72	6	1.6	2 est.	—	—	0.13	14	20	5	2
1,265	75	25	2.4	3 est.	—	—	0.50	13	ND	4	26
745	64	18	1.7	3 est.	—	—	ND	14	10	4	5
1,005	73	21	1.2	5	—	—	0.001	13	20	4	7
1,160	76	11	0.7	5	—	—	ND	15	20	13	11
1,170	73	8	1.7	5	—	—	ND	16	30	14	17
1,140	76	8	1.5	3 est.	—	—	0.002	15	20	12	11
1,470	7	6	0.3	2 est.	—	—	—	5	—	—	—
815	65	86	35	2 est.	—	—	0.07	10	110	14	7
870	66	72	19	4 est.	1.1	0.6	0.02	12	210	21	9
850	65	78	26	6	1.3	0.8	0.02	12	120	12	5
795	66	80	50	4 est.	1.0	0.8	0.02	9	440	8	5
825	69	84	44	ND	0.9	0.6	0.02	9	600	9	5
840	66	89	32	5	1.9	0.7	—	7	—	13	2
835	58	76	1.2	2	—	—	0.005	12	580	16	13
1,245	7	13	0.4	2	ND	0.03 est.	—	4	—	18	1
1,375	8	9	0.5	2	0.03 est.	0.03 est.	—	5	—	22	2
1,225	7	8	0.4	2	ND	0.06 est.	—	3	—	27	2
1,405	8	6	0.4	2	0.03 est.	0.06 est.	—	3	—	37	5
1,460	8	8	0.4	2	0.10	0.06 est.	—	4	—	25	3
1,465	8	10	1.9	2	0.13	0.15	—	5	—	13	ND
820	63	85	14.5	8	0.9	0.8	0.14	10	290	10	13
740	66	103	15.1	15	3.4	0.9	0.04	11	300	12	4
800	69	74	20.6	5	1.4	1.0	—	10	630	10	14
1,435	7	2	0.1 est.	2 est.	ND	0.03 est.	—	5	—	7	ND
1,380	7	6	0.2	2 est.	ND	0.06 est.	—	5	—	9	0.1
740	66	93	14.4	4 est.	1.7	1.0	—	11	390	11	8
775	66	90	5.3	7	1.6	0.7	0.03	11	530	14	3
1,665	65	88	14.8	3 est.	0.9	1.0	0.01	9	280	10	12
690	69	98	3.3	8	4.1	0.9	0.03	9	240	8	ND
820	60	78	18.8	4 est.	0.9	0.8	0.03	9	310	12	12
1,410	7	4	0.3	2	ND	<0.1	—	6	—	11	1
1,160	6	4	0.4	2	ND	<0.1	—	5	—	10	<1
840	63	81	18.8	7	1.4	0.9	0.05	12	390	10	3
840	66	81	15.8	3 est.	2.6	0.7	0.04	12	330	14	9
865	66	79	16.0	2 est.	1.4	0.9	0.04	11	350	12	7
800	69	81	21.5	3 est.	2.4	1.4	0.05	12	360	17	6.1
840	69	80	8.7	3 est.	2.6	0.7	0.05	11	320	11	ND
825	66	73	6.0	3 est.	2.3	1.4	0.05	10	350	8	<1
865	66	36	4.1	2 est.	1.7	0.5	—	10	320	6	ND
725	60	106	14.0	6	3.5	1.0	0.03	11	420	8	—
765	66	119	3.5	ND	3.4	0.8	§6.0	10	440	14	<1
725	66	101	14.0	6	3.5	1.1	0.08	11	400	15	8
725	69	134	7.1	15	—	—	0.04	11	400	19	3
800	65	225	3.0	3 est.	—	—	0.03	7	770	2	5
740	63	184	11.0	3 est.	1.4	0.4 est.	—	9	—	7	<1
740	60	205	10.1	2 est.	1.4	0.1 est.	—	6	210	5	ND
690	54	249	22.0	ND	2.7	1.1 est.	0.03	9	320	5	ND
725	66	274	18.4	8	3.4	1	0.03	9	260	7	ND

§ Possible contamination . . . not verified by rerun.
ND = Sought but not detected.
— = Not done.

Table 8

Sample Location	Fe^{++} HCl sol. (%)	Fe Total HCl sol. (%)	Fe Total (%)	Cu^+ (ppm)	Cu Total (%)	SO_4 H_2O sol. (%)	S Total calc. as SO_4 (%)
71 0–15	1.86	3.90	3.76	10	0.098	0.154	1.26
71 15–25	9.46	20.90	23.80	165	0.663		8.98
71 44–59	2.58	2.16	4.87	3	0.165	0.151	2.59
71(b) 59–74 pH 5.0	4.03	8.09	9.60	4	0.240	0.245	4.20
78C 0–15	1.14	5.52	6.54	8	0.061	0.137	1.02
			6.61		0.061		
[a]78C 15–25	2.56	12.40	12.50	11	0.113	0.169	1.79
78D 0–15	2.79	9.37	9.60	8	0.085	0.471	1.13
[a]78D 15–25	0.326	15.50	14.40	23	0.150	0.174	2.03
80P 0–15	0.79		5.22	0.5	0.021		—
					0.025		
80P 15–25	0.495	7.00	7.50	1	0.020	0.062	0.60
80P 100–120	1.10	4.25	5.00	0.3	0.014		—
80P 290–310	1.05	4.43	4.94	0.2	0.011	0.005	—
80P 310–320	0.71	2.20	5.56	0.1	0.015	0.024	0.09
80P 580–595	0.16		5.31	0.8	0.009	0.036	0.09
80P 595–615	0.96	4.32	4.99	0.5	0.016	0.018	0.08
80P 840–850	0.16	6.00	5.31	0.3	0.012	0.033	0.17
80P 850–870	1.16	2.64	4.99	0.7	0.016	0.010	0.04
81P 0–15	2.14	5.87	5.15	1	0.014	0.072	0.20
81P 15–25	0.47	3.54	3.13		0.007	0.028	0.03
81P 290–305	3.27	3.64	4.73	0.6	0.014	—	1.60
81P 315–325	2.13	6.23	5.00		0.010	0.018	1.85
							2.04
81P 590–605	2.78	4.39	4.62	0.7	0.016	—	1.65
			4.62		0.016		
81P 605–615	0.369	5.15	5.00	0.2	0.012	—	2.12
					0.012		
81P 855–865	2.17	4.81	5.00	0.1	0.012	—	1.87
							1.87
81P 865–880	2.67	3.71	4.53	0.3	0.018	—	1.79
82B 15–25	0.45	30.80	30.63	25	0.063	1.06	0.81
							0.80
82B 15–25 (f.f.)	0.772	17.30	17.50	6	0.488		8.62
82D Near catcher (f.f.)	0.34	27.30	34.70	0.7	0.150		0.04
84G 89–99	1.00	34.40	38.80	87	0.125		4.41
84G 99–114	3.28	5.83	6.25	5	0.088	0.354	2.20
84K 0–30	6.43	12.80	13.10	55	0.250	0.171	5.17
			13.10		0.250		5.63
84K 143	2.57	4.34	4.59	156	0.152	0.304	2.76
84K 206	3.85	7.05	6.56	0.5	0.225	0.625	6.02
			6.81				
84K 263	1.58	3.69	7.93	2	1.011	[b]2.93	54.91
84K 300	4.80	4.98	7.30	0.6	0.999	[c]3.54	31.69
84K 304	6.43	6.40	7.76	0.1	0.739	[d]1.99	43.52
84K 310–325	5.77	6.00	8.44	—	0.920	[e]3.60	39.22
			8.44		0.920		
84K 325–348	8.62	10.20	8.75	—	0.666	4.46	35.99
84K 325–348 L.P.	7.86	7.86	7.50	—	1.00		33.36

Second number for any one sample indicates repeat digestion.
— = Sought but not detected.
* Blank = Not run.

Chemistry of Sediments

S Total (%)	Cl (%)	F (ppm)	Na (%)	K (%)	Li (ppm)	Ca (%)	Mg (%)	Sr (ppm)	Ba (%)
0.42	51.9	*	24.6	1.01	6	1.75	0.288	130	0.069
	52.5								0.069
2.99	17.7	<100	11.2	0.75		1.94	0.375	75	0.044
0.86	50.5	<100	24.4	1.01	6	2.06	0.325	120	—
	50.5								
1.40	41.6	<100	19.6	0.88	5	1.66	0.344	110	0.031
0.34	49.5	<50	24.4	0.97	5	1.41	0.266	110	0.031
			24.6	0.98	5	1.41	0.266	110	
0.60	40.4		22.5	0.96		1.81	0.313	110	0.169
	41.4								
0.38	44.2		23.8	0.92	5	1.72	0.291	110	0.088
0.68	39.0		21.2	0.83		1.75	0.313	120	0.062
—	28.1		16.1	0.74	8	5.13	1.41	220	0.100
0.20	12.0		7.5	0.75		12.2	1.58	610	0.100
	11.7			0.79					
—	1.37	<100	8.1	0.86	8	16.9	1.82	660	0.119
—	14.4		8.4	0.83	8	16.9	1.79	560	0.119
0.03	7.6		4.1	0.87	9	20.0	2.02	680	0.056
0.03	7.0		4.4	0.81		18.1	1.95	700	0.138
0.03	14.0	180	8.0	1.12	11	18.9	1.72	640	0.050
0.06	6.5		4.4	0.88		17.2	2.08	680	0.088
				0.87					
0.01	13.2		7.9	1.04	11	18.1	1.75	700	0.062
0.07	17.4		9.5	0.79	11	14.4	1.99	440	0.081
0.01	6.7		5.0	0.96		20.5	1.88	700	0.094
	6.8		4.6						
0.53	16.8	170	10.1	0.82	8	15.0	1.60	420	0.300
	16.4								
0.62	10.4		6.0	0.87		17.5	1.88	620	0.081
0.68									0.081
0.55	17.7		10.2	1.04	10	14.8	1.47	440	0.088
			10.4		10	14.8	1.50	430	
0.71	10.5		6.2	0.96		17.8	1.56	860	0.050
			6.2			17.8			
0.62	9.8		5.6	0.87		17.5	1.25	840	0.099
0.62									
0.60	16.8	180	9.7	1.12	10	15.4	1.53	420	0.138
				1.08					
0.27	12.1		6.9	0.46		1.25	1.25	110	0.194
0.27	12.4								
2.88	22.8		13.1	0.75		11.6	0.56	100	0.194
0.01	5.3		2.9	0.17	2	0.81	2.06	30	0.088
1.47	3.1		1.2	0.21		0.56	0.75	180	0.200
0.73	47.7	<50	28.8	0.813		1.87	0.28	90	0.062
	47.2								
1.73	36.2		20.0	0.87		1.69	0.38	110	0.081
1.88			20.0	0.83		1.69		120	
0.92	46.9		26.2	1.04	5	1.28	0.48	120	0.025
2.01	44.0		27.3	1.42	5	1.44	1.19	100	0.038
18.32	12.3	160	7.1	0.27	5	8.90	1.56	140	0.031
9.83	27.9		16.9	0.79	8	4.00	2.00	120	0.125
14.52	21.1		11.9	0.54	5	6.56	1.25	150	0.019
13.08	19.5		11.9	0.44		8.40	1.58	160	0.038
			11.9	0.38		7.95	1.60	160	
12.01	16.1		9.4	0.50		5.00	0.62	120	0.188
11.13	24.7		13.1	0.70		3.13	1.25	40	0.031
	23.0								

[a] Squeeze and remainder.
[b] HCl soluble SO$_4$ in this sample was 21.8%.
[c] HCl soluble SO$_4$ in this sample was 9.91%.
[d] HCl soluble SO$_4$ in this sample was 20.6%.
[e] HCl soluble SO$_4$ in this sample was 20.3%.

Table 8

Sample Location	Zn (%)	Pb (ppm)	Mn (%)	Cr (ppm)	Ni (ppm)	Co (ppm)	Cd (ppm)
71 0–15	0.66	140	0.038	20	25	43	30
						38	26
71 15–25	4.13	880	0.120	18	25	195	198
71 44–59	1.42	240	0.032	10	16	62	53
71(b) 59–74 pH 5.0	1.72	320	0.051	12	19	68	87
78C 0–15	0.27	120	0.052	8	10	30	13
	0.28	130	0.052	8	10	31	13
a 78C 15–25	0.27	180	0.375	8	11	45	19
78D 0–15	0.55	180	0.150	9	12	41	26
							27
a 78D 15–25	1.05	260	0.155	8	9	51	58
80P 0–15	0.12	59	8.00	18	30	13	12
				20	25		
80P 15–25	0.29	120	7.44	3	50	35	12
80P 100–120	0.058	67	0.466	32	53	30	5
80P 290–310	0.049	65	0.455	32	53	30	4
80P 310–320	0.042	67	0.511	35	56	34	4
	0.040						
80P 580–595	0.046	50	0.500	53	60	35	3
80P 595–615	0.052	40	0.625	35	55	29	13
80P 840–850	0.053	55	0.750	60	55	35	2
80P 850–870	0.052	35	0.266	35	55	27	7
81P 0–15	0.036	59	0.944	42	60	34	5
81P 15–25	0.031	70	0.500	50	49	32	2
81P 290–305	0.046	65	0.278	29	63	28	4
						25	
81P 315–325	0.049	60	0.500	40	50	30	3
81P 590–605	0.055	35	0.276	35	75	24	7
	0.055	35	0.277	35		25	7
81P 605–615	0.049	60	0.375	60	50	29	2
	0.049	60	0.375	60	50	30	
81P 855–865	0.045	60	0.192	70	56	32	2
81P 865–880	0.050	40	0.279	35	50	26	9
82B 15–25	0.588	210	9.69	20	14	22	4
82B 15–25 (f.f.)	3.56	6.20	0.250	15	33	113	144
82D Near catcher (f.f.)	0.410	160	13.44	20	25	28	22
84G 89–99	0.970	500	10.31	30	14	22	4
84G 99–114	0.488	200	0.075	2	14	41	14
84K 0–30	1.92	370	0.065	35	17	91	69
	1.94	380	0.065		18	93	72
84K 143	0.860	170	0.097	8	15	48	39
84K 206	0.565	270	0.153	10	15	56	29
84K 263	10.06	450	0.014	10	49	198	313
84K 300	1.39	320	0.038	10	55	117	134
84K 304	4.66	400	0.032	15	35	136	130
84K 310–325	3.84	220	0.110	19	60	115	140
	3.86	220	0.120	20	60	110	143
84K 325–348	2.27	70	0.750	45	40	75	65
84K 325–348 L.P.	5.25	280	0.250	85	65	135	185
84K 365–370	0.055	90	0.125	30	14	29	4
84K 365–370 (352–400 same)	0.044	30	0.288	10	20	44	10

(continued)

Mo (ppm)	Al (%)	SiO$_2$ (%)	PO$_4$ (%)	B HCl sol. borate (ppm)	CO$_3$ (%)	H$_2$O (%)	Au (ppm)	Ag aqueous (ppm)	Ag extraction (ppm)
83	0.13	2.25	0.034	51	0.05	70.1	0.36	10	
100	0.72	5.75	0.120		3.37	46.1	1.3	54	65.8
90	0.19	3.75	0.040		0.04	63.4	0.28	12	
	0.22	6.50	0.089		0.04	65.6	0.59	23	
			0.077						
69	0.13	3.13	0.052		–	69.3	0.17	6	
	0.13	3.38				67.4			
59	0.24	6.25	0.084	64	1.39	65.5	0.33	10	
				63					
90	0.22	4.60	0.066		0.06	66.8	0.28	8	
53	0.25	7.25	0.109		0.85	63.7	0.15	15	
53	0.93	7.88	0.215		8.24	55.0	0.28	–	
47	1.70	12.50	0.279		12.57	40.1	0.083	5	
		12.50						6	
48	2.20	15.33	0.261		too high	38.1	0.047	–	
						41.9			
48	2.20	12.98	0.248		29.4	40.8	0.16	–	
23	2.40	3.25	0.304		too high	26.7	0.16	–	–
			0.313						
12	2.60	10.31	0.322			24.4	–	–	
83	2.20	10.25	0.258		too high	35.6	0.055	–	
75	2.00	7.38	0.279			25.9	–	–	
50	2.20	12.80	0.258		30.3	45.0	0.063	–	
56	2.20	11.34	0.227	59	too high	64.0	0.20	–	
13	2.80	7.00	0.304	37		24.4	0.083	–	
							0.035		
69	2.00	12.36	0.261		26.2	44.0	0.12	–	
					24.6				
17	2.20	8.00	0.313			36.6	0.021	–	
		8.00							
45	1.90	4.60	0.254		too high	44.0	0.079	–	
	1.90					41.5			
17	2.30	7.75	0.331			32.6	–	–	
13	1.80	8.13	0.314		19.57	31.1	–	–	
45	1.90	4.39	0.264		too high	42.9	0.16	–	
59	0.33	6.38	0.739		1.85	76.9	0.10	–	2.3
			0.744						
87	0.31	8.88	0.141		3.11	52.5	1.2	48	
60	0.28	4.49	0.736		2.60	63.5	0.28	–	
			0.744						
30	0.29	8.38	0.665	189	1.77	51.9	0.48	–	
			0.698						
40	0.11	3.06	0.037	38		69.3	0.13	15	
94	0.35	6.88	0.096			62.6	0.28	35	19
	0.35								25
48	0.19	5.88	0.034		0.31	67.6	0.40	16	
140	0.26	6.13	0.055		1.01	71.2	0.66	17	
			0.052						
1,290	0.72	9.50	0.074			34.9	5.6	109	
415	0.47	6.84	0.077	31	1.62	57.4	2.5	31	
570	0.25	6.25	0.043	19	5.72	46.4	3.3	31	
		5.75							
189	0.31	6.75	0.049		2.20	48.6	3.2	27	
225	0.33								
292	0.52	5.00	0.061	32		57.6	0.35	33	
							0.31		
458	0.72	8.25	0.089			50.0	1.9	71	
220	0.23	4.50	0.138			42.5	0.49	5	
280	0.22	2.85	0.080		0.70	51.4	0.35	5	4.8

Table 8

Sample Location	Fe^{++} HCl sol. (%)	Fe Total HCl sol. (%)	Fe Total (%)	Cu^{+} (ppm)	Cu Total (%)	SO$_4$ H$_2$O sol. (%)	S Total calc. as SO$_4$ (%)
84K 365–370	5.57	22.70	41.20	56	0.288	1.08	2.40
84K 365–370 (352–400 same)	3.64	28.00	27.20	0.5	0.20	2.48	6.65
[f](1) 84K 400 Catcher	5.45	21.10	20.30	0.2	0.132	0.117	
[f](2) 84K 400 Catcher	2.43	30.50	30.30	13	0.135	3.03	3.07
80P 120–130	1.56	4.90	5.00		0.013	0.010	0.12
84P 290–315	4.26	10.50	14.40		0.017		8.13
							7.88
84P 315–345	2.62	3.85	3.42	5	0.184	0.610	3.39
84P 580–595	4.03	4.44	6.75	4	0.470	3.06	21.12
					0.470		
84P 595–605	5.36	6.24	8.13	244	0.032	5.14	37.30
84P 850–865	1.56	31.90	43.10	–	0.310	0.288	1.51
84P 865–875	2.76	50.20	52.50	6	0.388	0.019	1.73
85K 30, Unit 6	1.06	43.10	40.00	205	0.255	0.666	0.97
85K 135	14.60	16.40	16.20	387	1.70	[g] 1.31	8.19
85K 170, Clinton type	2.24	38.50	35.00	431	0.563	0.758	1.91
85K 70, Unit 6	1.59		27.50	1,330	0.250	2.22	1.91
							2.00
85K 190, Unit 5	1.45	45.60	47.50	22	0.050		0.34
85K 200, Unit 5	0.629		28.70	40	0.059		1.42
85K 280	0.757	40.60	40.00	103	0.039	1.59	1.43
85K 300, Unit 3	17.20	42.00	40.00	73	0.183	1.39	1.50
					0.175		
85K 340, Unit 2	2.02	22.10	40.00	10	0.275		0.56
85K 360, Unit 2	7.00	33.80	33.10	42	0.310	0.777	1.56
85K 258, Unit 4	0.598	27.90	40.00	55	0.063	0.273	0.56
85K Unit 1, Core catcher	14.10	15.90	20.60	15	3.34	[h] 0.100	62.7
89C 0–10	4.50	6.72	7.81	393	0.17	0.208	2.78
			7.81		0.17		
89C 10–30	5.71	15.20	15.50	12	0.338	0.154	4.76
89C 0–20	1.16	2.74	2.75	24	0.050	2.35	33.20
							33.46
89C 20–30	5.51	15.00	16.00	1	0.175		22.60
94C 130–145	0.13	0.51	0.40	28	0.021	[i] 1.96	43.6
95K 0–10	2.72	8.68	8.75	0.3	0.025	0.05	0.02
			9.38		0.025		
95K 10–20	0.088	5.76	6.25	0.8	0.009	0.32	0.11
95K 40–45	3.11	7.01	7.50	0.6	0.007		0.14
95K 45–55	0.239	3.95	4.38	0.2	0.008	0.328	0.29
95K 55–60	0.458	14.30	23.10		0.006	0.566	0.79
95K 270–305 From bottom	2.00	45.40	45.00	29	0.075	0.497	2.15
95K From above catcher	0.965	28.80	37.00	57	0.100	1.33	3.00
[j] (A) 97G From bottom of core	0.284	3.32	13.10		0.025	0.087	1.21
[j] (B) 97G From bottom of core			3.13		0.025		1.06
106G 0–30	2.09	3.35	4.38	23	0.138	0.646	2.90
					0.138		
107G 0–30	0.58	3.83	3.88	3	0.050	0.655	1.68
108G 0–30	1.77	3.98	5.63	21	0.088	0.103	1.39
108G 173–193	1.26	1.46	3.75	5	0.125	0.189	3.83
							4.06
109G 5–35	0.93	3.97	5.63	5	0.025	0.232	0.20
109G 140–170	0.348	0.72	3.44	0.1	0.049	0.317	0.98
			3.75		0.049		
111C 122 Core catcher	1.62	4.98	5.63	3	0.063	0.164	0.73
111D 122 Core catcher	3.20	3.20	6.88	8	0.450	0.329	14.84
117K 200	1.32	32.60	33.80	1	0.004	0.901	0.92

[f] (1) and (2) were different jars with the same sample station and depth label.
[g] HCl soluble SO$_4$ in this sample was 4.56%.

(continued)

S Total (%)	Cl (%)	F (ppm)	Na (%)	K (%)	Li (ppm)	Ca (%)	Mg (%)	Sr (ppm)	Ba (%)
0.80	15.4		10.0	0.37		1.56	0.56	60	0.069
2.22	25.4		14.6	0.58	2	2.05	0.44	60	0.188
			14.4	1.29	–	6.56	0.50	1,250	0.438
1.02	20.5		12.6	0.42	2	2.13	0.84	440	–
0.04	7.9		5.0	0.87		18.1	1.94	740	0.062
2.71	26.7	90	15.6	1.33		1.63	0.63	240	0.275
2.62	26.3								
1.13	47.6		28.6	1.25	9	1.48	0.34	270	0.188
7.04	30.4		18.3	0.87	6	3.75	0.41	230	0.312
	30.3		18.1		6	3.75	0.38	230	
12.44	12.5		6.9	0.33		16.0	0.69	330	0.131
0.50	16.8		10.0	0.31		1.46	0.42	30	0.062
0.58	7.3	180	4.1	0.12		2.29	0.45	<20	0.112
	7.0								
0.32	7.7		4.4	0.25		0.63	0.80	420	0.431
									0.444
2.73	9.5		5.6	0.25		0.94	1.50	520	1.38
0.64	7.7		4.4	0.25		0.47	0.70	1,620	1.20
0.64	22.1		12.5	0.50		0.80	1.40	190	0.256
0.67									
0.11	4.2	280	2.5	0.12		0.23	0.50	80	0.400
0.47	16.8		8.8	0.54	2	0.44	1.00	130	0.250
			9.2	0.58	2	0.44	1.00	150	
0.48	15.3		8.0	0.71	3	0.50	0.81	60	0.062
	15.1								
0.50	14.2		7.6	0.50	2	0.39	0.81	140	0.412
			8.1						
0.19	2.8		1.5	0.25		0.94	0.83	140	0.469
0.52	7.4	180	3.8	0.50		0.96	1.50	180	0.312
	7.4								
0.18	5.6		3.1	0.125		0.39	0.48	130	0.588
20.92	3.0		0.6	0.063		1.53	1.06	40	0.125
0.92	41.8		23.8	1.00		2.19	0.44	110	0.012
			23.8	0.94		2.06	0.34	100	
1.58	25.4		15.3	0.92	5	0.44	0.47	100	0.048
11.08	26.0	<50	15.6	0.44		18.2	0.15	230	0.112
11.16									
7.54	13.0		8.1	0.34		14.6	0.30	120	0.162
14.53	17.9		8.8	0.31		13.6	0.14	160	0.019
1.9			1.2	0.38		21.9	1.80	880	0.044
			1.2			22.1	1.87	880	
4.9			3.8	0.83		18.5	1.83	420	0.019
0.05	3.3	300	1.9	0.47		25.0	1.87	1,050	0.075
0.10	2.6		1.9	0.75		22.4	2.08	1,000	0.069
0.26	11.0		6.2	0.62		7.08	1.67	210	0.262
0.72	5.3		3.0	0.22		2.50	1.18	180	0.094
1.00	9.7		5.2	0.31		2.08	1.46	140	0.650
0.40	29.1	120	16.2	0.62		2.08	0.42	2,120	2.00
0.35	50.0		29.4	1.4		1.63	0.38	1,880	0.688
0.97	49.1		28.8	0.78		2.08	0.30	130	0.225
			30.0	0.81			0.30	130	
0.56	49.7		32.5	0.81		1.87	0.30	120	0.044
0.46	47.2		28.8	0.56		1.87	0.28	120	0.019
1.28	44.3		28.1	1.10		1.94	0.25	120	0.044
1.35	45.2								
0.07	49.0	<50	30.6	1.00		1.66	0.19	120	–
0.33	45.7		27.5	1.50		1.89	0.36	130	–
			26.9	1.38		1.87	0.36	130	
0.24	47.2		28.1	1.1		1.56	0.25	110	–
4.98	7.7		3.8	0.9		9.38	1.06	150	0.144
0.31	10.0		4.4	0.2		2.88	0.94	120	–

[h] HCl soluble SO$_4$ in this sample was 0.16%.
[i] HCl soluble SO$_4$ in this sample was 43.3%.
[j] A and B were two different samples in one jar. No separate labels.

Table 8

Sample Location	Zn (%)	Pb (ppm)	Mn (%)	Cr (ppm)	Ni (ppm)	Co (ppm)	Cd (ppm)
[f] (1) 84K 400 Catcher	0.025	35	0.16	10	25	25	4
	0.042						
[f] (2) 84K 400 Catcher	0.040	65	0.130	14	18	40	5
	0.042	67	0.160	25	25		6
80P 120–130	0.045	70	0.030	45	55	33	3
84P 290–315	1.33	750	0.100	55	29	119	74
84P 315–345	1.22	160	0.080	10	15	38	82
84P 580–595	2.35	380	0.283	15	30	76	88
	2.35	390	0.279	15	30	74	89
84P 595–605	2.25	280	0.155	10	52	105	107
84P 850–865	0.025	65	0.150	23	15	65	4
84P 865–875	0.013	40	0.130	24	19	52	—
85K 30, Unit 6	0.813	620	0.588	17	14	26	2
85K 135	3.53	2,000	1.89	14	10	15	376
85K 170, Clinton type	0.750	880	1.50	7	27	26	24
85K 70, Unit 6	0.750	500	0.500	10	10	12	—
85K 190, Unit 5	0.875	250	4.13	19	9	29	3
85K 200, Unit 5	0.595	250	3.75	15	15	25	9
	0.595	240	3.75	15	15	25	9
85K 280	0.375	250	0.472	20	20	24	11
85K 300, Unit 3	0.344	55	0.192	20	15	25	6
85K 340, Unit 2	0.844	320	0.245	20	19	27	22
85K 360, Unit 2	0.400	150	4.21	29	25	35	4
85K 258, Unit 4	0.688	620	1.00	17	14	29	3
85K Unit 1, Core catcher	20.0	220	4.98	23	40	70	600
89C 0–10	1.54	220	0.066	15	15	70	50
	1.54	220	0.067		15	75	51
89C 10–30	2.13	260	0.162	15	25	139	86
89C 0–20	0.188	60	0.120	14	10	23	—
89C 20–30	0.719	240	0.145	17	51	104	26
94C 130–145	0.086	45	0.006	5	10	10	50
95K 0–10	0.281	54	6.25	39	77	32	—
		58	5.98	36	75	33	
95K 10–20	0.740	65	5.80	27	75	40	51
95K 40–45	0.063	30	0.875	46	58	27	2
95K 45–55	0.050	35	0.500	66	62	32	—
95K 55–60	0.063	23	0.375	49	36	26	—
95K 270–305 From bottom	0.375	47	0.150	17	35	62	—
95K From above catcher	0.175	63	0.140	22	36	65	2
[j] (A) 97G From bottom of core	0.138	22	4.00	4	13	15	2
[j] (B) 97G From bottom of core	0.023	50	0.040	25	20	20	5
106G 0–30	0.906	190	0.080	9	13	55	31
	0.906	190	0.040	10	13	58	31
107G 0–30	0.013	100	0.060	7	9	30	—
108G 0–30	0.500	130	0.065	4	9	33	23
108G 173–193	1.17	240	0.500	5	40	60	55
109G 5–35	0.038	65	0.065	5	20	20	8
109G 140–170	0.150	90	0.150	3	15	25	8
	0.150	90	0.140	3	20	25	8
111C 122 Core catcher	0.275	120	0.030	5	20	25	20
111D 122 Core catcher	1.88	480	1.75	50	90	65	75
117K 200	0.028	30	0.150	30	50	50	3

(continued)

Mo (ppm)	Al (%)	SiO₂ (%)	PO₄ (%)	B HCl sol. borate (ppm)	CO₃ (%)	H₂O (%)	Au (ppm)	Ag aqueous (ppm)	Ag extraction (ppm)
207	0.18	7.38				50.8			
332	0.25		0.104		1.85	49.4	0.32	—	
17	2.90	13.13	0.310 0.297			28.8	—	—	
58	0.34	7.50	0.086			55.6	0.85	42	
45	0.18	4.25	0.031		0.72 0.83	71.0	0.40	22	
120	0.24	8.50	0.061		2.47	59.1	1.0	37	
125	0.23								
147	0.29	9.75	0.074			34.8	1.0	42	
219	0.16	4.38	0.098		1.48	33.9	0.30	8	
136	0.22	3.38	0.212 0.208			25.4	0.47	8	
34	0.36	6.26 5.61	0.334	39	1.09	69.8	0.54	74	
83	1.10	11.13	0.028		9.23	79.2	5.0	250 225	260 295
57	0.32	6.95	0.325			72.9	0.89	30	
28	0.13	3.89	0.221		0.39	87.4	0.62	22	
82	0.22	8.13	0.500 0.475			58.2	0.36	—	
95	0.16	4.50	0.331		0.33	81.9	0.19	—	
100	0.17								
95	0.14	3.38	0.228	97	0.06	83.4	0.35	—	
75	0.21	4.70	0.521		—	81.2	0.44	16	
9	0.62	7.25	0.724			29.8	0.56	31	
81	1.90	3.68 3.78	0.454		H₂S inter.	70.4	0.61	10	10
37	0.08	3.38	0.368 0.337		—	44.4	0.18	—	
156	0.33	2.88	0.058	26	H₂S inter.	14.4	4.3 4.1	90	
113	0.31	7.63	0.067		0.93	66.7	0.40	18	
125	0.33								
140	0.55	4.00	0.159		2.57	56.1	0.68	30	32 35
—	0.69	1.94	0.15	27	0.69	53.4	0.035	—	
45	0.38	9.40	0.116		2.88	40.6	0.40	16	
63	0.015	1.00 1.50	—	14	—	37.4	0.14	6	
17	1.50 1.50	7.96	0.533		too high	36.7	0.057	—	
99	1.30	8.25 7.63	0.362		31.1	62.7	—	—	
9	2.20	11.54	0.408		too high	54.9	—	—	
9	3.00	5.31	0.276		too high	50.7	0.036	—	
62	0.32	9.80	0.331		4.86	64.4	—	—	
79	0.49	4.49	1.23		2.24	45.5	—	—	
88	0.51	9.73	1.43		3.11	80.0	0.036	—	0.5
9	0.20	5.93	0.043 0.043		0.10	67.8	—	—	
—	0.05	7.00		38		72.2	—	5	
22	1.90	3.48 3.78	0.040		0.43	67.8	0.36	13	
9	0.075	2.86	0.015		—	70.2	0.13	6	
20	0.16	3.99	0.049		—	57.5	0.22	10	
37	0.16	5.96	0.061		0.97	30.2	0.25	12	
21	0.038	4.24	0.245	250	—	71.3	0.11	6	4.9
—	0.14 0.16	8.85	0.043		0.38	30.9	0.069	5	
21	0.11	5.39	0.055		1.36	74.4	0.063	8	
141	1.90	6.84	0.279		too high	53.0	0.66	38	
50	0.36	4.43	0.472		3.90	71.3	—	—	

Table 8

Sample Location	Fe++ HCl sol. (%)	Fe Total HCl sol. (%)	Fe Total (%)	Cu+ (ppm)	Cu Total (%)	SO4 H2O sol. (%)	S Total calc. as SO4 (%)
117K 255	2.12	31.20	34.40	3	0.009	1.21	0.95
117K 322	0.638	23.50	34.40	3	0.002	0.816	0.77
117K 335	0.903	2.58	2.50		0.013	0.092	0.08
			2.50		0.012		
117K 290–400	1.64	3.72	7.50		0.013	0.084	0.30
118K 75	2.13	4.54	6.25	0.3	0.012	0.102	1.10
118K 220	1.31	2.54	3.75	0.2	0.011	0.108	4.04
118K 310	2.67	3.75	5.00	–	0.005	0.044	3.50
118K 360	1.25	18.10	16.90	0.8	0.014	0.037	1.10
119K 30–40	7.23	15.80	15.00	1	0.400		1.46
119K 50–60	2.69	6.03	5.63	1	0.075	0.208	0.43
119K 90	4.49	10.80	13.80	2	0.288		6.14
119K 100	1.05	1.15	3.13	2	0.063	0.163	1.25
			3.13		0.063		
119K 110	6.36	6.42	11.90	39	0.563	0.132	12.57
119K 115	0.536	0.65	1.88	42	0.088	0.120	0.29
119K 220	0.92	5.45	5.63	3	0.050	0.167	0.40
119K 240	1.76	5.64	14.40	–	0.238	0.012	1.99
119K 325–345	0.45	4.51	6.25	1	0.100	0.165	0.51
			6.25		0.100		
119K 345–360	1.63	12.30	15.60		0.150	0.061	0.81
119K 400–430	3.16	6.75	8.75	2	0.130	0.581	2.28
121G 0–30	0.051	3.47	7.50	3	0.050	0.041	0.86
121G 270–280	0.056	2.99	6.88	2	0.075	0.080	0.51
121G 270–290	0.831	12.31	13.80	2	0.075	0.058	0.37
			13.75		0.074		
126P 284–314	5.09	11.00	12.50	47	0.200	0.081	5.07
126P 33–63 Top	0.895	6.82	6.25	51	0.050	0.145	0.73
161K 0–70	2.65	25.50	26.20	97	0.213	–	1.25
161K 70–130	2.74	3.02	29.38	133	0.300	–	1.78
161K 130–180	0.487	25.60	24.40	268	0.01	0.013	0.14
			24.40		0.150		
[k] 84K 143	4.08	4.14	4.38	90	0.175	0.158	2.76
[k] 84P 315–345	3.20	3.38	3.75	125	0.188	0.311	
[k] 85K 70, Unit 6	2.20	37.20	25.00	1,893	0.263	1.380	2.00
[k] 85K 280	1.67	40.80	38.80	95	0.038	0.926	

[k] 84K 143, 84P 315–345, 70 Unit 6, and 85K 280 were repeated splittings made after all samples were split, to check splitting methods and stability of oxidation states.

(continued)

S Total (%)	Cl (%)	F (ppm)	Na (%)	K (%)	Li (ppm)	Ca (%)	Mg (%)	Sr (ppm)	Ba (%)
0.32	8.9		4.4	0.2		2.09	1.00	60	—
0.26	11.8		6.2	0.2		1.31	1.00	60	—
0.03	1.7		1.3	0.5		25.0	1.50	5,000	—
	1.8		1.4	0.6		25.0	1.56	5,620	
0.10	2.0		1.2	0.7		15.6	2.69	1,250	—
0.37	26.7		15.6	1.0		10.9	1.00	620	0.106
	26.7								
118K 220			13.8	0.7		14.1	0.94	620	0.019
1.17	21.8		13.1	1.0		9.38	1.50	1,250	0.081
0.37	27.1		15.6	0.7		7.81	0.75	210	—
0.98	20.7		13.1	0.6		6.25	0.69	144	0.081
0.14	46.8		28.1	0.9		2.44	0.38	180	0.144
2.05	23.6		15.0	0.8		2.56	0.50	620	0.456
0.42	48.3		29.4	1.3		1.63	0.25	280	0.125
			29.8	1.3		1.63	0.25	280	
4.19	15.7		9.4	0.6		1.75	0.38	3,750	0.350
0.10	52.0		32.5	1.0		1.63	0.25	190	0.019
0.13	44.1		29.4	0.9		1.69	0.25	130	0.044
	45.7								
0.66	38.4		17.5	1.3		1.31	0.56	150	0.125
0.17	46.1		29.4	1.1		1.44	0.25	90	0.025
			30.6	1.1		1.50	0.25	90	
0.27	37.6		16.9	1.9		1.00	0.44	60	—
			16.2						
0.76	37.8		23.9	1.0		1.69	0.38	110	0.050
	38.5		23.8	0.9		1.59	0.39	120	
0.29	47.2		29.4	0.90		1.69	0.19	80	—
0.17	46.1		28.1	1.1		1.63	0.50	80	—
0.12	36.9		23.8	1.06		1.50	0.50	80	—
			24.4	1.06		1.49	0.50		
1.90	35.6		22.5	1.3		1.44	0.31	70	—
0.24	47.8		30.6	1.1		1.94	0.38	80	—
0.42	21.4		12.5	0.7		2.00	0.31	110	0.062
	21.4			0.7					
0.59	19.2		11.2	0.7		1.13	0.38	80	0.062
0.05	9.4		6.2	0.5		2.00	0.56	170	0.125
			5.6	0.5		2.00	0.56	160	
0.92			28.8	1.3		1.88	0.44	90	0.025
			30.0	1.2		1.88	0.25	270	0.375
0.67			10.0	0.7		0.69	1.19	130	0.169
	14.2		10.0	0.5		0.50	0.81	50	0.062

Table 8

Sample Location	Zn (%)	Pb (ppm)	Mn (%)	Cr (ppm)	Ni (ppm)	Co (ppm)	Cd (ppm)
117K 255	0.038	30	1.13	20	35	65	3
117K 322	0.010	35	0.125	25	40	40	3
117K 335	0.016	50	0.375	30	50	25	3
	0.017		0.375	35	45	25	3
117K 290–400	0.030	45	0.250	167	70	40	3
118K 75	0.075	45	0.750	25	50	25	5
118K 220	0.038	30	0.250	20	55	30	3
118K 310	0.038	35	0.135	40	65	35	3
118K 360	0.025	45	0.195	20	65	50	5
119K 30–40	1.50	280	0.875	25	40	70	100
119K 50–60	0.163	75	0.100	5	30	20	10
119K 90	1.46	250	0.500	20	50	135	65
119K 100	0.250	140	0.035	4	20	35	15
	0.250		0.035	4	20	40	15
119K 110	1.42	1,120	2.00	40	80	105	70
119K 115	0.013	45	0.065	2	15	15	5
119K 220	0.016	70	0.375	45	25	20	3
119K 240	0.016	140	0.875	20	45	60	8
119K 325–345	0.016	75	0.115	10	30	25	5
	0.016	75	0.115	15	30	20	5
119K 345–360	0.023	115	0.165	15	35	30	5
119K 400–430	0.035	120	0.250	18	30	30	3
						30	3
121G 0–30	0.050	120	0.080	5	25	30	5
121G 270–280	0.016	85	0.250	5	40	25	5
121G 270–290	0.013	60	0.500	20	30	25	3
	0.013	65	0.530	13	30	25	5
126P 284–314	1.25	380	0.088	10	40	65	63
126P 33–63 Top	0.173	100	0.050	5	20	25	8
161K 0–70	1.10	380	2.00	20	50	35	48
161K 70–130	1.62	250	0.375	25	55	35	68
161K 130–180	1.10	750	12.50	40	40	40	6
	1.33	750	12.50		45	40	
*84K 143	0.541	180	0.090	10	25	55	58
*84P 315–345	1.67	210	0.075	5	20	45	93
*85K 70, Unit 6	0.625	620	0.625	50	50	20	8
*85K 280	0.350	250	1.00	45	80	25	8

Determinations of Au, aqueous Ag, and CO$_3$ by E. J. Mahaffey; all other data by R. L. Hendricks.

(continued)

Mo (ppm)	Al (%)	SiO$_2$ (%)	PO$_4$ (%)	B HCl sol. borate (ppm)	CO$_3$ (%)	H$_2$O (%)	Au (ppm)	Ag aqueous (ppm)	Ag extraction (ppm)
17	0.38	5.59	0.383		1.84	77.8	–	–	
		5.59	0.386					–	
33	0.24	4.14	1.01		0.68	71.4	–	–	
			0.929						
–	1.60	13.00	0.218			47.0	0.035	–	
	1.60								
–	4.50	7.99	0.307	66		42.3	–	–	
8	1.10	8.79	0.190		11.09	42.8	0.057	–	0.5
21	1.20	8.19	0.285			52.3	–	–	
		7.99							
50	2.30	10.01	0.218			49.0	–	–	
50	1.00	7.23	0.628		5.33	55.0	–	–	
46	0.56	11.36	0.242		6.59	48.4	0.37	47	35
–	0.12	5.96	0.061	41	0.66	69.5	0.13	9	
50	0.46	9.24	0.236		2.60	52.3	0.35	60	
–	0.63	6.35	0.024		6.30	70.9	0.071	7	
	0.69								
166	0.94	7.89	0.196		8.76	40.4	1.1	41	
			0.193						
–	0.10	4.63	0.064		0.87	71.4	0.063	–	
37	0.10	6.74	0.040		1.36	68.4	0.091	5	
125	0.47	7.00	0.116	53	2.39	56.1	0.40	26	
83	0.16	7.00	0.043		–	66.9	0.11	6	
83									
125	0.25	5.75	0.292		8.24	59.1	0.10	12	
42	0.23	10.63	0.126		3.45	65.2	0.23	–	
	0.22								
42	0.094	3.75	0.049		0.30	43.5	0.14	–	
29	0.16	8.38	0.043		0.29	46.2	0.16	9	
81	0.16	9.50	0.086		2.81	63.8	0.17	6	
83	0.16	10.13							
125	0.34	9.75	0.107	66	0.11	62.4	0.51	33	
			0.110						
42	0.063	3.75	0.043		–	70.2	0.13	6	
125	0.41	9.38	0.297	181	1.75	55.4	0.33	25	
125	0.34	9.25	0.279		0.56	52.2	0.58	40	
112	0.41	7.88	0.374		8.99	31.5	0.41	9	
	0.38								
42	0.22	5.25				68.9			
17	0.22	4.00				69.0			
83	0.20	5.13				89.0			
125	0.10	4.00				82.7			

Summary

The chemical and spectrographic analyses of sediments and interstitial waters from the Red Sea Deeps show a wide variation in total salts and in heavy metal content. The determinations of oxidation states of iron, copper and sulfur also show the wide variation in oxidation-reduction conditions in various parts of the deposits. The cross-checks between the chemical and spectrographic data, where applicable, show a general agreement but some discrepancies exist which have not been resolved, especially in minor elements. There are significant differences in the calcium figures by the wet and spectrographic methods, lesser differences in magnesium and strontium. On the elements of principal economic interest, the cross-checks between the two methods are generally good, at least in the sediments. Additional checks by other methods on copper, zinc, silver and gold also showed generally good agreement with the reported figures.

The moisture loss to 50°C varies widely from 25 per cent to 80 per cent (by weight) with an average of 55 per cent. The moisture content *in situ* was certainly higher since some of the sediments were squeezed. The additional weight loss from 50°C to 110°C was found to be insignificant so that the 50° dry weight figures can be considered essentially comparable to 110° dry weight values. The water soluble salt portion of the sediment was not determined separately in this study, but in many of the samples this material makes up over half the solid portion of the sediment. The oxidation state determinations on repeat splits of four samples indicate that the sediments are being slowly reduced while in storage under nitrogen at −6°C. It is not known whether this reduction started when the samples were first opened for initial splitting in the laboratory or when the samples were first packaged under inert atmosphere on board ship. If the latter case is true, the condition of the sediments in place may have been more oxidized than the reported data would indicate.

The highest concentrations of heavy metals coincide in general with the dark sulfide-rich sediments, as indicated by the high ratios of total sulfur to HCl-soluble sulfur. Some of the sulfide-rich samples contain up to 20 per cent zinc, 3 per cent copper and significant amounts of gold and silver (all on a dry weight basis). These samples, however, represent very small intervals, both vertically and laterally. The average of all the samples obtained is much lower in all the heavy metals. Because of the extreme variability in composition of the sediments, additional sampling on a systematic basis would be necessary to determine the average composition of the deposit with respect to the various elements of interest.

Reference

Ross, D. A., and E. T. Degens: Shipboard collection and preservation of sediment samples collected during **CHAIN 61** from the Red Sea. *In: Hot brines and recent heavy metal deposits in the Red Sea*, E. T. Degens and D. A. Ross (eds.). Springer-Verlag New York Inc., 363–367 (1969).

Microscopic and Electron Beam Microprobe Study of Sulfide Minerals in Red Sea Mud Samples

J. D. STEPHENS AND R. W. WITTKOPP

MMD Research
Kennecott Copper Corporation
Salt Lake City, Utah

Abstract

Base metal distribution in ten Red Sea mud samples recovered by Woods Hole Oceanographic Institution personnel was studied by microscopic and electron beam microprobe techniques.

Most of the base metals in the samples were present as sulfides. The most common sulfide mineral was marcasite, but chalcopyrite and marmatite were also identified in many of the samples. Marcasite occurred as euhedral plate or lath-shaped crystals and were frequently perched along the center line of euhedral anhydrite crystals. Chalcopyrite and marmatite occurred as irregular or subhedral particles. In several cases, marmatite rims completely surrounded chalcopyrite particles or chalcopyrite rims surrounded marmatite particles. Marcasite was observed filling the interiors of microfossils in several instances. Characteristic x-ray display photographs are presented to show the distribution of copper, zinc, sulfur and iron in a number of particles. Static electron beam analysis of the marmatite showed that it contained large amounts of iron. Some of the marmatite also contained substantial amounts of copper and traces of cadmium.

Copper and zinc values also occurred as a sub-micron dispersion in the mud matrix. The mineralogical nature of the dispersed copper and zinc could not be determined, but they were probably present as sulfides since microprobe analysis showed that sulfur had the same distribution pattern as the base metals.

Introduction

Since the discovery of hot, metal-bearing brines and muds in "holes" or basins in the Red Sea, the possibility of mining this mate-rial to recover its metal values has been considered. Many highly speculative articles and papers have been written about the potential value of the deposits and about how they could be mined and processed.

A group of ten Red Sea mud samples was studied to determine the mineralogy and mode of occurrence of base metals in the muds. The samples were collected by Woods Hole Oceanographic Institution during the R.V. **CHAIN** cruise in 1966. The samples were analyzed for Cu, Zn, Ag, Fe, Mn and S (Hendricks *et al.*, 1969), and an economic evaluation of deposits was also made (Bischoff and Manheim, 1969; Walthier and Schatz, 1969).

Sample Material

The samples studied were identified with the core station designations given them by the Woods Hole Oceanographic Institution personnel when they were taken (Ross and Degens, 1969). These identification data are presented in Table 1.

Table 1 Core Station Designation and Sample Depth of Red Sea Muds

Station Designation	Sample Depth	Color
71b	59–74cm	Dark brown
78c	0–15cm	Red-brown
80	595–615cm	Tan
80P	100–120cm	Tan
81P	290–305cm	Gray-brown
81P	590–605cm	Brown
84K	300cm	Black
84K	304cm	Brown-black
84K	365–379cm	Red
84K	400cm	Red

Two sample preparation techniques were used in studying the samples. First, a representative portion of the mud was placed in water and dispersed by an ultrasonic generator. The plus ten micron fraction from the sample was concentrated by elutriation, the process being repeated until all of the minus ten micron material was removed from the sample. Part of the plus 10 micron fraction from each sample was then mounted in leucite and a polished surface was prepared for microscopic study.

A second portion of the mud was prepared for microscopic and electron beam microprobe study. The mud was dried and stabilized by saturating it with epoxy resin which was diluted with an equal volume of acetone to decrease its viscosity. After saturation, the resin in the mud was set by oven curing at 100°C. A flat surface was then ground on the stabilized mud, and the mud fragment was set in a normal metallurgical mount. Polished surfaces were prepared on the stabilized mud fragments, and these were examined microscopically and analyzed by the electron beam microprobe.

Microscopic Study of the Samples

The plus ten micron materials which were recovered by elutriation of the samples were examined under a stereoscopic microscope. They contained mud agglomerates of several colors, microfossils, bladed transparent crystals and sulfide crystals. The sulfide crystals were generally euhedral but varied widely in morphology. The bladed transparent crystals and the sulfide crystals were quite large, sometimes exceeding two millimeters on their maximum dimensions.

The transparent bladed crystals were hand picked from the plus 10 micron products recovered from several of the mud samples and identified by both x-ray diffraction and infrared analysis. In each case they were anhydrite, $CaSO_4$. The euhedral sulfide crystals were also hand picked from these same samples and identified by x-ray diffraction analysis. In each case they were identified as marcasite, FeS_2.

Some of the large marcasite crystals that were hand sorted from sample 84K-304cm

are shown in Fig. 1. Examination of the irregular masses from which the marcasite crystals extend showed that they were also marcasite in some cases, but in other cases they appeared to be different sulfide species. Subsequent examination of such samples in polished section and analysis of the sulfides by the electron beam microprobe showed that chalcopyrite, $CuFeS_2$ and marmatite, $(Zn,Fe)S$, were often the principal constituents of these irregular sulfide masses.

Examination of the samples under a stereo-microscope showed that the marcasite crystals were often intergrown with the anhydrite crystals. Where such intergrowths were seen, the clusters of marcasite crystals occurred predominantly along the center lines of the bladed anhydrite crystals. This relation between the two minerals is shown in Fig. 2. When examined in polished section with vertically

Fig. 1. Photograph (10×) euhedral marcasite crystals hand picked from the plus 10 micron fraction of Red Sea mud sample 84K 304cm. The irregular masses attached to one end of the marcasite crystals are mixtures of marcasite, chalcopyrite and marmatite.

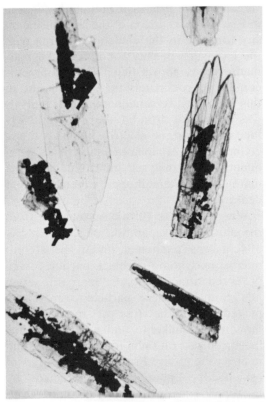

Fig. 2. Photomicrograph (20×) showing clusters of euhedral marcasite crystals perched along the center line of euhedral anhydrite crystals hand picked from the plus 10 micron fraction of Red Sea mud sample 84K 304cm.

reflected light, marcasite was observed filling the interior chambers of several of the microfossils. No alteration of the calcium carbonate shells could be seen or detected by subsequent microprobe analysis of these marcasite filled fossils.

Electron Beam Microprobe Studies

An Applied Research Laboratories model AMX electron beam microprobe was used to study the distribution of elements in the mud and in hand-picked individual mineral particles. An electron beam diameter of approximately one micron was used in all cases, and electron beam scanning techniques were used in studying most of the sample areas. Element distributions were determined by monitoring appropriate characteristic x-ray fluorescent radiation from the sample and recording the data by photographing the oscilloscope screen. In

some instances, semiquantitative analytical data were obtained for elements of particular interest by counting the quanta of characteristic x-rays emitted per unit time from the sample when excited by the static electron beam focused on the particle being analyzed.

An example of the association of chalcopyrite with marcasite is shown in Fig. 3. These particles were observed in the polished section prepared with epoxy stabilized mud from sample 84K-304cm. The mud matrix in which these particles occurred was brown-black colored. It consisted largely of silicates, and there was little iron or copper dispersed in the matrix at this point. Analysis of other areas in this sample showed that iron, copper and zinc all occurred as sub-micron dispersions in the matrix mud.

An interesting aspect of the sulfide mineralization in two of the samples from core 84K (300cm and 304cm) was the fact that in some particles chalcopyrite and marma-

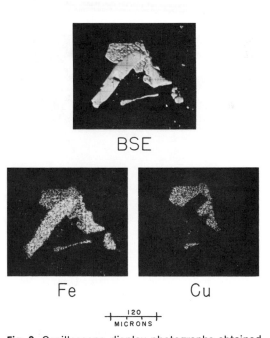

Fig. 3. Oscilloscope display photographs obtained by electron beam scanning a composite particle of marcasite and chalcopyrite. The particle was observed in the polished section prepared with epoxy stabilized mud from sample 84K 304cm. The photograph designated "BSE" was obtained with backscattered electrons. The photographs designated "Fe" and "Cu" were obtained by monitoring iron K alpha and copper K alpha radiation respectively.

tite were alternately deposited around the same nucleus. This type of deposition formed composite particles that were made up of successive rims of the two minerals. Some examples of such composite particles are shown in Figs. 4 and 5.

The photographic series in Fig. 4 was obtained by analysis of particles in the polished mud sample from 84K-300cm. Two composite particles can be seen in the series. The large, elongated particle is mostly marcasite, but approximately one-fourth of it at the left-hand end is marmatite. The other particle is roughly hexagonal in shape and has a chalcopyrite core which is surrounded by a thick rim of marmatite.

The areas occupied by chalcopyrite and marmatite are defined clearly by the distribution photographs for copper and zinc respectively. The photographs for these elements show that there are also numerous small particles of chalcopyrite and marmatite disseminated throughout the mud matrix.

A particularly interesting relation is

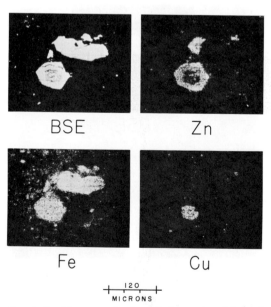

Fig. 4. Oscilloscope display photographs obtained by electron beam scanning two composite particles observed in the epoxy stabilized mud from sample 84K 300cm. The elongated particle is a composite of marcasite and marmatite. The roughly hexagonal particle is a composite with chalcopyrite in the center surrounded by a rim of marmatite. The photograph designated "BSE" was obtained with backscattered electrons. The photographs designated "Zn," "Fe" and "Cu" were obtained by monitoring zinc K alpha, iron K alpha and copper K alpha radiation respectively.

shown by the iron distribution photograph in Fig. 4. Comparison of the exposure density for iron in the chalcopyrite area with the exposure density for iron in the marmatite areas shows that the two minerals contain approximately the same amount of this element. A semiquantitative analysis was made for iron in the marmatite by focusing the static electron beam on it and counting iron K alpha radiation. This semiquantitative analysis indicated that the marmatite contained approximately 27 per cent iron.

When the plus 10 micron material from the elutriation separation of sample 84K-304cm was examined under the stereomicroscope, some jet black particles were observed attached to the marcasite crystals. Some of these particles were handpicked from the rest of the material, mounted in bakelite and a polished surface prepared on them for microscopic and microprobe study. Fig. 5 shows a series of oscilloscope photographs obtained by microprobe analysis of one of these handpicked particles.

The particle represented by the series of photographs in Fig. 5 is made up of marcasite, chalcopyrite and marmatite intergrown with each other. The center of the broadest portion of the particle is marmatite. The marmatite core is surrounded by a discontinuous rim of chalcopyrite, and this in turn is surrounded by another discontinuous rim of marmatite. All of these are partially surrounded or adjoined by marcasite.

The characteristic x-ray photographs for the distribution of iron, zinc and copper in the particle show several interesting characteristics. The iron distribution photograph shows that the iron contents of the marmatite and chalcopyrite are essentially the same, and that these minerals contain somewhat less iron than the marcasite which is attached to them.

The zinc distribution photograph shows that there were two periods of marmatite deposition, and comparison with the copper distribution photograph shows that they were separated by a period of chalcopyrite deposition. Close inspection of the zinc distribution photograph shows that the concentration of this element was somewhat

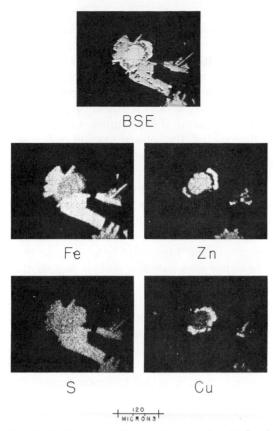

Fig. 5. Oscilloscope display photographs obtained by electron beam scanning a composite particle hand picked from sample 84K 304cm. The particle contains marcasite, chalcopyrite and marmatite. The photograph designated "BSE" was obtained with backscattered electrons. The photographs designated "Fe," "Zn," "S" and "Cu" were obtained by monitoring iron K alpha, zinc K alpha, sulfur K alpha and copper K alpha radiation respectively.

lower in the central marmatite core than in the rim of marmatite that partially surrounds it. This difference in zinc content of the two marmatite zones may also be partly explained by comparison with the copper distribution photograph. This comparison shows that the central marmatite area contains an appreciable amount of copper. A semiquantitative analysis was made by counting characteristic copper K alpha radiation when the static electron beam was focused on the central area. This analysis showed that there was approximately 2 per cent copper in the central marmatite area. From the distribution pattern for copper in this area, it appears to be rather uniformly distributed in the marmatite. A small amount of cadmium is also present in this area.

The central marmatite area was examined microscopically at 1000x magnification using vertically reflected light. A few particles of chalcopyrite were observed included in the marmatite, but not enough to account for the 2 per cent copper that was determined by the semiquantitative analysis. Inhomogeneity of the central marmatite zone was indicated by vague, ill-defined areas which tended to be elongated and which reflected more light than adjacent areas. The cause of these areas of high reflectance was not apparent from the microprobe analyses.

The previous illustrations have shown the occurrence of relatively large, well-crystallized mineral particles, but Fig. 6

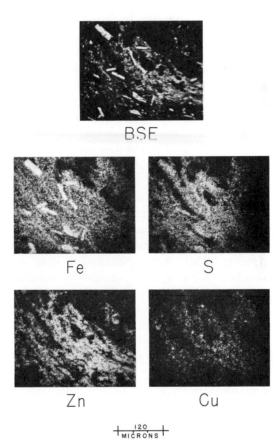

Fig. 6. Oscilloscope display photographs obtained by electron beam scanning an area in the epoxy stabilized mud section prepared from sample 84K 300cm. The photograph designated "BSE" was obtained with backscattered electrons and shows small marcasite crystals disseminated throughout the mud matrix. The photographs designated "Fe," "S," "Zn" and "Cu" were obtained by monitoring iron K alpha, sulfur K alpha, zinc K alpha and copper K alpha radiation respectively. These photographs show that the "mud" contains an extremely fine dispersion of copper and zinc sulfides.

shows that the metal values also occur as an extremely fine sulfide mud. While not as spectacular and photogenic as the larger sulfide particles, dispersions of this type contain a substantial portion of the base metal values in the samples. Because of this, and because of difficulties inherent in the handling of extremely fine aqueous dispersions, sulfide muds of this type will probably cause problems in any mining venture based on treating the Red Sea muds.

Summary and Discussion

The microscopic study of plus 10 micron elutriated products from ten Red Sea mud samples showed that euhedral marcasite crystals are common in samples that contained base metal values. The marcasite occurs as isolated single crystals and as interpenetrating clusters of crystals. Clusters of marcasite crystals often occur along the center lines of euhedral anhydrite crystals in the most strongly mineralized samples. Irregularly shaped masses of chalcopyrite and marmatite are frequently attached to ends of the elongated marcasite crystals.

The association of marcasite clusters along the center lines and to some extent intergrown with the anhydrite crystals suggests that the two minerals have a common genesis. This same conclusion may be reached with regard to the marcasite, chalcopyrite and marmatite. The basis for this latter conclusion lies in the fact that some of the marcasite clearly crystallized before some of the chalcopyrite and marmatite while other marcasite crystallized after these minerals. Fig. 3 showed an example where the euhedral marcasite clearly crystallized before the chalcopyrite, and Fig. 6 showed an example where the marcasite almost surrounds the chalcopyrite and marmatite. In this case the marcasite surely crystallized after these minerals.

The microprobe studies showed that several alternating periods of copper and zinc deposition occurred in the muds, and in some cases composite particles consisting of alternating concentric rims of chalcopyrite and marmatite were formed.

Semiquantitative analysis of the marmatite showed that it contained approximately 27 per cent iron. This is near the solubility limit for iron in zinc sulfide, and according to the proposed use of the sphalerite geologic thermometer such a high iron content should imply a high temperature of formation (Kullerud, 1959). Since the physical relations of the mineralization rules out really high temperatures, the sphalerite geologic thermometer must not apply to the Red Sea mud deposits. An interesting aspect of the Red Sea mineralization is the fact that the marmatite is associated with marcasite rather than the more common iron sulfides. Marcasite is less stable than the other iron sulfides, so it is logical that zinc sulfide associated with it might contain more iron than if it were associated with pyrite or pyrrhotite.

Some of the marmatite contained an appreciable amount of copper, as was seen in Fig. 5. Line profile scans were made for copper, zinc, iron and cadmium across the central marmatite area shown in Fig. 5. These profiles showed that iron and zinc behaved reciprocally to each other, iron being higher near its center. The copper profile suggests that this element was also most concentrated near the center of the inner marmatite particle. Some cadmium was detected in the marmatite, but no systematic distribution pattern was seen.

The close interlocking of the copper and zinc minerals in composite particles will make it impossible to obtain good separation of these metal values by conventional ore dressing techniques if mining and processing of the muds is attempted. In addition, the very high iron content of the marmatite will make it impossible to obtain a high grade zinc concentrate even if a successful method of physical beneficiation should be developed.

A further problem to be solved if physical beneficiation of the muds should be attempted would be the recovery of extremely fine, sub-micron dispersions of the copper and zinc sulfides in the mud matrix such as the material shown in Fig. 6. Extremely fine dispersions of this sort are impossible to concentrate by normal ore dressing methods. In fact, metal values of this sort would probably be diluted greatly during mining by muds containing little or no base metals. This dilution could very well lower the average metal content of the mud to the point where it would not be economically feasible to treat it by any method.

If direct chemical recovery of metals in the muds is considered without any physical treatment to concentrate the valuable constituents, their mineralogy and relations between themselves, the mud matrix, and included brines become even more important. Chemical treatment would probably involve either an acid or an alkaline leach to dissolve the sulfide minerals, followed by some process to separate and recover the valuable elements.

No serious consideration has been given chemical treatments, but they do not seem particularly promising. Calcium carbonate is a major constituent of some of the muds, and if an acid leach were used, reaction with this mineral would neutralize large amounts of the acid leach solution. If an alkaline leach were used, a large amount of alumina and silica would be dissolved from the mud and again a large amount of leach solution would be neutralized. In either case, recovery of the leach solution and

metal values from the extremely fine mud would be difficult.

References

Bischoff, J. L. and F. T. Manheim: Economic potential of the Red Sea heavy metal deposits. *In: Hot brines and recent heavy metal deposits in the Red Sea*, E. T. Degens and D. A. Ross (eds.). Springer-Verlag New York Inc., 535–541 (1969).

Hendricks, R. L., F. B. Reisbick, E. J. Mahaffey, D. B. Roberts, and M. N. A. Peterson: Chemical composition of sediments and interstitial brines from the Atlantis II, Discovery, and Chain Deeps. *In: Hot brines and recent heavy metal deposits in the Red Sea*, E. T. Degens and D. A. Ross (eds.). Springer-Verlag New York Inc., 407–440 (1969).

Kullerud, G.: Sulfide systems as geological thermometers. *In: Researches in Geochemistry*, P. H. Abelson (ed.). J. Wiley & Sons, Inc., New York, 301 (1959).

Ross, D. A. and E. T. Degens: Shipboard collection and preservation of sediment samples collected during CHAIN 61 from the Red Sea. *In: Hot brines and recent heavy metal deposits in the Red Sea*, E. T. Degens and D. A. Ross (eds.). Springer-Verlag New York Inc., 363–367 (1969).

Walthier, T. M. and C. F. Schatz: Economic significance of minerals deposited in the Red Sea Deeps. *In: Hot brines and recent heavy metal deposits in the Red Sea*, E. T. Degens and D. A. Ross (eds.). Springer-Verlag New York Inc., 542–549 (1969).

Mineralogy and Micropaleontology of a Goethite-Bearing Red Sea Core

YVONNE HERMAN AND P. E. ROSENBERG

Department of Geology
Washington State University
Pullman, Washington

Abstract

The mineralogy and micropaleontology of a goethite-bearing core from the northern Red Sea have been studied. Variations in planktonic microfaunal composition with depth in the core reveal three major climatic phases: Phase I, Postglacial, encompassing the last 11–12,000 years; Phase II, late glacial stages which began about 60–65,000 years ago; and Phase III, an interstadial, with a fauna similar to that of Phase I, commencing about 100,000 years ago.

The core is composed of sediments containing scattered, lithified, calcareous fragments and lumps, 3mm to 2cm in diameter. These lithified fragments which have a carbonate fraction containing principally high-magnesian calcite, low-magnesian calcite and aragonite are believed to have formed during recurrent evaporation cycles in the Red Sea.

Goethite is present in sediment samples at all depths and forms an almost monomineralic bed, deposited near the beginning of Phase III. Spectrochemical analyses of the goethite bed indicate a concentration of several heavy metals in excess of the average reported for marine sediments. The goethite is attributed to hot brine activity based on comparisons with the recent hot brine deposits of the central Red Sea. The goethite bed, which is underlain by carbonate-rich sediments, is thought to have been deposited after spreading from an undiscovered hot brine deep during a relatively short period of high activity. The presence of goethite in all samples implies continuous activity at the source during the deposition of the core sediments. Cross-correlations with ^{14}C dated Red Sea cores and extrapolations assuming constant rates of sedimentation, suggest that the history of hot brine deposition in the northern Red Sea extends back at least 100,000 years B.P.

Introduction

The discovery of active hot brines and associated iron deposits in a restricted portion of the Red Sea (Miller, 1964) suggested the possible existence of similar active or inactive areas elsewhere in the Red Sea. This speculation led to the re-examination of cores raised by the Lamont Geological Observatory and studied previously by Herman (1965, 1968).

In one of these cores, V14-122 (Fig. 1), an almost monomineralic bed of goethite was found which suggested a genetic relationship with the hydrous iron oxides now being deposited in the active hot brine area. A mineralogical and micropaleontological analysis of this core, with emphasis on the goethite bed, was undertaken to study this relationship and to extend present knowledge of hot brine deposits.

At the laboratory the core was extruded from the gutter pipe, cut in half longitudinally, photographed and described megascopically. Samples 1–2cm thick were taken at 10cm intervals for the upper (top) 50cm; below this level, samples were taken at 50cm intervals or at points where lithologic or faunal changes were observed. The dried and weighed material was washed through a 74μ or 62μ sieve which retained the pteropods and planktonic foraminifers as well as the skeletal fragments of larger organisms and mineral fragments. The coarse fraction was weighed for a second time and recorded as a percentage of the total. A curve constructed to represent

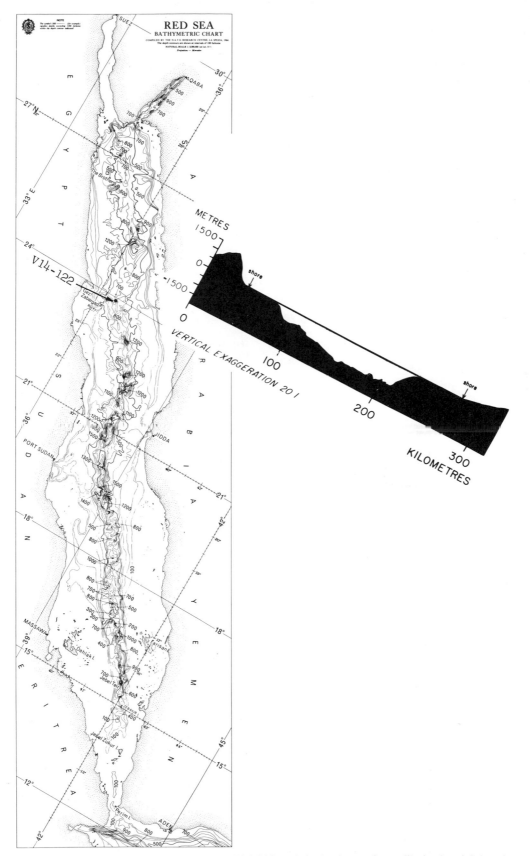

Fig. 1. Red Sea map showing location of core V14-122 and the bathymetric profile in the vicinity of core V14-122 (after Drake and Girdler, 1964).

449

these percentages (Fig. 2) was found to be useful in distinguishing between different types of sediments. For example, where abundant lithified calcareous fragments occur, the percentage of coarse fraction increases considerably (Fig. 2).

Micropaleontology

For micropaleontological analysis the washed sediment covering a tray of 50cm² was examined under a binocular microscope. Planktonic foraminifers and pteropods were determined specifically and counted; benthonic foraminifera were identified only in samples of particular interest. Estimates of abundance of tests are given in Table 2.

Table 1   Megascopic Sediment Characteristics

Core Location: Latitude − 23°55′N,
Longitude − 36°28′E.
Water Depth: 1,486m.
Core Length: 545cm.

Depth (cm)	
0–55	Gray, mottled silty-lutite rich in pteropods and foraminifers, base marked by 5mm thick dark gray layer containing organic detritus.
55.5–90	Gray layer with scattered lithified calcareous fragments (largest 2 cm × 1.5cm × 4mm). Porous, irregular matrix cements shells of *Creseis*.
90–290	Tan, mottled foraminiferal silty-lutite, lighter and coarser below 204cm. Scattered lithified calcareous grains 4–10 mm in diameter.
290–330	Brown foraminiferal silty-lutite.
330–460	Gray, mottled foraminiferal silty-lutite with scattered lithified fragments 4–7mm in diameter. A dark gray layer occurs between 455–456cm.
460–505	Brown lutite, no evidence of burrowing.
505–545	Gray foraminiferal silty-lutite, containing calcareous lithified lumps 3–4mm in diameter. Bottom 23–25cm are "flow in" sediments.

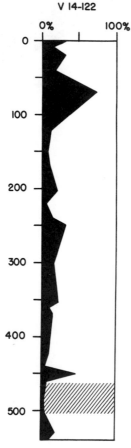

V 14-122

Fig. 2. Variation in the percentage of coarse fraction. The ordinate indicates depth in core in cms (after Herman, 1965). Stippled zone, goethite bed.

Faunal Analysis

The effect of water temperature, salinity, as well as other environmental factors, on the distribution and abundance of pteropods and planktonic foraminifers has long been known. More recently, as new sampling techniques have been developed, important quantitative and qualitative data concerning the relationship between planktonic organisms and their environment have become available. These data are essential for the interpretation of past climatic and hydrologic conditions. The composition of the preserved faunal assemblages is also the result of various post-depositionally operative factors, such as selective solution of limy shells (Berger, 1967), their removal by bottom currents and variations in the rates of detrital sediment accumulation.

In the Red Sea late Quaternary deepwater sediments, pteropods and planktonic foraminifers together constitute 90–95 per cent of the total faunal remains (in the fraction >74 or 62μ), the latter usually being more abundant. The remainder is made up of heteropods, other gastropods, embry-

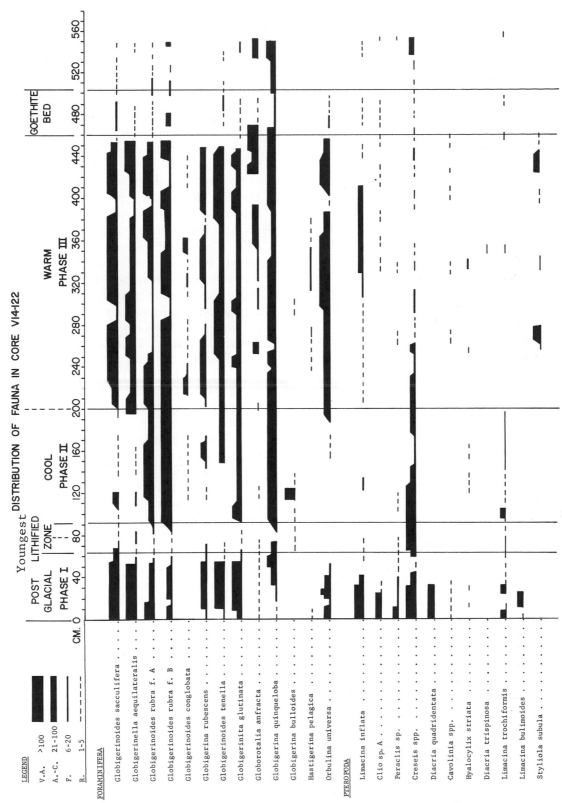

onic pelecypods, benthonic foraminifers, sponge spicules, alcyonarian and echinoid spines, ostracods and fish otoliths. Occasionally radiolarians, diatoms and bryozoan fragments occur.

In core V14-122 the Postglacial (Phase I) is contained in the topmost 60cm. It has a varied and abundant fauna dominated by *Globigerinoides sacculifera, Globigerinella aequilateralis, G. rubra, Limacina inflata* and *Creseis* spp. (Table 2). Between 55 and 55.5cm a sapropelitic layer is present; the faunal assemblage of this layer differs from that of the overlying and underlying beds in that there is an increase in the relative abundance of *Atlanta* spp. This genus generally constitutes 1–5 per cent of the planktonic remains, but here it makes up about 45 per cent of the fauna. ^{14}C age determination made on the coarse fraction ($>62\mu$) of the 30–50cm section gave an age of 7,935 ± 130 years, thus dating the time elapsed since the last stagnation of the basin water. Underlying the sapropelitic layer is a 30cm thick bed characterized by the presence of lithified calcareous fragments, deposited about 11,000–12,000 years ago (Herman, 1968). The fauna of this bed is dominated by epipelagic, eurytherm, euryhaline *Creseis acicula, Creseis conica* and *Creseis virgula.* The temperature and salinity ranges of *Creseis acicula* are 10°–27.9°C and 35.5–36.7 per mill, respectively; however, this species was found alive in a lagoon where temperatures range between 26° and 33°C and salinities between 25 and 45 per mill (van der Spoel, 1967). *Creseis virgula* is more stenotherm than the former species (van der Spoel, 1967) and the temperature and salinity ranges of *Creseis conica* are somewhat more restricted than those of *Creseis virgula* (van der Spoel, 1967). *Creseis* spp. constitutes >85 per cent of the total fauna in this section. The deposition and possible mode of formation of these lithified calcareous layers found in many Red Sea cores was discussed in earlier publications (Herman, 1965, 1968). They are thought to have been deposited during periods of extreme hydrologic conditions, possibly caused by temporary isolation or severe reduction of water exchange with the open ocean.

The 90–204cm bed was deposited during a cool climatic episode, here designated as Phase II. The fauna of this bed is characterized by the dominance of several epipelagic, eurythermal planktonic foraminifers and pteropods (Table 2). *Globigerinoides rubra* is one of the predominant species. Its temperature and salinity ranges in the Pacific are 10–27.3°C, and 34–36 per mill, respectively (Parker, 1960). *Globigerina quingueloba,* also present at high frequencies in this bed, is considered to be a typical cold water form by Bé and Hamlin (1967), who report its highest concentrations in north Atlantic waters with surface temperatures ranging from 12°–15°C. *Creseis* spp., whose temperature and salinity tolerance range are given above is the predominant pteropod. During this cool phase the relative abundance of several warm-water forms, namely *Globigerinoides sacculifera, Globigerinella aequilateralis, Limacina inflata, Diacria quadridentata* and *Clio* sp. A, was drastically reduced (Table 2 and Fig. 3). The scattered occurrence of lithified calcareous lumps at various horizons suggests recurrent episodes of restricted water exchange with the Indian Ocean. Biostratigraphic layer by layer correlation between core V14-122 and other northern Red Sea piston cores indicates that a 120–130cm thick section, from the basal part of this stratigraphic unit is missing (Herman, 1965). Based on cross-correlations with other cores collected from the proximity of V14-122 (Fig. 4) and on assumed constant rates of sedimentation of about 5cm/1,000 years in the northern sector of the sea, we obtain an age of about 60–65,000 years for the commencement of cool Phase II, which we equate with late Wisconsin (Würm) stades.

Phase III, a long warm interval with a fauna similar to that of the Postglacial (Phase I), underlies Phase II sediments, and the boundary with the latter is at about 204cm (Fig. 3 and Table 2). This long, warm phase is thought to represent an interstadial and is equated with the early Wisconsin (Würm) interstade which began about 100,000 years ago and lasted about 30,000 years (Ericson *et al.,* 1964). The basal section of the bed deposited during

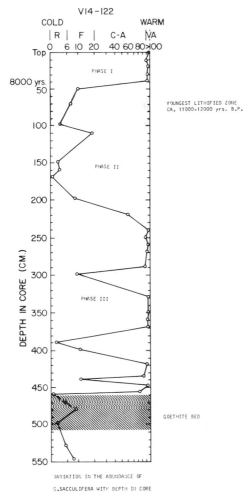

Fig. 3. Variation in abundance of Globigerinoides sacculifera. Abscissa numbers indicate actual specimen counts. R, rare—1 to 5 tests; F, frequent—6 to 20; C-A, common-abundant—21 to 100; VA, very abundant—>100.

this warm interval is marked by the presence of a goethite bed (460–505cm). Environmental conditions appear to have varied during the deposition of this goethite bed. In the 460cm layer the assemblage is monotonous, consisting of *Globigerina quinqueloba* and *Globorotalia anfracta,* which make up 50 per cent and 20 per cent of the total faunal remains, respectively. Unidentifiable juvenile globigerinids (0.10–0.15mm in diameter) constitute about 23 per cent of the skeletal remains. A few pteropod casts are also present. Benthonic foraminifers (1–2 per cent of the fauna) are represented by the following forms: *Bulimina marginata, Cassidulina laevigata, Cassidulina* sp., *Cibicides* sp., *Dentalina*

sp., *Textularia* sp., *Bolivina* spp., *Virgulina* sp., *Rotalia* cf. *beccarii, Lagena* sp. and several Miliolids. The tests show no evidence of solution such as pitting or broken shells in large quantities (Berger, 1967). The fauna of this layer resembles that contained in the 300, 440, 500, 530, 540 and 545cm samples and suggests relatively low surface water temperatures at the time of deposition.

The planktonic assemblage of the 470cm sample is varied; *Globigerinoides rubra* f.B is the dominant form (33 per cent of the fauna) followed by *G. sacculifera* (16 per cent), *Globigerina quinqueloba* (14 per cent), *G. rubra* f.A (9 per cent), *Orbulina universa* (8 per cent), *Globigerinita glutinata* (4 per cent), *Globigerinella aequilateralis* (3 per cent), *G. tenella* (1 per cent) and *Globorotalia anfracta* (1 per cent). The globigerinoids exhibit evidence of partial solution; about 50 per cent are broken and the percentage of test fragments is anomalously high. Unidentifiable juvenile globigerinids make up about 10 per cent of the assemblage. Few pteropod casts are preserved. Three specimens of benthonic foraminifers; *Quinqueloculina* sp., *Bolivina* sp. and a *Rotalid,* constituting about 2 per cent of the fauna were also observed.

Interestingly, the planktonic faunal composition of the 480cm sample closely resemble that in 470cm (Table 2). Furthermore, >50 per cent of the globigerinoids show the effect of solution described above. The benthonic foraminifers (1 per cent of the fauna) resemble those of 460cm. The following forms were observed: *Bulimina marginata, Cassidulina laevigata, Cassidulina* sp., *Bolivina* spp., *Cibicides* sp., *Quinqueloculina* sp., *Uvigerina* sp., *Virgulina* sp. and *Rotalia* cf. *beccarii.*

The 470–480cm section is thought to have been deposited during a relatively short, warm interval. The scarcity of tests (Table 2), may be due to dilution by the rapidly accumulating goethite and to the solution of the limy shells. Miller *et al.* (1966) report a pH as low as 5.3 in the active hot brines of the Red Sea. It is possible that similar, slightly acidic environments existed during the deposition of the 470–480cm section.

Correlation Between Seven Cores Based On Variation In Frequency Of
Globigerinoides sacculifera

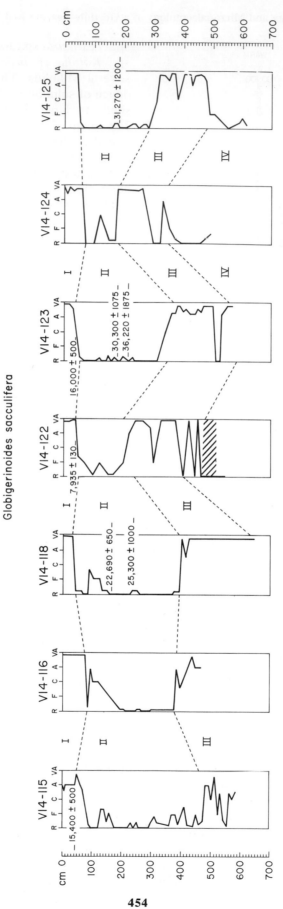

Fig. 4. For core location see Herman, 1968. Hatched zone in core V14-122 indicates goethite bed.

454

Underlying the goethite bed, between 505 and 545cm, cool water elements dominate the assemblages (Table 2). Inasmuch as the bottom 33–35 cm are "flow in" sediments, the *in situ* deposits terminate at about 510cm. Without absolute age determinations it is impossible to say whether the 505–510cm layer represents a relatively short, cool oscillation within the long interstadial (Phase III) or represents the last part of a preceding long, cool stade (the first Wisconsin (Würm) stade).

Discussion and Summary of Paleoclimatic Data

Fig. 3 shows the changes in abundance of warm water stenotherm *Globigerinoides sacculifera* in core V14-122. These variations are representative of the major climatic oscillations during the upper Quaternary in the northern Red Sea.

Based on data presently available, the following sequence of climatic successions is deduced:

Phase I, corresponding to the Postglacial and encompassing the last 11,000–12,000 years, has a fauna similar to that of the present. Lithified calcareous layers between 55.5–90cm (Fig. 2) mark the Postglacial-Last Glacial boundary.

During Phase II epipelagic, eurythermal planktonic foraminifers and pteropods dominated the faunal assemblages. This cool interval which began about 60–65,000 years ago is equated with late Wisconsin (Würm) stades. Hydrologic conditions varied, from fairly active water exchange with the ocean to more restricted conditions than at present as suggested by the occurrence of scattered lithified calcareous fragments.

Phase III, thought to represent a long interstadial (equated with the early Wisconsin (Würm) interstade), began about 100,-000 years ago and lasted approximately 30,000 years. A goethite bed 45cm thick occurs in the basal section of these deposits. It is impossible to assess precisely the time of deposition of the goethite bed; cross-correlation with other cores from the northern sector of the sea (Fig. 4) suggests

an age of about 90–100,000 years for the deposition of this bed.

Mineralogy and Chemistry

Mineralogy

The mineralogy of the goethite bed and of calcareous, lithified fragments and lumps (3mm to 2cm in diameter) found at various depths in the core were studied by means of x-ray diffractometry and optical microscopy. Although only the coarse fraction, $>62\mu$, was available for most lithified layers,* indurated fragments present in this fraction contain mainly fine materials which should approximate the mineral composition of the fraction, $<62\mu$. In the bulk sediment sample at 260cm the relative abundance of minerals in the fraction $<62\mu$ is not greatly different than that of the lithified fragments in the fraction $>62\mu$. Six whole sediment samples including three within the goethite bed were available for this study.

Samples from the goethite bed were washed with distilled water to remove NaCl and dried at room temperature. Larger shells and shell fragments were handpicked from all samples under a binocular microscope. The preparation of x-ray powder mounts was accomplished by drying the sample from a water slurry on a glass slide at room temperature. The compositions of magnesian calcites and dolomites were estimated by measurement of d_{211} (Harker and Tuttle, 1955) against an internal standard, CdF_2, which has a conveniently located x-ray reflection at $28.7°2\theta$ CuKα.

The mineralogy of the goethite bed and of the lithified fragments and lumps is summarized in Table 3. Approximate relative mineral abundances, categorized as major, minor or trace (Table 3), are based on intensities of x-ray reflections and visual estimates under the microscope. Although the larger tests were removed by washing

---

* Sediments containing scattered calcareous fragments and lumps (3mm to 2cm in diameter) are referred to as "lithified layers."

whole sediment samples through a 62μ sieve and by hand picking the coarse fraction, coccoliths, occasional foraminifera and fragments of larger shells were present in all samples and contribute significantly to the mineralogy of the lithified layers reported in Table 3. Detrital quartz usually accompanied by feldspar and mica, was observed in all samples.

Goethite, by far the most abundant iron oxide, is present in widely varying percentages in all samples examined and forms an essentially monomineralic bed between 460 and 505cm. Under the microscope goethite appears as yellow to brown irregular or roughly spherical (up to 50μ in diameter) aggregates. The intensities of goethite x-ray reflections are weaker in the fine fractions than in the coarse fractions of the goethite bed, suggesting the presence of considerable amounts of limonite. Amorphous appearing material has been observed under the microscope in the fine fractions. Goethite has also been found in appreciable amounts at depths of 70, 260, 330 and 456cm. Even where goethite is present only as a trace, extensive limonite staining of fine-grained carbonate aggregates, quartz and other minerals has been observed.

Lepidochrocite, a dimorph of goethite, was identified in two samples by its characteristic pleochroism (Deer *et al.*, 1966) and by x-ray diffraction analysis after careful hand picking in one of these samples. The presence of hematite and maghemite was inferred respectively from the deep reddish coloration of some goethite aggregates and from the distinct response of small fragments to a hand magnet. Small particles (<1mm in diameter) obtained from samples at 260 and 460cm after ultrasonic disaggregation were both distinctly reddish and magnetic. Hematite and maghemite, which are dehydration products of goethite and lepidochrocite respectively (Deer *et al.*, 1966), are thought to have formed from the hydrated iron oxides during the drying of the core.

Low-magnesian calcite (1–4 mole per cent $MgCO_3$) and high-magnesian calcite (9–14 mole per cent $MgCO_3$) make up the major portion of all lithified fragments and lumps. In some samples (e.g. 260cm) a complete series seems to exist between low- and high-magnesian calcites. Shell material was clearly subordinate to fine-grained magnesian calcite aggregates after hand picking of the larger tests. Aragonite, in the form of acicular and lath-shaped crystals, is an important mineral constituent occurring in virtually all samples in widely varying proportions. Rhombic crystals of dolomite averaging about 25μ in diameter are present in all lithified fragments and lumps. Dolomite x-ray reflections have also been identified in two samples from the goethite bed. Values of d_{211} for all dolomite observed are higher than those for the pure end-member, suggesting the presence of excess $CaCO_3$ or possibly the substitution of Fe^{+2} for Mg^{+2} (Rosenberg, 1967).

Chlorite, the only clay mineral found in the core, was identified in the sample at 456cm by weak 14Å and 7Å x-ray reflections and is believed to be detrital. Other detrital minerals include quartz, feldspar and mica. With the exception of weak unidentified x-ray reflections at approximately $24.8°2\theta$ CuKα in the samples at 390 and 530cm, all observed x-ray reflections are accounted for by the minerals reported in Table 3.

Chemistry

Two partial, quantitative spectrochemical analyses of whole sediment samples within the goethite bed at 470 and 480cm are given in Table 4. Values were determined by emission spectroscopy; several concentrations in the sample at 470cm were confirmed by atomic absorption techniques. Na is present in the sample at 470cm as NaCl; the sample at 480cm is NaCl-free.

Aside from the major component, iron, the goethite bed contains appreciable amounts of Si, Mg, Ca and Zn. With the exception of Zn these elements may be accounted for by the presence of small amounts of carbonates and detrital silicates. The percentages of Zn and a few other heavy metals (e.g. Mo and Sn) exceed their average concentrations in marine sediments. For example, the average concentration of Zn in marine sediments is on

Table 3 Mineralogy of Sediment Samples

Depth (cm)	Sediment Fraction	Minerals Present *		
		Major	Minor	Trace
0	>62μ	High-Mg Calcite	Quartz Aragonite	*Goethite*
10	>62μ	High-Mg Calcite Low-Mg Calcite	*Goethite*	Aragonite Quartz Mica
57	>62μ	High-Mg Calcite Quartz Feldspar Aragonite	Mica *Goethite* Opaque material	
70	>62μ	High-Mg Calcite Aragonite Low-Mg Calcite	Quartz Dolomite	Mica Feldspar *Goethite*
150	>62μ	High-Mg Calcite Low-Mg Calcite	Quartz Feldspar	Aragonite Dolomite *Goethite* Mica
250	>62μ	High-Mg Calcite Low-Mg Calcite	Quartz Dolomite	Aragonite Mica *Goethite*
260	Whole	High-Mg Calcite Low-Mg Calcite	Quartz Feldspar *Goethite* Dolomite	Lepidocrocite Hematite Maghemite
270	>62μ	High-Mg Calcite Low-Mg Calcite	Quartz Dolomite	Aragonite Feldspar Mica *Goethite*
290	>62μ	High-Mg Calcite Low-Mg Calcite	Quartz Aragonite *Goethite*	Dolomite Hematite
330	Whole	Low-Mg Calcite High-Mg Calcite Quartz	Dolomite Feldspar *Goethite*	Mica Aragonite Lepidocrocite Hematite
350	>62μ	High-Mg Calcite	Low-Mg Calcite Quartz	Aragonite Mica Feldspar *Goethite*
390	>62μ	High-Mg Calcite	Quartz Dolomite	Mica Aragonite *Goethite*
440	>62μ	Low-Mg Calcite High-Mg Calcite	Quartz	*Goethite*
456	Whole	High-Mg Calcite Low-Mg Calcite *Goethite*	Quartz Dolomite	Mica Feldspar Aragonite Chlorite
460	Whole	*Goethite*	Dolomite Low-Mg Calcite High-Mg Calcite	Quartz Feldspar Mica Hematite Maghemite
470	Whole	*Goethite*	Dolomite	Quartz Aragonite
480	Whole	*Goethite*		Mg Calcite Aragonite Quartz Mica
530	>62μ	High-Mg Calcite Low-Mg Calcite Aragonite	Quartz Dolomite Feldspar	Mica *Goethite*

* The relative abundances of minerals present in a sample are roughly estimated as major, minor and trace, based on x-ray and microscopic studies.

Table 4 Quantitative spectrochemical analyses of whole sediment samples from the goethite bed. The values are weight percentages ±15 per cent. Sample 470, unwashed, contains NaCl; sample 480, washed, NaCl-free. Percentages less than 0.001 are not reported

Element	470cm	480cm
Fe	major	major
Si	1.5	1.5
Mg	0.7	0.7
Mo	0.03	0.055
Na	1.2	—
Zn	0.23	0.23
Ca	0.6	0.5
Al	0.10	0.02
Sr	0.03	0.02
Mn	0.015	0.02
Ni	0.004	0.003
Sn	0.02	0.04
Cu	0.02	0.002
Pb	0.006	—

the order of 0.008 per cent (Miller et al., 1966) as compared with 0.2 per cent in samples from the goethite bed. Zn may be present as unidentified traces of a specific zinc mineral (e.g. sphalerite), or it may be adsorbed along with other heavy metals by the hydrous iron oxides. Although a few heavy metals are anomalous with respect to their average concentrations in marine sediments, heavy metal concentrations are much lower in the goethite bed than those in the recent iron deposits of the active hot brine area (Miller et al., 1966). Furthermore, heavy metal ratios are quite different in the two localities.

Discussion

Lithified layers with carbonate fractions composed of high-magnesian calcite, low-magnesian calcite and aragonite are not unusual in the Red Sea (Herman, 1965, 1968). To the authors' knowledge, dolomite has not been reported previously in these layers but its presence is not unexpected. The lithified layers are attributed to recurrent evaporation cycles caused by glacial lowering of sea levels (Deuser and Degens, 1969) and are widely distributed in the Red Sea (Gevirtz and Friedman, 1966).

The goethite bed described in this paper bears a close resemblance to the recent iron deposits of the Atlantis II Deep (Miller et al., 1966) in that both are composed predominantly of hydrous iron oxides and both contain heavy metals in excess of the average concentrations reported for marine sediments. The recent deposits are now forming in the presence of hot, weakly acid brines which are believed to be due to the submarine discharge of thermal waters (Miller et al., 1966).

Since the goethite bed is underlain by a typical lithified layer it cannot have formed within an active deep, but it could have been deposited beyond the margins of a deep after spreading from the active source area. A similar goethite bed has been found in a core taken approximately 6 miles beyond the edge of the Atlantis II Deep (Bischoff and Manheim, 1969).

This bed is interpreted as hot brine-derived, the goethite having spread from its source in the Atlantis II Deep. Like the goethite bed of core V14-122, it contains much lower but still anomalous concentrations of heavy metals.

The Atlantis II Deep, which lies approximately 150 miles to the south of the area under consideration in this paper, is too distant to be a likely source for the goethite in core V14-122. A much closer, undiscovered source, similar to the Atlantis II Deep, seems to be required. The presence of goethite in all samples (including the top) suggests that this source area was continuously active throughout the time span represented by the core and that it may still be active today.

The wide variation in the percentage of goethite with depth probably reflects changes in the activity at the source. During periods of low activity goethite production and spreading was restricted, only small amounts reaching the core location while during periods of high activity goethite production was increased and spreading was extensive, resulting in the deposition of the goethite bed. The corrosion of foraminiferal tests at 470 and 480cm suggests that acidic brines may have reached the vicinity of the core during the period of maximum activity. The fact that the minerals which occur as traces in the goethite bed are very similar to the major mineral constituents of the lithified layers supports the hypothesis of goethite spreading and rapid deposition in a "normal" Red Sea environment.

Conclusions

Mineralogical and micropaleontological analyses of core V14-122 from the northern Red Sea lead to the following conclusions:

1. Three major climatic phases are represented in the core: Phase I, Postglacial; Phase II, late glacial stades which began about 60–65,000 years ago; Phase III, an interstadial which began about 100,000 years ago.
2. Goethite was deposited after spreading from a nearby, undiscovered hot brine deep which was continuously active during the time span represented by the core and may still be active today.
3. An almost monomineralic goethite bed, 45cm thick, was deposited near the beginning of Phase II during a relatively short period of maximum activity in the hot brine deep.
4. The goethite bed of core V14-122 is considerably older than those reported from other parts of the Red Sea (Ku *et al.*, 1969); the history of hot brine activity and deposition extends back at least 100,000 years B. P. in the northern Red Sea.

Acknowledgments

Sediment samples were kindly made available by Lamont Geological Observatory which raised the core under the support of grants Nonr 266 (48) and NSF (GA-558). We are indebted to D. Lal, Tata Institute of Fundamental Research, India, for ^{14}C age determinations and to C. E. Harvey, Washington State University, and P. L. Siems, University of Idaho, for spectrochemical analyses. We also thank F. L. Parker, Scripps Institute of Oceanography, W. A. Berggren and E. T. Degens, Woods Hole Oceanographic Institution, for helpful discussions.

References

Bé, A. W. H. and W. H. Hamlin: Ecology of recent planktonic foraminifera. Micropaleontology, **13**, 87 (1967).

Berger, W. H.: Foraminiferal ooze: solution at depth. Science, **156**, 383 (1967).

Bischoff, J. L. and F. T. Manheim: Economic Potential of Red Sea heavy metal deposits. *In: Hot brines and recent heavy metal deposits in the Red Sea*, E. T. Degens and D. A. Ross (eds.). Springer-Verlag New York Inc., 535–541 (1969).

Deer, W. A., R. A. Howie, and J. Zussman: *An Introduction to the Rock-forming Minerals*, John Wiley and Sons, Inc., New York (1966).

Deuser, W. G. and E. T. Degens: O^{18}/O^{16} and C^{13}/C^{12} ratios of fossils from the hot-brine deep area of the central Red Sea. *In: Hot brines and recent heavy metal deposits in the Red Sea*, E. T. Degens and D. A. Ross (eds.). Springer-Verlag New York Inc., 336–347 (1969).

Ericson, D. B., M. Ewing, and G. Wollin: The Pleistocene Epoch in deep-sea sediments. Science, **146**, 723 (1964).

Gevirtz, J. L. and G. M. Friedman: Deep-sea carbonate sediments of the Red Sea and their implications on marine lithification. J. Sed. Pet., **36**, 143 (1966).

Harker, R. I. and O. F. Tuttle: Studies in the system CaO-MgO-CO$_2$. Part 2 Limits of Solid Solution along the binary join CaCO$_3$-MgCO$_3$. Amer. Jour. Sci., **253**, 274 (1955).

Herman, Y.: Etudes des sediments quaternaires de la Mer Rouge. Ann. Inst. Océan., **42**, 339 (1965).

———: Evidence of climatic changes in Red Sea cores. *In: Means of Correlation of Quaternary Successions*, R. B. Morrison and H. E. Wright (eds.), **8**, Univ. of Utah Press, 325 (1968).

Ku, T. L., D. L. Thurber, and G. Mathieu: Radiocarbon chronology of Red Sea sediments. *In: Hot brines and recent heavy metal deposits in the Red Sea*, E. T. Degens and D. A. Ross (eds.). Springer-Verlag New York Inc., 348–359 (1969).

Miller, A. R.: Highest salinity in the world ocean? Nature Lond. **203**, 590 (1964).

Miller, A. R., C. D. Densmore, E. T. Degens, J. C. Hathaway, F. T. Manheim, P. F. McFarlin, R. Pocklington, and A. Jokela: Hot brines and recent iron deposits in deeps of the Red Sea. Geochem. et Cosmochim. Acta, **30**, 341 (1966).

Parker, F. L.: Living planktonic foraminifera from the equatorial and southeast Pacific. Sci. Reports. Tohoku Univ., **4**, 71 (1960).

Rosenberg, P. E.: Subsolidus relations in the system CaCO$_3$-MgCO$_3$-FeCO$_3$ between 350° and 550°C. Amer. Mineral., **52**, 787 (1967).

Spoel, van der, S.: Euthecosomata. Noorduijn en Zoon, 375 p. (1967).

Magnetic Properties of Minerals from the Red Sea Thermal Brines *

D. W. STRANGWAY AND B. E. MC MAHON

Massachusetts Institute of Technology
Cambridge, Massachusetts

J. L. BISCHOFF

Woods Hole Oceanographic Institute
Woods Hole, Massachusetts

Abstract

Studies of the magnetic properties of sediment samples recovered from the thermal brine area of the Red Sea have been made. The minerals which have distinctive magnetic signatures, particularly on heating, are goethite, lepidocrocite, siderite, manganosiderite, pyrite, hematite and small quantities of a ferrimagnetic mineral, which is probably maghemite.

The following reactions take place during heating in air:

(a) goethite $\xrightarrow[300°C]{}$ hematite

The characteristic thermoremanent magnetization found in many goethites was not observed in these samples.

(b) lepidocrocite $\xrightarrow[250°C]{}$ maghemite $\xrightarrow[450°C]{}$ hematite

The development and breakdown of maghemite is very characteristic magnetically.

(c) siderite $\xrightarrow[500°C]{}$ magnetite

On cooling, the magnetite acquires a strong magnetization at its Curie temperature (580°C).

(d) manganosiderite $\xrightarrow[500°C]{}$ jacobsite

Jacobsite acquires a strong magnetization upon cooling. The Curie point is less than 580°C, depending upon the amount of substitution of manganese for iron in the magnetite structure.

(e) pyrite $\xrightarrow[400°C]{}$ maghemite $\xrightarrow[500°C]{}$ hematite

During heating a strong magnetic deflection is observed through the temperature range where maghemite is one of the mineral phases.

Hematite is present in two of the cores recovered from the Atlantis II Deep. The magnetic properties indicate that this material is coarse-grained ($>0.5\mu$d). By contrast the hematite formed chemically during the heating experiments is fine-grained ($<0.5\mu$d). A ferrimagnetic mineral, probably maghemite, is present in many of the samples examined. Most of this material is destroyed by heating above 500°C. The absence of an observable Neel temperature effect in the goethites at 120°C is attributed to the small grain sizes, making much of the material superparamagnetic at room temperature. Other data indicate that much of the goethite is in particles less than 100Å.

Introduction

Study of the magnetic properties of samples recovered from the Red Sea cores was undertaken to examine the mineralogical content by means of their magnetic properties. The results of the mineral identification by magnetic methods have been compared with identification based on x-ray analysis. Most of the materials examined in conjunction with x-ray analysis belong to the 2- to 62-micron fraction. In addition, a few samples of unfractionated material

* Woods Hole Oceanographic Institution Contribution No. 2200.

were studied. Examination of the magnetic properties does not indicate that a gradual mineralogical transformation of goethite to hematite takes place *in situ*. Goethite is the most abundant crystalline constituent of the Red Sea cores, but hematite has been reported in two cores in the Atlantis II Deep.

Types of Investigations

Saturation magnetization-versus-temperature (Js-T) measurements are the most commonly used procedures in the investigation of magnetic properties. In this approach a large magnetic field, usually several thousand oersteds (oe), is applied to a sample, and the changes in the magnetization of the samples with change in temperature are observed. At its Curie temperature a ferromagnetic mineral loses its magnetization and becomes paramagnetic. The Curie temperature is characteristic of the mineral and provides a useful means of identification. At very low temperatures some minerals also undergo characteristic changes in their magnetic properties (Fuller and Kobayashi, 1967). In addition to the characteristic temperatures, the reversibility of the heating and cooling curves also provides useful information. Failure of the cooling curve to follow the heating curve indicates that changes have taken place in the magnetic minerals on heating. These changes are due either to chemical reactions which change the mineralogy or to changes in grain size and grain perfection and can often be used in mineral identification.

A second type of thermomagnetic experiment is to heat and cool the sample in a weak magnetic field (usually less than 100oe). The remanence acquired by the sample in these weak fields is referred to as thermoremanent magnetism (TRM). Materials which are ferrimagnetic acquire a TRM upon cooling through their Curie temperatures. Materials which are antiferromagnetic, although essentially nonmagnetic, also have a characteristic temperature at which the ionic spins are ordered, the Neel temperature. Because there are as many spins oriented in one direction

as in the other, no net magnetic effect is to be expected. Recent work with antiferromagnetic materials (Smith and Fuller, 1967; Strangway *et al.*, 1967), however, shows that they often acquire a weak but very stable remanent magnetization upon cooling through the Neel temperature (the temperature of magnetic ordering in antiferromagnetic materials). This is believed to be due to spin imbalance in regions of irregularities and defects in the lattice. As the ordering temperature is characteristic of the mineral, the ability to detect this magnetization provides a convenient means of measuring the Neel temperature and identifying the mineral.

Another magnetic test involves the measurement of hysteresis loops or the magnetization of the mineral as the strength of the applied field is changed (B-H). The field strength necessary to saturate the mineral, the saturation magnetization, the remanence or magnetization remaining in the mineral after the applied field is returned to zero and the strength of the reversed field necessary to destroy the remanence yield useful information concerning the mineral.

Discussion

General

The oxides and oxhydroxides of iron, such as lepidocrocite, goethite, hematite, maghemite and magnetite, have characteristic magnetic properties or undergo chemical changes upon heating which also involve changes in magnetic properties. These changes permit identification of the minerals by means of thermomagnetic analysis (Huggett, 1929; Huggett and Chaudron, 1928). Other minerals, such as siderite, rhodochrosite and pyrite, oxidize to a magnetic phase during the heating process. The magnetic phase and conditions of the chemical changes are characteristic of the parent mineral and can be useful in mineral identification in a manner similar to the technique of differential thermal analysis (Kelly, 1956; Gheith, 1952; Kulp and Trites, 1951).

Magnetic analysis in conjunction with

x-ray examination has yielded useful results in the investigation of materials from the Red Sea thermal brines. The minerals encountered fall into natural mineralogical groups, such as oxhydroxides, oxides, carbonates and sulfides, and the results are discussed under these headings.

One difficulty with the use of magnetic properties for mineral identification is that many of these properties are dependent upon grain size. Because the Red Sea sediments contain colloidal material it is necessary to consider the effects of extremely small grain size on the magnetic properties.

Superparamagnetism

In very small particles the thermal disordering energy may equal the magnetic ordering energy, and many of the characteristic magnetic properties of large grains are lost or changed. Instead of behaving as ferrimagnetic and antiferromagnetic substances, they behave as paramagnetic substances and are therefore referred to as superparamagnetic. The critical size for the change to superparamagnetism has not been determined experimentally with great precision. It has been estimated, however, that the ferrimagnetic materials, magnetite and maghemite, will lose their ferrimagnetic character in grain sizes less than about 20Å. Creer (1961) calculated that the antiferromagnetic materials goethite and hematite should also lose their magnetic character in particle sizes less than 20Å. Experimental data suggest, however, that the effect of superparamagnetism is observable in grains as large as 250Å (Creer, 1962). Shinjo (1966) showed that goethite in particles less than 100Å is superparamagnetic. Hematite is peculiar in that it has, in addition to antiferromagnetism, a magnetization often referred to as parasitic ferromagnetism. Although it is very weak compared to ordinary ferrimagnetism, it is many times stronger than the small remanent magnetization associated with spin imbalance in the antiferromagnetism. This parasitic magnetization is lost in grain sizes smaller than 1000 to 5000Å (Strangway et al., 1967).

Between the grain sizes characterized by superparamagnetic behavior at low temperatures and the critical size at which magnetic ordering takes place at the Curie or Neel temperature, there is a spectrum of grain sizes which become ordered at progressively lower temperatures, often referred to as blocking temperatures. If none of the grains in a sample exceeds the critical size, no magnetic ordering will be possible at the characteristic Curie or Neel temperature. Furthermore, antiferromagnetic material in a superparamagnetic state shows a higher susceptibility than the same material in an ordered antiferromagnetic state. Superparamagnetic material is expected to show saturation when subjected to sufficiently large magnetic fields. Experimental observations have shown a field dependence in such materials (Creer, 1961; Bannerjee, 1968), particularly at cryogenic temperatures. It has not been determined if the effect is due entirely to superparamagnetism or to the presence of small quantities of ferrimagnetic material. Thus it is seen that in finely divided materials changes occur in the magnetic properties which affect the Curie or Neel temperature, the susceptibility and, possibly, the degree of field dependence.

Oxhydroxides

Goethite (αFeOOH). Goethite is the most abundant iron mineral identified in the sediments below the thermal brines of the Red Sea. Goethite is antiferromagnetic and has a Neel temperature at 120°C. Upon cooling from above 120°C, but from a temperature below that of its transition to hematite, goethite has been observed to acquire a weak thermoremanent magnetization (Strangway et al., 1967), which is explained as arising from spin imbalance.

Fig. 1 shows the Js-T curves and weight change curves obtained for several samples identified as containing goethite by x-ray analysis. Fig. 2 shows the character of the B-H curves seen on several samples of goethite.

The weight-change curves (Fig. 1) show a weight loss occurring between 250°C and 350°C. This is characteristic of all the Red Sea samples reported to contain goethite on the basis of x-ray examination and is caused by the transition from goethite to hematite. The transition temperature is slightly lower than that reported for well crystallized goethite, but differential thermal analysis has shown that poorly crystalline goethite has a lower temperature of transition than well crystallized material (MacKenzie, 1957). The weight change curves also show a weight loss at lower temperatures, particularly in the area of 100°C (Fig. 1b and 1c). This is believed to be due to the loss of water of hydration, an interpretation which is supported by the observation that the total weight loss is two to three times greater than can be accounted for by the dehydration of goethite, even assuming 100 per cent original goethite content in the sample.

Examination of the Js-T curves (Fig. 1)

shows no Curie temperature effect at 680°C although hematite is formed by the heating of goethite. The absence of the Curie temperature effect indicates that the hematite formed is in grain sizes smaller than 0.5μ, too small to carry the characteristic parasitic ferromagnetism.

Examination of Fig. 3 shows that there was no detectable change between heating and cooling curves when sample 85K (−160cm), weak goethite, was heated to 350°C although the weight-change curve indicates that goethite has already dehydrated to form hematite. It appears, therefore, that the effect of some included ferrimagnetic material dominates the Js-T curve up to 350°C.

The thermoremanent magnetization, which forms at 120°C and has been observed in many other goethites, was not

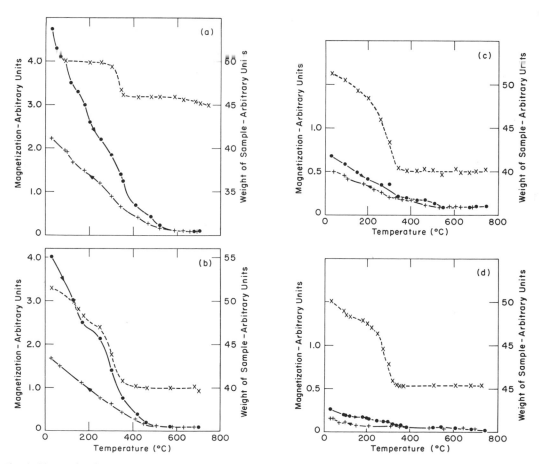

Fig. 1. Magnetization-versus-temperature curves for samples containing goethite by x-ray analysis. Dashed curve gives pattern of weight change with heating. Solid dots indicate the heating curve, the + signs the cooling curve. (a) Sample 161K (−325)—moderate goethite indication on x-ray. (b) Sample 161K (−360)—strong goethite indication on x-ray. (c) Sample 161K (−379)—strong goethite indication on x-ray. (d) Sample 127K (−455)—strong goethite indication on x-ray.

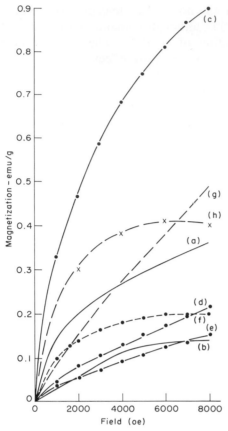

Fig. 2. Magnetization-versus-applied field. (a) Sample 161K (–325) before heating. (b) Sample 161K (–325) after heating. (c) Sample 85K (–160) before heating. (d) Sample 161K (–379) before heating. $x \approx 27 \times 10^{-6}$ emu/g. (e) Sample 127K (–455) before heating. $x \approx 19 \times 10^{-6}$ emu/g. (f) Curve (a) minus curve (e). Possible ferrimagnetic contribution to Sample 161K (–325). (g) Representation of a possible superparamagnetic contribution to curve (c). $x \approx 64 \times 10^{-6}$. (h) Representation of a possible ferrimagnetic contribution to curve (c).

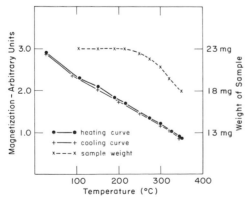

Fig. 3. Partial heating curve showing independence of magnetization and chemical change occurring between 250°C and 350°C. Sample 85K (–160).

present in these samples. This indicates that any thermoremanence formed on cooling was less than 3×10^{-6} emu/g, even in an applied field as high as 400oe. This may reflect the dilute nature of goethite in the samples measured although it is more likely that the grain size of the samples investigated is less than about 100Å.

Lepidocrocite (γFeOOH). Lepidocrocite, which is paramagnetic at room temperature, dehydrates to form maghemite, which has a large magnetic moment. Dehydration takes place between 260° and 300°C and is accompanied by a marked weight loss. The maghemite formed from lepidocrocite in turn breaks down to hematite. The temperature at which the inversion to hematite is complete varies with the sample and may occur anywhere between 400° and 600° although it is usually completed by 500°C. In all the heating experiments reported in the literature, the hematite formed is fine grained ($<0.5\mu$), since the end product does not possess the characteristic parasitic ferromagnetism. In fact, in a report of one of the earliest experiments with lepidocrocite, Huggett (1929) identified the product as a paramagnetic phase. The resulting magnetization-versus-temperature curve provides a means of identifying even very minute amounts of lepidocrocite (Fig. 4a).

Samples 119K (−27cm) and 127K (−762cm) contain lepidocrocite (Fig. 4b and Table 2) as shown by x-ray and Js-T analysis. Between 300° and 400°C a marked increase in magnetization shows the formation of maghemite. The decrease in magnetization at 400°C is the result of the breakdown of the maghemite to hematite. The curve displayed here is complicated above 480°C by the formation of magnetite through the breakdown and oxidation of siderite (see discussion under carbonates).

Carbonates

Siderite ($FeCO_3$). Upon heating, siderite breaks down at approximately 500°C with a marked weight loss as CO_2 is driven off. The FeO is then oxidized to an iron oxide which has an inverse spinel structure. The chemical changes have been observed and discussed in connection with mineral identification by differential thermal analysis (Kulp *et al.,* 1949; Rowland and Jonas, 1949). A brief summary of their observations follows:

Release of CO_2 is accompanied by an endothermic peak. The temperature of oxidation of the residual FeO, which is an exothermic reaction, depends upon the availability of oxygen to the system. Under well aerated conditions oxidation takes place concomitantly with release of CO_2, and the exothermic reaction (oxidation) masks the endothermic reaction (release of CO_2). According to Kulp *et al.* (1951), the product of the oxidation reaction is primarily maghemite (γFe_2O_3) with some hematite (αFe_2O_3). At 800°C there is a second exothermic peak, which is interpreted as the inversion of maghemite to hematite (Kulp *et al.,* 1951).

In our experiments the product of the first oxidation reaction appears to be magnetite rather than maghemite. The ability of the material to withstand heating to above 700°C and the Curie temperature of about 580°C (Fig. 4b) is typical of magnetite. For comparison, a thermomagnetic curve of a sample which contains only siderite is shown in Fig. 4c. Fig. 5 shows the B-H curve of the sample (119K −270) after heating to 735°C. It is seen that the material approaches saturation in an applied field of 1,000oe. Although finely divided (less than .5μ) hematite, formed from dehydration of lepidocrocite, is present, the

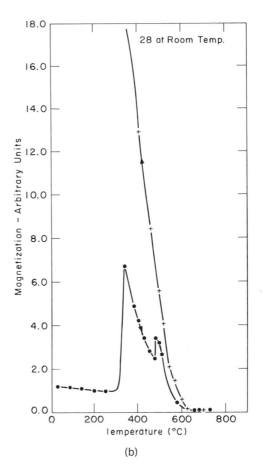

(b)

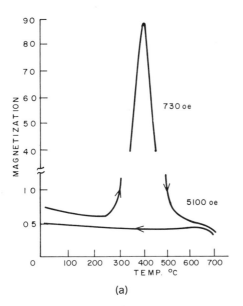

(a)

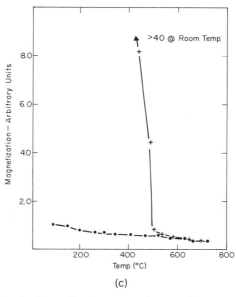

(c)

Fig. 4. Magnetization-versus-temperature curve; (a) Lepidocrocite; (b) Sample 119K (−270); (c) siderite.

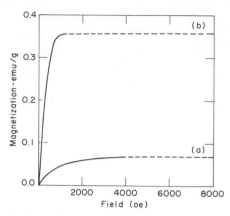

Fig. 5. Magnetization versus-applied field. Curves obtained from Sample 119K (–270) before (a) and after (b) heating to 735°C.

character of the B-H curve is dominated by magnetite formed from the breakdown of siderite. The saturation magnetization of 0.3 emu/g indicates that magnetite constitutes less than 0.4 per cent of the heated sample, and siderite in the original sample must have been equally dilute. A sample composed of siderite or siderite and lepidocrocite would have had a B-H curve similar to that displayed for curves a of Fig. 8. The curve obtained before heating (curve a, Fig. 5) indicates that the unheated sample also contains a small amount of ferrimagnetic material.

Rhodochrosite ($MnCO_3$). Rhodochrosite is the manganese carbonate. Ford (1917) suggested that there is a break in the series $FeCO_3$-$MnCO_3$ between Mn 50 per cent and 70 per cent. Kulp *et al.* (1951) and Rowland and Jonas (1949) report that complete substitution appears possible although the full sequence is not represented in nature. Upon heating rhodochrosite, CO_2 is driven off and the residual MnO is oxidized to hausmannite (Mn_3O_4), which is tetragonal at room temperature. Above 1160°C hausmannite has a cubic form and a complete solid solution series between magnetite and hausmannite is possible (Van Hook and Keith, 1958). Below 1160°C some exsolution occurs as high hausmannite (cubic) inverts to low hausmannite (tetragonal). At room temperature a two-phase mixture exists for compositions containing 54 mole per cent to 91 mole per cent Mn_3O_4 (Fig. 6). Within this compositional range, cubic

jacobsite forms intergrowths with tetragonal hausmannite.

As the siderite-rhodochrosite series oxidizes to the magnetite-jacobsite-hausmannite series within the temperature range of our Js-T experiments, the characteristics of the oxide series are important to an interpretation of the results of the experiment. Yun (1958), working with synthetic material which had been fused at 1100°C, has shown that there exists a relationship between the Curie temperature and the substitution of Mn for Fe in the magnetite-hausmannite solid solution series. The Curie temperature is progressively lowered by the substitution of Mn for Fe from 580°C for pure magnetite to 50°C for a Mn_3O_4/Fe_3O_4 ratio of 90 per cent. A reliable estimate of the Mn_3O_4 content probably requires heating to 1160°C to permit the development of a single phase. Exsolution is evidently too sluggish (Van Hook and Keith, 1958) to be accomplished in the time required for cooling through the critical temperature. In the normal heating range for Js-T experiments (up to 750°C) the two-phase region is not materially reduced. This limitation may have contributed to the high incidence of natural jacobsites with Mn_3O_4/Fe_3O_4 ratios of 60 per cent reported by Yun (1958). The same limitation applies to the jacobsite-hausmannite formed by heating manganosiderite. The temperatures of the experiment do not achieve a

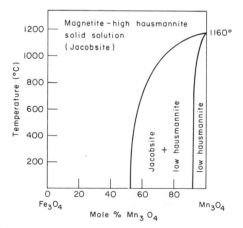

Fig. 6. Composition-temperature diagram for magnetite-hausmannite solid solution series, showing extent of two-phase region (Van Hook and Keith, 1958).

complete solid solution of the products of the chemical reaction for certain proportions of iron to manganese, and intergrowths of low hausmannite and jacobsite develop. For this reason it is expected that a disproportionate number of samples will indicate an Mn_3O_4 content of 55–60 mole per cent when substitution is estimated on the basis of Curie temperature data.

Samples 95G (−160cm), 95G (−190cm) and 127P (−514cm) contain manganosiderite. A characteristic curve obtained from the Js-T experiment is shown in Fig. 7. A sharp weight change occurs at 510°C where the chemical breakdown takes place. On cooling, the sample becomes magnetic at 210°C, the Curie temperature. For 2 of the samples the Curie temperature indicates an Mn_3O_4 content of approximately 50 mole per cent. Sample 95G (−190cm),

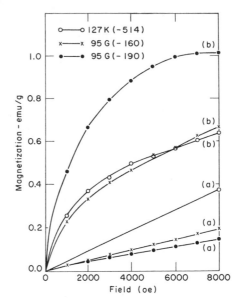

Fig. 8. Magnetization-versus-applied field before (a) and after (b) heating to above 700°C. Samples 127 (−514), 95G (−160), 95G (−190); (a) manganosiderite, (b) jacobsite.

with a Curie temperature of 300°C, is estimated to have an Mn_3O_4 content of 40 mole per cent. Assuming no substitution by other cations, such as Ca or Mg, a calculation of the Mn/Fe content based on x-ray data gives ratios showing slightly higher Mn content (Table 1). The B-H curves for samples 127K (−514cm) and 95G (−160cm) before heating show the linear relationship between magnetization and applied field expected for paramagnetic material (Fig. 8). The same samples after heating tend to saturate although saturation is not achieved in fields up to 8,000oe. After heating, sample 95G (−190cm) reaches saturation in a field of 8,000oe due to the presence of the magnetic oxide. Before heating, this sample shows only a very slight field dependence below 1,000oe.

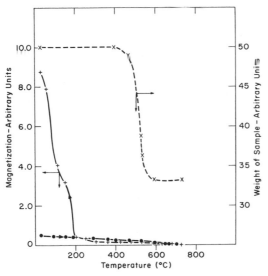

Fig. 7. Magnetization-versus-temperature curve for Sample 127 (−514) containing manganosiderite. Dashed curve shows pattern of weight change with heating.

Table 1

| | | Mole % Mn_3O_4 | |
| | | | |
Sample	Curie Temp. of Jacobsite Product	Tc Data (Jacobsite)	X-ray Data (Iron-Rich Rhodochrosite)
95G (−160cm)	280°	40	50% ± 5%
95G (−190cm)	220°	50	60% ± 5%
127K (−514cm)	210°	50	not run

Sulfides-pyrite (FeS$_2$). Pyrite is nonmagnetic, but, on heating, sulfur is driven off and the iron is oxidized. Kopp and Kerr (1958) have reported the chemical changes taking place and their effect upon the curves obtained for differential thermal analysis. The dominant peak is exothermic and signals the oxidation reaction. The character of the peak and the temperature

of the initial oxidation peak is influenced by grain size and sample weight. Kopp and Kerr report that after heating to 610°C, x-ray analysis showed the presence of hematite and a cubic iron oxide, whereas after heating to 1000°C only hematite lines are identifiable on the x-ray. The intermediate cubic phase they identified as either maghemite or magnetite. Because of the breakdown of this phase on heating, we believe it to be maghemite (see discussion below under oxides).

Experimental work based on thermomagnetic analysis is in good agreement with the observations made by Kopp and Kerr which are outlined above. The Js-T curve for a typical pyrite sample is shown in Fig. 9a. The peak on heating is due to the formation of maghemite at about 480°C and its subsequent breakdown. By x-ray analysis, sample 119K (−282cm) contains pyrite (Fig. 9b). Weight loss, caused by the release of SO$_2$, occurs between 360° and 410°C. The formation of maghemite is indicated by the large increase in magnetization at 410°C. Above 475°C the inversion of maghemite to hematite proceeds more rapidly than the formation of new maghemite. If this interpretation is correct, the oxidation peak is 40° to 60° lower than that observed by Kopp and Kerr. This difference is probably due to the very fine nature of the sample from the Red Sea or to the slow heating used in our experiment.

This sample also appears to contain a small quantity of ferrimagnetic material which breaks down after heating above 500°C. This is indicated by the tendency for saturation before heating (Fig. 10).

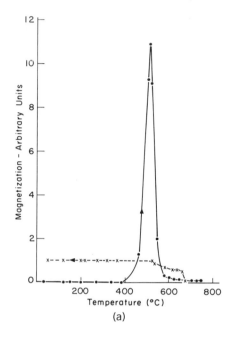

(a)

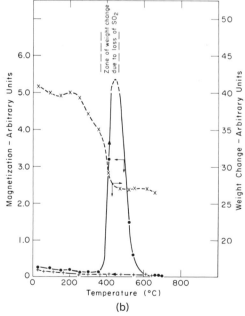

(b)

Fig. 9. Magnetization-versus-temperature curve: (a) Pyrite and (b) Sample 119K (−282) which contains pyrite. Dashed line shows pattern of weight change during heating.

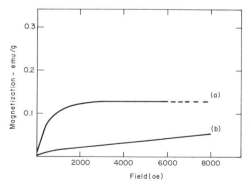

Fig. 10. Magnetization-versus-applied field before (a) and after (b) heating to 700°C. Sample 119K (−282).

Oxides

Magnetite (Fe₃O₄) and Maghemite (γFe₂O₃). Structurally, magnetite and maghemite are very similar. Maghemite commonly forms as a result of the oxidation of magnetite, and there is evidence to suggest that the two minerals can and do form a complete solid solution. Because of structural similarities and probably also because of the ability of the oxidation product to form a solid solution with the original mineral, it is difficult to distinguish between the two minerals by means of x-ray analysis. Their structural similarity also imparts to them very similar magnetic characteristics. Not only is it difficult to isolate maghemite from magnetite, but the thermal instability of maghemite (γFe₂O₃) has made it difficult to define its magnetic properties precisely. Because maghemite inverts so readily to hematite during heating, it has not been possible to measure the Curie temperature of maghemite directly. By stabilizing the mineral with Na ions, Michel and Chaudron (1935) extrapolated a value of 675°C for pure maghemite. Bannerjee (1965), by measuring properties of pure maghemite below the Curie temperature, obtained a value of 545°C by extrapolation.

Magnetite is more stable to heating, and complete oxidation to hematite is difficult to achieve. The oxidation reaction is sometimes quite complicated and probably varies with the sample. It seems to be partially controlled by grain size and the presence of associated water (Schmidt and Vermaas, 1955; Lepp, 1957; David and Welch, 1956). Under conditions of rapid oxidation, such as may occur in very small particles or in the presence of water, magnetite may be oxidized to maghemite at fairly low temperatures (~400°C). Continued heating causes an inversion of the maghemite to hematite. With coarse grained samples, rapid oxidation of a thin surface film to maghemite appears to occur at about 360°–375°C (Schmidt and Vermaas, 1955), but the bulk of the material does not oxidize until higher temperatures are reached. Some confirmation of this pattern is observed in magnetization-temperature experiments. The first time freshly crushed

material is tested the saturation magnetization of the cooling curve is somewhat less than that of the heating curve, suggesting that a minor amount of the sample has converted to hematite. Subsequent runs are reversible, indicating that no more chemical changes are taking place. Schmidt and Vermaas have reported tests in which the size of the endothermic peak at 360°–375°C is found to be directly proportional to the surface area of the sample tested. They also found that the low temperature peak did not reappear when the sample was reheated. At about 600°C an endothermic reaction is initiated which extends over a considerable temperature range (Schmidt and Vermaas, 1955; Gheith, 1952). The authors ascribe this reaction to the conversion of the bulk of the magnetite to hematite.

A number of the samples examined for their magnetic characteristics show a tendency to saturate in high fields (up to 8,000oe) although x-ray examination has not shown that any ferrimagnetic mineral is present. In fact, ferrimagnetic minerals are notably absent in all discussions of the minerals recovered from the Red Sea Deeps, with the exception of Miller et al. (1966), who reported the questionable identification of maghemite in a pipe dredge (station 543) of the Atlantis II Deep. Samples which show field dependence are 119K (−270cm) which contains lepidocrocite and siderite (Fig. 5), 119K (−282cm) which contains pyrite (Fig. 10) and some of the samples which contain goethite (Table 2, Fig. 2). The observed field dependence could be due to a superparamagnetic effect in small grains or it could be due to the presence of a small amount of ferrimagnetic material, such as maghemite or magnetite, which is too fine grained or too dilute to be identified by x-ray techniques.

The latter interpretation is preferred for the following reasons. There is no reason to suspect the presence of any ferrimagnetic or antiferromagnetic mineral that could contribute superparamagnetism in the samples containing lepidocrocite-siderite (119K −270cm) and pyrite (119K −282cm) because:

1. The dominant mineral content is paramagnetic above room temperature.

Table 2

Sample		Preliminary X-ray Determination	Determination by Magnetic Properties
Core	(Depth Below Surface, cm)		
127	(−380)	New clay	No distinguishing characteristics
127	(−455)	Goethite	Weight loss at 300°C characteristic of goethite otherwise indeterminate *
127	(−514)	Manganosiderite	Manganosiderite
127	(−762)	Lepidocrocite	Lepidocrocite, siderite
119K	(−270)	Lepidocrocite	Lepidocrocite, siderite, maghemite(?)
119K	(−282)	Pyrite	Pyrite, maghemite(?)
161K	(325)	Goethite	Weight loss at 300°C characteristic of goethite otherwise indeterminate,* maghemite(?)
161K	(−360)	Goethite	Weight loss at 300°C characteristic of goethite otherwise indeterminate,* maghemite(?)
161K	(−379)	Goethite	Weight loss at 300°C characteristic of goethite otherwise indeterminate *
85K	(−160)	Goethite	Weight loss at 300°C characteristic of goethite otherwise indeterminate,* maghemite(?)
95G	(−160)	Manganosiderite	Iron-rich rhodochrosite
95G	(−190)	Iron-rich rhodochrosite	Iron-rich rhodochrosite

* See text.

2. The roughly flat character of the Js-T curves between room temperature and 250°C indicates that there is no superparamagnetic effect at room temperature (Figs. 4b and 9b).

3. The form of the saturation-magnetization curves (Figs. 5a and 10a) suggests the effect of a single phase which saturates at very low field strengths, a characteristic of the two common ferrimagnetic minerals. Thus, in these samples the mineral which has a strong field dependence dominates the character of the B-H curve, and the effect of the paramagnetic materials, which are dominant by weight percent in the samples, are masked.

It is probable that some magnetite or maghemite is present. In this case, it is possible to calculate the approximate percentages of these minerals contained in the samples. The saturation magnetization of magnetite at room temperature is 93 emu/g and that of maghemite is approximately 83 emu/g. As sample 119K (−282cm) has a saturation magnetization of .127 emu/g, the portion of the ferrimagnetic mineral which is greater than the critical size constitutes no more than 0.15 per cent of the sample. Sample 119K (−270cm), with a saturation magnetization of .067 emu/g, contains a maximum of approximately 0.08 per cent ferrimagnetic material. The estimated content is far below the detection levels of x-ray analysis. It is noted that the B-H curve obtained on sample 119K (−282cm) after heating is linear and has a weaker response to high fields (Fig. 10b), suggesting that heating has destroyed the ferrimagnetic mineral.

By analogy with the samples containing pyrite and lepidocrocite-siderite discussed above, the field dependence observed in unheated samples containing goethite is probably also due to the presence of some ferrimagnetic material rather than to super-paramagnetic behavior in the fine grained, antiferromagnetic goethite (Fig. 2, curves a and c). The character of the heating curve below approximately 500°C in some of the samples (Fig. 1a and 1b) is also believed to be dominated by the ferrimagnetic material because heating through the goethite-to-hematite transition temperature produces no change in the Js-T curves (Fig. 3), whereas heating to above 500°C causes a decrease in the cooling curve (Fig. 1) and a more linear response to applied fields from 0 to 8,000 oe (Fig. 2, curve b). Again, it appears that a ferrimagnetic phase has been destroyed at about 500°C.

From the linear character of the B-H curve, sample 127 (−455cm) (Fig. 2e) appears to contain no ferrimagnetic material. X-ray peaks for this sample gave a strong goethite response, suggesting approximately 50 per cent finely crystalline goethite content. On this basis we assume that the B-H response observed for sample 127K (−455cm) is dominated by the goethite content. By contrast, curve a of Fig. 2 appears to reflect the response of a mixture of both goethite and the ferrimagnetic mineral. Curve f of Fig. 2, which is obtained by subtracting curve e from curve a, is very similar to curves a of Figs. 5 and 10 which are believed to be dominated by the ferrimagnetic minerals. Sample 161 (−325cm) had a moderate goethite indication on the basis of x-ray analysis, but, as shown in Fig. 2c, it appears to be a mixture of a small amount of ferrimagnetic material and a material with a linear field response and a much larger susceptibility than that indicated for the other samples containing goethite. X-ray analysis shows only weak goethite, indicating that the bulk of the sample is composed of amorphous goethite. The apparent high susceptibility observed in this sample (69×10^{-6} emu/g), which on the basis of x-ray analysis contains extremely finely divided material, compared to the relatively low susceptibility of sample 127 (−455cm) (19×10^{-6} emu/g), which contains slightly better crystalline material, suggests that the fine grained material is superparamagnetic. Curves g and h of Fig. 2 show a possible distribution between fer-rimagnetic and superparamagnetic contribution to the observed curve for sample 85K (−160cm), curve c of Fig. 2.

In some of the Red Sea sediments there appears to be a small amount of ferrimagnetic material which contributes to the saturation effect observed in some of the samples. There may also be a superparamagnetic effect contributed by very finely divided goethitic material as indicated by a high susceptibility in some of the samples.

On the basis of the available thermomagnetic data it is not possible to unequivocally identify the primary ferrimagnetic material. Its identification as maghemite is favored for a number of reasons: (1) The material breaks down upon heating to above 500°C. By contrast the material formed by the oxidation of siderite (magnetite) is unaffected by subsequent heating even above 700°C. (2) The material appears to be completely disseminated through the 2- to 62-micron fraction of the samples which show the ferrimagnetic component. This suggests that the material is a primary precipitate along with the goethite and lepidocrocite with which it most commonly occurs. Because of its association with oxhydroxides, it is probably more reasonable to assume that the ferrimagnetic material is maghemite rather than magnetite. Magnetite spherules have been found associated with slowly accumulating sediments (Mutch, 1966), and it is very probable that such materials have been deposited with the Red Sea sediments although they have not been identified to date. Because the ferrimagnetic material is so well distributed through the samples in which it occurs and because the oxhydroxide beds are believed to have accumulated at a fairly rapid rate, the ferrimagnetic effects observed in these samples are not believed to be due primarily to the inclusion of magnetite spherules.

Hematite (αFe_2O_3). In the Atlantis II Deep both core 84K and the adjacent core 84P contain hematite. A bulk sample from 350cm below the surface (core 84K), examined magnetically, was found to contain hematite. The magnetic determination has been supported by x-ray analysis. Fig. 11 displays the Js-T curve for 48K (−350cm) and Fig. 12 the B-H curves on

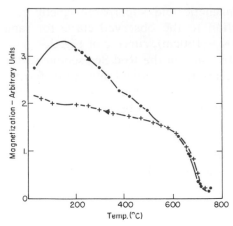

Fig. 11. Magnetization-versus-temperature curve-hematite. Sample 84K (–350).

the unheated sample (a), and the sample after it was heated to 700°C (b). The Js-T curve shows a Curie temperature effect at 680°C. The Curie point effect indicates that the hematite is coarse crystalline occurring in grain sizes that exceed $.5\mu$ diameter. During heating of the sample there was no weight change in the vicinity of 250°–350°C, the temperature zone that usually reflects the dehydration of goethite. No goethite was observed in the x-ray analysis of this sample or of selected samples higher in the core.

It is seen that the unheated sample has a

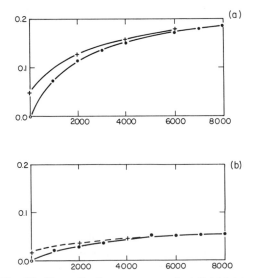

Fig. 12. Magnetization-versus-applied field (a) before heating and (b) after heating to 700°C. Sample 84K (–350).

tendency to saturate in high fields (Fig. 12a). In the heated sample the intensity of magnetization is decreased by more than 50 per cent, but the material still shows a field dependence, around 2,000 to 3,000oe, suggesting that the ferrimagnetic material has not been completely destroyed.

Conclusions

Table 2 compares the initial x-ray determinations of several samples taken from Red Sea cores with the determinations based on identification of the minerals by their magnetic properties. From the table it is seen that in several of the determinations the two methods are in complete agreement. On the other hand, there are several cases in which the two methods complement one another. The goethites could not have been identified by the magnetic method alone. The only observable indication of goethite is a weight loss in the vicinity of 300°C.

The saturation effect observed in many of the samples reveals the presence of a ferrimagnetic mineral, probably maghemite as indicated by its breakdown on heating to 500°C. It is present in quantities too dilute to be detected by x-ray analysis (0.1 per cent). Siderite as well as lepidocrocite was detected in samples 119K (−270cm) and 127K (−762cm).

The inability to observe a thermoremanent magnetization after cooling the Red Sea goethites through their Neel temperature at 120°C is probably due both to the fine grained character of the material and the dilute concentration of the material in the sediments. A significant proportion of the goethite probably falls within the size range of superparamagnetic behavior at room temperature so that the particles are around 100Å.

Acknowledgments

We would like to acknowledge the financial support of NSF on project G.P. 5341. We thank G. Olhoeft for his assistance in the laboratory.

References

Bannerjee, S. K.: On the transition of magnetite to haematite and its implication to rock magnetism. Journal Geomagnetism and Geoelectricity, **17**, 357 (1965).

————: Origin of weak ferromagnetism and remanence in natural cassiterite crystals. American Geophysical Union Transaction, **49**, 135 (1968).

Creer, K. M.: Superparamagnetism in red sandstones. Geophysical Journal, **5**, 16 (1961).

————: On the origin of the magnetization or red beds. Journal Geomagnetism and Geoelectricity, **13**, 68 (1962).

David, I. and A. J. E. Welch: The oxidation of magnetite and related spinels. Constitution of gamma ferric oxide. Trans. Faraday Soc., **52**, 1642 (1956).

Ford, W. E.: (1917) quoted in W. A. Deer, R. A. Howie, J. Zussman. *Rock-Forming Minerals.* John Wiley and Sons, Inc., New York, 371 p. (1962).

Fuller, M. D. and K. Kobayashi: Identification of magnetic phases in certain rocks by low-temperature analysis. *In: Methods in Paleomagnetism,* D. W. Collinson, K. M. Creer, S. K. Runcorn (eds.), 529 Elsevier Press, London (1967).

Gheith, M. A.: Differential thermal analysis of certain oxides and oxide hydrates. American Journal of Science, **250**, 677 (1952).

Huggett, J.: Application de l'analyse thermomagnétique à l'étude des oxydes et des mincrais de fer. Annal. de Chemie, **11**, 447 (1929).

———— and G. Chaudron: Etudes thermomagnétique de quelques minerais de fer. Comptes Rendus de l'Académie des Sciences, **186**, 694 (1928).

Kelly, W. C.: Applications of D.T.A. to identification of the natural hydrous ferric oxides. American Mineralogist, **41**, 353 (1956).

Kopp, O. C. and P. F. Kerr: Differential thermal analysis of pyrite and marcasite. American Mineralogist, **43**, 1079 (1958).

Kulp, J. L., P. Kent, and P. F. Kerr: Thermal study of the Ca-Mg-Fe carbonate minerals. American Mineralogist, **36**, 643 (1951).

Kulp, J. L. and A. F. Trites: Differential thermal analysis of natural hydrous ferric oxides. American Mineralogist, **36**, 23 (1951).

Kulp, J. L., H. D. Wright, and R. J. Holmes: Thermal study of rhodochrosite. American Mineralogist, **34**, 195 (1949).

Lepp, H.: Stages in the oxidation of magnetite. American Mineralogist, **42**, 679 (1957).

MacKenzie, P. C.: *The Differential Thermal Investigation of Clays.* Mineralogical Society, London (1957).

Mason, B.: The mineralogical aspects of the system FeO-Fe$_2$O$_3$-MnO-Mn$_2$O$_3$. Geol. För. Förh. Stockholm, **65**, 97 (1943).

Michel, A. and G. Chaudron: Etude du sesquioxyde de fer cubique stabilise. Comptes Rendus de l'Académie des Sciences, Paris, **201**, 1191 (1935).

Miller, A. R., D. C. Densmore, E. T. Degens, J. C. Hathaway, F. G. Manheim, P. F. McFarlin, R. Pocklington, and A. Jokela: Hot brines and recent iron deposits in deeps of the Red Sea. Geochimica et Cosmochimica Acta, **30**, 341 (1966).

Mutch, T. A.: Abundances of magnetic spherules in Silurian and Permian salt samples. Earth and Planetary Science Letters, **1**, 325 (1966).

Nicholls, G. D.: The mineralogy of rock magnetism. Phil. Mag. suppl., **4**, 113 (1955).

Rowland, R. A. and E. C. Jones: Variations in differtial thermal analysis curves of siderite. American Mineralogist, **34**, 550 (1949).

Schmidt, E. R. and F. H. S. Vermaas: Differential thermal analysis and cell dimensions of some natural magnetites. American Mineralogist, **40**, 422 (1955).

Shinjo, T.: Mössbauer effect in antiferromagnetic fine particles. Journal Physical Soc. Japan, **21**, 917 (1966).

Smith, R. W. and M. Fuller: Alpha hematite: stable remanence and memory. Science, **156**, 1130 (1967).

Strangway, D. W., B. E. McMahon, and R. M. Honea: Stable magnetic remanence in antiferromagnetic goethite. Science, **158**, 785 (1967).

Van Hook, H. J. and K. L. Keith: The system Fe$_3$O$_4$-Mn$_3$O$_4$. American Mineralogist, **43**, 69 (1958).

Yun, I.: Experimental studies on magnetic and crystallographic characters of Fe-bearing manganese oxides. Mem. Coll. Sci. Univ. of Kyoto, B, **25**, 125 (1958).

Sulfur Isotope Studies on Red Sea Geothermal Brines and Sediments *

I. R. KAPLAN, R. E. SWEENEY, AND ARIE NISSENBAUM

Department of Geology and Institute of Geophysics and Planetary Physics
University of California, Los Angeles

Abstract

S^{34}/S^{32} measurements and trace element analyses were performed on sediment from selected cores from the Red Sea geothermal deposit, the overlying brine and interstitial water.

The data show that δS^{34} falls into four general ranges: (1) >25 per mill, (2) +23 to +15 per mill, (3) +12 to +2 per mill, (4) <−25 per mill. Sulfate in the brine appears to have been derived from marine evaporites, whereas sulfide originates from two sources. In the Atlantis II Deep, it is derived from a hydrothermal process and introduced with the brine. In the other areas, biological sulfate reduction produces sulfide which precipitates metals from the brine originating in the Atlantis II Deep.

Introduction

The first description of core material taken by the Atlantis II expedition in hot, hypersaline brine basins from the median valley of the Red Sea (Miller *et al.,* 1966) indicated the presence of sulfur minerals. Sphalerite appeared to be the most abundant sulfide, but traces of pyrite (or marcasite) were also detected. Subsequent studies by Bischoff (1969) have demonstrated the presence of chalcopyrite and in a few isolated samples, greigite and possibly pyrrhotite. Anhydrite has also been identified in several cores.

Hartman and Nielsen (1966) analysed sulfur isotopes in soluble sulfate in normal Red Sea water and the hot brine from Atlantis II Deep. They also analysed sulfate from interstitial water and the total sulfide in a grab sample of sediment from the above basin. The four δS^{34} values they published are: +20.3, +20.3, +20.4 and +4.5 per cent, respectively. They concluded that the sulfate in the brine and interstitial water was probably derived from admixing of overlying Red Sea water and that the sulfide was introduced together with the hypersaline brine.

This chapter describes a more detailed study of the sulfur species dissolved in the brine and fixed in the sediment as sulfate, sulfide or elemental sulfur. Sulfur isotope ratios (δS^{34}), mineralogy and trace element concentrations have been measured in order to determine the origin of the ions and minerals in terms of the general geochemical history of the region.

Methods

In this study, water from seven hydrocasts (715, 717-P, 722-1, 722-2, 726, 727, 728-3) were analysed for soluble sulfate. In five of them (715, 717-P, 722-1, 726 and 727) δS^{34} measurements were made.

Detailed analysis from several depths were made on four cores, two from Atlantis II Deep (84K and 120K), one from Discovery Deep (119K) and one from the area outside the anomalous basins (118K). Additional samples were analysed from cores 127P (Atlantis II Deep), 85K (the sill between Atlantis and Chain Deeps) and 95K (an elevated knoll within the Atlantis II Deep).

* Contribution No. 680, Institute of Geophysics and Planetary Physics.

Cores 84K, 120K and 115K were stored at −20°C. Core 95K was stored at 2–4°C, whereas cores 119K and 85K were stored at ambient temperatures from time of collection until February, 1967, when samples were removed. All samples were subsequently stored at 4°C until analyzed (June, 1967–March, 1968). Interstitial water was removed by using a squeezer described by Presley *et al.* (1967; see also Brooks *et al.*, 1969).

Sulfur Species Separation

Soluble sulfate in the hydrocasts and interstitial waters was determined gravimetrically as $BaSO_4$. Elemental sulfur was extracted from the dry sediment by Soxhlet extraction or by refluxing with benzene. The sulfur was then oxidized to sulfate and determined gravimetrically as $BaSO_4$ above. Acid soluble sulfate was obtained by rapidly reacting the sediment with hot, dilute (1:10) hydrochloric acid. The sulfides in the non-soluble portion of the sediment were oxidized to sulfate by reacting the residue with boiling $1:1 HCl:HNO_3$.

Mineral Separation

Anhydrite, when occurring as small layers or as crystal aggregates, was separated by hand-picking. Otherwise, it was dissolved out with dilute acid as above.

Metal sulfide separation: the sediment was flushed with cold deoxygenated water, washed with acetone and dried in a vacuum desiccator at 30°C. This procedure minimized sulfide oxidation. The sediment was then gently heated in 25 per cent hydrofluoric acid (HF) for several hours. The acid was decanted and the residue washed and dried. X-ray analysis showed the only crystalline material left were chalcopyrite, pyrite and sphalerite. The residue was further separated by preferential dissolution of the sulfides. Moderate heating in HF slowly dissolved pyrite and chalcopyrite leaving the residue enriched in sphalerite. This treatment may also dissolve some fine-grained sphalerite.

Shaking the sulfides in 1:3 nitric acid for an hour at room temperature dissolved both sphalerite and pyrite, leaving chalcopyrite as the only crystalline material. Sphalerite dissolves when the sulfides are added to concentrated HCl. Upon heating, chalcopyrite dissolved but pyrite remained in the residue.

Attempts to isolate the sulfide minerals by usual physical methods, *e.g.* heavy liquids, centrifugation or magnetic separation, were not successful probably because of their small grain size.

All mineral identifications were made on a Norelco X-ray diffractometer, using high intensity $CuK\alpha$ radiation with a nickel filter.

Isotope Analysis

The sulfides were oxidized to sulfate and all sulfates were precipitated as $BaSO_4$. The sulfate was reduced to sulfide by graphite and then precipitated as silver sulfide. The sulfide was oxidized to sulfur dioxide by reacting it with copper oxide at 850°C. The sulfur dioxide was then analysed on a dual collection, 60°, 6-inch radius mass spectrometer manufactured by Nuclide Corporation.

Meteoritic sulfur (Canyon Diablo troilite) was used as the standard against which isotope measurements were made. δS^{34} is here defined in the usual manner:

$$\delta S^{34}\%_0 = \frac{(S^{34}/S^{32}\ \text{sample} - S^{34}/S^{32}\ \text{standard})}{S^{34}/S^{32}\ \text{standard}} \times 1{,}000$$

Chemical Analysis

Sediment and separated sulfide fractions were dissolved in 1:1 nitric acid. The solution was analysed for Ag, Cd, Mo, Cu, Co, Mn, Ni, Pb and Zn, using a Perkin-Elmer model 303 Atomic Absorption Spectrophotometer with acetylene-air fuel mixture. Analyses for As, Se and Te were performed by the U.S. Geological Survey.

Mineralogical Analysis

The relative abundance of the sulfide minerals was determined by X-ray diffraction after calibration with known amounts of pure sulfide minerals and aluminum as internal standard.

Several Red Sea sulfide concentrates were analysed with the internal Al standard, and relative amounts of each sulfide was determined from the peak areas.

Results

The location of the hydrocasts and cores are shown in Fig. 1. The sampling covers areas that are in the anomalous hot brine zones and those that are adjacent to them but outside these zones. Adequate discus-

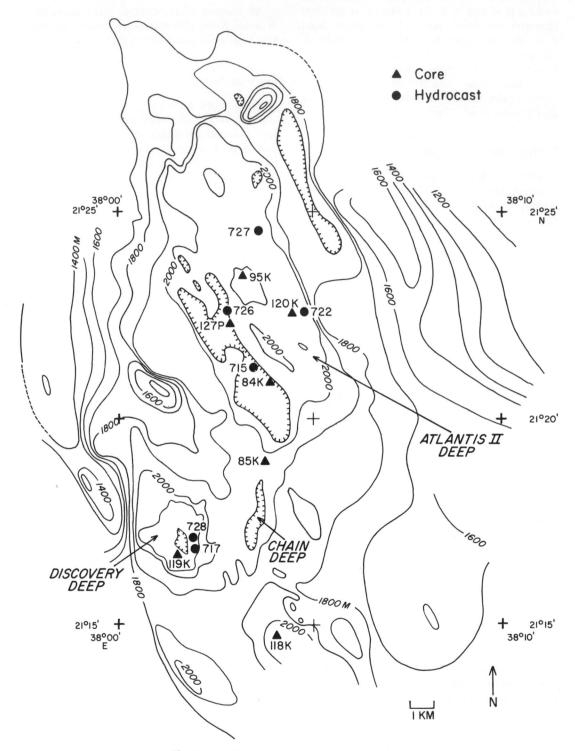

Fig. 1. Location map of cores and hydrocasts.

Table 1 Relative Abundance of Separated Sulfide Minerals
(In per cent of total sulfides)

	Depth	Sphalerite	Chalcopyrite	Pyrite
Core 118K	160–168cm	17%	15%	68%
	170–190cm	15%	15%	70%
	385cm	—	—	100%
	385–387cm	5%	—	95%
Core 119K	140–160cm	—	—	100%
	230–240cm	—	50%	50%
Core 84K	200–210cm	44%	10%	46%
	310–320cm	2%	—	98%
	350–360cm	70%	15%	15%
Core 120K	190–200cm	41%	16%	43%
	490–500cm	31%	23%	46%

sion has already appeared in the literature (Brewer *et al.*, 1965; Miller *et al.*, 1966; Riley, 1967; Hunt *et al.*, 1967; Degens and Ross, 1967), and will appear elsewhere in this volume, to serve as a background description of the environment.

Mineralogical Description

Sulfur exists in the sediment as three phases: sulfate, sulfide and elemental sulfur. The most abundant form of sulfate is anhydrite, where it is present in several cores from Atlantis II Deep (Bischoff, 1969). Fig. 2(3, 4) shows this anhydrite present as long euhedral prisms. This mineral is either intermixed with oxides or sulfides (especially sphalerite) or it is present as distinct layers of almost pure anhydrite, as in core 120K. X-ray diffraction analysis indicated that many of our samples studied were an admixture of anhydrite and some gypsum. This could have resulted from an inversion to gypsum during storage at low temperature,* which would shift the stability field away from anhydrite. Inspection of sediment mounted in a polyester polystyrene resin by optical microscopy and also scanning electron microscopy (Fig. 3) showed that some of the gypsum may be authigenic, rather than secondary replacement of anhydrite, which occurs as a thin

*According to Bischoff (personal communication) all gypsum in the sample formed upon storage at room temperature.

envelope surrounding anhydrite prisms. There is abundant evidence for well crystallized euhedral plates of gypsum, some crystals even show twinning as seen in Fig. 3. We did not find evidence of gypsum pseudomorphs after anhydrite, but found evidence for gypsum formation after anhydrite formation, since plates of the former were found enclosing well developed prisms of anhydrite. Further, a sample of anhydrite collected by R. Ginsburg in the Persian Gulf in 1963 and stored at ambient temperature until 1968, showed no detectable evidence of gypsum. Thus, inversion of anhydrite to gypsum is not rapid in all cases at lower temperatures and some gypsum in the Red Sea sediments may be authigenic. Some samples were washed with distilled water to remove the brine. This procedure, according to Hardie (1967), may cause anhydrite to revert to gypsum.

Barite has been reported by Bischoff (1969) from core 127P, 710cm. X-ray diffraction of the sample showed that pyrite was also present in the concentrate. Although the barite diffraction peak was strong, it appears to be only a minor component of the sediment, the bulk of which is an amorphous material. An attempt to separate the barite from the pyrite and other sediment yielded only trace quantities.

Three sulfide minerals have been identified, sphalerite (ZnS), chalcopyrite ($CuFeS_2$) and pyrite (FeS_2). Table 1 shows the relative proportions of each mineral. It is apparent that in the Atlantis II Deep

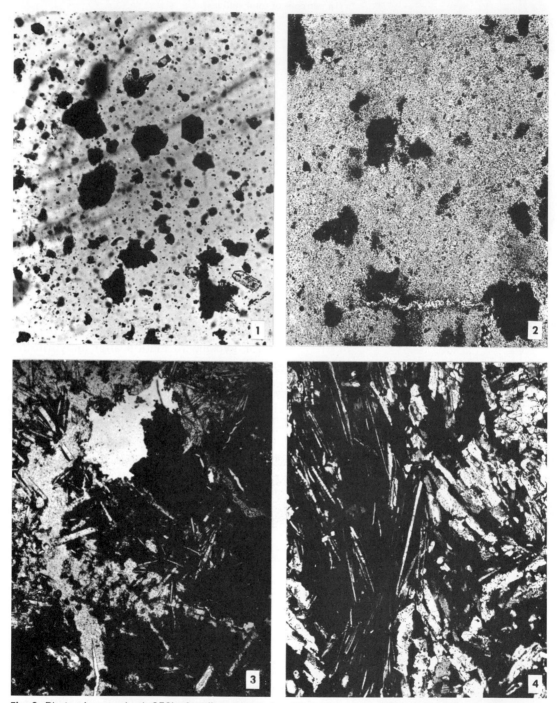

Fig. 2. Photomicrography (×250) of sediment samples. (1) Sulfide concentrate (after HF treatment); from core 84K, depth 350–360cm. (2) Iron oxide precipitate containing sulfide inclusions (dark minerals); core 120K, depth 455–460cm. (3) Anhydrite (elongated prisms), sphalerite (dark patches); assemblage from core 120K, depth 140–150cm. (4) Anhydrite crystals (polarized light) from anhydrite-rich layer; core 120K, depth 140–150cm.

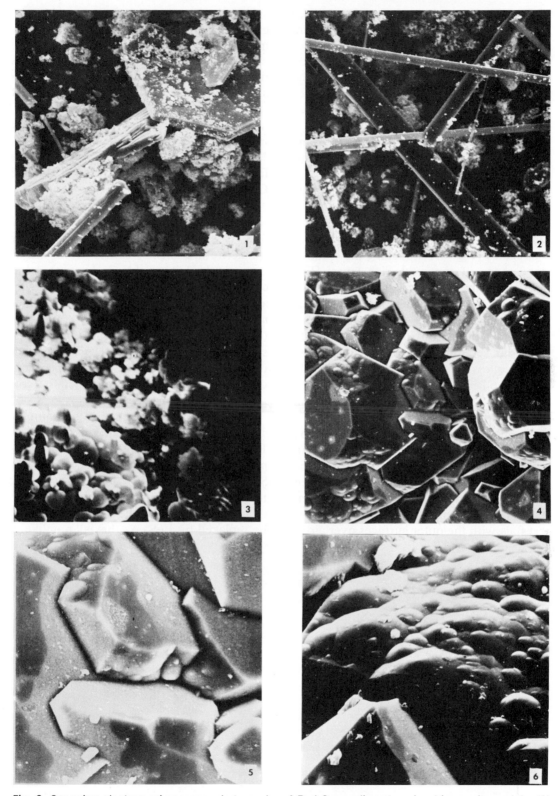

Fig. 3. Scanning electron microscope photographs of Red Sea sediment and marine pyrite nodule. (1) Anhydrite prisms and twinned gypsum crystals (×600); core 84K, 200–210cm. (2) Anhydrite prisms and fine-grained sulfides (×600); core 84K, 200–210cm. (3) Spheroidal sulfide (sphalerite?) grains (×4,000); core 120K, 490–500cm. Pyrite nodule from Long Basin, California, showing botryoidal growth (4) ×600, (5) ×2,000 (6) ×2,000.

(84K, 120K) sphalerite is an abundant or even dominant sulfide mineral. It occurs in much lower abundance outside the hot brine basins (core 118K) and appears to be absent in the single core studied from the Discovery Deep (119K). Pyrite becomes the dominant mineral in the areas removed from the Atlantis II Deep.

The sulfides were very fine-grained and generally intimately mixed with amorphous material, carbonates, oxides or clays. They were concentrated by treatment with dilute hydrofluoric acid, in order to separate them from the other material. Microscopic observation showed them to be opaque aggregates [Fig. 2(1)], lacking the granular appearance of the oxide-rich layers [Fig. 2(2)]. In some thin sections, opaque euhedral minerals tending to be hexagonal, were observed. An example of one such mineral is shown in the central right field of Fig. 2(1).

In thin sections the metal sulfides appear as opaque aggregates unresolved by the petrographical microscope [Fig. 2(2, 3)]. Investigation of sulfide concentrate from core 120K, 490–500cm by scanning electron microscopy showed the sulfide (possibly sphalerite) to be composed of aggregates of spherules. By comparison, a sample of marine pyrite nodule found in sediment of a core from Long Basin, off southern California, is composed mainly of euhedral crystals with the outer surfaces overlain by

spheroids [Fig. 3 (4, 5, 6)]. Presumably, the growth of the pyrite is by addition of new materials (amorphous or cryptocrystalline) in spheroids and subsequent recrystallization. Lebedev (1967a) gives numerous examples of sulfide aggregates which first form by packing of spheres and subsequently form euhedral crystals during aging. He gives one specific example (Lebedev, 1967b) where spherulites of sphalerite are forming in mixing tanks of the hypersaline Cheleken brine.

An example of the X-ray diffraction pattern from a sulfide concentrate in core 84K is shown in Fig. 4. The peaks for sphalerite are slightly shifted from that of a pure sphalerite. Table 2 shows that the cell edge (a), which was calculated using a modification of the least-square refinement program published by Burnham (1962), was larger in most samples than stoichiometric sphalerite. A significant difference was found in one sample from core 120K (#2). This indicated heterogeneity in the deposit.

The calculated cell edge in all samples shown in Table 2, with the exception of 120K #2, are too large for a simple replacement of FeS alone. Using the data of Skinner *et al.* (1959) and extrapolating to the calculated values of a in this study, over 50 per cent of the sphalerite should be replaced by FeS. As this is an unlikely,

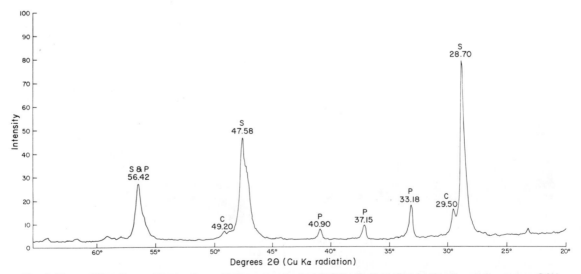

Fig. 4. X-ray diffraction pattern of a sulfide concentrate after hydrofluoric acid treatment from core 84K, depth 200–210cm. S = sphalerite, C = chalcopyrite, P = pyrite.

Table 2   Unit Cell Edges of Sphalerite Separates from Core Material

	Core 120K (490–500cm)			Core 127P (Depth 610cm)	Core 119K (Depth 110cm)	Crystal Plane
	#1	#2	#3			
2θ peak	28.75	28.76	28.66	28.44	28.65	111
	47.07	47.88	47.20	47.07	47.32	220
	56.06	56.14	55.98	55.77	56.14	311
a (cell edge in Å)	5.438	5.402	5.441	5.461	5.429	
σa	0.018	0.021	0.012	0.007	0.008	

if not a prohibitive situation, substitution of zinc in the sphalerite lattice is probably carried out by other metals. Table 10 illustrates the availability of several base metals potentially capable of performing this function.

Elemental sulfur was extracted from several cores. It could not be identified in the untreated sediment by X-ray diffraction, nor could it be observed by microscopy. In some instances, to be discussed later, the sulfur may have resulted from oxidation of sulfides during storage or during diagenesis. There is good evidence to believe, however, that most of the sulfur is authigenic and has formed at the same time as the sulfides with which it coexists. An attempt to extract elemental sulfur from oxide-rich sediment proved unsuccessful.

Sulfur Isotope Measurements

The distribution of S^{34}/S^{32} in the brine and sediments is shown in Fig. 5. It can be readily seen that the distribution falls into four distinct groups: (1) $\delta S^{34} > +251$ per mill dominated by interstitial sulfate, (2) $\delta S^{34} = +23$ to $+15$ per mill, controlled by dissolved sulfate in the brine, interstitial sulfate and sulfate minerals, (3) $\delta S^{34} = +2$ to $+12$ per mill representing sulfides, elemental sulfur and sulfate minerals, (4) $\delta S^{34} = -20$ to -37 per mill representing sulfides and elemental sulfur in cores 118K, 119K and 95K.

Dissolved Sulfate in Brine. The overall range for δS^{34} in the measured brines from Atlantis and Discovery Deeps (Table 3) is $+16.0$ to $+21.9$ per mill, giving an average δS^{34} of $+20.2$ per mill. If

the single low value from station 715, depth 2,008m is neglected, the spread is from $+18.2$ to $+21.9$ per mill with an average of $+20.3$ per mill. These values fall within the average spread for present day ocean sea water, and the latter average is identical to the single value measured by Hartmann and Nielsen (1966) for brine from the Atlantis II Deep. There is no significant difference between the overlying normal Red Sea water, the intermediate brine at 44°C. and the deeper Atlantis brine of 56°C.

An effort was made to identify the intermediate zone between the brine and normal (22°C) sea water in view of the detection of sulfate reducing bacteria in this zone by Trüper (1969) and by Watson and Waterbury (1969). The isotope results appear to indicate an enrichment of S^{32} in some of the water layers, particularly in the Atlantis Basin. These depletions in δS^{34} argue strongly against *in situ* bacterial sulfate reduction, but may constitute evidence for oxidation of sulfides (either biogenic or abiogenic). This is further indicated by the relatively high concentrations of sulfate in upper layers, such as #727, 2,015m and #722, 2,011m in Atlantis II Deep hydrocasts (Table 3).

Hydrogen sulfide has not been quantitatively measured in the brine, although Miller *et al.* (1966) made a passing remark of its "indicated presence." Evidently this was judged from the blackening of some messengers used to trigger the Nansen bottles (A. Jokela, personal communication). No sulfide was detected during the 1966 cruise by the **CHAIN**. Since, however, the enrichment of sulfate and S^{32} is observed in the transition zone, where

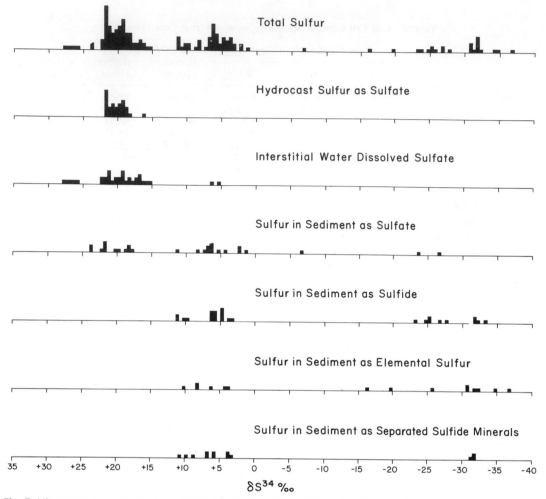

Fig. 5. Histogram of sulfur isotope (δS^{34}) distribution from different sulfur species in the brine, interstitial water and sediment from Atlantis II and Discovery Deeps and surrounding area.

the oxygen content increases, it is more logical to explain the biological activity as being due to sulfide oxidation by *Thiobacillus* sp. rather than sulfate reduction by *Desulfovibrio* sp.

Sulfur Species in Sediment. Values of δS^{34} from five cores (84K, 120K, 119K, 95K, 118K) are given in Tables 4 to 8 and Fig. 5. The Tables give data for dissolved sulfate in interstitial water, acid soluble sulfate in the sediment, metal sulfides and elemental sulfur (except for core 120K, Table 5, where no attempt was made to extract the elemental sulfur).

Core 84K (Table 4) and core 120K (Table 5) show a relatively uniform distribution for δS^{34} in the sulfides. With the exception of one sample (Table 4, depth 200–210cm),

the range of δS^{34} in the two cores is +3.1–+6.2 per mill. The elemental sulfur has a similar value to the sulfide for the two samples measured. In core 84K, the sulfur from depth 200–210cm is a little lighter than the sulfide. This may indicate a different origin of the two components, if not in terms of source then in terms of time.

The dissolved sulfate in the interstitial water of core 84K is on the average depleted in S^{34} relative to the interstitial water of core 120K and the overlying brine (#715, Table 3). The concentration of sulfate in the interstitial water is also higher than in the overlying hot brine. Certain layers, in particular at 200–210cm show a marked increase in concentration of soluble sulfate and a depletion of S^{34}. This relationship

Table 3 Concentration and S^{34}/S^{32} Ratios of Dissolved Sulfate in Brines from Atlantis II and Discovery Deeps

Depth (m)	SO_4 (gr/l)	δS^{34} (‰)
#726 (Atlantis II)		
1,990	2.4	+21.3
2,000	2.4	+21.9
2,010	1.9	+21.6
2,020	1.9	+21.8
2,025	1.9	+21.4
2,030	1.4	+20.6
2,035	1.2	+19.2
2,040	1.9	+21.2
2,045	1.7	+21.6
2,050	0.6	+21.8
2,055	0.7	+21.8
2,065	0.7	+19.5
2,075	0.7	+21.7
2,085	0.6	+21.0
#715 (Atlantis II)		
1,993	2.8	+19.7
2,008	2.9	+16.0
2,023	2.6	+20.2
2,038	2.5	+18.8
2,053	0.8	+18.8
core	0.5	+21.5
#727 (Atlantis II)		
2,015	3.3	+20.9
2,020	2.8	+20.7
2,040	2.9	+19.8
2,045	2.8	+18.6
2,055	2.7	+18.2
#722-1 (Atlantis II)		
2,011	3.2	+19.2
2,016	2.7	+19.4
2,030	2.5	+19.3
2,045	2.0	+19.8
#717-P (Discovery)		
1,985, 1,990, 2,024	1.9	+20.2
2,029	2.5	+19.0
2,039	2.3	+20.0
2,044	1.5	+20.5

between $^cSO_4^=$ and δS^{34} is also apparent in other cores as seen in Fig. 6. The most obvious explanation for it is the oxidation of sulfides. There is no evidence, however, to determine whether this occurs *in situ* at the time of deposition, during diagenesis or after sampling has occurred.

The acid soluble sulfate is somewhat more enriched in δS^{34}. The average δS^{34} for interstitial sulfate is +17.6 per mill and for acid soluble sulfate it is +20.2 per mill. This may indicate that oxidation occurred after sedimentation.

Core 120K (Table 5) shows a more uniform distribution of δS^{34} in the interstitial sulfate, with the exception of the sulfide-rich zone between 340–365cm. Here it appears that the interstitial sulfate as well as the acid soluble sulfate has been largely derived from sulfide. A marked increase of sulfate at depths of 170–180cm is reflected by a decrease in δS^{34} to +17.5 per mill. At depths 140–150cm, the sulfate concentration in the interstitial water is high due to a layer of anhydrite, which may have partially dissolved. The δS^{34} values for sulfides are very constant in this core, varying only between +4.5 and +6.3 per mill.

Core 119K, taken in Discovery Deep shows quite a different pattern from the two previous cores (Table 6). First, all samples of sulfate in the interstitial water, with the exception of 205–213cm depths, show an enrichment in δS^{34} relative to the brine with an average value of +26.8

Table 4 Concentration and S^{34}/S^{32} Ratios in Sulfur Species from Core 84K (Atlantis II Deep)

	Interstitial SO_4		Sediment		Elemental $S°$	
Depth (cm)	Conc. (g/l)	δS^{34} (‰)	SO_4 δS^{34} (‰)	Total Sulfide δS^{34} (‰)	Conc. (%)	δS^{34} (‰)
110–120	1.23	+17.1	+20.4	+3.8		
135–140			+21.8	+6.2		
150–160	1.15	+19.8		+3.1		
180–190	1.06	+18.1				
200–210	1.59	+15.1	+17.9	+9.8	0.32	+4.0
230–245	1.02	+16.7	+21.6	+4.8		
260–270	1.23	+18.5				
310–320	0.90		+19.6	+5.6		
340–350	0.76	+19.1				
350–360			+17.9	+6.2		
385–395	0.83	+16.3	+22.0	+5.8	0.16	+6.0

Table 5   Concentration and S^{34}/S^{32} Ratios in Sulfur Species from Core 120K (Atlantis II Deep)

| | Interstitial SO_4 | | Sediment | | | |
| | | | SO_4 | | Total $S^=$ | |
Depth (cm)	Conc. (g/l)	δS^{34} (‰)	Conc. (%)	δS^{34} (‰)	Conc. (%)	δS^{34} (‰)
100–110	0.5	+19.1				
120–130	0.3	+21.5				
130–140	0.2	+22.0				
140–150	2.0	+21.1				
170–180	1.1	+17.5				
210–220	0.6	+21.0				
220–230	0.6	+21.7				
235–240	0.5	+21.0				
300–310	0.6	+20.4				
330–340					0.07	+5.6
340–360	0.6	+6.0	0.20	+4.4	0.32	+4.5
360–365			0.22	+6.4	0.21	+5.0
365	0.6	+5.4	0.17	+11.1		
380–385	0.5	+19.9				
400–405	0.6	+21.1				
425–430	0.5	+20.9				
455–460	0.3	+20.4				
490–500	0.6	+17.0				+6.3

Table 6   Concentration and S^{34}/S^{32} Ratios in Sulfur Species from Core 119K (Discovery Deep)

| | Interstitial SO_4 | | Sediment | | | | | |
| | | | SO_4 | | Total Sulfide | | Elemental $S°$ | |
Depth (cm)	Conc. (g/l)	δS^{34} (‰)	Conc. (%)	δS^{34} (‰)	Conc. (%)	δS^{34} (‰)	Conc. (%)	δS^{34} (‰)
0–30	0.73		0.17	+6.1			0.08	−34.7
55–60	0.66	+25.8	0.15	+1.2	0.01	−24.8	0.19	−31.7
100–110	0.80	+27.1	0.32	+2.5	0.54	−23.3	0.93	−30.8
140–160	0.80	+26.9		+6.7		−25.1	0.19	−25.5
205–213	1.03	+17.1	0.05	+18.4			<0.01	
230–250	0.77	+26.3	0.20	−6.7	0.08	−26.7	0.08	−36.5
350–358	0.74	+27.9	0.13	+8.3	0.01	−32.1	0.06	−19.8
Sample from Bischoff 110cm						−25.0		

Table 7   Concentration and S^{34}/S^{32} Ratios in Sulfur Species from Core 118K

| | Interstitial SO_4 | | Sediment | | | | | |
| | | | SO_4 | | Total Sulfide | | Elemental $S°$ | |
Depth (cm)	Conc. (g/l)	δS^{34} (‰)	Conc. (%)	δS^{34} (‰)	Conc. (%)	δS^{34} (‰)	Conc. (%)	δS^{34} (‰)
20–40	3.42	+16.6	0.20	+18.9	—	—	None	
95–100	3.33	+19.2	—	—	—	—	0.10	−30.8
110–120	3.30	+19.1	—	+6.7	—	−27.7	0.19	−16.3
160–168	—	—	1.3	−26.9	0.28	−31.8	0.02	—
170–190	3.57	+18.1	0.54	−23.8	0.25	−31.8	0.02	—
381–385	—	—	—	—	—	—	0.45	+8.4
385	—	—	—	+2.3	—	+11.0	—	—
385–387	—	—	—	+5.2	—	+10.3	0.02	+10.2
387–400	3.96	+15.8	—	+6.1	—	+11.4	0.63	+8.4

Table 8   Concentration and S^{34}/S^{32} Ratios of Sulfide, Sulfur and Interstitial Sulfate from Core 95K

| | Interstitial SO$_4$ | | Sediment | | | |
| | | | Total Sulfide | | Elemental S° | |
Depth (cm)	Conc. (g/1)	δS^{34} (‰)	Conc. (%)	δS^{34} (‰)	Conc. (ppm)	δS^{34} (‰)
110–120	2.58	+17.3	Not detected		tr	—
180–185	6.36	−18.1	1.4	−33.4	30	−32.0
282–300	3.30					
390–395	2.55					

per mill. The sulfides, and even more so the elemental sulfur, are markedly depleted in δS^{34}. This strongly suggests microbiological sulfate reduction in a manner similar to that found in normal marine sediments. The fractionation factors are similar to those measured in sediments from marine basins in the continental shelf off southern California (Kaplan et al., 1963).

The sample from depth 205–213cm shows an increase in sulfate concentration relative to the other samples, but this was an oxidized layer where no sulfides were located. The δS^{34} in this layer probably signifies a mixture of sulfate in the brine and sulfate formed by oxidation of sulfide. This appears to represent an oxidation in situ and not after collection, since there appeared to be very little sulfur (reduced or oxidized) fixed in this sediment layer.

The results from core 118K (Table 7), which lies outside the deeps, show two unusual and interesting results. First, there is an abrupt change in the isotopic ratio of the sulfides between 190 and 381cm. Second, the δS^{34} of the samples near the base of the core fall between +10.3 and +11.4 per mill. This anomalously high ratio for the sulfide is reflected in the high δS^{34} of the elemental sulfur. With the exception of the top and bottom sample, δS^{34} of the interstitial water sulfate is somewhat lower than the average for sea water. The values for the acid soluble sulfate in this core (and also in core 119K) are probably not significant, and most likely are artifacts of experimental technique.

Core 95K was taken in a rise within the Atlantis II Deep. The upper sediment in this core appeared oxidizing, whereas the sediment in other sections of the core were

more reducing. Interstitial water analysis (Brooks et al., 1969) indicates that the surface 100cm of this core is enriched in trace elements and is characteristic of the brine-Red Sea water interface zone.

Sulfur analysis was undertaken on samples at two depths of this core. Sulfide and elemental sulfur was detected at 180–185cm depth (Table 8). The δS^{34} values of −33.5 and 32.0 per mill, respectively, are in close agreement and point to a relationship. The origin of the sulfide is biogenic. Interstitial sulfate at depth 180–185cm is a mixture of sea water sulfate and oxidized sulfide, since there is a significant enrichment both in concentration and in S^{32} over the dissolved sulfate at 110–120cm. The elemental sulfur was probably also derived during the oxidation, since it is present only in small amounts.

Of interest, is the presence of sphalerite in the sulfide, concentrated at depths 180–185cm. Coexisting with this sulfide enriched layer was fossiliferous material (pteropods). This organic material evidently provided the reducing matter needed for the bacterial conversion of the sulfate to sulfide. The source of the metals was probably the brine, to which may have been added metals adsorbed by the organic matter.

It is significant (Table 9) that in cores from the Atlantis II Deep and the bottom of core 118K, the fractionation factor, α, lies between 1.002 and 1.017, whereas in the upper half of core 118K and in core 119K, α lies between 1.042 and 1.060. The former value falls in the range of volcanic or hydrothermal deposits (Ault and Kulp, 1960; Gavelin et al., 1960; Steiner and Rafter, 1966), whereas the latter is in the

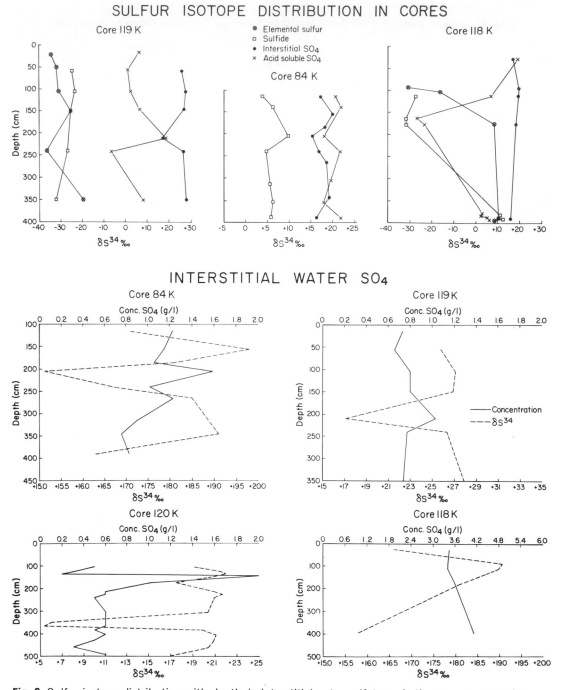

Fig. 6. Sulfur isotope distribution with depth, in interstitial water sulfate and other components of cores.

range measured in normal marine sediments (Kaplan *et al.*, 1963).

In order to learn about the sequential history of the sulfide formation, it was decided to measure separate minerals where possible. From Table 10 it can be seen that slight differences do exist. Pyrite generally appears to be somewhat more enriched in δS^{32} than sphalerite or chalcopyrite. Insufficient data exists, however, to determine whether this is a significant factor. After the mineral separations, sphalerite was left as a residue. There was, however, an insufficient quantity in most of the separates for isotopic measurement. The data clearly show the marked difference in the

Table 9   Enrichment Factors (α) of Total Sulfide-Interstitial Sulfate Pairs

$$[\alpha = 1 + (\delta S^{34}{}_{SO_4} - \delta S^{34}{}_{S^=})]$$

Core 118K		Core 120K	
cm	α	cm	α
110–120	1.047	340–360	1.002
170–180	1.050	490–500	1.011
387–400	1.004		

Core 119K		Core 84K	
cm	α	cm	α
55–60	1.051	110–120	1.014
100–110	1.050	150–160	1.017
140–160	1.052	200–210	1.005
230–250	1.053	230–245	1.012
350–358	1.060	385–395	1.011

values for pyrite and chalcopyrite in the top section of core 118K.

Anhydrite was hand-picked from three sections and analysed separately. The average value of δS^{34} is +2 to +3 per mill heavier than the average δS^{34} for brine and sea water. This is consistent with values obtained for gypsum precipitating out of modern evaporitic basins (Holser and Kaplan, 1966). By comparison, a single analysis of anhydrite from the Persian Gulf yielded $\delta S^{34} = +18.1$ per mill.

Trace Metal Analysis of Sediments

Selected samples from several cores were analysed for zinc, silver, cadmium, cobalt, copper, manganese, nickel, lead, molybdenum, arsenic, selenium and tellurium. In particular, an effort was made to compare the composition of the total sediment with the sulfide concentrate separated by hydrofluoric acid treatment of the sediment. For comparison, a pyrite nodule and the sediment from a core in Long Basin (100 miles west of San Diego, California) were also analysed.

Table 11 shows that generally there is a marked increase in the base metals within the sulfide phase. Zinc shows the highest concentration, reaching 4.4 and 4.5 per cent respectively in two sulfide concentrates from cores 85K and 127P. These values are considerably lower than the results for zinc obtained by Bischoff (1969) and by Bischoff and Manheim (1969) on sample 610cm, core 127P, where they measured a value of about 17 per cent Zn. Their analysis was made on a particular size fraction (2–62μ), whereas ours was carried out on a sample treated with hydrofluoric acid. The difference could therefore be explained either on the fact that Bischoff and Manheim had measured a particularly rich sediment fraction or else the trace metals are not entirely associated with the sulfide fractions, but with other minerals which are dissolved by the hydrofluoric acid treatment.*

Cadmium, silver, copper and lead generally follow zinc in concentration, pointing to a common origin and history of these metals. Cobalt shows a greater affinity for

---

* It has been noted by Hartmann that the bulk of the zinc is associated with the hydroxide rather than the sulfide portion (editors).

Table 10 δS^{34} (‰) of Separated Sulfur Minerals

Core	Depth (cm)	Total Sulfide δS^{34} (‰)	Sphalerite δS^{34} (‰)	Pyrite δS^{34} (‰)	Chalcopyrite δS^{34} (‰)	Anhydrite δS^{34} (‰)
Stat. 104						+23.7
127P	610	+4.6	—	+5.7	+5.9	—
85K	180	+3.4	—	—	+3.3	—
120K	140–150	—	—	—	—	+22.2
	490–500	+6.3	+6.5	+3.6	—	+23.5
84K	200–210	+9.8	—	+7.7	+8.9	—
	310–320	—	—	+4.6	—	—
118K	160–168	—	—	−31.7	—	—
	170–190	—	—	−31.6	−31.3	—
	385	—	—	+10.6	—	—
	385–387	—	—	+10.0	—	—
Long Basin, Calif.	~500	—	—	−52.2	—	—

the sulfide phase than for the total sediment. This also appears to be the case for nickel, although to a lesser extent. In core 118K, both cobalt and nickel are less enriched in the sulfide phase at the surface of the core than at depth. Further, in this same core there is a reversal of the Co/Ni ratio. Whereas in all Atlantis II Deep cores this ratio is >1, it is <1 for core 119K and the top of core 118K. In core 119K the average Co/Ni ratio is 0.43 and in core 95K it is 0.33 (for two samples measured) compared to 0.43 for deep sea red clays and 0.29 for the average near shore sediments (Chester, 1965). The Co/Ni ratios for the pyrite nodule from Long Basin and for the sediment from this basin are 0.13 and 0.30 respectively. The average Ni content in the sulfide-rich sediments of the Red Sea is appreciably lower than in deep-sea argillaceous sediments.

Manganese is not enriched in the sulfide zone and on the average appeared to be highest in the cores outside of Atlantis II

Deep (with the one exception of core 120K, 365cm). These values are probably not representative, however, since concentrations of manganese minerals are known at certain horizons.

Molybdenum was generally found to be in concentrations of less than 10ppm (at the limit of experimental sensitivity) for samples studied from the Red Sea. By comparison, a marine sediment from Long Basin, California, yielded 15ppm whereas organic-rich and sulfide-rich sediment from Saanich Inlet (off Vancouver Island, British Columbia) showed values of 40–70ppm. A surprisingly high value of 2,300ppm (average of 3 analyses) was measured in a pyrite nodule from Long Basin (Fig. 3). In comparison the Red Sea deposits are therefore very low in Mo.

Arsenic (Table 12) shows marked concentration in some sediment and sulfide concentrates of the Atlantis II Deep. It is most enriched in the sulfide concentrate, the highest content measured being

Table 11   Trace Element Concentrations in Selected Sediment and Sulfide Pairs (Sulfide Minerals Have Been Concentrated by Treating Sediment with Hydrofluoric Acid). All Values in ppm

Core	Depth (cm)		% S	Zn	Ag	Cd	Pb	Cu	Mn	Ni	Co	Co/Ni
118K	160–168	Sediment	0.28	391	2	13	65	39	1,432	117	57	.49
	160–168	Sulfide		513	12	26	103	103	496	51	37	.73
	170–190	Sediment	0.25	703	1	14	84	82	1,406	94	33	.35
	170–190	Sulfide		2,590	12	32	130	130	518	32	29	.91
	385	Sediment		1,070	3	36	107	178	1,961	18	10	.55
	385	Sulfide		1,071	7	36	214	3,213	321	89	104	1.24
	385–387	Sediment	1.8	757	3	15	151	252	504	177	179	1.01
	385–387	Sulfide		5,690	6	34	341	5,120	142	142	213	1.50
84K	200–210	Sediment	7.3	19,220	52	160	621	6,941	53	53	217	4.10
	200–210	Sulfide		20,744	105	192	768	11,371	576	58	280	4.83
	310–320	Sediment	0.54	341	14	<1	136	2,725	256	17	27	1.59
	350–360	Sulfide		31,181	9	289	481	5,295	770	112	199	1.77
119K	100–110	Sediment		1,540	0.3	2	69	94	2,225	34	14	.41
	140–160	Sediment		1,491	7	2	107	128	2,983	75	32	.43
	350–358	Sediment	0.01	983	3	1	49	49	1,721	62	27	.44
120K	365	Sediment		1,536	7	<1	142	496	10,620	18	20	1.11
	490–500	Sulfide		12,744	16	57	793	4,135	—	57	99	1.76
85K	180	Sulfide		44,380	140	247	1,315	15,615	123	82	289	3.53
127P	610	Sulfide		45,459	24	77	1,837	12,800	204	153	245	1.62
95K	110–120	Sediment		560	—	<1	44	28	3,880	38	9	.24
	180–185	Sediment	1.4	530	—	<1	53	73	2,900	64	26	.41
Long Basin,	220	Sediment *		73	2	<1	14	37	150	40	12	.30
California	500	Pyrite nodule *		500	52	<1	170	36	510	72	9	.13

* Mo for Long Basin sediment 15ppm and for pyrite nodule 2,300ppm. All Red Sea sediments <10ppm.

Table 12   Arsenic and Selenium in Brine, Interstitial Water, Sediment and Sulfide Concentrates

Sample	Depth	As (ppm)	Se (ppm)
Brines			
Discovery Deep (Hydrocast 728)	1,930m	<0.1	<0.05
Discovery Deep (Hydrocast 728)	2,100m	<0.1	<0.05
Atlantis II Deep (Hydrocast 727)	2,075m	<0.1	0.15
Interstitial Water			
Atlantis II Deep, core 84K	200–210cm	<0.1	<0.05
Sediment			
Atlantis II Deep, core 84K	200–210cm	400	<1
Atlantis II Deep, core 84K	310–320cm	15	2
Atlantis II Deep, core 120K	365cm	40	<1
Discovery Deep, core 119K	100–110cm	10	2
Discovery Deep, core 119K	140–160cm	30	4
Discovery Deep, core 119K	350–358cm	90	6
Outside of hot brine zone, core 118K	160–168cm	120	15
Outside of hot brine zone, core 118K	170–190cm	80	15
Outside of hot brine zone, core 118K	385cm	30	4
Outside of hot brine zone, core 118K	385–387cm	20	4
Sulfide Concentrates			
Atlantis II Deep, core 84K	350–360cm	250	6
Atlantis II Deep, core 120K	490–500cm	80	<1
Atlantis II Deep, core 127P	610cm	600	2
Outside of hot brine zone, core 118K	160–168cm	150	30

600ppm. The As content seems to correlate well with the lead content (Table 11), and was below the limit of detection in the brine overlying the sediment and in the interstitial water.

Selenium content was very low in the Atlantis II Deep and Discovery Deep samples with values falling from below the limit of detection (<1ppm) to 6ppm. Its highest concentration occurred in core 118K, where 30ppm was measured in the sulfide concentrate from 160–168cm.

An attempt to analyze for tellurium by H. Lakin (U.S. Geological Survey) showed this element to be in the range below the detection limit (0.05ppm).

Discussion

General Statement of Problem

The origin of base metal sulfide deposits is of major significance to economic geology. Several excellent reviews have recently been published on the mechanisms of formation of "strata-bound" metal sulfides (Dunham, 1964; Davidson, 1965; White, 1968) which have some bearing on interpreting the origin of Red Sea deposits. Of particular interest is the presence of a similar (but not identical) hypersaline brine environment near the Salton Sea, California, where sulfides are precipitating inside the casing of the brine-carrying pipe. It appears from the recent description of the sulfide scale deposits (Skinner et al., 1967) that copper sulfides (bornite and digenite) followed by copper-silver sulfides, are the most abundant sulfide minerals there. Another sulfide-mineral precipitating brine is the Cheleken brine, where sphalerite is forming in mixing tanks (Lebedev, 1967b).

Four aspects are considered by White (1968) to be most important in the generation of base metal sulfide ore deposits:

1. A source must exist for the ore constituents.
2. A fluid medium must exist for solubilization and transportation of metals.
3. An energy gradient must be present for physical movement of the fluid phase.
4. The ore deposits are formed by selective precipitation of certain constituents in response to physical and chemical changes as the fluid migrates into new environments.

These criteria will be referred to later in drawing a model for the Red Sea sulfide deposits.

Summary of Information on Sulfide-Rich Red Sea Sediments

The sediments in the median valley of the Red Sea are apparently different from normal deep sea marine sediments in that at some depths metal sulfides are present. Particularly high concentrations of metal sulfides are present in the Atlantis II Deep. In the Discovery Deep, and in the region south of the anomalous deeps (core 118K) pyrite is the most abundant sulfide mineral, although chalcopyrite and sphalerite were also detected. In the Atlantis II Deep sediments, sphalerite was highly abundant in the sulfide-rich layers, even within the sediment of the rise within this basin (core 95K). Elemental sulfur was found in all cores extracted and was present in most, but not all, sulfide-containing sediment. Calcium sulfate, mainly as anhydrite but possibly containing authigenic gypsum, was present in cores from the Atlantis II Deep. One sample of sediment found by Bischoff (core 120K, 710cm), contained small quantities of barite admixed with unidentified amorphous material and pyrite.

The S^{34}/S^{32} ratios are distributed in four ranges. Above +22 per mill, δS^{34} values are restricted to sulfates, either dissolved in the brine as in the case of core 119K, or the solid anhydrite phase. Around +20 per mill, δS^{34} values represent sulfate dissolved in sea water, the brine in the Discovery and Atlantis II Deeps and interstitial water. Between +3 to +11 per mill, δS^{34} values are restricted to sulfide and elementary sulfur from the Atlantis Deep and the bottom section of core 118K. δS^{34} values below −20 per mill are restricted to sulfides and elemental sulfur in the core from Discovery Deep (119K), the rise within the Atlantis Deep (95K) and the upper part of core 118K.

Interpretation of Isotope Data

δS^{34} measurements show that two distinct origins exist for the reduced sulfur in the sediments. Minerals with values of $\delta S^{34} < -25$ per mill must be interpreted as originating from microbiological sulfate reduction. Similar values have been measured in marine sediments (Kaplan et al. 1963; Nakai and Jensen, 1964). The fractionation factor between the sulfide pairs and the sulfate dissolved in interstitial water (cores 118K and 119K, Table 9) lies between 1.047 and 1.060. Theoretically, it is possible to assume an equilibrium exists between the dissolved sulfate and the sulfide (or elemental sulfur). Equilibrium constants based on partition functions have been calculated by Tudge and Thode (1950) and Sakai (1957), where $K = {}^{Q}SO_4/{}^{Q}S^{=}$. For fractionations with the observed range of α, the temperature range would be approximately 100–150°C. Thus, although it is possible for the sulfide to have arisen from the sulfate by an inorganic reduction, there are two factors arguing against such a formation.

First, there is no evidence for such high temperatures to have existed within the three cores studied (118K, 119K, 95K). Second, due to the large valence difference and the strength of the sulfur-oxygen bonds, no equilibrium between sulfide and sulfate has been measured at temperatures as low as 100°C.

On the other hand, fractionation factors as high as 1.047 have been measured in the laboratory with the bacterium *Desulfovibrio desulfuricans* (Kaplan and Rittenberg, 1964). This organism, or a related subspecies, has been detected in the sediments of the Discovery Deep (Watson and Waterbury, 1969) as well as in the interface between the hypersaline brine and the overlying normal sea water of the Atlantis II Deep (Trüper, 1969). By analogy with other marine environments, it would appear that the reduced sulfur has therefore arisen from sulfate by biological reduction. This appears to be an *in situ* process and does not occur in the water column. Evidence for this stems from the enrichment of δS^{34} in many of the interstitial sulfates of core 119K (Table 6) where values ranged from +25.8 to +27.9 per mill.

Sulfides in the Atlantis II Deep have a totally different range of δS^{34} (+3.1 to +9.8 per mill). The fractionation factor. α, between dissolved sulfate and sulfides from the same stratigraphic depth is 1.002 to 1.017. Three low values for α shown in

Table 9 (cores 84K, 120K and 118K) may not represent original conditions, since the δS^{34} of the sulfate was low and probably represents formation from sulfide (either in place or as an experimental artifact). Neglecting these values, the fractionation factor falls between 1.010 to 1.017.

These values are very close to published results of Ault and Kulp (1960), who gave average, α's, for barite-galena pairs of 1.024 and gypsum-galena pairs 1.014. Gavelin *et al.* (1960) measured average α's for barite-pyrite pairs equal to 1.012. Tatsumi (1965) gave a range of α for several barite-sulfide pairs as 1.017 to 1.026. Steiner and Rafter (1966) show average values of $\alpha = 1.011$ for coexisting pyrite and anhydrite in the New Zealand geothermal region and $\alpha = 1.019$ for coexisting dissolved hydrogen sulfide and sulfate in the fluid of the same region. Values in the Red Sea therefore fall in the median range of published data for sulfide-sulfate fractionation factors from known hydrothermal deposits.

The origin of the sulfide in such systems is a problem that has been frequently discussed in the literature. There is no direct evidence that the sulfide is magmatic, although it may well be. Generally, however, magmatic sulfur has a δS^{34} value close to 1 per mille (Ault and Kulp, 1959; Gross and Thode, 1965). It is also unlikely that the sulfide arises from leaching and transport of sedimentary sulfide, since the total spread in δS^{34} is narrow and not characteristic of most sedimentary sulfides. This narrow range in δS^{34} points to some relatively large and homogeneous reservoir of sulfur. Such a reservoir is the large sequence of Tertiary evaporites in the Red Sea Basin ringing north Africa and south-west Asia (Heybroek, 1965).

The source of the sulfide we wish to suggest here is evaporitic sulfate (dissolved gypsum or anhydrite) which has undergone reaction with organic compounds in shales to produce hydrogen sulfide and perhaps elemental sulfur. Toland (1960) has shown that reactions between organic compounds and aqueous sulfate (especially in the presence of sulfide) can occur at elevated temperatures under laboratory conditions. For

example, at temperatures between 300–350°C, quantitative oxidation of several hydrocarbons (saturated and unsaturated) was accomplished in periods of time from one to two hours at pressures of approximately 20 atmospheres. Even methane was partially oxidized under the experimental conditions used. The reaction products were organic carboxylic acids, hydrogen sulfide and often free sulfur.

Indirect evidence for such reactions having occurred under natural conditions is given by Germanov (1965). In the Kansay district (Karamazar, Tadjik, USSR) a metal sulfide ore body lies within a large carbonate deposit which has been bleached, apparently by sulfate-rich hydrothermal water, which also carries the metals. Barton (1967) also produces evidence from analysis of fluid inclusions within sphalerite, in which methane and barite apparently coexist out of equilibrium. Barton suggests that the sulfide in Mississippi Valley-type deposits may have originated from organic carbon reduction of dissolved sulfate.

The fractionation factors measured between sulfide and sulfate (1.011 to 1.017) in the Atlantis II Deep could be produced either by equilibrium exchange or by kinetic effects. In the former event isotopic exchange under equilibrium conditions ($S^{32}O_4^= + H_2S^{34} \rightleftarrows S^{34}O_4^= + H_2S^{32}$), with an equilibrium constant equal to the above fractionation factor, would represent a theoretical temperature range of 400–550°C (Sakai, 1957). Such temperatures may be feasible, because of the anomalous thermal gradient, but no experimental data exist to indicate whether, in fact, exchange can occur at these temperatures.

An alternative interpretation for the fractionation is a kinetic effect. The two isotopes in question will have different rates of reaction during the reduction. Thus, $S^{32}O_4^= \xrightarrow{k_1} 1H_2S^{32}$ and $S^{34}O_4^= \xrightarrow{k_2} 2H_2S^{34}$. The relative rates will depend on the nature of the activated complexes formed. In the simplest event the relative rates $\dfrac{k_1}{k_2} = \left(\dfrac{m_2}{m_1}\right)^{1/2}$. Harrison and Thode (1957) found that at temperatures between 18 and 50°C, $\dfrac{k_1}{k_2}$ (or α) = 1.022. This value is controlled by the

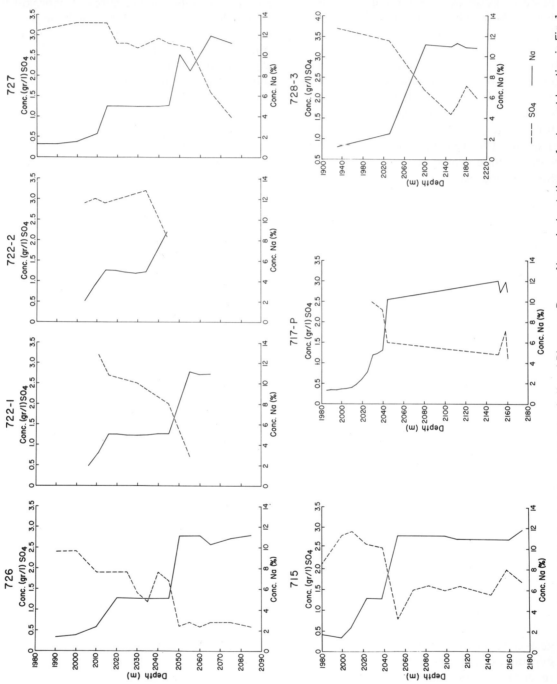

Fig. 7. Na$^+$ and SO$_4^=$ distribution in brines from Atlantis II and Discovery Deeps. Numerical notations refer to cast location in Fig. 1.

rate of reduction which is dependent on concentration, temperature and other factors. It also assumes reduction of a small proportion of the starting sulfate. Under the natural conditions obtaining in the Red Sea geothermal environment, it is very probable that fractionation factors lying between 1.01 and 1.02 can be attained.

The high temperature reduction of sulfate described above may be responsible for removing dissolved sulfate from the brine. Fig. 7 shows graphically that as the total salinity (or sodium concentration) rises, the concentration of dissolved sulfate decreases. This step may also produce isotopically light carbon dioxide. Deuser and Degens (1969) show that CO_2 dissolved in the brine of Atlantis II Deep has an average $\delta C^{13} = -3$ to -5 per mill. This range is typical of isotopic values measured on HCO_3^- in other geothermal areas (Hulston and McCabe, 1962).

Sulfide-rich layers with δS^{34} values of +9.8 per mill (core 84K, 200–210cm), +10.6 per mill (118K, 385cm) and +10.0 per mill (118K, 385–387cm) may represent conditions of high thermal activity. In such an event, unusual amounts of sulfide may form and rise to the surface with the brine. It is suggested here that such brine may have completely filled the Atlantis II Deep and overflowed into surrounding basins. One such event is observed in the sulfide at the base of core 118K.

Elemental sulfur was found in all the cores and was associated with the sulfides. Oxide layers did not contain any significant free sulfur. Since the δS^{34} values of the elemental sulfur was generally very similar to that of the associated sulfide, it would appear that they have a common origin. The origin suggested by us is the reduction of soluble sulfate by organic matter in the shales associated with evaporites. The thermal energy heating the fluid will drive the reaction and allow for volatilization of the sulfur and its dispersal either in a vapor or colloidal phase within the brine. Once this very finely dispersed sulfur is deposited in the sediment, it may undergo reactions with the metal phase. This would be enhanced during compaction through grain contact. Thermal heating would probably

raise the sulfur fugacity to a range where it will react with the metals.

Anhydrite and gypsum (in smaller amounts) are present in the sediments. The conditions prevailing at the sediment-brine interface are generally favorable for the precipitation of anhydrite (Hardie, 1967). Sulfates occur as admixtures with other sediments, especially metal sulfides, and also as enriched zones as almost pure calcium sulfate. Bischoff (1969) suggests that admixing of calcium-rich hot brines ($^cCa^{++} \approx 5,000$ppm in the Atlantis II Deep) with sulfate-rich overlying normal sea water would cause supersaturation. In such an event anhydrite (or gypsum) layers should also be present in the Discovery Deep, yet none have been detected.

Analysis of O^{18} in samples of anhydrite and gypsum were performed by Dr. R. M. Lloyd, Shell Development Co. For a sample of anhydrite-gypsum from core 120K, 140–150cm $\delta O^{18} = +10.07$ per mill, and in the brine of the Atlantis II Deep $\delta O^{18} = 7.21$ per mill, whereas dissolved sulfate in normal Red Sea water has $\delta S^{34} = 9.14$ per mille (Longinelli and Craig, 1967; Lloyd has measured 9.7 per mill for dissolved sulfate in Persian Gulf water).

On the basis of Lloyd's (1967) data, it would appear that the Atlantis II Deep anhydrite has precipitated out of the hypersaline brine rather than normal sea water, since the anhydrite-gypsum is between +2.5 to 2.9 per mill more enriched in δO^{18} than the dissolved sulfate of the brine. A similar fractionation was observed by Lloyd in comparing other marine evaporites with sea water sulfate.

It is therefore probable that the anhydrite did not form by evaporation processes, either from normal Red Sea water or from the brine. It could have formed, however, by reaction of calcium and sulfate ions. Such an event may have been rapid due to a sudden injection of brine, since undisturbed layers of almost pure calcium sulfate are observed. In such an event, fractionation between dissolved sulfate and precipitated sulfate would probably be very small.

Exchange between the sulfate and brine H_2O could, however, occur after deposition, since, under the conditions of preservation ($T \sim 60°C$ and pH ~ 6.3) the half-time of exchange is approximately 100 years (Lloyd, 1967). Such an exchange under equilibrium conditions would have a fractionation of 1.03, thus making the δO^{18} of the SO_4 about 31 per mill, since δO^{18} of the hypersaline brine H_2O is 1.2 per mill (Longinelli and Craig, 1966). It is apparent that no such exchange has occurred.

The present information on the calcium sulfate phase indicates that a precipitation occurred by supersaturation of calcium sulfate. It is insufficient to determine whether this occurred at the sediment water interface through injection of sulfate-rich brine, or whether it took place in the water column at the sea water-brine boundary as Bischoff (1969) suggests. In all probability both processes are operating. The first would account for rare occurrences of the beds of pure anhydrite, the second would explain the presence of calcium sulfate dispersed in other sediments.

Trace Metal Associations

The trace element contacts of the sediments and sulfide concentrates of several cores in the Red Sea area are listed in Table 11. It is evident that the sediments within the Atlantis II Deep are the most enriched. The sulfide fractions on the flanks of the basin (cores 120K and 85K) are also very rich in Zn, Cu and Pb compared to normal marine sediments. The sediment in the Discovery Deep shows less enrichment in these metals, yet their concentration in the brine appears to be the same (Brooks *et al.*, 1969) in both deeps. The rise above the Atlantis II Deep (core 95K) and the zone outside (core 118K) still show higher concentrations than normal marine sediments, but lower concentrations than the deeps. One exception appears to be molybdenum which was very highly enriched in one specimen of marine pyrite analyzed (from Long Basin, off California). We have no explanation for this, apart from the assumption that the geothermal fluids are incapable of either concentrating or transporting Mo. It is also possible that molybdenum was precipitated as sulfide at

the bottom of the "plumbing system," depleting the ascending brine.

Pb, Cu, Cd and Co are very closely correlated with zinc (Table 11). Ag shows concentrations in certain sulfide layers only, indicating that it is probably not derived from a constant source, or else that it is removed from the fluids at greater depth and does not rise to the surface except under conditions of low sulfide concentration. Mn appears to be most concentrated in the non-sulfide sediment fraction. It probably exists either as oxide or carbonate (Bischoff, 1969). Co and Ni show an interesting relationship to the total sediment and to the sulfide fraction. Co is concentrated in the sulfide fraction, whereas Ni appears to be concentrated in the non-sulfide fraction. With the exception of two layers (core 118K, 385–387cm, and 127P, 610cm), there is little variation in the nickel content of the Red Sea sediment. The Co-Ni content is highest in the Atlantis II Deep sediments and the bottom of core 118K, which we conclude arose from hydrothermal processes.

Although Pb shows relatively high concentrations both in the brine and sediment, no lead sulfide was detected by X-ray diffractometry. This suggests that the concentration of lead minerals is too low. An alternative suggestion is that Pb forms mixed sulfides with Zn and that the sphalerite contains admixtures of galena. It is entrance of Pb into the sphalerite structure that may be partially responsible for distortion of the crystal lattice (see Table 2).

Arsenic has an average concentration of 10ppm in deep sea sediments and 2ppm or less in most igneous rocks (Onishi and Sandell, 1955), and it is known to be highly concentrated in metal sulfides (Fleischer, 1955). Its low concentration in the brine and interstitial water indicates that it is not being continuously introduced and may have a history closely linked to the sulfide ion.

Selenium is generally found in very low abundance in marine sediments (<1ppm, Turekian and Wedepohl, 1961). It is normally most highly concentrated in chalcopyrite and to a lesser degree in sphalerite (Vlasov, 1966). Our data seem to show some concentration, but very small, with the highest concentration occurring in the sediment outside the hot brine deeps. The selenium may be concentrating close to the source and is not effectively transported by the brine.

Syngenesis Versus Epigenesis

The mechanism of sulfide formation has been given a great deal of attention. It is safe to assume that should the Red Sea metal sulfide-bearing sediments now be lithified and part of the continental stratigraphic sequence, they would be defined as "strata-bound" or "strati-form" deposits. Since we know that these deposits are forming authigenically, they must be termed "Sedimentary Exhalative" or "Sedimentary Hydrothermal" (Stanton, 1960; Dunham, 1964).

Davidson in many writings (e.g., 1962a, 1962b and 1965) generally assumes that these strata-bound deposits are epigenetic, formed through later mineralization of biogenic sedimentary sulfides captured in shales. Stanton (1960) and Dunham (1964) assume the metals to be contributed either by leaching of volcanics or igneous intrusions (Stanton) or by the introduction of metal enriched fluids along faults or fractures (Dunham) into epicontinental anoxic seas, where sulfide is being formed biogenically. Such origins have been ascribed, for example, to some of the Australian deposits and to the Zechstein Kupferschiefer deposits.

The Red Sea deposits show that Dunham's hypothesis (it is of course understood that others have postulated these ideas, also) does operate, but only in the areas adjacent to the area of maximum sulfide deposition, which must also be the source of the brine. The S^{34}/S^{32} data discussed in this paper shows two distinct sources for the sulfide, a biogenic origin for the areas outside the Atlantis II Deep and a hydrothermal source for the sulfides within this Deep. There is no evidence for replacement of authigenic sulfides (biogenic pyrite) by other metals in the Atlantis II sediments. In such an event, δS^{34}

would probably be very similar to values in Discovery Deep sediments. This, however, does not exclude the possibility of replacement of the metal sulfides by base metal enriched brines. Thus the metal sequence may be different at depths, containing, for example, higher concentrations of the less soluble copper and silver sulfides.

We propose here that the terms "syngenetic" and "epigenetic," as applied to strata-bound or stratiform sulfide deposits of the sediment-hydrothermal type, do not necessarily have two different operational definitions, but in fact may be contemporaneous steps of the overall "diagenetic" process. Deposition of one metal sulfide suite could be occurring at the sediment water interface, whereas replacement of sulfides by dissolved metals in interstitial brine may be taking place at depth. This reason may, for example, account for the irregular concentrations of Ag.

The Salton Sea sulfides and brines demonstrate another example of a metal-rich fluid capable of forming an ore deposit. According to Skinner et al. (1967), the brine (containing about 30ppm sulfur species) is in equilibrium with solid sulfide assemblages in the reservoir rock and there is no evidence of pyrite replacement. The most insoluble sulfides deposit first (Cu and Ag) when the temperature and pressure are lowered. Thus, rising fluid could deposit sulfides within lithified surface host rocks (not necessarily sediments) which would then be considered epigenetic. Alternatively, if it would reach the surface and deposit in a basin (e.g., the Salton Sea) the sulfide deposits would be considered syngenetic. In either case the sulfide molecule could or could not be derived from a common source.

A particularly interesting comparison is that of the Co/Ni ratio. Davidson (1962a) made a strong point of using this as an index to determine the origin of the fluids. He concluded that Co/Ni ratios >1 are characteristic of fluids liberated by or leaching granitic intrusions, since in normal sediments and in basic igneous rocks the ratio was always <1. Loftus-Hills and Solomon (1967) found high Co/Ni ratios (general spread of 0.2 to 4.5 with one value 8.15) in pyrite and chalcopyrite assemblages associated with Cambrian volcanics, whereas the sphalerite-galena assemblages in these volcanics show low Co/Ni ratios. They concluded that volcanic exhalations contributed the Co and Ni which coprecipitated with iron-copper sulfides. The present deposits demonstrate no direct connection with active volcanism, no evidence for replacement at the sediment surface and no preferential concentration of Co over Ni in pyrite or chalcopyrite over sphalerite.

Model for Sulfide-ore Formation in Red Sea Sediments

Subsequent to the previous discussion and following the criteria previously outlined, we suggest that the mechanism of formation of the sulfides is the following:

1. The source of the metal constituents are the shales and sediments underlying the Red Sea Basin. It is possible that igneous intrusives are an added source, although the trace element data argue against this being a major factor.

2. The fluid medium for solubilization and transport of the metal could come from the sinking of surface Red Sea water as suggested by Craig (1966) from $\delta O^{18}/\delta D$ evidence. It may also be possible to derive the water from dehydration of gypsum to anhydrite. According to Davidson (1965) $1m^3$ of gypsum can produce $0.486m^3$ of water, and Heybroek (1965) has shown that as much as 3km of Miocene evaporites exist in places. For example, the Maghersum formation, underlying Maghersum Island due west of the hot brine area, is "1,435m thick and is composed of 465m of salt overlain by a succession of gypsum and anhydrites interbedded with sandstone" (Heybroek, 1965). The Miocene *Clysmic Gulf* is thought to have extended from the Mediterranean Sea almost to the Indian Ocean; an evaporitic body covering an area of hundreds of thousands km². No doubt the high salt content of the brine was derived from this

source. Water derived from the gypsum anhydrite transition would be more than adequate to fill the approximately 10km³ volume of the Atlantis II Deep.

The review by White (1968) refers to recent studies by Helgeson and others to show that metal-chloride complexes are stable, even in the presence of 15ppm sulfides as in the Salton Sea brine, and can be transported in a brine solution. Such brine, especially when heated would be an excellent leaching agent.

3. Geothermal heating in the median valley caused elevation in temperature of the brine, enhanced its leaching abilities and reduced the density of the brine. Geophysical evidence (Drake and Girdler, 1964) indicates that separation of the African and Asian continents is occurring along the axis of the Red Sea. Fractures through which the brine can rise along a density gradient assisted by gas pressure are therefore available.

4. It is suggested that this brine rises from restricted vents into the Atlantis II Deep. Periodic injections occur of sulfide-rich brine, derived from reduction of sulfate with organic-rich shale during deep-seated hydrothermal activity. This sulfide will cause metal sulfide precipitation in this basin. On rare occasions the sulfide-rich brine may spill over into adjacent basins. Evidence for this seems to exist from S^{34}/S^{32} isotope data. Generally, sulfide does not escape from the Atlantis II Deep, but base metal-enriched brines do. In such an event they will react with hydrogen sulfide generated biogenically in the sediments under favorable conditions.

Acknowledgments

We wish to thank W. Dollase, UCLA, for calculations of the sphalerite cell dimension; J. T. Turner and E. P. Welsch, U.S. Geological Survey, Denver, for As, Se and Te analysis by courtesy of H. Lakin and J. B. McHugh; R. B. MacAdam and A. Loeblich, Chevron Research and Oil Field Company for the Scanning Electron Microscope photographs and P. Doose, UCLA, for mass spectrometer analysis and K. D. Watson, UCLA, for critical reading of the manuscript.

The study was carried out by support from American Chemical Society, Petroleum Research Fund (PRF #2815-A2) and the Atomic Energy Commission Contract No. AT (11-1)-34.

References

Ault, W. V. and J. L. Kulp: Isotopic geochemistry of sulphur. Geochim. et Cosmochim. Acta, **16**, 201 (1959).

——, ——: Sulfur isotopes and ore deposits. Econ. Geol., **55**, 73 (1960).

Barton, P. B., Jr.: Possible role of organic matter in the precipitation of the Mississippi Valley ores. Econ. Geol., **3**, 371 (1967).

Bischoff, J. L.: Red Sea geothermal deposits: their mineralogy, chemistry and genesis. *In: Hot brines and recent heavy metal deposits in the Red Sea*, E. T. Degens and D. A. Ross (eds.). Springer-Verlag New York Inc., 368–401 (1969).

Bischoff, J. L. and F. T. Manheim: Economic potential of the Red Sea heavy mineral deposits. *In: Hot brines and recent heavy metal deposits in the Red Sea*, E. T. Degens and D. A. Ross (eds.). Springer-Verlag New York Inc., 535–541 (1969).

Brewer, P. G., J. P. Riley, and F. Culkin: The chemical composition of the hot salty water from the bottom of the Red Sea. Deep Sea Res., **12**, 497 (1965).

Brooks, R. R., I. R. Kaplan, and M. N. A. Peterson: Trace element composition of the Red Sea geothermal brine and interstitial water. *In: Hot brines and recent heavy metal deposits in the Red Sea*, E. T. Degens and D. A. Ross (eds.). Springer-Verlag New York Inc., 180–203 (1969).

Burnham, C. W.: Lattice constant refinement. Carnegie Inst. Wash. Yearbook, **61**, 132 (1962).

Chester, R.: Elemental geochemistry of marine sediments. *In: Chemical Oceanography*, J. P. Riley and G. Skirrow (eds.). **2**, 23 (1965).

Craig, H.: Isotopic composition and origin of the Red Sea and Salton Sea geothermal brines. Science, **154**, 1544 (1966).

Davidson, C. F.: On the cobalt-nickel ratio in ore deposits. Mining Mag., **106**, 78 (1962a).

——: The origin of some strata-bound sulfide ore deposits. Econ. Geol., **57**, 265 (1962b).

——: A possible mode of origin of strata-bound copper ores. Econ. Geol., **60**, 942 (1965).

Degens, E. T. and D. A. Ross: Hot brines and heavy metals in the Red Sea. Oceanus, **13**, 24 (1967).

Drake, C. L. and R. W. Girdler: A geophysical study of the Red Sea. Geophys. J. Roy. Ast. Soc., **8**, 473 (1964).

Dunham, K. C.: Neptunist concepts in ore genesis. Econ. Geol., **59**, 1 (1964).

Fleischer, M.: Minor elements in some sulfide minerals. Econ. Geol., **50th anniv. vol.**, 970 (1955).

Gavelin, S., A. Parwel, and R. Ryhage: Sulfur isotope fractionation in sulfide mineralization. Econ. Geol., **55**, 510 (1960).

Germanov, A. I.: Geochemical significance of organic matter in the hydrothermal process. Geochemistry International, **2**, 643 (1965).

Gross, W. H. and H. G. Thode: Ore and the source of acid intrusives using sulfur isotopes. Econ. Geol., **60**, 576 (1965).

Hardie, L. A.: The gypsum-anhydrite equilibrium at one atmosphere pressure. Am. Min., **52**, 171 (1967).

Harrison, A. G. and H. G. Thode: The kinetic isotope effect in the chemical reduction of sulphate. Trans. Farad. Soc., **53**, 1648 (1957).

Hartmann, M. and H. Nielsen: Sulfur isotopes in the hot brine and sediment of Atlantis II Deep (Red Sea). Marine Geol., **4**, 305 (1966).

Heybroek, F.: The Red Sea Miocene Evaporite Basin. *In: Salt Basins Around Africa*, Inst. Petroleum, London, 17 (1965).

Holser, W. T. and I. R. Kaplan: Isotope geochemistry of sedimentary sulfates. Chem. Geol., **1**, 93 (1966).

Hulston, J. R. and W. J. McCabe: Mass spectrometer measurements in the thermal areas of New Zealand. Part 2: Carbon isotopic ratios. Geochim. et Cosmochim. Acta, **26**, 399 (1962).

Hunt, J. M., E. E. Hays, E. T. Degens, and D. A. Ross: Red Sea: detailed survey of hot brine area. Science, **156**, 517 (1967).

Kaplan, I. R., K. O. Emery, and S. C. Rittenberg: The distribution and isotopic abundance of sulphur in recent marine sediments off Southern California. Geochim. et Cosmochim. Acta, **27**, 297 (1963).

Kaplan, I. R. and S. C. Rittenberg: Microbial fractionation of sulphur isotopes. J. Gen. Microbiol., **34**, 195 (1964).

Lebedev, L. M.: Modern growth of sphalerite. Doklady Akad. Nauk USSR (English trans.), **175**, 196 (1967a).

——: *Metacolloids in Endogenetic Deposits*. Plenum Press, New York, 1 (1967b).

Lloyd, R. M.: Oxygen-18 composition of oceanic sulfate. Science, **156**, 1228 (1967).

Loftus-Hills, G. and M. Solomon: Cobalt, nickel and selenium in sulphides as indicators of ore genesis. Min. Deposita, **2**, 228 (1967).

Longinelli, A. and H. Craig: Oxygen-18 variations in sulfate ions in sea water and saline lakes. Science, **156**, 56 (1967).

Miller, A. R., C. D. Densmore, E. T. Degens, J. C. Hathaway, F. T. Manheim, P. F. McFarlin, R. Pocklington, and A. Jokela: Hot brines and recent iron deposits in deeps of the Red Sea. Geochim. et Cosmochim. Acta, **30**, 341 (1966).

Nakai, N. and M. L. Jensen: Sulfur isotope fractionation by bacterial oxidation and reduction of sulfur. Geochim. et Cosmochim. Acta, **28**, 1893 (1964).

Onishi, H. and E. B. Sandell: Geochemistry of arsenic. Geochim. et Cosmochim. Acta, **7**, 1 (1955).

Presley, B. J., R. R. Books, and H. M. Kappel: A simple squeezer for removal of interstitial water from ocean sediments. J. Marine Res., **25**, 355 (1967).

Riley, J. P.: The hot saline waters of the Red Sea bottom and their related sediments. *In: Oceanogr. Mar. Biol. Ann. Rev.,* H. Barnes (ed.). **5,** 141 (1967).

Sakai, H.: Fractionation of sulfur isotopes in nature. Geochim. et Cosmochim. Acta, **12,** 150 (1957).

Skinner, B. J., P. B. Barton, and G. Kullerud: Effect of FeS on the unit cell edge of sphalerite. A revision. Econ. Geol., **54,** 1040 (1959).

Skinner, B. J., D. E. White, H. J. Rose, and R. E. Mays. Sulfides associated with the Salton Sea geothermal brine. Econ. Geol., **62,** 316 (1967).

Stanton, R. L.: The application of sulphur isotope studies in ore genesis theory—a suggested model. N.Z.J. Geol. Geophys., **3,** 375 (1960).

Steiner, A. and T. A. Rafter: Sulfur isotopes in pyrite, pyrrhotite, alunite and anhydrite from steam wells in the Taupo volcanic zone, New Zealand. Econ. Geol., **61,** 1115 (1966).

Tatsumi, T.: Sulfur isotopic fractionation between coexisting sulfide minerals from Japanese ore deposits. Econ. Geol., **60,** 1645 (1965).

Toland, W. G.: Oxidation of organic compounds with aqueous sulfate. J. Am. Chem. Soc., **82,** 1911 (1960).

Trüper, H. G.: Bacterial sulfate reduction in the Red Sea hot brines. *In: Hot brines and recent heavy metal deposits in the Red Sea,* E. T. Degens and D. A. Ross (eds.). Springer-Verlag New York Inc., 263–271 (1969).

Trudge, A. P. and H. G. Thode: Thermodynamic properties of isotopic compounds of sulphur. Canad. J. Res., **28B,** 567 (1950).

Turekian, K. K. and K. H. Wedepohl: Distribution of the elements in some major units of the earth's crust. Bull. Geol. Soc. Amer., **72,** 175 (1961).

Vlasov, K. A.: Geochemistry of rare elements. Translated by Israel Program for Scientific Translations, Jerusalem, **1,** 552 (1966).

Watson, S. W. and J. B. Waterbury: The sterile hot brines of the Red Sea. *In: Hot brines and recent heavy metal deposits in the Red Sea,* E. T. Degens and D. A. Ross (eds.). Springer-Verlag New York Inc., 272–281 (1969).

White, D. E.: Ore forming fluids of diverse origins. Econ. Geol., **63,** 301 (1968).

Lead Isotope Measurements on Sediments from Atlantis II and Discovery Deep Areas

J. A. COOPER AND J. R. RICHARDS

Australian National University
Canberra, Australia

Abstract

In a study of the recent mineralized sediments from the Central Red Sea, a clear difference in the lead isotope pattern is observed in core sections of up to 5m in depth between those from Atlantis II Deep on the one hand and the Discovery Deep and barren sediments on the other. In general, low heavy-metal concentrations correlate with a low overall content of lead with variable isotopic ratios, a pattern observed in other areas of marine sedimentation. The more mineralized samples of Atlantis II Deep, with greater quantities (up to 500ppm) of lead, point to an ore source material with only small isotopic variation. The Discovery Deep samples, although mineralized, exhibit isotopic variation in the lead, suggesting significant contamination, in conformity with ideas that this "pool" is an overflow from the Atlantis II Deep.

Possible sources of the mineralization are considered. It is suggested that the mode of mineralization represented by this deposit is one very possible mechanism for the formation of "Pyritic Conformable" or "Stratiform" ore bodies.

These measurements also comprise the first lead isotope ratios for this laboratory that have been normalized by the double-spike technique.

Introduction

Stable lead isotope measurements have been used widely in attempts to establish the chronological relationships of lead ore deposits (Russell and Farquhar, 1960; Ostic *et al.*, 1967) and to relate such bodies to their local geological settings (Sinclair and Walcott, 1966). Although the logic used for lead orebodies could be applied to other types of mineral deposit in which lead plays a minor role, attempts to do this have been few because of serious technological problems in obtaining an accurate isotopic measurement upon small quantities of lead.

In recent times, interest has increased with improving techniques, principally in the trace quantities of lead in silicate systems. The most serious limitation upon interpretation has been the variable mass-dependent perturbations encountered in the single-filament, solid-source measurements needed for less than 50 microgram sample sizes. The double-spike method described below had its inspiration in the procedures used for strontium isotopes by Compston and his co-workers.* It does much to reduce these perturbations while permitting the use of much smaller samples than those required by other current methods. Details of this technique will be described more fully elsewhere.

Thus, the way is cleared for an accurate investigation of the newly discovered Red Sea mineralization in which lead occurs as a minor metal. This deposit is evidently associated with a magmatic heat source in its

---

* We understand (W. Compston, pers. comm.) that this technique was first used for Sr by G. W. Wetherill and L. T. Aldrich at some time prior to 1959. Application to the lead system was first advocated to our knowledge in about 1963 by W. Compston, who developed a successful normalization routine while on study leave at the Lamont Geological Observatory in 1966. The routine used in this work was developed independently at about the same time by one of us (J. R. R.).

early stages and has sedimentary control in the latter stages (Miller *et al.,* 1966). It is the first known orebody of zero age that has a relatively clear geological setting and is quite unaffected by post-depositional metamorphism. The results are of immediate interest in the interpretation of lead isotope systematics and ore genesis.

Sample Locations

The samples treated were those taken during the **CHAIN 61** cruise to the Red Sea (Ross and Degens, 1969). Sections were taken from five different Kasten Cores: two from mineralized areas of the Atlantis II Deep (84K and 120K), and one each from a local high projecting above the 2,000m contour within the Atlantis II Deep (95K),

from the Discovery Deep (119K) and from an apparently unmineralized area further south (118K). An additional sample, Ml, was obtained by the **METEOR** Expedition (Dietrich *et al.,* 1966) with a "Boomerang Corer." Sampling localities are shown in Fig. 1 and the depth of sections, where known, is indicated in Table 1.

Experimental

Sampling Details

The core sections of about 500g weight, including the occluded brine, were received doubly wrapped in thin plastic bags. No attempt was made to homogenize each section of core or to wash out the acidic brine before taking grab samples of 30–100g for chemical processing.

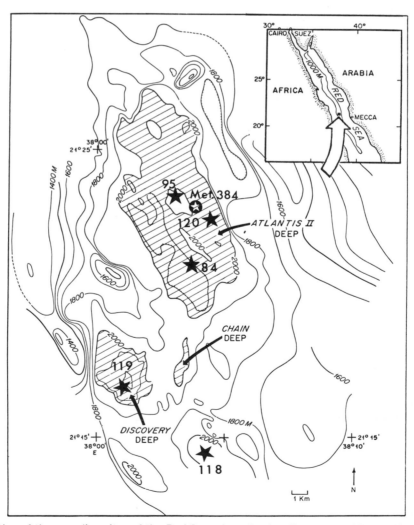

Fig. 1. Location of the sampling sites of the Red Sea mineralized sediments used for lead isotope studies.

Table 1   Cores Sampled for Pb Isotope Study

Cruise	Station	Lat. E	Long. N	Water Depth (m)	Core Section (cm)	A.N.U. No. M—	Notes
METEOR I	384	21° 23′	38° 04′	2,091	Top few cm	1	AT. II Deep
CHAIN 61	84K	21° 21.0′	38° 03.8′	2,164	120–130	2	
					180–190	3	
					260–270	4	Southern AT. II Deep
					340–350	5	
					390–400	6	
	95K	21° 23.4′	38° 03.3′	1,910	0–100	7	
					140–150	8	AT. II Deep on a high
					170–185	9	
					340–345 *	10	
	118K	21° 14.4′	38° 04.3′	1,989	100–120	11	
					250	12	Unmin. area
					310	13	
	120K	21° 22.6′	38° 04.5′	2,060	100–120	14	
					250–280	15	Eastern AT. II Deep
					350–370	16	
					490–500 *	17	
	119K	21° 16.8′	38° 01.7′	2,175	0–10	18	
					90–100	19	
					150	20	Discovery Deep
					230–250	21	
					290	22	
					350	23	
ATLANTIS II	543	21° 20.5′	39° 03.5′	2,064			AT. II Deep brine †

* Core sections marked on sample bags indicate a greater depth than the reported core length. Such samples were taken from the "core catcher" attachment (D. A. Ross, pers. comm.).

† Sample measured by Delevaux *et al.* (1967).

Metal Concentrations

For the metal concentration study 10g samples were dried, ground under A. R. acetone in an agate mortar, then dried again.

Metal Concentrations

The contents of Pb, Zn, Cu, Fe, Mn and Cl were measured for each of the sediment samples by x-ray fluorescence techniques. The method, involving peak to background ratios (Anderman and Kemp, 1958), was used for Mn, Cu, Zn and Pb; direct sample counts were used for Fe and Cl. Artificial standards were used for Zn, Cu and Cl; the results for Fe and Mn were based upon the rock Standards G1 and W1. The chlorine analyses were used to correct the other concentrations to a salt-free solid basis. These estimates, displayed diagrammatically in Figs. 2 and 3, should be regarded as no more than approximations which serve as a useful guide to the general state of mineralization of the samples. They could be in error by as much as 10 per cent of the reported value.

Chemical Separation of Lead

The first six samples were boiled with purified concentrated ammonium acetate solution, treated with H_2S gas and the saline liquid was discarded after centrifugation. The residue was washed several times, then heated first with HCl, then HCl + HNO_3, until no further breakdown occurred. The insoluble solids were removed by centrifuging and the lead was recovered from the dried supernate by standard dithizone techniques (Cooper and Richards, 1966b).

The first stage of this procedure was devised to ensure solution of any $PbSO_4$ which may have been present. It did not appear necessary for the remainder of the samples. Lead was extracted directly from these with 6N HCl (Chow and Patterson, 1962). This solution was then purified by anion exchange from 1N HCl and followed by a dithizone solvent extraction. All isotopic analyses were made with the lead as oxalate (Cooper and Richards, 1966a).

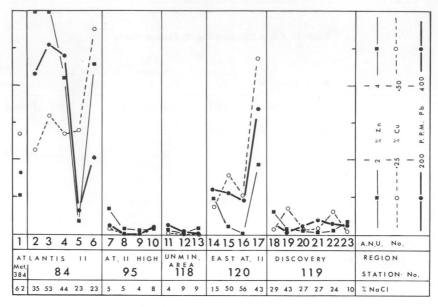

Fig. 2. Variation diagram of Pb, Cu and Zn content in dry salt-free Red Sea sediments.

Normalization by Double Spike

The major errors in the measurement of lead isotope ratios by modern solid-source techniques are due to isotopic fractionation of the lead sample as it evaporates from the mass-spectrometer filament. To correct for this, aliquots of a number of the samples were double-spiked with a previously calibrated mixture of enriched ^{207}Pb and ^{204}Pb isotopes. Isotopic fractionation was eliminated by an iterative adjustment of both spiked and unspiked data—sets

relative to the standard ratio of the spike. Calibration was effected against the Broken Hill Standard Galena UBC1, assuming values ^{207}Pb/^{204}Pb = 15.5423 and ^{206}Pb/^{204}Pb = 16.116 (Kollar *et al.*, 1960). The accepted ^{208}Pb/^{206}Pb of this standard was slightly adjusted so that the calculations might close for differing spike-sample ratios. While it is almost certain that this value is not absolutely correct, it permits the normalized ratios to be compared directly with the published "ore-lead growth curve" and with the many gas-source lead isotope measurements

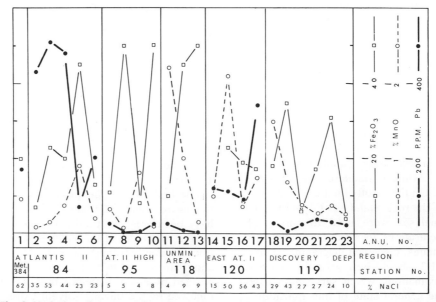

Fig. 3. Variation diagram of Pb, Mn and Fe content in dry salt-free Red Sea sediments.

on which it has been based, all of which have been adjusted relative to this value (Russell, 1963; Kanasewich and Farquhar, 1965; Ostic *et al.,* 1967). It is closely related to other oxalate solid source results (Cooper and Richards, 1966a).

Mass-Spectrometry

Lead isotope ratios were measured on a 12-inch radius of curvature, 60-degree sector, single-focusing mass-spectrometer, using a single-filament thermal-ionization source and a single-plate collector. Isotopic ratios were obtained by the usual magnetic peak-switching procedures, using voltage-to-frequency conversion and digital print-out (Compston *et al.,* 1965). Spiked ratios were obtained more quickly by switching the accelerating voltage about a mean reference voltage. Voltage discrimination was effectively removed during the fractionation correction (Arriens and Compston, 1968). Isotope ratios were measured by repetitive switching between pairs of peaks in the order 208/206, 206/204, 207/206, 208/206; closing on the 208/206 enabled correction of all ratios to a common time, thus allowing for any drift in the isotopic fractionation that may have occurred during a run. The internal measurement precision for the unspiked runs averaged 0.04 per cent for the $^{206}Pb/^{204}Pb$ ratio and 0.02 per cent for the $^{208}Pb/^{206}Pb$ and $^{207}Pb/^{206}Pb$ ratios (95 per cent confidence limits of the mean). The voltage-switched spiked runs showed slightly better statistics.

Discussion

Mineralization of Sediment Samples

The erratic concentration patterns of Figs. 2 and 3 are consistent with the reported fine layering in the cores and its implication of a complicated sequence of sedimentation conditions (Hunt *et al.,* 1967). The lead content varies over two orders of magnitude, but it nevertheless remains a comparatively minor component of the mineralization. While it shows a poor relationship with Fe and Mn abundances, it generally varies sympathetically with Cu and Zn. Hence, the samples with higher Pb content should give lead isotope ratios more closely representing the source of the Cu-Zn mineralization, whereas low and intermediate amounts of Pb, associated with

lower Cu and Zn, could permit the lead from local sediments of non-brine origin to contribute significantly to the isotopic pattern. Such rudimentary thinking is probably complicated by chemical fractionation processes when the sedimentary metallic elements are precipitated from solution (Miller *et al.,* 1966; Hunt *et al.,* 1967). Nevertheless, the above reasoning remains a useful guide.

In general, it would appear that samples (A.N.U. Nos.) M2 through M6 from station 84, M14 through M17 from station 120 and **METEOR** Sample M1 would probably contain mostly "hot-spot" lead. From station 95 on the small high within the confines of Atlantis II Deep, samples M8 and M9 should have a comparatively high "sedimentary" component, which may diminish as we pass from M10 to M7. Samples M18 through M23 from station 119 in the Discovery Deep are the most erratic of the suite. It appears that this deep has received metal-laden brine as an overflow from the Atlantis II Deep (Hunt *et al.* 1967). This would involve flow (possibly turbulent) of acidic brine over unconsolidated carbonate-rich sediment and so may be the basic cause of the differences noted here. Samples M11, M12 and M13 from the unmineralized station 118 show little heavy-metal content.

Presence of Carbonate in Sediment Samples

Another way of assessing the status of samples is to ascertain if other sediment has been added. The carbonate-rich nature of local deposition (Said, 1951; Gevirtz and Friedman, 1966) enables gross additions to be detected by a simple effervescence test with acid on dried sediment. Such testing indicated no carbonate from either of the Atlantis II stations 84 and 120. Cores from station 95, the local high in the Atlantis II Deep, yielded two samples M7 and M9 with a strong carbonate reaction and two, M8 and M10, with no carbonate reaction. From the unmineralized station 118 two samples, M11 and M12, yielded strong carbonate reactions, but, surprisingly, M13 did not. All the core samples from the Discovery Deep gave carbonate reactions: M18, M19, M22 and M23 were very strong while M19 and M21 were a little less so. Sample M1 did not react.

These results support the inference of a fluctuating brine level which results in intermittent metal deposition at station 95 and further indicate that considerable carbonate sediment has been caught up with the metal-rich material as it was transferred to the Discovery Deep. By way of supporting evidence, we note that samples M7, M8, M9, M11, M12, M18 and M20 all contained considerable gritty matter, indicating significant amounts of incorporated clastic material.

Status of Samples

Considering all these points, it seems that the lead of samples M2 through M6, and M14 through M17, could turn out to represent the "hot-spot" lead, while all

the others could be significantly contaminated by lead of origin similar to the lead in the barren sediments.

Lead Isotope Ratio Measurements

Isotope ratios were measured on unspiked extracts of lead from all twenty-three samples. The results are listed in Table 2 and are plotted diagramatically in Figs. 4a and 5a.

Three reference lines are also drawn on the $^{206}Pb/^{204}Pb$–$^{207}Pb/^{204}Pb$ (*i.e.*, x–y) plots (Fig. 4). These are as follows:

1. The "ore-lead growth" curve (Kanasewich and Farquhar, 1965; Ostic *et al.*, 1967), which corresponds to the progressive addition over a period of 4.55×10^9 years of radiogenic lead isotopes to lead of the meteoritic, troilite composition in an infinite reservoir with constant and homo-

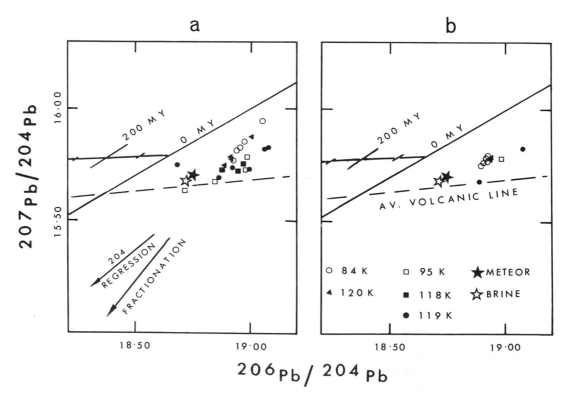

Fig. 4. $^{206}Pb/^{204}Pb$ v $^{207}Pb/^{204}Pb$ plot of Red Sea sediment samples: (a) Unspiked runs, (b) Runs normalized by double spiking for fractionation. Station 84K, Southern Atlantis II Deep; Station 95K, high within Atlantis II Deep; Station 120K, Eastern Atlantis II Deep; Station 119K, Discovery Deep; Station 118K, Unmineralized Area; Meteor, Atlantis II Deep, R/V **METEOR** station M384; Brine, R/V **ATLANTIS II** station 543. Horizontal solid line = Pb isotope growth curve for mu = 8.99. Dashed line = average regression of unspiked Pb from world-wide volcanics.

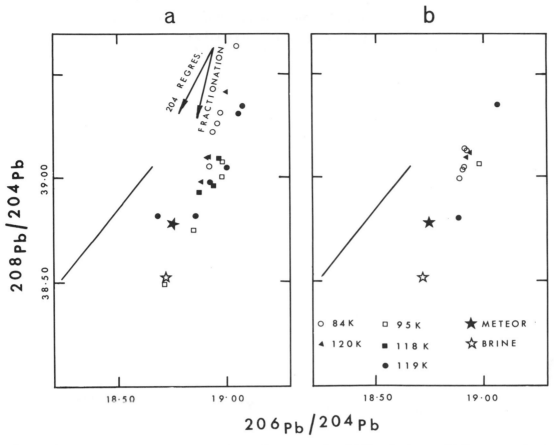

Fig. 5. $^{208}Pb/^{204}Pb$ v $^{206}Pb/^{204}Pb$ plot of Red Sea sediment samples: (a) Unspiked runs, (b) Runs normalized by double spiking. Solid line = growth curve for Th/U = 3.92, Co = 30.00. Other markings as in Fig. 4.

geneous $^{238}U/^{204}Pb = 8.99$. (This "single-stage" growth curve passes through the value of the Broken Hill standard used to calibrate the double spike.)

2. The "meteoritic geochron" or "zero isochron" (Murthy and Patterson, 1962), the family of zero age, single stage leads which differ from case (1) in that they have grown in individual regions of constant but different $^{238}U/^{204}Pb$ for 4.55×10^9 years.

3. A dashed line which approximates to the regression line of unspiked lead isotope measurements from many oceanic and continental volcanic samples (Cooper and Richards, 1966b). A lead isotope measurement on lead extracted from brine of the Atlantis II Deep (Delevaux *et al.*, 1967) is also plotted.

The rather complex pattern in Fig. 4a clarifies somewhat when the samples are subdivided, as discussed above, according to the probable amount of contamination by sediment of other than hot-brine origin. The samples with low-expected contamination lie very close to a line with slope appropriate for a mass-dependent perturbation. These "hot-spot" data group much more closely when this "fractionation" component is normalized against the corresponding spiked runs (Fig. 4b). The rather excessive degree of fractionation displayed by these samples was not altogether unexpected. Difficulties were encountered in obtaining absolutely pure lead from the highly mineralized samples. For example, submicrogram quantities of beam-quenching bismuth took additional time to burn off before measurements of a desirable precision could be made. Espe-

Table 2   Lead Isotope Measurements on Red Sea Sediments

(The numbers offset to the lower right are those corrected for
isotopic fractionation by the double-spike technique.)

Sample No.	206/204	207/206	208/206	207/204	208/204
M1a	18.75_3	0.8374	2.068_5	15.70_4	38.79
	18.75_0		2.068_1	15.70_0	38.78
M2	18.95_6	0.8349	2.070_9	15.82_6	39.26
	18.89_0	0.8335	2.063_7	15.74_4	38.98
M3	18.93_6	0.8350	2.071_0	15.81_2	39.22
	18.91_7	0.8346	2.068_9	15.78_8	39.14
M4	19.05_0	0.8370	2.080_7	15.94_5	39.64
	18.90_5	0.8338	2.064_7	15.76_3	39.03
M5	18.97_2	0.8356	2.072_2	15.85_3	39.31
	18.92_6	0.8345	2.067_2	15.79_5	39.12
M6	18.92_3	0.8333	2.063_7	15.76_8	39.05
	18.92_0	0.8332	2.063_0	15.76_4	39.04
M7	18.84_7	0.8316	2.055_8	15.67_3	38.75
M8	18.71_2	0.8355	2.056_8	15.63_4	38.49
M9	18.98_0	0.8287	2.054_9	15.72_7	39.00
M10	18.98_3	0.8316	2.058_0	15.78_6	39.07
	18.98_2	0.8315	2.057_8	15.78_4	39.06
M11	18.96_9	0.8307	2.060_6	15.75_8	39.09
M12	18.94_2	0.8299	2.056_8	15.72_0	38.96
M13	18.87_5	0.8333	2.062_3	15.72_8	38.93
M14	18.91_4	0.8343	2.067_0	15.78_0	39.10
M15	18.92_2	0.8341	2.066_5	15.78_2	39.10
	18.92_1	0.8340	2.066_3	15.78_1	39.10
M16	18.88_8	0.8338	2.063_9	15.74_9	38.98
M17	19.00_3	0.8356	2.074_4	15.87_8	39.42
	18.93_1	0.8340	2.066_4	15.78_7	39.12
M18	18.68_2	0.8430	2.078_0	15.74_9	38.82
M19	18.86_0	0.8320	2.058_4	15.69_1	38.82
M20	18.92_5	0.8311	2.059_6	15.72_8	38.98
	18.88_1		2.054_9	15.67_5	38.80
M21	19.07_3	0.8296	2.063_0	15.82_2	39.35
	19.07_4	0.8302	2.063_1	15.82_3	39.35
M22	19.05_9	0.8301	2.062_5	15.82_0	39.31
M23	18.99_0	0.8283	2.056_1	15.73_0	39.05

cially when single-filament emission is employed, all effects of this nature can cause variation in the amount of sample consumption prior to measurement and, thereby, fractionation (Eberhardt *et al.*, 1964).

The samples with probable sedimentary contamination exhibit a considerably wider range of ^{206}Pb/^{204}Pb (and of ^{208}Pb/^{204}Pb). This is particularly noticeable for the Discovery Deep samples, and it does not correspond to any simple expected instrumental perturbation pattern. The normalized results on three of these samples (M10, M20 and M21) demonstrate that this scatter is in fact "real," *i.e.*, geological, and not unlike the patterns observed for sediments in other areas (Chow and

Patterson, 1962). These three results demonstrate further that for chemically clean samples where "ideal" mass spectrometer runs are obtained, the results are much more reproducible. Each has a negligible ratio-shift upon normalization against the current set of spike calibration data. For the present, as stated above, these spike calibrations have been made against the conventional U.B.C. values (x, y) for the Broken Hill Standard Lead No. 1, irrespective of whether these are "true" values. This procedure has the significant virtue that it simultaneously minimizes the much-feared ^{207}Pb/^{204}Pb uncertainty and gives meaning to shifts in ^{207}Pb/^{204}Pb relative to the ore-lead growth curve (see under Normalization by Double Spike).

A closer look at the plotted distribution of the "hot-spot" lead data reveals a small linear scatter after fractionation correction, with amplitude around ±0.1 per cent. Comparison with unpublished data from this laboratory suggests that this is predominantly a residual instrumental effect. If there is a "geological" variation in these "hot-spot" leads, it is too small to be detected. With all this in mind, it is apparent that the mean $^{207}Pb/^{204}Pb$ ratio is approximately level with the end of the growth curve and is thus appropriate for a zero-age "primary" ore. It lies, however, significantly to the right of the meteoritic Geochron, thus indicating an excess of $^{206}Pb/^{204}Pb$ over that predicted by the simple model. It might be argued that the Geochron is based on measurements that have not been rigorously corrected for fractionation and is thus conceivably subject to possible error. To bring the Geochron and "hot-spot" lead average into line would demand, however, a general perturbation of the meteoritic results of about twice the magnitude of any which are apparent in this work. Alternatively a reconciliation between the growth-curve and these results could perhaps be achieved by assumption of an older age for the Earth (Tilton and Steiger, 1965; Murthy and Compston, 1965) while leaving the primordial values unaltered. This would, however, produce more discrepancies in other aspects than the one it solves. Thus, for the moment we must accept that these leads, while apparently close to being isotopically uniform, show some small amount of ^{206}Pb in excess of the composition appropriate for a zero age, single stage lead derived from the hypothetical infinite reservoir.

The $^{208}Pb/^{204}Pb$–$^{206}Pb/^{204}Pb$ plot of Fig. 5 shows a similar fractionation effect, and again the scatter of "hot-spot" leads reduces after double-spiking. The general state of the growth curve for a fixed U/Th/Pb source is less satisfactory (Kanasewich and Farquhar, 1965), particularly since these authors were forced to a slight adjustment of the primeval $^{208}Pb/^{206}Pb$ in this context. Allowing for this, we observe that the $^{208}Pb/^{204}Pb$ values are also close to that expected for a modern lead, but again in a "future" position relative to the accepted end of the growth curve.

The one mineralized mud which appears to differ from the close isotopic similarity of the remainder of this class is sample M1 from the **METEOR** expedition. It is isotopically similar to the soluble lead in the brine sample of Delevaux *et al.* (1967). Ross and Hunt (1967) have demonstrated a remarkable thermal layering in the brine pools. From this it can be deduced that at least two and probably three major volumes of brine have been injected into the bottom of the Atlantis II Deep within a period which has been short compared with the time scale of diffusion or heat-transfer processes. Watson and Waterbury (1969) report that anaerobic bacteria, including some sulfide reducers, have been detected in the transition water overlying the sterile brines in the Atlantis II Deep. Limited sulfur isotope measurements indicate that sulfide-reducing bacteria have not played a part in the sulfide precipitation (Hartmann and Nielsen, 1966). The sample studied came from the same core as our sample M1. On this basis metal precipitation is likely to be predominantly chemical, coming principally from the higher (earlier) brine layers by oxidation (Miller *et al.*, 1966; Hunt *et al.*, 1967) and alkalinity increase from contact with normal Red Sea water. Reaction with unconsolidated carbonate-rich sediments would also increase alkalinity and hence assist precipitation as would also reduction of solubility because of temperature decrease. In either case most of the precipitation so far analysed has probably come from what is now the upper brine volume or from even earlier injections for which no hydrographic evidence remains.* In contrast, sample M1, taken with a "Boomerang Corer," could well have come from the very top sediment layer in contact with the latest 56°C brine volume. Similarly, the brine sample measured by

---

* Different viewpoints are expressed by Watson and Waterbury (1969) who believe that sulfate-reducing bacteria in the transition layer are responsible for the sulfide sulfur in Atlantis II Deep.

Delevaux *et al.* (1967) was taken from 2,064m, a depth comparable to the M1 horizon and also well down into the 56°C brine. We are therefore led to invoke some mechanism which may have caused the latest brine to differ from previous injections. This may be difficult to substantiate. We note, however, that hot brines such as this are a proved metal solvent, and the present lead content of this latest pulse (0.5ppm Pb; Delevaux *et al.,* 1967) is far below that of similar, hotter brines (100ppm; White *et al.,* 1963; Doe *et al.,* 1966). Consequently, it is not inconceivable that some of this 56°C brine lead may represent the result of some contaminating process. Similar arguments could apply to the earlier brines although there is no supporting data available apart from the comparatively low Pb concentrations in the upper 5m of mineralized sediments (<500ppm Pb; Fig. 2). Thus, for the present, we can only acknowledge that we do not know the explanation for the isotopic difference between the 56° brine and M1 leads on the one hand, and the leads from stations 84K and 120K on the other.

Problems of Ore Genesis

If we ignore for the moment these two awkward results and consider only the data from the two mineralized cores 84K and 120K, it seems generally reasonable, on the available isotopic and chemical evidence, to equate the apparently uniform lead with the source of the mineralization and the variable lead in the other cores (including the Discovery Deep, 119K) with contamination by local sedimentary material. This may well be the correct interpretation but it cannot be substantiated until, at some time in the future, we observe a new pulse of brine containing lead in truly "ore" amounts and find that it corresponds isotopically with the apparently homogeneous "hot-spot" lead we have already analysed.

The next question is to inquire if there is any evidence which will help decide whether the heavy metals and the associated "hot-spot" lead have been leached from local country rocks or derived from a magmatic source. Craig (1966) has dem-

onstrated from $^{18}O-^{2}D$ measurements that water similar to normal Red Sea water is the major aqueous component of the brines (his reasoning which specifies water from near Bab al Mandab does not affect this discussion). The studies of Drake and Girdler (1964) and Knott *et al.* (1965, 1966) indicate that abundant local evaporite is available to produce the brine, and they also suggest the presence of basic rocks at nearly surface levels in the central axial trough along the rift. Hunt *et al.* (1967) report very high, localized heat flows in parts of the Atlantis II Deep. This demands some sort of magma very near to the surface of the sea floor in this, the deepest part of the Red Sea Rift. With an inferred magma so close, it is tempting to suppose it to be the source of the heavy metals. There is, however, a small amount of Pb isotope evidence which suggests that a country rock source, at least of the lead, cannot be overlooked. The limited analyses of Delevaux *et al.* (1967) indicate that lead minerals of comparable isotopic composition have been found both in Egypt and in the Arabian Peninsula. In a more extensive study, both of very hot brines from great depths near the Salton Sea and of the local rocks, Doe *et al.* (1966) deduced that considerable leaching of sedimentary and volcanic material could have supplied the dissolved lead, although their data did not permit a conclusive answer. These hot brines were also enriched in other heavy metals (White *et al.,* 1963).

Such data as is available suggests that heavily mineralized brines have occurred exclusively in "hot-spots" in contrast to other cold, but relatively unmineralized, brines. This tends to reinforce the prejudice that proximity to a heat source (*i.e.,* magma) is a necessary requirement for a mineralized brine.

With this in mind, it is of interest to compare the apparent processes and geological setting of this presently forming orebody with the observations and deductions of Stanton (1960) concerning the orebodies which he labelled as "Pyritic Conformable" but now prefers to call "Stratiform" (Stanton and Rafter, 1966). In brief, he emphasized the apparent sedimentary nature of ore deposition, the close association of volcanic lavas or tuffs or both with the mineralized sediments, the neighbouring limestone deposits suggesting shallow near-shore sedimentation and the ubiquitous presence of carbon. From this he suggested off-reef, lagoonal depressions in volcanic, probably andesitic, island arc regions as likely environments. The carbon content was thought to be the result of metamorphism of sediments of high organic content, the presence of which was due to conditions of slow sedimentation. Reducing conditions and anaerobic sulfate-reducing bacterial action were considered to play a significant role (this latter conclusion has since been revised, Stanton and Rafter, 1966). Stanton and Russell (1959) suggested that this type of sulfide deposit would have lead isotope ratios compatible with single-stage development in an infinite reservoir, *i.e.*, isotopic points would plot on a single growth-curve with constant U/Pb.

This Red Sea mineralization has been deposited in a sedimentary fashion. Association with igneous activity is suggested by the implied intrusive basic rocks (Drake and Girdler, 1964). Basaltic fragments have been observed in some cores (Chase, 1969), and an inferred active heat source very close to the surface almost heralds future volcanism. It exists in an area where calcareous sediments abound, and dense coral reefs rim the shores of the southern half of the Red Sea (Drake and Girdler 1964). Sedimentation is slow but the normal Red Sea water contains ample free oxygen (Miller *et al.*, 1966); however, the brines themselves are in a highly reduced state. The limited presence of anaerobic bacteria has been referred to earlier. Hartmann and Nielsen (1966) consider bacterial sulfate reduction as unlikely in the presence of so much oxidized iron. They regard the sulfide sulfur as a primary constituent of the brine. We know of no observations bearing on the role played by carbon in this deposit.

Stanton's observations fit this situation remarkably well. Even the apparent "^{206}Pb and ^{208}Pb excesses" have their parallels in some of the type examples selected by

him. Deposits at Rosebery, Tasmania, and Halls Peak, New South Wales (Ostic *et al.*, 1967), are both stratiform and associated with volcanic sequences. They both have "model ages" significantly younger by quite different amounts than their deduced geological ages—*i.e.*, they contain different proportions of "excess ^{206}Pb." Their "apparent $^{238}U/^{204}Pb$" or "mu-values," are both 8.93, at the lower extreme of the confidence limits quoted for this parameter by Kanasewich and Farquhar (1965). The mean "apparent mu" for the "hot-spot" lead is 8.93 (using the intercomparison spike calibration); its "model age"−240 m.y. This leads to the speculation that the geological environments have something in common; that the Red Sea geological environment may turn out to be the medium for the formation of more major sulfide bodies than has hitherto been recognized, although this has been obscured by one or more metamorphic events.

The Red Sea geological environment differs, however, from that deduced by Stanton. Instead of island arc volcanism this igneous activity is associated with continental rifting. For offshore lagoons and fringing reefs it substitutes a submerged rift valley, occasionally with two coral reefs, one parallel to each shore line. (It is noted that such reefs could be close at hand or far away depending on the width of the rift. Also they have not formed in the Northern half of the Red Sea. The lime-rich sediments in the median valley itself are the result of a local arid climate.) The function of hot saturated brine as a metal carrier was not visualized at that time (Wilson, 1967). Furthermore, Stanton (1960) observed that the volcanics often appeared to "have undergone a silica metasomatism" of some kind. The Red Sea brines carry abnormal quantities of silica, some of which has precipitated with the metals (Miller *et al.*, 1966). Its incorporation into tuffaceous material, followed by a metamorphic event, could well give an apparent silica enrichment in the resulting rock.

Thus the circumstantial evidence is strong; the median valley of a submerged continental rift is a hitherto unconsidered environment which could sometimes favour the concentration of metals into an orebody.

Conclusions

1. A clear isotopic distinction has been observed between mineralized Atlantis II Deep cores, on the one hand, and both the Discovery Deep cores and barren cores, on the other.
2. The isotopic variations within Discovery Deep appear to be the result of contamination of a fairly homogeneous lead associated with the mineralization by varying proportions of local sediment.
3. The fairly homogeneous lead may also have been slightly contaminated during upward transit in the brine solution, but this is not experimentally detectable with the present set of results, except for one near-surface sample which suggests possible variations.
4. The mineralization lead could be reasonably associated isotopically either (a) with the ore-leads classified by Stanton as "Stratiform" or (b) with the linear trend of modern volcanics. A local crustal average cannot be entirely excluded.
5. Some conditions of formation of the Red Sea orebody agree well with those observed by Stanton for certain stratiform lead ore-deposits. This modern deposit suggests that submerged continental rift areas, plus solution of minerals in locally produced hot brines and subsequent deposition from them, may be additional factors warranting further investigation in studies of old ore-deposits.

Acknowledgments

We wish to thank Dr. B. W. Chappell for the chloride analyses of the sediment samples and Drs. D. A. Ross and E. T. Degens for permitting us to take part in the investigation of this most interesting phenomenon.

References

Anderman, G. and J. W. Kemp: Scattered x-rays as internal standards in x-ray emission spectrometry. Analyt. Chem., **30**, 1306 (1958).

Arriens, P. A. and W. Compston: In preparation, (1968).

Chase, R. L. Basalt from the axial trough of the Red Sea. *In: Hot brines and recent heavy metal deposits in the Red Sea*, E. T. Degens and D. A. Ross (eds.). Springer-Verlag New York Inc., 122–128 (1969).

Chow, T. J. and C. C. Patterson: The occurrence and significance of lead isotopes in pelagic sediments. Geochim. et Cosmochim. Acta, **26**, 263 (1962).

Compston, W., J. F. Lovering, and M. J. Vernon: The rubidium-strontium age of the Bishopville aubrite and its component enstatite and feldspar. Geochim. et Cosmochim. Acta, **29**, 1085 (1965).

Cooper, J. A. and J. R. Richards: Solid source lead isotope measurements and isotopic fractionation. Earth and Plan. Sci. Let., **1**, 58 (1966a).

——, ——: Lead isotopes and volcanic magmas. Earth and Plan. Sci. Let., **1**, 259 (1966b).

Craig, H.: Isotopic composition and origin of the Red Sea and Salton Sea Geothermal Brines. Science, **154**, 1544 (1966).

Delevaux, M. H., B. R. Doe, and G. F. Brown: Preliminary lead isotope investigations of brine from the Red Sea, galena from the Kingdom of Saudi Arabia, and galena from United Arab Republic (Egypt). Earth and Plan. Sci. Let., **3**, 139 (1967).

Dietrich, G., G. Krause, E. Seibold, and K. Vollbrecht: Reisebericht der Indischen Ozean Expedition mit dem Forschungsschiff METEOR 1964–1965. *In: METEOR Forschungsergebnisse.* Gebrueder Borntraeger, Berlin (1966).

Doe, B. R., C. E. Hedge, and D. E. White: Preliminary investigation of the source of lead and strontium in deep geothermal brines underlying the Salton Sea geothermal area. Economic Geology, **61**, 462 (1966).

Drake, C. L. and R. W. Girdler: A geophysical study of the Red Sea. Geophys. J. Roy. Astr. Soc., **8**, 473 (1964).

Eberhardt, A., R. Delwiche, and J. Geiss: Isotopic effects in single filament thermal ion sources. Zeit. f. Naturforsch., **19a**, 736 (1964).

Gevirtz, J. L. and G. M. Friedman: Deep sea carbonate sediments of the Red Sea and their implications on marine lithification. J. Sed. Pet., **36**, 143 (1966).

Hartmann, M. and H. Nielsen: Sulfur isotopes in the hot brine and sediment of Atlantis II Deep (Red Sea). Marine Geol., **4**, 305 (1966).

Hunt, J. M., E. E. Hays, E. T. Degens, and D. A. Ross: Red Sea: detailed survey of hot brine areas. Science, **156**, 514 (1967).

Kanasewich, E. R. and R. M. Farquhar: Lead isotope ratios from the Cobalt-Noranda Area, Canada. Canadian Journal of Earth Sciences, **2**, 361 (1965).

Knott, S. T. and E. Bunce: Geophysical studies of the

Red Sea north of latitude 17°N. Trans. Amer. Geophys. Un., **46,** 102 (1965).

——, ——, and R. L. Chase: Red Sea seismic reflection studies. *In: The World Rift System,* T. N. Irvine (ed.). Geological Survey of Canada Paper 66-14, 33 (1966).

Kollar, F., R. D. Russell, and T. J. Ulrych: Precision intercomparisons of lead isotope ratios: Broken Hill and Mount Isa. Nature, **187,** 754 (1960).

Miller, A. R., C. D. Densmore, E. T. Degens, J. C. Hathaway, F. T. Manheim, P. F. McFarlin, R. Pocklington, and A. Jokela: Geochim. et Cosmochim. Acta, **30,** 341 (1966).

Murthy, V. R. and W. Compston: Rb-Sr ages of chondrules and carbonaceous chondrites. Journ. Geophys. Res., **70,** 5297 (1965).

—— and C. C. Patterson: Primary isochrons of zero age for meteorites and the earth. J. Geophys. Res., **67,** 1161 (1962).

Ostic, R. G., R. D. Russell, and R. L. Stanton: Additional measurements of the isotopic composition of lead from stratiform deposits. Canad. J. of Earth Sci., **4,** 245 (1967).

Ross, D. A. and E. T. Degens: Shipboard collection and preservation of sediment samples collected during CHAIN 61 from the Red Sea. *In: Hot brines and recent heavy metal deposits in the Red Sea,* E. T. Degens and D. A. Ross (eds.). Springer-Verlag New York Inc., 363–367 (1969).

—— and J. M. Hunt: Third brine pool in the Red Sea. Nature, **213,** 687 (1967).

Russell, R. D.: Some recent researches on lead isotope abundances. *In: Earth Science and Meteoritics,* J. Geiss and E. D. Goldberg (eds.). North Holland Publishing Company, Amsterdam, 44 (1963).

——, and R. M. Farquhar: Lead Isotopes in Geology, Interscience, New York (1960).

Said, R.: Organic origin of some calcareous sediments from the Red Sea. Science, **113,** 517 (1951).

Sinclair, A. J. and R. I. Walcott: The significance of Th/U ratios calculated from West Central New Mexico multistage lead data. Earth and Planetary Science Letters, **1,** 38 (1966).

Stanton, R. L.: General features of the conformable "Pyritic" orebodies. Canad. Min. and Met. Bull., **63,** 22 (1960).

—— and T. A. Rafter: The isotopic constitution of sulfur in some Stratiform Lead Zinc Sulfide Ores. Mineralium Deposita, **1,** 16 (1966).

—— and R. D. Russell: Anomalous leads and the emplacement of lead sulfide ores. Econ. Geol., **54,** 588 (1959).

Tilton, G. R. and R. H. Steiger: Lead isotopes and the age of the earth. Science, **150,** 1805 (1965).

Watson, S. W. and J. B. Waterbury: The sterile hot brines of the Red Sea. *In: Hot brines and recent heavy metal deposits in the Red Sea,* E. T. Degens and D. A. Ross (eds.). Springer-Verlag New York Inc., 272–281 (1969).

White, D. E., E. J. Anderson, and D. K. Grubbs: Geothermal brine well, mile deep drill hole may tap ore bearing magmatic water and rocks undergoing metamorphism. Science, **139,** 919 (1963).

Wilson, H. D. B.: Volcanism and ore deposits in the Canadian Archaean. Proc. Geol. Assoc. Canada, **18,** 11 (1967).

Uranium Series Isotopes in Sediments from the Red Sea Hot-Brine Area *

TEH-LUNG KU
Woods Hole Oceanographic Institution
Woods Hole, Massachusetts

Abstract

The distribution of the longer-lived uranium and thorium series isotopes in the iron-rich sediments of the Red Sea geothermal brine area is characteristic as compared to that of normal oceanic deposits. Concentrations of uranium on a salt- and water-free basis range from 6 to 30ppm, whereas those of thorium range from 0.1 to 0.5ppm. The low Th/U ratios are also reflected by the low activity ratios of Th^{230}/U^{234} and Pa^{231}/U^{235}, the latter ratios being considerably less than the secular equilibrium value of unity. Maximum ages of the deposits may be estimated from the growths of Th^{230} and Pa^{231}. These upper-limit ages indicate that the metalliferous sediments accumulated at a rate of at least $40cm/10^3$ yr and that the sampled four to eight meters of sediment columns in the Atlantis II Deep were deposited in less than 10,000 years. Rates higher than $100cm/10^3$ yr should not be considered as rare occurrences. While thorium, protactinium and radium in these sediments may originate from the extruding brines, an important fraction of uranium may have been scavenged out of the overlying normal Red Sea water through co-precipitation with the colloidal ferric hydroxide and silica.

The low specific activities of the uranium-unsupported Th^{230} (<1dpm/gr) and Pa^{231} (<0.1dpm/gr) associated with the calcareous lutite sediments impose limitations on the use of the decay of these nuclides for age-dating. They are, however, suggestive of a lutite accumulation rate of the order of $10cm/10^3$ yr in the Red Sea.

Introduction

Submarine discharge of geothermal brine in the central Red Sea deeps results in deposition of metal oxides and sulfides, silicates, carbonates and anhydrite (Miller *et al.*, 1966; Bischoff, 1969). With hydrous iron oxide as a dominant matrix, the deposits are characterized by an enrichment of heavy metals, notably zinc, manganese and copper.

The present study is primarily concerned with the distribution of the uranium and thorium series nuclides, particularly the longer-lived members: U^{238}, U^{234}, Th^{232}, Th^{230} and Pa^{231} in these precipitates. Since precipitation mechanisms of these metalliferous deposits involve cooling as well as interaction of the brine with the normal Red Sea water, the investigation is expected to yield information on the source and deposition of the radioelements associated with the geothermal activities. Furthermore, the radioactive disequilibrium relationships among these isotopes could also provide a basis for chronological studies of the sediments.

Methods of Investigation

Samples and Sample Preparation

Ten cores were chosen for the study. As shown in Fig. 1, cores 84K, 120K, 127P and 128P were taken from the Atlantis II Deep, 81G and 119K from the Discovery Deep, 85K and 161K from the sill area dividing the two deeps, and 118K and 154P from outside the hot-brine area.

General features in the lithologic variation of the cores have already been described (Bischoff,

* Woods Hole Oceanographic Institution Contribution No. 2201.

1969; Ku *et al.*, 1969). The sediments are mixtures of two end-member components from entirely different sources. The first is the normal pelagic sedimentary components: skeletons of microorganisms, detrital quartz and aluminosilicate minerals. The other is the iron-rich brine precipitates in the form of oxides, silicates, sulfides, and so forth. Since this latter end-member is a product of brine discharge in the Atlantis II Deep, its contribution to the sediments diminishes in areas further removed from the brine area. Thus the samples we analyzed can be divided into three categories:

Type I is calcareous lutite sediments as represented by core 154P located about 100 miles south of the brine area.

Type II is the brine precipitates. This is typified by samples in cores taken from the Atlantis II Deep and the sill area (e.g., 128P and 85K). According to Bischoff (1969), three major mineralogical facies prevail: amorphous-goethite, Fe-montmorillonite, and sulfide. The dominant chemical components are Fe_2O_3, FeO and SiO_2

for the first two facies and FeO, ZnO, CuO and S for the sulfide facies.

Type III is represented by sediments from the Discovery Deep and nearby cores outside the deeps (e.g., 118K) where intermixing of the Types I and II components occurs.

Samples of about $20cm^3$ size were taken soon after the cores were split open and then the material in contact with the coring tube was trimmed off. Except for samples of Type I, each sample was leached with distilled water and dried at 100°C. The weights of the original wet samples and those of the dried specimens allow interstitial brine contents to be estimated. The reported concentrations are on a water- and salt-free basis.

Analytical

About 3 grams of each sample were digested in a mixture of hot concentrated HCl and HNO_3 and then taken to fume with HF and $HClO_4$.

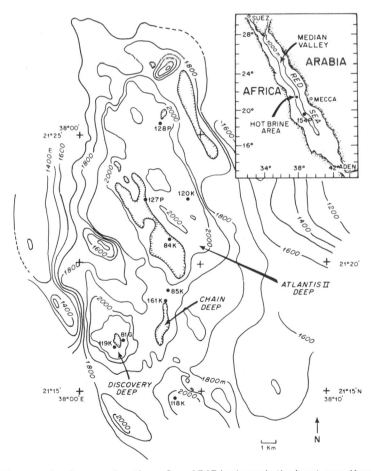

Fig. 1. Bathymetric map showing core locations. Core 154P is shown in the inset map. K refers to 4m Kasten cores, P to 8m piston cores, and G to gravity cores (see Ross and Degens, 1969).

This dissolution treatment was found to take up the majority of the radioisotopes of interest.

Methods of measurement of the uranium, thorium and protactinium isotopes have been published (Ku, 1965; 1968) and need no elaboration here. In brief, uranium and thorium were determined by alpha-spectrometry using U^{232} and Th^{228} as yield tracers. Pa^{231} was counted in a proportional counter of known efficiency. Its yield was monitored by β-counting of the Pa^{233} tracer added. The alpha-spectrometer consists of an "Ortec" silicon solid state detector coupled through a low-noise amplification system to a Nuclear Data 512-channel pulse-height analyzer. The alpha resolution (FWHM) at 5.5Mev is about 68Kev. The proportional counter used for Pa^{231} measurements is a windowless gas flow type (NMC model PCC-11T-DS-1T). The counter background in the alpha region was 0.018 ± 0.002cpm.

Ra^{226} was measured by the radon emanation method (Broecker, 1963).

The method for analyzing Po^{210} was developed by Dr. A. Kaufman. Po^{210} was self-plated from a warm ($\sim 75°C$) 0.6N HCl solution onto a copper disk and measured by alpha-spectrometry using Po^{209} as a tracer.

Total iron in the samples was determined by atomic absorption spectrophotometry.

EDTA titration of the acid soluble calcium was used to estimate the $CaCO_3$ content.

Results and Discussion

The analytical results are grouped according to the three categories of sediment types mentioned above and are listed in Tables 1, 2 and 3. The errors quoted are standard deviations derived from counting statistics.

Nature of Radioactive Inequilibrium Relationships

Presently available information on the distribution of the uranium series nuclides in the ocean has established that: (1) The sea is rather uniform in its uranium content (about $3\mu gU/l$; Rona et al. (1956); Torii and Murata, 1964; and others) and isotopic composition ($U^{234}/U^{238} = 1.15$; Thurber, 1962; Koide and Goldberg, 1965; Veeh,

1968; and others); (2) The precipitation of Th^{230} and Pa^{231} as they are formed from the dissolved oceanic uranium is essentially quantitative, less than 1% of the Th^{230} and Pa^{231} produced remaining in solution (Sackett et al., 1958; Moore and Sackett, 1964).

From the above, we may deduce that, by taking the average depth of the Red Sea as 1 km, under steady state conditions, about 2dpm Th^{230} and 0.2dpm Pa^{231} are precipitated per thousand years on a square centimeter of the sea floor. The concentration of these "authigenic" nuclides in the sediments will be roughly inversely proportional to the deposition rate of bulk sediments, i.e., $\frac{2}{R}$ dpm Th^{230}/gr sed. and $\frac{0.2}{R}$ dpm Pa^{231}/gr sed., where R is in units of gr sed./cm$^2 \cdot 10^3$ yr.

My interpretation of the data will be proceeded within the framework of the above background information.

Type I Sediments. Uranium and thorium in normal pelagic sediments are predominantly of detrital origin (as summarized by Ku, 1965). The evidence includes:

1. Foraminifera and coccoliths in globigerina oozes contain negligible amounts of uranium and thorium as compared to those of clays. The deep sea clays, furthermore, have contents of uranium (~ 2.6ppm) and thorium (~ 12-ppm) which are similar to those of the average igneous rocks and shales reported by Adams et al. (1959). The Th/U ratios are about 4 to 5.
2. The U^{234}/U^{238} ratios in the top layers of cores approach more closely the equilibrium value of unity than the oceanic value of 1.15.

The data in Table 1 conform in a general way with the above observations. A somewhat lower Th/U ratio (2–2.5) than that of deep sea clays may imply that the aragonitic pteropods, which occur commonly in these samples, contain an appreciable amount of uranium (as opposed to the $0.0x$ ppm for foraminifera and pteropods). There is little excess Th^{230} and Pa^{231} in the top layers of the cores. Two possible explanations for

Table 1   Data on Calcareous Lutite Sediments (Type I)

Core No.	Depth (cm)	CaCO$_3$ (%)	Fe (%)	U (ppm)	Th (ppm)	$\frac{Th}{U}$, wt.	$\frac{U^{234}}{U^{238}}$	$\frac{Th^{230}}{U^{234}}$	$\frac{Pa^{231}}{U^{235}}$
118K	0–10	78	1.9	0.76 ± .01	1.75 ± .07	2.3	1.10 ± .02	2.38 ± .07	—
154P	0–10	75	1.5	1.17 ± .04	2.67 ± .16	2.3	1.06 ± .05	1.10 ± .04	1.36 ± .27
	600–605	68	1.2	1.12 ± .02	2.11 ± .08	1.9	1.04 ± .03	1.10 ± .05	0.98 ± .19
	850–855	65	1.5	1.16 ± .03	2.34 ± .11	2.0	1.03 ± .04	1.07 ± .06	1.10 ± .20

this could be:

1. The pteropod components contribute low values (<1.0) of Th^{230}/U^{234} and Pa^{231}/U^{235} to the bulk sediments.
2. Radiocarbon dating (Ku *et al.,* 1969) shows the sedimentation rates (R) of these cores to be about 7gr/cm^2 · 10^3 yr. Hence the "authigenically precipitated" Th^{230} and Pa^{231} of the top layer are expected to have concentrations of the order of 0.3dpm Th^{230}/gr and 0.03dpm Pa^{231}/gr. These estimates are borne out by the data of Table 1.

Type II Sediments. When compared with data for the normal pelagic sediments, the results in Table 2 show:

1. Very low values of Th/U, Th^{230}/U^{234}, and Pa^{231}/U^{235}.
2. High uranium concentrations.
3. U^{234}/U^{238} ratios close to the oceanic value of 1.15.

4. High iron contents.

Bischoff (1969) suggests that formation of the deposits involves cooling of the subterranean brine as it discharges into the Atlantis II Deep and mixing of the brine with the overlying sea water. Most of the metals were transported as chloride complexes in the brine at depth, where physicochemical conditions are those of high temperature (>56°C), low Eh and near neutral (or slightly acidic) pH. The data in Table 2 indicate that the brine is not effective in transporting thorium and protactinium.

A positive correlation is seen between iron and uranium contents in the sediments (Fig. 2), giving a U/Fe ratio that varies from 2.5 × 10^{-5} to 7 × 10^{-5}. The correlation implies that the quantity of uranium precipitated is mainly related to the amounts of iron available.

I have analyzed iron and uranium in a 56°C brine sample of the Atlantis II Deep collected during the **CHAIN** 61 cruise. The

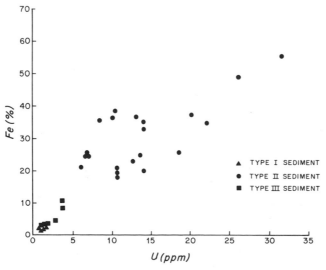

Fig. 2. Graph showing covariation of uranium with iron contents in sediments.

Table 2 Data on Iron-Rich Geothermal Deposits (Type II)

Core No.	Depth (cm)	Brine (%)	Fe (%)	U (ppm)	Th (ppm)	$\frac{Th}{U}$ wt.	$\frac{U^{234}}{U^{238}}$	$\frac{Th^{230}}{U^{234}}$	$\frac{Pa^{231}}{U^{235}}$	$\frac{Th^{230}}{Th^{232}}$	Mineral* Assemblage
84K	120–125	87	35.1	13.9 ± .3	0.19 ± .04	.013	1.13 ± .02	.025 ± .002	.037 ± .003	6.61 ± 1.4	Fe-m, sp, py
	350–355	70	26.0	6.60 ± .17	0.18 ± .04	.027	1.13 ± .03	.065 ± .004	.087 ± .010	8.27 ± 1.1	h, cal
85K	0–10	85	37.2	20.2 ± .2	0.46 ± .04	.023	1.14 ± .01	.039 ± .002	.051 ± .007	5.88 ± .36	g, cal
	30–35	90	25.6	18.5 ± .4	0.21 ± .03	.011	1.16 ± .02	.046 ± .002	.060 ± .004	14.7 ± 1.8	g
	90–95	91	34.9	22.0 ± .4	0.19 ± .03	.009	1.11 ± .02	.060 ± .002	.063 ± .004	22.8 ± 3.6	g
	290–295	85	48.8	26.0 ± .4	0.26 ± .03	.010	1.14 ± .01	.061 ± .003	.100 ± .006	21.1 ± 3.6	g
	320–325	75	56.5	31.3 ± .3	0.34 ± .03	.011	1.13 ± .01	.065 ± .002	.139 ± .010	19.3 ± 1.6	g
118K	90–95	80	36.6	10.1 ± .09	0.52 ± .05	.051	1.14 ± .01	.126 ± .005	—	8.57 ± 1.00	g, cal, qtz
	250–255	80	38.7	10.3 ± .3	0.33 ± .05	.032	1.12 ± .03	.176 ± .009	.314 ± .010	18.7 ± 2.4	g
	300–305	81	36.3	8.43 ± .18	0.10 ± .02	.012	1.11 ± .03	.196 ± .009	—	55.1 ± 15.5	g
120K	170–175 †	81	21.6	6.00 ± .09	0.19 ± .04	.031	1.14 ± .03	.030 ± .004	.034 ± .003	3.36 ± 0.8	Fe-m, mgs
	270–275	76	24.5	6.56 ± .09	0.11 ± .02	.017	1.16 ± .02	.058 ± .003	.084 ± .005	12.0 ± 2.2	Fe-m, mgs
	352–357	71	24.8	6.70 ± .14	0.19 ± .03	.028	1.12 ± .02	.090 ± .006	—	10.9 ± 2.0	Fe-m, mgs
127P	80–85	87	33.2	14.2 ± .4	0.25 ± .03	.017	1.15 ± .03	.030 ± .002	—	6.17 ± .87	Fe-m
	559–565 †	80	37.0	13.0 ± .3	0.36 ± .03	.027	1.14 ± .02	.096 ± .003	.150 ± .010	12.3 ± 1.8	g
128P	130–135	88	19.4	10.7 ± .2	0.35 ± .06	.030	1.14 ± .02	.050 ± .004	—	5.78 ± 1.19	g, qtz, cal, mgs
	440–445	85	25.0	13.6 ± .3	0.25 ± .04	.018	1.14 ± .02	.041 ± .002	—	7.90 ± 1.25	g, cal
	740–745	75	22.7	12.5 ± .3	0.21 ± .04	.017	1.16 ± .02	.060 ± .003	.105 ± .007	12.4 ± 1.8	Fe-m
161K	0–5 †	82	17.6	10.7 ± .4	0.12 ± .02	.011	1.13 ± .03	.025 ± .002	—	7.81 ± 1.50	g, ar, py
	180–185	86	19.9	14.1 ± .2	0.22 ± .04	.016	1.15 ± .01	.031 ± .002	—	6.78 ± 2.25	g, ar, py
	360–365	85	21.2	10.5 ± .2	0.15 ± .02	.014	1.14 ± .01	.042 ± .002	—	10.4 ± 1.5	g, ar, py

* Minerals identifiable from X-ray diffraction patterns. Comparisons with peak heights of pure standards of the minerals reveal that in all the samples more than 50% are X-ray amorphous materials. Abbreviations: g – goethite, Fe-m – Fe-montmorillonite, cal – calcite, qtz – quartz, sp – sphalerite, h – hematite, ar – aragonite, py – pyrite, mgs – manganosiderite. Minerals are listed in order of decreasing abundance.
† <16μ fraction.

results * are: 98 ± 2mgFe/l, $0.58 \pm .01\mu$g U/l, and $U^{234}/U^{238} = 1.12 \pm .02$. The iron content compares favorably with that reported by Brewer and Spencer (1969). The concentration of uranium is much lower than that of the normal sea water ($\sim3\mu$g U/l). The U/Fe ratio in the brine is about 0.6×10^{-5}, which is significantly lower than the ratio found in the sediments.

It would appear that the majority of the uranium in Type II sediments in the Atlantis II Deep comes from the normal Red Sea water. Uranium may be co-precipitated with the iron hydroxide which was formed by oxidation of the ferrous chloride complex in the brine as it comes in contact and mixes with the overlying Red Sea water. Since the iron-controlled precipitation processes as illustrated by the data of Fig. 2 are not expected to distinguish between the uranium from the two sources: brine and sea water, there must be a sizable amount of uranium from the latter source co-precipitating out at the brine-sea water interface. A one to one volume proportion in the mixing of brine with sea water is perhaps needed to yield the observed U/Fe ratios in sediments.

Can the amount of uranium in sediments be accounted for by the Red Sea water column above? If we take a uranium content of 15ppm (Table 2) and a sedimentation rate of 100cm/10^3 yr (see later discussion and Ku *et al.*, 1969), *i.e.*, 30gr/cm$^2 \cdot 10^3$ yr (bulk sediment density ~0.3gr/cm^3) for the Type II deposits, we estimate a uranium deposition rate of about 450μg/cm$^2 \cdot 10^3$ yr. Considering that the bottom water renewal time in the Red Sea is of the order of 200 years (Craig, 1969) and that even all the precipitated uranium comes from the Red Sea water, we require a supply of 90μg/cm$^2 \cdot 200$ yr. This is small compared to the value of about 300μg of uranium in a 1km water column above one square centimeter of the ocean floor in the Red Sea.

I should point out that the uranium "source" problem need be pursued only

when the time scale involved for the mixing between the brine and the sea water is slow, such that a significant amount of uranium is precipitated before complete mixing, or diffusing of uranium out from the brine into the overlying sea water takes place. The old C^{14} age of the brine (Craig, 1969) and the aforementioned uranium deposition rate justify our attempt put forth here. The picture presented above would also indicate that the Red Sea hot brine deeps act as a sink for uranium precipitation in the sea.

It is interesting to note that the data in Table 2 bear considerable analogy to a report by Kazachevskii *et al.* (1964, Table 2) for the Banu-Vukhu Volcano of Indonesia. These workers found that the hydrous iron and manganese oxide suspensions introduced by subaqueous exhalations of the volcano give values: $U^{234}/U^{238} = 1.14$, $Th^{230}/U^{238} = 0.001$, and Th/U = 0.003.

In terms of the co-precipitation of radioisotopes, the conditions of rapid accumulations of iron and manganese differ very much from the very slow formation of manganese concretions in the deep sea. The Th/U, Th^{230}/U^{238} and Pa^{231}/U^{235} ratios in the latter deposits are much higher (Tatsumoto and Goldberg, 1959; Ku and Broecker, 1967).

Type III Sediments. The nature of the sediments as mixtures of the above two types is essentially reflected by the results of Table 3. Compared with those of Tables 1 and 2, data in Table 3 all give intermediate values. The low iron and the ensuing low uranium contents of the samples from the Discovery Deep are presumably a consequence of the low iron contents (Miller *et al.*, 1966) in the Discovery Deep brine.

Age Estimates

The near equilibrium values for Th^{230}/U^{234} and Pa^{231}/U^{235} ratios in sediments of Types I and III render the use of decay of Th^{230} and Pa^{231} for age determinations impractical. This situation is created by the high total sediment accumulation rates (>10cm/10^3 yr) for the Type I sediments. The Type III sediments are complicated

---

* Uranium was measured by alpha spectrophotometry, iron by atomic absorption spectrometry. Errors indicated are standard deviations of duplicate runs.

Table 3   Data on Mixed Sediments (Type III)

Core No.	Depth (cm)	Brine (%)	Fe (%)	U (ppm)	Th (ppm)	$\frac{Th}{U}$, wt.	$\frac{U^{234}}{U^{238}}$	$\frac{Th^{230}}{U^{234}}$	$\frac{Pa^{231}}{U^{235}}$
81G	0–5	81	4.8	$2.72 \pm .10$	$1.84 \pm .15$	0.67	$1.13 \pm .05$	$0.73 \pm .04$	$0.89 \pm .06$
	30–35	80	3.6	$1.74 \pm .08$	$1.74 \pm .19$	1.0	$1.15 \pm .06$	$1.01 \pm .05$	$1.03 \pm .11$
118K	120–125	64	8.5	$3.53 \pm .09$	$0.33 \pm .04$	0.1	$1.14 \pm .03$	$0.30 \pm .01$	$0.40 \pm .05$
119K	30–35	50	3.3	$1.27 \pm .05$	$1.82 \pm .12$	1.1	$1.08 \pm .06$	$1.56 \pm .08$	—
	150–155	81	3.3	$1.23 \pm .05$	$0.98 \pm .10$	0.8	$0.95 \pm .06$	$2.06 \pm .13$	—
	390–395	75	10.7	$3.73 \pm .06$	$0.60 \pm .06$	0.16	$1.08 \pm .03$	$0.47 \pm .02$	—

due to mixing of two components of unknown proportions, hence unknown initial extents of disequilibrium.

Type II sediments contain high uranium contents and very low ratios of Th/U, Th^{230}/U^{234} and Pa^{231}/U^{235}. It may be possible to date these deposits by the extent of Th^{230} and Pa^{231} in return to secular equilibrium with their uranium parents. The main problem concerned is the amount of Th^{230} and Pa^{231} which could be initially incorporated into the sample during its formation. Sources of such contamination may be: (1) precipitation from the water column above, (2) brought in with the brine, (3) detrital minerals containing Th^{230} and Pa^{231}.

Since the water above contains $\sim 3\mu g$ U/l, the concentrations of the precipitating Th^{230} and Pa^{231} in freshly deposited sediments, as pointed out earlier, will be about $\frac{2}{R}$ dpm Th^{230}/gr and $\frac{0.2}{R}$ dpm Pa^{231}/gr, respectively. If R is of the order of

30 gr/cm² · 10³ yr (as will be shown to be the case), the level of contamination due to source (1) should be low in comparison with most of the measured Th^{230}/U^{234} and Pa^{231}/U^{235} ratios.

It was hoped that the extraneous Th^{230} and Pa^{231} due to sources (2) and (3) could be reduced by separating out the phases containing Th^{232} or high Th/U ratios. Several physical methods of separation (e.g., magnetic, heavy-liquid, size fractions, etc.) were employed, but clean separation was not achieved, presumably due to the very fine and aggregative nature of the material. Table 4, for example, gives some of the results based on the size separation. Four size separates were obtained for the sample 127P (559–565cm). The two smaller size fractions ($<2\mu$ and 2–16μ) contain X-ray amorphous material (~ 90 per cent) and goethite (~ 10 per cent). The 16–64μ fraction contains amorphous material (~ 90 per cent), traces of goethite, and about 10 per

Table 4   Comparison of Data for Different Size Fractions

Sample	Size Fraction	U (ppm)	Th (ppm)	$\frac{Th}{U}$, wt.	$\frac{U^{234}}{U^{238}}$	$\frac{Th^{230}}{U^{234}}$	$\frac{Pa^{231}}{U^{235}}$
127P (559–565cm)	$<2\mu$	$14.5 \pm .3$	$0.40 \pm .03$	27.6	$1.14 \pm .02$	$.095 \pm .003$	$.151 \pm .015$
	2–16μ	$11.5 \pm .3$	$0.32 \pm .03$	28.0	$1.13 \pm .02$	$.098 \pm .003$	$.140 \pm .010$
	16–64μ	$3.66 \pm .09$	$1.77 \pm .02$	0.48	$1.13 \pm .04$	$.230 \pm .020$	$.433 \pm .042$
	$>64\mu$	$0.58 \pm .03$	$0.10 \pm .02$	0.17	$1.11 \pm .06$	$.158 \pm .022$	$.387 \pm .045$
119K (390–395cm)	all sizes	$3.73 \pm .06$	$0.60 \pm .06$	0.16	$1.08 \pm .03$	$.467 \pm .022$	—
	$<16\mu$	$3.61 \pm .04$	$0.64 \pm .09$	0.18	$1.08 \pm .02$	$.430 \pm .050$	$.214 \pm .024$
118K (120–125cm)	all sizes	$3.53 \pm .09$	$0.33 \pm .04$	0.1	$1.14 \pm .03$	$.302 \pm .014$	$.398 \pm .050$
	$<16\mu$	$6.71 \pm .14$	$0.55 \pm .12$	.08	$1.15 \pm .03$	$.307 \pm .017$	$.415 \pm .051$
128P (130–135cm)	all sizes	$10.7 \pm .2$	$0.35 \pm .06$	.03	$1.14 \pm .02$	$.050 \pm .004$	—
	$<16\mu$	$10.1 \pm .2$	$0.10 \pm .07$	.01	$1.14 \pm .02$	$.045 \pm .008$	$.051 \pm .005$

cent manganosiderite. The $>64\mu$ size consists of about 30 per cent manganosiderite, anhydrite and todorokite(?), with the rest being amorphous. Table 4 shows that data for the two smaller size fractions are nearly identical but differ from those of the larger sizes. However, since the former fractions constitute more than 85 wt. per cent of the total sample, their compositions approximate those of the whole sample. This is also indicated by the comparisons listed in the same table.

Although in general the size separations allow minerals such as goethite and Fe-montmorillonite to be concentrated in the $<16\mu$ fraction, and quartz and carbonate minerals to appear in coarser fractions, the bulk sample appears to consist mainly of amorphous aggregates. My attempt to obtain mineral phases with low Th/U ratios from the latter aggregates must be regarded as unfeasible.

The ages presented in Table 5 are order-of-magnitude estimates only. They are calculated from the Th^{230}/U^{234} and Pa^{231}/U^{235} ratios assuming nil initial concentrations for Th^{230} and Pa^{231} and that samples act as a closed system for the nuclides of interest after deposition. These ages serve at their best only as upper limits.

One notable feature is that in almost all cases the apparent Th^{230}/U^{234} ages are larger than those based on the Pa^{231}/U^{235} ratios. This has several implications:

1. It strengthens the validity of the "closed system" assumption used. Extensive post-depositional migration of the nuclides concerned might have led to more spurious results; the apparent Th^{230}/U^{234} ages could be either older or younger than the Pa^{231}/U^{235} ages.
2. Some of the Th^{230} and Pa^{231} must have come from sources other than *in situ* decay of uranium in the samples. Since such sources are most likely minerals of detrital origin or derived from the chloride brine, they will introduce Th^{230} and Pa^{231} in a ratio close to 21.8, *i.e.*, the activity ratio of $U^{234}:U^{235}$. It can be shown (see Appendix) that these contaminations will then result

in apparent ages based on the Th^{230}/U^{234} ratios being older than those derived from the Pa^{231} growth. Theoretically, the ages in Table 5 may be "improved" by making reasonable corrections. For instance, as the Th^{230}/Th^{232} activity ratio of the average crustal rocks is near unity, a correction for the "external" Th^{230} using values comparable to the Th^{232} activities found in the samples may be applied. Also, as shown in the Appendix, it may be possible to derive the true ages from the differences between the observed Th^{230} ages and Pa^{231} ages. However, due to the large analytical uncertainties, these corrections are not applied to the present data.

3. The apparent Pa^{231}/U^{235} ages in all cases should be better estimates than the Th^{230}/U^{234} ages. A few available radiocarbon dates (Table 5) indicate that our Pa^{231} age estimates in general are not in serious error.

In Table 5, except for core 118K which is located about 6km south of the hot-brine area, the cores are located in or very near the Atlantis II Deep. These cores comprise brine-derived materials throughout their entire lengths of 4 to 8m. Only core 127P reaches a layer of calcareous lutite at its bottom part (about 8m deep) which has been dated by C^{14} (Ku *et al.*, 1969) at about 12,000 years B.P. Our data show that the ages of the sampled geothermal deposits from the Atlantis II Deep are all younger than 10,000 years. If core 127P represents an uninterrupted sedimentary record, then the 10,000 year age should approximate the onset of the most recent phase of the brine activity in the Atlantis II Deep. The Pa^{231} age of 18,600 years in core 118K is checked out by C^{14} dates. As pointed out by Ku *et al.* (1969), it may record an older brine extrusion event in the area.

The sedimentation rates can be estimated as follows. We may use the apparent age (Pa^{231} ages, wherever available) of the deepest measured section in each core and obtain a minimum average rate down to that depth. We see that the *minimum* value is about 40cm/10^3 yr. If we choose to divide

Table 5   Apparent Ages Derived from $\frac{Th^{230}}{U^{234}}$ and $\frac{Pa^{231}}{U^{235}}$ Ratios for the Iron-Rich Sediments (Type II)

Core No.	Depth (cm)	$\frac{Th^{230}}{U^{234}}$	Th^{230} Age (years)	$\frac{Pa^{231}}{U^{235}}$	Pa^{231} Age (years)
84K	120–125	.025 ± .002	2,700 ± 200	.037 ± .003	1,900 ± 200
	350–355	.065 ± .004	7,300 ± 500	.087 ± .010	4,600 ± 500
85K	0–10	.039 ± .002	4,400 ± 200	.051 ± .007	2,600 ± 400
	30–35	.046 ± .002	5,100 ± 300	.060 ± .004	3,100 ± 200
	90–95	.060 ± .002	6,700 ± 300	.063 ± .004	3,200 ± 200
	290–295	.061 ± .003	6,800 ± 400	.100 ± .006	5,200 ± 300
	320–325	.065 ± .002	7,300 ± 300	.139 ± .010	7,400 ± 500
118K *	90–95	.126 ± .005	14,600 ± 600	–	
	250±255	.176 ± .009	21,000 ± 1,200	.314 ± .010	18,600 ± 700
	300±305	.196 ± .009	23,600 ± 1,300	–	–
120K	170±175	.030 ± .004	3,300 ± 500	.034 ± .003	1,700 ± 200
	270±275	.058 ± .003	6,500 ± 400	.084 ± .005	4,400 ± 300
	352±357	.090 ± .006	10,200 ± 800	–	–
127P *	80–85	.030 ± .002	3,300 ± 200	–	–
	559–565	.096 ± .003	10,900 ± 400	.150 ± .010	8,000 ± 600
128P	130–135	.050 ± .004	5,500 ± 500	–	–
	440–445	.041 ± .002	4,500 ± 300	–	–
	740–745	.060 ± .003	6,700 ± 400	.105 ± .007	5,500 ± 400
161K	0–5	.025 ± .002	2,700 ± 300	–	–
	180–185	.031 ± .002	3,400 ± 300	–	–
	360–365	.042 ± .002	4,600 ± 300	–	–

* C^{14} ages (Ku et al., 1969):

118K	120–125cm:	8,350 ± 200 yrs.
	210–215cm:	17,300 ± 400 yrs.
	310–315cm:	20,000 ± 600 yrs.
127P	783–790cm:	12,390 ± 180 yrs.

the depth difference of any two measured sections by their age difference, we see that rates over 100cm/10^3 yr are common. Variations in the sedimentation rates both in space and time are expected, as they may well be a function of local bottom topography, distance from the locale of brine discharge, intermittent nature of the brine activities, and so forth. It should be mentioned that the lithological correlations of these cores (see Fig. 2 of Bischoff, 1969) further demonstrate such variations. Goll (1969) also finds the occurrence of Radiolaria in core 128P to be at a considerably greater depth than in core 127P.

As a concluding remark, I may say that since the errors due to the initial Th^{230} and Pa^{231} contaminations become less important in samples of older ages, the use of these nuclides should hold promise for the geochronological studies of the *older* geothermal brine events.

Some Results on Ra^{226} and Po^{210}

Ra^{226} and Pb^{210} have been noted to show enrichment in certain thermal waters, petroleum brines, and volcanic effusions (Begemann et al., 1954; Cherdyntsev, 1955). I made a few analyses of Ra^{226} and Po^{210} in top portions of two cores. The results are summarized in Table 6.

Po^{210} is seen essentially in secular equilibrium with Ra^{226}. Due to the short half-lives of Pb^{210} (22 years) and Po^{210} (138 days), ages of the samples would have

Table 6 Ra^{226} and Po^{210} Results

Core No.	Depth (cm)	$\frac{Ra^{226}}{Th^{230}}$	$\frac{Po^{210}}{Ra^{226}}$	$\frac{Ra^{226}}{U^{238}}$
81G	0–5	11.8 ± 2.8	0.9 ± .1	9.7 ± 1.2
	30–35	3.8 ± .8	1.4 ± .2	4.4 ± .5
85K	0–10	5.2 ± .8	1.5 ± .3	0.23 ± .03
	90–95	2.5 ± .4	1.2 ± .2	0.17 ± .02

presumably allowed initial disequilibrium (if any) among the Ra^{226}-Pb^{210}-Po^{210} chain to disappear. The data may also suggest that extensive post-depositional migrations of these elements and radon leakage on a time scale of tens of to a hundred years are absent.

Excess Ra^{226} over Th^{230} is noticeable. A large portion of the Ra^{226} must have been derived from the subterranean brine. However, Ra^{226} is deficient with respect to U^{238} in the Atlantis II Deep core 85K. Does this imply that, while radium may be present in notable quantities in the brine upon exhalation of the brine into the Atlantis II Deep, it escapes the rather immediate precipitations as iron and uranium do? It should be interesting to make radium measurements in the brines as well as in the Red Sea water in the future.

Appendix

To explain why the apparent Th^{230}/U^{234} ages are expected to be greater than the Pa^{231}/U^{234} ages, let us assume that the analyzed samples contain in addition Pa^{231} and Th^{230} derived from materials of old age. Assuming also that we are dealing with a chemically closed system, then the Pa^{231} and Th^{230} growths can be represented schematically by the graphs of Fig. 3.

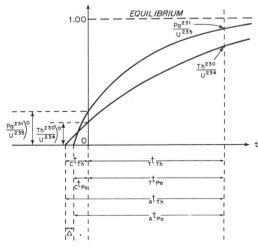

Fig. 3.

Here the notations are:

$\left.\dfrac{Th^{230}}{U^{234}}\right)^0, \left.\dfrac{Pa^{231}}{U^{235}}\right)^0$ = initial activity ratios in freshly deposited sediments.

$\dfrac{Th^{230}}{U^{234}}, \dfrac{Pa^{231}}{U^{235}}$ = measured activity ratios.

$A^t Th, A^t Pa$ = apparent ages derived from the measured ratios of $\dfrac{Th^{230}}{U^{234}}$ and $\dfrac{Pa^{231}}{U^{235}}$, respectively (assuming zero initial ratios)

$T^t Th, T^t Pa$ = true ages

$C^t Th, C^t Pa$ = age corrections due to contamination of the initially present Th^{230} and Pa^{231}, respectively.

$$\Delta = A^t Th \text{-} A^t Pa$$
$$\equiv C^t Th \text{-} C^t Pa$$

Let us further define:

$^\lambda Th$ = decay constant of Th^{230}
$$= \frac{0.693}{75,200} \text{ yr}^{-1}$$

$^\lambda Pa$ = decay constant of Pa^{231}
$$= \frac{0.693}{34,300} \text{ yr}^{-1}$$

$$r = \frac{^\lambda Pa}{^\lambda Th} = 2.19$$

$$S = \left.\frac{Th^{230}}{U^{234}}\right)^0 \bigg/ \left.\frac{Pa^{231}}{U^{235}}\right)^0 = \left.\frac{Th^{230}}{Pa^{231}}\right)^0 \bigg/ \left.\frac{U^{234}}{U^{235}}\right)^0$$

$$= \frac{21.8}{25} = 0.87$$

(We assume $\dfrac{U^{234}}{U^{238}}$ ratio in the freshly formed deposits to be 1.15, hence $\left.\dfrac{U^{234}}{U^{238}}\right)^0 = 21.8 \times$ 1.15 = 25. A choice of $\dfrac{U^{234}}{U^{238}}$ values to be <1.15 will not alter our argument presented below.)

Now,

$$\left.\frac{Th^{230}}{U^{234}}\right)^0 = 1 - e^{-\lambda Th \cdot C^t Th}$$

and

$$\left(\frac{Pa^{231}}{U^{235}}\right)^0 = 1 - e^{-\lambda_{Pa} \cdot C^t Pa}$$

$$\therefore \quad \Delta = C^t Th - C^t Pa$$

$$= -\frac{1}{\lambda_{Th}} \ln\left[1 - \left(\frac{Th^{230}}{U^{234}}\right)^0\right] + \frac{1}{\lambda_{Pa}}$$

$$\ln\left[1 - \left(\frac{Pa^{231}}{U^{235}}\right)^0\right]$$

$$= \ln\left\{ \frac{\left[1 - \left(\frac{Pa^{231}}{U^{235}}\right)^0\right]^{\frac{1}{\lambda_{Pa}}}}{\left[1 - \left(\frac{Th^{230}}{U^{234}}\right)^0\right]^{\frac{1}{\lambda_{Th}}}} \right\}$$

$$= \ln\left\{ \frac{\left[1 - \left(\frac{Pa^{231}}{U^{235}}\right)^0\right]}{\left[1 - S\left(\frac{Pa^{231}}{U^{235}}\right)^0\right]^r} \right\}^{\frac{1}{\lambda_{Pa}}} \qquad (1)$$

or:

$$\Delta = \ln\left\{ \frac{\left[1 - S\left(\frac{Th^{230}}{U^{234}}\right)^0\right]^{\frac{1}{r}}}{\left[1 - \left(\frac{Th^{230}}{U^{234}}\right)^0\right]} \right\}^{\frac{1}{\lambda_{Th}}} \qquad (2)$$

It is readily shown by substituting the numerical values of r and S into (1) that for

$$\left(\frac{Pa^{231}}{U^{235}}\right)^0 < 0.9 \left[i.e., \left(\frac{Th^{230}}{U^{234}}\right)^0 < 0.8\right]$$

$$\left[1 - \left(\frac{Pa^{231}}{U^{235}}\right)^0\right] > \left[1 - S\left(\frac{Pa^{231}}{U^{235}}\right)^0\right]^r$$

i.e., $\Delta > 0$, or $A^t Th > A^t Pa$.

Since measured values for $\frac{Pa^{231}}{U^{235}}$ are all much less than 0.9, the apparent $\frac{Th^{230}}{U^{234}}$ ages should all be older than the $\frac{Pa^{231}}{U^{235}}$ ages.

Let us go a step further. If this model is valid, it would enable us to make corrections for the initially present Th^{230} and Pa^{231} contaminations from the apparent age differences Δ, thus obtaining true ages for the samples.

Δ can be rewritten in the forms of

$$\Delta = C^t Th + \frac{1}{r\lambda_{Th}} \ln\left[1 - \frac{1}{S}\left(\frac{Th^{230}}{U^{234}}\right)^0\right]$$

and

$$\Delta = -\frac{r}{\lambda_{Pa}} \ln\left[1 - S\left(\frac{Pa^{231}}{U^{235}}\right)^0\right] - C^t Pa$$

i.e.,

$$C^t Th = \Delta - \frac{1}{r\lambda_{Th}} \ln\left[1 - \frac{1}{S}\left(\frac{Th^{230}}{U^{234}}\right)^0\right]$$

$$C^t Pa = -\Delta - \frac{r}{\lambda_{Pa}} \ln\left[1 - S\left(\frac{Pa^{231}}{U^{235}}\right)^0\right]$$

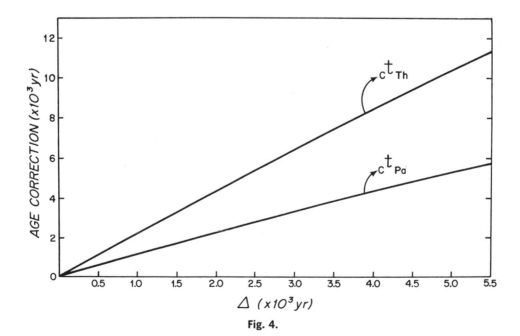

Fig. 4.

By substituting $\dfrac{Th^{230}}{U^{234}}\Big)^{0}$ and $\dfrac{Pa^{231}}{U^{235}}\Big)^{0}$ expressed in terms of Δ (from eqs. (1) and (2)) into the last two equations, we obtain C^tTh and C^tPa as functions of Δ alone. This is plotted in Fig. 4 for values of Δ up to 5,500 years.

Unfortunately, the large uncertainties involved for the data as given in Table 5 of the main text do not permit us to verify the validity of such an approach. A definite test can be made if samples of older age (hence the correction terms, C^tTh and C^tPa, are relatively small compared to the sample ages) are available, or if the analytical errors in the determination of Th^{230}/U^{234} and Pa^{231}/U^{235} are reduced. The suggested model may also be applicable to dating of marine carbonates, in which a certain amount of detrital silicate minerals is sometimes incorporated.

Acknowledgments

I am indebted to Dr. How-Kin Wong for the many fruitful discussions, to Dr. J. L. Bischoff and Mr. J. C. Hathaway for advice on X-ray analyses and size separations and to Mrs. Charlotte Lawson and Miss Barbara Gotthardt for laboratory assistance.

I also thank Dr. A. Kaufman of Lamont Geological Observatory for instruction on the Po^{210} analyses and for providing the brine sample from the A II Deep.

This work was supported by National Science Foundation Grant No. GA-584.

References

Adams, J. A. S., J. K. Osmond, and J. J. W. Rodgers: The geochemistry of uranium and thorium. *In: Physics and Chemistry of the Earth,* Pergamon Press, London, **3,** 298 (1959).

Begemann, F., J. Geiss, F. G. Houtermans, and W. Buser: Isotopenzusammensetzung und Radioaktivität von rezentem Vesurblei. Nuovo Cimento, **11,** 663 (1954).

Bischoff, J.: Red Sea geothermal brine deposits: their mineralogy, chemistry, and genesis. *In: Hot brines and recent heavy metal deposits in the Red Sea,* E. T. Degens and D. A. Ross (eds.). Springer-Verlag New York Inc., 368–401 (1969).

Brewer, P. G. and D. W. Spencer: A note on the chemical composition of the Red Sea brines. *In:*

Hot brines and recent heavy metal deposits in the Red Sea, E. T. Degens and D. A. Ross (eds.). Springer-Verlag New York Inc., 174–179 (1969).

Broecker, W. S.: A preliminary evaluation of uranium series inequilibrium as a tool for absolute age measurement on marine carbonates. J. Geophys. Res., **68,** 2817 (1963).

Cherdyntsev, V. V.: On the isotopic constitution of radioelements in natural objects with reference to problems of geochronology. Trudy III Sessii Komissii Opredelen. Absolyut. Vozrasta Geol. Formatsii, **1954,** 175 (1955).

Craig, H.: Geochemistry and origin of the Red Sea brine. *In: Hot brines and recent heavy metal deposits in the Red Sea,* E. T. Degens and D. A. Ross (eds.). Springer-Verlag New York Inc., 208–242 (1969).

Goll, R.: Radiolaria: the history of a brief invasion. *In: Hot brines and recent heavy metal deposits in the Red Sea,* E. T. Degens and D. A. Ross (eds.). Springer-Verlag New York Inc., 306–312 (1969).

Kazachevskii, I. V., V. V. Cherdyntsev, E. A. Kua'mina, L. D. Sulerzhitskii, V. F. Mochalova, and T. N. Kyuregyan: Isotopic composition of uranium and thorium in the supergene zone and in volcanic products. Geokhimiya, **11,** 1116 (1964).

Koide, M. and E. D. Goldberg: Uranium-234/uranium-238 ratios in sea water. *In: Progress in Oceanography,* M. Sears (ed.). Pergamon Press, New York, **3,** 177 (1965).

Ku, T. L.: An evaluation of the U^{234}/U^{238} method as a tool for dating pelagic sediments. J. Geophys. Res., **70,** 3457 (1965).

———: Protactinium-231 method of dating coral from Barbados Island. J. Geophys. Res., **73,** 2271 (1968).

———: and W. S. Broecker: Uranium, thorium, and protactinium in a manganese nodule. Earth and Planet. Sci. Letters, **2,** 317 (1967).

Ku, T. L., D. L. Thurber, and G. G. Mathieu: Radiocarbon chronology of Red Sea sediments. *In: Hot brines and recent heavy metal deposits in the Red Sea,* E. T. Degens and D. A. Ross (eds.). Springer-Verlag New York Inc., 348–359 (1969).

Miller, A. R., C. D. Densmore, E. T. Degens, J. C. Hathaway, F. T. Manheim, P. F. McFarlin, R. Pocklington, and A. Jokela: Hot brines and recent iron deposits in deeps of the Red Sea. Geochim. et Cosmochim. Acta, **30,** 341 (1966).

Moore, W. S. and W. M. Sackett. Uranium and thorium series inequilibrium in sea water. J. Geophys. Res., **69,** 5401 (1964).

Rona, E., L. O. Gilpatrick, and L. M. Jeffrey: Uranium determination in seawater. Trans. Amer. Geophys. Union, **37,** 697 (1956).

Ross, D. A. and E. T. Degens: Shipboard collection and preservation of sediment samples collected during CHAIN 61 from the Red Sea. *In: Hot brines and recent heavy metal deposits in the Red Sea,* E. T. Degens and D. A. Ross (eds.). Springer-Verlag New York Inc., 363–367 (1969).

Sackett, W. M., H. A. Potratz, and E. D. Goldberg: Thorium content of ocean water. Science, **128,** 204 (1958).

Tatsumoto, M. and E. D. Goldberg: Some aspects of the marine geochemistry of uranium. Geochim. et Cosmochim. Acta, **17,** 201 (1959).

Thurber, D. L.: Anomalous U^{234}/U^{238} in nature. J. Geophys. Res., **11,** 4518 (1962).

Torii, T. and S. Murata: Distribution of uranium in the Indian and the southern ocean waters. *In: Recent Researches in the Fields of Hydrosphere, Atmosphere, and Nuclear Geochemistry,* Y. Miyake and T. Koyama (eds.), Tokyo, Maruzen Co., 321 (1964).

Veeh, H. H.: U^{234}/U^{238} in the East Pacific sector of the Antarctic Ocean and in the Red Sea. Geochim. et Cosmochim. Acta, **32,** 117 (1968).

Comparison between Red Sea Deposits and Older Ironstone and Iron-Formation *

HAROLD L. JAMES

U.S. Geological Survey
Washington, D.C.

Abstract

The iron-rich sedimentary rocks of the geologic record are divided into two major types: ironstone, typically as oolitic limonite and chamosite in beds a few meters to a few tens of meters thick; iron-formation, which consists of alternating silica-rich and iron-rich layers and which occurs in units commonly hundreds of meters thick. Most, but not all, ironstone is post Precambrian in age, and most, but again not all, iron-formation is Precambrian. The great majority of ironstone and iron-formation deposits are the products of normal erosional and sedimentary processes, but a few are believed to be exhalative-sedimentary in origin, genetically related to seabottom hot springs and fumaroles. These latter deposits are marked by temporal and spatial association with volcanic rocks, rapid facies changes, and higher content of metals than that of "normal" ironstone and iron-formation.

Physically the Red Sea sediment is utterly unlike either ironstone or iron-formation, but chemically it has affinities to both except for its remarkable content of base and precious metals. If the Red Sea sediment were being deposited in an environment of strong current and wave action, the resulting product probably would be an ironstone of reasonably orthodox character. If it were to be buried in its present form, a remote possibility exists that it could become chemically differentiated through dewatering and diagenetic processes to yield an end product comparable to iron-formation, but more likely a substantial part of the iron and much or all of the base metals would migrate in hot brine solutions to be reprecipitated in laterally continuous carbonate host rocks.

Introduction

The discovery of iron-rich sediments in the Red Sea, and their relation to hot brines, raises the question as to their significance to the problem of older ironstones and iron-formations. It is worthwhile, therefore, to review the character and occurrence of these older deposits, the origin of which has remained speculative because of the absence of known counterparts in present environments, and to consider the implications of the Red Sea sediments.

Older Iron-Rich Sedimentary Rocks

Iron-rich sedimentary rocks have been defined as those containing 15 per cent or more Fe of depositional or diagenetic origin. Two major groups are recognized: *ironstones,* which are noncherty and mostly of Paleozoic or younger age; *iron-formations,* which typically are laminated with chert and are largely of Precambrian age (James, 1966). This nomenclature, which has arisen entirely from custom rather than prior definition, is utilized because it provides ready distinction between the quite different rock groups.

Of lesser quantitative importance among the iron-rich sedimentary rocks are the "blackband" (or "clayband") siderites

* Publication authorized by the Director, U.S. Geological Survey.

found particularly in coal-bearing strata, and the "bog ores," which are deposits of concretionary iron oxides found in swamps and small lakes in northern latitudes.

Both ironstone and iron-formation have a wide range in physical and chemical characteristics. Mineralogically they may be of oxide, silicate, carbonate, or sulfide facies, depending upon the dominant iron mineral (goethite, hematite, magnetite; chamosite, glauconite, greenalite, minnesotaite; siderite; pyrite). In general, the mineralogy of iron reflects differences in either sedimentary or diagenetic environment of origin, with the oxidation potential (Eh) as the dominant factor.

Ironstone

The typical ironstone consists of oolitic iron oxide (goethite or hematite) or chamosite in a matrix of calcite, chamosite or siderite. Commonly, the oolite contains some clastic material, and the ooliths show abundant evidence of reworking by current and wave action.

Ironstone, in beds as much as several tens of meters thick, is of fairly common occurrence in strata of lower and middle Paleozoic age in many parts of the northern hemisphere and is a particularly distinctive component of Jurassic strata in northern Europe, where it is mined as iron ore ("minette" of France, Belgium, and Luxembourg). The youngest beds of significant thickness are those of the Kerch Basin in Russia; these were deposited in an embayment of the Black Sea in Pliocene time (Markevich, 1960).

Iron-Formation

Chert, in layers less than a millimeter to a few tens of centimeters thick, is a distinguishing characteristic of iron-formation. The chert is interlayered with hematite, magnetite, iron silicates, siderite, or pyrite in laminae or layers of equivalent thickness.

Iron-formation is largely confined to strata of Precambrian age, in which it forms units as much as 600m thick. The major iron-formations of the world for which adequate geochronologic data are available all have an age of approximately 2 billion years (Lake Superior and Labrador in North America, Krivoi Rog and Kursk in Russia, Transvaal System of South Africa, Brockman Formation of western Australia). Iron-formation of lower Paleozoic age has been reported in the Snake River area of the Yukon and Northwest Territories, Canada, in the Lesser Khingan range of Manchuria and in a few other places, but the age assignments have not yet been fully verified.

Origin

Many aspects of origin of the iron-rich sedimentary rocks are fairly well understood. The rocks accumulated as chemical sediments in marine or brackish water basins of small to moderate extent—rarely greater than 100 miles in maximum dimension for ironstone, perhaps as much as 1,000 miles for iron-formation. The presence in most units of reduced facies, such as siderite, further suggests that the basins of deposition were of restricted circulation, with the bottom environment at least locally being depleted of oxygen. Recognizable clastic material, other than that locally derived from the iron sediments, is generally scarce or absent so that the deposits accumulated either during periods of structural quiescence or at sites well removed from the erosional products of uplift or structural disturbance. Diagenetic effects are evident in most rocks and in some are profound. These consist of formation of new minerals as replacements of primary materials, in consequence of the lower Eh of the burial environment, and of progressive changes in the mineralogy of oxides, silicates and sulfides.

The major unsolved problem is the source of the iron (and, for iron-formation, silica). In general, hypotheses fall into three general categories:

1. Iron is derived by deep chemical weathering of land masses (Taylor, 1949; James, 1954, 1966; Hough, 1958; Woolnough, 1941; and many others).

2. Iron is mobilized from iron-bearing clastic materials by sea-bottom reactions, transported in solution and selectively precipitated (Strakhov, 1959; Borchert, 1960).

3. Iron is contributed in solution by processes related to submarine or subbottom volcanic and igneous activity (Van Hise and Leith, 1911; Goodwin, 1956; Oftedahl, 1958).

For the latter process, Oftedahl has applied the general term "exhalative-sedimentary."

Most of the major ironstone and iron-formation deposits of the world occur in rock sequences in which evidence of contemporaneous volcanic or igneous activity is either wholly absent or very sparse. For a few, however, igneous processes can be shown to have been active during the general period of iron deposition, and a genetic relation can be reasonably inferred. The process of "exhalative-sedimentary" deposition can be summed up briefly as follows: magma rises to a relatively high level in the crust beneath the sea bottom, with or without extrusion of submarine lavas. During the emplacement and cooling process of the magma, hot, metal-bearing fluids are released, which rise along fissures to mingle with the bottom waters. Precipitation results from the cooling and mixing, with the materials precipitated being drawn in part from magmatic contributions, in part from subbottom strata altered by those solutions and in part from seawater. The following are examples of older iron-rich sedimentary rocks believed to have been produced by exhalative-sedimentary processes:

1. Iron-formation of Precambrian age in the Michipicoten area, Ontario (Goodwin, 1962).

2. Iron and manganese deposits of "Karadzhal-type," of Devonian age, central Kazakhstan, USSR (Maksimov, 1960).

3. Ironstone of Devonian age in the Lahn-Dill district, Germany (Borchert, 1960).

The process also has been invoked for many other types of metalliferous deposits (Oftedahl, 1958). A particularly strong case has been made by Ehrenberg *et al.* (1954) for the bedded pyrite-sphalerite-galena-barite ores of Meggen, Germany.

Red Sea Deposits

The occurrence, general character and chemical composition of the Red Sea deposits are summarized below in order to provide ready reference for comparison with older iron-rich sedimentary rocks. The deposits are bottom sediments in two or more brine pools centrally located in the median trough of the Red Sea. The brines, which extend about 150m above the bottom, have a range in temperature from about 45°C in the Discovery Deep to about 56°C in the Atlantis II Deep (Miller *et al.*, 1966). The iron-rich sediments are confined to bottom areas overlain by hot brine. The material is a well-layered, dark, gel-like ooze with a large amount of interstitial brine. Mineralogically, the material consists dominantly of amorphous to poorly crystallized iron oxide, amorphous silica and montmorillonite. It also contains variable to large amounts of Zn (as sphalerite), Cu and other metals. Four complete analyses of washed and airdried bottom material are presented in Table 1. The thickness and areal extent of the deposits are not yet completely established, but a preliminary estimate by F. T. Manheim (written communication April 27, 1967) is that the Atlantis II deposit covers an area of 53km² with an average thickness of at least 10m.

The median valley of the Red Sea, in which the brine pools and associated sediment occur, almost certainly reflects one or more faults of the African rift system. The strata beneath and adjacent to the trough and hot brine pools consist of coral limestone and unconsolidated sediments that overlie a sequence of carbonate and evaporite beds (Drake and Girdler, 1964, p. 489). The source of heat is assumed to be subjacent bodies of igneous rock. Origin of the brine is not known for certain, but a reasonable hypothesis is that its salt content is derived by circulation through the underlying or adjacent evaporite deposits.

Table 1 Chemical Analyses of 4 Samples of Bottom Sediment from the Atlantis II Deep, in Weight Per Cent

Major Elements	A	B	C	D
SiO_2	28.1	23.8	25.2	27.6
Al_2O_3	6.1	3.2	3.1	1.9
Fe_2O_3 (total iron)	18.6	43.8 [3]	17.0	36.5
Mn_3O_4	4.1	3.4	0.53	0.65
CaO	2.7	2.1	0.68	5.1
MgO	1.7	1.5	0.60	1.6
SrO	0.004	0.23	0.068	0.034
BaO	0.45	1.05	0.032	0.042
CuO [1]	3.7	0.35	3.5	0.58
ZnO [1]	11.2	0.19	21.2	2.1
PbO [1]	0.14	0.07	0.17	0.06
K_2O	0.89	0.55	0.58	0.61
Na_2O	1.4	1.2	1.8	2.8
P_2O_5	0.34	0.33	0.50	0.29
H_2O^-	(1.2)	(3.3)	(5.1)	(2.4)
Loss on ignition (1000°C)	20.8	18.2	24.1	20.0
Total [2]	100.2	100.0	99.1	99.8
Trace Elements				
Ti	0.02	0.02	0.02	0.04
Ni	0.039	0.025	0.033	0.014
Mo	0.025	0.02	0.067	0.007
Ag	0.017	0.0022	0.017	0.0051

[1] Cu, Zn, and Pb probably present as sulfides in original sample but converted to oxides in ignition.
[2] Excludes H_2O^- and trace element contents.
[3] Fe_2O_3 determined by difference (direct value 41.9).
Analyst, F. T. Manheim, U.S. Geological Survey.

Similar hot brine springs occur in nearby Ethiopia (Holwerda and Hutchinson, 1968), also in close association with structures related to the rift system and to evaporite-bearing strata. Here igneous activity is plainly evident, expressed at the surface as volcanic flows and ejecta of Quaternary age. No metalliferous deposits are reported, however, and the chemical data presented on brine composition do not contain any reference to metal content.

Comparative Chemistry

The average compositions of typical examples of the principal facies of ironstone and iron-formation are given in Table 2 together with the average of the four analyses of Red Sea sediment reduced to the same basis. The chemical differences between ironstone and iron-formation are readily apparent. Iron-formation is much higher in SiO_2, reflecting the presence of chert as an essential component, and much lower in Al_2O_3, alkalis and alkaline earths, TiO_2 and P_2O_5.

Table 2 Average Compositions of Representative Samples of Ironstone (A–D) and Iron-Formation (E–H), Compared to Red Sea Sediment (I)

(Weight per cent, recalculated to 100 per cent after subtraction of H_2O, CO_2, C, and base metals.)

	A	B	C	D	E	F	G	H	I
SiO_2	14.2	12.0	16.4	8.5	42.5	36.8	54.3	43.4	39.1
Al_2O_3	5.4	6.3	9.3	6.5	0.5	1.2	1.2	2.5	5.7
Fe_2O_3	49.2	50.1	14.0	2.1	52.5	36.2	5.3	4.3	43.2 [1]
FeO	11.9	11.5	35.6	70.0	4.2	22.1	32.5	39.8	—
MnO	0.5	0.3	—	0.3	<0.05	0.2	0.0	1.8	2.7
MgO	2.3	2.1	4.7	4.6	0.1	2.1	6.7	4.8	1.9
CaO	13.9	15.0	16.2	6.2	0.1	1.2	<0.05	2.3	3.7
$(Na,K)_2O$	—	0.2	0.4	0.9	0.1	0.1	0.0	0.4	3.3 [2]
TiO_2	0.2	0.3	0.7	0.3	<0.05	<0.05	—	0.2	<0.05
P_2O_5	2.3	1.8	2.1	0.4	<0.05	0.1	0.0	0.4	0.4
S	0.1	0.4	0.6	0.2	<0.05	<0.05	<0.05	0.1	nd
Total	100.0	100.0	100.0	100.0	100.0	100.0	100.0	100.0	100.0

[1] Total iron reported as Fe_2O_3.
[2] Partly occluded NaCl from brine not completely removed from sediment.
A – Goethitic ironstone (James, 1966, Table 8, A–G).
B – Hematitic ironstone (James, 1966, Table 9, A–D, G–I, N).
C – Chamositic ironstone (James, 1966, Table 13, E–G, L).
D – Sideritic ironstone (James, 1966, Table 16, A, D, E).
E – Hematitic iron-formation (James, 1966, Table 10, A–D).
F – Magnetitic iron-formation (James, 1966, Table 12, A–B).
G – Greenalite iron-formation (James, 1966, Table 17, A, D–G).
H – Sideritic iron-formation (James, 1966, Table 17, A, D–G).
I – Red Sea sediment. Average of 4 analyses given in Table 1.

The tabulation shows a curious similarity in contents of certain elements between Red Sea sediment and iron-formation, notably in silica and iron, but also in titania and phosphate. Alumina, on the other hand, is in the range of typical ironstone compositions. These relations between principal elements are expressed graphically in Fig. 1 as an ASF (Al + Si + Fe = 100 weight per cent) diagram.

Both ironstone and iron-formation are characteristically deficient in alkalis, values for which rarely are as much as one percent and commonly are virtually zero. The low contents of alkalis are in contrast to the average of 3.3 per cent in the Red Sea sediment, but the latter value includes an unknown amount of occluded NaCl from interstitial brine.

Contents of minor elements, including base metals, are very low in nearly all older iron-rich sedimentary rocks, particularly iron-formation (Landergren, 1948). Commonly they are less than crustal abundance (James, 1966).

Nevertheless there are exceptions, though somewhat difficult to document, particularly for ironstone and iron-formation that, on other criteria, have been classed as exhalative-sedimentary. In the iron-formation of the Michipicoten district of Ontario, concordant streaks of chalcopyrite and sphalerite are described by Good-

win (1962, p. 575). Goodwin (p. 576), drawing on earlier work, also notes that the pyritic member contains "persistent traces of nickel, arsenic, and gold." The Karadzhal-type iron and manganese ores of Kazakhstan, USSR, contain barite-rich lenses and locally pyrite, arsenopyrite, chalcopyrite, galena, sphalerite and tetrahedrite (Maksimov, 1960, p. 513). Hegemann and Albrecht (1954) show that at least some facies of the Lahn-Dill ironstone of Germany have a metal content significantly higher than normal sedimentary ironstone.

It should be noted that all of the deposits referred to above have undergone some metamorphism since deposition and diagenesis so that the present mineralogy and position of the metals may bear little relation to original character and distribution.

Speculations

If the Red Sea sediment were to be lithified and slightly metamorphosed without significant chemical modification, the resultant rock would consist of hematite, minor magnetite, quartz and mica, plus abundant sulfides of zinc, copper and lead. It would bear only superficial resemblance, physically or mineralogically, either to older iron-rich sedimentary rock or to any mineral deposit of orthodox type. The question may be asked, therefore, whether the deposit is in fact unique or of a type not present in the continental record, or whether it may be expected to undergo change in the burial environment that would convert it to a deposit of more familiar character.

Chemically the sediment has some similarities to both ironstone and iron-formation, provided the high content of base metals is disregarded. Could any process be envisaged that would result in reasonably normal ironstone or iron-formation, both of which are characterized by distinctive structure, mineralogy and low content of base metals?

Probably only few of the sedimentary accumulates that now form the stratigraphic sequence of geologic record are the result of single-stage processes; almost certainly

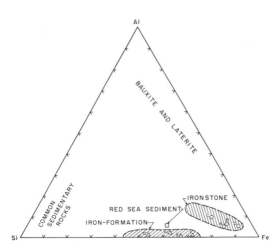

Fig. 1. ASF diagram showing compositional fields of typical ironstone (A–D) and iron-formation (E–G) compared with Red Sea sediments. Al + Si + Fe = 100 weight per cent.

most passed through at least several cycles of deposition and redistribution before becoming fixed in place. Evidence for this is particularly well preserved in some of the oolitic ironstones, in which individual ooliths are composed of chamosite, goethite, or alternating skins of chamosite and goethite, each of which reflects a different bottom environment, but which now coexist as a result of current sorting and transport from previous sites of deposition. If the Red Sea sediment were being precipitated in a shallower, more active environment or if the environment were to become more energetic, the sediment would undergo extensive reworking, probably with the formation of ooliths that in turn would be current-transported to an ultimate site of deposition. Furthermore, it is predictable that this process of physical reconstruction would also result in significant chemical differentiation. If the silica were in the form of clastic grains hydraulically equivalent to ooliths, it would continue to be associated with the iron, whereas if it were amorphous or in the form of very fine particles, it probably would be swept out of the system. Under conditions of continued working and reworking the present high content of metals might well be lost by resolution in sea water or be dispersed with and diluted by fine-grained, clastic material to a point where it was no longer a distinctive component.

The second set of conditions to appraise are those of the post-burial environment. If the Red Sea sediment were to be reworked as described in the previous paragraph, diagenetic processes would produce some marked changes in mineralogy, but the end product still would retain the distinctive aspects of ironstone. Let us consider what the effects might be if the material were to be buried in its present form.

The gross chemistry of the sediment, aside from content of metals (other than iron) and alkalis, is rudely similar to that of iron-formation. Are there any processes in diagenesis that might result in chemical differentiation of the principal components, that is, iron and silica, so as to produce a banded rock physically comparable to iron-formation? The answer probably is that there are not, yet the sensitivity of iron to environmental changes, particularly in Eh, raises some interesting possibilities. In a normal sea-bottom environment Eh and pH of the bottom waters are high, so that solubility of iron is virtually zero, and the stable precipitate is iron oxide. When the system becomes isolated from the oxygenated marine reservoir, Eh and pH can be expected to decrease progressively downward because of oxidation of trapped organic matter, which depletes the system of oxygen and contributes CO_2. In a normal diagenetic system, therefore, the solubility of iron increases markedly and interstitial waters will become enriched. As the overlying sedimentary load increases, these waters will move upward into environments of progressively higher Eh and pH in which dissolved iron again is precipitated. If the volume of interstitial water were very large and the process long continued, such a mechanism conceivably could result in a layered and chemically differentiated rock, by continuous transfer of iron to higher interfaces and residual accumulation of silica.

The discussion of the preceding paragraph focuses on the possible fate of the sediment if "normal" conditions prevailed. In fact, however, the prevailing physicochemical conditions doubtless depart radically from those of a normal burial environment. The metals that constitute much of the sediment can be assumed to have been transported in solution, via the medium of heated brine, to the sea bottom where they were precipitated as a result of temperature loss and reaction with sea water. Presumably, brines of the same character continue to move vertically and laterally through the deposits. Will these accumulated precipitates remain stable in the burial environment? This seems unlikely; indeed a fair possibility exists that at least part of the iron and much or all of the base metal content of the sediment eventually would return to solution and undergo further transport, either upward or laterally. The present deposit, therefore, may well be ephemeral in a geologic time sense.

The regional association of many "strata-bound" ores of lead, zinc and copper with evaporites has drawn considerable attention in the past few years (Davidson, 1966), and the fact that fluid inclusions in these ores are essentially chloride brines lends support to a concept of genetic relation. Jackson and Beales (1967) propose that lead-zinc ores of the Mississippi Valley type are derived by lateral migration of metal-bearing brines, with precipitation of the metals as sulfides taking place when the fluids encounter H_2S-bearing carbonate strata. The Red Sea environment can be readily fitted to this model: reef limestone known to be laterally continuous with the brine-charged, metal-rich ooze could be the ultimate trap; the resulting deposit would consist of disseminated pyrite, sphalerite, galena and other sulfides in a reef limestone or dolomite considerably removed from the initial site of accumulation of the metals.

Conclusions

Despite similarities in gross chemical composition on a water-free basis, the Red Sea sediment differs greatly from older, iron-rich sedimentary rocks in two ways: (1) structure and (2) high content of base and precious metals. Ironstones typically consist of oolitic goethite and chamosite, and iron-formation is laminated with chert. Both normally are markedly deficient in metals other than iron. We can conclude, therefore, that the Red Sea sediment cannot be considered a modern unlithified equivalent of either ironstone or iron-formation. Furthermore, if lithified without significant change in composition, it would bear little resemblance to any known base metal deposit of orthodox type.

The probability seems high that the existing deposit is likely to undergo extensive change before it finally is incorporated in the rock record. Three circumstances are considered:

1. *Change in physical environment.* If the relative sea level were to be lowered so that the deposit became subject to wave and current action,

the deposit would be dispersed and reworked. The base metals probably would be lost to the sea water, and the recognizable end product would be a thin layer of oolitic ironstone.

2. *Burial in a normal diagenetic environment.* If the present hot water plumbing system were to be shut off, so that the burial environment became more nearly normal, it is conceivable that profound chemical differentiation could take place during dewatering and diagenesis so as to yield a layered iron- and silica-rich material comparable to iron-formation.

3. *Burial under existing conditions.* The present deposits are precipitates resulting from interaction between emerging metal-charged hot brine and colder sea water. Upon burial, these precipitates will be subject to attack by solutions of the same character as those in which the metals were earlier transported. Although it is true that the base metal sulfides, which probably have drawn at least part of their sulfur from anaerobic sources near or at the sea water interface, will be only slightly soluble, nevertheless there will be a persistent tendency to return to solution. The same can be expected of the iron oxide precipitates, so that if active circulation of hot brines continues for a long period of time, the net result would be a argillaceous, siliceous residue from which a large part of the metals had been extracted.

It is this writer's view that the existing deposit is not in fact a progenitor of either ironstone or iron-formation, but that it may well be an initial stage in formation of base metal deposits of familiar type—that ultimately, through long-continued lateral movement of metal-charged brines, the deposit will be largely dispersed and the metals reprecipitated as sulfides in a favorable host rock some distance from the present site of deposition. The laterally continuous reef limestones of the Red Sea would provide an ideal trap. Perhaps a Mississippi Valley-type ore deposit is even now in the making.

References

Borchert, H.: Geosynklinale Lagerstätten, was dazu gehört und was nicht dazu gehört, sowie deren Beziehungen zu Geotektonik und Magmatismus. Freiberger Forschungshefte, H. C79, 8 (1960).

Davidson, C. F.: Some genetic relationships between ore deposits and evaporites. Inst. Mining and Metallurgy, Sec. B., Trans., 75, B-216–B-225 (1966).

Drake, C. L. and R. W. Girdler: A geophysical study of the Red Sea. Royal Astron. Soc. Geophys. Jour., 8, 473 (1964).

Ehrenberg, H., A. Pilger, and F. Schröder: Das Schwefelkies - Zinkblende - Schwerspatlager von Meggen (Westfalen). Geol. Jahrb. Beihefte, 12, 353 p. (1954).

Goodwin, A. M.: Facies relations in the Gunflint iron formation. Econ. Geology, 51, 565 (1956).

———: Structure, stratigraphy, and origin of iron-formations, Michipicoten area, Algoma district, Ontario. Geol. Soc. America Bull., 73, 561 (1962).

Hegemann, F. and F. Albrecht: Zur Geochemie oxydischer Eisenerze. Chemie Erde, 17, 81 (1954).

Holwerda, J. G. and R. W. Hutchinson: Potash-bearing evaporites in the Danakil area, Ethiopia. Econ. Geology, 63, 124 (1968).

Hough, J. L.: Fresh-water environment of deposition of Precambrian banded iron-formations. Jour. Sed. Petrology, 28, 414 (1958).

Jackson, S. A. and F. W. Beales: An aspect of sedimentary basin evolution: the concentration of Mississippi Valley-type ores during late stages of diagenesis. Can. Petrol. Geol. Bull., 15, 383 (1967).

James, H. L.: Sedimentary facies of iron formation. Econ. Geology, 49, 235 (1954).

———: Chemistry of the iron-rich sedimentary rocks. U.S. Geol. Survey, Prof. Paper, 440-W, 61 p. (1966).

Landergren, S. On the geochemistry of Swedish iron ores and associated rocks. Sveriges Geol. Undersökning Årsbok, 42, ser. C, 496, 182 p. (1948).

Maksimov, A. A.: Types of manganese and iron-manganese deposits in central Kazakhstan. Internat. Geology Rev., 2, 508 (1960).

Markevich, V. P.: The concept of facies, Part 1. Internat. Geology Rev., 2, 367 (1960).

Miller, A. R., C. D. Densmore, E. T. Degens, J. C. Hathaway, F. T. Manheim, P. F. McFarlin, R. Pocklington, and A. Jokela: Hot brines and recent iron deposits in deeps of the Red Sea. Geochim. et Cosmochim. Acta, 30, 341 (1966).

Oftedahl, C.: A theory of exhalative-sedimentary ores. Geol. fören. Stockholm Förh., 8, 1 (1958).

Strakhov, N. M.: Schema de la diagenese des depots marins. Eclogae Geol. Helvetiae, 51, 761 (1959).

Taylor, J. H.: Petrology of the Northampton sand ironstone formation. Great Britain Geol. Survey, Mem., 111 p. (1949).

Van Hise, C. R. and C. K. Leith: The geology of the Lake Superior region. U.S. Geol. Survey Mon., 52, 641 p. (1911).

Woolnough, W. G.: Origin of banded iron deposits — a suggestion. Econ. Geology, 36, 465 (1941).

Economic and
Legal Implications

Table 2 Partial Composition of Atlantis II and Discovery Deep Deposits

Numbers in parentheses are estimated standard deviation of core means. Au data is based on an assumed ratio of Ag:Au = 100, which frequently holds for sedimentary sulfide deposits. One sample of Red Sea material analyzed by the Washington Laboratories of the U.S. Geological Survey (Au by neutron activation) yielded an Ag:Au ratio of 80. All analyses are from Bischoff (1969a). Chemical components are based on brine-free, air-dry sediment.

	Atlantis II Deep			Discovery Deep
	"Goethite"	"Fe-Montmo-rillonite"	"Sulfide"	"Goethite-Detrital"
Fe_2O_3 (tot.)	49 (8.7)	35 (2)	33 (10)	20
Mn_3O_4 (tot.)	2.8 (2.1)	2 (2)	1.3 (.8)	1
ZnO	1.0 (0.9)	4.7 (3)	11.1 (4)	0.15
CuO	0.5 (0.5)	0.7 (0.4)	4.6 (3)	0.05
PbO	0.1 (0.1)	0.1 (0.1)	0.2 (0.2)	0.03
AgO	0.0033 (0.0015)	0.0062 (0.005)	0.013 (0.008)	<0.001
Au	0.00003	0.00006	0.00013	–
Percent brine	84 (4)	92 (2)	75 (15)	68
Area (m^2)	56×10^6	56×10^6	56×10^6	11×10^6
Thickness (m)	3.5	4	1	(10)
Volume (m^3)	196×10^6	224×10^6	56×10^6	110×10^6
Sp. G. sed. (wet)	1.33	1.26	1.39	1.49
Brine-free tonnage	42×10^6	22×10^6	19×10^6	52×10^6

Total Atlantis II Deep (brine-free) tonnage: 83×10^6.
Total Atlantis II Deep volume (m^3): 48×10^7.

solids.* For example, for the "Sulfide" layer:

Volume = area × thickness
$$= 56 \times 10^6 m^2 \times 1m = 56 \times 10^6 m^3$$

Sp. Gravity (natural sed.)

$$= \frac{1}{\text{volume brine} + \text{volume sediment}}$$

$$= \frac{1}{\frac{\text{wt. \% brine}}{\text{Sp. G. brine}} + \frac{\text{wt. \% sed.}}{\text{Sp. G. sed.}}}$$

$$= \frac{1}{\frac{75}{1.20} + \frac{25}{2.7}} = 1.39$$

Dry tonnage
$$= \text{Volume} \times \text{Sp. G. (nat.)} \times \text{wt. \% sed.}$$
$$= 56 \times 1.39 \times .25 = 19.4 \times 10^6$$

Values of the Atlantis II Deposits

Tonnages for individual metals in the 3 facies in the Atlantis II Deep may be calculated from the data in Table 2. Typical

* Determined by pycnometer on 10 representative samples.

United States smelter prices for 1967, adjusted for subsequent rises in prices for copper and silver, have been applied to these tonnages to give estimated metal values in Table 3. The limitations of this type of estimate are discussed later.

The main conclusion from the data in Table 3 is that the sulfide zone, although only 1m thick, contains more valuable metal than the other facies combined. In order of value, the metals are: Cu, Zn, Ag, Au and Pb. The value of other minor elements would probably add only slightly to the total. Semiquantitative and quantitative analyses have shown about 0.005 per cent tin, 0.03 per cent cadmium, 0.03 per cent cobalt and nickel, between 0.01 and 0.05 per cent arsenic, less than 0.005 per cent germanium, and less than 0.001 per cent bismuth, indium and mercury.

Fe and Mn are not usually recovered in heavy metal ores of this type since the separation costs are too great. Both of these metals are supplied to world markets by simple earth-moving processes. The ores are often, as is desired, quite pure.

The metal prices have been taken from Engineering and Mining Journal, Decem-

Table 3 Gross Value of Metals in Upper 10m of the Atlantis II Deep

Metal prices from *Engineering and Mining Journal,* December 1967. Value per ton brine-free material = \$30; per m³ in-place sediment = \$5.2.

	Price (dollars per ton)	"Goethite"		"Fe Silicate"		"Sulfide"		Totals		
		Tons	Value $\times 10^7$	Tons	Value $\times 10^7$	Tons	Value $\times 10^7$	Averaged Assay (%)	Tons	Value $\times 10^7$
Fe	—	143 $\times 10^5$	—	55 $\times 10^5$	—	45 $\times 10^5$	—	29	243 $\times 10^5$	—
Zn	300	3.3 $\times 10^5$	9.9	8.4 $\times 10^5$	25.2	16.9 $\times 10^5$	50.7	3.4	29 $\times 10^5$	86
Cu	1,200	2.1 $\times 10^5$	25.2	1.3 $\times 10^5$	15.6	7.2 $\times 10^5$	85.3	1.3	10.6 $\times 10^5$	127
Pb	300	0.3 $\times 10^5$	0.9	0.2 $\times 10^5$	0.6	0.3 $\times 10^5$	0.9	0.1	0.8 $\times 10^5$	2
Ag	62,400	12 $\times 10^2$	7.5	11.9 $\times 10^2$	7.4	21 $\times 10^2$	13.1	0.0054	45 $\times 10^2$	28
Au	1.2×10^6	12	1.3	11.9	1.4	21	2.3	0.5 ppm	45	5
Totals			44.8		50.2		152.3			248

ber, 1967. Prices may fluctuate considerably, as has been the case with the recent rise in silver price, following the U.S. Treasury move to discontinue silver support. We are also aware of the dangers of citing values based on total metal content in raw materials, which may be non-competitive for one or more reasons. The common example that ordinary shale or even soils would yield astronomic values for such metals as iron, aluminum and titanium make it clear that value estimates of this type are meaningful only if the deposits can be worked competitively with other ores. These conditions remain to be fulfilled for the Red Sea deposits. Since any other value estimate would be equally speculative at this stage, we have retained the smelter or refined price estimates, recognizing that many sulfide and heavy metal deposits yield a high percentage recovery of assayed metals.

Brooks (1966) has pointed out another pitfall in assaying new, large raw material supplies in evaluating Mero's (1965) discussion of sea floor manganese nodules. He noted that the tremendous potential input on the market of Co and Ni from nodules could hardly fail to bring prices down and, hence, render Mero's use of current prices too optimistic. The presently assayed metals in the Red Sea deposits, however, do not appear to reach such levels.

Deeper Deposits

Seismic reflection studies suggest that deposits in the brine deeps may, in places, extend to about 20m depth (Ross *et al.*, 1969) and in isolated places may extend to 100m. Brine-derived "goethite" has spread widely outside the source area in the Atlantis II Deep. Outlying sediments contain much lower, but still anomalous, concentrations of heavy metals. For example, in core 118K, approximately 6 miles beyond the edge of the Atlantis II Deep, "goethite" is found at a level at which Ku *et al.* (1969) determined a radio-carbon date of about 20,000 years. The deposits within the Atlantis II Deep at the bottom of the longest cores are no older than about 12,000 years (Ku *et al.,* 1969). Thus, the old "goethite" suggests a history of metalliferous brine deposition at least twice as long as represented by the length of the cores taken in Atlantis II Deep.

Near the bottom of core 127P detrital components were found abundantly. Their presence does not necessarily contradict the above lines of evidence, since elsewhere, such as core 128P, beds of detrital material are underlain by highly metalliferous deposits.

Finally, there is a possibility that local areas of the deeps through which brines have and are continuing to discharge may be recognizable by mineralogical and chemical features in the deposits. In particular, a "supergene" type of enrichment of the deposits may take place at such areas by replacement of the more electronegative elements such as iron and zinc in sulfides by copper and silver from the incoming brines. Such effects may help explain the locally higher Cu/Zn ratios in deposits near the presumed discharge zone, station 84K (Bischoff, 1969b).

Conclusion

The assayed sediments contain amounts and proportions of metals which would, no doubt, be economically exploitable on land. If means for their recovery and separation can be found, they represent an attractive prospect, especially since their full potential is still to be explored.

Acknowledgments

In addition to our colleagues in the Woods Hole Oceanographic Institution, and the collaborators in this volume, especially D. A. Ross, we wish to thank William Prinz, D. E. White and William Dempsey of the U.S. Geological Survey, J. W. Padan of the U.S. Bureau of Mines, G. Parks of Stanford University, T. Coleman Morton of International Geomarine Corporation and H. E. Harper of the Hecla Mining Company for useful discussion and various forms of assistance. The use of analytical data, especially semi-quantitative trace analyses, supplied by G. Sheary of the Bethlehem Steel Corporation and T. N. Walthier of the Bear Creek Mining Company is appreciated. The manuscript was improved by criticism from K. O. Emery and R. H. Meade. This work was supported in part by NSF Grant GA-584.

References

Bischoff, J. L.: Red Sea geothermal brine deposits: their mineralogy, chemistry, and genesis. *In: Hot brines and recent heavy metal deposits in the Red Sea,* E. T. Degens and D. A. Ross (eds.). Springer-Verlag New York Inc., 368–401 (1969a).

———: Goethite-Hematite stability relations with relevance to sea water and the Red Sea brine environment. *In: Hot brines and recent heavy metal deposits in the Red Sea,* E. T. Degens and D. A. Ross (eds.). Springer-Verlag New York Inc., 402–406 (1969b).

Brooks, D. B.: *Low Grade and Non-Conventional Sources of Manganese.* Johns Hopkins Press, Baltimore, 123 p. (1966).

Degens, E. T. and D. A. Ross: Hot brines and heavy metals in the Red Sea. Oceanus, **13,** 24 (1967).

Hunt, J. M., E. E. Hays, E. T. Degens, and D. A. Ross: Red Sea: detailed survey of hot brine areas, Science, **156,** 514 (1967).

Ku, T. L., D. L. Thurber, and G. Mathieu: Radiocarbon chronology of Red Sea sediments. *In: Hot brines and recent heavy metal deposits in the Red Sea,* E. T. Degens and D. A. Ross (eds.). Springer-Verlag New York Inc., 348–359 (1969).

Landergren, Sture, William Muld, and Rajandi Benita: Analytical methods. Landergren, Sture. On the Geochemistry of Deep Sea Sediments. Reports of the Swedish Deep Sea Expedition, **10,** Special Investigation No. 5, 148 (1964).

Manheim, F. T. In preparation (1968).

Mero, J. L.: *The Mineral Resources of the Sea.* Elsevier, Amsterdam, 312 p. (1965).

Miller, A. R., C. D. Densmore, E. T. Degens, J. C. Hathaway, F. T. Manheim, P. F. McFarlin, R. Pocklington, and A. Jokela: Hot brines and recent iron deposits in deeps of the Red Sea. Geochim. et Cosmochim. Acta, **30,** 341 (1966).

Ross, D. A., E. E. Hays, and F. C. Allstrom: Bathymetry and continuous seismic profiles of the hot brine region of the Red Sea. *In: Hot brines and recent heavy metal deposits in the Red Sea,* E. T. Degens and D. A. Ross (eds.). Springer-Verlag New York Inc., 82–97 (1969).

White, D. E.: Environments of generation of some base metal ore deposits. Econ. Geol. (In press, 1968).

Economic Significance of Minerals Deposited in the Red Sea Deeps

THOMAS N. WALTHIER

International Exploration, Occidental Minerals Corporation
Denver, Colorado

CLIFFORD E. SCHATZ

Bear Creek Mining Company
San Diego, California

Abstract

Chemical analyses of samples from cores recovered from the Red Sea show a high content of metals. Using current market prices for the metals, a high *in situ* gross value can be calculated. In making an economic analysis to determine whether a deposit represents an ore body, it is not normal industry practice to equate such a calculation with the economic worth of the deposit. An economic evaluation must necessarily consider only those constituents which hold reasonable expectation of commerical recovery: how the deposit can be mined and how the metals of interest can be extracted, as well as the capital and operating costs involved in the operation.

It would appear that the Red Sea oozes, as now known, are of more academic than economic value; however, the results of some assays are rich enough to warrant attention from the explorationist. They are of important theoretical value as a possible example of metal-rich sediments forming today which are akin to the important ancient sedimentary ores of iron, copper, lead and other metals.

Introduction

In this chapter we shall address ourselves to the potential commercial importance of the Red Sea oozes. For the oozes in the Red Sea Deeps to have commercial significance they must exist in large enough amounts and contain enough metal to be extracted and reduced into marketable products to justify developing a technology to exploit them. If projections of costs and revenues indicate profits, considerable capital must be invested to build the requisite facilities. What follows is a partial analysis of the critical problems that must be solved if this is to be accomplished.

Gross Deposit Value

What is the tonnage and grade of the mud? Is the average grade of the ooze attractive? If not, are portions of it rich enough, and, if so, how are the richer zones of ore distributed amidst the low grade waste? What tools and equipment are available to sample the ooze and outline the ore with enough precision to justify a sizable investment?

On land, ore bodies are usually sampled by drilling, prior to mining, with holes spaces 50 to 500 feet apart. Occasionally, some ore bodies are continuous and uniform enough that test holes 1,000 feet apart are adequate, but this is unusual. It may be shown that the oozes are so uniform that precision sampling is not necessary. Should early coring on widely spaced and poorly located sites indicate unusual continuity, the Red Sea mining investor might be willing to take a big gamble that the indicated continuity was a fact. Dry land ore deposits give ample evidence and experience that this is a very dangerous assumption.

The discussion on the value of the Red Sea muds is based upon "incomplete" analytical data (Hendricks *et al.,* 1969). The cores have not been correctly sampled for economic appraisal by splitting the cores so as to yield grade figures for ore reserve calculations. In this case, the limited supply of cores necessitated samples being distributed among a large number of research workers. Consequently, our knowledge of the average and total metal content of those muds comes from samples of the various lithologies cored.

Many examples can be drawn from the lithologic and assay logs to support the premise that the muds in the Red Sea Deeps have considerable lateral variability, at least when measured over distances of 1,000 to 2,000 feet.

Core 85K contains the highest metal contents found (copper, zinc, gold and silver). Elsewhere, a few other cores have high values, but they do not seem as continuously developed as at station 85K. This relationship, of course, could be due to the incomplete assaying of all the cores. Moreover, all assays may be misleading because they are reported on a dry weight basis, whereas the mud *in situ* contains appreciable water which adds considerably to the weight and also to the volume that the ooze occupies.*

The uppermost sediment zones are commonly 60–70 per cent water. Analyses conducted on board ship, immediately following recovery of the samples, indicated that the moisture content approached 80 per cent by weight. Later, Kennecott Exploration Services Laboratories in Salt Lake City measured the moisture and the range was

from 25 per cent to 80 per cent. The average was 55 per cent for all samples assayed. In the Atlantis II Deep, the measured water content of all samples from the top 200cm was 60 per cent; from 200–400cm, the water content was 50 per cent. Moreover, many of the drier samples included in this average were those which on shipboard had the interstitial water squeezed out for other studies. In spite of the precautions taken, there probably was some water loss between the time the samples were collected and when analyzed in Salt Lake City several months later. Therefore, the average Kennecott Exploration Services Laboratories water content values are probably too low. Samples collected in 1965 (Miller, *et al.,* 1966) were reported as containing 90 percent brine.

As a consequence of the water problem, the average *in situ* grade of the cores would be considerably less than the assays shown. If, for example, Core 85K contained 60 per cent water and the change in weight and volume were proportional to water content, which is generally true except for the carbonate-rich zones, the grade of the mud that would have to be handled could be calculated as follows:

Core 85K 0–175cm

	Dry Weight Assay *	*In Situ* Grade at 60% Water
Copper	1.1%	0.45%
Zinc	2.3%	0.90%
Silver/ton	7.25 oz	2.90 oz
Gold/ton	0.05 oz	0.02 oz

* Includes all samples collected by Schatz and Peterson and samples analyzed by J. L. Bischoff (personal communication).

Core 85K 0–400cm

	Dry Weight Assay *	*In Situ* Grade at 60% Water
Copper	1.2%	0.48%
Zinc	2.7%	1.08%
Silver/ton	1.75 oz	0.70 oz
Gold/ton	0.035 oz	0.014 oz

* Includes all samples collected by Schatz and Peterson and samples analyzed by J. L. Bischoff (personal communication).

* Editorial Note: The results of Walthier and Schatz differ considerably from those of Bischoff and Manheim (1969). Eliminating analytical errors, two main reasons exist for these differences:

1. Bischoff and Manheim (1969) performed their analyses on brine-free sediment, while Walthier and Schatz analyzed sediments that still contained their interstitial brine. Since the brines may be as much as 90 per cent of the total ooze, by weight, and have a salinity of 25 per cent, only ⅓ of the analyzed material of Walthier and Schatz (under the above conditions) is sediment, the rest mostly sodium chloride.

2. The analyses of Walthier and Schatz are confined mostly to the upper 4m of sediment while those of Bischoff and Manheim are over a 9-m interval.

Thus, all assays reported in this chapter for the oozes (unless specifically quoted on a wet basis) should be adjusted downward to obtain the wet, *in situ* grade, by the formula:

In situ grade = dry assay grade
$$(1.00 - X) \text{ where } X = \% \text{ water}$$

Not only does the water content have a profound effect on the grade in place, but it also affects the total tonnage of metal-bearing solids that are available for processing.

There has been serious argument whether it is correct to adjust the grade down because of the water content. Opposing views can be briefly summarized as follows:

PRO: 1. The water constitutes a large percentage of the natural material and causes the bulk of the mud to be greatly increased over the dried samples. Therefore, the volume of mineral matter available for exploitation is seriously misrepresented as well as the grade of the material, unless the water is considered as one of the gangue minerals and part of the ore.

2. The muds are more like saline brines or the dilute copper solutions draining copper mine drumps than they are like solid rock ores. The metal content of brines and the mine dump waters always takes the water content into account, and values are usually expressed as percent or ppm of solution.

CON: 1. The metals in the Red Sea oozes are NOT dissolved but occur as particulate matter. It may be easy by simple, physical means to remove most of the water. Therefore, the metal content of the solid portion is the significant factor. The analogy to brines and dump waters is a poor one. It is also incorrect to equate the water in the muds with a quartz-feldspar gangue in normal ore.

2. While conventional ores are reported on a dry weight basis, once they are mined they are mixed with water for the milling process. The slurry coming out of a ball mill ready for mineral concentration may be similar in grade and physical properties to the Red Sea muds;

hence, it is important to know what the dry weight assay is, just as the assay for normal mill heads is reported on a dry weight basis.

Rather than try to adjudicate this argument, we wish simply to point out that the water in the muds adds significantly to the weight and to the volume occupied by the oozes in place. This fact must be kept in mind when calculating tonnages and volumes of material to be moved and processed, as well as the effect of this on the value of the contained metals per unit of volume or ton handled.

What is the potential tonnage of metal-rich oozes in the Atlantis II Deep? The area is roughly 6.12×10^8 square feet. The volume is about 4×10^{10} cubic feet, assuming a depth of 20 to 30m. Assuming an ooze density about 1.6 (or 100 lbs/ft³), then 2 billion tons are indicated.

What is the potential grade of this nearly 2 billion tons of material? Taking every assay (55) made at the Kennecott Laboratories from the cores within the area (Cores 71, 78, 84, 94C, 95K, 97, 106, 107, 108, 109, 121 and 126) and giving them equal weight, the average dry weight assays are:

Iron	12.00%
Copper	0.19%
Zinc	0.93%
Silver	0.60 oz/ton
Gold	0.01 oz/ton

Bischoff (1969) selected other samples from the cores in this deep for assay. An average of 34 assays show the Atlantis II muds contain 0.74 per cent Cu and 2.1 per cent zinc. Assuming an average of 50 per cent water and 60 feet of sediment, the average grade, using all assays *in situ,* is:

Copper	0.2%
Zinc	0.7%
Silver *	0.3 oz/ton
Gold *	0.005 oz/ton

* Based on the analytical results obtained at Kennecott Exploration Services Laboratories only.

At current prices (38 cents/lb Cu, 13.5 cents/lb Zn, $1.86/oz Ag, and $35/oz Au), each ton in place contains approximately $4.15 of these metals, which amounts to 8 billion dollars in the Atlantis II Deep. Whether the Atlantis II Deep holds metals worth nearly 8 billion dollars is another question. These metals have no value unless they can be recovered and enter the market place profitably.

These average grades from Atlantis II Deep are far below those of the ores being exploited today on land. The Chain and Discovery Deeps have oozes of comparable grade. Thin zones, it is true, can be identified of higher grade although even most of these grades are not very exciting when adjusted for water content. Moreover, coupled with being thin, these zones are buried beneath lower grade material, and there is evidence to suggest they will be erratic in lateral extent. Other metals are present and might be recovered, but the four we have been discussing, in our opinion, have the main commercial potential.

An exception to the low average grades is the assays obtained for core 85K shown below. These grades, over a minimum thickness of ten feet, are comparable to the grades of some ores being mined today. Certainly, one big question is whether these assays are truly representative. Using the same metal price quotations as above and assuming the assays shown are representative, then the average value of the mud in core 85K is approximately as follows:

	Dry Weight	*In Situ* Grade at 60% Water
0–175cm	$30.00/ton	$12.00/ton
0–400cm	$20.00/ton	$8.00/ton

However, the values in 85K may be isolated and unique and of no tonnage significance. On the other hand, 85K may be representative of the sill between the Atlantis II and Chain Deeps. If we assume continuity over 1 square kilometer and thickness is ten feet, then 5 million tons of ooze would be present. If it is worth $8.00/ ton, as calculated above, then the gross value in place is $40,000,000.

Gold was detected in 15 samples by spectrographic analysis. Average content of four composite samples, measured by a chemical pre-concentration-spectrographic method, was 0.27ppm. The average gold content found in all samples in the Kennecott Laboratories, determined by atomic absorption, was 0.48ppm.

Mining

The ooze is very fluid in its natural state, and a little agitation should fluidize the material readily. The fluid ooze could then be pumped up a pipe much as crude oil is raised to the surface. The 6,000-foot depth should not be prohibitive. The ooze will be slightly heavier than oil and pumping charges higher, but the cost difference may be slight. In fact, the hydrostatic pressure of the 6,000 feet of Red Sea water will maintain the level of the ooze in the pipe stem well above the sea floor without pumping. We envision higher maintenance costs in the ocean system and possibly the development of special pipe and joints. In contrast, the high cost of drilling a 6,000-foot oil well through rock would be eliminated and help to offset the other costs. The ship, or surface platform for the Red Sea mine, should be no more expensive than a modern offshore oil well platform, and it could be much less. As a first approximation, then, one might use the cost of lifting oil from 2,000–3,000 feet (without any important natural gas reservoir pressure) as the direct cost of mining the ooze.

With the above described mining system, however, we cannot visualize how a mining pit could be maintained in the sea floor. The upper layers would flow freely and immediately into the depression created by sucking up the first amount of ooze. This means that there is no place to dispose of waste ooze even if a pit slope could be maintained. The cost of transporting waste to a neighboring deep for disposal several kilometers away would pose a very serious cost item as well as raise engineering problems on how to do so efficiently. Because of the fluid nature of the material, it would also present extremely difficult problems to try to mine a richer zone, selectively, whether localized at the surface or buried at depth. The probability of considerable dilution of the richer zones by lower grade material will seriously complicate any attempt to mine richer zones.

For a venture of this type, we believe that only a rather sizeable operation could hope to be successful. Assuming the venture to be profitable, optimum size can only be determined after many detailed engineering studies, but let us consider what a facility handling 5,000 tons per day (1,500,000 tons per year) of dry ooze would entail. Upon our premise that the fluid character of the ooze would preferentially be mined, an average water content of 75 per cent does not seem excessive. Each ton of dry ooze equivalent would involve mining 3 tons of brine as well.

Using a wet ooze, specific gravity of about 1.6, 20 ft³ of ooze equals one ton, but this is hardly a half ton of dry sediment equivalent. Therefore, about 50 ft³ of ooze are needed for 1 short ton of dry sediment. To produce 5,000 tons per day requires handling 250,000 ft³ of ooze per day. An interesting question arises whether any reduction in volume is practical on board ship. The fine grained nature of the ooze means that this will be complicated.

The very fact that some of the ooze "settles rapidly" means that it is important that dewatering be done at sea. Ships are stable when carrying an all-liquid cargo or a totally solid one, but two phase bulk cargoes (solids dispersed in a liquid) could cause serious instability situations.

If it turns out that dewatering at sea is not feasible and the whole mass must be transported to port, then the entire wet ooze volume must be transported to a land base for recovery of the metal values. Because of the low unit value and high water content, transit distance and time to port should be a minimum. This identifies Saudi Arabia or the Sudan as the first, and perhaps only, places to be considered.

Three transport methods can be considered:

1. The mining ship could be basically a hopper dredge. When the hold becomes filled with ooze it carries its cargo to port. To be efficient it should mine for a number of consecutive days. With one round trip per week and 5 to 6 days of mining the hold must be able to store from 1,250,000 ft³ to 1,500,000 ft³ of liquid ore if dewatering were not accomplished. This means a large ship with commensurate port facilities.

2. The mining ship could load ooze into a fleet of barge-tankers which, in turn, would carry the ooze to port. Mass transfer of material on a routine basis from one ship to another in open water can be successful only if a slight sea state is to be counted on. While the Red Sea is generally calm and would simplify the operation, we point out this has always been a thorny problem in maritime logistics.

3. A pipeline from the mining ship to land is a possibility. Pipelines have been successfully laid in the shallow waters of the continental shelves and telephone cables are laid across the ocean depths, but no one has attempted construction of a pipeline in these kinds of depths. It would have to be nearly 60 miles long to reach the nearest coast.

The hot brines and oozes will undoubtedly be highly corrosive to pumps, pipes and tanks. Material used in these units will require special treatment to limit maintenance, replacement and down-time costs.

Processing

We can approach the discussion in two ways, based upon two different basic assumptions:

1. The ooze is so fine grained that simple physical methods cannot be used to yield a high grade concentrate.
2. The metal-rich constituents are sufficiently large, discrete particles so that conventional ore dressing techniques can be used to yield a high grade concentrate.

The high water content suggests that the water should be extracted first to reduce the volume of material to be stored, handled, and processed. This may have been accomplished already at the mining site using centrifuges, thickeners or cyclones.

It could be reduced further by these means at a land based mill. If no dewatering were done at sea, yet the milling process required a lower water content, the cheapest way to do this may be by solar evaporation. The climate of the countries bordering the Red Sea is favorable. In the Kennecott Laboratories the muds were observed to crust over during drying, indicating that agitation would be required if a high degree of desiccation were to be achieved.

Low grade ores on land can be worked profitably because the ore minerals are distinct mineral grains. The grains are large enough so that, once they are liberated from the barren rock during grinding, they can be concentrated owing to differences in magnetic susceptibility, electrical conductivity, density, grain size or shape or by the manner in which liquids and air selectively coat the surface of ore and gangue minerals (flotation). Under our assumption 1., such efficient and inexpensive techniques are precluded for the Red Sea oozes. What techniques are left?

The answer will most probably be found in the fields of hydrometallurgy or pyrometallurgy. In the former, the ore is entirely or partially leached with a solvent, and then the valuable materials will be selectively precipitated or selectively collected and harvested on ion-exchange resins. In pyrometallurgy, the ore is smelted in a furnace; additives flux off the unwanted components as slag; the values remain as a metallic residue or matte, which undergoes further refining.

Both of these tend to be expensive processes and are commercial, in general, only if the majority of the material treated is eventually recovered and sold, if a solvent can be found that leaches only the worthwhile constitutents and is not consumed by leaching a lot of worthless material or if the solvent can be recovered and recycled.

The Red Sea oozes were originally reported to contain a high content of iron. Our assays do not substantiate this. On a dried basis, the average for the Atlantis II Deep is 12 per cent (Hendricks *et al.*, 1969). It is conceivable that, if the salt could be removed, smelting would yield pig iron, copper, silver and zinc products.

While this may prove to be technically feasible, it is very difficult to see how it could compete effectively against the high grade iron, copper and zinc ores of the world. Note, too, that the presence of copper and zinc could be detrimental to the quality of the iron.

There are valid reasons to discount the value of the iron oxides in the oozes. Too many high grade, pure, cheaply mined iron ore deposits exist around the world to visualize effective competition from a low grade, impure, non-conventional source.

In the last few years, smelting, using sodium chloride as a flux (TORCO process), has proven attractive in treating Zambian copper ores. In the furnace, the NaCl is decomposed to form hydrochloric acid which then forms volatile cuprous chloride. The cuprous chloride is reduced to metallic copper on carbon particles (*World Mining,* December 1966, p. 20). The Red Sea muds, if evaporated to dryness, would contain appreciable NaCl. This suggests that metallurgists might be able to take the concept of the TORCO process and adapt it to the Red Sea material.

Another development in recent years is the Imperial Smelting Process, a technique wherein complex ores of lead, zinc and copper are efficiently treated to recover a high percentage of all three metals (*World Mining,* August 1966, pp. 10–15). We are not suggesting an Imperial smelter could handle the Red Sea material, but certainly these new advancements in the smelting art suggest that the smelter people, faced with the Red Sea ooze, might be able with an expenditure of research and development money to learn how to handle it. The cost of doing so would have to be charged against what would otherwise be profits from the Red Sea venture.

Hydrometallurgy could be a more promising approach for beneficiating the Red Sea ooze. A solvent, such as nitric acid, might dissolve the copper, zinc and silver ions and leave the iron untouched, but nitric acid is expensive. Hydrochloric acid is cheaper but probably would not dissolve the silver or get good recovery of some of the other metals. Sulphuric acid is traditionally the cheapest commercial acid but currently is scarce and expensive. For the long term, its price and availability may permit its consideration. Silver would not, however, be readily leached in this acid. Neither is gold soluble in any of these acids.

Moreover, we believe that the calcium carbonate content of the ooze is too high to consider acid leaching. The carbonates would consume an inordinate amount of acid. More promising than acid digestion, therefore, would be leaching with alkaline solvents. To be efficient, autoclaving may have to be used during the leaching process, thereby raising capital and operating costs.

Dissolving the desired metals preferentially is only the first of several problems. Once in solution, the liquid must be separated from the non-dissolved worthless products. In this case, we must separate a solution from suspended colloids and mud-sized grains, and it must be done quickly and cheaply at the rate of 1,000 gallons daily.

It may be possible that additives would coagulate the colloids and not affect the leaching characteristics of the system, but the precipitate would be fine grained. Centrifuging would be more effective, but the power costs may be high. Proximity to the Near East oil and natural gas fields are a distinct plus factor in considering power requirements and costs.

Once the solutions are cleared of solid matter, the dissolved metals must be recovered, separately, if possible, and precipitated. We suspect the metallurgist would give careful consideration to separating and collecting the metals using selective ion-exchange. This technology is well developed for some metals, such as uranium, but for the metals with which we are concerned here the art is only now being developed. The possibilities are attractive, but research and development money will be needed by the Red Sea investor. Conceivably, an ion-exchange mill could extract other metals in addition to those we have been considering even though they are present in only minute amounts.

The authors cite the difficulties inherent in flotation where the particles are only a few microns in diameter. None of these difficulties are insurmountable. Some ores less than 10 microns in diameter are now being successfully floated. One of the main problems is reagent consumption. Very finely divided ores can develop exceedingly high charges for reagents.

Flotation is usually done with fresh water and the Red Sea oozes contain an unusually high saline brine. Sea water, however, was used to float chalcopyrite as early as 1930 by the Tocopilla Mining Company in Chile. The use of sea water in flotation of base metal ores was discussed by Rey and Raffinot (1966), who state that Penarroya has used sea water for flotation with good success at four mills in the Mediterranean basin: Carthagene in Spain, Aurium in Greece, Aïm Barbar in Algeria and Argentiera in Sardinia. The first two recover lead and zinc by selective flotation. The Aïm Barbar mill treats a lead-zinc-copper ore. From the literature, we might summarize by saying that sea water flotation, including differential flotation, yields results comparable to tests with fresh water, although reagent consumption is higher in sea water. Moreover, maintenance and capital costs are higher because of the corrosiveness of sea water.

Conventional ore must be ground to a fine mesh before flotation is possible. Next, water is added to prepare it for flotation. At Yerington, Nevada (Huttl, 1962) the ore entering the first stage flotation units has characteristics very similar to those possessed by the oozes. This makes for the intriguing speculation whether this material could, with only minimal preparation, be readied for selective flotation. No blasting of rock would be required; crushing, grinding and density control of the pulp is eliminated. Is this a case where significant capital and operating cost savings can be envisioned and would offset the difficulties of floating very finely divided ore in salt water? If only flotation is necessary, could this be done at sea and avoid the problems of transporting ore to port? We suspect the metallurgist studying recovery of the Red Sea oozes will want to give careful consideration to flotation.

The price a smelter pays for metal contained in a mill concentrate, naturally, is less than market price. The reduction covers:

1. Freight from mine-mill to smelter.
2. Smelting and refining costs.
3. Freight cost of the final metal to market.
4. Smelter's profit.
5. Penalty for the presence of undesirable constituents.

In the last few years a new smelting technique, called the Imperial Smelting Process, takes mixed concentrates of copper, lead and zinc and recovers a high percentage of all metals. If such a smelter were available for treatment of the Red Sea ooze concentrates, then revenue would be higher than if only a conventional smelter were available.

The above discussions are not a complete analysis of the processing of the oozes. They do, however, identify critical areas that have to be considered and resolved.

Summary

There are indications that the metal rich oozes in and near the Atlantis II Deep contain metals, in particular, copper, zinc, gold and silver, in sufficient enough quantities such that profitable extraction might be considered. Unfortunately, these muds have not been sampled and assayed rigorously enough to ascertain precisely how much metal is present. Regardless of how much metal is present, the muds are of no economic interest unless they can be mined and processed and the recovered metals delivered to the market place and sold profitably.

The ooze is very fluid in its natural state, and we feel that with a little agitation the whole mass could be readily fluidized and pumped up a pipe.

Little attention has been given to the extraction of the metal values after the muds arrive at the mill. Three broad approaches can be considered. None has been tested:

1. Dewater the ooze, employing centrifuging, cycloning and/or solar evaporation. Dissolve the metals contained in the residue in other acid or alkaline solvents and concentrate the individual metals by ion-exchange.
2. Dewater the ooze and smelt the entire residue.
3. Attempt conventional ore dressing techniques to yield high grade concentrates that can be sold to a custom smelter.

If flotation for concentrating the ore proves feasible, it might even be accomplished on board the mining ship at the mining site, eliminating the many problems involved in transporting this material to a land based mill.

It would seem that more sampling is critical before the economic significance of the Red Sea muds can be truly assessed. Obviously, additional study of the Red Sea basins may uncover richer material, semi-indurated and more coarsely crystalline in nature rather than colloidal soup. If so, the economic attraction would increase. Such material could occur deeper than yet penetrated by the cores or near the surface in those parts of the basins not yet sampled.

Acknowledgments

We would like to thank all the scientists at Woods Hole, especially Drs. J. L. Bischoff, E. T. Degens, J. M. Hunt and D. A. Ross, who generously cooperated with us during the **CHAIN** Cruise and our work on the cores. Our thanks are also due to Dr. Mel Peterson of Scripps for his active assistance and to Messrs. J. R. Atkinson, Les Camp, Blair Roberts and James Stephens for their fruitful criticism of this manuscript.

References

Bischoff, J. L. and F. T. Manheim: Economic potential of the Red Sea heavy metal deposits. *In: Hot brines and recent heavy metal deposits in the Red Sea*, E. T. Degens and D. A. Ross (eds.). Springer-Verlag New York Inc., 535–541 (1969).

Hendricks, R. L., F. B. Reisbick, E. J. Mahaffer, D. B. Roberts, and M. N. A. Peterson: Chemical composition of sediments and interstitial brines from the Atlantis II, Discovery, and Chain Deeps. *In: Hot brines and recent heavy metal deposits in the Red Sea*, E. T. Degens and D. A. Ross (eds.). Springer-Verlag New York Inc., 407–440 (1969).

Huttl, J.: Engineering and Mining Journal, **16**, no. 3 (1962).

Miller, A. R., C. D. Densmore, E. T. Degens, J. C. Hathaway, F. T. Manheim, P. F. McFarlin, R. Pocklington, and A. Jokela. Hot brines and recent iron deposits in deeps of the Red Sea. Geochim. et Cosmochim. Acta, **30**, 341 (1966).

Rey, M. and P. Raffinit: World Mining, June (1966).

International Legal Rights to Minerals in the Red Sea Deeps

WILLIAM L. GRIFFIN
Washington, D.C.

Abstract

The legal rights of the mineral deposits in the Red Sea are discussed. It is concluded that the Red Sea floor must be deemed to be adjacent to the nearest coastal state. These considerations place the Red Sea deeps and its ore deposits on the legal continental shelf of Sudan.

The Problem Situation

An American marine resources firm recently announced* that it had applied to the United Nations for a 38.5 square-mile exclusive mineral exploration lease to survey thermal sea floor springs in the middle of the Red Sea. According to the firm, no nation claims sovereignty over the area, and the firm wished to have United Nations' approval to sample and map mineral deposits over a three-year period to determine their economic significance. The United Nations replied that it has no authority to grant minerals rights in the Red Sea floor.† The purpose of this chapter is to examine the problem of who has what exploitation authority regarding minerals in the Red Sea deeps and how such authority is acquired.

Nature of International Law

International law, in its application to a given situation, may fairly be described as a prediction of the rights that a state may validly claim for itself and its nationals from other states and of the duties which in consequence it must observe towards them. Prediction of legal rights and duties pertaining to minerals of the Red Sea deeps requires exegesis of the physiographical and legal parameters of the problem situation.

Historical Background

The question of who has what rights, if any, in the sea bottom has ancient lineage. In Roman law the seabed was deemed *res nullius* in which rights could be acquired by occupation. For years international lawyers debated whether or not this Roman rule had been "received" into international law. Where actual exploitation has been practicable, states have not hesitated to claim rights in the seabed, *e.g.,* pearl fisheries and coal mine tunnels from shore.

The question assumed its contemporary form only when technology made substantial exploitation possible by vertical approach through the sea rather than by tunnelling from shore. It was first adumbrated by Britain and Venezuela in 1942 when they agreed to divide the "submarine areas of the Gulf of Paria" (Whiteman, 1965). The Gulf is narrow; only the littoral states could reasonably exploit its bed. No state protested.

In 1945 President Truman proclaimed that "the United States regards the natural resources of the subsoil and seabed of the continental shelf beneath the high seas but continuous to the coasts of the

---

* *Am. Metal Mkt.,* **15,** Feb. 15, 1968.
† *Wash. Eve. Star,* **D-10, col. 5,** Feb. 23, 1968.

United States as appertaining to the United States, subject to its jurisdiction and control" (Whiteman, 1965). The continental shelf was not defined.

The Truman Proclamation was not protested by any state; it was emulated. By the mid-1950's more than 40 more or less similar declarations had been published. International lawyers began to debate whether or not coastal states have *ipso jure,* without proclamation, sovereign rights over adjacent submarine areas as part of customary international law.

From 1951 to 1956 the International Law Commission (ILC), which is the United Nations organ responsible for progressive development of international law, debated and adopted draft articles on the law of the continental shelf. These draft articles became the foundation of the Convention on the Continental Shelf, a treaty adopted by the 1958 United Nations Geneva Conference on the Law of the Sea.

The Geological Continental Shelf

The term "continental shelf" is in origin a term used by geologists and physical oceanographers for a unique feature of the earth. The International Committee on Nomenclature of the Sea Floor, established by the Association of Physical Oceanography, agreed upon the following definitions:

"Continental Shelf, Shelf Edge . . . The zone around the continents, extending from the low water line to the depth at which there is a marked increase of slope to greater depth. Where this increase occurs the term shelf edge is appropriate. Conventionally its edge is taken at 100 fathoms (or 200m) but instances are known where the increase of slope occurs at more than 200 or less than 65 fathoms. . . .

"Continental Slope.—The declivity from the outer edge of the continental shelf . . . into great depths" (Whiteman, 1965).

These definitions were brought to the attention of the Geneva Conference on the Law of the Sea (UNESCO, 1957).

In structure and stratigraphy the continental shelf is a seaward continuation of the continent, directly facing the large ocean basins. Next beyond the continental shelf edge is the continental slope which is the general boundary between the light rocks of the continents and the denser rocks of the sea floor. Thus the continental slope is approximately the continental limit.

Around the world there are several large submarine areas, often less than 200m deep, which are submerged depressions largely surrounded by land or large islands. Such submarine areas are not part of a continental marginal zone directly facing the large ocean basins. Hence, some geologists do not consider them to be continental shelf and refer to them as "inner shelves." Examples are the floor of the North Sea and the Persian Gulf. In general all geologists agree that inner shelves and the continental shelf are parts of continental areas whose outer limit is the slope of the continental mass.

Physiography of the Red Sea

Physiographically, the bed of the Red Sea is an inner shelf because it is a fault depression in the Arabian-Nubian shield (Drake and Girdler, 1964). The length of the depression at sea level is about 1,875km from the Sinai Peninsula to the Strait of Bab-el-Mandeb. The width of the depression at sea level varies from about 200km in the north to about 360km in the south. The width of the Red Sea is substantially the same as the width of the depression, except at the extreme southern end where the Sea begins to funnel into the Strait. In the vicinity of the deeps, 21°15′–25′N latitude, the Red Sea is about 225 km wide. The predominant bathymetric features of the Red Sea are: (a) a shelf along the shore, (b) a main fault trough running almost the entire length, within which there is (c) a deep axial trough (Drake and Girdler, 1964).

The coastal shelf is everywhere very shallow (well under 200m). North of 20°N (which includes the deeps) it is 30–40km wide, but to the south it widens extensively. It slopes steeply to the bottom of the main trough.

The main trough runs from off Sinai to the Strait of Bab-el-Mandeb at the southern end of the Red Sea. The trough is steep-sided and ranges in depth from about 600 to 1,000m. North of 24°N it is about 150km wide. To the south it widens to about 200km in the vicinity of the deeps; at about 19°N it begins to narrow.

The deep axial trough within the main trough begins to develop at about 25°N, soon becoming about 10km wide and almost 2,000m deep. To the south it widens and deepens, becoming about 50km wide and over 2,000m deep near the brine area. The actual trough increases in width to about 70km at 16°N, further south it eventually becomes obscured by sedimentary matter. In addition to the foregoing fault pattern and bathymetry of the Red Sea, magnetic anomalies and seismic refraction profiles are additional evidence that the Red Sea depression is a tension feature in which the deep axial trough and probably the main trough marks the lateral separation of Arabia and Africa.

The Convention on the Continental Shelf

The Convention on the Continental Shelf is a repository of the minimum general principles upon which two thirds of the states at the Conference could agree. Its Articles relevant to the present Red Sea situation are:

ARTICLE 1

For the purpose of these articles, the term "continental shelf" is used as referring . . . to the seabed and subsoil of the submarine areas adjacent to the coast but outside the area of the territorial sea, to a depth of 200m or, beyond that limit, to where the depth of the superjacent waters admits of the exploitation of the natural resources of the said areas. . . .

ARTICLE 2

1. The coastal States exercise over the continental shelf sovereign rights for the purpose of exploring it and exploiting its natural resources.
2. The rights referred to in paragraph 1 of this article are exclusive in the sense that if the coastal State does not explore the continental

shelf or exploit its natural resources, no one may undertake these activities, or make a claim to the continental shelf, without the express consent of the coastal State.
3. The rights of the coastal State over the continental shelf do not depend on occupation, effective or notional, or on any express proclamation. . . .

ARTICLE 6

1. Where the same continental shelf is adjacent to the territories of two or more States whose coasts are opposite each other, the boundary of the continental shelf appertaining to such States shall be determined by agreement between them. In the absence of agreement, and unless another boundary line is justified by special circumstances, the boundary is the median line, every point of which is equidistant from the nearest points of the baselines from which the breadth of the territorial sea of each State is measured. . . .

The foregoing general principles are law between the 37 states who have ratified the Convention. They are not, however, necessarily the last word because they do not answer all questions and are subject to interpretation and formal amendment.

As to states who have not ratified the Convention, such as the Red Sea coastal states, the principles of the Convention are not without legal effect. Whether or not they be regarded as declaratory of customary international law, experience with older "law-making" treaties indicates that these general principles stand as authoritative bases for the prediction of coastal sea floor rights in specific problem situations.

The Convention text, however, does little more than recognize that a coastal state has some sort of special right over natural resources in an adjacent submarine area.* Such generalized language is intended to apply to the world ocean and is too inexact to be definitive. Its precise meaning for application to a specific problem situation such as the Red Sea deeps, requires identification of the theoretical basis and purpose of the coastal state's right, examination of the drafting history of

---

* The Convention also deals with other rights, as well as duties, of states regarding natural resources, which are beyond the scope of this chapter.

the Convention and description of the actual practice of states regarding other similar problem situations.

Purpose and Theoretical Basis of Coastal State's Rights

Both the purpose and the theoretical basis of the coastal state's special right are revealed throughout the drafting history of the Convention contained in the records of the ILC and the Geneva Conference. They were summarized by the ILC thus (Whiteman, 1965):

"The Commission does not deem it necessary to expatiate on the question of the nature and legal basis of the sovereign rights attributed to the coastal State. The considerations relevant to this matter cannot be reduced to a single factor. In particular, it is not possible to base the sovereign rights of the coastal State exclusively on recent practice, for there is no question in the present case of giving the authority of a legal rule to a unilateral practice resting solely upon the will of the States concerned. However, that practice itself is considered by the Commission to be supported by considerations of law and of fact. In particular, once the seabed and the subsoil have become an object of active interest to coastal States with a view to the exploration and exploitation of their resources, they cannot be considered as *res nullius, i.e.,* capable of being appropriated by the first occupier. It is natural that coastal States should resist any such solution. Moreover, in most cases the effective exploitation of natural resources must presuppose the existence of installations on the territory of the coastal State. Neither is it possible to disregard the geographical phenomenon whatever the term — propinquity, contiguity, geographical continuity, appurtenance or identity — used to define the relationship between the submarine areas in question and the adjacent nonsubmerged land. All these considerations of general utility provide a sufficient basis for the principle of the sovereign rights of the coastal State as now formulated by the Commission."

The theoretical basis, in a word, is adjacency. The purpose is to allocate those resources which are of immediate concern to coastal states because their adjacency to land makes their exploitation technologically possible or foreseeable.

Recognition that a coastal state has a special right regarding resources in a particular area raises the problems of defining the area, the nature and the manner of acquisition of the right. The concept of adjacency permeates the solution of these problems.

Outer Limit of Coastal State's Right

Convention Articles 1 and 6(1)

Articles 1 and 6(1) of the Convention on the Continental Shelf are verbal formulas defining the outer geographical limit of the coastal state's right. Article 1 contains two factors which circumscribe the outer limit. The first is the opening phrase of Article 1 — "For the purpose of these articles." This indicates that the right is confined to a "legal continental shelf" which may extend seaward either more or less than the geological continental shelf.

The second circumscribing factor is the phrase "adjacent to the coast." Article 1 would be grammatically complete without this phrase. Therefore this phrase is meaningless unless it qualifies what follows it in Article 1. Thus the area out to the 200m isobath is adjacent by definition and beyond that limit the area to the depth of exploitability is confined to the area "adjacent to the coast." Of course, an area "adjacent to the coast" is relative but that does not necessarily mean that limitless rights are recognized. What constitutes an area "adjacent to the coast" must be resolved with due regard to the circumstances of each problem situation.

Where the coasts of two states are opposite each other, *e.g.,* Sudan and Saudi Arabia, in the vicinity of the deeps, their legal continental shelf may take up the entire intervening submarine area. In that case the boundary between them is the median line defined in Article 6(1), absent agreement upon another boundary or another boundary justified by special circumstances.

The submarine area between Sudan and Saudi Arabia is more than 200m deep. Therefore, the question whether Article

6(1) is relevant depends on whether the entire submarine area is deemed "adjacent" to their coasts under the circumstances.

History of Article 1

The ILC had two reasons for defining the outer limit of the coastal state's right independently of the geological continental shelf. One reason is the disagreements among scientists as to whether an "inner shelf," such as the Red Sea floor, is a true geological shelf. The other reason is that where exploitation of an adjacent coastal submarine area is technologically possible at a depth exceeding 200m, exploitation should not be prohibited merely because the area is not a geological continental shelf (Whiteman, 1965).

The ILC, in view of its decision not to use the geological continental shelf to define the outer limit of the coastal state's right, considered whether it would be better to use the term "submarine areas." A majority decided to use "continental shelf" because it was in current use and because "submarine areas" standing alone would not give sufficient indication of the nature of the areas in question (Whiteman, 1965).

In 1951 the ILC considered adopting the 200m isobath as practically sufficient for the outer limit of the legal continental shelf. It decided not to adopt that limit because technology of the then near future might make possible exploitation beyond that depth and because geological continental shelf areas deeper than 200m might be exploitable from neighboring installations at less than 200m. The ILC adopted solely the exploitability limit in 1951 (Whiteman, 1965).

In 1953 the ILC reversed itself and adopted solely the 200m limit in view of comments by some governments that the exploitability limit lacks precision. The ILC considered that the 200m limit would be practically sufficient for a long time to come and that a stable limit is particularly desirable between states adjacent or opposite each other. It was aware that future technology might make exploitation possible beyond 200m, in which case the limit

would have to be revised (Whiteman, 1965).

In 1956 an Inter-American conference concluded that the outer limit of the coastal state's right should be 200m depth or beyond that limit, to the depth of exploitability. This disjunctive formula was designed to lessen the effect of inequality in the distribution of geological continental shelf areas throughout the world (Garcia-Amador, 1959). For example, off western South America the 200m isobath is within 5 to 30km of the coast, whereas it extends over 320km out to sea off Argentina and eastern North America.

Later in 1956 a majority of the ILC decided to adopt the Inter-American view on the outer limit of the legal continental shelf so that when exploitation at a depth greater than 200m became technologically possible, recognition of the coastal state's right would not have to await revision of the Convention. A minority of the ILC were of the opinion that addition of the exploitability limit unjustifiably impaired the stability of the outer limit of the legal continental shelf (Whiteman, 1965).

In 1958 the Geneva Conference redebated the ground covered by the ILC on the outer limit of the legal continental shelf. The conference recognized the advantages and the defects of the disjunctive 200m depth or exploitability depth limit but were unable to find a better formula to strike a balance between the rigidity of the 200m limit and the instability of the exploitability limit (Whiteman, 1965).

Practice of States

The final step in determining the outer limit of a coastal state's right is the practice of states in that regard. Two examples will suffice, the Persian Gulf and the North Sea.

The resources of the Persian Gulf floor are regarded by the coastal states, none being parties to the Convention on the Continental Shelf, as exclusively theirs without protest from other states. Saudi Arabia and Iran have agreed upon their median line boundary. However, because the Persian

Gulf is everywhere under 200m deep, it could be said not to be a conclusive example, but this is not true of the North Sea.

At the north end of the North Sea there is a trough in the sea floor about 7km wide and greatly in excess of 200m deep, which follows the southern coast of Norway just outside her territorial sea. Seaward of the trough the sea floor rises again to under 200m deep. Norway is not a party to the Convention on the Continental Shelf but claims exclusive rights to sea floor resources out to the median line, thus ignoring the trough as merely an irrelevant depression in the adjacent sea floor. No state has protested Norway's claim and she has made agreements with the United Kingdom and Denmark, both parties to the Convention, confirming the median line boundary between them. Similar boundary agreements also exist between the United Kingdom and Denmark and the Netherlands and Denmark (International and Comparative Law Quarterly, 1966). The Netherlands is a party to the Convention.

Conclusion and Application to Red Sea Deeps

In the light of the foregoing, it seems fair to conclude that the rights of a coastal state over sea floor resources extend to the 200m isobath or, beyond that depth, to a submarine area reasonably adjacent to the coast under the circumstances. It also seems fair to conclude that any given point on the Red Sea floor must be deemed to be adjacent to the nearest coastal state. The Red Sea floor is a geological inner shelf just as is the floor of the Persian Gulf and the North Sea. The Red Sea floor is slightly narrower than the Persian Gulf and much narrower than the North Sea. If an alien conducting exploration operations at any given point on the Red Sea floor without permission of the nearest coastal state were evicted by it, any protest by the alien's government seems most unreasonable and unlikely. These considerations place the Red Sea deeps on the legal continental shelf of Sudan since they are west of the median line (Fig. 1).

Mode of Acquisition Rights

Assuming adjacency places the minerals of the Red Sea deeps on the legal continental shelf of Sudan over which it has not proclaimed any rights, the question arises whether any proclamation is a necessary prerequisite to Sudanese acquisition of rights. With regard to determining the manner of acquisition of rights, adjacency as

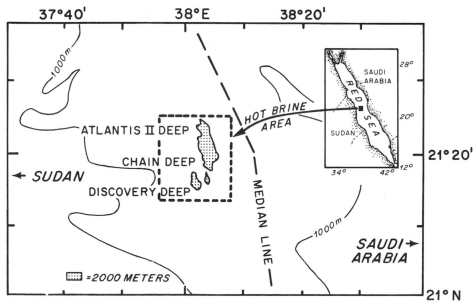

Fig. 1. Bathymetric map of brine area.

556   William L. Griffin

the theoretical basis for allocation of resources has consequences for states adjacent and nonadjacent to the resources.

With regard to states not adjacent to
the resources, the consequence is that they
cannot be deemed *res nullius, i.e.,* capable
of being appropriated by taking possession. As the ILC said, "It is natural that
coastal States should resist any such solution."

With regard to adjacent states there must
be a choice between alternative consequences. Adjacent resources could remain *res nullius* until the coastal state
takes possession or the resources could be
allocated to the coastal state without possession. The latter prevents the socially
undesirable result of the former, viz., that
if the coastal state did not have the technological ability to exploit adjacent resources neither would it have any right
which it could grant to others having the
technological ability. Under the former,
this socially undesirable result could also
be prevented by the notion that the coastal
state is deemed to have possession of adjacent resources, especially if it issued a
proclamation to that effect. But allocation
of adjacent resources by such a fiction of
possession is in substance an allocation
without possession. It is an allocation *ipso
jure, i.e.,* by operation of law. The latter
rule is the one recognized [Article 2 (3) of
the Convention]; hence, the lack of a formal
claim by Sudan of rights in the minerals of
the Red Sea deeps is immaterial.

Nature of Coastal State's Rights

Recognition that a coastal state has
rights regarding adjacent submarine resources raises the problem of determining
what content of the right and terms de-
scriptive thereof will promote resources
development while preserving freedom of
the superjacent high sea. With regard to
content of the right, adjacency, as its theoretical basis, has the consequence that
the resources must appertain solely to the
coastal state even if it does not have the
technological capability of developing
them, but in the latter event, their trans-
ferability is desirable [Article 2 (2) of the
Convention].

Both the ILC and the Geneva Conference carefully considered terms descrip-
tive of the right. The term "jurisdiction
and control" was rejected as too weakly
expressive of the content of the right. The
term "sovereignty" was rejected as lending itself to abuse of freedom of the high
seas. The language finally adopted [Article
2 (1) of the Convention] was "sovereign
rights." This was meant to confirm the exclusive nature of the right as well as to imply that the coastal state has all ancillary
rights reasonably necessary for and connected with exploration and exploitation,
including jurisdiction to prevent or punish
violations of its law regarding the resources
(Whiteman, 1965).

References

Drake, C. L. and R. W. Girdler: A geophysical study
of the Red Sea. Geophys. Jour. Roy. Astronom.
Soc., **8,** 473 (1964).

Garcia-Amador, F. V.: Exploitation and Conserva-
tion of the Resources of the Sea, 107 (1959).

UNESCO, Conference on the Law of the Sea, "Scientific Considerations Relating to the Continental
Shelf" (mimeo, 1957).

Whiteman, M. W.: Digest of International Law, **4,**
756, 789, 818, 829, 831, 843, 856 (1965).

Summary of Hot Brines and Heavy Metal Deposits in the Red Sea *

K. O. EMERY, J. M. HUNT, AND E. E. HAYS

Woods Hole Oceanographic Institution
Woods Hole, Massachusetts

Abstract

The chapters in this book record the results and ideas of many authors who studied different aspects of the Red Sea, its hot brines, and its metalliferous deposits. Separate chapters deal with regional structure and stratigraphy, local topography and rocks, measurements of geomagnetism and gravity, temperatures of water and sediments, movement of the water, general chemistry of water and sediments, distribution and significance of both stable and radioactive isotopes, flora and fauna of overlying water and bottom sediments, facies of sediments and their metal contents, comparison with ancient metalliferous strata, and finally economic and legal factors that will have to be considered by possible exploiters of the deposit. This last chapter in the book summarizes the previous ones and attempts to draw many related observations into a somewhat more condensed description of the hot brines and metal deposits.

Introduction

The Red Sea is one of the first large bodies of water recorded in human history. Its length was sailed by the famous expeditions of Egyptian Queen Hatshepsut about 1500 B.C., those of Israelite King Solomon and of Yemenese Queen Sheba about 950 B.C., and of the Phoenicians in their circumnavigation of Africa about 600 B.C. Many unrecorded trading ships traveled through it during this period and probably even earlier.

The sea has an area of about 438,000 square km, a length of 1,800km (2,100km if the Gulf of Suez is included), a maximum width of 270km, and it contains some of the hottest and saltiest ocean water of the world within its volume of about 215,000 cubic km. Its maximum recorded depth is 2,920m, but the sill at the southeastern end, opposite Aden, is only about 120m deep. At the northwestern end shallow canals were dug before the time of Christ to connect the Red Sea with the Nile River. A deeper canal to connect directly with the Mediterranean Sea was advocated by Caliph Haroun-al-Rachid of Baghdad about 800 A.D., but little was accomplished until Ferdinand de Lesseps with funding from an international company completed the precursor of the present Suez Canal in 1869.

Between the time of the opening of the canal in 1869 and its closing in 1967, the Red Sea was traversed probably at least 100 times by oceanographic research ships of many nations. Some of these ships paused briefly to make measurements and to collect samples of water or bottom sediment. Few ships obtained observations and samples from great depths; thus, the first indication of higher than normal water temperatures came only in 1948 by ALBATROSS, and of much higher than normal salinity in 1959 by ATLANTIS. Between 1963 and 1967 ATLANTIS II, DISCOVERY, METEOR, and OCEANOGRAPHER made eight visits to one or more of the areas of deep hot brine, and they accumu-

* Woods Hole Oceanographic Institution Contribution No. 2203.

lated information on the temperature (to about 56°C), the salinity (to about 7.5 times that of ordinary sea water), the depiction of the salt-water interfaces by echo sounding profiles, and the presence of several separate basins containing this unusual water. During October and November 1966 **CHAIN,** under the funding of the National Science Foundation, made the first cruise directly intended to investigate the special characteristics of the hot brine and its underlying sediments. The results of this cruise are the chief basis for this collection of 50 chapters written by 75 authors from several countries about many different aspects of the hot brine and associated sediment.

The contributions of the many authors help to answer such broad questions as: What characteristics of the regional and local geology are so unique as to indicate the origin of this unusual brine and deposit? Are the chemical, mineralogical, and physical properties of the deposit due to igneous or sedimentary processes, or both? When did the deposit begin to form, and is it still forming? Are other similar environments likely to occur elsewhere in the world, and in what kinds of areas should they be sought? What is the probable economic value of the deposit? And what are the legal implications of the deposit with respect to its exploitation?

This summary is intended to draw together the results that have been presented by the numerous authors, but in the interest of simplicity repeated references to the detailed presentations will be omitted. For specifics and details the reader should refer to appropriate chapters.

Geological and Geophysical Setting

The stratigraphy and structure of the Red Sea and its surroundings can be examined from outer space, on the ground, and aboard ship. Each vantage point provides information that is unique and supplementary to that from the other points.

Color photographs from manned **GEMINI** satellite capsules contain far more detail than the earlier televised imagery of the

TIROS and **NIMBUS** weather satellites. These photographs, even at casual inspection, confirm the general information given by navigational charts that the opposite sides of the Red Sea are sinuous and that the projections of one side fit the indentations of the other side. Closer inspection of the photographs show that the sinuosities of the shorelines are controlled by the intersections of several sets of faults that strike in slightly different directions. The similarities of the opposite shorelines suggest that the small faults mark a zone of weakness, and that the Arabian side was pulled away from the African side at a not-ancient time. Former continuity of Africa and Arabia is also indicated by faults which lie at an angle to the shore and whose broken continuation is far away on the opposite shore. The main obstacle to conformity of the two sides is opposite Aden, where a broad triangle is now filled with volcanic rocks. During the opening of the land to form the Red Sea, the Arabian side also moved laterally northwestward about 130km, relative to the African side. A similar left-lateral offset of about 100km occurred in the Gulf of Aqaba and its northerly extension past the Dead Sea, as indicated by displacement of rock areas having distinctive color and topographic texture.

Earlier studies of areal geology made on the ground had revealed most of the information that can be seen in so grand a manner on the photographs from outer space. Moreover, the ground observations included stratigraphy, by which events in the history of the Red Sea could be inferred and dated. These studies show that the Red Sea cuts through a huge central mass of Precambrian igneous and metamorphic rocks, whose outcrops form the more rugged and mountainous parts of the region. Surrounding much of the massif was a broad shelf on which Paleozoic marine sediments were deposited. Marine sedimentation continued during the Mesozoic, with continental sediments deposited atop the massif that lay astride the site of the middle part of the present Red Sea. Folding and block faulting that began in Late Paleozoic time continued, alternately causing the shore zone to advance and retreat, as shown by the shifting boundary between marine and continental strata. General submergence is indicated by thick accumulation of marine sediments, especially in the region of the Gulf of Suez.

Similar geological history continued into the Early Tertiary time with general emergence particularly in the northwest. During Late Oligocene time the Gulf of Suez took shape, followed during Early Miocene time by the main Red Sea, and during Late Pliocene time by the Gulf of Aqaba. As a result, the Gulf of Suez is underlain by very thick sediments, an almost continuous section from Paleozoic to Recent. The Red Sea has a bordering coastal plain of Middle Miocene to Recent continental and marine strata, and the Gulf of Aqaba has no coastal plain at all. The three areas contain coral reefs of Pleistocene age, many of which are now tens of meters above present sea level.

The ground studies suggest that during Late Eocene and certainly in Oligocene time compressional forces developed in the Gulf of Suez region and tensional forces developed at the southeastern end of the Red Sea. These forces gradually caused the crust to open from southeast to northwest. The accompanying volcanism in the southeast largely filled the triangular gap opposite Aden that is in line with the East African Rift Zone and its northeastward continuation in the Gulf of Aden. The opening of the Red Sea was relieved in the northwest by left-lateral movement along the Gulf of Aqaba-Dead Sea graben, as indicated by offsets of the northern limits of Precambrian, Cretaceous, and other mapped rock units. Temporary plugging of the early Red Sea by the volcanic rocks deposited opposite Aden during the opening stages of the sea is supported by the nature of strata that are visible along the sides of the Red Sea. Strata of Middle Miocene age consist mainly of basal clastics overlain by gypsum, probably continental in origin. Most of the overlying Upper Miocene strata are of brackish environment, although a few marine fossils indicate that ocean water penetrated into the Red Sea by the end of the Miocene Epoch. Marine sediments were widespread by Pliocene times.

Shipboard investigations aboard **CHAIN**

were concentrated in the general area of the hot brine and metalliferous sediment, near the midpoint of the Red Sea. Soundings revealed the presence of three separate basins that contain the hot brine, as indicated by acoustic reflections from a water interface at about 2,000m depth and by subsequent sampling of water and sediments. All three basins are enclosed by the 1,900m contour. The northernmost and largest by far of the basins is Atlantis II Deep, centering at North Latitude 21°23′ and East Longitude 38°04′ and reaching a maximum depth of 2,170m. The next basin to the south is Chain Deep, probably separated from Atlantis II Deep by a sill at about 2,009m. Chain Deep is the smallest of the three basins and its maximum depth is only 2,066m. West of Chain Deep and separated by a 1,980m sill is Discovery Deep, a subcircular basin having a maximum depth of 2,220m. The floors of Atlantis II and Discovery basins are generally flat, although Atlantis II Deep contains three central hills that lie at a slight angle to the general trend of the Red Sea. Side slopes of the basins range in steepness from 5 to 35 degrees. The total area of the three basins below the 2,000m contour is about 70 square km, only 0.016 per cent of the total area of the Red Sea, and only about 0.003 per cent of its total volume.

Continuous seismic reflection profiling showed the presence of a good acoustic reflecting horizon that ranged from 0 to about 350m below the floor of the Red Sea in the region surrounding the brine-filled basins, but it was not noted beneath the basins themselves. This horizon is common throughout the entire Red Sea. Its distribution and character is suggestive of an unconformity, perhaps dating from Late Miocene or Early Pliocene times. Seismic profiles within the basins show little detail, perhaps because of the peculiar nature of the sediments and the overlying brines. In addition, the low frequency used for profiling permits a maximum resolution of only about 30m. More precise acoustic measurements were provided by high-frequency pulses from an acoustic pinger that was lowered to near the bottom. The pinger records revealed several kinds of bottom layering and indicated general average thicknesses of sediment in the basins amounting to about 25m. Maximum thicknesses (including pelagic sediment) were estimated to be 100m in Atlantis II Deep, 50m in Chain Deep, and 200m in Discovery Deep.

Fragments of volcanic glass and basalt were found in six cores in or near the axis of the Red Sea within 160km of the hot brine basins. Chemical analyses indicate that at least two of the basalts resemble oceanic basalts in their low content of potassium and lack of nepheline and that they resemble African rift-valley basalts in their high contents of iron and phosphorus. Concentrations of rare-earth elements are more like those of oceanic basalts. The Red Sea basalts appear to be distinct from continental basalts and are comparable with those of oceanic ridges, the sites of sea-floor spreading.

Nearly a decade ago shipboard measurements of the earth's magnetic intensity indicated the presence of an axial belt of anomalies in the Red Sea relative to the regional field. The belt zig-zags in accordance with changes in the trend of the shorelines. New detailed measurements showed that the hot brine basins lie within this axial belt. A negative anomaly under Atlantis II Deep and positive ones under Chain Deep and Discovery Deep indicate that the magnetic materials that cause the anomalies are deep-seated ones. Moreover, laboratory measurements showed that the sediments, even the metalliferous ones, are so weakly magnetic, as well as being present in only a thin layer, that they are not the cause of the anomalies. The anomalies near the hot brine basins have short wave lengths, an amplitude of about 1,000 gammas, and gradients up to 200 gammas per km in the axial belt. A lower amplitude of anomalies occurs farther northwest and a higher amplitude farther southeast. The character of the anomalies is similar to that in belts parallel to the Atlantic and other rift zones of the earth that have been interpreted as due to sea-floor spreading away from a rift zone. New extrusive rocks have welled up at a more or less constant rate for millions of years in the zones of divergence. As they

cooled and hardened, their enclosed magnetic minerals took an orientation that was impressed by the earth's general magnetic field. For unknown reasons the polarity of the earth's field has reversed at irregular intervals whose dates during the past several tens of million years are approximately known from measurements of magnetic polarity of dated rocks on land.

The patterns of magnetic anomalies in the Red Sea fit very closely the pattern known elsewhere on land and at sea for the past five million years. When half the belt width is divided by the time span represented by all of the anomalies, the rate of divergence or spreading is found to be 1.5cm per year. Since geological considerations show that the movement is not a simple opening at right angles to the rift trend, but that the opening also has a left-lateral component, the rate of sea-floor spreading in the area of the hot brine basins must be about 1.6cm per year. Such a rate is about average for the other rift zones of the world. The seven-degree rotational opening of the Red Sea required by its geometry indicates a slightly faster rate to the southeast and a slower one to the northwest. When this rate near the hot brine basins is extrapolated to the entire distance of displacement between formerly adjacent shore points, one computes that rifting began 6 to 8 million years ago. Such a date is only about half as ancient as the date given by geological considerations, but the difference in the two results is probably due to periodic halts in the rifting.

Shipboard gravity measurements yield high positive (+150mgals) Bouguer anomalies along the axial belt of the Red Sea. These anomalies indicate a thinner crust or higher density rocks than at either side of the belt, thus confirming the interpretation derived from magnetic measurements that the axis has been intruded by dense basalts. Information about the deep structure underlying the Red Sea was provided by a model constructed to conform with the gravity data and with earlier seismic refraction velocities. Interpretation of the model suggests that the region near the hot brine basins is underlain by sedimentary rocks, evaporites, and pyroclastics having densities of 2.31 to 2.43g/cm³ and seismic velocities of 3.5 to 4.5km/sec; thicknesses range from 7km at either side of the Red Sea to only 1km along the axis. Flanking the sides of the Red Sea are deeper wedges that thicken landward to 15km and consist of rocks having a density of about 2.66g/cm³ and a seismic velocity of 5.5 to 6.4km/sec, presumably granitic shield rocks. Between the two wedges and lying directly beneath the blanket of sedimentary rocks is a broad ridge of rock having a density of 2.92gm/cm³ and seismic velocities of 6.8 to 7.3km/sec; this is inferred to represent basalt that flowed into the rift zone as the shield rocks parted along the axis of the Red Sea during rifting. The Mohorovičič discontinuity lies at a depth of about 22km beneath the axial belt of the Red Sea and more than 30km at either side, in response to the presence of dense rocks that lie above it and below the axial belt of the Red Sea. This intermediate depth to the Mohorovičič discontinuity is typical of newly formed ocean basins elsewhere in the world.

Measurements of thermal gradients in the sediments at fourteen stations in Atlantis II Deep revealed sediment temperatures as high as 62.3°C, in contrast with 56.5°C in the hot brine and 44°C in the overlying brine. Gradients ranged from +3.78°C/m (showing present heating of overlying water by the sediment) to −0.87°C/m (showing present heating of sediment by the overlying water). These data when combined with laboratory measurements of heat conductivity of the sediment, yield heat fluxes of 15 to 20μcal/cm²/sec in the Atlantis II Deep, with values of 1.5 to 8.0 in the axial belt of the Red Sea farther northwest and southeast. For comparison, the average for many hundred measurements of the ocean floor elsewhere in the world is about 1μcal/cm²/sec.

Long-term variation in the rate of heat loss from the floor of the Red Sea may be indicated by an observation that the temperature of the hot brine in Atlantis II Deep had increased by about 0.5°C in 20 months. Also, sediment similar to that underlying the hot brine of Atlantis II Deep occurs in Discovery Deep and Chain Deep, but beneath brines of lower temperature as

though brine had once flowed from Atlantis II Deep across the sills to the other basins. A former higher temperature and higher level of the top surface of the brine pool would have produced the observed negative gradients at depth in some sediments; subsequent cooling of both brine and sediment would then have developed the positive gradients nearer the surface in some areas.

The average measured heat flux from the bottom is inadequate to have caused the heating observed in 20 months or even to have replaced the heat lost upward to the water of the Red Sea. Computations of upward heat loss suggest that the present average heat flux in Atlantis II Deep may actually be 80μcal/cm²/sec or more, which means that most heating is due to forced discharge of very hot brine through local vents now as well as during irregular past intervals. Greater discharge during the past is suggested by the presence of the unique metalliferous layers in the basin sediment, and by past larger volume of hot brine in order for it to cross the sills into the other two basins, and to have periodically covered the tops of the hills that rise above the central part of Atlantis II Deep. Contribution of heat by convection of water through the sediment occurs, but heat from this source and especially from molecular conduction is probably relatively small. Heating by bulk convection or mixing of the sediments is of no consequence, as it would have destroyed their well developed layering.

Data from geology and geophysics produce a convincing picture of the environment of the hot brine basins and the source of the heat. The pulling apart of Arabia and Africa to form the Red Sea began in Miocene time about 25 million years ago, ceased for a period, and continues at present, as shown by stratigraphy along the sides of the sea and by magnetic reversals near the axis. Gravity and seismic refraction measurements indicate the presence of a layer of sediment that is thinnest along the axial belt, and is underlain by a broad ridge of dense rock, probably basalt, that has filled the region left by the pulling apart of a once continuous granitic shield. Volcanoes and massive flows on land, as well as fragments of glass and basalt in samples from the axial belt, support the concept of inflow of basalt during sea-floor spreading. Although the region has few earthquakes, secondary deformation is illustrated by unconformities, block faulting, folding, and raised coral reefs. Inflow of the basalt probably was and is accompanied by the injection of hot brines into the sea through the floor of the Red Sea, with the most recent injection through the southern part of the Atlantis II Deep. During times of greater than normal inflow, the brine pool became hotter and thicker than now so that it overflowed into adjacent basin areas, where it heated the underlying sediments. A subsequent period of lesser activity permitted some cooling of the brine and sediment, but renewed heating is indicated by the previously mentioned slight increase in temperature of the brine.

Water

Surface water in the Red Sea varies from less than 26°C in winter to more than 30°C in summer, and from less than 37 per mill salinity at the southeastern end to more than 40 per mill at the northwestern end. Below a transition zone at 100 to 400m depth and above the hot brines the water is reasonably uniform at 22°C, 40.6 per mill salinity, and 2.0ml/l oxygen. At the main point of exchange with the world ocean, opposite Aden at the southeastern end, the surface water flows inward from the Indian Ocean probably at all seasons; but the greatest flow is during winter and the least in summer due to reversal of wind directions. An outflow from the Red Sea through the strait occurs at depth. The total water budget of the Red Sea is estimated to be equivalent to a complete renewal of all its water in about 20 years.

Above the Atlantis II Deep the normal Red Sea water was noted to change through a depth range of 25m to an intermediate brine below about 2,009m; this intermediate brine was nearly uniform at 44.2°C, about 123 per mill salinity, and nearly zero oxygen content. Below a second interface layer less than 5m thick the main pool of hot brine began at 2,042m; it averaged 56.5°C, 257 per mill salinity, and zero oxygen. Active turbulence is indicated by the scatter of temperature observations below the interfaces averaging 0.25°C, about ten times the error of measurement with reversing thermometers; such is probably to be expected of convection over a non-uniform heat source on the bottom. Temperature variations along the length of a thermoprobe before it was driven into the bottom sediment amounted to 0.2°C with a period of about three minutes; these were deemed internal waves. Measurements of light transparency showed that the base of the normal 22°C water, the top of the 44°C brine, and the top of the 56.5°C brine had dense accumulations of suspended sediment. Most of it probably consisted of organic debris that floats atop the underlying denser brine. Instruments lowered through the interfaces also occasionally caught pieces of newspaper, rags, and other debris that had been thrown overside from passing ships. Probably the suspended material of whatever source serves as material for local concentrated bacterial activity.

Dense water in Discovery Deep lay below a 37m zone of transition. Between 2,023 and 2,027m was a layer of 36°C water. Another interface of 15m thickness separated the bottom water that began at 2,042, and averaged 44.7°C, 257 per mill salinity, and nearly zero oxygen. In Chain Deep a transitional zone began at about 2,010m and continued to the bottom (2,042m at that station) where the temperature was 34°C. In comparison, the water at the sills between the basins was less hot and salty than that above the basins at equal depth. This means that especially over Atlantis II Deep the hot brine forms a biconvex pool whose slope must be a function of the rate of production of new brine and the rate of cooling and dilution by overlying water. Periodic renewal of the brine may be indicated by the double layering observed in both Atlantis II Deep and Discovery Deep; however, theory and laboratory experiments suggest that well-mixed layers separated by sharp interfaces are characteristic of liquids that are stabilized with salt, but are made unstable by heating from below. Partial mixing of the hot brine with overlying less hot and salty water is also promoted by internal waves.

Cooling of the overflow into Discovery Deep and Chain Deep is especially well shown by the approximate similarity of the temperatures of their brines to that of the overlying water and the similarity of their salinities to that in Atlantis II Deep. Moreover, cooling of the brine that flowed into Discovery Deep is indicated by the strong negative temperature gradient ($-0.015°C/$m) in the depth range 20 to 40m above the bottom. Water that was within 20m of the bottom had a lesser negative gradient, as though it had settled to the lowest levels after partial cooling had increased its density. Water in Discovery Deep more than 60m above the bottom had a small positive gradient ($+0.0005°C/m$), equivalent to that expected from adiabatic considerations alone. In contrast, three stations in Atlantis II Deep exhibited a slight negative gradient ($-0.0003°C/m$), indicating stability.

The concentrations of major and minor ions in the bottom waters of the three hot brine pools differ from each other (Table 1). Except for sodium, potassium, and magnesium, all ions are slightly more concen-

trated in Atlantis II Deep than in Discovery Deep. Particularly noticeable in Discovery Deep is the lesser sulfate and trace metals, as though greater precipitation of metallic sulfides had occurred from that brine. In Chain Deep the major ions have concentrations intermediate between the concentrations in the brines from the other two basins and the concentration in ordinary ocean water, but the ratios of the ions to chlorine show that simple dilution is not the only factor. As a matter of fact, the concentration of most ions in all of the brines depart from their ordinary ocean-water ratios with chlorine. Most striking are the various trace metals which are concentrated about a thousand times over their abundance in ordinary ocean water, in contrast with only an eight-fold increase in chlorine. On the other hand, magnesium and sulfate ions are less concentrated in all of the brines than in ordinary ocean water. Also of interest is a comparison of the composition of the hot brines with the slightly saltier bottom water of the Dead Sea that is believed to owe its origin to long-term evaporation of ocean water. Major differences are the much higher concentrations in the Red Sea brines of sodium, iron, and manganese, and the much

Table 1   Major and Some Minor Ions in the Brines

(g/kg)

	Atlantis II Deep	Discovery Deep	Chain Deep	Dead Sea	Ocean Water
Na^+	92.60	93.05	24.0	32.30	10.76
K^+	1.87	2.14	0.78	6.17	0.39
Ca^{++}	5.15	5.12	1.18	13.98	0.41
Mg^{++}	0.76	0.81	1.42	34.45	1.29
Sr^{++}	0.04	0.04	0.01	0.20	0.008
Cl^-	156.03	155.3	41.9	178.20	19.35
Br^-	0.13	0.12	0.08	4.28	0.066
SO_4^{--}	0.84	0.70	2.81	0.34	2.71
HCO_3^-	0.14	0.03	–	0.18	0.72
Si	0.03	0.003	–	<0.01	0.004
Fe	0.08	0.003	–	<0.002	0.00002
Mn	0.08	0.05	0.005	0.004	0.00001
Zn	0.005	0.0008	–	<0.02	0.000005
Cu	0.0003	0.0001	–	<0.002	0.00001
Co	0.0002	0.0001	–	–	0.000004
Pb	0.0006	0.0002	–	<0.002	0.0000004
Ni	–	0.0003	–	<0.002	0.0000001
Salinity	257.76	257.37	72.19	270.11	35.71
Temperature	56.5°C	44.7°C	29.1°C	22.0°C	–
Density	1.178	1.183	–	1.233	1.03

lower concentrations of potassium, calcium, magnesium, strontium, chlorine, and bromine. The composition of the Red Sea brine rather strongly supports the other kinds of evidence that it is not a simple concentrate developed by extreme evaporation of surface water. Such evaporation should have increased, rather than decreased, the ratio to chlorine of the highly soluble ions, potassium, magnesium, and bromine, and it could not alone have caused the observed great increase in ratio of the many trace metals.

Expressed in ratio to the highly soluble potassium, the ions in the ocean and in the bottom brines of the Dead Sea and the Red Sea have an interesting grouping (Table 2). Comparison of the ratios indicates that sodium and chlorine are depleted in the Dead Sea with respect to the ocean, but enriched in the Red Sea; magnesium and bromine are enriched in the Dead Sea, but depleted in the Red Sea; and calcium is enriched in both the Dead Sea and the Red Sea. The changes between ocean water and Dead Sea water have been ascribed to deposition of sodium chloride and accumulation in the water of the more soluble potassium, magnesium, calcium, and bromine as bitters. The opposite ratios in the hot brine from the Red Sea accord with the concept of resolution of previously deposited halite and gypsum, somewhat like the deposits of the Dead Sea but long separated from the bitters that remained in solution. Thus the hot brines may consist of ground waters that obtained their major salts from resolution of bedded former evaporites and their trace elements from solution of interbedded shales. When the

ground water reached a mass of recently emplaced and still hot basalt, its dissolving power would have been enhanced, it may have received new materials dissolved in juvenile waters, and its decreased density would have caused it to rise, eventually to escape as hot brine at the bottom of the Red Sea, in Atlantis II Deep.

The stable isotopes of oxygen and hydrogen (deuterium) accord with the evidence provided by major ions. Their values ($\delta O^{18} = 1.21$ per mill; $\delta D = 7.5$ per mill) are nearly the same as for surface ocean water of the same salinity as that near the entrance of the Red Sea. These ratios mean that the brine could not have been formed by simple evaporation of ocean water, because such evaporation would have produced much higher concentrations of heavy oxygen and of deuterium; likewise, it could not have been formed through solution of salts by fresh water (as rain falling on outcrops), for this would have yielded lower concentrations of heavy oxygen and deuterium than are present. The observed concentrations are about what would be expected if ocean water like that near the southeastern end of the Red Sea passed through sediments or strata that cropped out in shallow water, dissolved buried evaporites (that are known to be present at depth in the Cenozoic strata), and eventually emerged on the floor of the Red Sea.

Emergence of brines at several different times is indicated by slightly different values of δO^{18}, δD and δS^{34}, as well as by differences in major ions and trace elements in the 56.5°C and 44°C brines of Atlantis II Deep, and the brine of Discovery Deep.

Sulfur isotopes proved to be especially useful in establishing the origin of the hot brines. Values for δS^{34} ranged around +29 per mill for dissolved sulfate in all of the brines and interstitial waters; this is the same as for the world ocean, indicating not only the absence of extreme evaporation to form the brines, but also the absence of extensive bacterial fractionation in the brines after they emerged on the floor of the Red Sea. Metallic sulfides in sediments of the Atlantis II Deep have δS^{34} of +3 to +11 per mill, a range that is suggestive of high

Table 2 Ion Ratios

	Ocean	Dead Sea	Red Sea	Ratios to Ocean Dead Sea	Ratios to Ocean Red Sea
Na/K	25.64	5.23	49.50	0.20	1.93
Ca/K	1.05	2.26	2.75	2.15	2.52
Mg/K	3.31	5.58	0.41	1.69	0.12
Cl/K	49.60	28.85	83.50	0.58	1.68
Br/K	0.17	0.69	0.07	4.06	0.41

temperature reduction of sulfate by organic matter in shales, not by sulfate-reducing bacteria.

Methane, ethane, and other light hydrocarbons in the brines probably were dissolved from strata by the brines. Although they are about a thousand times more concentrated than in near-surface ocean waters, these hydrocarbon gases totalled only about 0.06ml/l, and cannot have been responsible for effervescence of the brines as reported for some early water samples.

The chemistry of the present hot brines is a function not only of the composition of the original brine that was injected through the sea floor and mixed with some overlying ocean water, but it is also a function of later changes in the brine. Some of these changes resulted from continued mixing with overlying water, such as may have produced the intermediate water atop the hot brine. Other more interesting ones are due to chemical reactions within the brine itself. One that occurs near the top interface of the 44°C brine of Atlantis II Deep was shown by both field and laboratory experience, whereby a brown precipitate developed when the anaerobic brine came into contact with some dissolved oxygen; this precipitate consists mainly of iron deposited as ferric hydroxide (but it also contains traces of copper, nickel, cobalt, manganese, and lead scavenged by the precipitate).

Other chemical changes may take place within or a bit below the top of the anaerobic mass of the brine. Chief of these changes is the reduction of sulfate ion to sulfide by bacterial activity and the subsequent reaction of the sulfide ion with dissolved heavy metals, which are thus precipitated from the brine and added to the underlying sediments. The effects of these bacteria are well shown in Discovery Deep, where the metallic sulfides in the sediments have δS^{34} less than -20 per mill, in contrast to δS^{34} of $+20$ per mill in the sulfate of the overlying brine. Failure to detect hydrogen sulfide, even by its odor, in the brines probably means only that an excess of metals is present so that hydrogen sulfide is immediately taken up as soon as it was formed or became available. The availability of some

sulfide for reaction with copper and lead is indicated by the observed blackening of brass fittings and lead weights of sampling equipment that was lowered into the hot brines. Related information is provided by reported redox potentials of -100mv at a pH of 5.0. The anaerobic environment of the hot brine may also have led to bacterial denitrification of ammonia produced by decomposition of organic matter. This denitrification could account for the nearly 70 per cent excess of molecular nitrogen, relative to argon in the hot brine, as compared with the ratio of nitrogen to argon in ordinary ocean water.

Dissolved organic matter in the brines has about the same concentration as in ordinary ocean water, about 1.5mg carbon per liter. Some of this organic matter is probably scavenged from the top of the brine by the precipitated ferric hydroxide, as the precipitate was observed to remove 30 per cent of the dissolved organic matter from samples of the brine that were examined aboard ship. Some particulate organic matter undoubtedly is carried to the bottom by the weight of enclosing siliceous or calcareous tests, and by other inorganic particles. The few measurements of organic carbon in the sediments show the presence of 0.2 to 0.7 per cent organic carbon and 0.02 to 0.03 per cent nitrogen. These concentrations are in the range of most deep-sea sediments, or even greater if they are expressed on a heavy metal-free basis (about 4 per cent carbon).

Organisms

Organisms or their remains that may be found within the hot brines have two obvious possible sources: they inhabit the brines or their underlying sediments, or they have fallen from the overlying waters. Organisms of either source should change the composition of the brines or of sediment solutions; organisms of the second type should in addition provide most of the organic matter to the sediments, and their variation with time (depth of burial) should provide insight into the climatic history and changes in character of the waters of the Red Sea.

Search for organisms, chiefly bacteria, that might inhabit the hot brines is of especial interest because of the presumptive evidence of their local activity yielded by the sulfate reduction and denitrification in the brines. On the other hand, the high temperature, salinity, and concentration of heavy metals are formidable obstacles to life processes. Samples of ordinary Red Sea water, the hot brines, and underlying sediments were collected by sterile methods, inoculated into various media, and incubated at various temperatures. The only samples that yielded positive evidence of viable bacteria were those of the 22°C water, the interface between the 22°C water and the 44°C brine (note that this was one of only three very large samples), sediments that were shallower than the top of the hot brines (including ones in the sill south of Atlantis II Deep), and sediments of Discovery Deep.

The hard parts of many small pelagic organisms that lived in near-surface waters and sank to the bottom after death occur in cores from the hot brine pools and from surrounding areas. Chief of these are the calcareous tests of Foraminifera (unicellular protozoans). A latitudinal variation of the number and abundance of species plus a general affinity of the species to those of the Indian Ocean indicate that the principal or only source of Foraminifera was from the southeast. Most of these species occur in the sediments of the past 11,000 years, showing that this period had the most favorable living conditions. Two species, *Globigerinoides ruber* and *G. sacculifer,* were probably the most tolerant with *G. ruber* better able to withstand low water temperatures and high salinities. The ratio of these two species serve as a reasonable index of climatic conditions. Between the present and 11,000 years ago, *G. sacculifer* dominated. Sediments that were deposited 11,000 to 20,000 years ago contain both species with *G. sacculifer* gradually becoming dominant and indicating a warming trend with time. Between about 20,000 and 45,000 years ago *G. ruber* dominated and *G. sacculifer* commonly was absent; this presumably was a long cold period of high salinity during much of the Würm glacial epoch. Brief appearances of *G. sacculifer* in sediments deposited about 30,000 and about 37,000 years ago testify to brief warmer intervals. The ratios indicate a much longer warm interval 45,000 to 60,000 years ago, preceded by about 20,000 years of fluctuating climate. Benthic species reached only moderate abundance relative to pelagic species in the normal calcareous sediments, and they are virtually absent in cores from the hot brine areas.

The remains of other pelagic organisms in the sediments are less abundant and less diagnostic of environmental conditions than are the foraminiferal tests. Nearly all of them can be zoned with depth in the sediments, with zonations attributed to changes in water temperatures, salinities, or depths. Many of the zonations for the various organisms, however, cannot be correlated with certainty chiefly because the studies were made on different cores, mostly outside the hot-brine area. Most abundant by far are Coccolithophorida (unicellular brown algae that secrete small calcareous plates). The species in a single long core that probably spans about 80,000 years are tropical to sub-tropical, except for ones characteristic of cooler water from a zone that probably corresponds to about 13,000 to 24,000 years ago, the latest time of glacially low sea level in the open ocean. The siliceous tests of Radiolaria (unicellular protozoans) were sought in many cores but were found in only a few sections of three cores in Atlantis II Deep; sections at depths corresponding to about 9,000 to 12,000 years ago. They are of especial interest owing to their much closer affinity to species in the Indian Ocean than in the Mediterranean Sea. Pteropoda (pelagic gastropods) are represented by their conical shells in many of the cores. Most striking is a reduction in number of species, and the dominance of a shallow-water form in the 13,000 to 24,000 year zone that corresponds to low sea level in the open ocean. The chitinous plates of Dinoflagellata (a flagellated protozoan), studied in two cores, exhibited a cyclical variation in concentration akin to that of Foraminifera, supporting the concept that

chemical precipitation or ion exchange, (b) smelting of the whole sediment or some concentrate of it, (c) flotation of the metal-bearing minerals, (d) some other techniques.

4. Related studies of whether the sediment must first be leached, dried, or treated in bulk. The method of treatment also controls whether the sediment can be processed aboard ship (with related disposal of wastes to avoid recycling them) or whether it must be processed ashore (with added costs of bulk transportation and probable transfer to shuttle carriers from the mining ship).

5. Cost analysis of the operation to determine what is the minimum economic rate of mining in order to offset the large capital investment for the final evaluation and the initiation of the operation, plus the regular operating costs.

Solution of these mining, metallurgical, and cost-analysis problems is likely to be expensive and time-consuming, and it is only a precursor to the building and assembly of the necessary hardware. Yet, experience with large-scale mining on land has shown that unless such studies are thorough, the cost of mining and extraction of metals may exceed the value of the metals as determined by the demand for them and the cost of production from alternate cheaper sources. Obviously, a time will come that production of the metals from other deposits will increase in cost, but most likely the cost of sea-floor mining will also have increased. Thus, the question of the economic value of the Red Sea deposits remains open, as does the date at which the exploitation may become profitable.

Exploitation probably will also await a decision on the question of ownership of the deposit. This is a difficult question because ownership presumably involves not only the right to lease and tax, but also the obligation to protect the lease-holder's right to exploit without interference by others. A second uncertainty results from the present unsatisfactory status of the definition of sovereign rights. Stripped of legal jargon, the Geneva Convention says that the adjacent country's sovereign rights extend to the edge of the continental shelf or to any depth that can be exploited. The exploitation term overrides the restriction to continental shelf, but some legal talent has advocated redefining the term continental shelf to include any desired portion (even all) of the deep-sea floor. If the decision as to sovereign rights is based only upon nearness to coastal state (law of median lines), then the deposit clearly belongs to Sudan. At least one group has applied to Sudan for mining rights. Two others have considered the hot brine areas to be international territory, beyond the sovereignty of any nation. One of these groups has applied to the United Nations and the other has incorporated in Lichtenstein on the theory that no country owns these deep-water areas. Complications arise about whether the deep water in the Red Sea is legally a deep-sea region, because it is a local trough that is separated from the deep-sea floor of the world ocean by a shallow strait at one end and by a canal at the other. Also, there is a question about whether a nation that claims the right to lease and tax has an obligation to legally or militarily protect the lease holder against other claimants. Where all of this maneuvering will end is unknown, but there is a distinct possibility that lawyers will profit more from the Red Sea deposits than will scientists or the metal industry.

Indices

Index of Names

The page numbers in lightface indicate mentions in text; in italics,
reference listings; and in boldface, chapters (inclusive pages).

Index of Latin Names

A dagger before a page number indicates a table or a figure.

Index of Subjects

*Italic page numbers indicate chapters on the given subject;
a dagger before a page number indicates a table or a figure.*